MEMS

Design and
Fabrication

Mechanical Engineering Series
Frank Kreith and Roop Mahajan - Series Editors

MEMS

Design and Fabrication

Edited by
Mohamed Gad-el-Hak

CRC Press
Taylor & Francis Group
Boca Raton London New York

CRC Press is an imprint of the
Taylor & Francis Group, an **informa** business

A TAYLOR & FRANCIS BOOK

Foreground: A 24-layer rotary varactor fabricated in nickel using the Electrochemical Fabrication (EFAB*) technology. See Chapter 6 for details of the EFAB* technology. Scanning electron micrograph courtesy of Adam L. Cohen, Microfabrica Incorporated (www.microfabrica.com), U.S.A.

Background: A two-layer, surface micromachined, vibrating gyroscope. The overall size of the integrated circuitry is 4.5 × 4.5 mm. Sandia National Laboratories' emblem in the lower right-hand corner is 700 microns wide. The four silver rectangles in the center are the gyroscope's proof masses, each 240 × 310 × 2.25 microns. See Chapter 4, *MEMS: Applications* (0-8493-9139-3), for design and fabrication details. Photography courtesy of Andrew D. Oliver, Sandia National Laboratories.

Published in 2006 by CRC Press
Taylor & Francis Group
6000 Broken Sound Parkway NW, Suite 300
Boca Raton, FL 33487-2742

ISBN-13: 9780849391385 (hbk)
ISBN-13: 9780367391638 (pbk)

Library of Congress Card Number 2005050109

Library of Congress Cataloging-in-Publication Data

MEMS : design and fabrication / edited by Mohamed Gad-el-Hak.
   p. cm. -- (Mechanical engineering series (Boca Raton, Fla.))
  Includes bibliographical references and index.
  ISBN 0-8493-9138-5 (alk. paper)
  1. Microelectromechanical systems. 2. Microelectromechanical systems--Design and construction. 3. Microfabrication. I. Gad-el-Hak, M. II. Series.

TK7875.M46 2005
621.381--dc22
                                                            2005050109

Visit the Taylor & Francis Web site
at http://www.taylorandfrancis.com

and the CRC Press Web site at http://
www.crcpress.com

Preface

In a little time I felt something alive moving on my left leg, which advancing gently forward over my breast, came almost up to my chin; when bending my eyes downward as much as I could, I perceived it to be a human creature not six inches high, with a bow and arrow in his hands, and a quiver at his back. ... I had the fortune to break the strings, and wrench out the pegs that fastened my left arm to the ground; for, by lifting it up to my face, I discovered the methods they had taken to bind me, and at the same time with a violent pull, which gave me excessive pain, I a little loosened the strings that tied down my hair on the left side, so that I was just able to turn my head about two inches. ... These people are most excellent mathematicians, and arrived to a great perfection in mechanics by the countenance and encouragement of the emperor, who is a renowned patron of learning. This prince has several machines fixed on wheels, for the carriage of trees and other great weights.

(From *Gulliver's Travels—A Voyage to Lilliput*, by Jonathan Swift, 1726.)

In the Nevada desert, an experiment has gone horribly wrong. A cloud of nanoparticles — micro-robots — has escaped from the laboratory. This cloud is self-sustaining and self-reproducing. It is intelligent and learns from experience. For all practical purposes, it is alive.

It has been programmed as a predator. It is evolving swiftly, becoming more deadly with each passing hour.

Every attempt to destroy it has failed.

And we are the prey.

(From Michael Crichton's techno-thriller *Prey*, HarperCollins Publishers, 2002.)

Almost three centuries apart, the imaginative novelists quoted above contemplated the astonishing, at times frightening possibilities of living beings much bigger or much smaller than us. In 1959, the physicist Richard Feynman envisioned the fabrication of machines much smaller than their makers. The length scale of man, at slightly more than 10^0 m, amazingly fits right in the middle of the smallest subatomic particle, which is approximately 10^{-26} m, and the extent of the observable universe, which is of the order of 10^{26} m. Toolmaking has always differentiated our species from all others on Earth. Close to 400,000 years ago, archaic *Homo sapiens* carved aerodynamically correct wooden spears. Man builds things consistent with his size, typically in the range of two orders of magnitude larger or smaller than himself. But humans have always striven to explore, build, and control the extremes of length and time scales. In the voyages to Lilliput and Brobdingnag in *Gulliver's Travels*, Jonathan Swift speculates on the remarkable possibilities which diminution or magnification of physical dimensions provides. The Great Pyramid of Khufu was originally 147 m high when completed around 2600 B.C., while the Empire State Building constructed in 1931 is presently 449 m high. At the other end of the spectrum of manmade artifacts, a dime is slightly less than 2 cm in diameter. Watchmakers have practiced the art of miniaturization since the 13th century. The invention of the microscope in the 17th century opened the way for direct observation of microbes and plant and animal cells. Smaller things were manmade in the latter half of the 20th century. The transistor in today's integrated circuits has a size of 0.18 micron in production and approaches 10 nanometers in research laboratories.

Microelectromechanical systems (MEMS) refer to devices that have characteristic length of less than 1 mm but more than 1 micron, that combine electrical and mechanical components, and that are fabricated using integrated circuit batch-processing technologies. Current manufacturing techniques for MEMS include surface silicon micromachining; bulk silicon micromachining; lithography, electrodeposition, and plastic molding; and electrodischarge machining. The multidisciplinary field has witnessed explosive growth during the last decade and the technology is progressing at a rate that far exceeds that of our understanding of the physics involved. Electrostatic, magnetic, electromagnetic, pneumatic and thermal actuators, motors, valves, gears, cantilevers, diaphragms, and tweezers of less than 100 micron size have been fabricated. These have been used as sensors for pressure, temperature, mass flow, velocity, sound and chemical composition, as actuators for linear and angular motions, and as simple components for complex systems such as robots, lab-on-a-chip, micro heat engines and micro heat pumps. The lab-on-a-chip in particular is promising to automate biology and chemistry to the same extent the integrated circuit has allowed large-scale automation of computation. Global funding for micro- and nanotechnology research and development quintupled from $432 million in 1997 to $2.2 billion in 2002. In 2004, the U.S. National Nanotechnology Initiative had a budget of close to $1 billion, and the worldwide investment in nanotechnology exceeded $3.5 billion. In 10 to 15 years, it is estimated that micro- and nanotechnology markets will represent $340 billion per year in materials, $300 billion per year in electronics, and $180 billion per year in pharmaceuticals.

The three-book *MEMS set* covers several aspects of microelectromechanical systems, or more broadly, the art and science of electromechanical miniaturization. MEMS design, fabrication, and application as well as the physical modeling of their materials, transport phenomena, and operations are all discussed. Chapters on the electrical, structural, fluidic, transport and control aspects of MEMS are included in the books. Other chapters cover existing and potential applications of microdevices in a variety of fields, including instrumentation and distributed control. Up-to-date new chapters in the areas of microscale hydrodynamics, lattice Boltzmann simulations, polymeric-based sensors and actuators, diagnostic tools, microactuators, nonlinear electrokinetic devices, and molecular self-assembly are included in the three books constituting the second edition of *The MEMS Handbook*. The 16 chapters in *MEMS: Introduction and Fundamentals* provide background and physical considerations, the 14 chapters in *MEMS: Design and Fabrication* discuss the design and fabrication of microdevices, and the 15 chapters in *MEMS: Applications* review some of the applications of micro-sensors and microactuators.

There are a total of 45 chapters written by the world's foremost authorities in this multidisciplinary subject. The 71 contributing authors come from Canada, China (Hong Kong), India, Israel, Italy, Korea, Sweden, Taiwan, and the United States, and are affiliated with academia, government, and industry. Without compromising rigorousness, the present text is designed for maximum readability by a broad audience having engineering or science background. As expected when several authors are involved, and despite the editor's best effort, the chapters of each book vary in length, depth, breadth, and writing style. These books should be useful as references to scientists and engineers already experienced in the field or as primers to researchers and graduate students just getting started in the art and science of electromechanical miniaturization. The Editor-in-Chief is very grateful to all the contributing authors for their dedication to this endeavor and selfless, generous giving of their time with no material reward other than the knowledge that their hard work may one day make the difference in someone else's life. The talent, enthusiasm, and indefatigability of Taylor & Francis Group's Cindy Renee Carelli (acquisition editor), Jessica Vakili (production coordinator), N. S. Pandian and the rest of the editorial team at Macmillan India Limited, Mimi Williams and Tao Woolfe (project editors) were highly contagious and percolated throughout the entire endeavor.

Mohamed Gad-el-Hak

Editor-in-Chief

Mohamed Gad-el-Hak received his B.Sc. (summa cum laude) in mechanical engineering from Ain Shams University in 1966 and his Ph.D. in fluid mechanics from the Johns Hopkins University in 1973, where he worked with Professor Stanley Corrsin. Gad-el-Hak has since taught and conducted research at the University of Southern California, University of Virginia, University of Notre Dame, Institut National Polytechnique de Grenoble, Université de Poitiers, Friedrich-Alexander-Universität Erlangen-Nürnberg, Technische Universität München, and Technische Universität Berlin, and has lectured extensively at seminars in the United States and overseas. Dr. Gad-el-Hak is currently the Inez Caudill Eminent Professor of Biomedical Engineering and chair of mechanical engineering at Virginia Commonwealth University in Richmond. Prior to his Notre Dame appointment as professor of aerospace and mechanical engineering, Gad-el-Hak was senior research scientist and program manager at Flow Research Company in Seattle, Washington, where he managed a variety of aerodynamic and hydrodynamic research projects.

Professor Gad-el-Hak is world renowned for advancing several novel diagnostic tools for turbulent flows, including the laser-induced fluorescence (LIF) technique for flow visualization; for discovering the efficient mechanism via which a turbulent region rapidly grows by destabilizing a surrounding laminar flow; for conducting the seminal experiments which detailed the fluid–compliant surface interactions in turbulent boundary layers; for introducing the concept of targeted control to achieve drag reduction, lift enhancement and mixing augmentation in wall-bounded flows; and for developing a novel viscous pump suited for microelectromechanical systems (MEMS) applications. Gad-el-Hak's work on Reynolds number effects in turbulent boundary layers, published in 1994, marked a significant paradigm shift in the subject. His 1999 paper on the fluid mechanics of microdevices established the fledgling field on firm physical grounds and is one of the most cited articles of the 1990s.

Gad-el-Hak holds two patents: one for a drag-reducing method for airplanes and underwater vehicles and the other for a lift-control device for delta wings. Dr. Gad-el-Hak has published over 450 articles, authored/edited 14 books and conference proceedings, and presented 250 invited lectures in the basic and applied research areas of isotropic turbulence, boundary layer flows, stratified flows, fluid–structure interactions, compliant coatings, unsteady aerodynamics, biological flows, non-Newtonian fluids, hard and soft computing including genetic algorithms, flow control, and microelectromechanical systems. Gad-el-Hak's papers have been cited well over 1000 times in the technical literature. He is the author of the book *"Flow Control: Passive, Active, and Reactive Flow Management,"* and editor of the books *"Frontiers in Experimental Fluid Mechanics," "Advances in Fluid Mechanics Measurements," "Flow Control: Fundamentals and Practices," "The MEMS Handbook,"* and *"Transition and Turbulence Control."*

Professor Gad-el-Hak is a fellow of the American Academy of Mechanics, a fellow and life member of the American Physical Society, a fellow of the American Society of Mechanical Engineers, an associate fellow of the American Institute of Aeronautics and Astronautics, and a member of the European Mechanics

Society. He has recently been inducted as an eminent engineer in Tau Beta Pi, an honorary member in Sigma Gamma Tau and Pi Tau Sigma, and a member-at-large in Sigma Xi. From 1988 to 1991, Dr. Gad-el-Hak served as Associate Editor for *AIAA Journal*. He is currently serving as Editor-in-Chief for *e-MicroNano.com*, Associate Editor for *Applied Mechanics Reviews* and *e-Fluids*, as well as Contributing Editor for Springer-Verlag's *Lecture Notes in Engineering* and *Lecture Notes in Physics*, for McGraw-Hill's Year Book of Science and Technology, and for CRC Press' *Mechanical Engineering Series*.

Dr. Gad-el-Hak serves as consultant to the governments of Egypt, France, Germany, Italy, Poland, Singapore, Sweden, United Kingdom and the United States, the United Nations, and numerous industrial organizations. Professor Gad-el-Hak has been a member of several advisory panels for DOD, DOE, NASA and NSF. During the 1991/1992 academic year, he was a visiting professor at Institut de Mécanique de Grenoble, France. During the summers of 1993, 1994 and 1997, Dr. Gad-el-Hak was, respectively, a distinguished faculty fellow at Naval Undersea Warfare Center, Newport, Rhode Island, a visiting exceptional professor at Université de Poitiers, France, and a Gastwissenschaftler (guest scientist) at Forschungszentrum Rossendorf, Dresden, Germany. In 1998, Professor Gad-el-Hak was named the Fourteenth ASME Freeman Scholar. In 1999, Gad-el-Hak was awarded the prestigious Alexander von Humboldt Prize — Germany's highest research award for senior U.S. scientists and scholars in all disciplines — as well as the Japanese Government Research Award for Foreign Scholars. In 2002, Gad-el-Hak was named ASME Distinguished Lecturer, as well as inducted into the Johns Hopkins University Society of Scholars.

Contributors

Gary M. Atkinson
Department of Electrical and
    Computer Engineering
Virginia Commonwealth
    University
Richmond, Virginia, U.S.A.

Christopher A. Bang
Microfabrica Inc.
Burbank, California, U.S.A.

Glenn M. Beheim
NASA Glenn Research Center
Cleveland, Ohio, U.S.A.

Gary H. Bernstein
Department of Electrical
    Engineering
University of Notre Dame
Notre Dame, Indiana, U.S.A.

Liang-Yu Chen
OAI/NASA Glenn Research
    Center
Cleveland, Ohio, U.S.A.

Todd Christenson
HT MicroAnalytical Inc.
Albuquerque, New Mexico,
    U.S.A.

Adam L. Cohen
Microfabrica Inc.
Burbank, California, U.S.A.

Laura J. Evans
NASA Glenn Research Center
Cleveland, Ohio, U.S.A.

Mohamed Gad-el-Hak
Department of Mechanical
    Engineering
Virginia Commonwealth
    University
Richmond, Virginia, U.S.A.

Holly V. Goodson
Department of Chemistry and
    Biochemistry
University of Notre Dame
Notre Dame, Indiana, U.S.A.

Gary W. Hunter
NASA Glenn Research Center
Cleveland, Ohio, U.S.A.

Jaesung Jang
School of Electrical and Computer
    Engineering
Purdue University
West Lafayette, Indiana, U.S.A.

Guangyao Jia
Department of Mechanical and
    Aerospace Engineering
University of California, Irvine
Irvine, California, U.S.A.

Ezekiel J. J. Kruglick
Microfabrica Inc.
Burbank, California, U.S.A.

Sang-Youp Lee
School of Veterinary Medicine
Purdue University
West Lafayette, Indiana, U.S.A.

Jih-Fen Lei
NASA Glenn Research Center
Cleveland, Ohio, U.S.A.

Chung-Chiun Liu
Electronics Design Center
Case Western Reserve
    University
Cleveland, Ohio, U.S.A.

Marc J. Madou
Department of Mechanical and
    Aerospace Engineering
University of California, Irvine
Irvine, California, U.S.A.

Darby B. Makel
Makel Engineering, Inc.
Chico, California, U.S.A.

Mehran Mehregany
Electrical Engineering and
    Computer Science
    Department
Case Western Reserve University
Cleveland, Ohio, U.S.A.

Jill A. Miwa
National Institute
    of Scientific Research
University of Quebec
Varennes, Quebec, Canada

Robert S. Okojie
NASA Glenn Research Center
Cleveland, Ohio, U.S.A.

Zoubeida Ounaies
Department of Aerospace
    Engineering
Texas A&M University
College Station, Texas, U.S.A.

Federico Rosei
National Institute
    of Scientific Research
University of Quebec
Varennes, Quebec, Canada

Gregory L. Snider
Department of Electrical
    Engineering
University of Notre Dame
Notre Dame, Indiana, U.S.A.

Steven T. Wereley
School of Mechanical Engineering
Purdue University
West Lafayette, Indiana, U.S.A.

Jennifer C. Xu
NASA Glenn Research Center
Cleveland, Ohio, U.S.A.

Christian A. Zorman
Electrical Engineering and
    Computer Science Department
Case Western Reserve University
Cleveland, Ohio, U.S.A.

Table of Contents

The farther backward you can look,
the farther forward you are likely to see.

(Sir Winston Leonard Spencer Churchill, 1874–1965)

Janus, Roman god of
gates, doorways and all
beginnings, gazing both
forward and backward.

As for the future, your task is not to foresee, but to enable it.

(Antoine-Marie-Roger de Saint-Exupéry, 1900–1944,
in Citadelle [*The Wisdom of the Sands*])

1

Introduction

Mohamed Gad-el-Hak
Virginia Commonwealth University

How many times when you are working on something frustratingly tiny, like your wife's wrist watch, have you said to yourself, "If I could only train an ant to do this!" What I would like to suggest is the possibility of training an ant to train a mite to do this. What are the possibilities of small but movable machines? They may or may not be useful, but they surely would be fun to make.

(From the talk "There's Plenty of Room at the Bottom," delivered by Richard P. Feynman at the annual meeting of the American Physical Society, Pasadena, California, December 1959.)

Toolmaking has always differentiated our species from all others on Earth. Aerodynamically correct wooden spears were carved by archaic *Homo sapiens* close to 400,000 years ago. Man builds things consistent with his size, typically in the range of two orders of magnitude larger or smaller than himself, as indicated in Figure 1.1. Though the extremes of length-scale are outside the range of this figure, man, at slightly more than 10^0 m, amazingly fits right in the middle of the smallest subatomic particle, which is

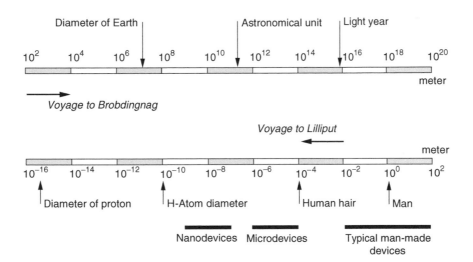

FIGURE 1.1 Scale of things, in meters. Lower scale continues in the upper bar from left to right. One meter is 10^6 microns, 10^9 nanometers, or 10^{10} Angstroms.

approximately 10^{-26} m, and the extent of the observable universe, which is of the order of 10^{26} m (15 billion light years); neither geocentric nor heliocentric, but rather egocentric universe. But humans have always striven to explore, build, and control the extremes of length and time scales. In the voyages to Lilliput and Brobdingnag of *Gulliver's Travels*, Jonathan Swift (1726) speculates on the remarkable possibilities which diminution or magnification of physical dimensions provides.[1] The Great Pyramid of Khufu was originally 147 m high when completed around 2600 B.C., while the Empire State Building constructed in 1931 is presently — after the addition of a television antenna mast in 1950 — 449 m high. At the other end of the spectrum of manmade artifacts, a dime is slightly less than 2 cm in diameter. Watchmakers have practiced the art of miniaturization since the 13th century. The invention of the microscope in the 17th century opened the way for direct observation of microbes and plant and animal cells. Smaller things were man-made in the latter half of the 20th century. The transistor — invented in 1947 — in today's integrated circuits has a size[2] of 0.18 micron (180 nanometers) in production and approaches 10 nm in research laboratories using electron beams. But what about the miniaturization of mechanical parts — machines — envisioned by Feynman (1961) in his legendary speech quoted above?

Manufacturing processes that can create extremely small machines have been developed in recent years (Angell et al., 1983; Gabriel et al., 1988, 1992; O'Connor, 1992; Gravesen et al., 1993; Bryzek et al., 1994; Gabriel, 1995; Ashley, 1996; Ho and Tai, 1996, 1998; Hogan, 1996; Ouellette, 1996, 2003; Paula, 1996; Robinson et al., 1996a, 1996b; Tien, 1997; Amato, 1998; Busch-Vishniac, 1998; Kovacs, 1998; Knight, 1999; Epstein, 2000; O'Connor and Hutchinson, 2000; Goldin et al., 2000; Chalmers, 2001; Tang and Lee, 2001; Nguyen and Wereley, 2002; Karniadakis and Beskok, 2002; Madou, 2002; DeGaspari, 2003; Ehrenman, 2004; Sharke, 2004; Stone et al., 2004; Squires and Quake, 2005). Electrostatic, magnetic, electromagnetic, pneumatic and thermal actuators, motors, valves, gears, cantilevers, diaphragms, and tweezers of less than 100 μm size have been fabricated. These have been used as sensors for pressure, temperature, mass flow, velocity, sound, and chemical composition, as actuators for linear and angular motions, and as simple components for complex systems, such as lab-on-a-chip, robots, micro-heat-engines and micro heat pumps (Lipkin, 1993; Garcia and Sniegowski, 1993, 1995; Sniegowski and Garcia, 1996; Epstein and Senturia, 1997; Epstein et al., 1997; Pekola et al., 2004; Squires and Quake, 2005).

Microelectromechanical systems (MEMS) refer to devices that have characteristic length of less than 1 mm but more than 1 micron, that combine electrical and mechanical components, and that are fabricated using integrated circuit batch-processing technologies. The books by Kovacs (1998) and Madou (2002) provide excellent sources for microfabrication technology. Current manufacturing techniques for MEMS include surface silicon micromachining; bulk silicon micromachining; lithography, electrodeposition, and plastic molding (or, in its original German, *Lithographie Galvanoformung Abformung, LIGA*); and electrodischarge machining (EDM). As indicated in Figure 1.1, MEMS are more than four orders of magnitude larger than the diameter of the hydrogen atom, but about four orders of magnitude smaller than the traditional manmade artifacts. Microdevices can have characteristic lengths smaller than the diameter of a human hair. Nanodevices (some say NEMS) further push the envelope of electromechanical miniaturization (Roco, 2001; Lemay et al., 2001; Feder, 2004).

The famed physicist Richard P. Feynman delivered a mere two, albeit profound, lectures[3] on electromechanical miniaturization: "There's Plenty of Room at the Bottom," quoted above, and "Infinitesimal Machinery," presented at the Jet Propulsion Laboratory on February 23, 1983. He could not see a lot of use for micromachines, lamenting in 1959 that "(small but movable machines) may or may not be useful, but they surely would be fun to make," and 24 years later said, "There is no use for these machines, so I still don't

---

[1]*Gulliver's Travels* were originally designed to form part of a satire on the abuse of human learning. At the heart of the story is a radical critique of human nature in which subtle ironic techniques work to part the reader from any comfortable preconceptions and challenge him to rethink from first principles his notions of man.

[2]The smallest feature on a microchip is defined by its smallest linewidth, which in turn is related to the wavelength of light employed in the basic lithographic process used to create the chip.

[3]Both talks have been reprinted in the *Journal of Microelectromechanical Systems*, vol. 1, no. 1, pp. 60–66, 1992, and vol. 2, no. 1, pp. 4–14, 1993.

understand why I'm fascinated by the question of making small machines with movable and controllable parts." Despite Feynman's demurring regarding the usefulness of small machines, MEMS are finding increased applications in a variety of industrial and medical fields with a potential worldwide market in the billions of dollars.

Accelerometers for automobile airbags, keyless entry systems, dense arrays of micromirrors for high-definition optical displays, scanning electron microscope tips to image single atoms, micro heat exchangers for cooling of electronic circuits, reactors for separating biological cells, blood analyzers, and pressure sensors for catheter tips are but a few of the current usages. Microducts are used in infrared detectors, diode lasers, miniature gas chromatographs, and high-frequency fluidic control systems. Micropumps are used for ink jet printing, environmental testing, and electronic cooling. Potential medical applications for small pumps include controlled delivery and monitoring of minute amount of medication, manufacturing of nanoliters of chemicals, and development of artificial pancreas. The much sought-after lab-on-a-chip is promising to automate biology and chemistry to the same extent the integrated circuit has allowed large-scale automation of computation. Global funding for micro- and nanotechnology research and development quintupled from $432 million in 1997 to $2.2 billion in 2002. In 2004, the U.S. National Nanotechnology Initiative had a budget of close to $1 billion, and the worldwide investment in nanotechnology exceeded $3.5 billion. In 10 to 15 years, it is estimated that micro- and nanotechnology markets will represent $340 billion per year in materials, $300 billion per year in electronics, and $180 billion per year in pharmaceuticals.

The multidisciplinary field has witnessed explosive growth during the past decade. Several new journals are dedicated to the science and technology of MEMS; for example *Journal of Microelectromechanical Systems, Journal of Micromechanics and Microengineering, Microscale Thermophysical Engineering, Microfluidics and Nanofluidics Journal, Nanotechnology Journal,* and *Journal of Nanoscience and Nanotechnology.* Numerous professional meetings are devoted to micromachines; for example Solid-State Sensor and Actuator Workshop, International Conference on Solid-State Sensors and Actuators (Transducers), Micro Electro Mechanical Systems Workshop, Micro Total Analysis Systems, and Eurosensors. Several web portals are dedicated to micro- and nanotechnology; for example, <http://www.smalltimes.com>, <http://www.emicronano.com>, <http://www.nanotechweb.org/>, and <http://www.peterindia.net/NanoTechnologyResources.html>.

The three-book *MEMS set* covers several aspects of microelectromechanical systems, or more broadly, the art and science of electromechanical miniaturization. MEMS design, fabrication, and application as well as the physical modeling of their materials, transport phenomena, and operations are all discussed. Chapters on the electrical, structural, fluidic, transport and control aspects of MEMS are included in the books. Other chapters cover existing and potential applications of microdevices in a variety of fields, including instrumentation and distributed control. Up-to-date new chapters in the areas of microscale hydrodynamics, lattice Boltzmann simulations, polymeric-based sensors and actuators, diagnostic tools, microactuators, nonlinear electrokinetic devices, and molecular self-assembly are included in the three books constituting the second edition of *The MEMS Handbook.* The 16 chapters in *MEMS: Introduction and Fundamentals* provide background and physical considerations, the 14 chapters in *MEMS: Design and Fabrication* discuss the design and fabrication of microdevices, and the 15 chapters in *MEMS: Applications* review some of the applications of microsensors and microactuators.

There are a total of 45 chapters written by the world's foremost authorities in this multidisciplinary subject. The 71 contributing authors come from Canada, China (Hong Kong), India, Israel, Italy, Korea, Sweden, Taiwan, and the United States, and are affiliated with academia, government, and industry. Without compromising rigorousness, the present text is designed for maximum readability by a broad audience having engineering or science background. As expected when several authors are involved, and despite the editor's best effort, the chapters of each book vary in length, depth, breadth, and writing style. The nature of the books — being handbooks and not encyclopedias — and the size limitation dictate the noninclusion of several important topics in the MEMS area of research and development.

Our objective is to provide a current overview of the fledgling discipline and its future developments for the benefit of working professionals and researchers. The three books will be useful guides and references to the explosive literature on MEMS and should provide the definitive word for the fundamentals and applications of microfabrication and microdevices. Glancing at each table of contents, the reader may rightly sense an overemphasis on the physics of microdevices. This is consistent with the strong conviction of the Editor-in-Chief that the MEMS technology is moving too fast relative to our understanding of the unconventional physics involved. This technology can certainly benefit from a solid foundation of the underlying fundamentals. If the physics is better understood, less expensive, and more efficient, microdevices can be designed, built, and operated for a variety of existing and yet-to-be-dreamed applications. Consistent with this philosophy, chapters on control theory, distributed control, and soft computing are included as the backbone of the futuristic idea of using colossal numbers of microsensors and microactuators in reactive control strategies aimed at taming turbulent flows to achieve substantial energy savings and performance improvements of vehicles and other manmade devices.

I shall leave you now for the many wonders of the small world you are about to encounter when navigating through the various chapters of these volumes. May your voyage to Lilliput be as exhilarating, enchanting, and enlightening as Lemuel Gulliver's travels into "Several Remote Nations of the World." *Hekinah degul!* Jonathan Swift may not have been a good biologist and his scaling laws were not as good as those of William Trimmer (see Chapter 2 of *MEMS: Introduction and Fundamentals*), but Swift most certainly was a magnificent storyteller. *Hnuy illa nyha majah Yahoo!*

References

Amato, I. (1998) "Formenting a Revolution, in Miniature," *Science* **282**, no. 5388, 16 October, pp. 402–405.

Angell, J.B., Terry, S.C., and Barth, P.W. (1983) "Silicon Micromechanical Devices," *Faraday Transactions I* **68**, pp. 744–748.

Ashley, S. (1996) "Getting a Microgrip in the Operating Room," *Mech. Eng.* **118**, September, pp. 91–93.

Bryzek, J., Peterson, K., and McCulley, W. (1994) "Micromachines on the March," *IEEE Spectrum* **31**, May, pp. 20–31.

Busch-Vishniac, I.J. (1998) "Trends in Electromechanical Transduction," *Phys. Today* **51**, July, pp. 28–34.

Chalmers, P. (2001) "Relay Races," *Mech. Eng.* **123**, January, pp. 66–68.

DeGaspari, J. (2003) "Mixing It Up," *Mech. Eng.* **125**, August, pp. 34–38.

Ehrenman, G. (2004) "Shrinking the Lab Down to Size," *Mech. Eng.* **126**, May, pp. 26–29.

Epstein, A.H. (2000) "The Inevitability of Small," *Aerospace Am.* **38**, March, pp. 30–37.

Epstein, A.H., and Senturia, S.D. (1997) "Macro Power from Micro Machinery," *Science* **276**, 23 May, p. 1211.

Epstein, A.H., Senturia, S.D., Al-Midani, O., Anathasuresh, G., Ayon, A., Breuer, K., Chen, K.-S., Ehrich, F.F., Esteve, E., Frechette, L., Gauba, G., Ghodssi, R., Groshenry, C., Jacobson, S.A., Kerrebrock, J.L., Lang, J.H., Lin, C.-C., London, A., Lopata, J., Mehra, A., Mur Miranda, J.O., Nagle, S., Orr, D.J., Piekos, E., Schmidt, M.A., Shirley, G., Spearing, S.M., Tan, C.S., Tzeng, Y.-S., and Waitz, I.A. (1997) "Micro-Heat Engines, Gas Turbines, and Rocket Engines — The MIT Microengine Project," AIAA Paper No. 97-1773, AIAA, Reston, Virginia.

Feder, T. (2004) "Scholars Probe Nanotechnology's Promise and Its Potential Problems," *Phys. Today* **57**, June, pp. 30–33.

Feynman, R.P. (1961) "There's Plenty of Room at the Bottom," in *Miniaturization*, H.D. Gilbert, ed., pp. 282–296, Reinhold Publishing, New York.

Gabriel, K.J. (1995) "Engineering Microscopic Machines," *Sci. Am.* **260**, September, pp. 150–153.

Gabriel, K.J., Jarvis, J., and Trimmer, W., eds. (1988) *Small Machines, Large Opportunities: A Report on the Emerging Field of Microdynamics, National Science Foundation*, published by AT&T Bell Laboratories, Murray Hill, New Jersey.

Gabriel, K.J., Tabata, O., Shimaoka, K., Sugiyama, S., and Fujita, H. (1992) "Surface-Normal Electrostatic/Pneumatic Actuator," in *Proc. IEEE Micro Electro Mechanical Systems '92*, pp. 128–131, 4–7 February, Travemünde, Germany.

Garcia, E.J., and Sniegowski, J.J. (1993) "The Design and Modelling of a Comb-Drive-Based Microengine for Mechanism Drive Applications," in *Proc. Seventh International Conference on Solid-State Sensors and Actuators (Transducers '93),* pp. 763–766, Yokohama, Japan, 7–10 June.

Garcia, E.J., and Sniegowski, J.J. (1995) "Surface Micromachined Microengine," *Sensor. Actuator. A* **48,** pp. 203–214.

Goldin, D.S., Venneri, S.L., and Noor, A.K. (2000) "The Great out of the Small," *Mech. Eng.* **122,** November, pp. 70–79.

Gravesen, P., Branebjerg, J., and Jensen, O.S. (1993) "Microfluidics — A Review," *J. Micromech. Microeng.* **3,** pp. 168–182.

Ho, C.-M., and Tai, Y.-C. (1996) "Review: MEMS and Its Applications for Flow Control," *J. Fluids Eng.* **118,** pp. 437–447.

Ho, C.-M., and Tai, Y.-C. (1998) "Micro–Electro–Mechanical Systems (MEMS) and Fluid Flows," *Annu. Rev. Fluid Mech.* **30,** pp. 579–612.

Hogan, H. (1996) "Invasion of the Micromachines," *New Sci.* **29,** June, pp. 28–33.

Karniadakis, G.E., and Beskok A. (2002) *Microflows: Fundamentals and Simulation,* Springer-Verlag, New York.

Knight, J. (1999) "Dust Mite's Dilemma," *New Sci.* **162,** no. 2180, 29 May, pp. 40–43.

Kovacs, G.T.A. (1998) *Micromachined Transducers Sourcebook,* McGraw-Hill, New York.

Lemay, S.G., Janssen, J.W., van den Hout, M., Mooij, M., Bronikowski, M.J., Willis, P.A., Smalley, R.E., Kouwenhoven, L.P., and Dekker, C. (2001) "Two-Dimensional Imaging of Electronic Wavefunctions in Carbon Nanotubes," *Nature* **412,** 9 August, pp. 617–620.

Lipkin, R. (1993) "Micro Steam Engine Makes Forceful Debut," *Sci. News* **144,** September, p. 197.

Madou, M. (2002) *Fundamentals of Microfabrication,* second edition, CRC Press, Boca Raton, Florida.

Nguyen, N.-T., and Wereley, S.T. (2002) *Fundamentals and Applications of Microfluidics,* Artech House, Norwood, Massachusetts.

O'Connor, L. (1992) "MEMS: Micromechanical Systems," *Mech. Eng.* **114,** February, pp. 40–47.

O'Connor, L., and Hutchinson, H. (2000) "Skyscrapers in a Microworld," *Mech. Eng.* **122,** March, pp. 64–67.

Ouellette, J. (1996) "MEMS: Mega Promise for Micro Devices," *Mech. Eng.* **118,** October, pp. 64–68.

Ouellette, J. (2003) "A New Wave of Microfluidic Devices," *Ind. Phys.* **9,** no. 4, pp. 14–17.

Paula, G. (1996) "MEMS Sensors Branch Out," *Aerospace Am.* **34,** September, pp. 26–32.

Pekola, J., Schoelkopf, R., and Ullom, J. (2004) "Cryogenics on a Chip," *Phys. Today* **57,** May, pp. 41–47.

Robinson, E.Y., Helvajian, H., and Jansen, S.W. (1996a) "Small and Smaller: The World of MNT," *Aerospace Am.* **34,** September, pp. 26–32.

Robinson, E.Y., Helvajian, H., and Jansen, S.W. (1996b) "Big Benefits from Tiny Technologies," *Aerospace Am.* **34,** October, pp. 38–43.

Roco, M.C. (2001) "A Frontier for Engineering," *Mech. Eng.* **123,** January, pp. 52–55.

Sharke, P. (2004) "Water, Paper, Glass," *Mech. Eng.* **126,** May, pp. 30–32.

Sniegowski, J.J., and Garcia, E.J. (1996) "Surface Micromachined Gear Trains Driven by an On-Chip Electrostatic Microengine," *IEEE Electron Device Lett.* **17,** July, p. 366.

Squires, T.M., and Quake, S.R. (2005) "Microfluidics: Fluid Physics at the Nanoliter Scale," *Rev. Mod. Phys.* **77,** pp. 977–1026.

Stone, H.A., Stroock, A.D., and Ajdari, A. (2004) "Engineering Flows in Small Devices: Microfluidics Toward a Lab-on-a-Chip," *Annu. Rev. Fluid Mech.* **36,** pp. 381–411.

Swift, J. (1726) *Gulliver's Travels,* 1840 reprinting of *Lemuel Gulliver's Travels into Several Remote Nations of the World,* Hayward & Moore, London, Great Britain.

Tang, W.C., and Lee, A.P. (2001) "Military Applications of Microsystems," *Ind. Phys.* **7,** February, pp. 26–29.

Tien, N.C. (1997) "Silicon Micromachined Thermal Sensors and Actuators," *Microscale Thermophys. Eng.* **1,** pp. 275–292.

2

Materials for Microelectromechanical Systems

Christian A. Zorman and
Mehran Mehregany
Case Western Reserve University

2.1 Introduction

Without question, one of the most exciting technological developments during the last decade of the 20th century was the field of microelectromechanical systems (MEMS). MEMS consists of microfabricated mechanical and electrical structures working in concert for perception and control of the local environment. It was no accident that the development of MEMS accelerated rapidly during the 1990s, as the field was able to take advantage of innovations created during the integrated circuit revolution of the 1960s–80s in terms of processes, equipment, and materials. A well-rounded understanding of MEMS requires a mature knowledge of the materials used to construct the devices, as the material properties of each component can influence device performance. Because the fabrication of MEMS structures often depends on the use of structural, sacrificial, and masking materials on a common substrate, issues related to etch selectivity, adhesion, microstructure, and a host of other properties are important design considerations. A discussion of the materials used in MEMS is really a discussion of the material systems used in MEMS, as the fabrication technologies rarely utilize a single material but rather a collection of materials, each serving a critical function. It is in this light that this chapter is constructed. The chapter does not attempt to present a comprehensive review of all materials used in MEMS because the list of materials is just too long. It does, however, detail a selection of material systems that illustrate the importance of viewing MEMS in terms of material systems as opposed to individual materials.

2.2   Single-Crystal Silicon

Use of silicon (Si) as a material for microfabricated sensors can be traced to 1954, when the first paper describing the piezoresistive effect in germanium (Ge) and Si was published [Smith, 1954]. The results of this study suggested that strain gauges made from these materials could be 10 to 20 times larger than those for conventional metal strain gauges, which eventually led to the commercial development of Si strain gauges in the late 1950s. Throughout the 1960s and early 1970s, techniques to mechanically and chemically micromachine Si substrates into miniature, flexible mechanical structures on which the strain gauges could be fabricated were developed and ultimately led to commercially viable high-volume production of Si-based pressure sensors in the mid 1970s. These lesser known developments in Si microfabrication technology happened concurrently with more popular developments in the areas of Si-based solid-state devices and integrated-circuit (IC) technologies that have revolutionized modern life. The conjoining of Si IC processing with Si micromachining techniques during the 1980s marked the advent of MEMS and positioned Si as the primary material for MEMS.

There is little question that Si is the most widely known semiconducting material in use today. Single-crystal Si has a diamond (cubic) crystal structure. It has an electronic band gap of 1.1 eV, and like many semiconducting materials, it can be doped with impurities to alter its conductivity. Phosphorus (P) is a common dopant for n-type Si and boron (B) is commonly used to produce p-type Si. A solid-phase oxide (SiO_2) that is chemically stable under most conditions can readily be grown on Si surfaces. Mechanically, Si is a brittle material with a Young's modulus of about 190 GPa, a value that is comparable to steel (210 GPa). Being among the most abundant elements on earth, Si can be refined readily from sand to produce electronic-grade material. Mature industrial processes exist for the low-cost production of single-crystal Si wafered substrates that have large surface areas (>8 in diameter) and very low defect densities.

For MEMS applications, single-crystal Si serves several key functions. Single-crystal Si is perhaps the most versatile material for bulk micromachining, owing to the availability of well-characterized anisotropic etches and etch-mask materials. For surface micromachining applications, single-crystal Si substrates are used as mechanical platforms on which device structures are fabricated, whether they are made from Si or other materials. In the case of Si-based integrated MEMS devices, single-crystal Si is the primary electronic material from which the IC devices are fabricated.

Bulk micromachining of Si uses wet and dry etching techniques in conjunction with etch masks and etch stops to sculpt micromechanical devices from the Si substrate. From the materials perspective, two key capabilities make bulk micromachining a viable technology: (1) the availability of anisotropic etchants such as ethylene–diamine pyrocatecol (EDP) and potassium hydroxide (KOH), which preferentially etch single-crystal Si along select crystal planes, and (2) the availability of Si-compatible etch-mask and etch-stop materials that can be used in conjunction with the etch chemistries to protect select regions of the substrate from removal.

One of the most important characteristics of etching is the directionality (or profile) of the etching process. If the etch rate in all directions is equal, the process is said to be *isotropic*. By comparison, etch processes that are *anisotropic* generally have etch rates perpendicular to the wafer surface that are much larger than the lateral etch rates. It should be noted that an anisotropic sidewall profile could also be produced in virtually any Si substrate by deep reactive ion etching, ion beam milling, or laser drilling.

Isotropic etching of a semiconductor in liquid reagents is commonly used for removal of work-damaged surfaces, creation of structures in single-crystal slices, and patterning single-crystal or polycrystalline semiconductor films. For isotropic etching of Si, the most commonly used etchants are mixtures of hydrofluoric (HF) and nitric (HNO_3) acid in water or acetic acid (CH_3COOH), usually called the HNA etching system.

Anisotropic Si etchants attack the (100) and (110) crystal planes significantly faster than the (111) crystal planes. For example, the (100)–to–(111) etch-rate ratio is about 400:1 for a typical KOH/water etch solution. Silicon dioxide (SiO$_2$), silicon nitride (Si$_3$N$_4$), and some metallic thin films (e.g., Cr, Au) provide good etch masks for most Si anisotropic etchants. In structures requiring long etching times in KOH, Si$_3$N$_4$ is the preferred masking material due to its chemical durability.

In terms of etch stops, heavily B-doped Si ($>7 \times 10^{19}$/cm^3), commonly referred to as a p+ etch stop, is effective for some etch chemistries. Fundamentally, etching is a charge transfer process, with etch rates dependent on dopant type and concentration. Highly doped material might be expected to exhibit higher etch rates because of the greater availability of mobile carriers. This is true for isotropic etchants such as HNA, where typical etch rates are 1 to 3 mm/min for p- or n-type dopant concentrations greater than 10^{18}/cm^3 and essentially zero for concentrations less than 10^{17}/cm^3. On the other hand, anisotropic etchants such as EDP and KOH exhibit a much different preferential etching behavior. Si that is heavily doped with B ($>7 \times 10^{19}$/cm^3) etches at a rate that is about 5 to 100 times slower than undoped Si when etched in KOH and 250 times slower when etched in EDP. Etch stops formed by the p+ technique are often less than 10 μm thick, as the B doping is often done by diffusion. Using high diffusion temperatures (e.g., 1175°C) and long diffusion times (e.g., 15 to 20 hours), thick (~20 μm) p+ etch stop layers can be created. It is also possible to create a p+ etch stop below the Si surface using ion implantation; however, the implant depth is limited to a few microns and a high-energy/high-current ion accelerator is required for implantation. While techniques are available to grow a B-doped Si epitaxial layer on top of a p+ etch stop to increase the thickness of the final structure, this is seldom utilized due to the expense of the epitaxial process step.

Due to the high concentration of B, p+ Si has a high density of defects. These defects are generated as a result of stresses created in the Si lattice because B is a smaller atom than Si. Studies of p+ Si report that stress in the resultant films can either be tensile [Ding et al., 1990] or compressive [Maseeh and Senturia, 1990]. These variations may be due to postprocessing steps. For instance, thermal oxidation can significantly modify the residual stress distribution in the near-surface region of p+ Si films, thereby changing the overall stress in the film. In addition to the generation of crystalline defects, the high concentration of dopants in the p+ etch stops prevents the fabrication of electronic devices in these layers. Despite some of these shortcomings, the p+ etch-stop technique is widely used in Si bulk micromachining due to its effectiveness and simplicity.

A large number of dry etch processes are available to pattern single-crystal Si. The process spectrum ranges from physical etching via sputtering and ion milling to chemical plasma etching. Two processes, reactive ion etching (RIE) and reactive ion beam etching (RIBE), combine aspects of both physical and chemical etching. In general, dry etch processes utilize a plasma of ionized gases along with neutral particles to remove material from the etch surface. Details regarding the physical processes involved in dry etching can be found elsewhere [Wolfe and Tauber, 1999].

Reactive ion etching is the most commonly used dry etch process to pattern Si. In general, fluorinated compounds such as CF$_4$, SF$_6$, and NF$_3$ or chlorinated compounds such as CCl$_4$ or Cl$_2$ sometimes mixed with He, O$_2$ or H$_2$ are used. The RIE process is highly directional, thereby enabling direct pattern transfer from the masking material to the etched Si surface. The selection of masking material is dependent on the etch chemistry and the desired etch depth. For MEMS applications, photoresist and SiO$_2$ thin films are often used. Si etch rates in RIE processes are typically less than 1 mm/min, so dry etching is mostly used to pattern layers on the order of several microns in thickness. The plasmas selectively etch Si relative to Si$_3$N$_4$, or SiO$_2$, so these materials can be used as etch masks or etch-stop layers. Development of deep reactive ion etching processes has extended Si etch depths well beyond several hundred microns, thereby enabling a multitude of new designs for bulk micromachined structures.

2.3　Polysilicon

Without doubt the most common material system for the fabrication of surface micromachined MEMS devices utilizes polycrystalline Si (polysilicon) as the primary structural material, SiO$_2$ as the sacrificial material, and Si_3N_4 for electrical isolation of device structures. Heavy reliance on this material system stems in part from the fact these three materials find uses in the fabrication of ICs, and as a result, film deposition and etching technologies are readily and widely available. Like single-crystal Si, polysilicon can be doped during or after film deposition using standard IC processing techniques. SiO_2 can be grown or deposited over a broad temperature range (e.g., 200 to 1150°C) to meet various process and material requirements. SiO_2 is readily dissolvable in hydrofluoric acid (HF), an IC-compatible chemical, without etching the polysilicon structural material [Adams, 1988]. HF does not wet bare Si surfaces; as a result, it is automatically rejected from microscopic cavities between polysilicon layers after a SiO_2 sacrificial layer is completely dissolved.

For surface micromachined structures, polysilicon is an attractive material because it has mechanical properties comparable to single-crystal Si, because the required processing technology has been developed for IC applications, and because it is resistant to SiO_2 etchants. In other words, polysilicon surface micromachining leverages on the significant capital investment made by the IC industry in the important areas of film deposition, patterning, and material characterization.

For MEMS and IC applications, polysilicon thin films are commonly deposited by a process known as low-pressure chemical vapor deposition (LPCVD). This deposition technique was first commercialized in the mid-1970s [Rosler, 1977] and has since been a standard process in the microelectronics industry. The typical polysilicon LPCVD reactor (or furnace) is based on a hot-wall resistance-heated horizontal fused-silica tube design. The temperature of the wafers in the furnace is maintained by heating the tube using resistive heating elements. The furnaces are equipped with quartz boats that have closely spaced vertically oriented slots that hold the wafers. The close spacing requires that the deposition process be performed in the reaction-limited regime to obtain uniform deposition across each wafer surface. In the reaction-limited deposition regime, the deposition rate is determined by the reaction rate of the reacting species on the substrate surface, as opposed to the arrival rate of the reacting species to the surface (which is the diffusion-controlled regime). The relationship between the deposition rate and the substrate temperature in the reaction-limited regime is exponential; therefore, precise temperature control of the reaction chamber is required. Operating in the reaction-limited regime facilitates conformal deposition of the film over the substrate topography, an important aspect of multilayer surface micromachining. Commercial equipment is available to accommodate furnace loads exceeding 100 wafers.

Typical deposition conditions utilize temperatures from 580 to 650°C and pressures ranging from 100 to 400 mtorr. The most commonly used source gas is silane (SiH_4), which readily decomposes into Si on substrates heated to these temperatures. Gas flow rates depend on the tube diameter and other conditions. For processes performed at 630°C, the polysilicon deposition rate is about 100 Å/min. The gas inlets are typically at the load door end of the tube, with the outlet to the vacuum pump located at the opposite end. For door injection systems, depletion of the source gas occurs along the length of the tube. To keep the deposition rate uniform, a temperature gradient is maintained along the tube so that the increased deposition rate associated with higher substrate temperatures offsets the reduction due to gas depletion. Typical temperature gradients range from 5 to 15°C along the tube length. Some systems incorporate an injector inside the tube to allow for the additional supply of source gas to offset depletion effects. In this case, the temperature gradient along the tube is zero. This is an important modification, as the microstructure and physical properties of the deposited polysilicon are a function of the deposition temperature.

Polysilicon is made up of small single-crystal domains called *grains*, whose orientations and/or alignment vary with respect to each other. The roughness often observed on polysilicon surfaces is due to the granular nature of polysilicon. The microstructure of the as-deposited polysilicon is a function of the deposition conditions [Kamins, 1998]. For typical LPCVD processes (e.g., 100% SiH_4 source gas, 200 mtorr deposition pressure), the amorphous-to-polycrystalline transition temperature is about 570°C, with amorphous films deposited below this temperature (Figure 2.1) and polycrystalline films above this temperature (Figure 2.2). As the deposition temperature increases significantly above 570°C, the grain structure of the as-deposited polysilicon films changes dramatically. For example, at 600°C, the grains are very fine and equiaxed, while at 625°C, the grains are larger and have a columnar structure that is aligned

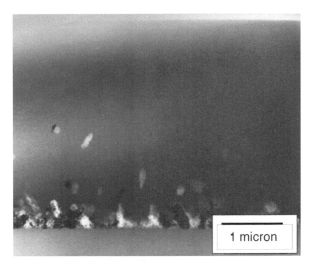

FIGURE 2.1 TEM micrograph of an amorphous Si film deposited at 570°C.

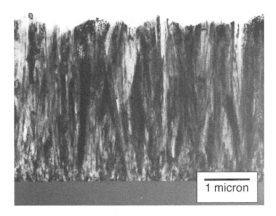

FIGURE 2.2 TEM micrograph of a polysilicon film deposited at 620°C.

perpendicular to the plane of the substrate [Kamins, 1998]. In general, the grain size tends to increase with film thickness across the entire range of deposition temperatures. As with grain size, the crystalline orientation of the polysilicon grains is dependent on the deposition temperature. For example, under standard LPCVD conditions (100% SiH_4, 200 mtorr), the crystal orientation of polysilicon is predominantly (110) for substrate temperatures between 600 and 650°C. In contrast, the (100) orientation is dominant for substrate temperatures between 650 and 700°C.

During the fabrication of micromechanical devices, polysilicon films typically undergo one or more high-temperature processing steps (e.g., doping, thermal oxidation, annealing) after deposition. These high-temperature steps can cause recrystallization of the polysilicon grains leading to a reorientation of the film and a significant increase in average grain size. Consequently, the polysilicon surface roughness increases with the increase in grain size, an undesirable outcome from a fabrication point of view because surface roughness limits pattern resolution. Smooth surfaces are desired for many mechanical structures, as defects associated with surface roughness can act as initiating points of structural failure. To address these concerns, chemical–mechanical polishing processes that reduce surface roughness with minimal film removal can be used.

Three phenomena influence the growth of polysilicon grains, namely strain-induced growth, grain-boundary growth, and impurity drag [Kamins, 1998]. If the dominant driving force for grain growth is the release of stored strain energy caused by such things as doping or mechanical deformation (wafer warpage), grain growth will increase linearly with increasing annealing time. To minimize the energy associated with grain boundaries, the gains tend to grow in a way that minimizes the grain boundary area. This driving force is inversely proportional to the radius of curvature of the grain boundary, and the growth rate is proportional to the square root of the annealing time. Heavy P-doping causes significant grain growth at temperatures as low as 900°C because P increases grain boundary mobility. If other impurities are incorporated in the gain boundaries, they may retard grain growth, which then results in the growth rate's being proportional to the cube root of the annealing time.

Thermal oxidation of polysilicon is carried out in a manner essentially identical to that of single-crystal Si. The oxidation rate of undoped polysilicon is typically between that of (100)- and (111)-oriented single-crystal Si. Heavily P-doped polysilicon oxidizes at a rate significantly higher than undoped polysilicon. However, this impurity-enhanced oxidation effect is smaller in polysilicon than in single-crystal Si. The effect is most noticeable at lower oxidation temperatures (<1000°C). Like single-crystal Si, oxidation of polysilicon can be modeled by using process simulation software. For first-order estimates, however, the oxidation rate of (100) Si can be used to estimate the oxidation rate of polysilicon.

The resistivity of polysilicon can be modified by impurity doping using the methods developed for single-crystal doping. Polysilicon doping can be achieved during deposition (called in situ doping) or after film deposition either by diffusion or ion implantation. In situ doping is achieved by adding reaction gases such as diborane (B_2H_6) and phosphine (PH_3) to the Si-containing source gas. The addition of dopants during the deposition process not only affects the conductivity of the as-deposited films, but also affects the deposition rate. Relative to the deposition of undoped polysilicon, the addition of P reduces the deposition rate, while the addition of B increases the deposition rate. In situ doping can be used to produce conductive films with uniform doping profiles through the film thickness without the need for high-temperature steps commonly associated with diffusion or ion implantation. Nonuniform doping through the thickness of a polysilicon film can lead to microstructural variations in the thickness direction that can result in stress gradients in the films and subsequent bending of released structural components. In addition, minimizing the maximum required temperature and duration of high-temperature processing steps is important for the fabrication of micromechanical components on wafers that contain temperature-sensitive layers.

The primary disadvantage of in situ doping is the complexity of the deposition process. The control of film thickness, deposition rate, and deposition uniformity is more complicated than the process used to deposit undoped polysilicon films, in part because a second gas with a different set of temperature- and pressure-related reaction parameters is included. Additionally, the cleanliness standards of the reactor are more demanding for the doped furnace. Therefore, many MEMS fabrication facilities use diffusion-based doping processes. Diffusion is an effective method for doping polysilicon films, especially for very heavy doping (e.g., resistivities of 10^{-4} Ω-cm) of thick (>2 µm) films. However, diffusion is a high-temperature process, typically from 900 to 1000°C. Therefore, fabrication processes that require long diffusion times to achieve uniform doping at significant depths may not be compatible with pre-MEMS, complementary metal-oxide-semiconductor (CMOS) integration schemes. Like in situ doping, diffusion processes must be performed properly to ensure that the dopant distribution through the film thickness is uniform, so that dopant-related variations in the mechanical properties through the film thickness are minimized. As will be discussed below, the use of doped oxide sacrificial layers relaxes some of the concerns associated with doping the film uniformly by diffusion because the sacrificial doped SiO_2 can also be used as a diffusion source. Phosphorous, which is the most commonly used dopant in polysilicon MEMS, diffuses significantly faster in polysilicon than in single-crystal Si, due primarily to enhanced diffusion rates along grain boundaries. The diffusivity in polysilicon thin films (i.e., small equiaxed grains) is about 1×10^{12} cm²/s.

Ion implantation is also used to dope polysilicon films. The implantation energy is typically adjusted so that the peak of the concentration profile is near the midpoint of the film. When necessary, several implant steps are performed at various energies in order to distribute the dopant uniformly through the thickness of the film. A high-temperature anneal step is usually required to electrically activate the implanted dopant, as well as to repair implant-related damage in the polysilicon film. In general, the resistivity of implanted polysilicon films is not as low as films doped by diffusion. In addition, the need for specialized implantation equipment limits the use of this method in polysilicon MEMS.

The electrical properties of polysilicon depend strongly on the grain structure of the film. The grain boundaries provide a potential barrier to the moving charge carriers, thus affecting the conductivity of the films. For P-doped polysilicon, the resistivity decreases as the amount of P increases for concentrations up to about $1 \times 10^{21}/cm^3$. Above this value, the resistivity reaches a plateau of about $4 \times 10^4 \Omega$-cm after a 1000°C anneal. The maximum mobility for such a highly P-doped polysilicon is about 30 cm^2/Vs. Grain boundary and ionized impurity scattering are important factors limiting the mobility [Kamins, 1988].

The thermal conductivity of polysilicon is a strong function of the grain structure of the film [Kamins 1998]. For fine-grain films, the thermal conductivity is about 0.30 to 0.35 W/cm-K, which is about 20 to 25% of the single-crystal value. For thick films with large grains, the thermal conductivity ranges between 50 and 85% of the single-crystal value.

In general, thin films are generally under a state of stress commonly referred to as residual stress, and polysilicon is no exception. In polysilicon micromechanical structures, the residual stress in the films can greatly affect the performance of the device. Like the electrical and thermal properties of polysilicon, the as-deposited residual stress in polysilicon films depends on microstructure. In general, as-deposited polysilicon films have compressive residual stresses, although reports regarding polysilicon films with tensile stress can be found in the literature [Kim et al., 1998]. The highest compressive stresses are found in amorphous Si films and polysilicon films with a strong columnar (110) texture. For films with fine-grained microstructures, the stress tends to be tensile. For the same deposition conditions, thick polysilicon films tend to have lower residual stress values than thin films; this is especially true for films with a columnar microstructure. Annealing can be used to reduce the compressive stress in as-deposited polysilicon films. For polysilicon films doped with phosphorus by diffusion, a decrease in the magnitude of compressive stress has been correlated with grain growth [Kamins, 1998]. For polysilicon films deposited at 650°C, the compressive residual stress is typically on the order of 5×10^9 to 10×10^9 dyne/cm^2. However, these stresses can be reduced to less than 10^8 dyne/cm^2 by annealing the films at high temperature (1000°C) in a N$_2$ ambient [Guckel et al., 1985; Howe and Muller, 1983]. Compressive stresses in fine-grained polysilicon films deposited at 580°C (100-Å grain size) can be reduced from 1.5×10^{10} to less than 10^8 dyne/cm^2 by annealing above 1000°C, or even can be made to be tensile (5×10^9 dynes/cm^2) by annealing at temperatures between 650 and 850°C [Guckel et al., 1988]. Advances in the area of rapid thermal annealing (RTA) as applied to polysilicon indicate that RTA is a fast and effective method of stress reduction in polysilicon films. For polysilicon films deposited at 620°C with compressive stresses of about 340 MPa, a 10 sec anneal at 1100°C was sufficient to completely relieve the stress [Zhang et al., 1998].

A second approach called the multipoly process has been developed to address issues related to residual stress [Yang et al., 2000]. As the name implies, the multipoly process is a deposition method to produce a polysilicon-based multilayer structure where the composite has a predetermined stress level. The multilayer structure is comprised of alternating tensile and compressive polysilicon layers deposited sequentially. The overall stress of the composite is simply the superposition of the stress in each individual layer. The tensile layers consist of fine-grained polysilicon grown at a temperature of 570°C, while the compressive layers are made up of polysilicon deposited at 615°C and having a columnar microstructure. The overall stress in the composite film depends on the number of alternating layers and the thickness of each layer. With the proper set of parameters, a composite polysilicon film can be deposited with a near-zero residual stress. Moreover, despite the fact that the composite has a clearly changing microstructure through the thickness of the film, the stress gradient is also nearly zero. The clear advantage of the multipoly process is that stress reduction can be achieved without the need for high-temperature annealing, a considerable advantage for polysilicon MEMS processes with on-chip CMOS integration. A transmission electron microscopy (TEM) micrograph of a multipoly structure is shown in Figure 2.3.

Conventional techniques to deposit polysilicon films for MEMS applications utilize LPCVD systems with deposition rates that limit the maximum film thickness to roughly 5 μm. Many device designs, however, require thick structural layers that are not readily achievable using LPCVD processes. For these devices,

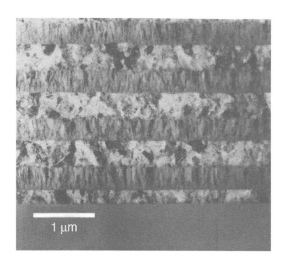

FIGURE 2.3   TEM micrograph of a polysilicon multilayer film created using the multipoly process.

wafer bonding and etchback techniques are often used to produce thick (>10 μm) single-crystal Si films on sacrificial substrate layers. There is, however, a deposition technique to produce thick polysilicon films on sacrificial substrates. These thick polysilicon films are called epi-poly films because epitaxial Si reactors are used to deposit them using a high-temperature process. Unlike conventional LPCVD polysilicon deposition processes, which have deposition rates of 100 Å/min, epi-poly processes have deposition rates on the order of 1 μm/min [Gennissen et al., 1997]. The high deposition rates are a result of the deposition conditions used — specifically, much higher substrate temperatures (>1000°C) and deposition pressures (>50 torr). The polysilicon films are usually deposited on SiO_2 sacrificial substrate layers and have been used in the fabrication of mechanical properties test structures [Lange et al., 1996; Gennissen et al., 1997; Greek et al., 1999], thermal actuators [Gennissen et al., 1997], electrostatically actuated accelerometers [Gennissen et al., 1997], and gryoscopes [Funk et al., 1999]. An LPCVD polysilicon seed layer is used to control nucleation, grain size, and surface roughness. In general, the microstructure and residual stress of epi-poly films is related to deposition conditions, with compressive films having a mixture of (110) and (311) grains [Lange et al., 1996; Greek et al., 1999] and tensile films having a random mix of (110), (100), (111), and (311) grains [Lange et al., 1996]. The Young's modulus of epi-poly measured from micromachined test structures is comparable to LPCVD polysilicon [Greek et al., 1999].

Porous Si is a "type" of Si finding applications in MEMS technology. Porous Si is made by room-temperature electrochemical etching of Si in HF. Under normal conditions, Si is not etched by HF, hence its widespread use as an etchant of sacrificial oxide in polysilicon surface micromachining. In an electrochemical circuit using an HF-based solution, however, positive charge carriers (holes) at the Si surface facilitate the exchange of F atoms with the H atoms terminating the Si surface bonds. The exchange continues with the exchange of subsurface bonds, leading to the eventual removal of the fluorinated Si. The quality of the etched surface is related to the density of holes at the surface, which is controlled by the applied current density. For high current densities, the density of holes is high and the etched surface is smooth. For low current density, the density of holes is low and they are clustered in highly localized regions associated with surface defects. The surface defects become enlarged by etching, leading to the formation of pores. Pore size and density are related to the type of Si used and the conditions of the electrochemical cell. Both single-crystal and polycrystalline Si can be converted to porous Si, with porosities of up to 80% possible.

The large surface-to-volume ratios make porous Si attractive for many MEMS applications. As one might expect, use of porous Si has been proposed for a number of gaseous and liquid applications including filter membranes and absorbing layers for chemical and mass sensing [Anderson et al., 1994]. The large surface-to-volume ratio also permits the use of porous Si as the starting material for the formation of thick thermal oxides, as the proper pore size can be selected to account for the volume expansion of the thermal oxide. When single-crystal substrates are used in the formation of porous Si films, the unetched material remains single crystalline, thus providing the appropriate template for epitaxial growth. It has been shown that CVD coatings will not penetrate the porous regions but, rather, overcoat the pores at the surface [Lang et al., 1995]. The formation of localized surface-micromachinable Si on insulator structures is possible by simply combining electrochemical etching, epitaxial growth, dry etching (to create access holes), and thermal oxidation. A third MEMS-related application is the direct use of porous Si as a sacrificial layer in polysilicon and single-crystalline Si surface micromachining. The process involves the electrical isolation of the structural Si layer either by the formation of pn-junctions through selective doping or by use of electrically insulating thin films [Lang, 1995]. In essence the formation of pores occurs only on electrically charged surfaces. A weak Si etchant aggressively attacks the porous regions with little damage to the structural Si layers. Porous Si may be an attractive option for micromachining processes that are chemically stable in HF but are tolerant of high-temperature processing steps.

With the possible exception of porous Si, all of these processes to prepare polysilicon for MEMS applications utilize substrate temperatures in excess of 570°C, either during film deposition or in subsequent stress-relieving annealing steps. Such high-temperature processing restricts the use of non-Si derivative materials, such as aluminum for metallization and polymers for sacrificial layers, both of which are relatively straightforward to deposit and pattern and would be of great benefit to polysilicon micromachining if they could be used throughout the process. Work in developing low-temperature deposition processes for polysilicon has focused on sputter deposition techniques [Abe and Reed, 1996; Honer and Kovacs, 2000]. Early work [Abe and Reed, 1996] emphasized the ability to deposit very smooth (25-Å roughness average) films at reasonable deposition rates (191 Å/min) and with low residual compressive stresses. The process involved DC magnitron sputtering from a Si target using an Ar sputtering gas, a chamber pressure of 5 mtorr, and a power of 100 W. The substrates consisted of thermally oxidized Si wafers. The authors reported that a postdeposition anneal at 700°C in N_2 for 2 hr was performed to crystallize the deposited film and perhaps lower the stress. A second group [Honer and Kovacs, 2000] sought to develop a polymer-friendly Si-based surface-micromachining process. The Si films were sputter-deposited on polyimide sacrificial layers. To improve the conductivity of the micromachined Si structures, the sputtered Si films were sandwiched between two TiW cladding layers. The device structures were released by etching the polyimide in a O_2 plasma. The processing step with the highest temperature was the polyimide cure, which was performed for 1 hr at 350°C. To test the robustness of the process, sputter-deposited Si microstructures were fabricated on substrates containing CMOS devices. As expected from thermal budget considerations, the authors reported no measurable degradation of device performance.

2.4 Silicon Dioxide

SiO_2 can be grown thermally on Si substrates as well as deposited using a variety of processes to satisfy a wide range of different requirements. In polysilicon surface micromachining, SiO_2 is used as a sacrificial material, as it can be dissolved easily using etchants that do not attack polysilicon. In a less prominent role, SiO_2 is used as an etch mask for dry etching of thick polysilicon films because it is chemically resistant to dry polysilicon etch chemistries.

The SiO_2 growth and deposition processes most widely used in polysilicon surface micromachining are thermal oxidation and LPCVD. Thermal oxidation of Si is performed at high temperatures (e.g., 900 to 1000°C) in the presence of oxygen or steam. Because thermal oxidation is a self-limiting process (i.e., the oxide growth rate decreases with increasing film thickness), the maximum practical film thickness that can be obtained is about 2 μm, which for many sacrificial applications is sufficient.

SiO_2 films for MEMS applications can also be deposited using an LPCVD process known as low-temperature oxidation (LTO). In general, LPCVD provides a means for depositing thick (>2 μm) SiO_2

films at temperatures much lower than thermal oxidation. Not only are LTO films deposited at low temperatures, but the films also have a higher etch rate in HF than thermal oxides, which results in significantly faster releases of polysilicon surface-micromachined devices. An advantage of the LPCVD processes is that dopant gases can be included in the flow of source gases in order to dope the as-deposited SiO_2 films. One such example is the incorporation of P to form phosphosilicate glass (PSG). PSG is formed using the same deposition process as LTO, with PH_3 added to dope the glass with a P content ranging from 2 to 8 wt%. PSG has an even higher etch rate in HF than LTO, further facilitating the release of polysilicon surface-micromachined components. PSG flows at high temperatures (e.g., 1000 to 1100°C), which can be exploited to create a smooth surface topography. Additionally, PSG layers sandwiching a polysilicon film can be used as a P-doping source, improving the uniformity of diffusion-based doping.

Phosphosilicate glass and LTO films are deposited in hot-wall low-pressure fused-silica reactors in a manner similar to the systems for polysilicon. Typical deposition rates are about 100 Å/min. Precursor gases include SiH_4 as a Si source, O_2 as an oxygen source, and in the case of PSG, PH_3 as a source of phosphorus. Because SiH_4 is pyrophoric (i.e., spontaneously combusts in the presence of O_2), door injection of the deposition gases would result in a large depletion of the gases at deposition temperatures of 400 to 500°C and nonuniform deposition along the tube. Therefore, the gases are introduced in the furnace through injectors distributed along the length of the tube. The wafers are placed vertically in caged boats; this is to ensure uniform gas transport to the wafers. In the caged boats, two wafers are placed back to back in each slot, thus minimizing the deposition of SiO_2 on the wafers' backs. The typical load of an LTO system is over 100 wafers.

Low-temperature oxidation and PSG films are typically deposited at temperatures of 425 to 450°C and pressures ranging from 200 to 400 mtorr. The low deposition temperatures result in LTO and PSG films that are slightly less dense than thermal oxides due to incorporation of hydrogen in the films. LTO films can, however, be densified by an annealing step at high temperature (1000°C). The low density of LTO and PSG films is partially responsible for the increased etch rate in HF, which makes them attractive sacrificial materials for polysilicon surface micromachining. LTO and PSG deposition processes are not typically conformal to nonplanar surfaces because the low substrate temperatures result in low surface migration of reacting species. Step coverage is, however, sufficient for many polysilicon surface-micromachining applications, although deposited films tend to thin at the bottom surfaces of deep trenches and therefore must be thoroughly characterized for each application.

The dissolution of the sacrificial SiO_2 to release free-standing structures is a critical step in polysilicon surface micromachining. Typically, 49% (by weight) HF is used for the release process. To pattern oxide films using wet chemistries, etching in buffered HF (28 ml 49% HF, 170 ml H_2O, 113 g NH4F), also known as buffered oxide etch (BOE), is common for large structures. A third wet etchant, known as P-etch, is traditionally used to selectively remove PSG over undoped oxide (e.g., to deglaze a wafer straight from a diffusion furnace).

Thermal SiO_2, LTO, and PSG are electrical insulators suitable for many MEMS applications. The dielectric constants of thermal oxide and LTO are 3.9 and 4.3, respectively. The dielectric strength of thermal SiO_2 is 1.1×10^6 V/cm, and for LTO it is about 80% that of thermal SiO_2 [Ghandhi, 1983]. Thermal SiO_2 is in compression with a stress level of about 3×10^9 dyne/cm² [Ghandhi, 1983]. For LTO, however, the as-deposited residual stress is tensile, with a magnitude of about 1 to 4×10^9 dyne/cm² [Ghandhi, 1983]. The addition of phosphorous to LTO (i.e., PSG) decreases the tensile residual stress to about 10^8 dyne/cm² for a phosphorus concentration of 8% [Pilskin, 1977]. These data are representative of oxide films deposited directly on Si substrates under typical conditions; however, the final value of the stress in an oxide film can be a strong function of the process parameters as well as any postprocessing steps.

One report documents the development of another low-pressure process, known as plasma-enhanced chemical vapor deposition (PECVD), for MEMS applications. The objective was to deposit low-stress, very thick (10 to 20 μm) SiO_2 films for insulating layers in micromachined gas turbine engines [Zhang et al., 2000]. PECVD was selected in part because it offers the possibility to deposit films of the desired thickness at a reasonable deposition rate. The process used a conventional parallel plate reactor with tetraethylorthosilicate (TEOS), a commonly used precursor in LPCVD processes, as the source gas.

As expected, the authors found that film stress is related to the concentration of dissolved gases in the film and that annealed films tend to suffer from cracking. By using a thin Si_3N_4 film in conjunction with the thick SiO_2 film, conditions were found where a low-stress, crack-free SiO_2 film could be produced.

Two other materials in the SiO_2 family are receiving increasing attention from MEMS fabricators, especially now that the material systems have expanded beyond conventional Si processing. The first of these is crystalline quartz. The chemical composition of quartz is SiO_2. Quartz is optically transparent and, like its amorphous counterpart, quartz is electrically insulating. However, the crystalline nature of quartz gives it piezoelectric properties that have been exploited for many years in electronic circuitry. Like single-crystal Si, quartz substrates are available as high-quality large-area wafers. Also like single-crystal Si, quartz can be bulk micromachined using anisotropic etchants based on heated HF and ammonium fluoride (NH_4F) solutions, albeit the structural shapes that can be etched into quartz do not resemble the shapes that can be etched into Si. A short review of the basics of quartz etching and its applications to the fabrication of a micromachined acceleration sensor can be found in Danel et al. (1990).

A second SiO_2-related material that has found utility in MEMS is spin-on-glass (SOG), which is used in thin-film form as a planarization dielectric material in IC processing. As the name implies, SOG is applied to a substrate by spin coating. The material is polymer based with a viscosity suitable for spin-coating, and once dispensed at room temperature on the spinning substrate, it is cured at elevated temperatures to form a solid thin film. Two publications illustrate the potential uses of SOG in MEMS. In the first example, SOG was developed as a thick-film sacrificial molding material to pattern thick polysilicon films [Yasseen et al., 1999]. The authors reported a process to produce SOG films that were 20 μm thick, complete with a chemical–mechanical polishing (CMP) procedure and etching techniques. The thick SOG films were patterned into molds that were filled with 10 μm thick LPCVD polysilicon films, planarized by selective CMP and subsequently dissolved in a $HCl:HF:H_2O$ wet etchant to reveal the patterned polysilicon structures. The cured SOG films were completely compatible with the polysilicon deposition process, indicating that SOG could be used to produce MEMS devices with extremely large gaps between structural layers. In the second example, high-aspect-ratio channel-plate microstrucures were fabricated from SOG [Liu et al., 1999]. The process required the use of molds to create the structures. Electroplated nickel (Ni) was used as the molding material, with Ni channel plate molds fabricated using a conventional Lithographie, Galvanoformung, Abformung (LIGA) process. The Ni molds were filled with SOG, and the sacrificial Ni molds were removed in a reverse electroplating process. In this case, the fabricated SOG structures were over 100 μm tall, essentially bulk micromachined structures fabricated using a sacrificial molding material system.

2.5 Silicon Nitride

Si_3N_4 is widely used in MEMS for electrical isolation, surface passivation, and etch masking and as a mechanical material. Two deposition methods are commonly used to deposit Si_3N_4 thin films: LPCVD and PECVD. PECVD Si_3N_4 is generally nonstoichiometric and may contain significant concentrations of hydrogen. Use of PECVD Si_3N_4 in micromachining applications is somewhat limited because its etch rate in HF can be high (e.g., often higher than that of thermally grown SiO_2) due to the porosity of the film. However, PECVD offers the potential to deposit nearly stress-free Si_3N_4 films, an attractive property for many MEMS applications, especially in the area of encapsulation and packaging. Unlike its PECVD counterpart, LPCVD Si_3N_4 is extremely resistant to chemical attack, thereby making it the material of choice for many Si bulk and surface micromachining applications. LPCVD Si_3N_4 is commonly used as an insulating layer to isolate device structures from the substrate and from other device structures because it is a good insulator with a resistivity of 10^{16} Ω-cm and a field breakdown limit of 10^7 V/cm.

The LPCVD Si_3N_4 films are deposited in horizontal furnaces similar to those used for polysilicon deposition. Typical deposition temperatures and pressures range between 700 and 900°C and 200 to 500 mtorr respectively. A typical deposition rate is about 30 Å/min. The standard source gases are dichlorosilane (SiH_2Cl_2) and ammonia (NH_3). SiH_2Cl_2 is used in place of SiH_4 because it produces films with a higher degree of thickness uniformity at the required deposition temperature and it allows the wafers to be spaced close together, thus increasing the number of wafers per furnace load. To produce stoichiometric Si_3N_4, a NH_3–to–SiH_2Cl_2 ratio of 10:1 is commonly used. The standard furnace configuration uses door injection of the source gases with a temperature gradient along the tube axis to accommodate for the gas depletion effects. LPCVD Si_3N_4 films deposited between 700 and 900°C are amorphous; therefore, the material properties do not vary significantly along the length of tube despite the temperature gradient. As with polysilicon deposition, a typical furnace can accommodate over 100 wafers. Because Si_3N_4 is deposited in the reaction-limited regime, film is deposited on both sides of each wafer with equal thickness.

The residual stress in stochiometric Si_3N_4 is large and tensile, with a magnitude of about 10^{10} dyne/cm^2. Such a large residual stress limits the practical thickness of a deposited Si_3N_4 film to a few thousand angstroms because thicker films tend to crack. Nevertheless, stoichiometric Si_3N_4 films have been used as mechanical support structures and electrical insulating layers in piezoresistive pressure sensors [Folkmer et al., 1995]. To reduce the residual stress, thus enabling the use of thick Si_3N_4 films for applications that require durable, chemically resistant membranes, nonstoichiometric silicon nitride (SixNy) films can be deposited by LPCVD. These films, often referred to as Si-rich or low-stress nitride, are intentionally deposited with an excess of Si by simply decreasing the NH_3–to–SiH_2Cl_2 ratio in the reaction furnace. For a NH_3–to–SiH_2Cl_2 ratio of 1:6 at a deposition temperature of 850°C and pressure of 500 mtorr, the as-deposited films are nearly stress free [Sekimoto et al., 1982]. The increase in Si content not only leads to a reduction in tensile stress but also decreases the etch rate of the film in HF. As a result, low-stress silicon nitride films have replaced stoichiometric Si_3N_4 in many MEMS applications and even have enabled the development of fabrication techniques that otherwise would not be feasible with stoichiometric Si_3N_4. For example, low-stress silicon nitride has been successfully used as a structural material in a surface micromachining process that uses polysilicon as the sacrificial material [Monk et al., 1993]. In this case, Si anisotropic etchants such as KOH and EDP were used for dissolving the sacrificial polysilicon. A second low-stress nitride surface micromachining process used PSG as a sacrificial layer, which was removed using a HF-based solution [French et al., 1997]. Of course, wide use of Si_3N_4 as a MEMS material is restricted by its dielectric properties; however, its Young's modulus (146 GPa) is on par with Si (~190 GPa), making it an attractive material for mechanical components.

The essential interactions among substrate, electrical isolation layer, sacrificial layers, and structural layers are best illustrated by examining the critical steps in a multilevel surface micromachining process. The example used here (shown in Figure 2.4) is the fabrication of a Si micromotor using a technique called the rapid prototyping process. The rapid prototyping process utilizes three deposition and three photolithography steps to implement flange-bearing side-drive micromotors such as in the SEM of Figure 2.5. The device consists of heavily P-doped LPCVD polysilicon structural components deposited on a Si wafer using LTO both as a sacrificial layer and as an electrical isolation layer. Initially, a 2.4 µm thick LTO film is deposited on the Si substrate. A 2 µm thick doped polysilicon layer is then deposited on the LTO film. Photolithography and RIE steps are then performed to define the rotor, stator, and rotor–stator gap. To fabricate the flange, a sacrificial mold is created by etching into the LTO film with an isotropic etchant and then partially oxidizing the polysilicon rotor and stator structures to form what is called the bearing clearance oxide. This oxidation step also forms the bottom of the bearing flange mold. A 1 to 2 µm thick, heavily doped polysilicon film is then deposited and patterned by photolithography and RIE to form the bearing. At this point, the structural components of the micromotor are completely formed, and all that remains is to release the rotor by etching the sacrificial oxide in HF and performing an appropriate drying procedure (detailed later in this chapter). In this example, the LTO film serves three purposes: it is the sacrificial underlayer for the free-spinning rotor; it comprises part of the flange mold; and it serves as an insulating anchor for the stators and bearing post. Likewise, the thermal oxide serves as a mold and electrical isolation layer. The material properties of LTO and thermal oxide allow for these films to be used as they are in the rapid prototyping process, thus enabling the fabrication of multilayer structures with a minimum of processing steps.

Without question, SiO_2 is an excellent sacrificial material for polysilicon surface micromachining; however, other materials could also be used. In terms of chemical properties, aluminum (Al) would

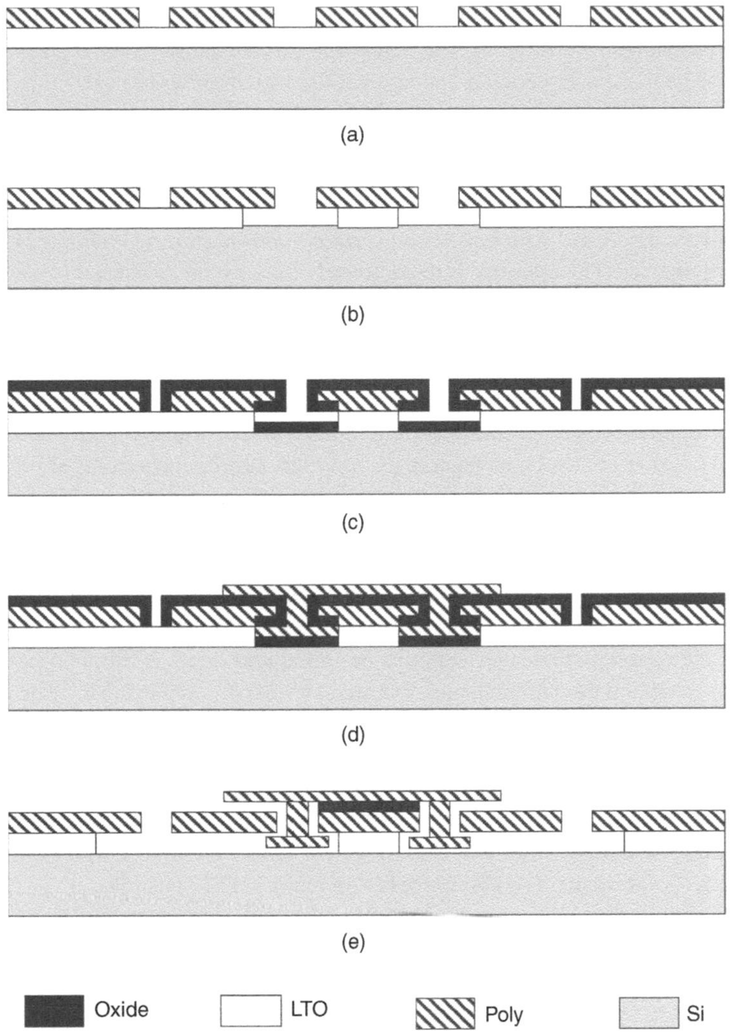

FIGURE 2.4 Cross-sectional schematics of the rapid prototyping process used to fabricate polysilicon micromotors by surface micromachining: (a) after the rotor–stator etch, (b) after the flange mold etch, (c) after the bearing clearance oxidation step, (d) after the bearing etch, and (e) after the release step.

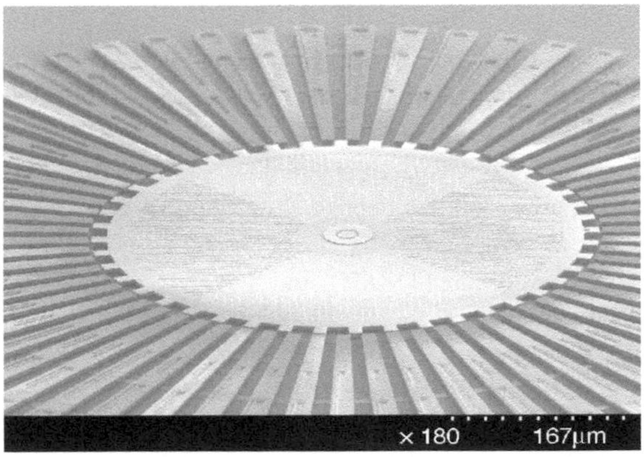

FIGURE 2.5 SEM micrograph of a polysilicon micromotor fabricated using the rapid prototyping process.

certainly be a satisfactory candidate as a sacrificial layer, as it can be dissolved in acidic-based Al etchants that do not etch polysilicon. However, the thermal properties tell a different story. LPCVD polysilicon is often deposited at temperatures between 580 and 630°C, which are excessively close to the Al melting temperature at the deposition pressure. Independent of the temperature incompatibility, polysilicon is often used as the gate material in MOS processes. As a result, for MEMS and IC processes that share the same LPCVD polysilicon furnace, as might be the case for an integrated MEMS process, putting Al-coated wafers in a polysilicon furnace would be inadvisable due to cross-contamination considerations.

The release process associated with polysilicon surface micromachining is simple in principle but can be complicated in practice. The objective is to completely dissolve the sacrificial oxide from beneath the freestanding components without etching the polysilicon structural components. The wafers or dies are simply immersed in the appropriate solution for a period of time sufficient to release all desired parts. This is done with various concentrations of electronic-grade HF, including BOE, as the etch rates of SiO_2 and polysilicon are significantly different. It has been observed, however, that during the HF release step, the mechanical properties of polysilicon including residual stress, Young's modulus, and fracture strain can be affected [Walker et al., 1991]. In general, the modulus and fracture strain of polysilicon decreases with increasing time of exposure to HF and with increasing HF concentration. This decrease in the modulus and fracture strain indicates a degradation of the film mechanical integrity. To minimize the HF release time, structures are designed with access holes and cuts of sufficient size to facilitate the flow of HF to the sacrificial oxide. In this manner, polysilicon structures can be released without appreciable degradation to film properties and hence device performance.

Following the HF release step, the devices must be rinsed and dried. A simple process includes rinses in deionized (DI) water and then in methanol, followed by a drying step using N_2. The primary difficulty with the wet release process is that surface tension forces, which are related to the surface properties of the material, tend to pull the micromechanical parts toward the substrate as the devices are immersed and pulled out of the solutions. Release processes that avoid the surface tension problem by using frozen alcohols that are sublimated at the final rinse step have been developed [Guckel et al., 1990]. Processes based on the use of supercritical fluids [Mulhern et al., 1993], such as CO_2 at 35°C and 1100 psi, to extinguish surface tension effects vanish are now commonplace in many MEMS facilities.

2.6   Germanium-Based Materials

Germanium (Ge) has a long history in the development of semiconducting materials, dating to the development of the earliest transistors. The same is true in the development of micromachined transducers and the early work on the piezoresistive effect in semiconducting materials [Smith, 1954]. Development of Ge for microelectronic devices might have continued if only a water-insoluble oxide could be formed on Ge surfaces. Nonetheless, there is a renewed interest in Ge for micromachined devices, especially for devices that require use of low-temperature processes.

Thin polycrystalline Ge (poly-Ge) films can be deposited by LPCVD at temperatures much lower than polysilicon, namely 325°C at a pressure of 300 mtorr on Si, Ge, or SiGe substrates [Li et al., 1999]. Ge does not nucleate on SiO_2 surfaces, which prohibits use of thermal oxides and LTO films as sacrificial substrate layers but does enable use of these films as sacrificial molds, as selective growth using SiO_2 masking films is possible. Residual stress in poly-Ge films deposited on Si substrates is about 125 MPa compressive, which can be reduced to nearly zero after a 30 s anneal at 600°C. Poly-Ge is essentially impervious to KOH, tetramethyl ammonium hydroxide (TMAH), and BOE, making it an ideal masking and etch-stop material in Si micromachining. In fact, the combination of low residual stress and inertness to Si anisotropic etches enables the fabrication of Ge membranes on Si substrates [Li et al., 1999]. The mechanical properties of poly-Ge are comparable with polysilicon, with a Young's modulus measured at 132 GPa and a fracture stress ranging between 1.5 GPa and 3.0 GPa [Franke et al., 1999]. Poly-Ge can also be used as a sacrificial layer. Typical wet etchants are based on mixtures of HNO_3, H_2O, and HCl and of H_2O, H_2O_2 and HCl as well as the RCA SC-1 cleaning solution. These mixtures do not etch Si, SiO_2, Si_3N_4, or SixNy, thereby enabling the use of poly-Ge as a sacrificial substrate layer in polysilicon surface micro-machining. Using the above-mentioned techniques, poly-Ge-based thermistors and Si_3N_4-membrane-based pressure sensors using poly-Ge sacrificial layers have been fabricated [Li et al., 1999]. In addition, poly-Ge microstructures, such as lateral resonant structures, have been fabricated on Si substrates containing CMOS structures with no process-related degradation in performance, thus showing the advantages of low deposition temperatures and compatible wet chemical etching techniques [Franke et al., 1999].

SiGe is an alloy of Si and Ge and has recently received attention for its usefulness in microelectronics; therefore, deposition technologies for SiGe thin films are readily available. While the requirements for SiGe-based electronic devices include single-crystal material, the requirements for MEMS are much less restrictive allowing for the use of polycrystalline material in many applications. Polycrystalline SiGe (poly-SiGe) films retain many properties comparable to polysilicon but can be deposited at lower sub-strate temperatures. Deposition processes include LPCVD, atmospheric pressure chemical vapor deposi-tion (APCVD), and RTCVD (rapid thermal CVD) using SiH_4 and GeH_4 as precursor gases. Deposition temperatures range from 450°C for LPCVD [Franke et al., 2000] to 625°C for RTCVD [Sedky et al., 1998]. The LPCVD processes can be performed in horizontal furnace tubes similar in configuration and size to those used for the deposition of polyslicon films. In general, the deposition temperature is related to the concentration of Ge in the films, with higher Ge concentration resulting in lower deposition tempera-tures. Like polysilicon, poly-SiGe can be doped with B and P to modify its conductivity. In fact, it has been reported that as-deposited in situ B-doped poly-SiGe films have a resistivity of 1.8 mΩ-cm [Franke et al., 2000].

Poly-SiGe can be deposited on a number of sacrificial substrates, including SiO_2 [Sedky et al., 1998], PSG [Franke et al., 1999], and poly-Ge [Franke et al., 1999], which as already detailed can also be deposited at relatively low processing temperatures. For films rich in Ge, a thin polysilicon seed layer is sometimes used on SiO_2 surfaces, as Ge does not readily nucleate on oxide surfaces. Because poly-SiGe is an alloy, variations in film stoichiometry can result in changes in physical properties. For instance, attack of poly-SiGe by H_2O_2, a main component in some Ge etchants, becomes problematic for Ge concentra-tions over 70%. As with most CVD thin films, residual stress is dependent on the substrate used and the deposition conditions; however, for in situ B-doped films, the as-deposited stresses are quite low at 10 MPa compressive [Franke et al., 2000].

In many respects, fabrication of devices made from poly-SiGe thin films follows processing methods used in polysilicon micromachining as Si and Ge are quite compatible. The poly-SiGe/poly-Ge material system is particularly attractive for surface micromachining, as it is possible to use H_2O_2 as a release agent. It has been reported that in H_2O_2, poly-Ge etches at a rate of 0.4 mm/min, while poly-SiGe with Ge con-centrations below 80% have no observable etch rate after 40 hr [Heck et al., 1999]. The ability to use H_2O_2 as a sacrificial etchant makes the poly-SiGe and poly-Ge combination perhaps the ideal material system for surface micromachining. To this end, several interesting devices have been fabricated from poly-SiGe. Due to the conformal nature of the poly-SiGe coating, poly-SiGe-based high-aspect-ratio structural ele-ments, such as gimbal/microactuator structures made using the Hexil process [Heck et al., 1999], can readily be fabricated. Capitalizing on the low substrate temperatures associated with the deposition of poly-SiGe and poly-Ge thin films, an integrated MEMS fabrication process on Si wafers has been demon-strated [Franke et al., 2000]. In this process, CMOS structures are first fabricated into standard Si wafers. Poly-SiGe thin-film mechanical structures are surface micromachined atop the CMOS devices using a poly-Ge sacrificial layer and H_2O_2 as an etchant. A significant advantage of this design lies in the fact that the MEMS structure is positioned directly above the CMOS structure, thus significantly reducing the par-asitic capacitance and contact resistance characteristic of interconnects associated with the side-by-side integration schemes often used in integrated polysilicon MEMS. Use of H_2O_2 as the sacrificial etchant means that no special protective layers are required to protect the underlying CMOS layer during release. Clearly, the unique properties of the poly-SiGe/poly-Ge material system, used in conjunction with the Si/SiO_2 material system, enable fabrication of integrated MEMS that minimizes interconnect distances and potentially increases device performance.

2.7   Metals

Metals are used in many different capacities ranging from hard etch masks and thin film conducting interconnects to structural elements in microsensors and microactuators. Metallic thin films can be deposited using a wide range of deposition techniques, the most common being evaporation, sputtering, CVD, and electroplating. Such a wide range of deposition methods makes metal thin films one of the most versatile classes of materials used in MEMS devices. A complete review would constitute a chapter in itself; the following illustrative examples are included to give the reader an idea of how different metal thin films can be used.

Aluminum (Al) is probably the most widely used metal in microfabricated devices. In MEMS, Al thin films can be used in conjunction with polymers such as polyimide because the films can be sputter-deposited at low temperatures. In most cases, Al is used as a structural layer; however, Al can be used as a sacrificial layer as well. The polyimide/aluminum combination as structural and sacrificial materials, respectively, has also been demonstrated to be effective for surface micromachining [Schmidt et al., 1988; Mahadevan et al., 1990]. In this case, acid-based Al etchants can be used to dissolve the Al sacrificial layer. A unique feature of this material system is that polyimide is significantly more compliant than polysilicon and silicon nitride (e.g., its elastic modulus is nearly 50 times smaller). At the same time, polyimide can withstand large strains (up to 100% for some chemistries) before fracture. Finally, because both polyimide and Al can be processed at low temperatures (e.g., below 400°C), this material system can be used subsequent to the fabrication of ICs on the wafer. A drawback of polyimide is its viscoelasticity (i.e., it creeps).

Tungsten (deposited by CVD) as a structural material and silicon dioxide as a sacrificial material have also been used for surface micromachining [Chen and MacDonald, 1991]. In this case, HF is used for removing the sacrificial oxide. In conjunction with high-aspect-ratio processes, nickel and copper are being used as structural layers with polyimide and other metals (e.g., chromium) as the sacrificial layers. The study of many of these material systems has been either limited or is in the preliminary stages; as a result, their benefits are yet to be determined.

Metal thin films are among the most versatile MEMS materials, as alloys of certain metallic elements exhibit a behavior known as the shape-memory effect. The shape-memory effect relies on the reversible transformation from a ductile martensite phase to a stiff austenite phase upon the application of heat. The reversible nature of this phase change allows the shape-memory effect to be used as an actuation mechanism. Moreover, it has been found that high forces and strains can be generated from shape-memory thin films at reasonable power inputs, thus enabling shape memory actuation to be used in MEMS-based microfluidic devices such as microvalves and micropumps. Alloys of Ti and Ni, collectively known as TiNi, are among the most popular shape-memory alloys owing to their high actuation work densities (reported to be up to $50 \, MJ/m^3$) and large bandwidth (up to 0.1 kHz) [Shih et al., 2001]. TiNi is also attractive because conventional sputtering techniques can be employed to deposit thin films of the alloy, as detailed in a recent report [Shih et al., 2001]. In this study, TiNi films deposited by two methods — cosputtering elemental Ti and Ni targets and cosputtering TiNi alloy and elemental Ti targets — were compared for use in microfabricated shape-memory actuators. In each case, the objective was to establish conditions so that films with the proper stoichiometry, and hence phase transition temperature, could be maintained. The sputtering tool was equipped with a substrate heater in order to deposit films on heated substrates as well as to anneal the films in vacuum after deposition. It was reported that cosputtering from TiNi and Ti targets produced better films than cosputtering from Ni and Ti targets, due to process variations related to roughening of the Ni target. The TiNi/Ti cosputtering process has been successfully used as an actuation material in a silicon spring-based microvalve [Hahm et al., 2000].

Use of thin-film metal alloys in magnetic actuator systems is yet another example of the versatility of metallic materials in MEMS. From a physical perspective, magnetic actuation is fundamentally the same in the microscopic and macroscopic domains, with the main difference being that process constraints limit the design options of microscale devices. Magnetic actuation in microdevices generally requires the magnetic layers to be relatively thick (tens to hundreds of microns), so as to create structures that can be used to generate magnetic fields of sufficient strength to generate the desired actuation. To this end, magnetic materials are often deposited by thick-film methods such as electroplating. The thicknesses of these layers often exceeds what can feasibly be patterned by etching, so plating is often conducted in microfabricated molds usually made from X-ray-sensitive materials such as polymethylmethacrylate (PMMA). The PMMA mold thickness can exceed several hundred microns, so X-rays are used as the exposure source. In some cases, a thin-film seed layer is deposited by sputtering or other conventional means before the plating process begins. At the completion of the plating process, the mold is dissolved, freeing the metallic component. This process, commonly known as LIGA, has been used to produce high-aspect-ratio structures such as microgears from NiFe magnetic alloys [Leith and Schwartz, 1999]. LIGA is not restricted to the creation of magnetic actuator structures and, in fact, has been used to make such structures as Ni fuel atomizers [Rajan et al., 1999]. In this application, Ni was selected for its desirable chemical, wear, and temperature properties, not its magnetic properties.

2.8   Silicon Carbide

Use of Si as a mechanical and electrical material has enabled the development of MEMS for a wide range of applications. Of course, use of MEMS is restricted by the physical properties of the material, which in the case of Si-based MEMS limits the devices to operating temperatures of about 200°C in low-wear and benign chemical environments. Therefore, alternate materials are necessary to extend the usefulness of MEMS to areas classified as harsh environments. In a broad sense, harsh environments include all conditions where use of Si is prohibited by its electrical, mechanical, and chemical properties. These would include high-temperature, high-radiation, high-wear, and highly acidic and basic chemical environments. To be a direct replacement for Si in such applications, the material would have to be a chemically inert, extremely hard, temperature-insensitive, micromachinable semiconductor. These requirements pose significant fabrication challenges, as micromachining requires the use of chemical and mechanical processes to remove unwanted material. In general, a class of wide bandgap semiconductors that includes silicon carbide (SiC) and diamond embodies the electrical, mechanical, and chemical properties required for many harsh environment applications, but until recently these materials found little usefulness in MEMS because the necessary micromachining processes did not exist. The following two sections review the development of SiC and diamond for MEMS applications.

SiC has long been recognized as a semiconductor with potential for use in high-temperature and high-power electronics. SiC is polymorphic, meaning that it exists in multiple crystalline structures, each sharing a common stoichiometry. SiC exists in three main polytypes: cubic, hexagonal, and rhombehedral. The cubic polytype, called 3C-SiC, has an electronic bandgap of 2.3 eV, which is over twice that of Si. Numerous hexagonal and rhombehedral polytypes have been identified, the two most common being the 4H-SiC and 6H-SiC hexagonal polytypes. The electronic bandgap of 4H- and 6H-SiC is even higher than 3C-SiC, being 2.9 and 3.2 eV respectively. SiC in general has a high thermal conductivity, ranging from 3.2 to 4.9 W/cm-K, and a high breakdown field (30×10^5 V/cm). SiC films can be doped to create n- and p-type material. The stiffness of SiC is quite large relative to Si; with measured Young's modulus values in the range of 300 to 700 GPa, it is very attractive for micromachined resonators and filters, as the resonant frequency increases with increasing modulus. SiC is not etched in any wet chemistries commonly used in Si micromachining. SiC can be etched in strong bases like KOH, but only at temperatures in excess of 600°C. SiC does not melt but rather sublimes at temperatures in excess of 1800°C. Single-crystal 4H- and 6H-SiC wafers are commercially available, although they are smaller (3 in diameter) and much more expensive than Si. With this list of properties, it is little wonder why SiC is being actively researched for MEMS applications.

SiC thin films can be grown or deposited using a number of different techniques. For high-quality single-crystal films, APCVD and LPCVD processes are most commonly employed. The high crystal quality is achieved by homoepitaxial growth of 4H- and 6H-SiC films on substrates of like crystal type. These processes usually employ dual precursors to supply Si and C, with the common sources being SiH_4 and

C_3H_8. Typical epitaxial growth temperatures range from 1500 to 1700°C. Epitaxial films with p- or n-type conductivity can be grown using such dopants as Al and B for p-type films and N or P for n-type films. In fact, doping with N is so effective at modifying the conductivity that growth of undoped SiC is virtually impossible because the concentrations of residual N in these deposition systems can be quite high. At these temperatures, the crystal quality of the epilayers is sufficient for the fabrication of electronic device structures.

Both APCVD and LPCVD can be used to deposit the only known polytype to grow epitaxially on a non-SiC substrate, namely 3C-SiC on Si. Heteroepitaxy is possible because 3C-SiC and Si have similar lattice structures. The growth process involves two key steps. The first step, called carbonization, involves converting the near-surface region of the Si substrate to 3C-SiC by simply exposing it to a propane/hydrogen mixture at a substrate temperature of about 1300°C. The carbonized layer forms a crystalline template on which a 3C-SiC film is grown by adding silane to the hydrogen/propane mix. A 20% lattice mismatch between Si and 3C-SiC results in the formation of crystalline defects in the 3C-SiC film. The density is highest in the carbonization layer, but it decreases with increasing thickness, although not to a level comparable with epitaxial 6H- and 4H-SiC films. Regardless, the fact that 3C-SiC does grow on Si substrates enables the use of Si bulk micromachining techniques to fabricating a host of SiC-based MEMS structures such as pressure sensors and resonant structures.

Polycrystalline SiC, hereafter referred to as poly-SiC, has proven to be a very versatile material for SiC MEMS. Unlike single-crystal versions of SiC, poly-SiC can be deposited on a variety of substrate types, including common surface micromachining materials such as polysilicon, SiO_2 and Si_3N_4. Moreover, poly-SiC can be deposited using a much wider set of processes than epitaxial films; LPCVD, APCVD, PECVD, and reactive sputtering have all been used to deposit poly-SiC films. The deposition of poly-SiC requires much lower substrate temperatures than epitaxial films, ranging from roughly 500 to 1200°C. The microstructure of poly-SiC films is temperature and substrate dependent [Wu et al., 1999]. In general, grain size increases with increasing temperature. For amorphous substrates such as SiO_2 and Si_3N_4, poly-SiC films tend to be randomly oriented with equiaxed grains, with larger grains deposited on SiO_2 substrates. In contrast, for oriented substrates such as polysilicon, the texture of the poly-SiC film matches that of the substrate as a result of grain-to-grain epitaxy [Zorman et al., 1996]. This variation in microstructure suggests that device performance can be tailored by selecting the proper substrate and deposition conditions.

Direct bulk micromachining of SiC is very difficult due to its outstanding chemical durability. Conventional wet chemical techniques are not effective; however, several electrochemical etch processes have been demonstrated. These techniques are selective to certain doping types, so dimensional control of the etched structures depends on the ability to form doped layers, which can only be formed by in situ or ion implantation processes, as solid source diffusion is not possible at reasonable processing temperatures. This constraint limits the geometrical complexity of fabricated devices. To fabricate thick (hundreds of microns), three-dimensional, high-aspect-ratio SiC structures, a molding technique has been developed [Rajan et al., 1999]. The molds are fabricated from Si substrates using deep reactive ion etching, a dry etch process that has revolutionized Si bulk micromachining. The micromachined Si molds are then filed with SiC using a combination of thin epitaxial and thick polycrystalline film CVD processes. The thin-film process, which produces 3C-SiC films with a featureless SiC/Si interface, is used to ensure that the molded structure has smooth outer surfaces. The mold-filling process coats all surfaces of the mold with a very thick SiC film. To remove the mold and free the SiC structure, the substrate is first mechanically polished to expose sections of the mold and then the substrate is immersed in a Si etchant to completely dissolve the mold. Because SiC is not attacked by Si etchants, the final SiC structure is released without the need of any special procedures. This process has been successful in the fabrication of solid SiC fuel atomizers, and a variant has been used to fabricate SiC structures in Si-based micro-gas-turbines [Lohner et al., 1999]. In both cases, the process capitalizes on the chemical inertness of SiC in conjunction with the reactivity of Si to create structures that could otherwise not be fabricated with existing technologies.

Although SiC cannot be etched using conventional wet etch techniques, thin SiC films can be patterned using conventional dry etching techniques. RIE processes using fluorinated compounds such as CHF_3 and

SF$_6$ combined with O$_2$ and sometimes with an inert gas or H$_2$ are used. The high oxygen content in these plasmas generally prohibits the use of photoresist as a masking material; therefore, hard masks made of metals such as Al and Ni are often used. RIE processes are generally effective patterning techniques; however, a phenomenon called micromasking, which results in the formation of etch-field grass, can sometimes be a problem. Nonetheless, RIE-based SiC surface-micromachining processes using polysilicon and SiO$_2$ sacrificial layers have been developed [Fleischman et al., 1996, 1998]. These processes are effective means to fabricate single-layer SiC structures, but multilayer structures are very difficult to fabricate because the etch rates of the sacrificial layers are much higher than the SiC structural layers. The lack of a robust etch stop makes critical dimensional control in the thickness direction unreliable, thus making RIE-based SiC multilayer processes impractical.

To address the materials compatibility issues facing RIE-based SiC surface micromachining in the development of a multilayer process, a micromolding process for SiC patterning on sacrificial layer substrates has been developed [Yasseen et al., 2000]. In essence, the micromolding technique is the thin film analog to the molding technique presented earlier. The cross-sectional schematic shown in Figure 2.6 illustrates the steps to fabricate a SiC lateral resonant structure. The micromolding process utilizes polysilicon and SiO$_2$ films as sacrificial molds, Si$_3$N$_4$ as an electrical insulator, and SiO$_2$ as a sacrificial substrate. These films are deposited and patterned by conventional methods, thus leveraging the well characterized and highly selective processes developed for polysilicon MEMS. Poly-SiC films are deposited into and onto the micromolds. Mechanical polishing with a diamond-based slurry is used to remove poly-SiC from atop the molds, then the appropriate etchant is used to dissolve the molds and sacrificial layers. An example of device structure fabricated using this method is shown in Figure 2.7. The

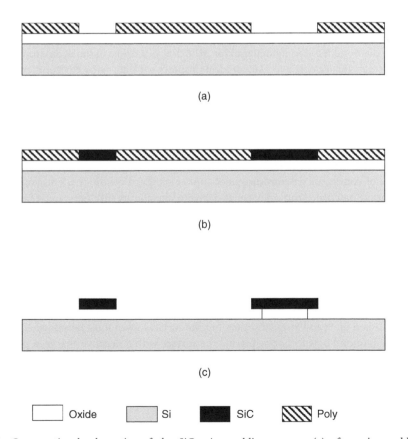

FIGURE 2.6 Cross-sectional schematics of the SiC micromolding process: (a) after micromold fabrication, (b) after SiC deposition and planarization, and (c) after mold and sacrificial layer release.

FIGURE 2.7   SEM micrograph of a SIC lateral resonant structure fabricated using the micromolding process.

micromolding method clearly utilizes the differences in chemical properties of the three materials in this system in a way that bypasses the difficulties associated with chemical etching of SiC.

2.9   Diamond

Along with SiC, diamond is a leading material for MEMS applications in harsh environments. It is commonly known as nature's hardest material, an ideal property for high-wear environments. Diamond has a very large electronic bandgap (5.5 eV) that is well suited for stable high-temperature operation. Diamond is a high-quality insulator with a dielectric constant of 5.5; however, it can be doped with B to create p-type conductivity. In general, diamond surfaces are chemically inert in the same environments as SiC. Diamond has a very high Young's modulus (10^{35} GPa), making it the ideal material for high-frequency micromachined resonators. Perhaps diamond's only disadvantage from a materials properties perspective is that a stable oxide cannot be grown on its surface. Thermal oxidation results in the formation of CO and CO_2, which, of course, are gaseous substances under standard conditions. This complicates the fabrication of diamond-based electronic devices as deposited insulating thin films must be used. Operation of diamond-based sensors at high temperatures requires the use of passivation coatings to protect the diamond structures from oxidation. These limitations, however, can be overcome and do not severely restrict the use of diamond films in harsh environment applications.

Unlike SiC, fabrication of diamond MEMS structures is restricted to polycrystalline and amorphous material. Although diamond epitaxy has been demonstrated, the epi films were grown on small, irregular, single-crystalline pieces because single-crystalline diamond wafers are not yet available. 3C-SiC thin films have been used to deposit highly oriented diamond films on Si substrates. Polycrystalline diamond films can be deposited on Si and SiO_2 substrates, but the surfaces often must be seeded either by damaging the surface with diamond powders or by biasing the surface with a negative charge, a process called bias enhanced nucleation. In general, diamond nucleates much more readily on Si surfaces than on SiO_2 surfaces, and this fact can be exploited to pattern diamond films into microstructures, such as a micromachined atomic force microscope (AFM) cantilever probe, using a selective growth process in conjunction with SiO_2 molding masks [Shibata et al., 2000]. As mentioned previously, diamond can be made insulating or semiconducting, and it is relatively straightforward to produce both types in polycrystalline diamond. This capability enables the fabrication of all-diamond microelectromechanical structures, thus eliminating the need for Si_3N_4 as an insulating layer.

Bulk micromachining of diamond is more difficult than SiC because electrochemical etching techniques have not been demonstrated. Using a strategy similar to that used in SiC, bulk micromachined diamond structures have been fabricated using bulk micromachined Si molds [Bjorkman et al., 1999]. The Si molds were fabricated using conventional micromachining techniques and filed with polycrystalline diamond deposited by HFCVD. The HFCVD process uses hydrogen as a carrier gas and methane as the carbon source. A hot tungsten wire is used to crack the methane into reactive species as well as to heat the substrate. The process was performed at a substrate temperature of 850 to 900°C and a pressure of 50 mtorr. The Si substrate was seeded with diamond particles suspended in an ethanol solution prior to deposition. After diamond deposition, the top surface of the diamond structure was polished using a hot iron plate. The material removal rate was reported to be around 2 mm/hr. After polishing, the Si mold was removed in a Si etchant, leaving behind the micromachined diamond structure. This process was used to produce all-diamond high-aspect-ratio capillary channels for microfluidic applications [Rangsten et al., 1999].

Surface micromachining of polycrystalline diamond thin films requires modifications of conventional micromachining practices to compensate for the nucleation and growth mechanisms of diamond thin films on sacrificial substrates. Early work in this area focused on developing thin-film patterning techniques. Conventional RIE methods are generally ineffective, so effort was focused on developing selective growth methods. One early method used selective seeding to form patterned templates for diamond nucleation. The selective seeding process was based on the lithographic patterning of photoresist mixed with diamond powders [Aslam and Schulz, 1995]. The diamond-loaded photoresist was deposited onto a Cr-coated Si wafer, exposed and then developed, leaving a patterned structure on the wafer surface. During the diamond deposition process, the photoresist rapidly evaporates, leaving behind the diamond seed particles in the desired structural shapes, which then serve as a template for diamond growth.

A second process has been developed for selective deposition directly on sacrificial substrate layers. This process combines a conventional diamond seeding technique with photolithographic patterning and etching to fabricate micromachined diamond structures using SiO_2 sacrificial layers [Ramesham, 1999]. The process can be executed in one of two approaches. The first approach begins with the formation of a SiO_2 layer by thermal oxidation on a Si wafer. The wafer is then seeded with diamond particles, coated with photoresist, and photolithographically patterned to form a mask for SiO_2 etching. Unmasked regions of the seeded SiO_2 film are then partially etched in BOE to form a surface unfavorable for diamond growth. The photoresist is then removed and a diamond film is selectively deposited. The second approach begins with an oxidized Si wafer, which is coated with photoresist. The resist is photolithographically patterned, and the wafer is then seeded, with the photoresist protecting select regions of the SiO_2 surface from the damage caused by the seeding process. The photoresist is removed, and selective diamond deposition is performed. In each case, once the diamond film is patterned, the structures can be released using conventional means. These techniques have been used to fabricate cantilever beams and bridge structures.

A third method to surface-micromachine polycrystalline diamond films follows the conventional approach of film deposition, dry etching, and release. The chemical inertness of diamond renders most conventional plasma chemistries useless for etching diamond films. Oxygen-based ion beam plasmas, however, can be used to etch diamond thin films [Yang et al., 1999]. The oxygen ion beam prohibits the use of photoresist masks, so hard masks made from metals such as Al are required. A simple ion-beam, etching-based, surface-micromachining process begins with the deposition of a Si_3N_4 film on a Si wafer and is followed by the deposition of a polysilicon sacrificial layer. The polysilicon layer is seeded with a diamond slurry, and a diamond film is deposited by HFCVD. To prepare the diamond film for etching, an Al masking film is deposited and patterned. The diamond films are then etched in the O_2 ion beam plasma, and the structures are released by etching the polysilicon with KOH. This process has been used to create lateral resonant structures; although the patterning process was successful, the devices were not operable because of a significant stress gradient in the film. With a greater understanding of the structure–property relationships of diamond thin films, such problems with surface micromachined structures should be solvable, thus enabling the successful fabrication of a new class of highly functional devices.

2.10   III–V Materials

Galium arsenide (GaAs), indium phosphide (InP), and other III–V compounds are attractive electronic materials for various types of sensors and optoelectronic devices. In general, III–V compounds have favorable piezoelectric and optoelectric properties, high piezoresistive constants, and wide electronic bandgaps (relative to Si). In addition, III–V materials can be deposited as ternary and quaternary alloys that have lattice constants closely matched to the binary compounds from which they are derived (e.g., $Al_xGa_{1-x}As$ and GaAs), thus permitting the fabrication of a wide variety of heterostructures that facilitate device performance. Although the III–V class of materials is quite large, this section of the chapter will focus on GaAs and InP for MEMS applications.

Crystalline GaAs has a zinc-blend crystal structure. It has an electronic bandgap of 1.4 eV, enabling GaAs electronic devices to function at temperatures as high as 350°C [Hjort et al., 1994]. High-quality single-crystal wafers are commercially available, as are well developed metallorganic chemical vapor deposition (MOCVD) and molecular beam epitaxy (MBE) growth processes for epitaxial layers of GaAs and its alloys. GaAs does not outperform Si in terms of mechanical properties; however, its stiffness and fracture toughness are still suitable for micromechanical devices. A favorable combination of mechanical and electrical properties makes GaAs attractive for certain MEMS applications.

Micromachining of GaAs is relatively straightforward, as many of its lattice-matched ternary and quaternary alloys have sufficiently varying chemical properties to allow their use as sacrificial layers. For example, the most common ternary alloy for GaAs is $Al_xGa_{1-x}As$. For values of $x \leqslant 0.5$, etchants containing mixtures of HF and H_2O etch $Al_xGa_{1-x}As$ without attacking GaAs. In contrast, etchants consisting of NH_4OH and H_2O_2 mixtures attack GaAs isotropically but do not etch $Al_xGa_{1-x}As$, thereby enabling the bulk micromachining of GaAs wafers with lattice-matched etch stops. An extensive review of III–V etch processes can be found in Hjort (1996). By taking advantage of the single-crystal heterostructures that can be formed on GaAs substrates, both surface micromachined and bulk micromachined devices can be fabricated from GaAs. The list of devices is widely varying and includes comb-drive lateral resonant structures [Hjort, 1996], pressure sensors [Fobelets et al., 1994; Dehe et al., 1995b], thermopile sensors [Dehe et al., 1995a], and Fabry–Perot detectors [Dehe et al., 1998].

Micromachining of InP closely resembles the techniques used for GaAs. Many of the properties of InP are similar to GaAs in terms of crystal structure, mechanical stiffness, and hardness; however, the optical properties of InP make it particularly attractive for micro-optomechanical devices to be used in the 1.3 to 1.55 µm wavelengths [Seassal et al., 1996]. Like GaAs, single-crystal wafers of InP are readily available. Ternary and quaternary lattice-matched alloys of InP include InGaAs, InAlAs, InGaAsP, and InGaAlAs compounds; like GaAs, some of these can be used as either etch stop and/or sacrificial layers, depending on the etch chemistry. For instance, InP structural layers deposited on In0.53Al0.47As sacrificial layers can be released using $C_6H_8O_7$:H_2O_2:H_2O etchants. At the same time, InP films and substrates can be etched in HCl:H_2O-based solutions with In0.53Ga0.47As films as etch stops. A comprehensive list of wet chemical etches for InP and related alloys is reviewed in Hjort (1996). InP-based micromachining techniques have been used to fabricate multi-air-gap filters [Leclerq et al., 1998], bridge structures [Seassal et al., 1996], and torsional membranes [Dehe et al., 1998] from InP and its related alloys.

2.11   Piezoelectric Materials

Piezoelectric materials play an important role in MEMS technology, mainly for mechanical actuation but also to a lesser extent for sensing applications. In a piezoelectric material, mechanical stress polarizes the material, which results in the production of an electric field. The effect also works in reverse; that is, an applied electric field acts to produce a mechanical strain. Many materials retain some sort of piezoelectric behavior, such as quartz, GaAs, and ZnO, to name a few. Recent work in MEMS has focused on the development of the compound lead zirconate titanate, $Pb(Zr_xTi_{1-x})O_3$ (PZT). PZT is attractive because it has high piezoelectric constants that lead to high mechanical transduction.

PZT can be deposited by a wide variety of methods including cosputtering, CVD, and sol–gel processing. Sol–gel processing has been receiving attention lately due to its being able to control the composition and the homogeneity of the deposited material over large surface areas. The sol–gel process uses a liquid precursor containing Pb, Ti, Zr, and O to create PZT solutions [Lee et al., 1996]. The solution is then deposited on the substrate using a spin-coating process. The substrates in this example consist of a Si wafer with a Pt/Ti/SiO$_2$ thin-film multilayer on its surface. The deposition process produces a PZT film in multi-layer fashion, with each layer consisting of a spin-coated layer that is dried at 110°C for 5 min and then heat treated at 600°C for 20 min. After building up the PZT layer to the desired thickness, the multilayer was heated at 600°C for up to 6 hr. Prior to this anneal, a PbO top layer was deposited on the PZT surface. A Au/Cr electrode was then sputter-deposited on the surface of the piezoelectric stack. This process was used to fabricate a PZT-based force sensor. Like Si, PZT films can be patterned using dry etch techniques based on chlorine chemistries, such as Cl$_2$/CCl$_4$, as well as ion beam milling using inert gases such as Ar.

2.12 Conclusions

The early development of MEMS can be attributed to the recognition of silicon as a mechanical material. The rapid expansion of MEMS over the last decade is due in part to the inclusion of new structural materials that have expanded the functionality of microfabricated devices beyond what is achievable in silicon. As shown by the examples in this chapter, the materials science of MEMS is not only about the structural layers, but also about the associated sacrificial and masking layers and how these layers interact during the fabrication process in order to realize the final device. In simple terms, MEMS is about material systems; therefore, analysis of what makes MEMS devices work (or, in many cases, not work) relies on a thorough understanding of this fact.

References

Abe, T., and Reed, M.L. (1996) "Low Strain Sputtered Polysilicon for Micromechanical Structures," in *Proc. 9th Int. Workshop on Microelectromechanical Systems*, 11–15 February, San Diego, pp. 258–62.

Adams, A.C. (1983) "Dielectric and Polysilicon Film Deposition," *VLSI Technology*, 2nd ed., McGraw-Hill, New York, pp. 93–129.

Alley, R.L., Cuan, C.J., Howe, R.T., and Komvopoulos, K. (1992) "The Effect of Release Etch Processing on Surface Microstructure Stiction," in *Technical Digest: Solid-State Sensor and Actuator Workshop*, 4–8 June, Hilton Head, SC, pp. 202–7.

Anderson, R., Muller, R.S., and Tobias, C.W. (1994) "Porous Polycrystalline Silicon: A New Material for MEMS," *J. MEMS* 3, pp. 10–18.

Aslam, M., and Schulz, D. (1995) "Technology of Diamond Microelectromechanical Systems," in *Proc. 8th Int. Conf. Solid-State Sensors and Actuators*, 25–29 June, Stockholm, pp. 222–24.

Bjorkman, H., Rangsten, P., Hollman, P., and Hjort, K. (1999) "Diamond Replicas from Microstructured Silicon Masters," *Sensor. Actuator. A* 73, pp. 24–29.

Chen, L.Y., and MacDonald, N. (1991) "A Selective CVD Tungsten Process for Micromotors," in *Technical Digest: 6th Int. Conf. on Solid-State Sensors and Actuators*, June, San Francisco, June 24–27, 1991, pp. 739–42.

Danel, J.S., Michel, F., and Delapierre, G. (1990) "Micromachining of Quartz and Its Application to an Acceleration Sensor," *Sensor. Actuator. A* 21–23, pp. 971–77.

Dehe, A., Fricke, K., and Hartnagel, H.L. (1995a) "Infrared Thermopile Sensor Based on AlGaAs-GaAs Micromachining," *Sensor. Actuator. A* 46–47, pp. 432–36.

Dehe, A., Fricke, K., Mutamba, K., and Hartnagel, H.L. (1995b) "A Piezoresistive GaAs Pressure Sensor with GaAs/AlGaAs Membrane Technology," *J. Micromech. Microeng.*, 5, pp. 139–42.

Dehe, A., Peerlings, J., Pfeiffer, J., Riemenschneider, R., Vogt, A., Streubel, K., Kunzel, H., Meissner, P., and Hartnagel, H.L. (1998) "III–V Compound Semiconductor Micromachined Actuators for Long Resonator Tunable Fabry–Perot Detectors," *Sensor. Actuator. A* 68, pp. 365–71.

Ding, X., Ko, W.H., and Mansour, J. (1990) "Residual and Mechanical Properties of Boron-Doped p+ Silicon Films," *Sensor. Actuator. A* **21–23**, pp. 866–71.

Fleischman, A.J., Roy, S., Zorman, C.A., and Mehregany, M. (1996) "Polycrystalline Silicon Carbide for Surface Micromachining," in *Proc. 9th Int. Workshop on Microelectromechanical Systems*, 11–15 February, San Diego, pp. 234–38.

Fleischman, A.J., Wei, X., Zorman, C.A., and Mehregany, M. (1998) "Surface Micromachining of Polycrystalline SiC Deposited on SiO$_2$ by APCVD," *Mater. Sci. Forum* **264–68**, pp. 885–88.

Fobelets, K., Vounckx, R., and Borghs, G. (1994) "A GaAs Pressure Sensor Based on Resonant Tunnelling Diodes," *J. Micromech. Microeng.* **4**, pp. 123–28.

Folkmer, B., Steiner, P., and Lang, W. (1995) "Silicon Nitride Membrane Sensors with Monocrystalline Transducers," *Sensor. Actuator. A* **51**, pp. 71–75.

Franke, A., Bilic, D., Chang, D.T., Jones, P.T., King, T.J., Howe, R.T., and Johnson, C.G. (1999) "Post-CMOS Integration of Germanium Microstructures," in *Proc. 12th Int. Conf. on Microelectromechanical Systems*, 17–21 January, Orlando, pp. 630–37.

Franke, A.E., Jiao, Y., Wu, M.T., King, T.J., and Howe, R.T. (2000) "Post-CMOS Modular Integration of Poly-SiGe Microstructures Using Poly-Ge Sacrificial Layers," in *Technical Digest: Solid-State Sensor and Actuator Workshop*, 4–8 June, Hilton Head, SC, pp. 18–21.

French, P.J., Sarro, P.M., Mallee, R., Fakkeldij, E.J.M., and Wolffenbuttel, R.F. (1997) "Optimization of a Low-Stress Silicon Nitride Process for Surface Micromachining Applications," *Sensor. Actuator. A* **58**, pp. 149–57.

Funk, K., Emmerich, H., Schilp, A., Offenberg, M., Neul, R., and Larmer, F. (1999) "A Surface Micromachined Silicon Gyroscope Using a Thick Polysilicon Layer," in *Proc. 12th Int. Conf. on Microelectromechanical Systems*, 17–21 January, Orlando, pp. 57–60.

Gennissen, P., Bartek, M., French, P.J., and Sarro, P.M. (1997) "Bipolar-Compatible Epitaxial Poly for Smart Sensors: Stress Minimization and Applications," *Sensor. Actuator. A* **62**, pp. 636–45.

Ghandhi, S.K. (1983) *VLSI Fabrication Principles: Silicon and Gallium Arsenide*, John Wiley & Sons, New York.

Greek, S., Ericson, F., Johansson, S., Furtsch, M., and Rump, A. (1999) "Mechanical Characterization of Thick Polysilicon Films: Young's Modulus and Fracture Strength Evaluated with Microstructures," *J. Micromech. Microeng.* **9**, pp. 245–51.

Guckel, H., Randazzo, T., and Burns, D.W. (1985) "A Simple Technique for the Determination of Mechanical Strain in Thin Films with Application to Polysilicon," *J. Appl. Phys.* **57**, pp. 1671–75.

Guckel, H., Burns, D.W., Visser, C.C.G., Tilmans, H.A.C., and Deroo, D. (1988) "Fine-Grained Polysilicon Films with Built-In Tensile Strain," *IEEE Trans. Electron. Devices* **ED-35**, pp. 800–1.

Guckel, H., Sniegowski, J.J., Christianson, T.R., and Raissi, F. (1990) "The Applications of Fine-Grained Polysilicon to Mechanically Resonant Transducers," *Sensor. Actuator. A* **21–23**, pp. 346–51.

Hahm, G., Kahn, H., Phillips, S.M., and Heuer, A.H. (2000) "Fully Microfabricated Silicon Spring Biased Shape Memory Actuated Microvalve," in *Technical Digest: Solid-State Sensor and Actuator Workshop*, 4–8 June, Hilton Head, SC, pp. 230–33.

Heck, J.M., Keller, C.G., Franke, A.E., Muller, L., King, T.-J., and Howe, R.T. (1999) "High Aspect Ratio Polysilicon–Germanium Microstructures," in *Proc. 10th Int. Conf. on Solid-State Sensors and Actuators*, 7–10 June, Sendai, Japan, pp. 328–34.

Hjort, K. (1996) "Sacrificial Etching of III–V Compounds for Micromechanical Devices," *J. Micromech. Microeng.* **6**, pp. 370–75.

Hjort, K., Soderkvist, J., and Schweitz, J.-A. (1994) "Galium Arsenide as a Mechanical Material," *J. Micromech. Microeng.* **4**, pp. 1–13.

Honer, K., and Kovacs, G.T.A. (2000) "Sputtered Silicon for Integrated MEMS Applications," in *Technical Digest: Solid-State Sensor and Actuator Workshop*, 4–8 June, Hilton Head, SC, pp. 308–11.

Howe, R.T., and Muller, R.S. (1983) "Stress in Polysilicon and Amorphous Silicon Thin Films," *J. Appl. Phys.* **54**, pp. 4674–75.

Kamins, T. (1998) *Polycrystalline Silicon for Integrated Circuits and Displays*, 2nd ed., Kluwer Academic, Berlin.

Kim, T.W., Gogoi, B., Goldman, K.G., McNeil, A.C., Rivette, N.J., Garling, S.E., and Koch, D.J. (1998) "Substrate and Annealing Influences on the Residual Stress of Polysilicon," in *Technical Digest: Solid-State Sensor and Actuator Workshop*, 8–11 June, Hilton Head, SC, pp. 237–40.

Lang, W., Steiner, P., and Sandmaier, H. (1995) "Porous Silicon: A Novel Material for Microsystems," *Sensor. Actuator. A* **51**, pp. 31–36.

Lange, P., Kirsten, M., Riethmuller, W., Wenk, B., Zwicker, G., Morante, J.R., Ericson, F., and Schweitz, J.A. (1996) "Thick Polycrystalline Silicon for Surface-Micromechanical Applications: Deposition, Structuring, and Mechanical Characterization," *Sensor. Actuator. A* **54**, pp. 674–78.

Leclerq, J., Ribas, R.P., Karam, J.M., and Viktorovitch, P. (1998) "III–V Micromachined Devices for Microsystems," *Microelectron. J.* **29**, pp. 613–19.

Lee, C., Itoh, T., and Suga, T. (1996) "Micromachined Piezoelectric Force Sensors Based on PZT Thin Films," *IEEE Trans. Ultrason. Ferroelectr. Freq. Control* **43**, pp. 553–59.

Leith, S.D., and Schwartz, D.T. (1999) "High-Rate Through-Mold Electrodeposition of Thick (>200 micron) NiFe MEMS Components with Uniform Composition," *J. MEMS* **8**, pp. 384–92.

Li, B., Xiong, B., Jiang, L., Zohar, Y., and Wong, M. (1999) "Germanium as a Versatile Material for Low-Temperature Micromachining," *J. MEMS* **8**, pp. 366–72.

Liu, R., Vasile, M.J., and Beebe, D.J. (1999) "The Fabrication of Nonplanar Spin-On Glass Microstructures," *J. MEMS* **8**, pp. 146–51.

Lohner, K., Chen, K.S., Ayon, A.A., and Spearing, M.S. (1999) "Microfabricated Silicon Carbide Microengine Structures," *Mater. Res. Soc. Symp. Proc.* **546**, pp. 85–90.

Mahadevan, R., Mehregany, M., and Gabriel, K.J. (1990) "Application of Electric Microactuators to Silicon Micromechanics," *Sensor. Actuator. A* **21–23**, pp. 219–25.

Maseeh, F., and Senturia, D.D. (1990) "Plastic Deformation of Highly Doped Silicon," *Sensor. Actuator. A* **21–22**, pp. 861–65.

Monk, D.J., Soane, D.S., and Howe, R.T. (1993) "Enhanced Removal of Sacrificial Layers for Silicon Surface Micromachining," in *Technical Digest: The 7th Int. Conf. on Solid-State Sensors and Actuators*, June, Yokohama, June 7–10, 1993, pp. 280–83.

Mulhern, G.T., Soane, D.S., and Howe, R.T. (1993) "Supercritical Carbon Dioxide Drying of Microstructures," in *Technical Digest: 7th Int. Conf. on Solid-State Sensors and Actuators*, June, Yokohama, June 7–10, 1993, pp. 296–99.

Pilskin, W.A. (1977) "Comparison of Properties of Dielectric Films Deposited by Various Methods," *J. Vac. Sci. Technol.* **21**, pp. 1064–81.

Rajan, N., Mehregany, M., Zorman, C.A., Stefanescu, S., and Kicher, T. (1999) "Fabrication and Testing of Micromachined Silicon Carbide and Nickel Fuel Atomizers for Gas Turbine Engines," *J. MEMS* **8**, pp. 251–57.

Ramesham, R. (1999) "Fabrication of Diamond Microstructures for Microelectromechanical Systems (MEMS) by a Surface Micromachining Process," *Thin Solid Films* **340**, pp. 1–6.

Rangsten, P., Bjorkman, H., and Hjort, K. (1999) "Microfluidic Components in Diamond," in *Proc. of the 10th Int. Conf. on Solid-State Sensors and Actuators*, 7–10 June, Sendai, Japan, pp. 190–93.

Rosler, R.S. (1977) "Low Pressure CVD Production Processes for Poly, Nitride, and Oxide," *Solid State Technol.* **20**, pp. 63–70.

Schmidt, M.A., Howe, R.T., Senturia, S.D., and Haritonidis, J.H. (1988) "Design and Calibration of a Microfabricated Floating-Element Shear-Stress Sensor," *Trans. Electron. Devices* **ED-35**, pp. 750–57.

Seassal, C., Leclercq, J.L., and Viktorovitch, P. (1996) "Fabrication of InP-Based Freestanding Microstructures by Selective Surface Micromachining," *J. Micromech. Microeng.* **6**, pp. 261–65.

Sedky, S., Fiorini, P., Caymax, M., Loreti, S., Baert, K., Hermans, L., and Mertens, R. (1998) "Structural and Mechanical Properties of Polycrystalline Silicon Germanium for Micromachining Applications," *J. MEMS* **7**, pp. 365–72.

Sekimoto, M., Yoshihara, H., and Ohkubo, T. (1982) "Silicon Nitride Single-Layer X-ray Mask," *J. Vac. Sci. Technol.* **21**, pp. 1017–21.

Shibata, T., Kitamoto, Y., Unno, K., and Makino, E. (2000) "Micromachining of Diamond Film for MEMS Applications," *J. MEMS* **9**, pp. 47–51.

Shih, C.L., Lai, B.K., Kahn, H., Phillips, S.M., and Heuer, A.H. (2001) "A Robust Co-Sputtering Fabrication Procedure for TiNi Shape Memory Alloys for MEMS," *J. MEMS* **10**, pp. 69–79.

Smith, C.S. (1954) "Piezoresistive Effect in Germanium and Silicon," *Phys. Rev.* **94**, pp. 1–10.

Walker, J.A., Gabriel, K.J., and Mehregany, M. (1991) "Mechanical Integrity of Polysilicon Films Exposed to Hydrofluoric Acid Solutions," *J. Electron. Mater.* **20**, pp. 665–70.

Wolfe, S., and Tauber, R. (1999) *Silicon Processing for the VLSI Era*, 2nd ed., Lattice Press, Sunset Beach, CA.

Wu, C.H., Zorman, C.A., and Mehregany, M. (1999) "Growth of Polycrystalline SIC Films on SiO_2 and Si_3N_4 by APCVD," *Thin Solid Films* **355–56**, pp. 179–83.

Yang, Y., Wang, X., Ren, C., Xie, J., Lu, P., and Wang, W. (1999) "Diamond Surface Micromachining Technology," *Diamond Relat. Mater.* **8**, pp. 1834–37.

Yang, J., Kahn, H., He, A.-Q., Phillips, S.M., and Heuer, A.H. (2000) "A New Technique for Producing Large-Area As-Deposited Zero-Stress LPCVD Polysilicon Films: The Multipoly Process," *J. MEMS* **9**, pp. 485–94.

Yasseen, A., Cawley, J.D., and Mehregany, M. (1999) "Thick Glass Film Technology for Polysilicon Surface Micromachining," *J. MEMS* **8**, pp. 172–79.

Yasseen, A., Wu, C.H., Zorman, C.A., and Mehregany, M. (2000) "Fabrication and Testing of Surface Micromachined Polycrystalline SiC Micromotors," *Electron. Device Lett.* **21**, pp. 164–66.

Zhang, X., Zhang, T.Y., Wong, M., and Zohar, Y. (1998) "Rapid Thermal Annealing of Polysilicon Thin Films," *J. MEMS* **7**, pp. 356–64.

Zhang, X., Ghodssi, R., Chen, K.S., Ayon, A.A., and Spearing, S.M. (2000) "Residual Stress Characterization of Thick PECVD TEOS Film for Power MEMS Applications," in *Technical Digest: Solid-State Sensor and Actuator Workshop*, 4–8 June, Hilton Head, SC, pp. 316–19.

Zorman, C.A., Roy, S., Wu, C.H., Fleischman, A.J., and Mehregany, M. (1996) "Characterization of Polycrystalline Silicon Carbide Films Grown by Atmospheric Pressure Chemical Vapor Deposition on Polycrystalline Silicon," *J. Mater. Res.* **13**, pp. 406–12.

For Further Information

A comprehensive review of polysilicon as a material for microelectronics and MEMS is presented in *Polycrystalline Silicon for Integrated Circuits and Displays*, 2nd ed., by Ted Kamins. The Materials Research Society holds an annual symposium on the materials science of MEMS at its fall meetings. The proceedings from these symposia have been published as volumes 546B, 605B and 657B of the Materials Research Society Symposium Proceedings. Several regularly published journals contain contributed and review papers concerning materials aspects of MEMS, including: (1) the *Journal of Microelectromechanical Systems*, (2) *Journal of Micromachining and Microengineering*, (3) *Sensors and Actuators*, and (4) *Sensors and Materials*. These journals are carried by most engineering and science libraries and may be accessible online.

3

MEMS Fabrication

Guangyao Jia and
Marc J. Madou
University of California, Irvine

3.1   Wet Bulk Micromachining: Introduction

In wet bulk micromachining, features are sculpted in the bulk of materials such as silicon, quartz, SiC, GaAs, InP, and Ge by orientation-independent (isotropic) or orientation-dependent (anisotropic) wet etchants. The technology employs pools of liquid as tools [Harris, 1976] rather than dry etching plasmas. A vast majority of wet bulk micromachining work is based on single-crystal silicon and glass as a companion material. There has been some work on quartz, crystalline Ge, SiC, and GaAs, and a minor amount on GaP and InP.

Wet bulk micromachining, along with surface micromachining, form the principal commercial Si micromachining tool sets used today. Micromolding — from a lithography-defined master — has only

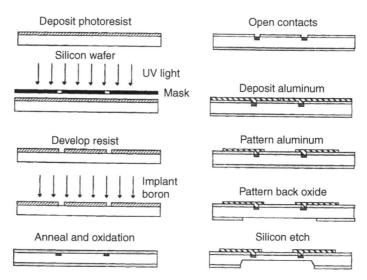

FIGURE 3.1   A wet bulk micromachining process is used to craft a membrane with piezoresistive elements. Silicon micromachining selectively thins the double-sided polished silicon wafer from a starting thickness of about 425 μm. A diaphragm having a typical thickness of 20 μm or less with precise lateral dimensions and vertical thickness control results.

more recently become commercially viable. A typical structure fashioned in a Si bulk micromachining process is shown in Figure 3.1. It was this type of piezoresistive membrane structure, a likely base for a pressure sensor, that demonstrated that batch fabrication of miniature components need not be limited to integrated circuits (ICs). This chapter's emphasis is on the wet etching process itself. Two other machining steps typically used in conjunction with wet bulk micromachining, additive processes and bonding processes, are covered in Madou (2002).

After a short historical note on wet bulk micromachining, we will begin with an introduction to the crystallography of single-crystal Si and a listing of its properties clarifying the importance of Si as a sensor material. Some empirical data on wet etching will be reviewed, and different models for anisotropic and isotropic etching behavior will follow. Etch stop techniques, which catapulted micromachining into an industrial manufacturing process, are then discussed. Subsequently, a discussion of problems associated with bulk micromachining, such as IC incompatibility, extensive real-estate usage, and issues involving corner compensation, are presented. We conclude with some example applications of wet bulk micromachining.

3.2   Historical Note

The earliest instance of wet etching of a substrate using a mask (wax) and etchants (acid-base) appears in the late 15th or early 16th century for decorating armor [Harris, 1976; Durant, 1957] (Inset 3.1). Engraving hand tools were not hard enough to work the armor, and more powerful acid-based processes took over. By the early 17th century, etching to decorate arms and armor was a well established process. Some pieces stemming from that period have been found in which the chemical milling was accurate to within 0.5 mm. The masking in traditional chemical milling was accomplished by cutting the maskant with a scribing tool and peeling off the maskant where etching was desired. Harris (1976) describes in detail the improvements that by the mid-1960s made this type of chemical milling a valuable and reliable method of manufacturing especially popular in the aerospace industry. The method enabled manufacturers to produce many parts more easily and cheaply than by other means and, in many cases, provided design and production configurations not previously possible. Through the introduction of photosensitive

INSET 3.1   Decorating armor.

masks by Niépce in 1822, chemical milling in combination with lithography became a reality, and a new level of tolerances were within reach. The major modern applications of lithography-based chemical milling are the manufacture of printed circuit boards, started during the Second World War, and by 1961 the fabrication of Si-integrated circuitry. Photochemical machining is also used for such precision parts as color television shadow masks, integrated circuit lead frames, light chopper and encoder discs, and decorative objects such as costume jewelry [Allen, 1986]. The geometry of a "cut" produced when etching silicon integrated circuits is similar to the chemical-milling cut of the aerospace industry, but the many orders of magnitude difference in size and depth of the cut account for a major difference in achievable accuracy. Accordingly, the tolerances for fashioning integrated circuitry are many orders of magnitude more precise than in the chemical milling industry.

In this chapter, we are concerned with lithography and chemical machining used in the IC industry and in microfabrication. A major difference between these two fields is in the aspect ratio (height-to-width ratio) of the features crafted. The IC industry deals with mostly very small, flat structures with aspect ratios of 1 to 2. In the microfabrication field, structures typically are somewhat larger, and aspect ratios might be as high as 400.

Isotropic etching has been used in silicon semiconductor processing since its introduction in the early 1950s. Representative work from that period is the impressive series of papers by Robbins and Schwartz [Robbins and Schwartz, 1959, 1960; Schwartz and Robbins, 1961, 1976] on chemical isotropic etching and Uhlir's paper on electrochemical isotropic etching [Uhlir, 1956]. The usual chemical isotropic etchant used for silicon was HF in combination with HNO_3 with or without acetic acid or water as diluent [Robbins and Schwartz, 1959, 1960; Schwartz and Robbins, 1961, 1976]. The early work on isotropic etching in an electrochemical cell (i.e., electropolishing) was carried out mostly in nonaqueous solutions to avoid black or red deposits that formed on the silicon surface in aqueous solutions [Halloas, 1971]. Turner showed that if a critical current density is exceeded, silicon can be electropolished in aqueous HF solutions without the formation of those deposits [Turner, 1958].

In the mid-1960s, Bell Telephone Laboratories started work on anisotropic Si etching in mixtures at first of KOH, water, and alcohol and later of KOH and water. The need for high-aspect-ratio cuts in silicon arose in the fabrication of dielectrically isolated structures in integrated circuits such as those for beam leads. Both chemical and electrochemical anisotropic etching methods were pursued [Stoller and Wolff, 1966; Stoller, 1970; Forster and Singleton, 1966; Kenney, 1967; Lepselter, 1966, 1967; Waggener, 1970; Kragness and Waggener, 1973; Waggener et al., 1967; Waggener et al., 1967; Bean and Runyan, 1977]. In the mid-1970s, a new surge of activity in anisotropic etching was associated with the work on V-groove and U-groove transistors [Rodgers et al., 1977; Rodgers et al., 1976; Ammar and Rodgers, 1980; Smith, 1954]. The first use of Si as a micromechanical element can be traced to a discovery and an idea from the mid-1950s and early 1960s respectively. The discovery was the large piezoresistance in Si and Ge by Smith in 1954 [Smith, 1954]. The idea stemmed from Pfann et al. (1961), who proposed a diffusion technique for the fabrication of Si piezoresistive sensors for stress, strain, and pressure. As early as 1962, Tufte et al. (1962) at Honeywell followed up on this suggestion. By using a combination of a wet isotropic etch, dry etching, and oxidation processes, they made the first thin Si piezoresistive diaphragms, of the type shown in Figure 3.1, for pressure sensors [Tufte et al., 1962]. In 1972, Sensym became the first to make stand-alone Si sensor products. By 1974, National Semiconductor Corporation in California carried an extensive line of Si pressure transducers in the first complete silicon pressure transducer catalog [National Semiconductor, 1974]. Other early commercial suppliers of micromachined pressure sensor products were Foxboro/ICT, Endevco, Kulite, and Honeywell's Microswitch. Innovative nonpressure-sensor micromachined structures began to be explored by the mid- to late 1970s. Texas Instruments produced a thermal print head in 1977 [Texas Instruments, 1977]. IBM produced ink-jet nozzle arrays in 1977 [Bassous et al., 1977]. Western Electric made fiber optic alignment structures in 1974 [Boivin, 1974], and Hewlett Packard made thermally isolated diode detectors in 1980 [O'Neill, 1980]. Many Silicon Valley and California based microsensor companies (Inset 3.2) played and continue to play a pivotal role in the development of the market for Si sensor products. However, from the mid-1990s on, many of the Silicon Valley MEMS companies became BIOMEMS oriented, and Si was no longer the preferred substrate.

Some Private California MEMS Companies

1972 Foxboro ICT (called SenSym ICT since 1999)
1972 Sensym (called SenSym ICT since 1999)
1975 Endevco
1975 IBM Micromachining
1976 Cognition (sold to Rosemount in 1978)
1980 Irvine Sensors Corp.
1982 IC Sensors (sold to EG&G in 1994)
1985 NovaSensor (sold to Lucas in 1990)
1988 Nanostructures
1988 Redwood Microsystems
1988 TiNi Alloys
1989 Abaxis
1989 Advanced Recording Technologies
1991 Incyte Genomics
1991 Sentir
1992 Silicon Microstructures
1993 Affymetrix
1993 Nanogen
1995 Aclara Biosciences
1995 Integrated Micromachines
1995 MicroScape
1996 Caliper
1996 Cepheid
1997 Microsensors
1998 Quantum Dot
1998 Zyomyx
1999 Symyx

INSET 3.2   Some private California MEMS companies.

European and Japanese companies followed the United States' lead more than a decade later; for example, Druck, Ltd., in the U.K. started exploiting Greenwood's micromachined pressure sensor in the mid-1980s [Greenwood, 1984].

Petersen's 1982 paper extolling the excellent mechanical properties of single-crystalline silicon helped galvanize academia's involvement in Si micromachining in a major way [Peterson, 1982]. Before that time, timid efforts had played out in industry, and practical needs (market pull) were driving the technology. The new generation of micromachined devices explored in academia often constituted gadgetry only, and as a consequence, the field is still perceived by many as a technology looking for applications (technology push). It has been estimated that by 1994 more than 10,000 scientists worldwide were involved in Si sensor research and development [Middlehoek and Dauderstadt, 1994]. To justify the continued high investments by government and industry, it became an absolute priority to understand the intended applications better, to be able to select an optimum micromachining tool set intelligently, and to identify more large market applications — "killer applications," or "killer aps." Some of these killer aps materialized only toward the end of the 20th century and are mainly to be found in information technology (IT) and biotechnology.

3.3   Silicon Crystallography

3.3.1   Introduction

Crystalline silicon substrates are available as circular wafers of 100 mm (4 in) diameter and 525 μm thickness or 150 mm (6 in) diameter with a thickness of 650 μm. Larger 200 mm and 300 mm diameter wafers are currently not economically justified for MEMS and are used only in the integrated circuit industry

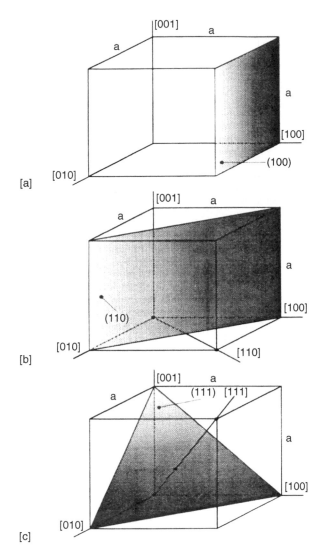

FIGURE 3.2   Miller indices in a cubic lattice: planes and axes. Shaded planes are (a) (100), (b) (110), (c) (111).

[Maluf, 2000]. Wafers polished on both sides, often used in MEMS, are about $100\,\mu m$ thinner than standard thickness substrates (see the $425\,\mu m$ thick Si substrate in Figure 3.1).

3.3.2   Miller Indices

The periodic arrangement of atoms in a crystal is called the lattice. The unit cell in a lattice is a segment that is representative of the entire lattice. For each unit cell, basis vectors (a_1, a_2, and a_3) can be defined such that if the unit cell is translated by integral multiples of these vectors a new unit cell identical to the original is obtained. A simple cubic-crystal unit cell for which $a_1 = a_2 = a_3$ and the axes angles are $\alpha = \beta = \gamma = 90°$ is shown in Figure 3.2. In this figure, the dimension a is known as the lattice constant. To identify a plane or a direction, a set of integers h, k, and l, called the Miller indices, are used. To determine the Miller indices of a plane, one takes the intercept of that plane with the axes and expresses these intercepts as multiples of the basis vectors a_1, a_2, a_3. The reciprocal of these three integers is taken, and to obtain whole numbers the three reciprocals are multiplied by the smallest common denominator. The resulting set of numbers is written down as (hkl). By taking the reciprocal of the intercepts, infinities (∞)

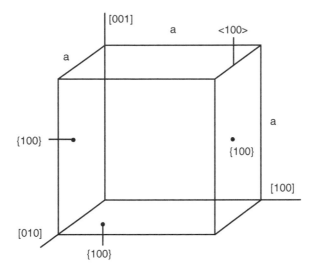

FIGURE 3.3   Miller indices for some of the planes of the {100} family of planes.

are avoided in the plane identification. Parentheses or braces are used to specify planes. A direction in a lattice is expressed as a vector with components as multiples of the basis vectors. The rules for determining the Miller indices of an orientation translate the orientation to the origin of the unit cube and take the normalized coordinates of its other vertex. For example, the body diagonal in a cubic lattice as shown in Figure 3.2 is $1a$, $1a$, and $1a$ or a diagonal along the [111] direction. Directions [100], [010], and [001] are all crystallographically equivalent and are jointly referred to as the family, form, or group of <100> directions. Brackets or carets specify directions. A form, group, or family of faces that bear like relationships to the crystallographic axes — for example, the planes (001), (100), (010), (001), (100), and (010) — are all equivalent, and they are marked as {100} planes. Figure 3.3 shows some of the planes of the {100} family of planes.

3.3.3   Crystal Structure of Silicon

Crystalline silicon forms a covalently bonded structure, the diamond-cubic structure, which has the same atomic arrangement as carbon in diamond form and belongs to the more general zinc-blend classification [Kittel, 1976]. Silicon, with its four covalent bonds, coordinates itself tetrahedrally, and these tetrahedrons make up the diamond-cubic structure. This structure can also be represented as two interpenetrating face-centered cubic lattices, one displaced $(1/4,1/4,1/4)$ times a with respect to the other, as shown in Figure 3.4. The structure is face-centered cubic (fcc), but with two atoms in the unit cell. For such a cubic lattice, direction [hkl] is perpendicular to a plane with the three integers (hkl), simplifying further discussions about the crystal orientation; that is, the Miller indices of a plane perpendicular to the [100] direction are (100). The lattice parameter a for silicon is 5.4309 Å, and silicon's diamond-cubic lattice is surprisingly wide open with a packing density of 34% compared with 74% for a regular fcc lattice. The {111} planes present the highest packing density, and the atoms are oriented such that three bonds are below the plane. In addition to the diamond-cubic structure, silicon is known to have several stable high-pressure crystalline phases [Hu et al., 1986], and a stress-induced metastable phase with a wurtzite-like structure referred to as diamond-hexagonal silicon. The latter has been observed after ion implantation [Tan et al., 1981] and hot indentation [Eremenko and Nikitenko, 1972].

When ordering silicon wafers, the crystal orientation must be specified. The most common orientations used in the IC industry are [100] and [111]; in micromachining, [110] wafers are used quite often as well. The [110] wafers break or cleave much more cleanly than other orientations. In fact, this is the

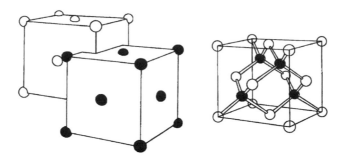

FIGURE 3.4   The diamond-type lattice can be constructed from two interpenetrating face-centered cubic unit cells. Si forms four covalent bonds making tetrahedrons.

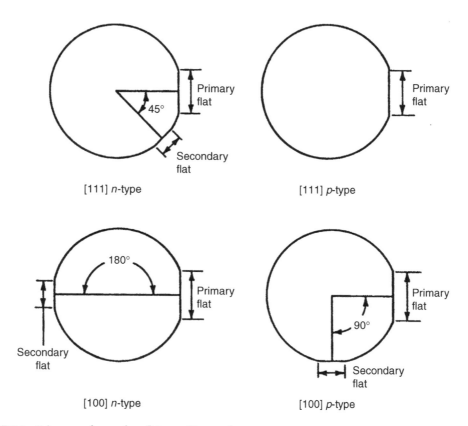

FIGURE 3.5   Primary and secondary flats on silicon wafers.

only major plane that can be cleaved with exactly perpendicular edges. The [111] wafers are used less often as they cannot be easily etched by wet anisotropic etchants except when using special techniques such as laser-assisted etching [Alavi et al., 1992]. On a [100] wafer, the [110] direction is often made evident by a flat segment, also called an orientation flat. The precision on the flat is about 3°. The flat's position on [110]–oriented wafers varies from manufacturer to manufacturer but often parallels a [111] direction. Flat areas help orientation determination and placement of slices in cassettes and fabrication equipment (large primary flat). They also help identify orientation and conductivity type (smaller secondary flat). Primary and secondary flats on [111] and [100] silicon wafers are indicated in Figure 3.5.

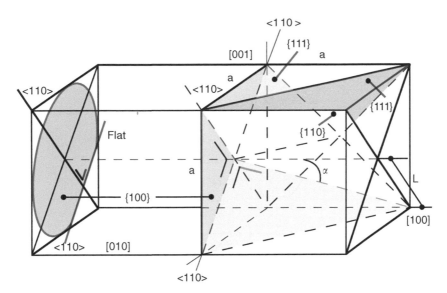

FIGURE 3.6 A (100) silicon wafer with reference to the unity cube and its relevant planes. (Reprinted with permission from Peeters, E. [1994] Process Development for 3D Silicon Microstructures, with Application to Mechanical Sensor Design, Ph.D. thesis, KUL, Belgium.)

3.3.4 Geometric Relationships among Some Important Planes in the Silicon Lattice

3.3.4.1 Introduction

To appreciate the different three-dimensional shapes resulting from anisotropically etched single-crystal Si (SCS) and to better understand the section below on corner compensation requires further clarification of some of the more important geometric relationships between different planes within the Si lattice. We will consider only silicon wafers with a (100) or a (110) as the surface planes. We will also accept, for now, that in anisotropic alkaline etchants, the {111} planes, which have the highest atom-packing density, are nonetching compared to the other planes. As the {111} planes are essentially not attacked by the etchant, the sidewalls of an etched pit in SCS will ultimately be bounded by this type of plane, given that the etch time is long enough for features bounded by other planes to be etched away. The types of planes introduced initially depend on the geometry and the orientation of the mask features.

We will clarify in the sections below how simple vector algebra proves that the angles between {100} and {110} planes and between {100} and {111} planes are 45° and 54.74° respectively and, similarly, that the {111} and {110} planes can intersect each other at 35.26°, 90°, or 144.74°.

3.3.4.2 [100]–Oriented Silicon

In Figure 3.6, the unity cell of a silicon lattice is shown along with the correct orientation of a [100]–type wafer relative to this cell [Peeters, 1994]. Intersections of the nonetching {111} planes with the {100} planes (e.g., the wafer surface) are mutually perpendicular and lying along the <110> orientations. Provided that a mask opening (say, a rectangle or a square) is accurately aligned with the primary orientation flat (that is, the [110] direction), only {111} planes will be introduced as sidewalls from the very beginning of the etch. Since the nonetching character of the {111} planes renders an exceptional degree of predictability to the recess features, this is the mask arrangement most often utilized in commercial applications.

During etching, truncated pyramids (square mask) or truncated V-grooves (rectangular mask) deepen but do not widen (Figure 3.7). The edges in these structures are <110> directions, the ribs are <211>

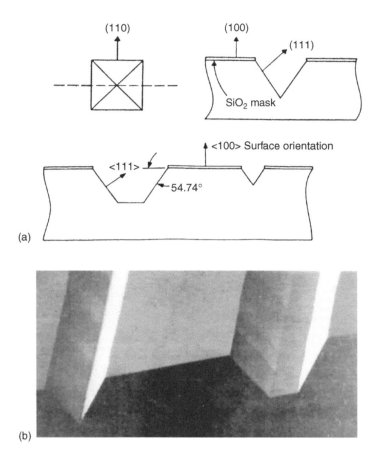

FIGURE 3.7 Anisotropically etched features in a (100) wafer with (a) square mask (schematic) and (b) rectangular mask (scanning electron microscope micrograph of resulting actual V- and U-grooves).

directions, the sidewalls are {111} planes, and the bottom is a (100) plane parallel with the wafer surface. After prolonged etching, the {111} family planes are exposed down to their common intersection, and the (100) bottom plane disappears creating a pyramidal pit (square mask) or a V-groove (rectangular mask). As shown in this figure, no underetching of the etch mask is observed due to the perfect alignment of the concave oxide mask opening with the [110] direction. Misalignment still results in pyramidal pits, but the mask will be undercut. For a mask opening with arbitrary geometry and orientation — a circle, for example — and for sufficiently long etch times, the anisotropically etched recess in a {100} wafer is pyramidal with a base perfectly circumscribing the circular mask opening [Peeters, 1994]. Convex corners ($>180°$) in a mask opening will always be completely undercut by the etchant after sufficiently long etch times. This can be disadvantageous (e.g., when attempting to create a mesa rather than a pit), or it can be advantageous for undercutting suspended cantilevers or bridges. The section on corner compensation will address the issue of undercutting in more detail. In corner compensation, the convex corner undercutting is compensated by clever layout schemes. The slope of the sidewalls in a cross-section perpendicular to the wafer surface and to the wafer flat is determined by the angle a as in Figure 3.6 depicting the off-normal angle of the intersection of a (111) sidewall and a (110) cross-secting plane; it can be calculated from:

$$\tan \alpha = \frac{L}{a} \tag{3.1}$$

with $L = a \times \frac{\sqrt{2}}{2}$ or $\alpha = \arctan \frac{\sqrt{2}}{2} = 35.26°$, or $54.74°$ for the complementary angle. The tolerance on this slope is determined by the alignment accuracy of the wafer surface with respect to the (100) plane. Wafer manufacturers typically specify this misalignment to $1°$ ($0.5°$ in the best cases).

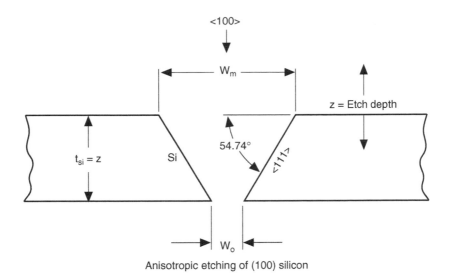

FIGURE 3.8   Relation of bottom cavity plane width with mask opening width.

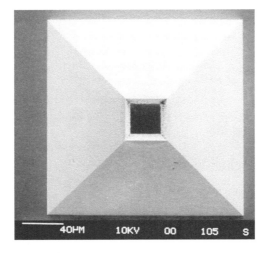

INSET 3.3   Orifice (a via through Si wafer).

The width of the rectangular or square cavity bottom plane, W_0, in Figure 3.8, aligned with the <110> directions is defined completely by the etch depth z the mask opening W_m and the above-calculated sidewall slope:

$$W_0 = W_m - 2\cot(54.74°)z$$

or:

$$W_0 = W_m - \sqrt{2}z \tag{3.2}$$

The larger the opening in the mask, the deeper the point at which the {111} sidewalls of the pit intersect. The etch stop at the {111} sidewalls' intersection occurs when the depth is about 0.7 times the mask opening. If the oxide opening is wide enough, $W_m > 849\,\mu m$ (for a typical 6 in wafer with thickness $t_{si} = z = 600\,\mu m$), the {111} planes do not intersect within the wafer. The etched pit in this particular case extends all the way through the wafer creating a small orifice, or via (Inset 3.3). If a high density of such vias through the Si is required, the wafer must be made very thin.

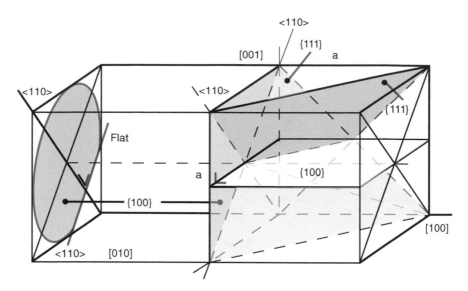

FIGURE 3.9   [100] silicon wafer with [100] mask-aligned features introduces vertical sidewalls. (Reprinted with permission from Peeters, E. [1994] Process Development for 3D Silicon Microstructures, with Application to Mechanical Sensor Design, Ph.D. thesis, KUL, Belgium.)

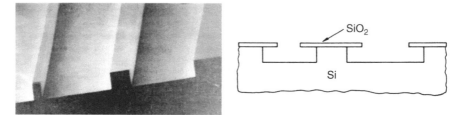

FIGURE 3.10   Vertical sidewalls in a (100) wafer.

Corners in an anisotropically etched recess are defined by the intersection of crystallographic planes, and the resulting corner radius is essentially zero. This implies that the size of a silicon diaphragm is very well defined, but it also introduces a considerable stress concentration factor. The influence of the zero corner radius on the yield load of diaphragms can be studied with finite element analysis (FEA).

One way to obtain vertical sidewalls instead of 54.7° sidewalls using a [100]–oriented Si wafer may be understood from Figure 3.9. There are {100} planes perpendicular to the wafer surface, and their intersections with the wafer surface are <100> directions. These <100> directions enclose a 45° angle with the wafer flat (i.e., the <110> directions). By aligning the mask opening with these <100> orientations, {100} facets are initially introduced as sidewalls. The {110} planes etch faster than the {100} planes and are not introduced. As the bottom and sidewall planes are all from the same {100} group, lateral under-etch equals the vertical etch rate, and rectangular channels bounded by slower etching {100} planes result (Figure 3.10).

Since the top of the etched channels is exposed to the etchant longer than the bottom, one might have expected the channels in Figure 3.10 to be wider at the top than at the bottom. We use Peeters' derivation to show why the sidewalls stay vertical [Peeters, 1994]. Assume the width of the mask opening to be W_m. At a given depth z into the wafer, the underlying Si is no longer masked by W_m but rather by the intersection of the previously formed {100} facets with the bottom surface. The width of this new mask is larger than the lithography mask W_m by the amount the latter is being undercut. Let us call the new mask width W_z, the effective mask width at a depth z. The relation between W_m and W_z is given by the lateral etch rate of a {100} facet and the time that facet was exposed to the etchant at depth z:

$$W_z = W_m + 2R_{xy}\Delta t_z \tag{3.3}$$

where R_{xy} = the lateral underetch rate (i.e., etch rate in the xy plane)
Δt_z = the etch time at depth z

The underetching, U_z, of the effective mask opening W_z is given by:

$$U_z = TR_{xy} - R_{xy}\Delta t_z \tag{3.4}$$

where T is the total etch time so far. The width of the etched pit W_{tot} at depth z is further given by the sum of W_m and twice the underetching for that depth:

$$W_{tot} = W_z + 2U_z = W_m + 2TR_{xy} \tag{3.5}$$

Or, because T can also be written as the measured total etch depth z divided by the vertical etch rate R_z, Equation (3.5) can be rewritten as:

$$W_{tot} = W_m + 2z(R_{xy}/R_z) \qquad \text{or since } R_{xy} = R_z$$

$$W_{tot} = W_m + 2z \tag{3.6}$$

The width of the etched recess is therefore equal to the photolithographic mask width plus twice the etch depth — independent of that etch depth, in other words. The walls remain vertical independent of the depth z.

For sufficiently long etch times, {111} facets eventually take over from the vertical {100} facets. These inward sloping {111} facets are first introduced at the corners of a rectangular mask and grow larger at the expense of the vertical sidewalls until the latter ultimately disappear altogether. Alignment of mask features with the <100> directions so as to obtain vertical sidewalls in [100] wafers, therefore, is not very useful for the fabrication of diaphragms. However, it can be very effective for anticipating the undercutting of convex corners on [100] wafers. This useful aspect will be revisited when discussing corner compensation.

3.3.4.3 [110]–Oriented Silicon

In Figure 3.11, we show a unit cell of Si properly aligned with the surface of a [110] Si wafer. This drawing will enable us to predict the shape of an anisotropically etched recess on the basis of elementary geometric crystallography. Four of the eight equivalent {111} planes are perpendicular to the (110) wafer surface. The remaining four are slanted at 35.26° with respect to the surface. Whereas the intersections of the four vertical {111} planes with the (100) wafer surface are mutually perpendicular, they enclose an angle $\gamma + 90°$ in the (110) plane. Moreover, the intersections are not parallel or perpendicular to the main wafer flat ([110] in this case) but rather enclose angles δ or $\delta + \gamma$ with this direction. It follows that a mask opening that will not be undercut (i.e., oriented such that resulting feature sidewalls are exclusively made up by {111} planes) cannot be a rectangle aligned with the wafer flat but must be a parallelogram skewed by $\gamma - 90°$ and δ degrees off-axis. The angles γ and δ are calculated by calculating β first [Peeters, 1994] (see Figures 3.11 and 3.12):

$$\tan \beta = \frac{\frac{1}{2} a \frac{\sqrt{2}}{2}}{\frac{a}{2}} = \frac{\sqrt{2}}{2} \tag{3.7}$$

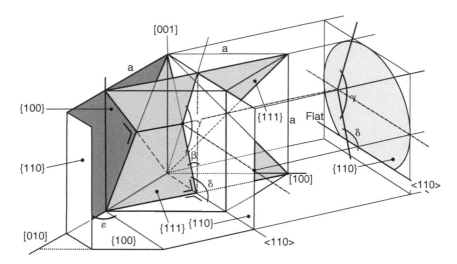

FIGURE 3.11 A (110) silicon wafer with reference to the unity cube and its relevant planes. The wafer flat is in a <110> direction. (Reprinted with permission from Peeters, E. [1994] Process Development for 3D Silicon Microstructures, with Application to Mechanical Sensor Design, Ph.D. thesis, KUL, Belgium.)

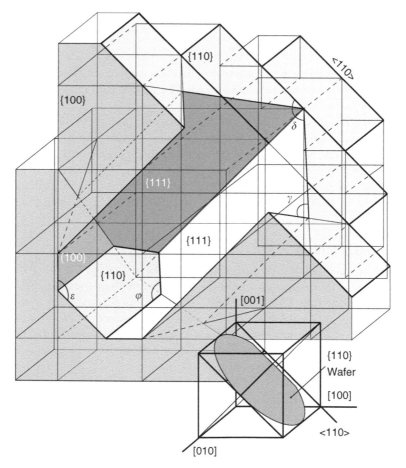

FIGURE 3.12 A (110) silicon wafer with anisotropically etched recess inscribed in the Si lattice. $\gamma = 109.47°$; $\delta = 125.26°$; $\varphi = 144.74°$; and $\varepsilon = 45°$. (Reprinted with permission from Peeters, E. [1994] Process Development for 3D Silicon Microstructures, with Application to Mechanical Sensor Design, Ph.D. thesis, KUL, Belgium.)

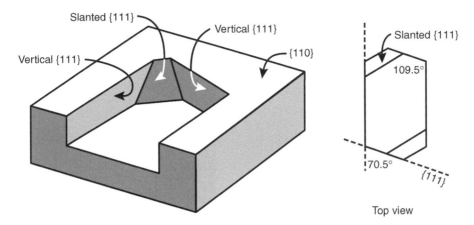

INSET 3.4   Illustration of anisotropic etching in {110} oriented silicon. Etched structures are delineated, by four vertical {111} lanes and two slanted {111} planes. The vertical {111} planes intersect at an angle of 70.5°.

and from β we can deduce these other pertinent angles:

$$\gamma = 180° - 2\beta = 180° - 2\arctan\left(\frac{\sqrt{2}}{2}\right) = 109.47°$$

$$\delta = 90° + \beta = 90° + \arctan\left(\frac{\sqrt{2}}{2}\right) = 125.26°$$

(3.8)

$$\varphi = 270° - \delta = 144.74°$$

The large inside angle of the parallelogram γ is 109.47°, and the small angle is thus 70.5°. A groove etched in [110] wafers has the appearance of a complex polygon delineated by six {111} planes, four vertical and two slanted (Inset 3.4). The bottom of the etch pit shown in Figure 3.12 is bounded at first by {110} and/or {100} planes, depending on the etch time. At short etch times, one mainly sees a flat {110} bottom (Inset 3.4). As the {110} planes are etching slightly faster than the {100} planes, the flat {110} bottom gets progressively smaller, and a V-shaped bottom bounded by {100} planes results. The angle ε as shown in Figure 3.12 equals 45°, being the angle enclosed by the intersections of a {100} and a {110} bottom plane. At even longer etch times, shallow {111} planes form the bottom, and they eventually stop the etching. For [110] etching, an arbitrary window opening is circumscribed by a parallelogram with the given orientation and skewness for sufficiently long etching times. Another difference between [100]– and [110]–oriented Si wafers is that on the [110] wafers it is possible to etch under microbridges crossing at a 90° angle a shallow V-groove formed by (111) planes. To undercut a bridge on a (100) plane, the bridge cannot be perpendicular to the V-groove; it must be oriented slightly off normal [Elwenspoek et al., 1994].

3.3.4.4   Selection of [100]– or [110]–Oriented Silicon

In Table 3.1, we compare the main characteristics of etched features in [100]– and [110]–oriented wafers. This guide can help decide which orientation to use for a specific microfabrication application at hand. From this table, it is obvious that for membrane-based sensors [100] wafers are preferred. An understanding of the geometric considerations with [110] wafers is important, however, to fully appreciate all the possible SCS micromachined shapes, and it is especially helpful to understand corner compensation schemes. Moreover, all processes for providing dielectric isolation require that the silicon be separated into discrete regions. To achieve a high component density with anisotropic etches on (100) wafers, the silicon must be made very thin because of the aspect ratio limitations due to the sloping walls. With vertical sidewall etching in a (100) wafer, the etch mask is undercut in all directions to a distance approximately equal to the depth of the etching. Vertical etching in (110) surfaces relaxes the etching requirement dramatically and enables more densely

TABLE 3.1   Selection of Wafer Type

[100] Orientation	[110] Orientation
Inward sloping walls (54.74°)	Vertical {111} walls
The sloping walls cause a lot of lost real-estate	Narrow trenches with high aspect ratio are possible
Flat bottom parallel to surface is ideal for membrane fabrication	Multifaceted cavity bottom ({110} and {100} planes) makes for a poor diaphragm
Bridges perpendicular to a V-groove bound by (111) planes cannot be underetched	Bridges perpendicular to a V-groove bound by (111) planes can be undercut
Shape and orientation of diaphragms are convenient and simple to design	Shape and orientation of diaphragms are awkward and more difficult to design
Diaphragm size, bounded by nonetching {111} planes, is relatively easy to control	Diaphragm size is difficult to control (the <100> edges are not defined by nonetching planes)

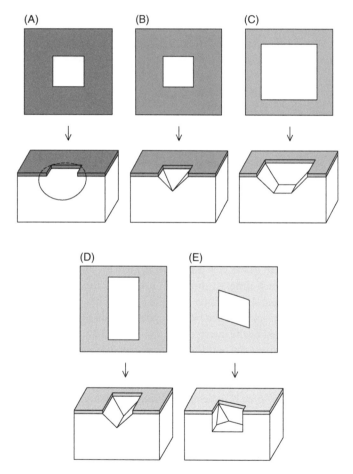

FIGURE 3.13   Isotropic and anisotropic etched features in [100] and [110] wafers. (A) isotropic etch, [100]-oriented wafer; (B) to (D) anisotropic etch, [100]-oriented wafers; and (E): anisotropic etch, [110]-oriented wafer.

packed structures such as beam leads or image sensors. Kendall describes and predicts a wide variety of applications for (110) wafers such as fabrication of trench capacitors, vertical multijunction solar cells, diffraction gratings, infrared interference filters, large area cathodes, and filters for bacteria [Kendall, 1975, 1979].

3.3.4.5   Examples of Wet Etched Structures in Si

In Figure 3.13, we compare a wet isotropic etch (A) with examples of anisotropic etch (B to E). In the anisotropic etching examples, a square (B and C) and a rectangular pattern (D) are defined in an oxide

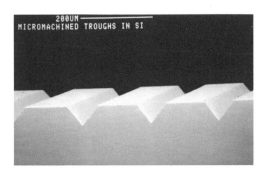

INSET 3.5   Long V-shaped grooves in a (100) Si wafer.

mask with sides aligned along the family of <110> directions on a [100]–oriented silicon surface. The square openings are precisely aligned (within one or two degrees) with the [110] directions on the (100) wafer surface to obtain pits that conform exactly to the oxide mask rather than undercutting it. Most [100] silicon wafers have a main flat parallel to a [110] direction in the crystal, allowing for a rough alignment of the mask (see Figure 3.5). Etching with the square pattern result s in a pit with well defined {111} sidewalls (at angles of 54.74° to the surface) and a (100) bottom.

The dimensions of the hole at the bottom of the pit, as we have seen, are given by Equation (3.2). The larger the square opening in the mask, the deeper the point where the {111} sidewalls of the pit intersect. If the oxide opening is wide enough, that is, $W_m > \sqrt{2}z$ (with $z = 600\,\mu$m for a typical 6 inch wafer, this means $W_m > 849\,\mu$m), the {111} planes do not intersect within the wafer. The etched pit in this particular case extends all the way through the wafer creating a small square opening on the bottom surface. As shown in Figure 3.13 (B to D), no underetching of the etch mask is observed due to the perfect alignment of the concave oxide mask opening with the [110] direction. Figure 3.13A shows an undercutting isotropic etch (acidic). Misalignment in the case of an anisotropic etch results in pyramidical pits, but the mask will be severely undercut. A rectangular pattern aligned along the <110> directions on a <100> wafer leads to long V-shaped grooves (see Figure 3.13D and Inset 3.5) or an open slit, depending on the width of the opening in the oxide mask.

Using a properly aligned mask (see above) on a [110] wafer, grooves with four vertical and two slanted {111} planes result (see Figures 3.13E and 3.14A). Even a slight mask misorientation leads to all skewed sidewalls instead (Figure 3.14B). Before emergence of slanted {111} bottom planes, the groove is defined by four vertical (111) planes and a horizontal (110) bottom (see also Figure 3.14C and Inset 3.4); in between, the {110} and {100} bottom planes are in competition. Self-stopping occurs when the tilted end {111} planes intersect at the bottom of the groove (Figure 3.14D). It is easy to etch very long, narrow grooves deeply into a [110] silicon wafer (Figure 3.15). However, it is impossible to etch a short, narrow groove deeply into it [Kendall, 1979] because the narrow dimension of the groove is quickly limited by slow-etching {111} planes that subtend an angle of 35.26° to the surface and cause etch termination. At a groove of length $L = 1$ mm on the top surface, etching will stop when it reaches a depth of 0.289 mm, that is, $z_{max} = L/(2\sqrt{3})$. For very long grooves, the tilted end planes are too far apart to intersect in practical cases making the end effects negligible compared to the remaining trench-shaped part of the groove. Tuckerman and Pease demonstrated the use of such trenches as liquid cooling fins for integrated circuits [Tuckerman and Pease, 1981].

A laser can be used to melt, or "spoil," the shallow {111} surfaces, making it possible to etch deep vertical-walled holes through a [110]–oriented wafer as shown in Figure 3.16A [Schumacher et al., 1994; Barth et al., 1985; Seidel and Csepregi, 1988]. The technique is illustrated in Figure 3.16B. The absorbed energy of an Nd:YAG laser beam causes a local melting or evaporation zone, enabling etchants to etch the shallow {111} planes in the line-of-sight of the laser. Etching proceeds until "unspoiled" (111) planes are encountered. Some interesting resulting possibilities including partially closed microchannels are shown in Figure 3.16C [Schumacher et al., 1994; Alavi et al., 1991]. With this method, it is possible to use [111]

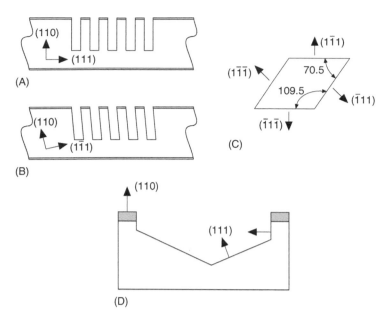

FIGURE 3.14   Anisotropic etching of [110] wafers. (A) Closely spaced grooves on correctly oriented (110) surface. (B) Closely spaced grooves on misoriented [110] wafer. (C) Orientations of the {111} planes looking down on a (110) wafer. (D) Shallow slanted (111) planes eventually form the bottom of the etched cavity.

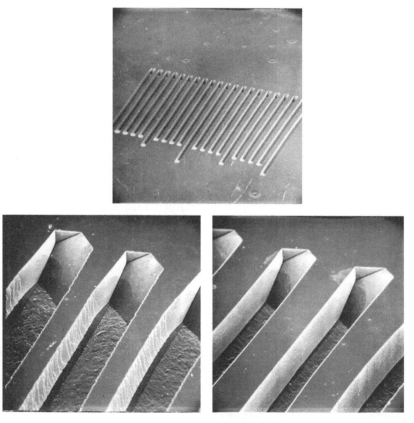

FIGURE 3.15   Test pattern of U-grooves in a <110> wafer to help in the alignment of the mask; final alignment is done with the groove that exhibits the most perfect long perpendicular walls.

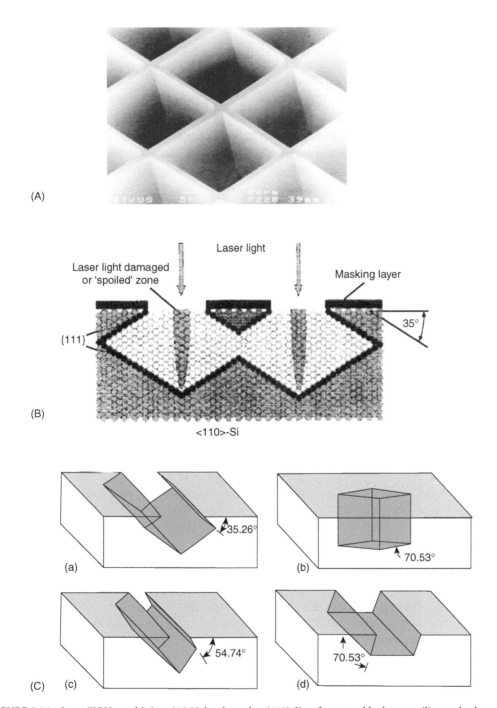

FIGURE 3.16 Laser/KOH machining. (A) Holes through a (110) Si wafer created by laser spoiling and subsequent KOH etching. The two sets of (111) planes making an angle of 70° are the vertical walls of the hole. The (111) planes making a 35.26° angle with the surface tend to limit the depth of the hole, but laser spoiling enables one to etch all the way through the wafer. (B) After laser spoiling, the line-of-sight (111) planes are spoiled, and etching proceeds until unspoiled (111) planes are reached. (C) Some of the possible features rendered by laser spoiling. (Reprinted with permission from Schumacher, A. et al. [1994] "Mit Laser und Kalilauge," *Technische Rundschau*, **86**, pp. 20–23.)

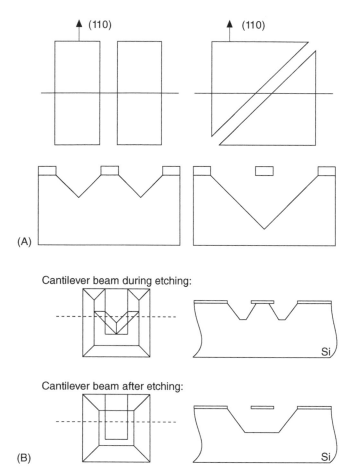

FIGURE 3.17 How to make (A) a suspension bridge from a [100] Si wafer and (B) a diving board from a [110] Si wafer. (Reprinted with permission from Barth, P.W., Shlichta, P.J., and Angel, J.B. [1985] "Deep Narrow Vertical-Walled Shafts in <110> Silicon," in *3rd International Conference on Solid-State Sensors and Actuators*, Philadelphia, pp. 371–73.)

wafers for micromachining. The light of the Nd:YAG laser is very well suited for this micromachining technique due to the 1.17 eV photon energy just exceeding the band gap energy of Si. Details on this laser machining process can be found, for example, in [Alavi et al., 1991, 1992].

The orientation of the wafer is of extreme importance, especially when machining surface structures by undercutting. Consider, for example, the formation of a bridge in Figure 3.17A [Barth et al., 1985]. When using a (100) surface, a suspension bridge cannot form across the etched V-groove; two independent truncated V-grooves flanking a mesa structure result instead. To form a suspended bridge, the V-groove must be oriented away from the [110] direction. Contrast this with a (110) wafer where a microbridge crossing a V-groove with a 90° angle will be undercut. Convex corners will be undercut by etchant, allowing formation of cantilevers as shown in Figure 3.17B. The diving board shown forms by undercutting starting at the convex corners.

3.4   Silicon as Substrate

For many mechanical sensor applications, single-crystal Si presents an excellent substrate choice based on its intrinsic mechanical stability and the feasibility of integrating sensing and electronics on the same

TABLE 3.2   Performance Comparison of Substrate Materials

Substrate	Cost	Metallization	Machinability
Ceramic	Medium	Fair	Poor
Plastic	Low	Poor	Fair
Silicon	High	Good	Very good
Glass	Low	Good	Poor

substrate. For chemical sensors, on the other hand, Si with few exceptions is merely the substrate, and as such it is not necessarily the most attractive option.

Table 3.2 shows a performance comparison of substrate materials in terms of cost, metallization ease, and machinability. Both ceramic and glass substrates are difficult to machine, and plastic substrates are not readily amenable to metallization. Silicon has the highest material cost per unit area, but this cost can often be offset by the small feature sizes possible in a silicon implementation. Silicon with or without passivating layers is often the preferred substrate due to its extreme flatness, relative low cost, and well established coating procedures; this is especially so for thin films. A lot of thin film deposition equipment is built to accommodate Si wafers, and as other substrates are harder to accommodate, this lends Si a convenience advantage. Silicon technology also affords more flexibility in design and manufacturing than other substrates. In addition, the initial capital equipment investment, although much more expensive, is not product specific. Once a first product is on line, a next generation of new products will require changes in masks and process steps, but not in the equipment itself.

The disadvantages of using Si are usually most pronounced with increasing device size and low production volumes and when electronics do not need to be, or cannot be, incorporated on the same Si substrate. The latter could be either for cost reasons (e.g., in the case of disposables such as glucose sensors) or for technological reasons (e.g., the devices will be immersed in conductive liquids or must operate at temperatures above 150°C).

An overwhelming determining factor for substrate choice is the final package of the device. A chemical sensor on an insulating substrate is almost always easier to package than a piece of Si with conductive edges in need of insulation.

Packaging is so important in sensors that, as a rule, sensor design should start from the package rather than from the sensor. In this context, an easier-to-package substrate has a huge advantage. The latter is the most important reason why chemical sensor development in industry changed direction from an all-out move toward integration on silicon in the 1970s and early 1980s to a hybrid thick film on ceramic approach in the late 1980s and early 1990s. In United States academic circles, chemical sensor integration with electronics continued until the late 1980s, stopped for a while, and then re-emerged with the successful introduction of a commercial ion sensitive field effect transistor (ISFET) [Sandifer and Voycheck, 1999] and DNA arrays with integrated electronics [Heller et al., 2000]. In Europe and Japan, such efforts were never completely abandoned [Madou, 1994].

Mechanical and material properties of Si are compared with other important sensor substrate materials in Tables 3.6 and 3.31. A good engineering guideline to determine if silicon is suitable for a given mechanical application is that at least two benefits should arise from the use of silicon over other substrates (aside from the possibility of integrating the electronics on the same substrate). For example, in the case of a torsional micromirror (Figure 3.18) fabricated based on an SOI (silicon on insulator) approach, the two Si-derived benefits are (1) excellent mirror-like surface and (2) superior torsional behavior of the silicon mirror hinges. Based on these two important properties, Si has become a formidable contender for making arrays of micromirrors for optical switching. Note that the mirror in Figure 3.18 is not optimized for a high-density mirror switch array. For such an application, the hinges should be very short or hidden behind the mirror because the mirror surface exposed to the incoming light should be maximized. The design as shown could be used, for example, for a linear scanner.

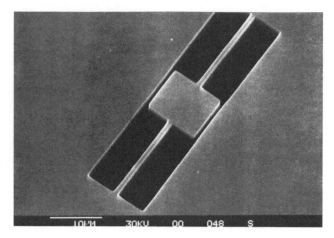

FIGURE 3.18   Silicon on insulator (SOI) micromirror.

3.5   Silicon as a Mechanical Element in MEMS

3.5.1   Introduction

In mechanical sensors, the active structural elements convert a mechanical external input signal (force, pressure, acceleration, etc.) into an electrical signal output (voltage, current, or frequency). The transfer functions in mechanical devices describing this conversion are mechanical, electromechanical, and electrical.

In mechanical conversion, a given external load is concentrated and maximized in the active member of the sensor. Structurally active members are typically high-aspect-ratio elements such as suspended beams or membranes. Electromechanical conversion is the transformation of the mechanical quantity into an electrical quantity such as capacitance, resistance, or charge. Often, the electrical signal needs further electrical conversion into an output voltage, frequency, or current. For electrical conversion into an output voltage, a Wheatstone bridge may be used, as in the case of a piezoresistive sensor, and a charge amplifier may be used, as in the case of a piezoelectric sensor. To optimize all three transfer functions, detailed electrical and mechanical modeling is required. One of the most important inputs required for mechanical models is the experimentally determined independent elasticity constants or moduli. In what follows, we describe what makes Si such an important structural element in mechanical sensors, and we present its elasticity constants.

3.5.2   Stress–Strain Curve and Elasticity Constants

Yield, tensile strength, hardness, and creep of a material all relate to the elasticity curve; that is, the stress–strain diagrams. Example stress–strain curves for several types of materials are shown in Figure 3.19. For small strain values, Hooke's (1635–1702) law applies; that is, stress (force per unit area, N/m^2) and strain (displacement per unit length, dimensionless) are proportional, and the stress–strain curve is linear with a slope corresponding to the elastic modulus E (Young's modulus, N/m^2). This regime, as marked in Figure 3.19A, is the elastic deformation regime (typically valid for $\varepsilon < 10^{-4}$). The magnitude of the Young's modulus ranges from 4.1×10^4 MPa (Pascal, Pa = N/m^2) for magnesium to 40.7×10^4 MPa for tungsten and 144 GPa for Invar, while concrete is 45 GPa, aluminum is about 70 GPa, and elastomeric materials are as low as 10^{-3} to 10^{-2} GPa. Silicon is a stiff and brittle material with a Young's modulus of 160 GPa (Figure 3.19B). With increasing temperature, the elastic modulus diminishes.

For isotropic media such as amorphous and polycrystalline materials, the applied axial force per unit area or tensile stress, σ_a, and the axial or tensile strain, ε_a, are related as:

$$\sigma_a = E\varepsilon_a \tag{3.9}$$

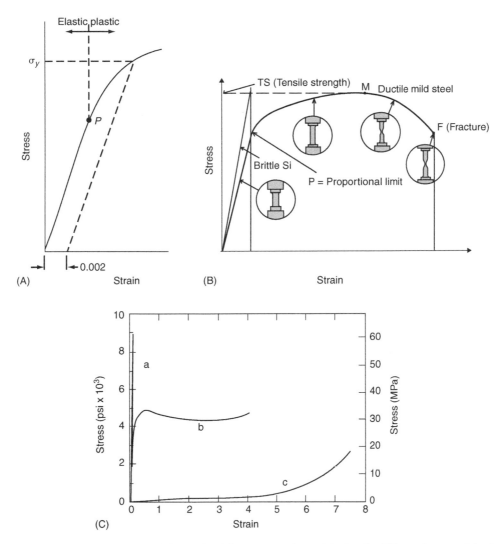

FIGURE 3.19 Typical stress–strain behavior for different types of materials showing different degrees of elastic and plastic deformations. (A) Stress–strain curve for a typical material with the proportional limit P and the yield strength σ_Y, as determined using the 0.002 strain offset method. (B) Stress–strain curve for Si and a ductile mild steel. The tensile strength (TS) of a metal is the stress at the maximum of the curve shown (M). Fracture is indicated by F. The various stages of a test metal sample are indicated as well. A high modulus material such as Si exhibits abrupt breakage; it is brittle with no plastic deformation region at all. (C) Stress–strain behavior for brittle (curve A), plastic (curve B), and highly elastic (elastomeric) (curve C) polymers.

with ε_a given by the dimensionless ratio of $(L_2 - L_1)/L_1$; that is, the ratio of the wire's elongation to its original length. The elastic modulus may be thought of as stiffness, or a material's resistance to elastic deformation. The greater the modulus, the stiffer the material. A tensile stress usually also leads to a lateral strain or contraction (Poisson effect) ε_1 given by the dimensionless ratio of $(D_2 - D_1)/D_1$ $(\Delta D/D_1)$, where D_1 is the original wire diameter, and ΔD is the change in diameter under axial stress (see Figure 3.20). The Poisson ratio is the ratio of lateral over axial strain:

$$v = -\frac{\varepsilon_1}{\varepsilon_a} \tag{3.10}$$

The minus sign indicates a contraction of the material. For most materials, v is a constant within the elastic range and fluctuates for different types of materials over a relatively narrow range. Generally, it is on

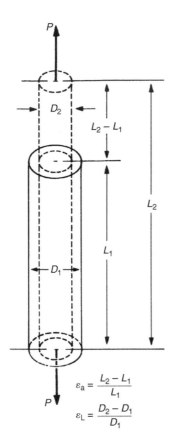

$$\varepsilon_a = \frac{L_2 - L_1}{L_1}$$

$$\varepsilon_L = \frac{D_2 - D_1}{D_1}$$

FIGURE 3.20   Metal wire under axial or normal stress; normal stress creates both elongation and lateral contraction.

the order of 0.25 to 0.35, and a value of 0.5 is the largest value possible. The latter is attained by materials such as rubber and indicates a material in which only the shape changes while the volume remains constant. Normally, some slight volume change does accompany the deformation, and, consequently v is smaller than 0.5. The Poisson ratios for metals are typically around 1/3; for example, aluminum and cast steel are 0.34 and 0.28, respectively. For ceramics, it is around 0.25, and for polymers it is typically between 0.4 and 0.5. In extreme cases, values as low as 0.1 (certain types of concrete) and as high as 0.5 (rubber) occur. For an elastic isotropic medium subjected to a triaxial state of stress, the resulting strain component in the x direction, ε_x, is given by the summation of elongation and contraction:

$$\varepsilon_x = \frac{1}{E}[\sigma_x - v(\sigma_y + \sigma_z)] \tag{3.11}$$

and so on for the y and z directions (three equations in total).

For an analysis of mechanical structures, we must consider not only compressional and tensile strains but also shear strains. Whereas normal stresses create elongation plus lateral contraction with accompanying volume changes, shear stresses (e.g., by twisting a body) create shape changes without volume changes, that is, shear strains. The one-dimensional shear strain, γ, is produced by the shear stress, τ (N/m^2). For small strains, Hooke's law may be applied again:

$$\gamma = \frac{\tau}{G} \tag{3.12}$$

where G is called the elastic shear modulus or the modulus of rigidity. For any three-dimensional state of shear stress, three equations of this type will hold. Isotropic bodies are characterized by two independent elastic constants only, because the shear modulus G can be shown [Chous and Pagano, 1967] to relate the Young's modulus and the Poisson ratio as:

$$G = \frac{E}{2(1 + v)} \tag{3.13}$$

Crystal materials, whose elastic properties are anisotropic, require more than two elastic constants, the number increasing with decreasing symmetry. Cubic crystals (body-centered cubic (bcc), for example, require three elastic constants; hexagonal crystals require five; and materials without symmetry require 21 [Kittel, 1976; Chou and Pagano, 1967]. The relation between stresses and strains is more complex in this case and depends greatly on the spatial orientation of these quantities with respect to the crystallographic axes. Hooke's law in the most generic form is expressed in two formulas:

$$\sigma_{ij} = E_{ijkl} \cdot \varepsilon_{kl} \quad \text{and} \quad \varepsilon_{ij} = S_{ijkl} \cdot \sigma_{kl} \tag{3.14}$$

where σ_{ij} and σ_{kl} are stress tensors of rank 2 expressed in N/m^2; ε_{ij} and ε_{kl} are strain tensors of rank 2 and are dimensionless; E_{ijkl} is a stiffness coefficient tensor of rank 4 expressed in N/m^2 with at the most $3 \times 3 \times 3 \times 3 = 81$ elements; and S_{ijkl} is a compliance coefficient tensor of rank 4 expressed in m^2/N with at the most $3 \times 3 \times 3 \times 3 = 81$ elements. The first expression is analogous to Equation (3.9), and the second expression is the inverse giving the strains in terms of stresses. The tensor representations in Equation (3.14) can also be represented as two matrices:

$$\sigma_m = \sum_{n=1}^{6} E_{mn}\varepsilon_n \quad \text{and} \quad \varepsilon_m = \sum_{n=1}^{6} S_{mn}\sigma_n \tag{3.15}$$

Components of tensors E_{ijkl} and S_{ijkl} are substituted by elements of the matrices E_{mn} and S_{mn}, respectively. To abbreviate the ij indices to m and the kl indices to n, the following scheme applies:

> $11 \rightarrow 1$, $22 \rightarrow 2$, $33 \rightarrow 3$, 23 and $32 \rightarrow 4$, 13 and
> $31 \rightarrow 5$, 12 and $21 \rightarrow 6$, $E_{ijkl} \rightarrow E_{mn}$ and $S_{ijkl} \rightarrow S_{mn}$
> when m and $n = 1,2,3$; $2S_{ijkl} \rightarrow S_{mn}$ when m or $n = 4,5,6$;
> $4S_{ijkl} \rightarrow S_{mn}$ when m and $n = 4,5,6$; $\sigma_{ij} \rightarrow \sigma_m$ when $m = 1,2,3$;
> and $\varepsilon_{ij} \rightarrow \varepsilon_m$ when $m = 4,5,6$

With these reduced indices, there are thus six equations of the type:

$$\sigma_x = E_{11}\varepsilon_x + E_{12}\varepsilon_y + E_{13}\varepsilon_z + E_{14}\gamma_{yz} + E_{15}\gamma_{zx} + E_{16}\gamma_{xy} \tag{3.16}$$

and hence 36 moduli of elasticity or E_{mn} stiffness constants. There are also six equations of the type:

$$\varepsilon_x = S_{11}\sigma_x + S_{12}\sigma_y + S_{13}\sigma_z + S_{14}\tau_{yz} + S_{15}\tau_{zx} + S_{16}\tau_{xy} \tag{3.17}$$

defining 36 S_{mn} constants, which are called the compliance constants. It can be shown that the matrices E_{mn} and S_{mn}, each composed of 36 coefficients, are symmetrical ($E_{mn} = E_{nm}$ and $S_{mn} = S_{nm}$); hence, a material without symmetrical elements has 21 independent constants or moduli. Due to symmetry of crystals, several more of these may vanish until for an isotropic medium they number two only (E and v). The stiffness coefficient and compliance coefficient matrices for cubic-lattice crystals with the vector of stress oriented along the [100] axis are given as:

$$E_{mn} = \begin{bmatrix} E_{11} & E_{12} & E_{12} & 0 & 0 & 0 \\ E_{12} & E_{11} & E_{12} & 0 & 0 & 0 \\ E_{12} & E_{12} & E_{11} & 0 & 0 & 0 \\ 0 & 0 & 0 & E_{44} & 0 & 0 \\ 0 & 0 & 0 & 0 & E_{44} & 0 \\ 0 & 0 & 0 & 0 & 0 & E_{44} \end{bmatrix} \quad S_{mn} = \begin{bmatrix} S_{11} & S_{12} & S_{12} & 0 & 0 & 0 \\ S_{12} & S_{11} & S_{12} & 0 & 0 & 0 \\ S_{12} & S_{12} & S_{11} & 0 & 0 & 0 \\ 0 & 0 & 0 & S_{44} & 0 & 0 \\ 0 & 0 & 0 & 0 & S_{44} & 0 \\ 0 & 0 & 0 & 0 & 0 & S_{44} \end{bmatrix} \tag{3.18}$$

In cubic crystals, the three remaining independent elastic moduli are usually chosen as E_{11}, E_{12}, and E_{44}. The S_{mn} values can be calculated simply from the E_{mn} values. Expressed in terms of the compliance constants, one can show that $1/S_{11} = E =$ Young's modulus, $-S_{12}/S_{11} = \nu =$ Poisson's ratio, and $1/S_{44} = G =$ shear modulus. In the case of an isotropic material, such as a metal wire, there is an additional relationship:

$$E_{44} = \frac{E_{11} - E_{12}}{2} \tag{3.19}$$

reducing the number of independent stiffness constants to two. The anisotropy coefficient α is defined as:

$$\alpha = \frac{2E_{44}}{E_{11} - E_{12}} \tag{3.20}$$

making $\alpha = 1$ for an isotropic crystal. For an anisotropic crystal, the degree of anisotropy is given by the deviation of α from 1. Single-crystal silicon has moderately anisotropic elastic properties [Brantley, 1973; Nikanorov et al., 1972], with $\alpha = 1.57$, and very anisotropic crystals may have a value larger than 8. Brantley (1973) gives the nonzero Si stiffness components, referred to the [100] crystal orientation, as $E_{11} = E_{22} = E_{33} = 166 \times 10^9$ N/m², $E_{12} = E_{13} = E_{23} = 64 \times 10^9$ N/m², and $E_{44} = E_{55} = E_{66} = 80 \times 10^9$ N/m²:

$$\begin{array}{c} \sigma_x \\ \sigma_y \\ \sigma_z \\ \tau_{xy} \\ \tau_{xz} \\ \tau_{yz} \end{array} = \begin{vmatrix} 166(E_{11}) & 64(E_{12}) & 64(E_{12}) & 0 & 0 & 0 \\ 64(E_{12}) & 166(E_{11}) & 64(E_{12}) & 0 & 0 & 0 \\ 64(E_{12}) & 64(E_{12}) & 166(E_{11}) & 0 & 0 & 0 \\ 0 & 0 & 0 & 80(E_{44}) & 0 & 0 \\ 0 & 0 & 0 & 0 & 80(E_{44}) & 0 \\ 0 & 0 & 0 & 0 & 0 & 80(E_{44}) \end{vmatrix} \times \begin{array}{c} \varepsilon_x \\ \varepsilon_y \\ \varepsilon_z \\ \gamma_{xy} \\ \gamma_{xz} \\ \gamma_{yz} \end{array} \tag{3.21}$$

with σ normal stress, τ shear stress, ε normal strain, and γ shear strain. The values for E_{mn} in Equation (3.21), compare with a Young's modulus of 207 GPa for a low-carbon steel. Variations on the values of the elastic constants on the order of 30%, depending on crystal orientation, must be considered; doping level and dislocation density have minor effects as well. From the stiffness coefficients, the compliance coefficients of Si can be calculated as $S_{11} = 7.68 \times 10^{-12}$ m²/N, $S_{12} = -2.14 \times 10^{-12}$ m²/N, and $S_{44} = 12.6 \times 10^{-12}$ m²/N [Khazan, 1994]. A graphical representation of elastic constants for different crystallographic directions in Si and Ge is given in Worthman et al. (1965) and is reproduced in Figure 3.21. Figure 3.21A to d displays E and ν for Ge and Si in planes (100) and (110) as functions of direction. Calculations show that E, G, and ν are constant for any direction in the (111) plane. In other words, a plate lying in this plane can be considered as having isotropic elastic properties. A review of independent determinations of the Si stiffness coefficients with their respective temperature coefficients is given in Metzger et al. (1970). Some of the values from that review are reproduced in Table 3.3. Values for Young's modulus and the shear modulus of Si can also be found in Greenwood (1988) and are reproduced in Table 3.4 for the three technically important crystal orientations.

3.5.3 Residual Stress in Si

Most properties, such as the Young's modulus, for lightly and highly doped silicon are identical. Residual stress and associated stress gradients in highly boron-doped single-crystal silicon present some controversy. Highly boron-doped membranes, which are usually reported to be tensile, have also been reported as compressive [Huff and Schmidt, 1992; Maseeh and Senturia, 1990]. From a simple atom-radius argument, a large number of substitutional boron atoms would be expected to create a net shrinkage of the lattice as compared with pure silicon and that the residual stress would be tensile with a stress gradient corresponding to the doping gradient. For example, an etched cantilever would be expected to bend up out of the plane of the silicon wafer. Maseeh et al. (1990) believe that the appearance of compressive behavior in heavily boron-doped single-crystal layers results from the use of an oxide etch mask. They suggest that plastic deformation of the p$^+$ silicon beneath the compressively stressed oxide can explain the observed behavior. Ding et al. (1991) also found compressive behavior for nitride-covered p$^+$ Si thin

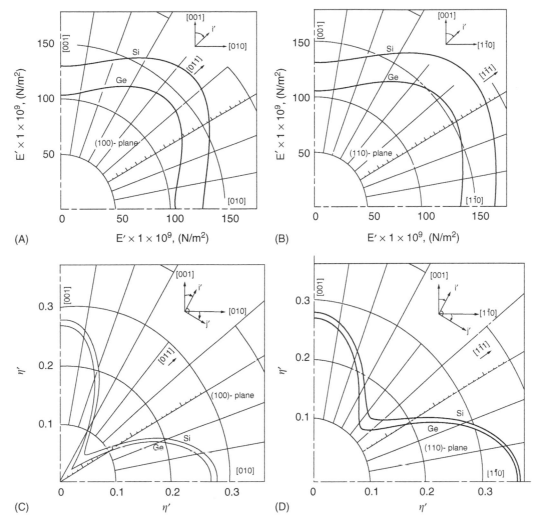

FIGURE 3.21 Elasticity constants for Si and Ge. (A) Young's modulus as a function of direction in the (100) plane. (B) Young's modulus as a function of direction in the (110) plane. (C) Poisson ratio as a function of direction in the (100) plane. (D) Poisson's ratio as a function of direction in the (110) plane. (Reprinted with permission from Worthman, J.J., and Evans, R.A. [1965] "Young's Modulus, Shear Modulus, and Poisson's Ratio in Silicon and Germanium," *J. Applied Physics* **36**, 153–56.)

TABLE 3.3 Stiffness Coefficients and Temperature Coefficient of Stiffness for Si

Stiffness Coefficients in Gpa (Value in GPa = 10^9 N/m²)	Temp. Coefficient of Young's Modulus (10^{-6} K⁻¹)
$E_{11} = 164.8 \pm 0.16$	−122
$E_{12} = 63.5 \pm 0.3$	−162
$E_{44} = 79.0 \pm 0.06$	−97

Source: Reprinted with permission from Metzger, H., and Kessler, F.R. (1970) "Der Debye-Sears Effect zur Bestimmung der Elastischen Konstanten von Silicium," *Z. Naturf.* A25, pp. 904–6.

TABLE 3.4   Derived Values for Young's Modulus and Shear Modulus
for Si

Crystal Orientation	Young's Modulus E (GPa)	Shear Modulus G (GPa)	Temp. Coeff. $(10^{-6}\,K^{-1})$
[100]	129.5	79.0	−63
[110]	168.0	61.7	−80
[111]	186.5	57.5	−46

Source: Reprinted with permission from Greenwood, J.C. (1988) "Silicon
in Mechanical Sensors," *J. Phys. E, Sci. Instrum.* **21**, 1114–28.

membranes; they believe that the average stress in p^+ silicon is tensile, but great care is required to establish this fact because the combination of heavy boron doping and a high-temperature drive-in under oxidizing conditions can create an apparent reversal of both the net stress (to compressive) and of the stress gradient (opposite to the doping gradient). A proposed explanation is that, at the oxide–silicon interface, a thin compressively stressed layer is formed during the drive-in that is not removed in buffered HF. It can be removed by reoxidation and etching in HF or by etching in KOH.

3.5.4   Yield, Tensile Strength, Hardness, and Creep

As a material is deformed beyond its elasticity limit, yielding or plastic deformation (permanent, nonrecoverable deformation) occurs. The point of yielding in Figure 3.19A is the point of initial departure from linearity of the stress–strain curve and is sometimes called the proportional limit indicated by a letter P. The Young's modulus of mild steel is $\pm 30{,}000{,}000$ psi, and its proportional limit (highest stress in the elastic range) is approximately 30,000 psi. Thus, the maximum elastic strain in mild steel is about 0.001 under a condition of uniaxial stress. This gives an idea as to the magnitude of the strains we are dealing with. A convention has been established wherein a straight line is constructed parallel to the elastic portion of the stress–strain curve at some specified strain offset, usually 0.002. The stress corresponding to the intersection of this line and the stress–strain curve as it bends over in the plastic region is defined as the yield strength, σ_y (see Figure 3.19A). The magnitude of the yield strength of a material is a measure of its resistance to plastic deformation. Yield strengths may range from 35 MPa (5000 psi) for a soft and weak aluminum to over 1400 MPa (200,000 psi) for high strength steels. The tensile strength is the stress at the maximum of the stress–strain curve. This corresponds to the maximum stress that can be sustained by a structure in tension; if the stress is applied and maintained, fracture will result. Crystalline silicon is a hard and brittle material deforming elastically until it reaches its yield strength, at which point it breaks. For Si, the yield strength is 7 GPa, equivalent to a 700 kg weight suspended from a 1 mm^2 area. Both tensile strength and hardness are indicators of a metal's resistance to plastic deformation. Consequently, they are roughly proportional [Callister, 1985]. Material deformation occurring at elevated temperatures and static material stresses is termed creep. It is defined as a time-dependent and permanent deformation of materials when subjected to a constant load or stress.

Silicon exhibits no plastic deformation or creep below 500°C; therefore, Si sensors are inherently very insensitive to fatigue failure when subjected to high cyclic loads. Silicon sensors have actually been cycled in excess of 100 million cycles with no observed failures. This ability to survive a very large number of duty cycles exists because there is no energy absorbing or heat generating mechanism due to intergranular slip or movement of dislocations in silicon at room temperature. However, single-crystal Si, as a brittle material, will yield catastrophically when stress beyond the yield limit is applied rather than deform plastically as metals do (see Figure 3.19B). At room temperature, high-modulus materials such as Si, SiO_2, and Si_3N_4 often exhibit linear-elastic behavior at lower strain and transition abruptly to brittle-fracture behavior at higher strain. Plastic deformation in metals is based on stress-induced dislocation generation in the grain boundaries and a subsequent dislocation migration that results in a macroscopic deformation from intergrain shifts in the material. No grain boundaries exist in SCS, and plastic deformation can occur only through migration of the defects originally present in the lattice or of those generated at the surface. As the number of these defects is very low in SCS, the material can be considered a perfect elastic material at normal temperatures. Perfect elasticity implies proportionality between stress and strain (i.e., load and flexure) and the absence of irreversibilities or mechanical hysteresis. The absence of plastic behavior also accounts for the extremely low mechanical losses in SCS, which enable the fabrication of resonating structures that exhibit exceptionally high Q-factors. Values of up to 10^8 in vacuum have been reported. At elevated temperatures, and with metals and polymers at ordinary temperatures, complex behavior in the stress–strain curve can occur. Considerable plasticity can be induced in SCS at elevated temperatures (>800°C), when silicon softens appreciably and the mobility of defects in the lattice is substantially increased. Huff and Schmidt (1992) actually report a pressure switch exhibiting hysteresis based on buckling of plastically deformed silicon membranes. To eliminate plastic deformation of Si wafers, it is important during high-temperature steps to avoid the presence of films that could stress or even warp the wafer asymmetrically, typically oxides or nitrides.

3.5.5 Piezoresistivity in Silicon

Piezoresistance is the fractional change in bulk resistivity induced by small mechanical stresses applied to a material. Lord Kelvin discovered the effect in 1856. Most materials exhibit piezoresistivity, but the effect, Smith found in 1954, is particularly important in some semiconductors (more than an order of magnitude higher than that of metals) [Smith, 1954]. Monocrystalline silicon has a high piezoresistivity that, combined with its excellent mechanical and electronic properties, makes it a superb material for the conversion of mechanical deformation into an electrical signal. Actually, the history of silicon-based mechanical sensors started with the discovery of the piezoresistance effect in Si (and Ge) more than four decades ago [Smith, 1954]. The piezoresistive effect in semiconductor materials originates in the deformation of the energy bands as a result of the applied stress. The deformed bands change the effective mass and mobility of the charge carriers (electrons and holes), hence modifying the resistivity. The two main classes of piezoresistive semiconductor sensors are membrane-type structures (typically pressure and flow sensors) and cantilever beams (typically acceleration sensors) with in-diffused resistors (boron, arsenic, or phosphorus) strategically placed in zones of maximum stress (Inset 3.6).

For a three-dimensional anisotropic crystal, the electrical field vector is related to the current vector i by a 3-by-3 resistivity tensor [Khazan, 1994]. Experimentally, the nine coefficients are always found to reduce to six, and the symmetric tensor is given by:

$$\begin{bmatrix} E_1 \\ E_2 \\ E_3 \end{bmatrix} = \begin{bmatrix} \rho_1 & \rho_6 & \rho_5 \\ \rho_6 & \rho_2 & \rho_4 \\ \rho_5 & \rho_4 & \rho_3 \end{bmatrix} \begin{bmatrix} i_1 \\ i_2 \\ i_3 \end{bmatrix} \tag{3.22}$$

For the cubic Si lattice, with the axes aligned with the <100> axes, ρ_1, ρ_2, and ρ_3 define the dependence of the electric field on the current along the same direction (one of the <100> directions). And, ρ_4, ρ_5, and ρ_6 are cross-resistivities, relating the electric field to the current along a perpendicular direction.

The six resistivity components in Equation (3.22) depend on the normal (σ) and shear (τ) stresses in the material as defined earlier in this chapter. Smith (1954) was the first to measure the resistivity coefficients for Si at room temperature. The coefficients are dependent on crystal orientation, temperature, and dopant concentration. Table 3.5 lists Smith's results for the three remaining independent resistivity coefficients; that is, π_{11}, π_{12}, and π_{44} of [100] oriented Si at room temperature (Smith, 1954). The piezoresistance coefficients are largest for π_{11} in n-type silicon and π_{44} in p-type silicon, about -102×10^{-11} and $138 \times 10^{-11}\,\text{Pa}^{-1}$, respectively. The piezoresistivity coefficients are related to the gauge factor (G_f) by the Young's modulus. (The gauge factor, the relative resistance change divided by the applied strain, is defined in Equation [3.28].)

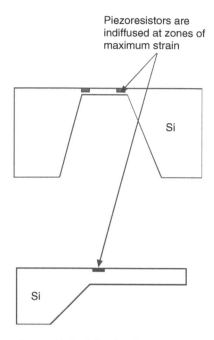

INSET 3.6   Si membrane and Si cantilever with in-diffused resistors.

TABLE 3.5   Resistivity and Piezoresistivity Coefficients at Room Temperature, <100> Si Wafers and Doping Levels below 10^{18} cm^{-3} in 10^{-11} Pa^{-1}

	ρ (Ω cm)	Direction	π_{11}	π_{12}	π_{44}	π_t	π_l
p-Si	7.8	<100>				−1.1	+6.6
		<110>	+6.6	−1.1	+138.1	−66	72
n-Si	11.7	<100>				+53.4	−102.2
		<110>	−102.2	+53.4	−13.6	−18	−31

Source: Smith (1954) and Khazan (1994).

Resistance change can be calculated as a function of the membrane or cantilever beam stress. The contribution to resistance changes from stresses that are longitudinal (σ_l) and transverse (σ_t) with respect to the current flow is given by:

$$\frac{\Delta R}{R} = \sigma_l \pi_l + \sigma_t \pi_t \tag{3.23}$$

where σ_l = longitudinal stress component (i.e., stress component parallel to the direction of the current);

σ_t = transverse stress component (i.e., the stress component perpendicular to the direction of the current);

π_l = longitudinal piezoresistance coefficient; and

π_t = transverse piezoresistance coefficient.

The piezoresistance coefficients π_l and π_t for (100) silicon as a function of crystal orientation are reproduced from Kanda in Figure 3.22A (for p-type) and B (for n-type) [Kanda, 1982]. For lightly doped silicon (n- or p-type $< 10^{18}$ cm^{-3}), the temperature coefficient of resistance (TCR) for π_l and π_t is approximately 0.25% per °C. It decreases with dopant concentration to about 0.1% per °C at 8×10^{19} cm^{-3}. By maximizing the expression for the stress-induced resistance change in Equation (3.23), the achievable sensitivity in a piezoresistive silicon sensor is optimized.

The orientation of a sensing membrane or beam is determined by its anisotropic fabrication. The surface of the silicon wafer is usually a (100) plane; the edges of the etched structures are intersections

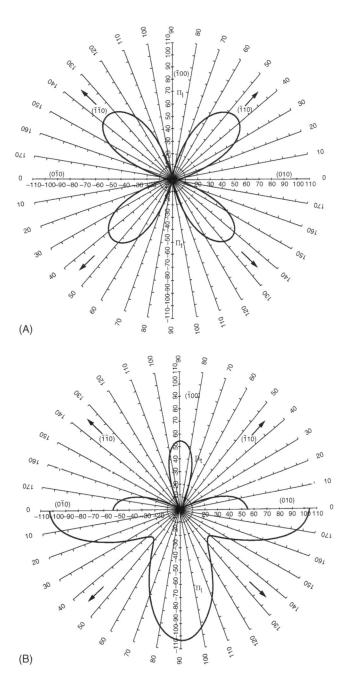

FIGURE 3.22 Piezoresistance coefficients π_l and π_t for (100) silicon in the (001) plane in $10^{-12}\,\text{cm}^2\,\text{dyne}^{-1}$ or $10^{-11}\,\text{Pa}^{-1}$. (A) For p-type. (B) For n-type. (Reprinted with permission from Kanda, Y. [1982] "A Graphical Representation of the Piezoresistance Coefficients in Silicon," *IEEE Trans. Electr. Dev.*, ED-29, pp. 64–70.)

of (100) and (111) planes and are thus <110> directions. For pressure sensing, p-type piezoresistors are most commonly used. This is because the orientation of maximum piezoresistivity (<110>) happens to coincide with the edge orientation of a conventionally etched diaphragm and because the longitudinal coefficient is roughly equal in magnitude but opposite in sign as compared with the transverse coefficient (see Table 3.5 and Figure 3.22A) [Peeters, 1994]. Piezoresistors oriented at 45° with respect to the primary flat, that is, in the <100> direction, are insensitive to applied stress, which provides an inexpensive way

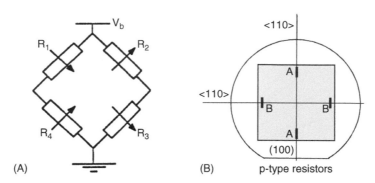

FIGURE 3.23 Measuring on a membrane with piezoresistors. (A) Wheatstone-bridge configuration of four in-diffused piezoresistors. The arrows indicate resistance changes when the membrane is bent downward. (B) Maximizing the piezoresistive effect with p-type resistors. The A resistors are stressed longitudinally and the B resistors are stressed transversally. (Based with permission on Peeters, E. [1994] Process Development for 3D Silicon Microstructures, with Application to Mechanical Sensor Design, Ph.D. thesis, KUL, Belgium, and Maluf, N. [2000] *An Introduction to Microelectromechanical Systems Engineering*, Artech House, Boston.)

to incorporate stress-independent diffused temperature sensors. With the values in Table 3.5, π_l and π_t now can be calculated numerically for any orientation. The longitudinal piezoresistive coefficient in the <110> direction is $\pi_l = 1/2(\pi_{11} + \pi_{12} + \pi_{44})$. The corresponding transverse coefficient is $\pi_t = 1/2(\pi_{11} + \pi_{12} - \pi_{44})$. From Table 3.5, we know that for p-type resistors π_{44} is more important than the other two coefficients, and Equation (3.23) is approximated by:

$$\frac{\Delta R}{R} = \frac{\pi_{44}}{2}(\sigma_l - \sigma_t) \tag{3.24}$$

For n-type resistors, π_{44} can be neglected, and we obtain:

$$\frac{\Delta R}{R} = \frac{\pi_{11} + \pi_{12}}{2}(\sigma_l + \sigma_t) \tag{3.25}$$

Equations (3.24) and (3.25) are valid only for uniform stress fields or if the resistor dimensions are small compared with the membrane or beam size. When stresses vary over the resistors, they have to be integrated, which is done most conveniently by computer simulation programs. More details on the underlying physics of piezoresistivity and its dependence on crystal orientation can be found in Kanda (1982) and Middlehoek and Audet (1989).

To convert the piezoresistive effect into a measurable electrical signal, a Wheatstone bridge is often used. A balanced Wheatstone bridge configuration is constructed as in Figure 3.23A by locating four p-piezoresistors midway along the edges of a square diaphragm as in Figure 3.23B (location of maximum stress). Two resistors are oriented so that they sense stress in the direction of their current axes, and two are placed to sense stress perpendicular to their current flow. Two longitudinally stressed resistors (A) are balanced against two transversely stressed resistors (B); two of them increase in value, and the other two decrease in value upon application of a stress. In this case, from Equation (3.24),

$$\frac{\Delta R}{R} = 70 \times 10^{-11}\,(\sigma_l - \sigma_t) \tag{3.26}$$

with σ in Pa. For a realistic stress pattern where $\sigma_l = 10\,\text{MPa}$ and $\sigma_t = 50\,\text{MPa}$, Equation (3.26) gives us a $\Delta R/R \approx 2.8\%$ [Peeters, 1994].

By varying the diameter and thickness of the silicon diaphragms, piezoresistive sensors in the range of 0 to 200 MPa have been made. The bridge voltages are usually between 5 and 10 V, and the sensitivity may vary from 10 mV/kPa for low-pressure sensors to 0.001 mV/kPa for high-pressure sensors.

A schematic illustration of a pressure sensor with diffused piezoresistive sense elements is shown in Figure 3.24. In this case, p-type piezoresistors are diffused into a thin epitaxial n-type Si layer on a p-type Si substrate [Maluf, 2000].

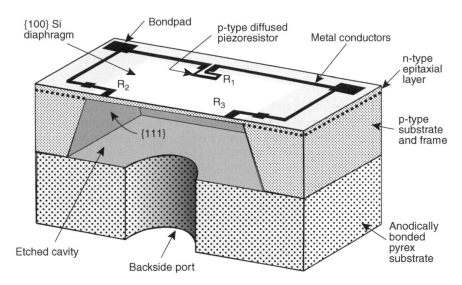

FIGURE 3.24 Schematic illustration of a pressure sensor with p-type diffused resistor in an n-type epitaxial layer (Based with permission on Maluf, N. [2000] *An Introduction to Microelectromechanical Systems Engineering*, Artech House, Boston.)

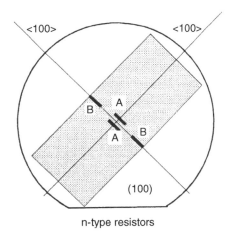

n-type resistors

FIGURE 3.25 Higher pressure sensitivity by strategic placement of in-diffused piezoresistors proposed by Peeters [Peeters, 1994]. The n-type resistors are stressed longitudinally with the A resistors under tensile stress and the B resistors under compressive stress. (Reprinted with permission from Peeters, E. [1994] Process Development for 3D Silicon Microstructures, with Application to Mechanical Sensor Design, Ph.D. thesis, KUL, Belgium.)

Peeters (1994) shows how a more sensitive device could be based on n-type resistors when all the n-type resistors oriented along the <100> direction are subjected to a uniaxial stress pattern in the longitudinal axis as shown in Figure 3.25. The overall maximum piezoresistivity coefficient (π_l in the <100> direction) is substantially higher for n-type silicon than it is for p-type silicon in any direction (maximum π_t and π_l in the <100> direction, Figure 3.22B). Exploitation of these high piezoresistivity coefficients is less obvious, though, since the resistor orientation for maximum sensitivity (<100>) is rotated over 45° with respect to the <110> edges of an anisotropically etched diaphragm. Also evident from Figure 3.22B is that a transversely stressed resistor cannot be balanced against a longitudinally stressed resistor. Peeters has circumvented these two objections by a uniaxial longitudinal stress pattern in the rectangular diaphragm represented in Figure 3.25. With a (100) substrate and a <100> orientation (45° to wafer flat), we obtain:

$$\frac{\Delta R}{R} = (53\sigma_t - 102\sigma_l) \times 10^{-11} \tag{3.27}$$

with σ in Pa. Based on Equation (3.27), with $\sigma_t = 10\,\text{MPa}$ and $\sigma_l = 50\,\text{MPa}$, $\Delta R/R \approx -4.6\%$. In the proposed stress pattern, it is important to minimize the transverse stress by making the device truly uni-axial, as the longitudinal and transverse stress components have opposite effects and can even cancel out. In practice, a pressure sensor with an estimated 65% gain in pressure sensitivity over the more traditional configurations could be made in the case of an 80% uniaxiality [Peeters, 1994].

The piezoresistive effect is often described in terms of the gauge factor, G_f, defined as:

$$G_f = \frac{1}{\varepsilon}\frac{\Delta R}{R} \tag{3.28}$$

which is the relative resistance change divided by the applied strain. The gauge factor of a metal strain gauge is typically around 2; for single-crystal Si it is 90; and for polycrystalline and amorphous Si, it is between 20 and 40.

3.5.6 Bending of Thin Si Plates

Above, we learned that silicon pressure sensors work on the principle of converting strains in a deformed thin silicon diaphragm (due to an applied pressure) to a desired form of electronic output. In most cases, these silicon diaphragms can be treated as thin plates subjected to bending by a uniformly applied pressure.

The governing differential equation of a rectangular plate subject to lateral bending can be expressed as [Timoshenko and Woinowsky-Krieger, 1959]:

$$\left(\frac{\partial^2}{\partial x^2} + \frac{\partial^2}{\partial y^2}\right)\left(\frac{\partial^2 w}{\partial x^2} + \frac{\partial^2 w}{\partial y^2}\right) = \frac{p}{D} \tag{3.29}$$

where $w = w(x, y)$ is the lateral deflection of a flat plate due to the applied pressure, p. The x–y plane defines the plate as shown in Figure 3.26. The parameter D is the flexural rigidity of the plate, which can be expressed as:

$$D = \frac{Et^3}{12(1 - v^2)} \tag{3.30}$$

where E and v are respectively the Young's modulus and Poisson's ratio of the plate material, and t is the thickness of the plate. (See also Equation [3.51].) The components of the bending moments M_x and M_y,

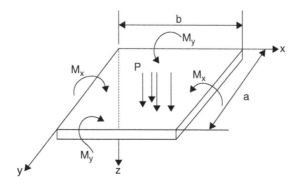

FIGURE 3.26   Bending of a rectangular plate.

and the twisting moment M_{xy}, can be computed from the solution of Equation (3.29) as:

$$M_x = -D\left(\frac{\partial^2 w}{\partial x^2} + v\frac{\partial^2 w}{\partial y^2}\right)$$

$$M_y = -D\left(\frac{\partial^2 w}{\partial y^2} + v\frac{\partial^2 w}{\partial x^2}\right) \qquad (3.31)$$

$$M_{xy} = D(1 - v)\left(\frac{\partial^2 w}{\partial x \partial y}\right)$$

and the maximum bending stresses derived from Equation (3.29) are:

$$(\sigma_{xx})_{max} = \frac{6(M_x)_{max}}{h^2}$$

$$(\sigma_{yy})_{max} = \frac{6(M_y)_{max}}{h^2} \qquad (3.32)$$

$$(\sigma_{xy})_{max} = \frac{6(M_{xy})_{max}}{h^2}$$

To solve Equation (3.29) for plate deflections under various boundary conditions, software packages such MATLAB or MATHCAD may be used. We return to the topic of stress and bending in thin films later in this chapter.

3.5.7 Silicon as a Mechanical MEMS Material: Summary

Mechanical stability is crucial for mechanical sensing applications. Any sensing device must be free of drift to avoid recalibration at regular intervals. Part of the drift in mechanical sensors may be associated with movement of crystal dislocations in the loaded mechanical part. In ductile materials such as metals dislocations move readily. By contrast, in brittle materials such as semiconductors dislocations hardly move. Mechanical engineers often avoid using brittle materials and opt for ductile materials, even though these plastically deform, meaning that they are subject to mechanical hysteresis. Single-crystal Si can be made virtually without defects and under an applied load no dislocation lines can move. This means that at room temperature Si can be deformed only elastically. This last property, coupled with an extremely high yield strength (comparable to steel), makes Si a material superior to any metal in most applications. As a consequence, Si has been quite successful as a structural element in mechanical sensors, particularly over the last 15 years. Pressure and acceleration sensors based on simple piezoresistive elements embedded in a silicon movable mechanical member have turned into major commercial applications. Besides the desirable mechanical and known electrical properties, this success also can be attributed to the available fabrication technology of integrated circuits.

The significant properties that have made Si a successful material, not only for electronic applications but also for mechanical applications, are reviewed in Tables 3.6 and 3.7. From these tables, we reiterate some reasons behind the success of Si as a mechanical sensor element:

- Silicon surpasses stainless steel in yield strength and also displays a density lower than that of aluminum. In fact, silicon's specific strength, defined as the ratio of yield strength to density, is significantly higher than for most common engineering materials.
- The hardness of Si is slightly better than that of stainless steel; it approaches that of quartz and is higher than most common glasses.
- The Young's modulus of Si has a value approaching that of stainless steel and is well above that of quartz. From Tables 3.6 and 3.7, we also note that Si_3N_4, a coating often used on silicon, has a hardness topped only by a material such as diamond. The combination of Si with Si nitride coatings, therefore, can be used for highly wear-resistant components as required in micromechanisms such as micromotors.

TABLE 3.6   Mechanical Properties of Single Crystal Silicon (SCS) among Other Technological Materials

	Yield Strength $(10^9 \text{N/m}^2 =$ GPa)	Specific Strength $[10^3 \text{m}^2\text{s}^{-2}]$	Knoop Hardness (kg/mm^2)	Young's Modulus $(10^9 \text{N/m}^2 =$ Gpa)	Density (10^3kg/m^3)	Thermal Conductivity at 300 K (W/cmK)	Thermal Expansion $(10^{-6}/°\text{C})$
Diamond (SC)	53	15000	7000	10.35	3.5	20	1.0
Si(SCS)	2.8–6.8	3040	850–1100	190(111)	2.32	1.56	2.616
GaAs(SC)	2.0			0.75	5.3	0.81	6.0
Si_3N_4	14	4510	3486	323	3.1	0.19	2.8
SiO_2 (fibers)	8.4		820	73	2.5	0.014	0.4–0.55
SiC (6H-SiC)	21	6560	2480	448	3.2	5	4.2
Iron	12.6		400	196	7.8	0.803	12
Tungsten (W)	4	210	485	410	19.3	1.78	4.5
Al	0.17	75	130	70	2.7	2.36	25
AlN	16			340		1.60	4.0
Al_2O_3	15.4		2100	275	4.0	0.5	5.4–8.7
Stainless steel	0.5–1.5		660	206–235	7.9–8.2	0.329	17.3
Quartz values							
// Z	9		850	107	2.65	0.014	7.1
⊥ Z							13.2 (increases with T)
Polysilicon	1.8 (annealed)			161			2.8

Note: SC = single crystal; SCS = single crystal silicon; T = temperature; ‖ Z = parallel with Z-axis; ⊥ Z = perpendicular to Z-axis.

TABLE 3.7   Silicon Single-Crystal Material Characteristics

Si Parameter	Value and Comment
Atomic weight	28.1
Atoms/cm^3	5×10^{22}
Band gap at 300 K	1.12 eV. Silicon has a high band gap, making it useful electrically at high temperatures. Indirect band-gap in the near infrared. It is opaque to ultraviolet and transparent to IR
Chemical resistance	High. Silicon is resistant to most acids, except combinations of HF/HNO_3 and certain bases
Density (gr/cm^3)	2.4 Si has a lower density than aluminum (2.7)
Dielectric constant	11.9 vs. 13.1 for GaAs
Dielectric strength (V/cm10^6)	3
Dislocation density	$<100 \text{ cm}^2$ IC grade silicon contains virtually no imperfections; thus it is relatively insensitive to cycling and fatigue failure
Electron mobility (cm^2/Vs)	1500
Hole mobility (cm^2/Vs)	400
Intrinsic carrier concentration (cm^{-3})	1.45×10^{10}
Intrinsic resistivity (Ω cm)	2.3×10^5 vs. 10^8 for GaAs
Knoop hardness (kg/mm^2)	850 (stainless steel is 820). Si is harder than steel and can readily be coated with silicon nitride, providing high abrasion resistance
Lattice constant (Å)	5.43
Linear coefficient of thermal expansion at 300 K $(10^{-6}/°\text{C})$	2.6. The low expansion coefficient of Si is closer to quartz than to metal, making it insensitive to thermal shock
Melting point	1415°C. Silicon is a high melting material making it suitable for high temperature applications
Minority carrier lifetime (s)	2.5×10^{-3}
Oxide growth	Si grows a dense, strong, chemically resistant, passivating layer of SiO_2. This oxide is an excellent thermal insulator with a low expansion coefficient
Poisson ratio	0.22
Relative permittivity	11.8

(Continued)

TABLE 3.7   (*Continued*)

Si Parameter	Value and Comment
Silicon nitride	A typical coating for Si with a hardness and wear resistance topped only by diamond
Specific heat at 300 K (J/gK)	0.713
Thermal conductivity at 300 K (W/cmK)	1.56 Si has a high thermal conductivity comparable to metals such as carbon steel (0.97) and Al (2.36)
Temperature coefficient of Young's modulus (10^{-6} K^{-1}) at 300 K	−90
Temperature coefficient of piezoresistance (10^{-6} K^{-1}) at 300 K (doping $<10^{18}$ cm^{-3})	−2500
Temperature coefficient of permittivity (10^{-6} K^{-1}) at 300 K	1000
Thermal diffusivity (cm²/s)	0.9
Yield strength (GPa)	7 (steel is 2.1). IC grade Si is stronger than steel
Young's modulus E (GPa)	190 [111] direction. The elastic modulus is similar to that of steel (steel is 200)

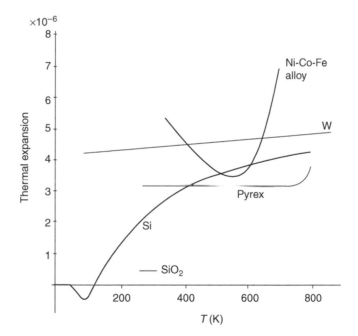

FIGURE 3.27   Thermal expansion coefficient versus absolute temperature. (Reprinted with permission from Greenwood, J.C. [1988] "Silicon in Mechanical Sensors," *J. Phys. E: Sci. Instrum.*, 21, pp. 1114–28.)

3.6   Other Si Sensor Properties

3.6.1   Thermal Properties of Silicon

In Figure 3.27, the thermal expansion coefficient of Si, W, SiO$_2$, Ni-Co-Fe alloy, and Pyrex® is plotted against absolute temperature. Single-crystal silicon has a high thermal conductivity (comparable to metals such as steel and aluminum and approximately 100 times larger than that of glass) and a low thermal expansion coefficient. Because of its thermal conductivity, Si is used in some devices as an efficient heat sink. Its thermal expansion coefficient is closely matched to Pyrex® glass but exhibits considerable temperature dependence. A good match in thermal expansion coefficient between the device wafer and the support substrate is required. A poor match introduces stress, which degrades the device performance. This makes it difficult to fabricate composite structures of Pyrex® and Si that are stress-free over a wide range of temperatures. Drift in silicon sensors often stems from packaging. In this respect, several types of stress-relief (subassemblies for stress-free mounting of the active silicon parts) play a major role; using silicon as the support for silicon sensors is highly desirable. Anodic bonding of Pyrex® glass to Si in this respect is inferior to fusion bonding of Si to Si [Shimbo et al., 1986; Tong et al., 1994; Barth, 1990].

Although the Si band gap is relatively narrow, by employing silicon-on-insulator (SOI) wafers (see Section 3.18.1), high-temperature sensors can be fashioned. For the latter application, relatively highly doped Si, which is relatively linear in its temperature coefficient of resistance and sensitivity over a wide range, typically is employed.

When fabricating thermally isolated structures on Si, the large thermal conductivity of Si poses a considerable problem, as the major heat leak occurs through the Si material. For thermally isolated structures, machining in glass, quartz, or ceramics, with their lower thermal conductivity, represents an important alternative.

3.6.2 Silicon Optical Properties

Silicon features an indirect band gap and is not an active optical material. As a consequence, silicon-based lasers do not exist. Silicon is effective only in detecting light as the indirect band gap makes emission of light difficult. Above 1.1 μm, silicon is transparent, but at wavelengths shorter than 0.4 μm (in the blue and ultraviolet portions of the visible spectrum), it reflects over 60% of the incident light. One of the most established applications for silicon sensors, although not often typified as MEMS, is visual imaging with charge-coupled devices (CCDs). In a CCD imager, each element in a two-dimensional array generates an electrical charge in proportion to the amount of light it receives. The charge, stored by CCD elements along a row, is subsequently transferred to the next element in bucket-brigade fashion as the light input is read out line by line. The number of picture elements (pixels) on the CCD determines the resolution. In the top-of-the-line CCD cameras (<http://www.pctechguide.com/19digcam.htm>), a 3.34 megapixel CCD is capable of delivering a maximum image size of 2048 × 1536 pixels. Si as the pixel semiconductor can be used for a wide variety of electromagnetic radiation wavelengths, from gamma rays to infrared. There is a trend now to configure pixels in clever ways to make novel optical sensors feasible.

One example of a smart pixel configuration is embodied in the retina chip shown in Figure 3.28 [IMEC, 1994]. This is an integrated circuit chip working like the human retina to select out only the necessary information from a presented image to greatly speed up image processing. The chip features 30 concentric circles

FIGURE 3.28   Photo of the retina sensor. (Courtesy of Dr. Lou Hermans, IMEC, Belgium.)

with 64 pixels each. The pixels increase in size from $30 \times 30\,\mu m$ for the inner circle to $412 \times 412\,\mu m$ for the outer circle. The circle radius increases exponentially with eccentricity. The center of the chip, called the fovea, is filled with 104 pixels measuring $30 \times 30\,\mu m$ placed on an orthogonal pattern. The total chip area is 11×11 mm. The chip is designed for those applications in which real-time coordination between the sensory perception and the system control is of prime concern. The main application area is active vision, and its potential application is expected in robot navigation, surveillance, recognition, and tracking. The system can cover a wide field of view with a relatively low number of pixels without sacrificing the overall resolution and leads to a significant reduction in required image processing hardware and calculation time. The fast, but rather insensitive, large pixels on the rim of the retina chip pick up a sudden movement in the scenery swiftly (peripheral vision), prompting the robot equipped with this "eye" to redirect itself in the direction of the movement in order to focus better on the moving object with the more sensitive fovea pixels.

The area in which MEMS has seen one of its biggest commercial breakthroughs lately is micromachined mirrors for optical switching in both fiber optic communications and data storage applications. Optical switches are to optical communications what transistors are to electronic signaling. What makes Si single-crystal attractive in this case is the optical quality of the Si surface (see example 4 in section 3.12, "Wet Bulk Micromachining Examples"). The quality of the mirror surface is primordial to obtain very low insertion loss even after multiple reflections. Wet bulk micromachining has an advantage here over deep reactive ion etching (DRIE), as the latter leaves lossier Si mirror surfaces due to the inevitable ripples [Hélin et al., 2000].

3.6.3 Biocompatibility of Si

Biocompatibility has both a mechanical and a chemical component. For example, it is well known that for better biocompatibility a smooth surface and rounded corners are preferable to a rough surface, sharp corners, and small crevices; that for a given surface finish and shape, carbon is preferable to copper. The research community remains somewhat divided on the relative contribution of mechanical effects (e.g., surface roughness and shape) vs. chemical effects (e.g., Cu vs. C) to the overall bioresponse of an implant. von Recum's work seems to demonstrate that micromachining (microtexturing) of surfaces of different materials may help deconvolute these relative contributions [von Recum and Cooke, 1988].

Some specific examples of substrate microtexturing effects follow. An increase in biosynthetic activity and mobility of bone cells on polymer films cast from micromachined molds has been identified (this might have applications for bone regeneration and bioengineering of prosthetics). The regeneration of severed tendons only on textured surfaces has been demonstrated and might lead to the use of textured bandages to accelerate healing of tissue. Many examples in the MEMS literature also illustrate directed growth of nerve cells on micromachined structures. In a typical procedure, a hydrophobic silane pattern on a glass slide is produced by "lift-off," where the hydrophobic silane (dimethyl-trichlorosilane) replaces the more familiar metallization step. Hydrophobic patterning is followed by the deposition of an amino silane (3-aminoethylaminopropyltrimethoxysilane) in the remaining areas, the amino group forming an attachment site for proteins [Cooper et al., 1993; Connolly et al., 1992]. Successful chemical patterning of glass slide surfaces is demonstrated by the growth pattern of cells. Some cells, such as fibroblasts, grow in hydrophilic areas but have difficulty adhering and spreading in hydrophobic areas. Sharp edge definition is thus obtained along edges of hydrophilic/hydrophobic interfaces. Using this technique, neuronal growth can be controlled, and neuronal processes may be induced to follow the straight lines of the original mask pattern. It has been suggested that this type of approach could be applied to nerve repair techniques. The method and its range of applications are described in Britland et al. (1990). Surface characteristics of BIOMEMS materials that should be considered include not only chemistry and surface texture but also hydrophilicity, charge, polarity and energy, heterogeneous distribution of functional groups, water absorption, and chain mobility (for a polymer).

The above background on the many aspects determining biocompatibility may explain why literature on biocompatibility of Si remains somewhat an enigma; it is often hard to understand what is meant when a Si device is proclaimed to be biocompatible. Maluf, without providing references, claims that preliminary medical evidence indicates that silicon even without protein adsorption control is benign in the body [Maluf, 2000]. In reality, it remains of considerable interest to improve the biocompatibility of silicon microdevices and engineer surfaces with reduced protein and cell adherence for the prevention of fouling and fibrotic response — especially when dealing with active silicon devices in contact with blood (e.g., a Si-based glucose sensor) [Zhang et al., 1998]. Even if significant control over protein adsorption and cell adhesion is achieved, Si remains a brittle material that may shard in the body and, just like glass, is to be avoided in any load-bearing application in living subjects or foodstuffs (e.g., as a micromachined needle). Although the motivation for most MEMS devices (especially active ones) is usually to eliminate any adhesion between proteins and surfaces, in some cases, the ability to enable protein adhesion locally but not in other areas and thus provide for a background signal (e.g., for cell-based assays) clearly requires two different surface chemistries. Studies at the University of Michigan and MIT, according to Issys's Website, demonstrated that boron-doped single-crystal Si has a superior biocompatibility when compared with standard single-crystal Si or polysilicon-based devices. Some quite definitive work has been done with Si for cortical implants. Si electrode arrays were proposed as permanent cortical implants with the purpose of restoring useful vision, and their biocompatibility with neural tissue was investigated. In this context, silicon electrode arrays were left in place within a rabbit cortex for six months. Afterward, the neuron density as a function of radius from the center of the electrode shafts was used as the means of assessing damage. The researchers concluded that the electrode array showed long-term biocompatibility because normal brain tissue was observed within microns of the surface of the shafts [Clark et al., 1992; Horch et al., 1993].

3.7   Wet Isotropic and Anisotropic Etching

3.7.1   Introduction

Wet etching is used for cleaning, shaping 3-D structures, removing surface damage, polishing, and characterizing structural and compositional features [Uhlir, 1956]. The materials etched include semiconductors, conductors, and insulators. The most important parameters in chemical etching are bias (undercut), tolerance, etch rate, anisotropy, selectivity, overetch, feature size control, and loading effects. The emphasis in this chapter is on etching Si. Wet chemical etching of Si provides a higher degree of selectivity than dry etching techniques. Wet etching usually is also faster: a few microns to tens of microns per minute for isotropic etchants and about 1 μm/min for anisotropic wet etchants vs. 0.1 μm/min in typical dry etching. More recently, though, with inductively coupled plasma (ICP) dry etching of Si, rates of up to 6 μm/min are achieved. Modification of wet etchant and/or temperature can alter the selectivity to silicon dopant concentration and type and, especially when using alkaline etchants, to crystallographic orientation. Etching proceeds by (1) reactant transport to the surface, (2) surface reaction, and (3) reaction product transport away from the surface. If either (1) or (3) is rate determining, etching is diffusion limited and may be increased by stirring. If (2) is the rate-determining step, etching is reaction-rate limited and depends strongly on temperature, etching material, and solution composition. Diffusion-limited processes have lower activation energies (on the order of a few Kcal/mol) than reaction-rate-controlled processes and therefore are relatively insensitive to temperature variations. In general, reaction-rate limitation is preferred, as it is easier to reproduce a temperature setting than a stirring rate. The best generic etching apparatus has both a good temperature controller and a reliable stirring facility [Kaminsky, 1985; Stoller et al., 1970].

Isotropic etchants (also polishing etchants) etch in all crystallographic directions at the same rate; they usually are acidic, such as $HF/HNO_3/CH_3COOH$ (HNA), and lead to rounded isotropic features in single-crystalline Si. The HNA etch is also known as the poly-etch because in the early days of the integrated circuit industry it was used as an etchant for polysilicon [Maluf, 2000]. Isotropic etchants are used at room temperature or slightly above (<50°C). Historically, they were the first Si etchants introduced [Robbins and Schwartz, 1959; 1960; Schwartz and Robbins, 1961; 1976; Uhlir, 1956; Hallas, 1971; Turner, 1958; Kern, 1978; Klein and D'Stefan, 1962]. Later, it was discovered that some alkaline chemicals will etch anisotropically; that is, they etch away crystalline silicon at different rates depending on the orientation of the exposed crystal plane. Typically, the pH stays above 12, while more elevated temperatures are used for these slower-type etchants (>50°C). The latter type of etchants surged in importance in the late 1960s for the fabrication of dielectrically isolated structures in silicon [Stoller and Wolff, 1966; Stoller, 1970; Forster and Singleton, 1966; Kenney, 1967; Lepselter, 1966, 1967; Waggener, 1970; Kragness and Waggener, 1973; Waggener et al., 1967a; Waggener et al., 1967b; Bean and Runyan, 1977; Rodgers et al., 1977; Rodgers et al., 1976; Ammar and Rodgers, 1980; Schnable and Schmidt, 1976]. Isotropic etchants typically show diffusion limitation, while anisotropic etchants are reaction rate limited. In both cases, the two principal reactions are oxidation of the silicon followed by dissolution of the hydrated silicate.

Preferential or selective etching (also structural etching) is usually isotropic but exhibits some anisotropy [Kern and Deckert, 1978]. These etchants are used to produce a difference in etch rate between different materials or between compositional or structural variations of the same material on the same crystal plane. These types of etches are often the fastest and simplest techniques to delineate electrical junctions and to evaluate the structural perfection of a single crystal in terms of, for example, slip and stacking faults. The artifacts introduced by the defects etch into small pits of characteristic shape. Most of the etchants used for this purpose are acids with an oxidizing additive [Yang, 1984; Chu and Gavaler, 1965; Archer, 1982; Schimmel and Elkind, 1973; d'Aragona, 1972].

3.7.2 Isotropic Etching

3.7.2.1 Usage of Isotropic Etchants

When etching silicon with aggressive acidic etchants, rounded isotropic patterns form. This method is widely used for

1. Removal of work-damaged surfaces
2. Rounding of sharp anisotropically etched corners (to avoid stress concentration)
3. Removing of roughness after dry or anisotropic etching
4. Creating structures or planar surfaces in single-crystal slices (thinning)
5. Patterning single-crystal, polycrystalline, or amorphous films
6. Delineation of electrical junctions and defect evaluation (with preferential isotropic etchants)

For isotropic etching of silicon, the most commonly used etchants are mixtures of nitric acid (HNO_3) and hydrofluoric acids (HF). Water can be used as a diluent, but acetic acid (CH_3COOH) is preferred because it prevents the dissociation of the nitric acid better and so preserves the oxidizing power of HNO_3, which depends on the undissociated nitric acid species for a wide range of dilution [Robbins and Schwartz, 1960]. The etchant is called the HNA system; we will return to this etch system below.

3.7.2.2 Simplified Reaction Scheme

In acidic media, the Si etching process involves hole injection into the Si valence band by an oxidant, an electrical field, or photons. Nitric acid in the HNA system acts as an oxidant; other oxidants such as H_2O_2 and Br_2 also work [Tuck, 1975]. The holes attack the covalently bonded Si, oxidizing the material, followed by a reaction of the oxidized Si fragments with OH^- and subsequent dissolution of the silicon oxidation products in HF. The following reactions describe these processes.

The holes are, in the absence of photons and an applied field, produced by HNO_3, along with water and trace impurities of HNO_2:

$$HNO_3 + H_2O + HNO_2 \rightarrow 2HNO_2 + 2OH^- + 2H^+ \qquad \text{(Reaction 3.1)}$$

The holes in reaction (3.1) are generated in an autocatalytic process; HNO_2 generated in the above reaction re-enters into the further reaction with HNO_3 to produce more holes. With a reaction of this type, an induction period begins before the oxidation reaction takes off and continues until a steady-state concentration of HNO_2 has been reached. This has been observed at low HNO_3 concentrations [Tuck, 1975].

After hole injection, OH^- groups attach to the oxidized Si species to form SiO_2, liberating hydrogen in the process:

$$Si^{4+} + 4OH^- \rightarrow SiO_2 + H_2 \qquad \text{(Reaction 3.2)}$$

Hydrofluoric acid (HF) dissolves the SiO_2 by forming the water-soluble H_2SiF_6. The overall reaction of HNA with Si looks like:

$$Si + HNO_3 + 6HF \rightarrow H_2SiF_6 + HNO_2 + H_2O + H_2(\text{bubbles}) \qquad \text{(Reaction 3.3)}$$

The simplification in the above reaction scheme is that only holes are taken into account. In the actual Si acidic corrosion reaction, both holes and electrons are involved. The question of hole and/or electron participation in Si corrosion will be considered after we introduce the model for the Si/electrolyte inter-facial energetics. We will learn from that model that the rate-determining step in acidic etching involves hole injection in the valence band, whereas in alkaline anisotropic etching it involves electron injection in the conduction band by surface states. The reactivity of a hole injected in the valence band is signifi-cantly greater than that of an electron injected in the conduction band. The observation of isotropy in acidic etchants and anisotropy in alkaline etchants centers on this difference in reactivity.

3.7.2.3 Iso-Etch Curves

By the early 1960s, the isotropic HNA silicon etch was well characterized. Schwartz and Robbins pub-lished a series of four very detailed papers on the topic between 1959 and 1976 [Robbins and Schwartz, 1959, 1960; Schwartz and Robbins, 1961; 1976]. Most of the material presented below is based on their work.

HNA etching results represented as iso-etch curves for various weight percentages of the constituents are shown in Figure 3.29. For this work, normally available concentrated acids of 49.2 wt% HF and 69.5 wt% HNO_3 are used. Water as diluent is indicated by dashed-line curves and acetic acid by solid-line curves. Acetic acid is less polar than water and helps prevent the dissociation of HNO_3, thereby allowing the formation of more of the species directly responsible for the oxidation of Si. A typical formulation for HNA is 250 mL HF, 500 mL HNO_3, and 800 mL CH_3COOH. When used at room temperature, this for-mulation results in an etch rate of about 4 to 20 μm/min, increasing with agitation [Kovacs et al., 1998]. In Figure 3.29, as in Wong's representation [Wong, 1990], we have recalculated the curves from Schwartz et al. to express the etch rate in μm/min and divided the authors' numbers by 2 because we are consider-ing one-sided etching only. The highest etch rate is observed around a weight ratio HF-HNO_3 of 2:1 and is nearly 100 times faster than anisotropic etch rates. Adding a diluent slows the etching. From these curves, the following characteristics of the HNA system can be summarized:

1. At high HF and low HNO_3 concentrations, the iso-etch curves describe lines of constant HNO_3 concentrations (parallel to the HF-diluent axis); consequently, the HNO_3 concentration controls the etch rate. Etching at those concentrations tends to be difficult to initiate and exhibits an uncer-tain induction period. In addition, it results in relatively unstable silicon surfaces proceeding to slowly grow a layer of SiO_2 over time. The etch is limited by the rate of oxidation so that it tends to be orientation dependent and affected by dopant concentration, defects, and catalysts (sodium nitrate often is used). In this regime, the temperature influence is more pronounced, and activa-tion energies for the etching reaction of 10 to 20 Kcal/mol have been measured.

2. At low HF and high HNO_3 concentrations, iso-etch curves are lines parallel to the nitric-diluent axis; that is, they are at constant HF composition. In this case, the etch rate is controlled by the abil-ity of HF to remove the SiO_2 as it is formed. Etches in this regime are isotropic and truly polish-ing, producing a bright surface with anisotropies of 1% or less (favoring the <110> direction) when used on <100> wafers [Wise et al., 1981]. An activation energy of 4 Kcal/mol is indicative of the diffusion-limited character of the process; consequently, in this regime, temperature changes are less important.

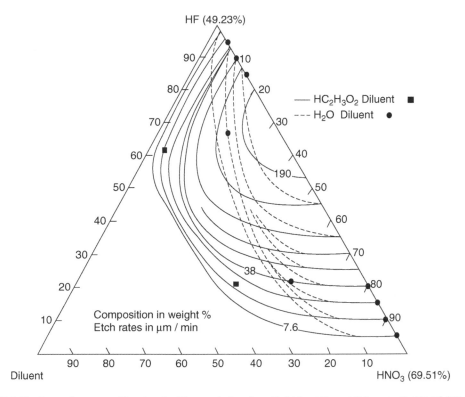

FIGURE 3.29 Iso-etch curves. (Reprinted with permission from Robbins, H., and Schwartz, B. [1960] "Chemical Etching of Silicon-II: The System HF, HNO_3, H_2O, and $HC_2C_3O_2$," *J. Electrochem. Soc.* **107**, pp. 108–11; recalculated for one-sided Si etching and expressed in μm/min.

3. In the region of maximal etch rate, both reagents play an important role. The addition of acetic acid, as opposed to the addition of water, does not reduce the oxidizing power of the nitric acid until a fairly large amount of diluent has been added. Therefore, the rate contours remain parallel with lines of constant nitric acid over a considerable range of added diluent.

4. In the region around the HF vertex, the surface reaction rate-controlled etch leads to rough, pitted Si surfaces and sharply peaked corners and edges. In moving toward the HNO_3 vertex, the diffusion-controlled reaction results in the development of rounded corners and edges, and the rate of attack on (111) planes and (110) planes becomes identical in the polishing regime (anisotropy less than 1%; see point 2).

Figure 3.30 summarizes how the topology of the Si surfaces depends strongly on the composition of the etch solution. Around the maximum etch rates, the surfaces appear quite flat with rounded edges, and very slow etching solutions lead to rough surfaces [Schwartz and Robbins, 1976].

3.7.2.4 Arrhenius Plot for Isotropic Etching

The effect of temperature on the reaction rate in the HNA system was studied in detail by Schwartz and Robbins (1961). An Arrhenius plot for etching Si in 45% HNO_3, 20% HF, and 35% $HC_2H_3O_2$ culled from their work is shown in Figure 3.31. Increasing the temperature increases the reaction rate. The graph shows two straight-line segments indicating a higher activation energy below 30°C and a lower one above this temperature. In the low temperature range, etching is preferential, and the activation energy is associated with the oxidation reaction. At higher temperatures, the etching leads to smooth surfaces, and the activation energy is lower and associated with diffusion-limited dissolution of the oxide [Schwartz and Robbins, 1961].

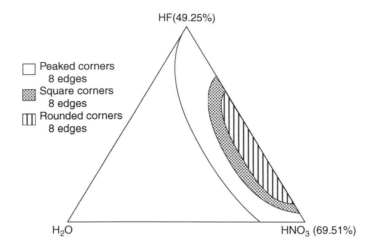

FIGURE 3.30 Topology of etched Si surfaces. (Reprinted with permission from Schwartz, B., and Robbins, H. [1976] "Chemical Etching of Silicon-IV: Etching Technology," *J. Electrochem. Soc.* **123**, 1903–9.)

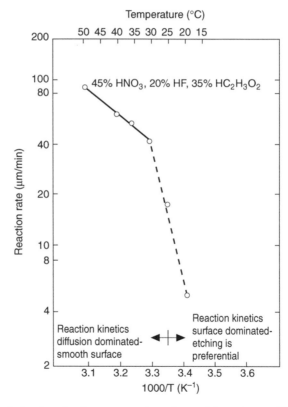

FIGURE 3.31 Etching an Arrhenius plot. Temperature dependence of the etch rate of Si in HF:HNO$_3$:CH$_3$:COOH (1:4:3). (Reprinted with permission from Schwartz, B., and Robbins, H. [1961] "Chemical Etching of Silicon-III: A Temperature Study in the Acid System," *J. Electrochem. Soc.* **108**, pp. 365–72.)

With isotropic etchants, the etchant moves downward and outward from an opening in the mask, undercuts the mask, and enlarges the etched pit while deepening it (Figure 3.32). The resulting isotropically etched features show more symmetry and rounding when agitation accompanies the etching (the process is diffusion limited). This agitation effect is illustrated in Figure 3.32. With agitation, the etched

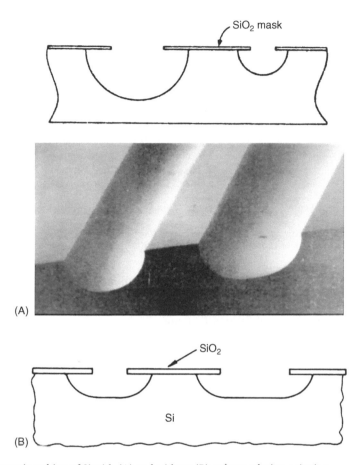

FIGURE 3.32   Isotropic etching of Si with (A) and without (B) etchant solution agitation.

feature approaches an ideal round cup; without agitation, the etched feature resembles a rounded box [Petersen, 1982]. The flatness of the bottom of the rounded box generally is poor because the flatness is defined by agitation.

3.7.2.5   Masking for Isotropic Silicon Etchants

Acidic etchants are very fast; for example, an etch rate for Si of up to $50\,\mu m\,min^{-1}$ can be obtained with 66% HNO_3 and 34% HF (volumes of reagents in the normal concentrated form) [Kern, 1978; Kern and Deckert, 1978]. Isotropic etchants are so aggressive that the activation barriers associated with etching the different Si planes are not differentiated; all planes etch equally fast, making masking a real challenge.

Although SiO_2 has an appreciable etch rate of 300 to 800 Å/min in the HF:HNO_3 system, thick layers of SiO_2 are often used as a mask, especially for shallow etching because the oxide is so easy to form and pattern. A mask of nonetching Au or Si_3N_4 is needed for deeper etching. Photoresists do not stand up to strong oxidizing agents such as HNO_3, and neither does Al.

Silicon itself is soluble to a small extent in pure HF solutions; for a 48% HF at 25°C, a rate of 0.3 Å/min was observed for n-type, 2-Ω-cm (111)-Si. It was established that Si dissolution in HF is not due to oxidation by dissolved oxygen. Diluted HF etches Si at a higher rate because the reaction in aqueous solutions proceeds by oxidation of Si by OH^- groups [Hu and Kerr, 1967]. A typically buffered HF (BHF) solution has been reported to etch Si at radiochemically measured rates of 0.23 to 0.45 Å/min, depending on doping type and dopant concentration [Hoffmeister, 1969].

By reducing the dopant concentration (n or p) to below 10^{17} atoms/cm^3, the etch rate of Si in HNA is reduced by ~ 150 [Muraoka et al., 1973]. The doping dependence of the etch rate provides yet another

TABLE 3.8 Masking Materials for Acidic Etchants

Masking	Piranha (4:1, H_2O_2: H_2SO_4)	Buffered HF (5:1NH_4F: conc.HF)	HNA
Thermal SiO_2		0.1 µm/min	300-800 Å/min. Limited etch time; thick layers often are used due to ease of patterning
CVD (450°C) SiO_2		0.48 µm/min	0.44 µm/min
Corning 7740 glass		0.063 µ/min	1.9 µ/min
Photoresist	Attacks most organic films	OK for short while	Resists do not stand up to strong oxidizing agents like HNO_3 and are not used
Undoped Si polysilicon	Forms 30 Å of SiO_2	0.23 to 0.45 Å/min	0.7 to 40 µm/min at RT (at a dopant concentration $<10^{17}$ cm^{-3} [n or p])
Black wax			Usable at room temperature
Au/Cr	OK	OK	OK
LPCVD Si_3N_4		1 Å/min	Etch rate is 10-100 Å/min. Preferred masking material

Note: The many variables involved necessarily mean that the given numbers are approximate only.

means of patterning a Si surface (see next section). Table 3.8 summary of masks that can be used in acidic etching is presented.

3.7.2.6 Dopant Dependence of Silicon Isotropic Etchants

The isotropic etching process is fundamentally a charge-transfer mechanism. This explains the etch rate dependence on dopant type and concentration. Typical etch rates with an HNA system (1:3:8) for n- or p-type dopant concentrations above 10^{18} cm^{-3} are 1 to 3 µm/min. As presented in the preceding section, a reduction of the etch rate by 150 times is obtained in n- or p-type regions with a dopant concentration of 10^{17} cm^{-3} or smaller [Muraoka et al., 1973]. This presumably is due to the lower mobile carrier concentration available to contribute to the charge transfer mechanisms. In any event, heavily doped silicon substrates with high conductivity can be etched more readily than lightly doped materials. Dopant-dependent isotropic etching can also be exploited in an electrochemical setup as described in the next section. Although doping does change the chemical etch rate, attempts to exploit these differences for industrial production have failed [Seidel, 1989]. This situation is different in electrochemical isotropic etching (see next section).

3.7.2.7 Electrochemical Isotropic Silicon Etch–Etch Stop

Sometimes, a high-temperature or extremely aggressive chemical etching process can be replaced by an electrochemical procedure utilizing a much milder solution, thus allowing a simple photoresist mask to be employed [Kern and Deckert, 1978]. In electrochemical acidic etching with or without illumination of the corroding Si electrode, an electrical power supply is employed to drive the chemical reaction by supplying holes to the silicon surface (W-EL, Figure 3.33A). A voltage is applied across the silicon wafer and a counter electrode (C-EL, usually platinum) arranged in the same etching solution. Oxidation is promoted by a positive bias applied to the silicon, causing an accumulation of holes in the silicon at the silicon/electrolyte interface. Under this condition, oxidation at the surface proceeds rapidly while the oxide is readily dissolved by the HF solution. No oxidant such as HNO_3 is needed to supply the holes; excess electron-hole pairs are created by the electrical field at the surface and/or by optical excitation, thereby increasing the etch rate. This technique proved successful in removing heavily doped layers leaving behind the more lightly doped membranes in all possible dopant configurations: p on p$^+$, p on n$^+$, n on p$^+$, and n on n$^+$ [Theunissen et al., 1970; Meek, 1971]. This electrochemical etch-stop technique is demonstrated in Figure 3.33B [Dijk and Jonge, 1970]. A 5% HF solution is used, the electrolyte cell is kept in the dark at room temperature, and the distance between the Si anode and the Pt cathode in the electrochemical cell is 1 to 5 cm. Instead of using HF, one can substitute NH_4F (5 wt%) for the electrochemical etching as described by Shengliang [Shengliang et al., 1987]. Shengliang reports a selectivity of n-type silicon to n$^+$-type silicon (0.001 Ω-cm) of 300 with the latter etchant. In Figure 3.33B, the current density vs. applied voltage across the anode and cathode during dissolution is plotted (I_d/η). The current density is related

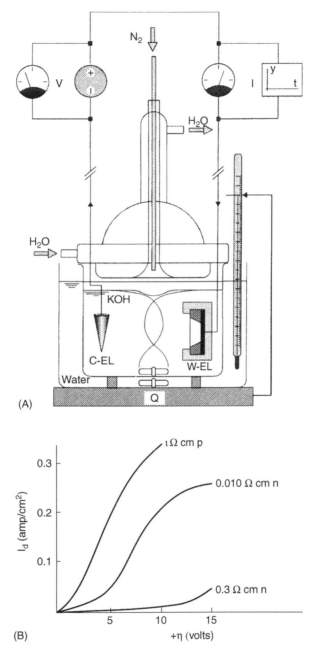

FIGURE 3.33 Electrochemical etching apparatus. (A) W-EL: working electrode (Si), C-EL: counter electrode (e.g., Pt), Q = heat supplied. (B) Current-voltage (I_d/η) curves in electrochemical etching of Si of various doping. Etch rate dependence on dopant concentration and dopant type for HF-anodic etching of silicon. (Reprinted with permission from van Dijk, H.J.A., and de Jonge, J. [1970] "Preparation of Thin Silicon Crystals by Electrochemical Thinning of Epitaxially Grown Structures," *J. Electrochem. Soc.* 117, pp. 553–54.)

to the dissolution rate of silicon. It can be seen that p-type and heavily doped n-type materials can be dissolved at relatively low voltages, whereas n-type silicon with a lower doping level does not dissolve at the same low voltages. Experiments in this same setup with homogeneously doped silicon wafers show that n-type silicon of about 3×10^{18} cm^{-3} (<0.01 Ω-cm) completely dissolves in these etching conditions, whereas n-type silicon of donor concentrations lower than $2 \times 10^{16}\,cm^{-3}$ ($>0.3\,\Omega$-cm) barely dissolves. For p-type silicon, dissolution is initiated when the acceptor concentration is higher than $5 \times 10^{15}\,cm^{-3}$ ($<3\,\Omega$-cm), and the dissolution rate further increases with increasing acceptor concentration. Under specific circumstances, namely high HF concentrations and low etching currents, porous Si may form [Bomchil et al., 1986].

The acidic electrochemical technique has not been used much in micromachining and is primarily used to polish surfaces. Since the etching rate increases with current density, high spots on the surface are more rapidly etched and very smooth surfaces result. This method of isotropic electrochemical etching has some major advantages that could make it a more important micromachining tool in the future. The etched surfaces are very smooth (say with an average roughness R_a of 7 nm); the process is room temperature and IC compatible; simpler resists schemes can be used because the process is much milder than etching in HNA; and etching can be controlled simply by switching a voltage on or off. We will resume the discussion of anodic polishing, photoetching, and formation of porous silicon in HF solutions after gathering more insight into various etching models.

3.7.2.8 Preferential Etching

A variety of additives to the HNA system, mainly oxidants, can be included to modify the etch rate, surface finish, or isotropy, rendering the etching baths preferential. It is clear that the effect of these additives will show up only in the reaction-controlled regime. Only additives that change the viscosity of the solution can modify the etch rate in the diffusion-limited regime, thereby changing the diffusion coefficient of the reactants [Tuck, 1975; Bogenschutz et al., 1967]. We will not review the effect of these additives any further; refer to Table 3.9 and the cited literature for more information [Yang, 1984; Chu and Gavaler, 1965; Archer, 1982; Schimmel and Elkind, 1973; d'Aragona, 1972].

3.7.2.9 Problems Associated with Isotropic Etchants

Several problems are associated with isotropic etching of Si. First, it is difficult to mask with high precision using a desirable and simple mask such as SiO_2 (etch rate is 2 to 3% of the silicon etch rate). Second, the etch rate is very agitation sensitive in addition to being temperature sensitive. This makes it difficult to control lateral as well as vertical geometries. Electrochemical isotropic etching and the development of anisotropic etchants in the late 1960s were able to overcome many of these problems.

A comprehensive review of isotropic etchants solutions can be found in Kern et al. (1978), including a review of different techniques practiced in chemical etching such as immersion etching, spray etching, electrolytic etching, gas-phase etching, and molten salt etching (fusion techniques). Table 3.9 lists some isotropic and preferential etchants and their specific applications.

3.7.3 Anisotropic Etching

3.7.3.1 Introduction

Anisotropic etchants shape, or machine, desired structures in crystalline materials by etching much faster in one direction than another. When carried out properly, anisotropic etching results in geometric shapes bounded by the slowest etching and perfectly defined crystallographic planes. Anisotropic wet etching techniques dating to the 1960s at Bell Laboratories, were developed mainly by trial and error. It seems fitting to go over experimental etch data before embarking upon the models especially because most models for higher index planes fail.

Figure 3.1 shows a cross section of a typical shape formed using anisotropic etching. The thinned membrane with diffused resistors is used for a piezoresistive pressure sensor. In the usual application, the wafer is selectively thinned from a starting thickness of 300 to 500 µm to form a diaphragm having a final thickness of 10 to 20 µm with precisely controlled lateral dimensions and a thickness control on the order of 1 µm or better. A typical procedure involves the steps summarized in Table 3.10 [Elwenspoek et al., 1994].

The development of anisotropic etchants solved the lateral dimension control lacking in isotropic etchants. Lateral mask geometries on planar photoengraved substrates can be controlled with an accuracy

TABLE 3.9   Isotropic and Preferential Defect Etchants and Their Specific Applications

Etchant	Application/Material	Remark/Reference
HF; 8 vol%, HNO_3; 75 vol% and CH_3COOH; 17 vol%	n- and p-type Si, all planes, general etching	Planar etch; e.g., 5 μm/min at 25°C
1 part 49% HF, 1 part of (1.5M-CrO_3) (by volume)	Delineation of defects on (111), (100), and (110) Si without agitation	Yang etch [Yange, 1984]
5 vol parts nitric acid (65%), 3 vol parts HF (48%), 3 vol parts acetic acid (96%), 0.06 parts bromine	Polishing etchant used to remove damage introduced during lapping	So-called CP_4 etchant; Heidenreich. US Patent 2619414
HF 48%	SiO_2	Etch rate is 20–2,000 nm/min. Etch rate for Si is 0.3 Å/min for n-type 2 ohm cm (111) Etch rate for Al is 5 nm/min
HF:NH_4F (buffered HF 28 ml HF,170 ml H_2O,113 g NH_4F)	SiO_2	Etch rate 100–500 nm/min at 25°C
1HF, 3HNO_3, 10 CH_3COOH	Delineates defects in (111) silicon. Etches p^+ or n^+ and stops at p^- or n^-	Dash etch [Dash, 1956]; p- and n-Si at 1300 Å/min in the [100] direction and 46 Å/min in the [111] direction at 25°C
1HF, 1(5M-CrO_3)	Delineates defects in (111); needs agitation; does not reveal etch pits on (100) well	Sirtl etch [Sirtl and Adler, 1961]
HF:H_2O_2	Titanium	880 nm/min
2HF,1 (0.15M-$K_2Cr_2O_7$)	Yields circular (100) Si dislocation etch pits; agitation reduces etch time	Secco etch [d'Aragona, 1972]
60 ml HF, 30 ml HNO_3, 30 ml (5M-CrO_3), 2 grams $Cu(NO_3)_2$, 60 ml CH_3COOH, 60 ml H_2O	Delineates defects in (100) and (111) Si; requires agitation	Jenkins etch [Jenkins, 1977]
H_2O_2	Tungsten	20–100 nm/min
34 g KH_2PO_4, 13.4 g KOH, 33 g $K_3FE(CN)_6$ and H_2O to make up 1 liter	Tungsten	160 nm/min
1 ml HCl, 9 ml saturated $CeSO_4$ solution	Chromium	80 nm/min
1 ml HCl, 1 ml glycerine	Chromium	80 nm/min
2HF,1(1M-CrO_3)	Delineates defects in (100) Si without agitation; works well on resistivities 0.6–15.0 Ωcm n and p-types)	Schimmel etch [Schimmel, 1979]
2HF, 1(1M-CrO_3), 1.5 H_2O	Works well on heavily doped (100) silicon	Modified Schimmel
HF/$KMnO_4$/CH_3COOH	Epitaxial Si	
3 ml HCl, 1 ml HNO_3	Gold	25–50 μm/min Aqua regia
4 g KI, 1 g I_2 and 40 ml H_2O	Gold	0.5–1 μm/min
H_3PO_4	Si_3N_4	Etch rate is 5 to 10 nm/min 160–180°C.
KOH + alcohols	Polysilicon	85°C
H_2SO_4/H_2O_2	Organic layers	>1,000 nm/min
Acetone	Organic layers	>4,000 nm/min
H_3PO_4/ HNO_3/$HC_2H_3O_2$	Al	Etch rate is 660 nm/min 40–50°C
HNO_3/BHF/water	Si and polysilicon	0.1 μm/min for single crystal Si

and reproducibility of 0.5 μm or better, and the anisotropic nature of the etchant allows this accuracy to be translated into control of the vertical etch profile. Different etch stop techniques, needed to control the membrane thickness, are available. The invention of these etch stop techniques truly made such applications as shown in Figure 3.1 manufacturable.

While anisotropic etchants solve problems of lateral control associated with isotropic etching, they are not problem free. These etchants are slow — even in the fast etching <100> direction — with etch rates

TABLE 3.10   Summary of the Process Steps Required for Anisotropic Etching of a Membrane

Process	Duration	Process Temperature (°C)
Oxidation	Variable (hr)	900–1200
Spinning resist at 5000 rpm	20–30 s	Room temperature
Prebake	10 min	90
Exposure	20 s	Room temperature
Develop	1 min	Room temperature
Postbake	20 min	120
Stripping of oxide (BHF:1:7)	±10 min	Room temperature
Stripping resist (acetone)	10–30 s	Room temperature
RCA1 [NH_3(25%) + H_2O + H_2O_2:1:5:1]	10 min	Boiling
RCA2 (HCl + H_2O + H_2O_2:1:6:1)	10 min	Boiling
HF-dip (2% HF)	10 sec	Room temperature
Anisotropic etch	From minutes up to one day	70–100

Source: Reprinted with permission from Elwenspoek, M. et al. (1994) "Micromechanics," Report No. 122830, University of Twente.

of about 1 μm/min or less. That means that etching through a wafer is a time-consuming process: to etch through a 300 μm thick wafer requires 5 hr. Anisotropic etchants also must be run hot to achieve these etch rates (80 to 115°C) precluding many simple masking options. Like the isotropic etchants, their etch rates are temperature sensitive; however, they are not particularly agitation sensitive, which is considered a major advantage.

3.7.3.2 Anisotropic Etchants

A wide variety of etchants have been used for anisotropic etching of silicon, including alkaline aqueous solutions of KOH, NaOH, LiOH, CsOH, RbOH, NH_4OH and quaternary ammonium hydroxides, with the possible addition of alcohol. Alkaline organics such as ethylenediamine, choline (trimethyl-2-hydroxyethyl ammonium hydroxide), hydrazine and sodium silicates with additives such as pyrocatechol and pyrazine are employed as well. Etching of silicon is possible without the application of an external voltage and is dopant insensitive over several orders of magnitude, but in a curious contradiction to its suggested chemical nature, it has been shown to be bias dependent [Allongue et al., 1993; Palik et al., 1987]. This contradiction will be explained with the help of the etching models presented below.

Alcohols such as propanol and isopropanol butanol typically slow the attack on Si [Linde and Austin, 1992; Price, 1973]. The role of pyrocatechol [Finne and Klein, 1967] is to speed up the etch rate through complexation or chelation of the reaction products. Additives such as pyrazine and quinone have been described as catalysts by some [Reisman et al., 1979], but this is contested by others [Seidel et al, 1990]. The etch rate in anisotropic etching is reaction rate controlled and thus temperature dependent. The etch rate for all planes increases with temperature, and the surface roughness decreases with increasing temperature, so etching at higher temperatures gives the best results. In practice, etch temperatures of 80 to 85°C are used to avoid solvent evaporation and temperature gradients in the solution.

3.7.3.3 Arrhenius Plots For Anisotropic Etching

A typical set of Arrhenius plots for <100>, <110>, and <111> silicon etching in an anisotropic etchant (EDP, or ethylenediamine pyrocatechol) is shown in Figure 3.34 [Seidel et al., 1990b]. The temperature dependence of the etch rate is large, and the slope differs for the different planes; that is, (111) > (100) > (110). Lower activation energies in Arrhenius plots correspond to higher etch rates. The anisotropy ratio (AR) derived from this figure is:

$$AR = \frac{(hkl)_1 \text{ etch rate}}{(hkl)_2 \text{ etch rate}} \tag{3.33}$$

The AR is approximately 1 for isotropic etchants and can be as high as 400/200/1 for (110)/(100)/(111) in 50 wt% KOH/H_2O at 85°C. Generally, the activation energies of the etch rates of EDP are smaller than those of KOH. Price found that (111) planes always etch slowest, and the selectivity with respect to (100) in KOH

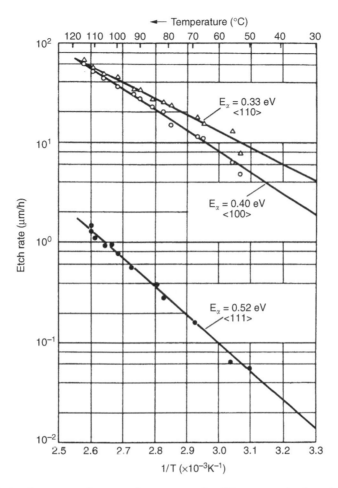

FIGURE 3.34 Vertical etch rates as a function of temperature for different crystal orientations: (100), (110), and (111). Etch solution is EDP (133 ml H_2O, 160 g pyrocatechol, 6 g pyrazine, and 1 l ED). (Reprinted with permission from Seidel, H., et al. [1990b] "Anisotropic Etching of Crystalline Silicon in Alkaline Solutions: Part 1. Orientation Dependence and Behavior of Passivation Layers," *J. Electrochem. Soc.* 137, 3612–26.)

etching can be greatly increased by adding isopropyl alcohol (IPA), a less polar diluent, to saturate the solution [Price, 1973]. The sequence for the (100) and (110) etching rate can be reversed, for example, 50/200/8 in 55 vol% ethylenediamine ED/H_2O, also at 85°C. The (110) Si plane etches eight times slower and the (111) eight times faster in KOH/H_2O than in ED/H_2O, while the (100) etches at the same rate [Kendall and Guel, 1985]. Working with alcohols and other organic additives often changes the relative etching rate of the different Si planes. Along this line, Seidel et al. (1990a, 1990b) found that the decrease in etch rate after adding isopropyl alcohol to a KOH solution was 20% for <100> but almost 90% for <110>. As a result of the much stronger decrease of the etch rate on a (110) surface, the etch ratio of (100):(110) is reversed.

Misalignment will change the etch rate greatly; a one-degree misalignment on the [111] direction may increase the etch rate on (near) the (111) surface by 300% [Seidel, 1987].

3.7.3.4 Selected Anisotropic Etchant Systems

Overview. In choosing an etchant, a variety of issues must be considered.

- Ease of handling
- Toxicity

- Etch rate
- Desired topology of the etched bottom surface
- IC-compatibility
- Etch stop
- Etch selectivity over other materials
- Mask material and thickness of the mask

The principal characteristics of four different anisotropic etchants are listed in Table 3.11. The most commonly used are KOH [Stoller and Wolff, 1966; Stoller, 1970; Forster and Singleton, 1966; Kenney, 1967; Lepselter, 1966; 1967; Waggener, 1970; Kragness and Waggener, 1973; Waggener et al., 1967; Waggener et al., 1967; Bean and Runyan, 1977; Rodgers et al., 1976, 1977; Ammar and Rodgers, 1980; Seidel et al., 1990a, 1990b; Kendall and Guel, 1985; Lee, 1969; Noworolski et al., 1995; Waggener and Dalton, 1972; Weirauch, 1975; Clemens, 1973; Bean et al., 1974; Declercq et al., 1975] and ethylenediamine pyrocatechol + water (EDP) [Finne and Klein, 1967; Reisman et al., 1979; Wu et al., 1986]; hydrazine[Declercq et al., 1975; Mehregany and Senturia, 1988]. More recently, quaternary ammonium hydroxide solutions such as tetramethyl ammonium hydroxide-water (TMAHW) and tetraethyl ammonium hydroxide-water (TEAHW) have become more popular [Asano et al., 1976; Tabata et al., 1990]. Each has its advantages and problems. NaOH is not used much anymore [Pugacz-Muraszkiewicz and Hammond, 1977].

Potassium Hydroxide (KOH). The simple KOH water system is the most popular Si anisotropic etchant. A KOH etch in near saturated solutions (1:1 in water by weight) at 80°C produces a uniform and bright surface. Williams and Muller used 50 wt% KOH at 80°C for a reported (100) etch rate of 1.4 μm/min [Williams and Muller, 1996]. Nonuniformity of etch rate becomes considerably worse above 80°C. An abundance of bubbles are seen emerging from the Si wafer while etching in KOH. The etching selectivity between Si and SiO_2 is not very good in KOH, as it etches SiO_2 too fast. KOH is also incompatible with the IC fabrication process (e.g., aluminum bond pads are quickly attacked and damaged) and can cause blindness when it comes in contact with the eyes. The etch rate for low index planes is maximal at around $4M$ (Figure 3.35A and C [Peeters, 1994] and Lambrechts et al. [1992]).

Herr (1991) found that the high-index crystal planes exhibit the highest etch rates for $6M$ KOH and that for lower concentrations the etch bottoms disintegrate into micro facets. In $6M$ KOH, the etch-rate order is (311) > (144) > (411) > (133) > (211) > (122). These authors could not correlate the particular etch rate sequence with the measured activation energies. This is in contrast to lower activation energies corresponding to higher etching rates for low index planes as shown in Figure 3.34. Their results obtained on large open-area structures differ significantly from previous results obtained by underetching special mask patterns. The vertical etching rates obtained here are substantially higher than the underetching rates described elsewhere, and the etch rate sequence for different planes is also significantly different. These results suggest that crevice effects may play an important role in anisotropic etching.

To create vertical (100) faces, as shown earlier in Figure 3.10, in general only KOH works (not EDP or TMAHW), and it has to be carried out in high-selectivity conditions (low temperature, low concentration: 25 wt% KOH, 60°C). Interestingly, high-concentration KOH (45 wt%) at higher temperatures (80°C) produces a smooth sidewall, controllable and repeatable at an angle of 80°. EDP produces 45° angled planes, and TMAHW usually makes a 30° angle [Palik et al., 1985].

Besides KOH [Clark and Edell, 1987], other hydroxides have been used, including NaOH [Allongue et al., 1993; Pugacz-Muraszkiewicz and Hammond, 1977], CsOH [Clark and Edell, 1987], and NH_4OH [Schnakenberg et al., 1990]. A major disadvantage of KOH is the presence of alkali ions, which are detrimental to the fabrication of sensitive electronic parts.

Ethylenediamine Pyrocatechol (EDP). With EDP (sometimes referred to as EPW, for ethylenediamine, pyrocatechol, and water), a variety of masking materials (SiO_2, Si_3N_4, Au, Cr, Cu, Ag, Ta, etc.) can be used, and though the etchant is toxic it is less so than hydrazine (see below). No sodium or potassium contamination occurs, and the etch rate of SiO_2 is much slower than with KOH. The ratio of etch rates of Si and SiO_2 using EDP can actually be as large as 5000:1, corresponding to about 2 Å/min of SiO_2 compared

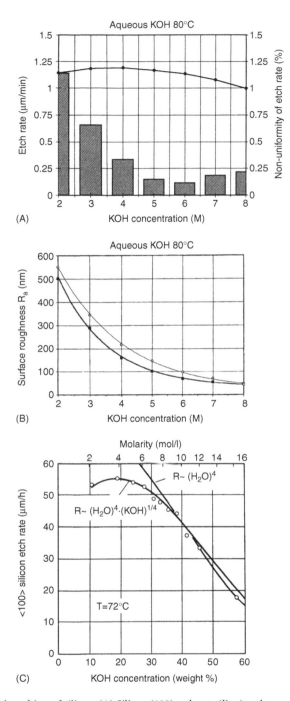

FIGURE 3.35 Anisotropic etching of silicon. (A) Silicon (100) etch rate (line) and nonuniformity of etch rate (column) in KOH at 80°C as a function of KOH concentration. The etch rate for all low index planes is maximal at around 4 *M*. (B) Silicon (100) surface roughness (Ra) in aqueous KOH at 80°C as a function of concentration for a 1-hour etch time (thin line) and for an etch depth of 60 μm (thick line). (C) Silicon (100) etch rate as a function of KOH concentration at a temperature of 72°C. ([A] and [B] reprinted with permission from Peeters, E. [1994] Process Development for 3D Silicon Microstructures, with Application to Mechanical Sensor Design, Ph.D. thesis, KUL, Belgium; [C] reprinted with permission from Seidel, H. et al. [1990] "Anisotropic Etching of Crystalline Silicon in Alkaline Solutions: Part 1. Orientation Dependence and Behavior of Passivation Layers," *J. Electrochem. Soc.* 137, 3612–26.)

with 1 µm/min of Si, which is much larger than the ratio for Si to SiO_2 in KOH, which is at the highest 400:1 [Bean, 1978]. Importantly, the etch rate slows at a lower boron concentration than with KOH. A typical fastest-to-slowest hierarchy of Si etch rates with EDP at 85°C according to Barth (1984) is (110) > (411) > (311) > (511) > (211) > (100) > (331) > (221) > (111). Petersen uses 750 mL ethylenediamine, 120 g pyrocatechol, and 100 mL water. This cocktail at 115°C results in an etch rate of 0.75 µm/min, with an (100):(111) etch-rate ratio of 35:1 [Petersen, 1982].

Ethylenediamine in EDP reportedly causes allergic respiratory sensitization, and pyrocatechol is described as a toxic corrosive. The material is also optically dense, making end-point detection harder, and it ages quickly; if the etchant reacts with oxygen, the liquid turns to a reddish brown, and it loses its useful properties. If EDP is cooled after etching, precipitation of silicates in the solution will occur. Sometimes precipitation during etching can happen, spoiling the results. The last ingredient added to the solution should be the water, because the water's addition causes the oxygen sensitivity. All of these characteristics make the etchant quite difficult to handle.

In terms of etching and masking layers, amine gallates are similar to EDP but perhaps safer [Linde and Austin, 1992]. Amine gallate etchants are not used much but appear promising. They are composed of a mixture of ethanolamine, gallic acid, water, pyrazine, hydrogen peroxide, and a surfactant. Etch rates as high as 2.3 µm/min have been measured on a (100) Si plane, and etch stops at lower boron concentration than it takes to stop EDP have been observed ($>3 \times 10^{19}$ cm^{-3}). Pyrazine and peroxide can be added to increase the etch rate, but they affect the surface roughness negatively.

Ammonium Hydroxide-Water (AHW) and Tetramethyl Ammonium Hydroxide-Water (TMAHW). Efforts continue to find anisotropic etchants that are more compatible with CMOS processing than alkali hydroxides and that are neither toxic nor harmful. Two examples are AHW mixtures [Schnakenberg et al., 1990] and TMAHW mixtures [Tabata et al., 1990] and Schnakenberg et al. [Schnakenberg et al., 1990]. Kern used AHW (9.7% in water) and achieved 0.11 µm/min etch rates on (100) Si at temperatures of 85 to 92°C [Kern, 1978]. Schnakenberg et al. reported their best AHW results with a 3.7 wt% solution at 75°C for well stirred etching baths [Schnakenberg et al., 1990]. For the same solution, these authors demonstrated a boron-dependent etch-stop at 1.3×10^{20} cm^{-3} with a selectivity of 1:8000 (see also under Etch-Stop Techniques). Ammonia-based etchants have not become widely used for several reasons, including their slow etch rate, tendency to lead to rough surfaces (hillocks), and rapid evaporative losses [Kovacs et al., 1998]. A TMAHW [$(CH_3)_4NOH$] solution, on the other hand, is one of the more useful wet etchants for silicon. TMAHW does not decompose at temperatures below 130°C, a very important feature from the viewpoint of production; it is nontoxic, not expensive, and can be handled easily. TMAHW solutions also exhibit excellent selectivity to silicon oxide and silicon nitride masks. The etchant is actually so selective for Si over SiO_2 that it is advisable to remove the thin native oxide of Si in HF prior to attempting a TMAHW etch. The solution is often already present in the clean room because it is used in many positive photoresist developers. At a solution temperature of 90°C and 22 wt% TMAH, a maximum (100) silicon etch rate of 1.0 µm/min is observed, 1.4 µm/min for (110) planes (higher than those observed with EDP, AHW, hydrazine water, and tetraethyl ammonium hydroxide [TEA], but slower than those observed for KOH) and an anisotropy ratio, AR(100)/(111), of between 12.5 and 50 [Tabata et al., 1992]. From the viewpoint of fabricating various silicon sensors and actuators, a concentration above 22 wt% is preferable, since lower concentrations result in more pronounced roughness on the etched surface. However, higher concentrations give a lower etch rate and lower etch ratio (100)/(111). Tabata (1995) also studied the etching characteristics of pH-controlled TMAHW. To obtain a low aluminum etching rate of 0.01 µm/min, pH values below 12 for 22 wt% TMAHW were required. At those pH values, the Si (100) etching rate is 0.7 µm/min. The aluminum etch rate can also be reduced by adding silicon powder to the etchant [Reay et al., 1994]. The etch rate for TMAHW begins to slow for boron doping levels above approximately 1×10^{19} cm^{-3} and falls down by a factor of 40 for 2×10^{20} cm^{-3} [Steinsland et al., 1995].

Hydrazine. Etch rates with hydrazine-water mixtures are on the order of 2 µm/min, and similar masking materials can be used as with EDP [Petersen, 1982]. The (100):(111) etch ratios are lower than those for KOH or EDP. Hydrazine-water is explosive at high hydrazine concentrations (rocket fuel) and is a suspected carcinogen, so its use should be avoided for safety reasons. A 50% hydrazine/water solution is stable, though, and according to Mehregany (1988), excellent surface quality and sharply defined corners are obtained in Si. Also on the positive side, the etchant has a very low SiO_2 etch rate and will not attack most metal masks except for Al, Cu, and Zn. According to Wise, on the other hand, Al does not etch in hydrazine either, but the etch produces rough Si surfaces [Wise et al., 1985].

3.7.3.5 Si Surface Roughness

Anisotropic etchants frequently leave too rough a Si surface behind, and a slight isotropic etch is used to touch up the surface. The average surface roughness R_a of Si continuously decreases with increasing KOH concentration, as can be gleaned from Figure 3.35B. The silicon etch rate as a function of KOH concentration is shown in Figure 3.35A and C [Seidel et al., 1900]. Since the difference in etch rates for different KOH concentrations is small, a highly concentrated KOH (e.g., 7 M) is preferred in obtaining a smooth surface on low index planes. Except at very high concentrations of KOH, the etched (100) plane becomes rougher the longer one etches. This is thought to be due to the development of hydrogen bubbles, which hinder the transport of fresh solution to the silicon surface, causing *micromasking*, which results in hillock formation [Kovacs et al., 1998]. Average roughness R_a is influenced strongly by fluid agitation. Stirring can reduce the R_a values by over an order of magnitude, probably caused by the more efficient removal of hydrogen bubbles from the etching surface when stirring [Gravesen, 1986]. Ohwada et al. noted that the use of ultrasonic agitation essentially eliminated surface roughness in KOH etching [Ohwada et al., 1995]. Baum et al. investigated the surface finish of (100) Si in KOH etching with an atomic force microscope and confirmed that mild ultrasonic agitation improved the surface finish considerably [Baum and Schriffrin, 1997]. Hillock formation may also be suppressed by the addition of a suitable oxidizing agent that competes with hydrogen evolution such as by adding ferricyanide or peroxydisulfate ions. Bressers et al. report a drastic reduction in hillock formation when adding 18 mM of $K_3Fe(CN)_6$ to a 4 M KOH solution at 70°C [Bressers et al., 1996], and Klaassen et al. accomplish the same by adding 5 g/L ammonium peroxydisulfate to a 5 wt% TMAHW solution [Klassen et al., 1996]. Baum et al. found that the inclusion in the KOH bath of oxygen and/or isopropanol results in root mean square roughness values smaller than 20 nm. The effectiveness of these additives has been related to changes of the contact angle between the liquid/gas/etching interface [Baum and Schiffrin, 1997].

A distinction must be made between macroscopic and microscopic roughness (Inset 3.7). Macroscopic roughness, also referred to as notching or pillowing, results when centers of exposed areas etch with a

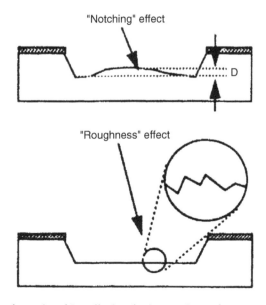

INSET 3.7   Macroscopic roughness (notching effect) and microscopic roughness.

seemingly lower average speed than the borders of the areas, so that the corners between sidewalls and (100) ground planes are accentuated. Membranes or double-sided clamped beams (microbridges) therefore tend to be thinner close to the clamped edges than in the center of the structure. This difference can be as large as 1 to 2 μm, which is quite considerable if one is etching 10 to 20 μm thick structures. Notching increases linearly with etch depth but decreases with higher concentrations of KOH. The microscopic smoothness of originally mirror-like polished wafers can also be degraded into microscopic roughness. This is the type of short-range roughness to which we referred in discussing Figure 3.35B.

3.7.3.6 Masking for Anisotropic Etchants

Etching through a whole wafer thickness (400 to 600 μm) takes several hours (a typical wet anisotropic etch rate being 1.1 μm/min), definitely not a fast process. When using KOH, SiO_2 cannot be used as a masking material for features requiring that long an exposure to the etchants. The SiO_2 etch rate as a function of KOH concentration at 60°C is shown in Figure 3.36. There is a distinct maximum at 35 wt% KOH of nearly 80 nm/hr. The shape of this curve will be explained further below on the basis of Seidel et al.'s model. Experiments have shown that even a 1.5 μm thick oxide is not sufficient for the complete etching of a 380 μm thick wafer (6 hr) because of pinholes in the oxide [Lambrechts and Sansen, 1992]. The etch rate of thermally grown SiO_2 in KOH-H_2O somewhat varies and apparently depends not only on the quality of the oxide but also on the etching container and the age of the etching solution, as well as other factors [Kendall, 1979]. The Si/SiO_2 selectivity ratio at 80°C in 7 M KOH is 30 ± 5. This ratio increases

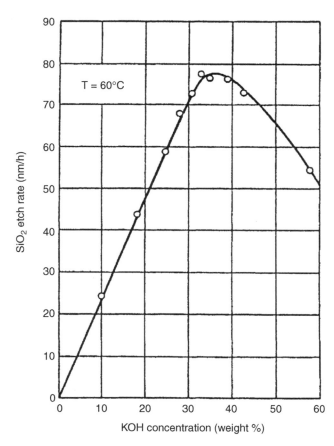

FIGURE 3.36   The SiO_2 etch rate in nm/hr as a function of KOH concentration at 60°C. (From Seidel, H., et al. [1990] "Anisotropic Etching of Crystalline Silicon in Alkaline Solutions: Part 1. Orientation Dependence and Behavior of Passivation Layers," *J. Electrochem. Soc.* 137, 3612–26.)

with decreasing temperature; reducing the temperature from 80 to 60°C increases the selectivity ratio from 30 to 95 in 7 M KOH [Kendall, 1975]. Thermal oxides are under strong compressive stress because in the oxide layer one silicon atom takes nearly twice as much space as in single-crystalline Si. This might have severe consequences; for example, if the oxide mask is stripped on one side of the wafer, the wafer will bend. Atmospheric pressure chemical vapor deposited (APCVD) SiO_2 tends to exhibit pinholes and etches much faster than a thermal oxide. Annealing of APCVD oxide removes the pinholes, but the etch rate in KOH remains greater by a factor of 2 to 3 than that of thermal oxide. Low-pressure chemical vapor deposited (LPCVD) oxide is a mask material of comparable quality to thermal oxide. The etch rate of SiO_2 in EDP is smaller by two orders of magnitude than in KOH.

For prolonged KOH etching, a high-density silicon nitride mask has to be deposited. An LPCVD nitride generally serves better for this purpose than a less-dense plasma deposited nitride [Puers, 1991]. With an etch rate of less than 0.1 nm/min, a 400-Å layer of LPCVD nitride suffices to mask against KOH etchant. The etch selectivity Si/Si_3N_4 was found to be better than 10^4 in 7 M KOH at 80°C. The nitride also acts as a good ion-diffusion barrier protecting sensitive electronic parts. Nitride can easily be patterned with photoresist and etched in a CF_4/O_2-based plasma or, in a more severe process, in H_3PO_4 at 180°C (10 nm/min) [Buttgenbach and Mikromechanik, 1991]. Nitride films are typically under a tensile stress of about 1×10^9 Pa. If in the overall processing of the devices nitride deposition does not pose a problem, KOH emerges as the preferential anisotropic wet etchant. For dopant-dependent etching, EDP is the better etchant and generally is better suited for deep etching since its oxide etch rate is negligible (<5 Å/min).

Summarizing oxide and nitride, oxide and nitride mask for anisotropic etchants to varying degrees with both mask types being used. A KOH solution etches SiO_2 at a relatively fast rate of 1.4 to 3 nm/min so that Si_3N_4 or Au/Cr must be used as a mask against KOH for deep and long etching. When these layers are used to terminate an etch in the [100] direction, a low etch rate of the mask layer allows overetching of silicon to compensate for wafer thickness variations.

3.7.3.7 Back-Side Protection

In many cases, it is necessary to protect the back of a wafer from an isotropic or anisotropic etchant. The back is either mechanically or chemically protected. In the mechanical method, the wafer is held in a holder, often made from Teflon®. The wafer is fixed between Teflon-coated O-rings that are carefully aligned to avoid mechanical stress in the wafer. In the chemical method, waxes or other organic coatings are spun onto the back of the wafer. Two wafers may be glued back to back for faster processing.

3.7.3.8 Etch Rate and Dopant Concentration Dependency

The Si etch rate R as a function of KOH concentration at 72°C, was shown in Figure 3.35C. The etch rate has a maximum at about 10% KOH of about 0.9 μm/min. The best fit for this experimentally determined etch rate, for most KOH concentrations, is [Seidel et al., 1990a, 1990b]:

$$R = k[H_2O]^4[KOH]^{\frac{1}{4}} \tag{3.34}$$

Any model of anisotropic etching will have to explain this peculiar dependency on the water and KOH concentration, as well as the fact that all anisotropic etchant systems of Table 3.11 exhibit drastically reduced etch rates for high boron concentrations in silicon ($\geqslant 5 \times 10^{19}$ cm^{-3} solid solubility limit). Other impurities (P, Ge) also reduce the etch rate but at much higher concentrations (see Figure 3.37 [Seidel et al., 1990]). Boron typically is incorporated using ion implantation (thin layers) or liquid/solid source deposition (thick layers >1 μm). These doped layers are used as very effective etch stop layers (see below). Hydrazine or EDP, which display a smaller (100) to (111) etch rate ratio (~35) than KOH, exhibit a stronger boron concentration dependency. The etch rate in KOH is reduced by a factor of 5 to 100 for a boron concentration larger than 10^{20} cm^{-3}. When etching in EDP, the factor climbs to 250 [Bogh, 1971]. With TMAHW solutions, the Si etch rate decreases to 0.01 μm/min for boron concentrations of about 4×10^{20} cm^{-3} [Steinsland et al., 1995]. The mechanism eludes us, but Seidel et al.'s model (see below)

TABLE 3.11   Principal Characteristics of Four Different Anisotropic Etchants

Etchant/Diluent/ Additives/ Temperature	Etch Stop	Etch Rate (100) μm/min	Etch Rate Ratio	Remarks	Mask (Etch Rate)
KOH (water) 85°C 44 g/100 ml	B $>10^{20}$ cm^{-3} reduces etch rate by 20	1.4	400 for (100)/(111) and 600 for (110)/(111)	IC incompatible, avoid in eyes, etches oxide fast, lots of H$_2$ bubbles	Photoresist (shallow etch at room temperature); Si$_3$N$_4$ (<1 nm /min) SiO$_2$ (28 Å/min)
Ethylenediamine pyrocatechol (water) 115°C 750 ml/ 120 g/240 ml	$\geqslant 7 \times 10^{19}$ cm^{-3} reduces the etch rate by 50	1.25	35 for (100)/(111)	Toxic, ages fast, O$_2$ must be excluded, few H$_2$ bubbles, silicates may precipitate	SiO$_2$ (2–5 Å/min) Si$_3$N$_4$ (1 Å/min) Ta, Au, Cr, Ag, Cu are not attacked Al at a 0.33 μm/min.
Tetramethyl ammonium hydroxide (TMAH) (water) 90°C	$>4 \times 10^{20}$ cm^{-3} reduces etch rate by 40	1	from 12.5 to 50 (100)/(111)	IC compatible, easy to handle, smooth surface finish, few studies	SiO$_2$ etch rate is four orders of magnitude lower than (100)LPCVD Si$_3$N$_4$
N$_2$H$_4$ (water, isopropyl alcohol) 100°C 100 ml/100 ml	$>1.5 \times 10^{20}$ cm^{-3} practically stops the etch	2.0	10(100)/(111)	Toxic and explosive, OK at 50% water	SiO$_2$ (<2 Å/min) and most metallic films; does not attack Al

Note: Given the many possible variables, the data in the table are only typical examples.

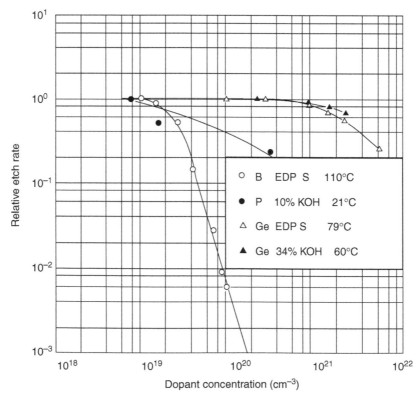

FIGURE 3.37   Relative etch rate for (100) Si in EDP and KOH solutions as a function of concentration of boron, phosphorus, and germanium. (Reprinted with permission from Seidel, H. et al. [1990] "Anisotropic Etching of Crystalline Silicon in Alkaline Solutions: Part 2. Influence of Dopants," *J. Electrochem. Soc.* 137, 3626–32.)

gives the most plausible explanation for now. Some of the different mechanisms to explain etch stop effects follow.

1. Several observations suggest that doping leads to a more readily oxidized Si surface. Highly boron- or phosphorus-doped silicon in aqueous KOH spontaneously can form a thin passivating oxide layer [Palik et al., 1985, 1982]. The boron-oxides and hydroxides initially generated on the silicon surface are not soluble in KOH or EDP etchants [Petersen, 1982]. The substitutional boron creates local tensile stress in the silicon, increasing the bond strength so that a passivating oxide might be more readily formed at higher boron concentrations. Boron-doped silicon has a high defect density (slip planes), encouraging oxide growth.

2. Electrons produced during oxidation of silicon are needed in a subsequent reduction step (hydrogen evolution in Reaction 3.2). When the hole density passes $10^{19} \, cm^{-3}$, these electrons combine with holes instead, thus stopping the reduction process [Palik et al., 1982]. Seidel et al.'s model follows this explanation (see below).

3. Silicon doped with boron is under tension as the smaller boron atoms enter the lattice substitutionally. The large local tensile stress at high boron concentration makes it energetically more favorable for the excess boron (above $5 \times 10^{19} \, cm^{-3}$) to enter interstitial sites. The strong B-Si bonds bind the lattice rigidly. With high enough doping, the high binding energy can stop etching [Petersen, 1982]. This hypothesis is similar to item 1, except that no oxide formation is invoked.

3.7.3.9 Alignment Patterns

When alignment of a pattern is critical, preetch alignment targets become useful to delineate the planes of interest because the wafer flats often are aligned to $\pm 1°$ only. To find the proper alignment for the mask, a test pattern of closely spaced lines can be etched (see Figure 3.15). The groove with the best vertical walls determines the proper final mask orientation. Along this line, Ciarlo (1987) made a set of lines 3 mm long and 8 μm wide, fanned out like spokes in a wagon wheel at angles 0.1° apart. This target was printed near the perimeter of the wafer and then etched 100 μm into the surface. Again, by evaluating the undercut in this target, the correct crystal direction could be determined. This method accomplished alignment with better than 0.05° accuracy. Similarly, to obtain detailed experimental data on crystal orientation dependence of etch rates, Seidel et al. (1990) used a wagon-wheel- or star-shaped mask (e.g., made from CVD-Si_3N_4; SiH_4 and NH_3 at 900°C), consisting of radially divergent segments with an angular separation of 1°. Yet finer 0.1° patterns were made around the principal crystal directions. The etch pattern emerging on a <100>-oriented wafer covered with such a mask is shown in Figure 3.38A. The blossom-like figure is due to the total underetching of the passivation layer in the vicinity of the center of the wagon wheel, leaving an area of bare, exposed Si. The radial extension of the bare Si area depends on the crystal orientation of the individual segments, leading to a different amount of total underetching. The observation of these blossom-like patterns was used for qualitative guidance of etching rates only. To establish quantitative numbers for the lateral etch rates, the width w of the overhanging passivation layer was measured with an optical line-width measurement system (see Figure 3.38B). Laser beam reflection was used to identify the crystal planes, and ellipsometry was used to monitor the etching rate of the mask itself. Lateral etch rates determined in this way on <100>- and <110>-oriented wafers at 95°C in EDP (470 mL water, 1 L ethylenediamine, 176 g pyrocatechol) and at 78°C in a 50% KOH solution are shown in Figure 3.39. Etch rates shown are normal to the actual crystal surface and are conveniently described in a polar plot in which the distance from the origin to the polar plot surface (or curve in two dimensions) indicates the etch rate for that particular direction. Note the deep minima at the {111} planes. It can also be seen that in KOH the peak etch rates are more pronounced. A further difference is that with EDP the minimum at {111} planes is steeper than with KOH. For both EDP and KOH, the etch rate depends linearly on misalignment. All the above observations have important consequences for the interpretation of anisotropy of an etch (see below). The difference between KOH and EDP etching behavior around the {111} minima has the direct practical consequence that it is more important for etching in EDP to align the crystallographic direction more precisely than in KOH [Elwenspoek et al., 1994].

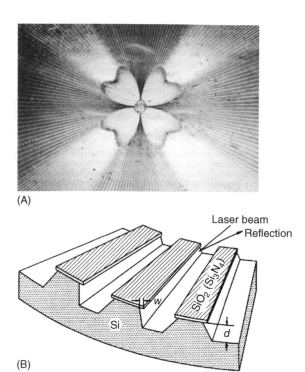

FIGURE 3.38 (A) Etch pattern emerging on a wagon-wheel-masked <100>–oriented Si wafer after etching in an EDP solution. (B) Schematic cross-section of a silicon test chip covered with a wagon-wheel-shaped masking pattern after etching. The measurement of *w* is used to construct polar diagrams of lateral underetch rates as shown in Figure 3.39. (Reprinted with permission from Seidel, H., et al. [1990] "Anisotropic Etching of Crystalline Silicon in Alkaline Solutions: Part 1. Orientation Dependence and Behavior of Passivation Layers," *J. Electrochem. Soc.* 177, 3612–26.)

When determining etch rates without using underetching masks but using vertical etching of beveled silicon samples, results are quite different from those obtained when working with masked silicon [Herr and Baltes, 1991]. The etch rates on open areas of beveled structures are much larger than in underetching experiments with masked silicon, and different crystal planes develop. Herr et al. conclude that crevice effects may play an important role in anisotropic etching. Elwenspoek et al.'s model [Elwenspoek et al., 1993, 1994], analyzed below, is the only model that predicts such a crevice effect. The authors explain why, when etchants are in a small restricted crevice area and are not refreshed fast enough, etching rates slow down and increase anisotropy.

3.7.4 Chemical Etching Models

3.7.4.1 Introduction

Conflicting data exist in the literature on the anisotropic etch rates of the different Si planes, especially for the higher index planes. This is not surprising, given the multiple parameters influencing individual results: temperature, stirring, size of etching feature (i.e., crevice effect), KOH concentration, addition of alcohols and other organics, surface defects, complexing agents, surfactants, pH, cation influence, and so forth. More rigorous experimentation and standardization will be needed, as well as better etching models, to better understand the influence of all these parameters on etch rates.

Several chemical models explaining the anisotropy in etching rates for the different Si orientations have been proposed. Presently, we will list all of the proposed models and compare the two most recent and detailed, one by Seidel et al. (1990a, 1990b) and another by Elwenspoek et al. (1993, 1994). Different Si crystal properties have been correlated with the anisotropy in silicon etching.

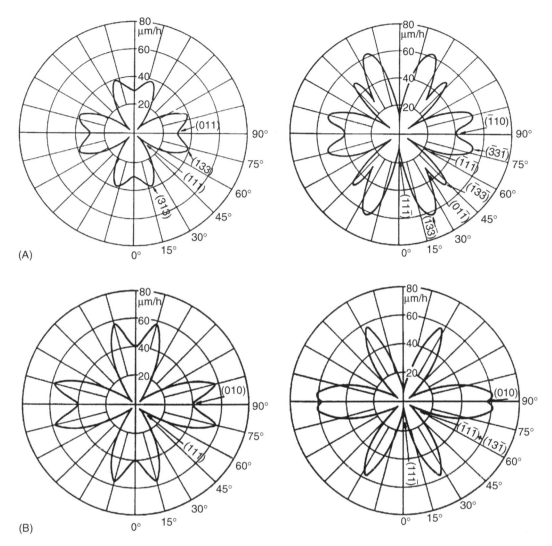

FIGURE 3.39 Lateral underetch rates as a function of orientation for (A) EDP (470 ml water, 1 l ED, 176 g pyrocatechol) at 95°C. (B) KOH (50% solution) at 78°C. Left, <100>– and right, <110>–oriented Si wafers. (Reprinted with permission from Seidel, H., et al. [1990] "Anisotropic Etching of Crystalline Silicon in Alkaline Solutions: Part 1. Orientation Dependence and Behavior of Passivation Layers," *J. Electrochem. Soc.* 177, 3612–26.)

1. It has been observed that the {111} Si planes present the highest density of atoms per cm^2 to the etchant and that the atoms are oriented such that three bonds are below the plane (Inset 3.8). It is possible that these bonds become chemically shielded by surface bonded OH or oxygen, thereby slowing the etch rate.

2. It also has been suggested that etch rate correlates with available bond density so that the surfaces with the highest bond density etch faster [Price, 1973]. The available bond densities in Si and other diamond structures follow the sequence 1:0.71:0.58 for the {100}:{110}:{111} surfaces. However, Kendall (1979) commented that bond density alone is an unlikely explanation because of the magnitude of etching anisotropy (e.g., a factor of 400), compared with the bond density variations of at most a factor of two.

3. Kendall (1979) explains the slow etching of {111} planes on the basis of their faster oxidation during etching; this does not happen on the other faces, due to greater distance of the atoms on planes other than (111). Since they oxidize faster, these planes may be better protected against etching.

INSET 3.8   On {111} planes three backbonds are below the plane.

The oxidation rate in particular follows the sequence {111} > {110} > {100}, and the etch rate Loften follows the reverse sequence. In the most used KOH-H₂O, however, the sequence is {110} > {100} > {111}.

4. In yet another model, it is assumed that the anisotropy is due to differences in activation energies and backbond geometries on different Si surfaces [Glembocki et al., 1985].

5. Seidel et al.'s model [Seidel et al., 1990a, 1990b] supports the previous explanation. They detail a process to explain anisotropy based on the difference in energy levels of backbond-associated surface states for different crystal orientations.

6. Finally, Elwenspoek (1993) and Elwenspoek et al. (1993, 1994) propose that it is the degree of atomic smoothness of the various surfaces that is responsible for the anisotropy of the etch rates. Basically, this group argues that the kinetics of smooth faces — the (111) plane is atomically flat — is controlled by a nucleation barrier that is absent on rough surfaces. The latter, therefore, would etch faster by orders of magnitude.

Until recently, the reason why acidic media lead to isotropic etching and alkaline media to anisotropic etching was not addressed in any of the models surveyed. In the following, we will give our own model as well as Elwenspoek et al.'s to explain isotropic vs. anisotropic etching behavior.

It is our hope that reading this section will inspire more detailed electrochemistry work on Si electrodes. The refining of an etching model will be of invaluable help in writing more predictive Si etching software code.

3.7.4.2   Seidel et al.'s Model

Seidel et al.'s model is based on the fluctuating energy level model of the silicon/electrolyte interface and assumes the injection of electrons in the conduction band of Si during the etching process. Consider the situation of a piece of Si immersed in a solution without applied bias at open circuit. After immersion of the silicon crystal into the alkaline electrolyte, a negative excess charge builds up on the surface due to the higher original Fermi level of the H_2O/OH^- redox couple as compared to the Fermi level of the solid; that is, the work function difference is equalized. This leads to a downward bending of the energy bands on the solid/liquid interface for both p- and n-type silicon (Figure 3.40A and B). The downward bending is more pronounced for p-type than for n-type due to the initially larger difference of the Fermi levels between the solid and the electrolyte.

Next, hydroxyl ions cause the Si surface to oxidize, consuming water and liberating hydrogen in the process. The detailed steps, based on suggestions by Palik (1985), are:

$$Si + 2OH^- \rightarrow Si(OH)_2^{2+} + 2e^-$$

$$Si(OH)_2^{2+} + 2OH^- \rightarrow Si(OH)_4 + 2e^-$$

$$Si(OH)_4 + 4e^- + 4H_2O \rightarrow Si(OH)_6^{2-} + 2H_2 \qquad \text{(Reaction 3.4)}$$

Silicate species were observed by Raman spectroscopy [Palik et al., 1983].

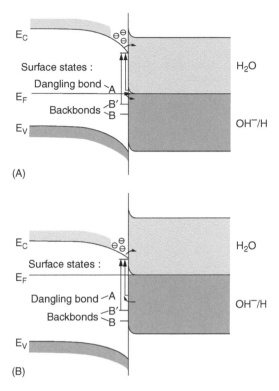

FIGURE 3.40 Band model of the silicon/electrolyte interface for moderately doped Si (electrolyte at pH > 12): (A) p-type Si and (B) n-type Si. We assume no applied bias and no illumination. The energy scale functions in respect to the saturated calomel electrode (SCE), an often-used reference in electrochemistry. Notice that p-type Si exhibits more band bending as its Fermi level is lower in the band gap. For simplicity, we show only one energetic position for surface states associated with dangling bonds and backbonds; in reality, there will be new surface states arising during reactions as the individual dangling bonds and backbonds are taking on different energies as new Si-OH bonds are introduced.

The overall silicon oxidation reaction consumes four electrons first injected into the conduction band, where they stay near the surface due to the downward bending of the energy bands. Evidence for injection of four electrons rather than a mixed hole and electron mechanism was first presented by Raley et al. (1984). The authors could explain the measured etch-rate dependence on hole concentration only by assuming that the proton or water reduction reaction is rate determining and that a four-electron injection mechanism with the conduction band is involved. These injected electrons are highly *reducing* and react with water to form hydroxide ions and hydrogen:

$$4H_2O + 4e^- \rightarrow 4H_2O^- \qquad \text{(Reaction 3.5)}$$

$$4H_2O^- \rightarrow 4OH^- + 4H^+ + 4e^- \rightarrow 4OH^- + 2H_2 \qquad \text{(Reaction 3.6)}$$

It is thought that the hydroxide ions in Reaction 3.6, generated directly at the silicon surface, react in the oxidation step. The hydroxide ions from the bulk of the solution may not play a major role, as they will be repelled by the negatively charged Si surface, whereas the hydroxide ions formed in situ do not need to overcome this repelling force. This would explain why the etch rates for an EDP solution with an OH$^-$ concentration of 0.034 mol/L are nearly as large as those for KOH solutions with a hundred-fold higher OH$^-$ concentration of 5 to 10 mol/L [Seidel et al., 1990]. The hydrogen formed in Reaction 3.6 can inhibit the reaction, and surfactants may be added to displace the hydrogen (IBM, U.S. Patent 4,113,551, 1978). Additional support for the involvement of four water molecules (Reaction 3.5) comes from the

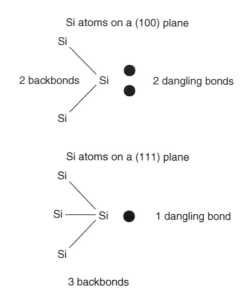

INSET 3.9   Dangling bonds.

experimentally observed correlation between the fourth power of the water concentration and the silicon etch rate for highly concentrated KOH solutions (Equation [3.34] and Figure 3.35C). The weak dependence of the etching curve on the KOH concentration (~1/4 power) supports the assumption that the hydroxide ions involved in the oxidation reactions are mostly generated from water. A strong influence of water on the silicon etch rate was also observed for EDP solutions. In molar water concentrations of up to 60%, a large increase of the etch rate occurs [Finne and Klein, 1967]. The driving force for the overall Reaction 3.4 is given by the larger Si–O binding energy of 193 kcal/mol as compared to a Si–Si binding energy of only 78 kcal/mol. The role of cations, K^+, Na^+, Li^+, and even complicated cations such as $NH_2(CH_2)_2NH_3^+$ can probably be neglected [Seidel et al., 1990].

The four electrons in Reaction 3.5 are injected into the conduction band in two steps. In the case of {100} planes, there are two dangling bonds per surface atom for the first two of the four hydroxide ions to react with, injecting two electrons into the conduction band in the process. As a consequence of the strong electronegativity of the oxygen atoms, the two bonded hydroxide groups on the silicon atom reduce the strength of the two silicon backbonds (Inset 3.9). With two new hydroxide ions approaching, two more electrons (now stemming from the Si–Si backbonds) are injected into the conduction band, and the silicon-hydroxide complex reacts with the two additional hydroxide ions. Seidel et al. (1990) claim that the step of activating the second two electrons from the backbonds into the conduction band is the rate-limiting step, with an associated thermal activation energy of about 0.6 eV for {100} planes.

The electrons in the backbonds are associated with surface states within the band gap (see Figure 3.40). The energy level of these surface states is assumed to vary for different surface orientations, being lowest for {111} planes. The thermal activation of the backbonds corresponds to an excitation of the electrons out of these surface states into the conduction band. Since the energy for the backbond surface state level is the lowest within the band gap for {111} planes, these planes will be hardest to etch. The {111} planes have only one dangling bond for a first hydroxide ion to react with. The second rate-limiting step involves breaking three lower energy backbonds. The lower energy of the backbond surface states for {111} Si atoms can be understood from the simple argument that their energy level is raised less by the electronegativity of a single binding hydroxide ion than by that of two in the case of the silicon atoms in {100} planes. The high etch rate generally observed on {110} surfaces is similarly explained by a high energy level of the backbond-associated surface states for these planes. Elwenspoek et al. (1993, 1994) do not accept this "two-vs.-three backbonds" argument. They point out that the silicon atoms in the {110} planes also have three backbonds and activation energy in these crystallographic directions should be comparable to that of {111} planes in contrast to experimental evidence. Seidel et al. would probably counter here that the backbonds and the energy levels of the associated surface states are not necessarily the same for {111} and {110} planes, as that energy also will be influenced by the effect of the orientation of these bonds. Another argument in favor of the high etching rates of {110} planes is the easier penetrability of {110} surfaces for water molecules along channels in that plane.

The final step in the anisotropic etching is the removal of the reaction product $Si(OH)_4$ by diffusion. If the production of $Si(OH)_4$ is too fast for solutions with a high water concentration, the $Si(OH)_4$ leads to the formation of a SiO_2-like complex before $Si(OH)_4$ can diffuse away. This might be observed experimentally as a white residue on the wafer surface [Wu et al., 1985]. The high pH values in anisotropic etching are required to obtain adequate solubility of the $Si(OH)_4$ reaction product and to remove the native oxide from the silicon surface. From silicate chemistry it is known that for pH values above 12 the $Si(OH)_4$ complex will undergo the following reaction by the detachment of two protons:

$$Si(OH)_4 \rightarrow SiO_2(OH)_2^{2-} + 2H^+ \qquad \text{(Reaction 3.7)}$$

$$2H^+ + 2OH^- \rightarrow 2H_2O \qquad \text{(Reaction 3.8)}$$

Pyrocatechol in an ethylenediamine etchant acts as a complex-forming agent for reaction products such as $Si(OH)_4$, converting these products into more complex anions:

$$Si(OH)_4 + 2OH^- + 3C_6H_4(OH)_2 \rightarrow Si(C_6H_4O_2)_3^{2-} + 6H_2O \qquad \text{(Reaction 3.9)}$$

There is evidence by Abu-Zeid et al. (1985) of diffusion control contribution to the etch rate in EDP, probably because the hydroxide ion must diffuse through the layer of complex silicon reaction products (see Reaction 3.9). The same authors also found that the etch rate depends on the effective Si area being exposed and on its geometry (crevice effect). That is why the silicon wafer is placed in a holder and the solution is vigorously agitated to minimize the diffusion layer thickness. For KOH solutions, no effect of stirring on etching rate was noticed. Stirring here is mainly used to decrease the surface roughness, probably through removal of hydrogen bubbles.

The influence of alcohol on the KOH etching rate in Equation (3.34) mainly is due to a change in the relative water concentration and its concomitant pH change; it does not participate in the reaction (this was confirmed by Raman studies by Palik et al. (1983). The reversal of etch rates for {110} and {100} planes through alcohol addition to KOH/water etchants can be understood by assuming that the alcohol covers the silicon surface [Palik et al., 1983], thus canceling the channeling advantage of the {110} planes. In the case of EDP, alcohol has no effect as the water concentration can be freely adjusted without significantly influencing the pH value due to the incomplete dissociation of EDP.

For the etching of SiO_2 shown in Figure 3.36, Seidel et al. propose the following reaction:

$$SiO_2 + 2OH^- \rightarrow SiO_2(OH)_2^{2-} \qquad \text{(Reaction 3.10)}$$

At KOH concentrations up to 35%, a linear correlation occurs between etch rate and KOH concentration. The SiO_2 etch rate in KOH solutions exceeds those in EDP by close to three orders of magnitude. For higher concentrations, the etch rate decreases with the square of the water concentration, indicating that water plays a role in this reaction. Seidel et al. speculate that at high pH values the silicon electrode is highly negatively charged (the point of zero charge of SiO_2 is 2.8), repelling the hydroxide ions while water takes over as reaction partner. An additional reason for the decrease is that the hydroxide concentration does not continue to increase with increasing KOH concentration for very concentrated solutions. The decrease of the Si/SiO_2 etch rate ratio with increasing temperature and pH value of the solution follows out of the larger activation energy of the SiO_2 etch rate (0.85 eV) and its linear correlation with the hydroxide concentration, whereas the silicon etch rate mainly depends on the water concentration.

TABLE 3.12   Effect of Water Concentration and pH Value on the Characteristics of
Silicon Etching

	$- H_2O+$	$- pH+$
SiO_2 etch rate	No effect	$- \Leftrightarrow +$
Si etch rate	$- \Leftrightarrow +$	Little effect
Solubility	No effect	$- \Leftrightarrow +$
Si/SiO_2 ratio	$- \Leftrightarrow +$	$+ \Leftrightarrow -$
Diffusion effects	$- \Leftrightarrow +$	$+ \Leftrightarrow -$
Residue formation	$- \Leftrightarrow +$	$+ \Leftrightarrow -$
p^+ etch stop	$- \Leftrightarrow +$	$+ \Leftrightarrow -$
p, n etch stop	$- \Leftrightarrow +$	$+ \Leftrightarrow -$

Source: Reprinted with permission from Seidel, H. (1990) "The Mechanism of Anisotropic
Electrochemical Silicon Etching in Alkaline Solutions," in *Technical Digest: 1990 Solid State
Sensor and Actuator Workshop*, Hilton Head Island, SC, 1990, pp. 86–91.

The effect of water concentration and pH value on the etching process in the Seidel et al. model is summarized in Table 3.12 [Seidel, 1990]. For aqueous KOH solutions within a concentration range from 10 to 60%, the following empirical formula for the calculation of the silicon etch rate R (see Equation [3.34]) proved to be in close agreement with the experimental data:

$$R = k_0 [H_2O]^4 [KOH]^{\frac{1}{4}} e^{-\frac{E_a}{kT}} \tag{3.35}$$

Etch rates for Si in μm/hr and for thermally grown SiO_2 in nm/hr for various KOH concentrations and etch temperatures are given in Madou (2002, Appendix C [Seidel et al., 1990]). The values for the fitting parameters used were $E_a = 0.595$ eV and $k_0 = 2480$ μm/hr $(mol/L)^{-4.25}$ for <100> Si, and $E_a = 0.6$ eV and $k_0 = 4500$ μm/hr $(mol/L)^{-4.25}$ for <110> Si. For the SiO_2 etch, an activation energy, E_a, of 0.85 eV was used.

In the section on etch stop techniques, we will see that the Seidel et al. model also nicely explains why all alkaline etchants exhibit a strong reduction in etch rate at high boron dopant concentration of the silicon; at high doping levels the conduction band electrons for the rate-determining reduction step are not confined to the surface anymore, and the reaction basically stops.

The key points of the Seidel et al. model can be summarized as follows (see also Table 3.12) [Seidel, 1990]:

1. The rate-limiting step is the water reduction.
2. Hydroxide ions required for oxidation of the silicon are generated through reduction of water at the silicon surface. The hydroxide ions in the bulk do not contribute to the etching, since they are repelled from the negatively charged surface. This implies that the silicon etch rate will depend on the molar concentration of water and that cations will have little effect on the silicon etch rate.
3. The dissolution of silicon dioxide is assumed to be purely chemical with hydroxide ions. The SiO_2 etch rate depends on the pH of the bulk electrolyte.
4. For boron concentrations in excess of 3×10^{19} cm^{-3}, the silicon becomes degenerate, and the electrons are no longer confined to the surface. This prevents the formation of the hydroxide ions at the surface and thus causes the etching to stop.
5. Anodic biases will prevent the confinement of electrons near the surface as well and lead to etch stop as in the case of a p^+ material.

Points 4 and 5 will become clearer when we discuss the workings of etch-stop techniques. This model applies well for lower index planes (i.e., {nnn} with n < 2) where high etch rates always correspond to low activation energies. But for higher index planes (i.e., {n11} and {1nn} with n = 2, 3, 4), Herr et al. (1991) found no correlation between activation energies and etch rates. For higher index planes, we must rely mainly on empirical data.

The Si etching reactions suggested by Seidel et al. are only the latest; in earlier proposed schemes according to Gandhi (1968) and Kern (1978), the silicon oxidation reaction steps suggested were injection of holes into the Si (raising the oxidation state of Si), hydroxylation of the oxidized Si species, complexation of the silicon reaction products, and dissolution of the reaction products in the etchant solution. In this reaction scheme, etching solutions must provide a source of holes as well as hydroxide ions, and they must contain a complexing agent with soluble reacted Si species in the etchant solution; for example, pyrocatechol forming the soluble $Si(C_6H_4O_2)_3^{2-}$ species. This older model still seems to be guiding the current thinking of many micromachinists, although Seidel et al.'s energy-level-based model of the silicon/electrolyte interface proves more satisfying.

3.7.4.3   Elwenspoek et al.'s Model

Elwenspoek et al. (1993, 1994) introduced an alternative model for anisotropic etching of Si, a model built on theories derived from crystal growth. According to these authors, the Seidel et al. model does not clearly explain the fast etching of {110} planes. Those planes, having three backbonds like the {111} planes, should etch equally slowly. The activation energy of the anisotropic etch rate depends on the etching system used; for example, etching in KOH is faster than in EDP, even when the pH of the solution is the same. Seidel et al. attribute this dependence to diffusion that plays a greater role in EDP than in KOH solutions. But Elwenspoek et al. point out that at least for slow etching the etch rate should not be diffusion controlled but governed by surface reactions. With surface reactions, diffusion should have a minor effect, analogous to growth at low pressure in an LPCVD reactor. Another comment focuses on the lack of understanding why certain etchants etch isotropically and others etch anisotropically.

Elwenspoek et al. note the parallels in the process of etching and growing of crystals; slowly growing crystal planes also etch slowly. A key to understanding both processes, growing or dissolution (etching), pertains to the concept of the energy associated with the creation of a critical nucleus on a single crystalline smooth surface — that is, the free energy associated with the creation of an island (growth) or a cavity (etching). Etching or growing of a material starts at active kink sites on steps. Kink sites are atoms with as many bonds to the crystal as to the liquid. Kinetics depend critically on the number of such kink sites. This aspect has remained neglected in the discussion of etch rates of single crystals up to now.

The free energy change ΔG involved in creating an island or digging a cavity (of circular shape in an isotropic material) of radius r on or in an atomically smooth surface is given by:

$$\Delta G = -N\Delta\mu + 2\pi r\gamma \tag{3.36}$$

where N is the number of atoms forming the island or the number of atoms removed from the cavity, $\Delta\mu$ is the chemical potential difference between silicon atoms in the solid state and in the solution, and γ is the step free energy. The step free energy in Equation (3.36) will be different at different crystallographic surfaces. This can easily be understood from the following example. A perfectly flat {111} surface in the Si diamond lattice has no kink positions (three backbonds, one dangling bond per atom), while on the {001} face every atom has two backbonds and two dangling bonds; that is, every position is a kink position. Consequently, creating an adatom-cavity pair on {111} surfaces costs energy: three bonds must be broken and only one is reformed. In the case of {001} faces, the picture is quite different. Creating an adatom-cavity pair now costs no energy because two bonds must be broken to remove an atom from the {001} face, but they can be returned by placing the atom back on the surface. The binding energy ΔE of an atom in a crystal slice with orientation (hkl) divided by kT (Boltzmann's constant times absolute temperature) is known as the α factor of Jackson of that crystal face [Jackson, 1966] or:

$$\alpha = \frac{\Delta E}{kT} \tag{3.37}$$

At sufficiently low temperature, where entropy effects can be ignored, $kT\alpha$ is proportional to the step free energy γ, and the number of adatom-cavity pairs is proportional to $\exp(-\alpha)$. This number is very small on the {111} silicon faces at low temperature, but 1 on the {001} silicon faces at any temperature. The

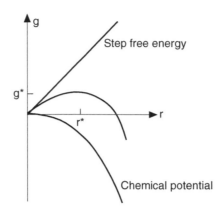

FIGURE 3.41 A plot of ΔG versus r based on Equation (3.39) exhibits a maximum.

consequence for {111} and {001} planes is that at sufficiently low temperatures the first are atomically smooth, and the latter are atomically rough. N in Equation (3.36) can be further written out as:

$$N = \pi r^2 h \rho \qquad (3.38)$$

where h is the height of the step, r is the diameter of the hole or island, and ρ is the density (atoms per cm^3) of the solid material. The result is:

$$\Delta G = -\pi r^2 h \rho \Delta \mu + 2\pi r \gamma \qquad (3.39)$$

where $\Delta \mu$ is counted positive, and γ is positive in any case. In Figure 3.41, we show a plot of ΔG vs. r. Equation (3.39) exhibits a maximum at:

$$r^* = \frac{\gamma}{h \rho \Delta \mu} \qquad (3.40)$$

At r^*, the free energy is:

$$\Delta G^* = \Delta G(r^*) = \frac{\pi \gamma^2}{h \rho \Delta \mu} \qquad (3.41)$$

Consequently, an island or an etch cavity of critical size exists on a smooth face. If by chance a cavity is dug into a crystal plane smaller than r^*, it will be filled rather than allowed to grow, and an island that is too small will dissolve rather than continue to grow because that is the easy way to decrease the free energy. With $r = r^*$, islands or cavities do not have any course of action, but with $r > r^*$, the islands or cavities can grow until the whole layer is filled or removed. In light of the above nucleation barrier theory, to remove atoms directly from flat crystal faces such as the {111} Si faces seems very difficult because the created cavities increase the free energy of the system and filling of adjacent atoms is more probable than removal; in other words, a nucleation barrier has to be overcome. The growth and etch rates R of flat faces are proportional to:

$$R \sim \exp\left(-\frac{\Delta G^*}{kT}\right) \qquad (3.42)$$

Since ΔG^* is proportional to γ^2, the activation energy is different for different crystallographic faces, and both the etch rate and the activation energy are anisotropic. If $\Delta G^*/kT$ is large, the etch rate will be very slow, as is the case for large step free energies and for small undersaturation (i.e., the "chemical drive" or $\Delta \mu$ is small) (see Equation [3.41]). Both $\Delta \mu$ and γ depend on the temperature and type of etchant, and these parameters might provide clues to understanding the variation of etch rate, degree of anisotropy, temperature dependence, etc., giving this model more bandwidth than Seidel et al.'s model. According to Elwenspoek et al., the chemical reaction energy barrier and the transport in the liquid are isotropic, and the most prominent anisotropy effect is due to the step free energy (absent on rough surfaces) rather than to the surface free energy. The surface free energy and the step free energies are related, though: when comparing flat faces, those having a large surface free energy have a small step free energy and vice versa. The most important difference in these two parameters is that the step free energy is zero for a rough surface, whereas the surface energy remains finite.

Flat faces grow and etch with a rate proportional to ΔG^*, which predicts that faces with a large free energy associated with forming a step will grow and etch much slower than faces with smaller free energy. Elementary analysis indicates that the only smooth face of the diamond lattice is the (111) plane. There may be other flat faces, but with lower activation energies due to reconstruction and/or adsorption, prominent candidates in this category are {100} and {110} planes. On the other hand, a rough crystal face grows and etches with a rate directly proportional to $\Delta \mu$. The temperature at which γ vanishes and a face transitions from smooth to rough is called the roughening transition temperature T_R [Elwenspoek and Weerden, 1987; Bennema, 1984]. Above T_R, the crystal is rough on the microscopic scale. Because the step free energy is equal to zero, new Si units may be added or removed freely to the surface without changing the number of steps. Rough crystal faces grow and dissolve with a rate proportional to $\Delta \mu$ and therefore proceed faster than flat surfaces. Imperfect crystals (for example, surfaces with screw dislocations) etch even faster with R proportional to $\Delta \mu^2$.

For the state of a surface slightly above or below the roughening temperature, T_R, thermal equilibrium conditions apply. Etching, in most practical cases, is far from equilibrium, and kinetic roughening might occur. Kinetic roughening [Bennema, 1984] occurs if the super- or undersaturation of the solution is so large that the thermally created islands or cavities are the size of the critical nucleus. One can show that if the super- or undersaturation is larger than $\Delta \mu_c$, given by:

$$\Delta \mu_c = \frac{\pi f_0 \gamma^2}{kT} \qquad (3.43)$$

(f_0 being the area one atom occupies in a given crystal plane), the growth and etch mechanism changes from a nucleation-barrier-controlled mechanism to a direct growth/etch mechanism. The growth rate and etch rate again become proportional to the chemical potential difference. It can thus be expected that, if the undersaturation becomes high enough, even the {111} faces could etch isotropically, as they indeed do in acidic etchants. If the undersaturation becomes so large that $\Delta \mu^* \ll kT$, the nucleation barrier breaks down. Each single-atom cavity acts as a nucleus made in vast numbers by thermal fluctuations. The face in question etches with a rate comparable to the etch rate of a rough surface. This situation is called kinetic roughening. If all faces are kinetically rough, the etch rate becomes isotropic.

Isotropic etching requires conditions of kinetic roughening because the etch rate is no longer dominated by a nucleation barrier but by transport processes in solution and the chemical reaction. To test this aspect of the model, Elwenspoek et al. show that there is a transition from isotropic to anisotropic etching if the undersaturation becomes too small. This can occur if one etches with an acidic etchant very long or if one etches through very small holes in a mask (crevice effect). In both cases, anisotropic behavior becomes evident as aging or limited transport of the solution causes the undersaturation to become very small. No proof is available to indicate that acidic etchants are much more undersaturated than the alkaline etchants. Still, the above explains some phenomena that Seidel et al.'s model fails to address. Another nice confirmation of Elwenspoek et al.'s model is in the effect of misalignment on etch rate. A misalignment of the mask close to smooth faces implies steps; there is no need for nucleation to etch. Since the density of steps is proportional to the angle of misalignment, the etch rate should be proportional to the misalignment angle, provided the distance between steps is not too large. Nucleation of new cavities becomes very probable. This has indeed been observed for the etch rate close to the <111> directions [Seidel et al., 1990]. Where Elwenspoek et al.'s model becomes a bit murky is in the classification of which surfaces are smooth and which are rough. Elementary analysis classifies only the {111} planes as smooth at low temperatures. At this stage, the model does not explain anything more than other models; every model has an explanation for the slower {111} etch rate. But these authors invoke the possibility of surface reconstruction and/or adsorption of surface species, which by decreasing the surface-free energy, could make faces such as {001} and {110} flat as well but with lower activation energies. They also take heart in the fact that CVD experiments often end up showing flat {110}, {100}, {331}, and strongest of all {111} planes. Especially where the influence of the etchant is concerned, more convincing thermodynamic data to estimate $\Delta\mu$ and γ are needed.

3.7.4.4 Isotropic vs. Anisotropic Etching of Silicon

In contrast to alkaline etching, with an acidic etchant such as HF, holes are needed for etching Si. An n-type Si electrode immersed in HF in the dark will not etch due to lack of holes. The same electrode in an alkaline medium etches readily. A p-type electrode in an HF solution, where holes are available under the proper bias, will etch even in the dark. For HF etchants, one might assume that the Ghandi and Kern model [Kern, 1978; Ghandi, 1968] relying on the injection of holes applies. In terms of the band model, this must mean that the silicon/electrolyte interface in acidic solutions exhibits quite different energetics from the alkaline case. It is not directly obvious why the energetics of the silicon/electrolyte interface would be pH dependent. To the contrary, since the flat-band potential of most oxide semiconductors as well as most oxide-covered semiconductors (such as Si) and the Fermi level of the H_2O/OH^- redox couple in an aqueous solution are both expected to change by 59 mV for each pH unit change [Madou and Morrison, 1989], one would expect the energetics of the interface to be pH independent. Since the electronegativity in a Si-F bond is higher than for a Si-OH bond, one might even expect the backbond surface states to be raised higher in an HF medium, making an electron injection mechanism even more likely than in alkaline media. To clarify this contradiction, we will analyze the band model of a Si electrode in an acidic medium in more detail. The band model shown in Figure 3.42 was constructed on the basis of a set of impedance measurements on an n-type Si electrode in a set of aqueous solutions at different pH values. From the impedance measurements Mott–Schottky plots were constructed to determine the pH dependency of the flat-band potential. From that, the position of the conduction band E_{cs} and valence band edges E_{vs} of the Si electrode in an acidic medium at a fixed pH of 2.2 (the point of zero charge) was calculated at 0.74 eV vs. SCE (saturated calomel electrode) for E_{cs}, and -0.36 eV vs. SCE for E_{vs} [Madou et al., 1980, 1981].

We have assumed in Figure 3.42 that the bands are bent upward at open circuit (see below for justification) so that holes in the valence band are driven to the interface where they can react with Si atoms or with competing reducing agents from the electrolyte. Since we want to etch Si, we are interested only in the reactions where Si itself is consumed. Reactions of holes with reducing agents are of great importance in photoelectrochemical solar cells [Madou et al., 1981]. For n-type Si, holes can be (1) injected by oxidants from the solution (e.g., by adding nitric acid to the HF solution); (2) supplied at the electrolyte/

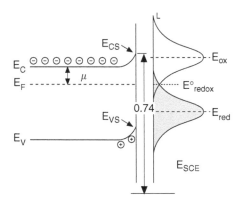

FIGURE 3.42 Band diagram for n-type Si in pH = 2.2 (no bias or illumination). Reference is the SCE. In Figure 4.40 no energy values were given; here we provide actual positions of the conduction band edge, E_{cs} = 0.74 eV versus SCE, and the valence band edge, E_{vs} = 0.74 eV − 1.1 eV = −0.36 eV versus SCE (1.1 eV is the band gap of Si). These values were determined by means of Mott–Schottky plots [Madou et al., 1980]. The separation between the Fermi energy and the bottom of the conduction band is indicated by μ.

semiconductor interface by shining light on a properly biased n-type Si wafer; or (3) created by impact ionization, that is, zener breakdown, of a sufficiently high reverse-biased n-type Si electrode [Levy-Clement et al., 1992]. With a p-type Si wafer, a small forward bias supplies all the holes needed for the oxidation of the lattice even without light, as the conduction happens via a hole mechanism. An important finding, explaining the different reaction paths in acidic and alkaline media, comes from plotting the flat band potential as a function of pH. It was found that the band diagram of Si shifts with less than 59 mV per pH unit. Actually the shift is only about 30 mV per pH unit [Madou et al., 1981]. As shown in Figure 3.43, with increasing pH, the energy levels of the solution rise faster than the energy levels in the semiconductor. As a consequence, it is more likely that electron injection takes place in alkaline media as the filled levels associated with the OH$^-$ are closer to the conduction band, whereas in acidic media, the filled levels of the redox system overlap better with the valence band, favoring a hole reaction. A lower position of the redox couple with respect to the conduction band edge, E_{cs}, in acidic media explains the upward bending of the bands as drawn in Figure 3.43. With isotropic etching in acidic media, the reaction starts with a hole in the valence band, equivalent to a broken Si-Si bond. In this case, the relative position of backbond-related surface states in different crystal orientations is of no consequence, and all planes etch at the same rate. A study of the interfacial energetics helps to understand why isotropic etching occurs in acidic media and anisotropic etching in alkaline media.

A few words of caution regarding our explanation for the reactivity difference between acidic and alkaline media are in order. Little is known about the width of the bell-shaped curves describing the redox levels in solution [Madou and Morrison, 1989]. Not knowing the surface concentrations of the reactive redox species involved in the etching reactions further hinders a better understanding of the surface energetics, as the bell-shaped curves for oxidant and reductant will only be the same height (as drawn in Figures 3.42 and 3.43) if the concentration of oxidant and reductant are the same. Clearly, the above picture is oversimplified; several authors have found that the dissolution of Si in HF might involve both the conduction and valence band, a claim confirmed by photocurrent multiplication experiments [Matsumura and Morrison, 1983; Lewerenz et al., 1988]. These photocurrent multiplication experiments showed that one or two holes generated by light in the Si valence band were sufficient to dislodge one Si [Brantley, 1973] unit, meaning that the rest of the charges were injected into the conduction band. Our contention here is only that the low pH dependence of the flat-band potential of a silicon electrode makes

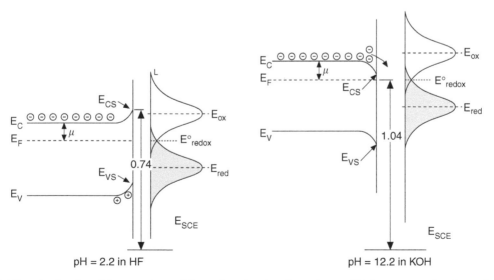

FIGURE 3.43 Band model comparison of the Si/electrolyte interface at low and high pH. Increasing the pH by 10 units shifts the redox-levels up by 600 mV, whereas the Si bands only move up by 300 mV. This leads to a different band-bending and a different reaction mechanism — that is, electron injection in the alkaline media (anisotropic) and hole injection in acidic media (isotropic).

a conduction band mechanism more favorable with alkaline-type etchants and a hole mechanism more favorable in acidic media.

Continued attempts at modeling the etch rates of all Si planes are under way. For example, Hesketh et al. (1993) attempted to model the etch rates of the different planes developing on silicon spheres in etching experiments with KOH and CsOH by calculating the surface free energy. The number of surface bonds per centimeter square on a Si plane is indicative of the surface free energy, which can be estimated by counting the bond density and multiplying by the bond energy. Using the unit cell dimension a of Si of 5.431 Å, and a silicon-to-silicon bond energy of 42.2 kcal/mol, the surface free energy, ΔG, can be related to N_B, the bond density, by the following expression:

$$\Delta G = \frac{N_B}{2} \times 2.94 \times 10^{-19} \frac{J}{m^2} \tag{3.44}$$

Although Hesketh et al. could not explain the etching differences observed between CsOH and KOH (these authors identified a cation effect on the etch rate!), a plot of the calculated surface free energy vs. orientation yielded minima for all low index planes such as {100}, {110}, and {111}, as well as for the high index {522} planes. Fewer bonds per unit area on the low index planes produce a lower surface energy and lower etching rate. When Hesketh et al. added the in-plane bond density to the surface bond density, producing a total bond density, a correlation with the hierarchy of etch rates in CsOH and KOH was found; that is, {311}, {522} > {100} > {111}. The surfaces with the higher bond density etched faster, suggesting that the etch rate might be a function of the number of electrons available at the surface. Hesketh et al. imply that their result falls in line with the Seidel et al.'s model, although it is unclear how the total bond density relates to surface state energies of backbonds. Moreover, Hesketh's model does not take into account the angles of the bonds, and in Elwenspoek's view, the surface free energy actually does not determine the anisotropy.

More research could focus on the modeling of Si etch rates. The semiconductor electrochemistry of corroding Si electrodes will be a major tool in further developments. Consult Sundaram et al. (1993) on Si etching in hydrazine and Palik et al. (1985) on the etch-stop mechanism in heavily doped silicon; both explain in some detail the silicon/electrolyte energetics. A more generic treatise on semiconductor electrochemistry can be found in Morrison (1980). A free etch simulator from the University of Illinois can be found at <http://mass.micro.uiuc.edu/research/completed/aces/index.html.>

3.8 Etching with Bias and/or Illumination of the Semiconductor

3.8.1 Introduction

The isotropic and anisotropic etchants discussed so far require neither a bias nor illumination of the semiconductor. The etching in such cases proceeds at open circuit, and the semiconductor is shielded from light. In a cyclic voltammogram as shown in Figure 3.44, the operational potential is the rest potential V_r, where anodic and cathodic currents are equal in magnitude and opposite in sign resulting in the absence of flow of current in an external circuit. This does not mean that macroscopic changes do not occur at the electrode surface, since the anodic and cathodic currents may be part of different chemical reactions. Consider isotropic etching of Si in an HF/HNO$_3$ etchant at open circuit where the local anodic reaction is associated with corrosion of the semiconductor:

$$\text{Anode: Si} + 2H_2O + nh^+ \rightarrow SiO_2 + 4H^+ + (4 - n)e^- \tag{Reaction 3.11}$$

$$SiO_2 + 6HF \rightarrow H_2SiF_6 + 2H_2O \tag{Reaction 3.12}$$

(the involvement of holes h$^+$ in the acidic reaction was discussed in the preceding section), while the local cathodic reaction could be associated with reduction of HNO$_3$:

$$\text{Cathode: HNO}_3 + 3H^+ \rightarrow NO + H_2O + 3H^+ \tag{Reaction 3.13}$$

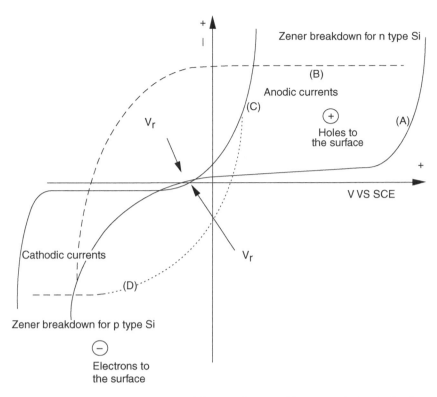

FIGURE 3.44 Basic cyclic voltammograms (*I* vs. *V*) for n- and p-type Si in an HF solution, in the absence of a hole injecting oxidant: (A) n-type Si without illumination; (B) n-type Si under illumination; (C) p-type Si without illumination; (D) p-type Si under illumination. If the reactions on the dark Si electrode determining the rest potential V_r for n- and p-type are the same, then V_r is expected to be the same as well. For clarity of the figure we have chosen V_rs different here; in practice, V_r for n- and p-type Si in HF are found to be identical.

We will now explore Si etching while illuminating and/or applying a bias to the silicon sample. To simplify the situation, we will first consider the case of an oxidant-free solution so that all the holes must come from within the semiconductor. Anodic dissolution of n-type Si in an HF-containing solution requires a supply of holes to the surface. For an n-type wafer under reverse bias, very few holes will show at the surface unless the high reverse (anodic) bias is sufficient to induce impact ionization or zener breakdown (Figure 3.44A). Alternatively, the interface can be illuminated, creating holes in the space charge region that the field pushes toward the semiconductor/electrolyte interface (Figure 3.44B). In the forward direction, electrons from the Si conduction band (majority carriers) reduce oxidizing species in the solution (e.g., reduction of protons to hydrogen). A p-type Si sample exhibits high anodic currents even without illumination at small anodic (forward) bias (Figure 3.44C). Here, the current is carried by holes. A p-type electrode illuminated under reverse bias gives rise to a cathodic photocurrent (Figure 3.44D). At relatively low light intensities, the photocurrent plateaus for both n- and p-types (Figure 3.44B and 3.44D, respectively) depend linearly on light intensity. The photocurrent is cathodic for p-type Si (species are reduced by photoproduced electrons at the surface, e.g., hydrogen formation) and anodic for n-type Si (species are oxidized by photoproduced holes at the surface; either the lattice itself is consumed or reducing compounds in solution are). In Figure 3.45, we show the cyclic voltammograms of n-type and p-type Si in the presence of a hole-injecting oxidant. The most obvious effect is on the dark p-type Si electrode. The injection of holes in the valence band increases the cathodic dark current dramatically. The current level measured in this manner for varying oxidant concentration or different oxidants could be used to estimate the efficiency of different isotropic Si etchants; a pointer to the fact that semiconductor electrochemistry has been underutilized as a tool to study Si etching. When n-type Si is

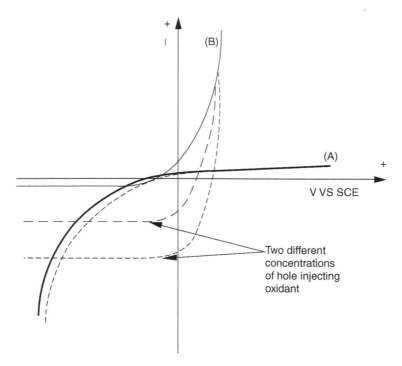

FIGURE 3.45 Basic cyclic voltammograms for n- and p-type Si in an HF solution and in the presence of a hole injecting oxidant, for example, HNO_3: (A) = n-type Si in the dark; (B) = p-type Si in the dark. An increase of the cathodic dark current on the p-type electrodes is most obvious. The current level is proportional to the oxidant concentration.

consumed under illumination, we experience photocorrosion (see Figure 3.44B). This photocorrosion phenomenon has been a major barrier to the long-term viability of photoelectrochemical cells [Madou et al., 1980]. In the following discussion, photocorrosion is put to use for electropolishing and formation of microporous and macroporous layers [Levy-Clement et al., 1992].

3.8.2 Electropolishing and Microporous Silicon

3.8.2.1 Electropolishing

Photoelectrochemical etching (PEC etching) involves photocorrosion in an electrolyte in which the semiconductor is generally chemically stable in the dark; that is, no hole-injecting oxidants are present, as in the case of an HF solution (see Figure 3.44). For carrying out the experiments, a setup as shown in Figure 3.33 may be used with a provision to illuminate the semiconductor electrode. At high light intensities, the anodic curves for n- and p-type Si are the same except for a potential shift of a few hundred millivolts (see Figure 3.46). Because of this equivalence, several of the etching processes described apply for both forward biased p-type and n-type Si under illumination. The anodic curves in Figure 3.46 present two peaks, characterized by i_{CRIT} and i_{MAX}. At the first peak, i_{CRIT}, partial dissolution of Si in reactions such as:

$$Si + 2F^- + 2h^+ \rightarrow SiF_2 \qquad \text{(Reaction 3.14)}$$

$$SiF_2 + 2HF \rightarrow SiF_4 + H_2 \qquad \text{(Reaction 3.15)}$$

leads to the formation of porous Si while hydrogen formation occurs simultaneously. The porous Si typically forms when using low current densities in a highly concentrated HF solution — in other words, by limiting the oxidation of silicon due to a hole and OH^- deficiency. Above i_{CRIT}, the transition from the charge-supply-limited to the mass-transport-limited case, the porous film delaminates and bright electropolishing

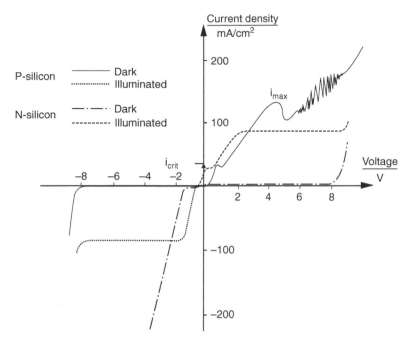

FIGURE 3.46 Cyclic voltammograms identifying porous Si formation regime and electropolishing regime. (Reprinted with permission from Levy-Clement, C. et al. [1992] "Photoelectrochemical Etching of Silicon," *Electrochim. Acta* 37, 877–88.)

occurs at potentials positive of i_{MAX}. With dissolution of chemical reactants in the electrolyte rate limiting, HF is depleted at the electrode surface, and a charge of holes builds up at the interface. Hills on the surface dissolve faster than depressions because the current density is higher on high spots. As a result, the surface becomes smoother; that is, electropolishing takes place [Lehmann, 1993]. Electropolishing in this regime can be used to smooth silicon surfaces or to thin epitaxially grown silicon layers. The peak and oscillations in Figure 3.46 are explained as follows: at current densities exceeding i_{MAX}, an oxide grows first on top of the silicon, leading to a decrease of the anodic photocurrent (explaining the i_{MAX} peak), until a steady state is reached in which dissolution of the oxide by HF through formation of a fluoride complex in solution (SiF_6^{2-}) equals the oxide growth rate. The oscillations observed in the anodic curve in Figure 3.46 can be explained by a nonlinear correlation between formation and dissolution of the oxide [Levy-Clement et al., 1992].

3.8.2.2 Porous Silicon

Introduction. The formation of porous Si was first discovered by Uhlir in 1956 [Uhlir, 1956]. His discovery has led to all types of interesting new devices from quantum structures, permeable membranes, and photoluminescent and electroluminescent devices to a basis for making thick SiO_2 and Si_3N_4 films [Bomchil et al., 1986; Smith and Collins, 1990]. Two types of pores exist: micropores and macropores. Their sizes can differ by three orders of magnitude, and the underlying formation mechanism is quite different. Some important features of porous Si, as detailed later, are

- Pore sizes in a diameter range from 20 Å to 10 μm
- Pores that follow crystallographic orientation [Chuang and Smith, 1988]
- Very high aspect ratio (~250) pores in Si maintained over several millimeters' distance [Lehman, 1993]
- Porous Si is highly reactive, oxidizes and etches at very high rate
- Porosity varies with the current density.

These important attributes contribute to the essential role porous Si plays in both micromachining [Barret et al., 1992] and the fashioning of quantum structures [Canham, 1990].

Microporous Silicon. Whether one is in the regime of electropolishing or porous silicon formation depends on both the anodic current density and the HF concentration. The surface morphology produced by the Si dissolution process critically depends on whether mass transport or hole supply is the rate-limiting step. Porous silicon formation is favored for high HF concentrations and low currents (weak light intensities for n-type Si), where the charge supply is limiting, while etching is favored for low HF concentration and high currents (strong light intensity for n-type Si), where mass transport is limiting. For current densities below i_{MAX} (Figure 3.46) holes are depleted at the surface, and HF accumulates at the electrode/electrolyte interface. As a result, a dense network of fine holes forms [Unagami, 1980; Watanabe et al., 1975]. The formation of a porous silicon layer (PSL) in this regime has been explained as a self-adjusting electrochemical reaction due to hole depletion by a quantum confinement in the microporous structure [Lehman and Gosele, 1991].The structure of the pores in PSL can best be observed by transmission electron microscopy (TEM), and its thickness can be monitored with an infrared (IR) microscope. The structure of the porous layer primarily depends on the doping level of the wafer and on the illumination during etching. Pore sizes decrease when etching occurs under illumination [Levy-Clement et al., 1992]. Pores formed in p-type Si show much smaller diameters than n-type ones for the same formation conditions. It is believed that the PSL consists of silicon hydrides and oxides. The pore diameter typically ranges from 40 to 200 Å [Lehmann, 1996; Yamana et al., 1990; Arita and Sunohara, 1977]. This very reactive porous material etches or oxidizes rapidly. Heat treatment in an oxidizing atmosphere (1100°C in oxygen for 30 min is sufficient to make a 4-μm thick film) leads to oxidized porous silicon (OPS). The oxidation occurs throughout the whole porous volume, and SiO_2 layers several micrometers thick can be obtained in times that correspond to the growth of a few hundred nanometers on regular Si surfaces. Porous silicon is low density and remains single crystalline, providing a suitable substrate for epitaxial Si film growth. These properties have been used to obtain dielectric isolation in ICs and to make SOI wafers [Watanabe et al., 1975]. Porous Si can also be formed chemically through Reactions 3.11 and 3.13. In this case, the difference between chemical polishing and porous Si formation conditions is more subtle. In the chemical polishing case, all reacting surface sites switch constantly from being local anode (Reaction 3.11) to being local cathode (Reaction 3.13) resulting in nonpreferential etching. If surface sites do not switch quickly between being local anode and cathode, charges have time to migrate over the surface. In this case, the original local cathode site remains a cathode for a longer time, and the corresponding local anode site, somewhere else on the surface, also remains an anode to keep the overall reaction neutral. A preferential etching results at the localized anode sites, making the surface rough and causing a porous silicon to form [Jung et al., 1993]. Any inhomogeneity, for example, some oxide or a kink site at the surface, might increase such preferential etching [Jung et al., 1993]. Unlike PSLs fabricated by electrochemical means, chemically etched porous Si film thickness is self-limiting.

Besides its use for dielectric isolation and the fabrication of SOI wafers, porous silicon has been introduced in a wide variety of other applications: Luggin capillaries for electrochemical reference electrodes, high surface area gas sensors, humidity sensors, sacrificial layers in silicon micromachining, and more. Recently, PSL was shown to exhibit photoluminescent and electroluminescent behavior; light-emitting porous silicon (LEPOS) was demonstrated. Visible light emission from regular Si is very weak due to its indirect band gap. Pumping porous Si with a green light laser (argon) caused it to emit a red glow. If a LEPOS device could be integrated monolithically with other structures on silicon, a big step in micro-optics, photon data transmission, and processing would be achieved. To explain the blue shift of the absorption edge of LEPOS of about 0.5 eV compared to bulk silicon [Lehmann and Gosele, 1991] and room temperature photoluminescence [Canham, 1990], Searson et al. (1993) have proposed an energy-level diagram for porous silicon in which the valence band is lowered with respect to bulk silicon to give a band gap of about 1.8 eV. Not only may PSL formation be explained invoking quantum structures as seen above, but its remarkable optical properties may also be explained this way. Canham believes that the thin Si filaments may act as quantum wires. Significant quantum effects require structural sizes below 5 nm, and porous Si definitely can have structures of that size. By treatment of the porous Si with NH_3 at high temperatures, it is possible to make thick Si_3N_4 films. Even at 13 µm, these films show little evidence of stress in contrast to stoichiometric LPCVD nitride films [Smith and Collins 1990].

Porous Si might represent a simple way of making the quantum structures of the future. Pore size of PSL can be influenced by both light intensity and current density. The quantum aspect has added significantly to the continued high interest in porous Si. For example, an optical biosensor based on porous Si has been demonstrated. Porous Si samples were prepared in such a way that the porous silicon films displayed Fabry-Perot fringes in their white-light reflection spectrum. When biological molecules were then chemically attached as recognition elements to the porous silicon surface and exposed to the appropriate complementary binding pair, binding occurs and is observed as a shift in the Fabry–Perot fringes [Doan and Sailor, 1992; Lin et al., 1997]. Lammel et al. (2000) fabricated tunable optical filters based on porous Si that can be used in reflective or transmission mode. The ratio of voids to total volume of porous Si determines its refractive index, and because porosity can be adjusted by varying the current density, these authors were able to micromachine a multilayer stack of porous Si of different indices by modulating the current density during porosification (in principle indices between 1.6 and 2.1 can be obtained this way). The process was applied to an area of single-crystal Si delineated by a silicon nitride window. The porosified Si plate was released from the substrate by subsurface electropolishing. During chemical release, two suspension arms lift the plate out of the plane automatically by internal stress release. The suspension arms are also provided with Cr/Ni heater wires to actuate the filters. By tilting the freestanding plate of porous Si with the thermal bimorph suspension arms, a wavelength scan is possible. Using this process, Lammel et al. achieved a 20 nm wavelength resolution in the visible part of the spectrum. Figure 3.47 is an SEM of the described porous Si plate optical filter.

Macroporous Silicon. In addition to micropores, well defined macropores can also be made in Si by photo and/or bias etching in HF solutions. Macropores have sizes as large as 10 µm visible with a scanning electron microscope (SEM) rather than TEM. The two types of pores often coexist, with micropores covering the walls of macropores. Sizes can differ by three orders of magnitude. This is not a matter of a broad fractal-type distribution of pores, but the formation mechanism is quite different.

Electrochemical etching of macropores or macroholes has been reported for n-type silicon in 2.5 to 5% HF under high voltage (>10 V), low current density (10 mA cm^{-2}), under illumination, and in the dark [Levy-Clement et al., 1992]. In the latter case, zener breakdown in silicon (electric field strength in excess

FIGURE 3.47 Optical filter of porous Si. SEM of a free-standing porous Si microplate containing a multilayered optical interference filter. The wavelength can be tuned by tilting the microplate, using the integrated thermal bimorph microactuator. The tilt angle is a function of the actuator DC voltage. This filter element can be used to build a microspectrometer. (Courtesy of Dr. Gerhard Lammel, EPFL, Lausanne, Switzerland.

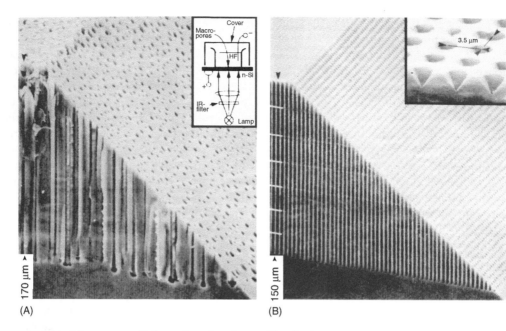

FIGURE 3.48 Macroporous Si; formation of random and localized macropores or macroholes. (A) Random: surface, cross-section, and a 45° bevel of an n-type sample (10^{15}/cm^3 phosphorus-doped) showing a random pattern of macropores. Pore initiation was enhanced by applying 10 V bias in the first minute of anodization followed by 149 minutes at 3 V. The current density was kept constant at 10 mA/cm^2 by adjusting the back-side illumination. A 6% aqueous solution of HF was used as an electrolyte. The setup used for anodization is sketched in the upper right corner. (B) Localized: surface, cross-section, and a 45° bevel of an n-type sample (10^{15}/cm^3 phosphorus-doped) showing a predetermined pattern of macropores (3 V, 350 min, 2.5% HF). Pore growth was induced by a regular pattern of pits produced by standard lithography and subsequent alkaline etching (inset upper right). To measure the depth dependence of the growth rate, the current density was kept periodically at 5 mA/cm^2 for 45 minutes and then reduced to 3.3 mA/cm^2 for 5 minutes. This reduction resulted in a periodic decrease of the pore diameter as marked by white labels in the figure. (Reprinted with permission from Lehmann, V. [1993] "The Physics of Macropore Formation in Low Doped n-type Silicon," *J. Electrochem. Soc.* 140, 2836–43.)

of 3×10^5 V/cm) causes the hole formation. The macropores are formed only with lightly doped n-type silicon at much higher anodic potentials than those used for micropore formation [Zhang, 1991]. By using a pore initiation pattern, the macropores actually can be localized at any desired location. This dramatic effect is illustrated by comparing Figure 3.48A and Figure 3.48B [Lehmann, 1993, 1996]. Pores orthogonal to the surface with depths up to a whole wafer thickness can be made, and aspect ratios as large as 250 become possible. The formation mechanism in this case cannot be explained on the basis of depletion of holes due to quantum confinement in the fine porous structures, given that these macropores exhibit sizes well beyond 5 nm. As with microporous Si, the surface morphology produced by the dissolution process depends critically on whether mass transport or charge supply is the rate-limiting step. For pore formation, one must work again in a charge depletion mode. Macropore formation, as micropore formation, is a self-adjusting electrochemical mechanism. In this case, the limitation is due to the depletion of the holes in the pore walls in n-type Si wafers causing them to passivate. Holes continue to be collected by the pore tip, where they promote dissolution. No passivating layer is involved to protect the pore wall. The only decisive differences between pore tips and pore walls are their geometry and their location. Holes generated by light or zener breakdown are collected at pore tips. Every depression or pit in the surface initiates pore growth because the electrical field at a curved pore bottom is much larger than that of a flat surface due to the effect of the radius of curvature. The latter leads to higher current and enhances local etching [Zhang, 1991]. Zener breakdown and illumination of n-type Si lead to different types of pore geometry [Lehmann, 1993; Theunissen et al., 1972; Lehmann and Foll, 1990]. Branched pores with

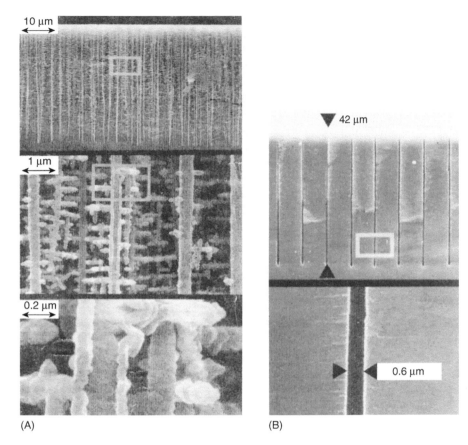

(A)   (B)

FIGURE 3.49   Comparison of macropores made with breakdown holes (A) and macropores made with light created holes (B). (A) An oxide replica of pores etched under weak back-side illumination visualizes the branching of pores produced by generation of charge carriers due to electrical breakdown (5 V, 3% HF, room temperature, $10^{15}/cm^3$ phosphorus-doped). (Reprinted with permission from Lehmann, V. [1993] "The Physics of Macropore Formation in Low Doped n-type Silicon," *J. Electrochem. Soc.* 140, 2836–43.) (B) Single pore associated with KOH pit. (Reprinted with permission from Lehmann, V., and Foll, H. [1990] "Formation Mechanism and Properties of Electrochemically Etched Trenches in n-type Silicon," *J. Electrochem. Soc.* 137, 653–59.)

sharp tips form if holes are generated by breakdown (Figure 3.49A) [Lehmann, 1993]. Unbranched pores with larger tip radii result from holes created by illumination (Figure 3.45B) [Lehmann and Foll, 1990]. The latter difference can be understood as follows: the electric field strength is a function of bias, doping density, and geometry. High doping level density or sharp pore tips will lower the required bias for breakdown, so macropores will tend to follow pores with the sharpest tips. Since every tip causes a new breakdown and hole generation, the position of the original pore tip becomes independent of the other pores, branching of the pores is possible, and fir-tree-type pores can be observed. With illumination, the pore radius may be larger, as the breakdown field strength is not necessary to generate charge carriers, so the pores remain unbranched.

The Si anisotropy common with alkaline etchants surprisingly shows up here with an isotropic etchant such as HF. For example, with breakdown-supplied holes, <100>-directed macroholes with <110> branches form (see also Figure 3.49A) [Lehmann, 1993], leading to a complex network of caverns beneath the silicon surface. Pyramidal pore tips [Lehmann, 1993] also were observed when the current density was limited by the bias ($i < i_{CRIT}$). Isotropic pore pits form when the current is larger than the critical current density, that is, isotropy in HF etching can be changed into anisotropy when the supply of holes is limited. We refer here to the Elwenspoek et al. model (see above), which predicts that in confined spaces etching will tend to be more anisotropic, even when using a normally isotropic etchant such as HF. It was also determined that macrohole formation depends on the wavelength of the light used. No hole formation occurs below about 800 nm. Depending on the wavelength, the shape of the hole can be manipulated as well [Lehmann and Foll, 1990]. For wavelengths above 867 nm, the depth profile of the holes changed from conical to cylindrical. The latter was interpreted in terms of the influence of the local minority carrier generation rate. Carriers generated deep in the bulk would promote the hole growth at the tips, whereas near-surface generation would lead to lateral growth.

In 1993, Cahill et al. (1993) reported the creation of 1 to 5 μm diameter pores with pore spacings (center to center) from 200 to 1000 μm. Until this finding, the pores typically formed were spaced in the range of 4 to 30 μm center-to-center and 0.6 to 10 μm in diameter. Making highly isolated pores presents quite another challenge. In the previous work, the relatively close spacing of the pores allowed the authors to conclude that the regions between the pores were almost totally depleted and that practically all carriers were collected by the pore tips. In such a case, neither the pore sidewalls nor the wafer surface etched, as all holes were swept to the pore tips. Since the surface was not attacked by pore-forming holes, the quality of the pore initiation mask lost its relevance. In Cahill et al.'s case, on the other hand, a long-lived mask (>20 hr) needed to be developed to help prevent pore formation everywhere except at initiation pits. The mask eventually used was a SiC layer sandwiched between two layers of silicon nitride. The silicon nitride directly atop the silicon serves to insulate the silicon carbide from the underlying substrate. As the silicon carbide proves very resistant to HF, loss of thickness does not show during the procedure. The top nitride serves to protect the carbide during anisotropic pit formation. By lowering the bias to less than 2 V with respect to SCE, side-branching is avoided.

Nano- and micropore formation phenomena are not limited to Si and extend to many types of semiconductors. Some preliminary evidence was collected with InP and GaAs [Lehmann, 1993].

3.9   Etch-Stop Techniques

3.9.1   Introduction

In many cases, it is desirable to stop etching in silicon when a certain cavity depth or a certain membrane thickness is reached. Nonuniformity of etched devices due to nonuniformity of the silicon wafer thickness can be quite high. Taper of double-polished wafers, for example, can be as high as 40 μm [Lambrechts and Sansen, 1992]! Even with the best wafer quality, the wafer taper is still around 2 μm. The taper and variation in etch depth lead to intolerable thickness variations for many applications. Etch-rate control typically requires monitoring and stabilization of

- Etchant composition
- Etchant aging
- Stabilization with N_2 sparging (especially with EDP and hydrazine)
- Taking account of the total amount of material etched (loading effects)
- Etchant temperature
- Diffusion effects (constant stirring is required, especially for EDP)
- Stirring also leads to a smoother surface through bubble removal
- Trenching (also pillowing) and roughness decrease with increased stirring rate
- Light may affect the etch rate (especially with n-type Si)
- Surface preparation of the sample can have a significant effect on etch rate (the native oxide retards etch start; a dip in dilute HF is recommended)

With good temperature, etchant concentration, and stirring control, the variation in etch depth typically is 1% (see Figure 3.50) [Lambrechts and Sansen, 1992]. A good pretreatment of the surface to be etched is a standard RCA clean combined with a 5% HF dip to remove the native oxide immediately prior to etching in KOH.

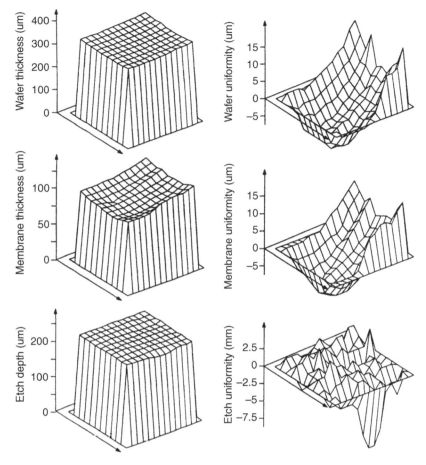

FIGURE 3.50 Map of wafer thickness, membrane thickness, and etch depth. (Reprinted with permission from Lambrechts, M., and Sansen, W. [1992] *Biosensors: Microelectrochemical Devices,* IPP, Philadelphia.)

In the early days of micromachining, one of the following techniques was used to etch a Si structure anisotropically to a predetermined thickness. In the simplest mode, the etch time was monitored (Table 3.11 lists some etch rates for different etchants) or a bit more sophisticated; the infrared transmittance through the etching membrane was followed. For thin membranes, the etch stop cannot be determined by a constant etch time method with sufficient precision. The spread in etch rates becomes critical if one etches membranes down to thicknesses of less than $20 \, \mu m$; it is almost impossible to etch structures down to less than $10 \, \mu m$ with a timing technique. In the V-groove technique, V-grooves with precise openings (see Equation [3.2]) were used such that the V-groove stopped etching at the exact moment a desired membrane thickness was reached (see Figure 3.51) [Samaun et al., 1973].

One can also design wider mask openings on the wafer's edge so that the wafer is etched through at those sites at the moment the membrane has reached the appropriate thickness. Although Nunn and Angell (1975) claimed that an accuracy of about $1 \, \mu m$ could be obtained using the V-groove method, none of the mentioned techniques are found to be production worthy. Nowadays, the above methods are almost completely replaced by etch-stop techniques based on a change in etch rate that depends on doping level or the chemical nature of a stopping layer. High-resolution silicon micromachining relies on the availability of effective etch-stop layers. It is actually the existence of impurity-based etch stops in silicon that has allowed micromachining to become a high-yield production process.

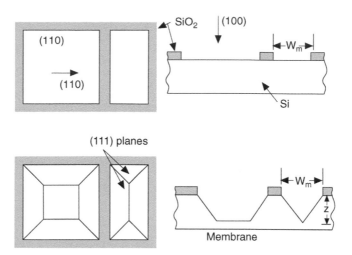

FIGURE 3.51 V-groove technique to monitor the thickness of a membrane. At the precise moment the V-groove is developed, the membrane has reached the desired thickness.

3.9.2 Boron Etch Stop

The most widely used etch-stop technique is based on the fact that anisotropic etchants, especially EDP, do not attack heavily boron-doped (p^{++}) Si layers. Selective p^{++} doping is typically implemented using a gaseous or solid boron diffusion source with a mask (such as silicon dioxide). The maximum practical depth achievable is $15\,\mu m$. The boron etch stop effect was first noticed by Greenwood in 1969 [Greenwood, 1969]. He assumed that the presence of a p-n junction was responsible. Bogh in 1971 [Bogh, 1971] found that an impurity concentration of about $7 \times 10^{19}/cm^3$ resulted in the etch rate of Si in EDP dropping sharply (see also Table 3.11) but without any requirement for a p-n junction. For KOH-based solutions, Price (1973) found a significant reduction in etch rate for boron concentrations above $5 \times 10^{18}\,cm^{-3}$. The model by Seidel et al. discussed above provides an elegant explanation for the etch stop at high boron concentrations. At moderate dopant concentration, we saw that the electrons injected into the conduction band stay localized near the semiconductor surface due to the downward bending of the bands (Figure 3.40). The electrons there have a small probability of recombining with holes deeper into the crystal even for p-type Si. This situation changes when the doping level in the silicon increases further. At a high dopant concentration, silicon degenerates and starts to behave like a metal. For a degenerate p-type semiconductor, the space charge thickness shrinks and the Fermi level drops into the valence band as indicated in Figure 3.52.

The injected electrons shoot (tunnel) directly through the thin surface charge layer into deeper regions of the crystal where they recombine with holes from the valence band. Consequently, these electrons are not available for the subsequent reaction with water molecules (Reaction 3.5), the reduction of which is necessary for providing new hydroxide ions in close proximity to the negatively charged silicon surface. These hydroxide ions are required for the dissolution of the silicon as $Si(OH)_4$. The remaining etch rate observed within the etch stop region is then determined by the number of electrons still available in the conduction band at the silicon surface. This number is assumed to be inversely proportional to the number of holes and thus the boron concentration. Experiments show that the decrease in etch rate is nearly independent of the crystallographic orientation, and the etch rate is proportional to the inverse fourth power of the boron concentration in all alkaline etchants.

From the above, it follows that a simple boron diffusion or implantation, introduced from the front of the wafer, can be used to create beams and diaphragms by etching from the back. A boron etch-stop technique is illustrated in Figure 3.53A for the fabrication of a micromembrane nozzle [Brodie and Muray, 1982]. The SiO_2 mesa in Figure 3.53A.b leads to the desired boron p^{++} profile. The anisotropic etch from the back clears the lightly doped p-type Si (Figure 3.53A[d]). By stripping and reoxidation an orifice is

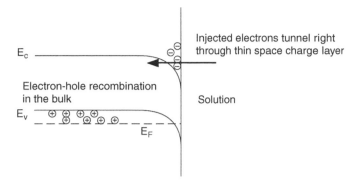

FIGURE 3.52 Si/electrolyte interface energetics at high doping level explaining etch-stop behavior.

produced in the suspended membrane (Figure 3.53A[e]). Two alternative ways of fabricating nozzles — one square and one circular — are illustrated in Figure 3.53B. The top method shown uses silicon nitride as an etch stop layer to etch a square nozzle. The side of the backside opening in the silicon nitride must be larger than 71% of the wafer thickness in order to etch all the way through the wafer (Equation [3.2]). The bottom approach shown in Figure 3.53B is again based on a p^{++} etch stop layer. The difference with the approach in Figure 3.53A is that a uniform p^{++} doping profile is first established here and that layer is subsequently etched through in a circular pattern. Layers of p^{++} silicon having a thickness of 1 to 20 μm can easily be fabricated and the boron etch stop is very effective; when the operator takes the wafer out of the etchant is not critical. One important practical note is that the boron etch stop may become badly degraded in EDP solutions that are allowed to react with atmospheric oxygen. Since boron atoms are smaller than silicon, a highly doped, freely suspended membrane or diaphragm will be stretched; the boron-doped silicon is typically in tensile stress and the microstructures are flat and do not buckle. While doping with boron decreases the lattice constant, doping with germanium increases the lattice constant. A membrane doped with B and Ge still etches much slower than undoped silicon, and the stress in the layer is reduced. A stress-free, dislocation-free, and slow etching layer (± 10 nm/min) is obtained at doping levels of 10^{20} cm^{-3} boron and 10^{21} cm^{-3} germanium [Seidel et al., 1990; Heuberger, 1989].

One disadvantage with the boron etch-stop technique is that the extremely high boron concentrations are not compatible with standard CMOS or bipolar techniques, so they can only be used for microstructures without integrated electronics. Another limitation of this process is the fixed number and angles of (111) planes one can accommodate. The etch stop is less effective in KOH compared to EDP. Besides boron, other impurities have been tried for use in an etch stop in anisotropic etchants. Doping Si with germanium has hardly any influence on the etch rate of either the KOH or EDP solutions. At a doping level as high as 5×10^{21} cm^{-3}, the etch rate is barely reduced by a factor of two [Seidel et al., 1990].

By burying the highly doped boron layer under an epitaxial layer of lighter doped Si, the problem of incompatibility with active circuitry can be avoided. A $\pm 1\%$ thickness uniformity is possible with modern epilayer deposition equipment (see, e.g., *Semiconductor International*, July 1993, pp. 80–83). A widely used method of automatically measuring the epi thickness is with IR instruments, especially Fourier transform infrared (FTIR) [Rehrig, 1990].

3.9.3 Electrochemical Etch Stop

For the fabrication of piezoresistive pressure sensors, the doping concentration of the piezoresistor must be kept smaller than 1×10^{19} cm^{-3} because the piezoresistive coefficients drop considerably above this value and reverse breakdown becomes an issue. Moreover, high boron levels compromise the quality of the crystal by introducing slip planes and tensile stress and thus prevent the incorporation of integrated electronics. As a result, a boron stop often cannot be used to produce well controlled thin membranes unless, as suggested above, the highly doped boron layer is buried underneath a lighter doped Si epilayer.

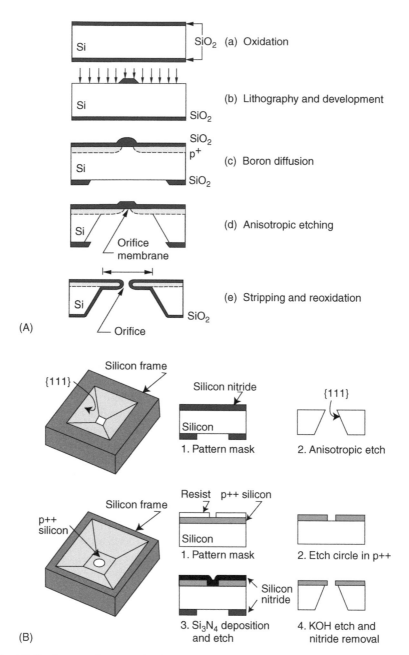

FIGURE 3.53 (A) The boron etch stop in the fabrication of a membrane nozzle. (Reprinted with permission from Brodie, I., and Muray, J.J. [1992] *The Physics of Microfabrication,* Plenum Press, New York.) (B) Two alternate methods to fabricate nozzles. Top: etch stop is based on silicon nitride. Bottom: etch stop is based on boron etch stop.

Alternatively, a second etch-stop method, an electrochemical technique, can be used. In this method, a lightly doped p-n junction is used as an etch stop by applying a bias between the wafer and a counter-electrode in the etchant. This technique was first proposed by Waggener in 1970. Other early work on electrochemical etch stops with anisotropic etchants such as KOH and EDP was performed by Palik et al. (1982), Jackson et al. (1981), Faust and Palik (1983), and Kim and Wise (1983). In electrochemical anisotropic etching, a p-n junction is made, for example, by the epitaxial growth of an n-type layer (phosphorus-doped, 10^{15} cm^{-3}) on a p-type substrate (boron-doped, $30\,\Omega$-cm). This p-n junction forms a

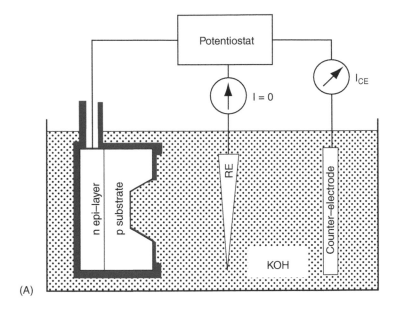

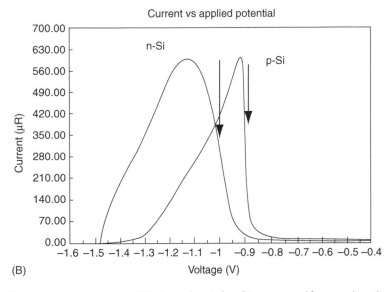

FIGURE 3.54 Electrochemical etch stop. (A) Electrochemical etching setup with potentiostatic control (three-electrode system). Potentiostatic control, mainly used in research studies, enables better control of the potential as it is referenced now to a reference electrode such as an SCE. In industrial settings, electrochemical etching is often carried out in a simpler two-electrode system; that is, a Pt counter electrode and Si working electrode [Kloeck et al., 1989]. (B) Cyclic voltammograms of n- and p-type silicon in an alkaline solution at 60°C. Flade potentials are indicated with an arrow.

large diode over the whole wafer. The wafer is usually mounted on an inert substrate, such as a sapphire plate, with an acid-resistant wax and is partly or wholly immersed in the solution. An ohmic contact to the n-type epilayer is connected to one pole of a voltage source and the other pole of the voltage source is connected via a current meter to a counterelectrode in the etching solution (see Figure 3.54A). In this arrangement, the p-type substrate can be selectively etched away, and the etching stops at the p-n junction leaving a membrane with a thickness solely defined by the thickness of the epilayer. The incorporation of a third electrode (a reference electrode) in the three-terminal method depicted in Figure 3.54A allows for a more precise determination of the silicon potential with respect to the solution than a two-terminal setup as we illustrated in Figure 3.33.

At the Flade potential in Figure 3.54B, the oxide growth rate equals the oxide etch rate; a further increase of the potential results in a steep fall of the current due to complete passivation of the silicon surface. At potentials positive of the Flade potential, all etching stops. At potentials below the Flade potential, the current increases as the potential becomes more positive. This can be explained by the formation of an oxide that etches faster than it forms; that is, the silicon is etched away. Whereas electrochemists like to talk about the Flade potential, physicists like to discuss matters in terms of the flat-band potential. Flat-band potential is the applied potential at which there are no more fields within the semiconductor; that is, the energy band diagram is flat throughout the semiconductor. The passivating SiO_2 layer is assumed to start growing as soon as the negative surface charge on the silicon electrode is cancelled by the externally applied positive bias, a bias corresponding to the flat band potential. At these potentials, the formation of $Si(OH)_x$ complexes does not lead to further dissolution of silicon because two neighboring Si-OH HO-Si groups will react by splitting off water, leading to the formation of Si-O-Si bonds. As can be learned from Figure 3.54B, the value of the Flade potential depends on the dopant type. Consequently, if a wafer with both n- and p-type regions exposed to the electrolyte is held at a certain potential in the passive range for n-type and the active range for p-type Si, the p-type regions are etched away, whereas the n-type regions are retained. In the case of the diode shown in Figure 3.54A, where at the start of the experiment only p-type Si is exposed to the electrolyte, a positive bias is applied to the n-type epilayer (V_n). This reverse biases the diode, and only a reverse bias current can flow. The potential of the p-type layer in this regime is negative to the flat-band potential, and active dissolution takes place. At the moment that the p-n junction is reached, a large anodic current can flow, and the applied positive potential passivates the n-type epilayer. Etching continues on the areas where the wafer is thicker until the membrane is reached there, too. The thickness of the silicon membrane is thereby solely defined by the thickness of the epilayer; neither the etch uniformity nor the wafer taper will influence the result. A uniformity of better than 1% can be obtained on a $10\,\mu m$ thick membrane. The current vs. time curve can be used to monitor the etching process; at first, the current is relatively low, limited by the reverse bias current of the diode. Then as the p-n junction is reached, a larger anodic current can flow until all the p-type material is consumed and the current falls again to a plateau. The plateau indicates that r, the current associated with a passivated n-type Si electrode, has been reached. The etch procedure can be stopped at the moment that the current plateau has been reached, and because the etch stop is thus basically one of anodic passivation, it is sometimes called an anodic oxidation etch stop. Registering an I vs. V curve as in Figure 3.54B will establish an upper limit on the applied voltage V_n [McNeil et al., 1990], and such curves are used for in situ monitoring and controlling of the etch stop. A crude endpoint monitoring can be accomplished by the visual observation of cessation of hydrogen bubble formation accompanying Si etching. Palik et al. (1988) presented a detailed characterization of Si membranes made by electrochemical etch stop.

Hirata et al. (1988), using the same anodic oxidation etch stop in a hydrazine-water solution at $90 \pm 5°C$ and a simple two-terminal electrochemical cell, obtained a pressure sensitivity variation of less than 20% from wafer to wafer (pressure sensitivity is inversely proportional to the square of membrane thickness). A great advantage of this etch-stop technique is that it works with Si at low doping levels of the order of $10^{16}\,cm^{-3}$. Due to the low doping levels, it is possible to fabricate structures with a very low, or controllable, intrinsic stress. Moreover, active electronics and piezomembranes can be built into the Si without problems. Reay et al. [1984] used TMAH with silicon dissolved in it. A disadvantage is that the back of the wafer with the aluminum contact has to be sealed hermetically from the etchant solution, which requires complex fixturing and manual wafer handling. The fabrication of a suitable etch holder is no trivial matter. The holder (1) must protect the epi-contact from the etchant, (2) must provide a low-resistance ohmic contact to the epi, and (3) must not introduce stress into wafers during etching [Peeters, 1994; Elwenspoek et al., 1994]. Stress introduced by etch holders easily leads to diaphragm or wafer fracture and etchant seepage through to the epi-side.

Using a four-electrode electrochemical cell, controlling the potentials of both the epitaxial layer and the silicon wafer as shown in Figure 3.55 can further improve the thickness control of the resulting

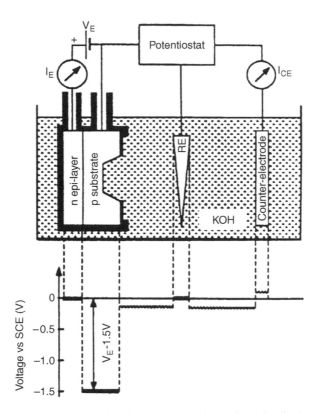

FIGURE 3.55 Four-electrode electrochemical etch-stop configuration; voltage distribution with respect to the SCE reference electrode (RE) for the four-electrode case. The fourth electrode enables an external potential to be applied between the epitaxial layer and the substrate, thus maintaining the substrate at etching potentials. (Reprinted with permission from Kloeck, et al. [1989] "Study of Electrochemical Etch-Stop for High-Precision Thickness Control of Silicon Membranes," *IEEE Trans. Electron Devices* **36**, 663–69.)

membrane by directly controlling the p/n bias voltage. The potential required to passivate n-type Si can be measured using the three-electrode system in Figure 3.54A, but this system does not take into account the diode leakage. If the reverse leakage is too large, the potential of the n-type Si, V_n, will approach the potential of the p-type Si, V_p. If there is a large amount of reverse diode leakage, the p-type region may passivate prior to reaching the n-type region, and etching will cease. In the four-electrode configuration, the reverse leakage current is measured separately via a p-type region contact, and the counterelectrode current may be monitored for endpoint detection. The four-electrode approach allows etch stopping on lower quality epis (larger leakage current) and should also enable etch stopping of p-type epi on n-type substrates. Kloeck et al. (1989) demonstrated that, using such an electrochemical etch-stop technique with current monitoring, the sensitivity of pressure sensors fabricated on the same wafer could be controlled to within 4% standard deviation. These authors used a 40% KOH solution at 60 ± 0.5°C. Without the etch stop, the sensitivity from sensor to sensor on one wafer varied by a factor of two.

The etching solution used in electrochemical etching can be either isotropic or anisotropic. Electrochemical etch stop in isotropic media was discussed above. In this case, HF/H_2O mixtures are used to etch the highly doped regions of p^+p, n^+n, n^+p, and p^+n systems [Theunissen et al., 1970; Meek, 1971; Theunissen et al., 1972; Wen and Weller, 1972; Dijk, 1972]. The rate-determining step in etching with isotropic etchants does not involve reducing water with electrons from the conduction band as it does in anisotropic etchants, and the etch stop mechanism, as we learned earlier, is obviously different. In isotropic media, the etch stop is simply a consequence of the fact that higher conductivity leads to higher corrosion currents and the etch slows down on lower conductivity layers. A major advantage of the KOH

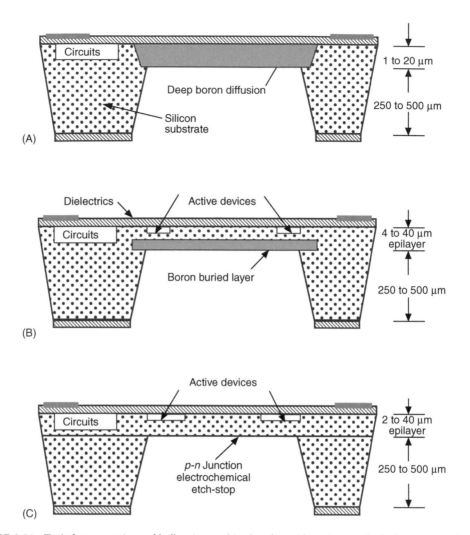

FIGURE 3.56 Typical cross sections of bulk micromachined wafers with various methods for etch-stop formation shown. (A) Diffused boron etch stop. (B) Boron etch stop as a buried layer. (C) Electrochemical etch stop. (Reprinted with permission from Wise, K.D. et al. [1985] "Silicon Micromachining and Its Applications to High Performance Integrated Sensors," in *Micromachining and Micropackaging of Transducers,* Fung, C.D. et al., eds., Elsevier, New York, pp. 3–18.)

electrochemical etch is that it retains all of the anisotropic characteristics of KOH without needing a heavily doped p[+] layer to stop the etch [Saro and Herwaarden, 1986].

In Figure 3.56, we review the etch-stop techniques discussed so far: diffused boron etch stop, buried boron etch stop, and electrochemical etch stop [Wise et al., 1985]. A comparison of the boron etch stop and the electrochemical etch stop reveals that the IC compatibility and the absence of built-in stresses, both due to the low dopant concentration, are the main assets of the electrochemical etch stop.

3.9.4 Photo-Assisted Electrochemical Etch Stop (for n-Type Silicon)

A variation on the electrochemical diode etch-stop technique is the photo-assisted electrochemical etch-stop method illustrated in Figure 3.57 [Mlcak et al., 1993]. An n-type silicon region on a wafer may be selectively etched in an HF solution by illuminating and applying a reverse bias across a p-n junction, driving the p-type layer cathodic and the n-type layer anodic. Etch rates up to 10 μm/min for the n-type material

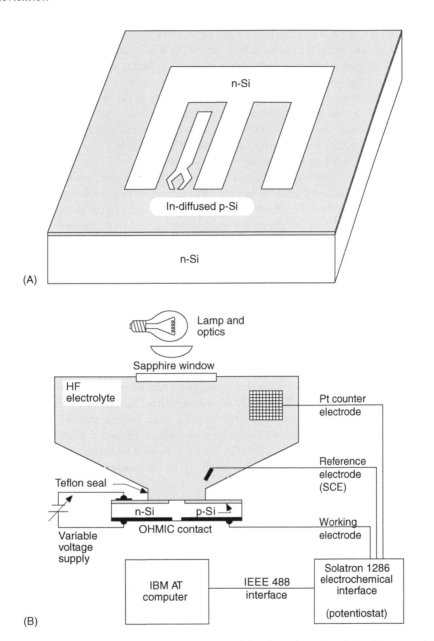

FIGURE 3.57 Photoelectrochemical etching. (A) Schematic of the photoelectrochemical etching experiment apparatus. (B) Schematic of the spatial geometry of the diffused p-type Si layer into n-type Si used to form cantilever beam structures. (Reprinted with permission from Mlcak, R. et al. [1993] "Photo-Assisted Electromechanical Machining of Micromechanical Structures," in *Proceedings: IEEE Micro Electro Mechanical Systems [MEMS '93]*, Fort Lauderdale, pp. 225–29.)

and a high resolution etch stop render this an attractive potential micromachining process. Advantages also include the use of lightly doped n-type Si, bias- and illumination-intensity-controlled etch rates, in situ process monitoring using the cell current, and the ability to spatially control etching with optical masking or laser writing. Using this method, Mlcak et al. (1993) prepared stress-free cantilever beam test samples. They diffused boron into a $10^{15}\,cm^{-3}$ (100) n-type Si substrate through a patterned oxide mask leaving exposed a small n-type region that defines two p-type cantilever beams (see Figure 3.57).

The boron diffusion resulted in a junction 3.3 μm underneath the surface. An ohmic contact on the back of the wafer was used to apply a variable voltage across the p-n junction, and both p and n areas were exposed to the HF electrolyte. The exposed n-type region was etched to a depth of 150 μm by shining light on the whole sample. The resulting n-type Si surface was at first found to be rough because porous Si up to 5 μm in height forms readily in HF solutions. The Si surface could be made smoother by etching at higher bias (4.3 V vs. SCE) and higher light intensity (2 W/cm²) to a finish with features of the order of 0.4 μm in height. Smoothing could also be accomplished by a 5 s dip in HNO3:HF:H₃COOH or a 30 s dip in 25 wt%, 25°C, KOH. Yet another way of removing unwanted porous Si is a 1000°C wet oxidation to make OPS followed by an HF etch (see next section) [Yoshida et al., 1992].

The gauge factor of a piezoresistor at the base of a thin rectangular cantilever as shown in Figure 3.57 can be calculated from Equation (3.28). To make a very high-sensitivity cantilever with a piezoresistor at its base, cantilevers under 1000Å thick have been produced. Example 3 of the section "Polysilicon Surface Micromachining Examples," explores the design and fabrication of such a cantilever are explored in more detail.

3.9.5   Photo-Induced Preferential Anodization (for p-Type Silicon)

Electrochemical etching requires the application of a metal electrode to apply the bias. The application of such a metal electrode often induces contamination and constitutes at least one extra process step, and extra fixturing is needed. With photoinduced preferential anodization (PIPA), it is not necessary to deposit metal electrodes. Here one relies on the illumination of a p-n junction to bias the p-type Si anodically, and the p-type Si converts automatically into porous Si while the n-type Si acts as a cathode for the reaction. The principle of photobiasing for etching purposes was known and patented by Shockley as far back as 1963 [Shockley, 1963]. In U.S. patent 3,096,262, he writes, "light can be used in place of electrical connections … for biasing of the sample.… This means a small isolated area of p-type material on an n-type body may be preferentially biased for removal of material beyond the junction by etching." The method was reinvented by Yoshida et al. (1992) in 1992 and in 1993 by Peeters et al. (1993a, 1993b). The latter group called the method PHET, for photovoltaic electrochemical etch-stop technique, and the former group coined the PIPA acronym.

In PIPA, etch rates of up to 5 μm/min result in the formation of porous layers that are readily removed with Si etching solutions. An important advantage of the technique is that very small and isolated p-type islands can be anodized at the same time. The method also lends itself to fabrication of three-dimensional structures using p-type Si as sacrificial layers [Yoshida et al., 1992]. A disadvantage of the technology is that one cannot control the process very well because the current cannot be measured for endpoint detection. Application of PIPA to form a micro-bridge is shown in Figure 3.58. First, a buried p-type layer doped to 10^{18} cm^{-3} and an n-type layer doped to 10^{15} cm^{-3} are formed on an n-type substrate using epitaxy (Figure 3.58A). Then, the p-type layer is preferentially anodized in 10% HF solution under 30 mW/cm² light intensity for 180 min (Figure 3.58B) forming porous Si. The porous Si is then oxidized in wet oxygen at 1000°C (Figure 3.58C). Finally, the sacrificial layer of oxidized porous silicon (OPS) is etched and removed with an HF solution (Figure 3.58D). The resulting surfaces of the n-type silicon are very smooth. It is interesting to consider making complicated three-dimensional structures by going immediately to the electropolishing regime instead. Yoshida et al. (1992) believe that porous silicon as a sacrificial intermediate is more suitable for fabricating complicated structures.

The authors are probably referring to the fact that electropolishing is much more aggressive and could not be expected to lead to the same retention of the shape of the buried, sacrificial p layers. Peeters et al. (1993) carry out their photovoltaic etching in KOH, thus skipping the porous Si stage of Yoshida et al. They, like Yosida et al., stress the fact that in one single etch step this technique can make a variety of complex shapes that would be impossible with electrochemical etching techniques. These authors found it necessary to coat the n-type Si part of the wafer with Pt to get enough photovoltaic drive for the anodic dissolution; this metallization step makes the process more akin to photoelectrochemical etching, and some advantages of the photo-biasing process are lost.

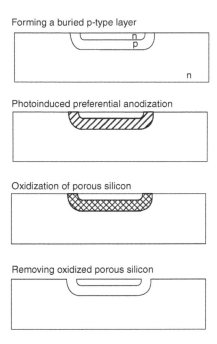

FIGURE 3.58   Photo-assisted electrochemical etch stop (for n-type Si). Fabrication process for a microbridge and SEM picture of Si structure before and after PIPA. (Reprinted with permission from Yoshida, T. et al. [1992] "Photo-Induced Preferential Anodization for Fabrication of Monocrystalline Micro-Mechanical Structures," in *Proceedings: IEEE Micro Electro Mechanical Systems [MEMS '92]*, Travemunde, Germany, pp. 56–61.)

3.9.6   Etch Stop at Thin Insoluble Films

Yet another distinct way (the fourth) to stop etching is by employing a change in composition of material. An example is an etch stopped at a Si_3N_4 diaphragm (see Figure 3.53B, top, for example). Silicon nitride is very strong, hard, and chemically inert, and the stress in the film can be controlled by changing the Si/N ratio in the LPCVD deposition process. The stress turns from tensile in stoichiometric films to compressive in silicon-rich films (for details, see later in this chapter). A great number of materials are not attacked by anisotropic etchants. Hence, a thin film of such a material can be used as an etch stop.

Another example is the SiO_2 layer in an SOI structure. A buried layer of SiO_2 sandwiched between two layers of crystalline silicon forms an excellent etch stop because of the good selectivity of many etchants of Si over SiO_2. The oxide does not exhibit the good mechanical properties of silicon nitride and consequently rarely is used as a mechanical member in a micro-device. As with PIPA, no metal contacts are needed with an SOI etch stop, greatly simplifying the process over an electrochemical etch-stop technique.

We have classified SOI micromachining under surface micromachining later in this chapter, where more details about this promising micromachining alternative are presented.

3.10   Issues in Wet Bulk Micromachining

3.10.1   Introduction

Despite the introduction of more controllable etch-stop techniques, bulk micromachining remains a difficult industrial process to control. It is also not an applicable submicron technology because wet chemistry cannot etch reliably on that scale. For submicron structure definition, dry etching is required (dry etching is also more environmentally safe). We will now look into some of the other problems associated with bulk micromachining — such as the extensive real estate consumption and difficulties in etching at convex corners — and detail current solutions to avoid, control, or alleviate them.

3.10.2   Extensive Real Estate Consumption

3.10.2.1   Introduction

Bulk micromachining involves extensive real estate consumption. This quickly becomes a problem in making arrays of devices. Consider the diagram in Figure 3.59, illustrating two membranes created by etching through a <100> wafer from the back until an etch stop, say a Si_3N_4 membrane, is reached. In creating two of these small membranes, a large amount of Si real estate is wasted, and the resulting device becomes quite fragile.

3.10.2.2   Real Estate Gain by Etching from the Front

One solution to limiting the amount of Si to be removed is to use thinner wafers, but this solution becomes impractical below 200 μm, as such wafers often break during handling. A more elegant solution is to etch from the front rather than from the back. Anisotropic etchants will undercut a masking material by an amount dependent on the orientation of the wafer with respect to the mask. Such an etchant will etch any <100> silicon until a pyramidal pit is formed, as shown in Figure 3.60. These pits have sidewalls with a characteristic 54.7° angle with respect to the surface of a <100> silicon wafer, since the delineated planes are {111} planes. This etch property makes it possible to form cantilever structures by etching from the front, as the cantilevers will be undercut and eventually will be suspended over a pyramidal pit in the silicon. Once this pyramidal pit is completed, the etch rate of the {111} planes exposed is extremely slow and practically stops. Process sequences, which depend on achieving this type of a final structure, are therefore very uniform across a wafer and very controllable. The upper drawings in Figure 3.60 represent patterned holes in a masking material: a square, a diamond, and a square with a protruding tab. The drawings immediately below represent the etched pit in the silicon produced by anisotropic etching. In the first drawing on the left a square mask aligned with the <110> direction produces a four-sided pyramidal pit. In the second drawing, a similar mask shape oriented at 45° produces an etched pit that is oriented parallel to the

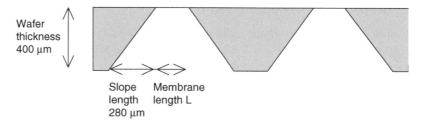

FIGURE 3.59   Two membranes formed in a <100>–oriented silicon wafer.

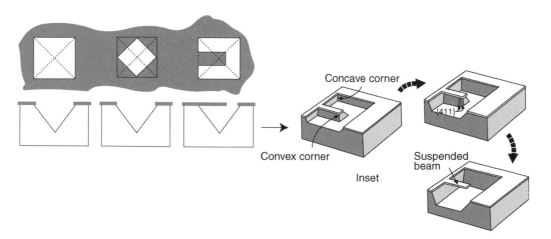

FIGURE 3.60   Three anisotropically etched pits etched from the front in a <100>–oriented silicon wafer. Inset: Illustration of etching at convex corners for the formation of suspended beams.

pit etched in the first drawing independent of the mask orientation. In the second drawing, the corners of the diamond are undercut by the etchant as it produces the final etch pit. The third drawing illustrates that any protruding member is eventually undercut by the anisotropic etchant leaving a cantilever structure suspended over an identical etch pit. The inset detailing the third drawing shows how undercutting at the convex corners of the cantilever in pure KOH is determined by {411} planes [Mayer et al., 1990].

3.10.2.3 Real Estate Gain by Using Silicon Fusion-Bonded Wafers

Using silicon fusion-bonded (SFB) wafers rather than conventional wafers also makes it possible to fabricate much smaller microsensors. The process, introduced in the 1980s, is clarified in Figure 3.61 for the fabrication of an absolute pressure sensor [Bryzek et al., 1990]. The bottom handle wafer has a standard thickness of 525 μm and is anisotropically etched with a square cavity pattern. Next, the etched handle wafer is fusion bonded to a top sensing wafer (see SFB and surface micromachining later in this chapter, and packaging in

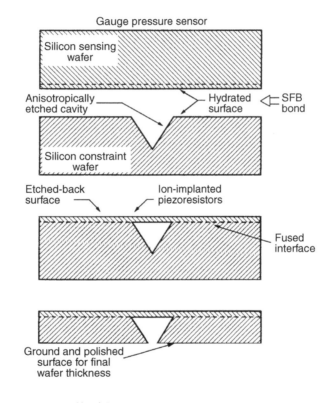

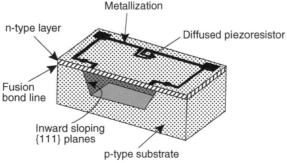

FIGURE 3.61 Fabrication process of an SFB-bonded gauge pressure sensor. (Reprinted with permission from Bryzek, J. et al. [1990] *Silicon Sensors and Microstructures,* NovaSensor, Fremont, California.)

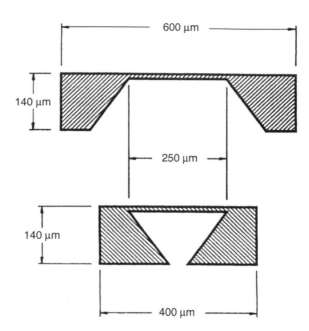

FIGURE 3.62   Comparison of conventional and SFB processes. The SFB process results in a chip that is at least 50% smaller than the conventional chip. (Reprinted with permission from Bryzek, J. et al. [1990] *Silicon Sensors and Microstructures,* NovaSensor, Fremont, California.) Inset: A miniature silicon-fusion bonded absolute pressure sensor made with Si fusion bonding. (Courtesy Lucas NovaSensor, Fremont, California.)

Madou [2002, Chapter 8]. The sensor wafer consists of a p-type substrate with an n-type epilayer corresponding to the required thickness of the pressure-transducing membrane. The sensing wafer is thinned all the way to the epilayer by electrochemical etching and resistors are ion implanted. The handle wafer is ground and polished to the desired thickness. For gauge measurement, the anisotropically etched cavity is truncated by the polishing operation, exposing the back of the diaphragm. For an absolute pressure sensor, the cavity is left enclosed. With the same diaphragm dimensions and the same overall thickness of the chip, an SFB device, because of the inward sloping {111} walls, is almost 50% smaller than a conventional machined device (Figure 3.62). Lucas NovaSensor, Fremont, California, for example, manufactures a sensor that is 400 µm wide, 800 µm long, and 150 µm thick and fits inside the tip of a catheter (Figure 3.61, inset).

3.10.3   Corner Compensation

3.10.3.1   Underetching

Underetching of a mask that contains no convex corners (that is, corners turning outside in) in principle stems from mask misalignment and/or from a finite etching of the {111} planes. Peeters measured the widening of {111}–walled V-grooves in a (100) Si wafer after etching in 7 M KOH at 80 ± 1° over 24 hr as 9 ± 0.5 µm [Peeters, 1994]. The sidewall slopes of the V-groove are a well defined 54.74°, and the actual etch rate R_{111} is related to the rate of V-groove widening R_v through:

$$R_{111} = \frac{1}{2}\sin(54.74°)R_v \quad \text{or} \quad R_{111} = 0.408 \cdot R_v \tag{3.45}$$

with R_{111} the etch rate in nm/min and R_v the groove widening, also in nm/min. The V-groove widening experiment then results in an R_{111} of 2.55 ± 0.15 nm/min. In practice, this etch rate implies a mask underetching of only 0.9 µm for an etch depth of 360 µm. For a 1 mm long V-groove and a 1° misalignment angle, a total underetching of 18 µm is theoretically expected, with 95% due to misalignment and only 5% due to etching of the {111} sidewalls [Peeters, 1994]. The total underetching will almost always be determined by misalignment rather than by etching of {111} walls.

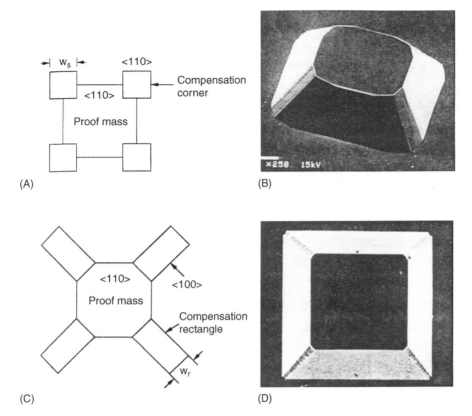

FIGURE 3.63 Formation of a proof mass by silicon bulk micromachining. (A, B) Square corner compensation method using EDP as the etchant. (C, D) Rotated rectangle corner compensation method using KOH as the etchant.

Mask underetching with masks that do include convex corners is much larger than the underetching just described, as the etchant tends to circumscribe the mask opening with {111} walled cavities. This is usually called undercutting rather than underetching. It is advisable to avoid mask layouts with convex corners. Often, mesa-type structures are essential, though, and in that case there are two possible ways to reduce the undercutting. One is by chemical additives, reducing the undercut at the expense of a reduced anisotropy ratio, and the other is by a special mask compensating the undercut at the expense of more lost real estate.

3.10.3.2 Undercutting

When etching rectangular convex corners, deformation of the edges occurs due to undercutting. This is often an unwanted effect, especially in the fabrication of, say, acceleration sensors, where total symmetry and perfect 90° convex corners on the proof mass are mandatory for good device prediction and specification. The undercutting is a function of etch time and thus directly related to the desired etch depth. An undercut ratio is defined as the ratio of undercut to etch depth (δ/H).

Saturating KOH solutions with IPA reduces the convex corner undercutting; unfortunately, this happens at the cost of the anisotropy of the etchant. This additive also often causes the formation of pyramidical or cone-shaped hillocks [Peeters, 1994; Gravesen, 1986]. Peeters claims that these hillocks are due to carbonate contamination of the etchant, and he advises etching under inert atmosphere and stockpiling all etchant ingredients under an inert nitrogen atmosphere [Peeters, 1994].

Undercutting can also be reduced or even prevented by "corner compensation structures," which are added to the corners on the mask layout so as to be undercut during etching. Depending on the etching solution, different corner compensation schemes are used, among them square corner compensation (EDP or KOH) and rotated rectangle corner compensation (KOH) as illustrated in Figure 3.63. In square corner compensation, the square of SiO_2 in the mask outlining the square proof mass feature for an accelerometer is enhanced by adding an extra SiO_2 square to each corner (Figure 3.63A). Both the proof mass and the compensation squares are aligned with their sides parallel to the <110> direction. In this way, two concave corners are created at the convex corner to be protected. Thus, direct undercutting is prevented. The three "sacrificial" convex corners at the protective square are undercut laterally by the fast etching planes during the etch process. The dimension of the compensation square w_s depends on the depth of the required cavity; for example, for a 300 μm deep cavity, a square with a side length of 300 μm is used. The resulting mesa structure after EDP or KOH etching is shown in Figure 3.63B. In rotated rectangle corner compensation, shown in Figure 3.63C, a properly scaled rectangle (w_r should be twice the depth of etching) is added to each of the mask corners. The four sides of the mesa square are still aligned along the <110> direction, but the compensation rectangles are rotated (45°) with their longer sides along the <100> directions. Using KOH as an etchant reveals the mesa shown in Figure 3.63D. A proof mass is frequently dislodged by simultaneously etching from the front and the back. Corner compensation requires a significant amount of real estate in the mask layout around the corners, making the design less compact, and the method is often only applicable for simple geometries.

Different groups using different corner compensation schemes all claim to have optimized spatial requirements. For an introduction to corner compensation, refer to Puers et al. (1990). Sandmaier et al. (1991) used <110>-oriented beams, <110>-oriented squares, and <010>-oriented bands for corner compensation during KOH etching. They found that spatial requirements for compensation structures could be reduced dramatically by combining several of these compensation structures. Figure 3.64 shows

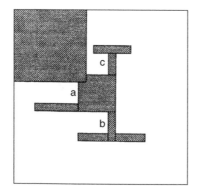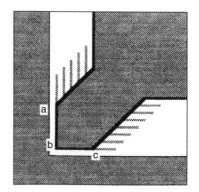

FIGURE 3.64   Mask layout for various convex corner compensation structures. (Reprinted with permission from Sandmaier, H. et al. [1991] "Corner Compensation Techniques in Anisotropic Etching of (100) Silicon Using Aqueous KOH," in *6th International Conference on Solid-State Sensors and Actuators [Transducers '91]*, San Francisco, pp. 456–59.)

the mask layouts for some of the different compensation schemes they used. To understand the choice and dimensioning of these compensation structures, as well as those in Figure 3.63, we will first look at emerging planes at convex corners during KOH etching.

There is some disagreement in the literature about the exact nature of planes that emerge as etching progresses at convex corners. Long et al. (1999) in their work assumed {310} to be the dominant etch plane. Mayer et al. (1990) found that the undercutting of convex corners in pure KOH etch is determined exclusively by {411} planes. The {411} planes of the convex underetching corner as shown in Figure 3.65 are not entirely laid free, however; rugged surfaces where only fractions of the main planes can be detected overlap the {411} planes under a diagonal line shown as AB in this figure. The ratio of {411} to {100} etching does not depend on temperature between 60 and 100°C. The value declines with increasing KOH concentration from about 1.6 at 15% KOH to 1.3 at above 40% where the curve flattens out [Mayer et al., 1990]. Ideally, one avoids rugged surfaces and searches for well defined planes bounding the convex corner. Figure 3.66 shows how a <110> beam is added to the convex corner to be etched. Fast-etching {411} planes, starting at the two convex corners, laterally underetch a <110>-oriented beam (dashed lines in Figure 3.66). The longer this <110>-oriented beam is, the longer the convex corner will be protected from undercutting. It is essential that the beam disappear by the end of the etch to maintain a minimum of rugged surface at the convex edge. On the other hand, as is obvious from Figure 3.66, a complete disappearance of the beam leads to a beveling at the face of the convex corner. The dimensioning of the compensating <110> beam then works as follows: the length of the compensating beam is calculated primarily from the required etch depth (H) and the etch rate ratio R{411}/R{100} ($\approx \delta/H$), at the concentration of the KOH solution used:

$$L = L_1 - L_2 = 2H \frac{R\{411\}}{R\{100\}} - \frac{B_{<110>}}{2\tan(30.9°)} \tag{3.46}$$

where H = etch depth, $B_{<110>}$ = width of the <110>-oriented beam, and $\tan(30.9°)$ = geometry factor [Mayer et al., 1990].

Factor 2 in the first term of this equation results as the etch rate of the {411} plane is determined normal to the plane and has to be converted to the <110> direction. The second term in Equation (3.46) takes into account that the <110> beam needs to disappear completely by the time the convex corner is

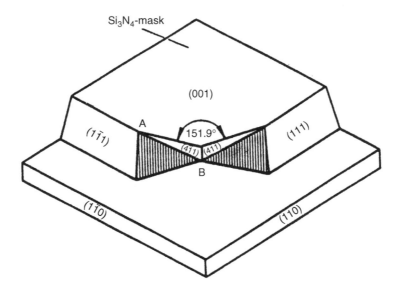

FIGURE 3.65 Planes occurring at convex corners during KOH etching. (Reprinted with permission from Mayer, G.K. et al. [1990] "Fabrication of Non-Underetched Convex Corners in Anisotropic Etching of (100)-Silicon in Aqueous KOH with Respect to Novel Micromechanic Elements," *J. Electrochem. Soc.* 137, 3947–51.)

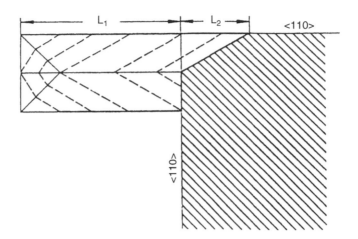

FIGURE 3.66 Dimensioning of the corner compensation structure with a <110>–oriented beam. (Reprinted with permission from Sandmaier, H. et al. "Corner Compensation Techniques in Anisotropic Etching of (100) Silicon Using Aqueous KOH," in *6th International Conference on Solid-State Sensors and Actuators [Transducers '91],* San Francisco, pp. 456–59.)

reached. The resulting beveling in Figure 3.66 can be reduced by further altering the compensation structures. This is done by decelerating the etch front, which largely determines the corner undercutting. One way to accomplish this is by creating more concave shapes right before the convex corner is reached. In Figure 3.64A, splitting of the compensation beam creates such concave corners, and by arranging two such double beams a more symmetrical final structure is achieved. By using these split beams, the beveling at the corner is reduced by a factor of 1.4 to 2 and leads to bevel angles under 45°.

Corner compensation with <110>-oriented squares (as shown in Figure 3.63A) features considerably higher spatial requirements than the <110>-oriented beams. Since these squares are again undercut by {411} planes that are linked to the rugged surfaces described above, the squares do not easily lead to sharp {111}–defined corners. Dimensioning of the compensation square is done by using Equation (3.46) again, where L_1 is half the side length of the square; for $B_{<110>}$, the side length is used. All fast-etching planes have to reach the convex corner at the same time. The spatial requirements of this square compensation structure can be reduced if it is combined with <110>-oriented beams. Such a combination is shown in Figure 3.64B. The three convex corners of the compensation square are protected from undercutting by the added <110> beams. During the first etch step, the <110>-oriented side beams are undercut by the etchant. Only after the added beams have been etched does the square itself compensate the convex corner etching. The dimensioning of this combination structure is carried out in two steps. First, the <110>-oriented square is selected with a size that is permitted by the geometry of the device to be etched. From these dimensions, the etch depth corresponding to this size is calculated from Equation (3.46). For the remaining etch depth, the <110>-oriented beams are dimensioned like any other <110>-oriented beam. If the side beam on corner b is made about 30% longer than the other two side beams, the quality of the convex corner can be further improved. In this case, the corner is formed by the etch fronts starting at the corners a and b (Figure 3.64B).

A drawback to all of the above proposed compensation schemes is the impossibility of obtaining a sharp corner in both the top and the bottom of a convex edge due to the rugged surfaces associated with {411} planes. Buser et al. (1988) introduced a compensation scenario in which a convex corner was formed by two {111} planes that were well defined all the way from the mask to the etch bottom. No rugged, undefined planes show in this case. The mask layout to create such an ideal convex corner uses <010> bands that are added to convex corners in the <100> direction (Figure 3.64C). These bands will be underetched by vertical {100} planes from both sides. With suitable dimensioning of such a band, a vertically oriented membrane results, thinning and eventually freeing the convex edge shortly before the final intended etch

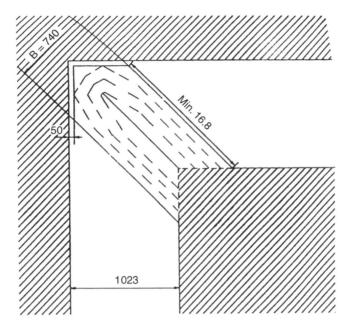

FIGURE 3.67 Beam structure open on one side. The beam is oriented in the <010> direction. The dimensions are in microns and B is the width of the beam.

depth is reached. In contrast to compensation structures undercut by {411} planes, posing problems with undefined rugged surfaces (see above), this compensation structure is mainly undercut by {100} planes. Over the temperature range of 50 to 100°C and KOH concentrations ranging from 25 to 50 wt%, no undefined surfaces could be detected in the case of structures undercut by {100} planes [Sandmaier et al., 1991]. The width of these <010>-oriented compensation beams, which determines the minimum dimension of the structures to etch, has to be twice the etching depth. These beams can either connect two opposite corners and protect both from undercutting simultaneously, or they can be added to the individual convex corners (open beam). With an open-beam approach (see Figure 3.67), it is extremely important that the {100} planes reach the corner faster than the {411} planes. For that purpose, the beams have to be wide enough to avoid complete underetching by {411} planes moving in from the front before they are completely underetched by {100} planes moving in from the side. For instance, in a 33% KOH etchant, a ratio between beam length and width of at least 1.6 is required. To make these compensation structures smaller while simultaneously maintaining {100} undercutting to define the final convex corner, Sandmaier et al. remarked that the shaping {100} planes do not need to be present at the beginning of the etching process. These authors implement delaying techniques by adding fan-like <110>-oriented side beams to a main <100>-oriented beam (see Figure 3.64C). As described above, these narrow beams are underetched by {411} planes and the rugged surfaces they entail until reaching the <100>-oriented beam. Then, the {411} planes are decelerated in the concave corners between the side beams by the vertical {100} planes with slower etching characteristics. The length of the <110>-oriented side beams is calculated from:

$$L_{<110>} = \left(H - \frac{B_{<010>}}{2}\right)\frac{R\{411\}}{R\{100\}} \tag{3.47}$$

with H being the etching depth at the deepest position of the device.

The width of the side beams does not influence the calculation of their required length. To avoid the rugged surfaces at the convex corner, the width of the side beams as well as the spaces between them should be kept as small as possible. For an etching depth of 500 μm, a beam width of 20 μm and a space width of 2 μm are optimal [Sandmaier et al., 1991]. The biggest drawback of this compensation strategy is that the mask layout is rather complicated, and it takes a lot of time to generate the pattern to fabricate the masks themselves.

Modifications of the compensation scheme that use <010> bands have been proposed to simplify the mask design. Zhang et al. (1996) introduced a modified method that uses a <100> bar with a width greater than twice the etching depth, so that lateral etching would stop at <410> sidewalls, and improved results were obtained. Another novel structure was described by Enoksson (1997); it involves a layout that consists of a <010> diagonal band connected to a concave corner on the opposite side of the groove (similar to the situation in Figure 3.67). The band has a slit in the middle of the concave corner and, according to Enoksson, this gives a near perfect square corner.

Depending on the etchant, different planes are responsible for undercutting. From the above, we learned that, in pure KOH solutions, undercutting — according to Mayer et al. (1977) and Sandmaier et al. (1991) — mainly proceeds through {411} planes or {100} planes. That the {411} planes are the fastest undercutting planes was confirmed by Seidel (1986). At the wafer surface, the sectional line of a (411) and a (111) plane point in the <410> direction, forming an angle of 30.96° with the <110> direction, and it was in this direction that he found a maximum in the etch rate. In KOH and EDP etchants, Bean (1978) identified the fast undercutting planes as {331} planes. Puers et al. (1990) for alkali/alcohol/water identified the fast underetching planes as {331} planes, as well. Mayer et al. (1990) working with pure KOH could not confirm the occurrence of such planes. Lee indicated that, in hydrazine-water, the fastest underetching planes are {211} planes (1969). Abu-Zeid (1984) reported that the main beveling planes are {212} planes in ethylenediamine-water solution (no added pyrocatechol). Wu et al. (1989) found the main beveling planes at undercut corners to be {212} planes whether using KOH, hydrazine, or EPW solutions. In view of our earlier remarks on the sensitivity of etching rates of higher index rates to a wide variety of parameters (temperature, concentration, etching size, stirring, cation effect, alcohol addition, complexing agent, etc.), these contradictory results are not too surprising. Along the same line, Wu et al. (1987) and Puers et al. (1990) have suggested triangles to compensate for underetching, but Mayer et al. (1990) found them to lead to rugged surfaces at the convex corner. Combining a chemical etchant with more limited undercutting (IPA in KOH) with Sandmaier's reduced compensation structure schemes could further decrease the required size of the compensation features while retaining an acceptable anisotropy.

Corner compensation for <110>-oriented Si was explored by Ciarlo (1987), who comments that both corner compensation and corner rounding can be minimized by etching from both surfaces so as to minimize the etch time required to achieve the desired features. This requires accurate front-to-back alignment and double-sided polished wafers.

Employing corner compensation offers access to completely new applications such as rectangular solids, orbiting V-grooves, truncated pyramids with low cross-sections on the wafer surface, bellows structures for decoupling mechanical stresses between micromechanical devices and their packaging, etc. [Sandmaier et al., 1991].

Some additional important references on corner compensation are Ted Hubbard's Ph.D. thesis (MEMS Design, Caltech 1994), Nikpour et al. (1998), and Kim et al. (1998).

3.11 Computer Simulation Software

Design of masking patterns in MEMS requires visualization in three dimensions. ACESTM, from the University of Illinois at Urbana–Champaign, stands for Anisotropic Crystalline Etch Simulation, and it is the first PC-based 3-D etch simulator to offer high computational speed. AnisE® from IntelliSense (<http://www.intellisensesoftware.com/datasheets/AnisE.pdf>) helps the user to simulate 3-D bulk silicon anisotropic etching processes and is capable of predicting the effects of etchant temperature, concentration, and etch time on the final 3-D device geometry. Corner compensation and process tolerances can be modeled and visualized. SEGS is an interactive on-line wet etch simulator that predicts the etched shape as a function of time for arbitrary isotropic or anisotropic etchants and any initial mask shape [Li et al., 1998]. Using any Java enabled Web browser, users can draw initial masks, choose the etchant, simulate the etching, and view animation results.

3.12   Wet Bulk Micromachining Examples

Example 1. Dissolved Wafer Process

Figure 3.68 illustrates the dissolved wafer [Cho, 1995]. This process, used by Draper Laboratory [Greiff, 1995; Weinberg et al., 1996] for the fabrication of low-cost inertial instruments, involves a sandwich of a silicon sensor anodically bonded to a glass substrate. The preparation of the silicon part requires only two masks and three processing steps. A recess is KOH-etched into a p-type (100) silicon wafer (step 1 with mask 1), followed by a high-temperature boron diffusion (step 2 with no mask). In step one, RIE may be used as well. Cho claims that by maintaining a high-temperature uniformity in the KOH etching bath ($\pm 0.1°C$) the accuracy and absolute variation of the etch across the wafer, wafer-to-wafer, and lot-to-lot, can be maintained to <0.1 μm using premixed, 45 wt% KOH. Cho is also using low-defect oxidation techniques (e.g., nitrogen annealing and dry oxidation) to form defect-free silicon surfaces. In the boron diffusion, the key is optimizing the oxygen content. In general, the optimal flow of oxygen is on the order of 3 to 5% of the nitrogen flow, in which case the doping uniformity is on the order of ± 0.2 μm. Varying the KOH etch depth and the shallow boron diffusion time, a wide variety of operating ranges and sensitivities for sensors can be obtained. Next, the silicon is patterned for a reactive ion etching (RIE) etch (step 3 with mask 2). Aspect ratios above 10 are accessible. Using some of the newest dry etching techniques, depths in excess of 500 μm at rates above 4 μm/min (with an SF6 chemistry) are now possible [Craven et al., 1995]. The glass substrate (#7740 Corning glass) preparation involves etching a recess, depositing, and in a one-mask step, patterning a multi-metal system of Ti/Pt/Au. The electrostatic bonding of glass to silicon takes place at 335°C with a potential of 1000 V applied between the two parts. Commercial bonders have alignment accuracies on the order of <1 μm. The lightly doped silicon is dissolved in an EDP solution at 95°C. The keys to uniform EDP etching are temperature uniformity and suppression of etchant depletion through chemical aging or restricted flow (e.g., through bubbles). These effects can be minimized by techniques that optimize temperature control and reduce bubbling (e.g., proper wafer spacing, lower temperature, large bath). The structures are finally rinsed in deionized (DI) water and a hot methanol bath.

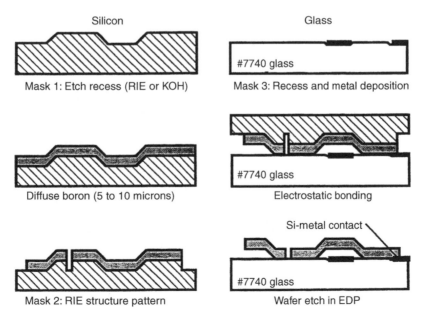

FIGURE 3.68   Dissolved wafer process. (Reprinted with permission from Greiff, P. [1995] "SOI-Based Micromechanical Process," in *Micromachining and Microfabrication Process Technology (Proceedings of the SPIE)*, Austin, TX, pp. 74–81.)

Draper Laboratory, although obtaining excellent device results with the dissolved wafer process, is now exploring an SOI process as an alternative. The latter yields an all-silicon device while preserving many of the dissolved wafer process advantages (see SOI Surface Micromachining later in this chapter).

Example 2. An Electrochemical Sensor Array Measuring pH, CO_2, and O_2 in a Dual Lumen Catheter

An electrochemical sensor array developed by the author is shown in Figure 3.69, packaged and ready for in vivo monitoring of blood pH, CO_2, and O_2. The linear electrochemical array fits inside a 20-gauge catheter (750 μm in diameter) without taking up so much space as to distort the pressure signal monitored by a pressure sensor outside the catheter. A classical (macro) reference electrode making contact with the blood through the saline drip was used for the pH signal, while the CO_2 and O_2 had their own internal reference electrodes. The high impedance of the small electrochemical probes makes a close integration of the electronics mandatory; otherwise, the high impedance connector leads in a typical hospital setting act as antennas for the surrounding electromagnetic noise. As can be seen from the computer-aided design (CAD) picture in Figure 3.70, the thickness of the sensor comes from two silicon pieces, the top piece containing electrochemical cells and the bottom piece containing the active electronics. Each wafer is 250 μm thick. The individual electrochemical cells are etched anisotropically into the top silicon wafer. The bottom piece is fabricated in a custom IC housing using standard IC processes.

The process sequence to build a generic electrochemical cell in a 250 μm thick Si wafer with one or more electrodes at the bottom of each well is illustrated in Figure 3.71. Electrochemical wells are etched from the front of the wafer, and after an oxidation step, access cavities for the metal electrodes are also etched from the back. The etching of the vias stops at the oxide-covered bottoms of the electrochemical wells (Figure 3.71A). Next, the wafers are oxidized a second time, with the oxide thickness doubling everywhere except in the suspended window areas where no Si can feed further growth (Figure 3.71B). The desired electrode metal is subsequently deposited from the back of the wafer into the access cavities and against the oxide window (Figure 3.71C). Finally, a timed oxide etch removes the sacrificial oxide window from

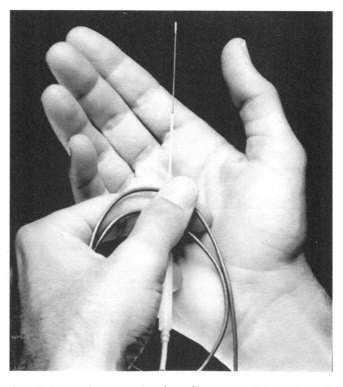

FIGURE 3.69   An *in vivo* pH, CO_2, and O_2 sensor based on a linear array of electrochemical cells.

above the underlying metal while preserving the thicker oxide layer in the other areas on the chip (Figure 3.71D) [Joseph et al., 1989; Madou and Otagawa, 1989; Holland et al., 1988]. An SEM micrograph demonstrating step D in Figure 3.71 is shown in Figure 3.72. The electrodes in the electrochemical cells of the top wafer are further connected to the bottom wafer electronics by solder balls in the access vias of the top

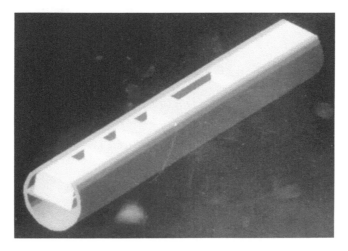

FIGURE 3.70 CAD of the electrochemical sensor array showing two pieces of Si (each 250-μm thick) on top of each other, mounted in a dual lumen catheter. The bottom part of the catheter is left open so pressure can be monitored and blood samples can be taken.

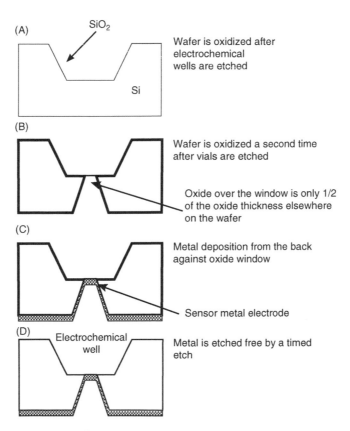

FIGURE 3.71 Fabrication sequence for a generic electrochemical cell in Si. Depth of the electrochemical well, number of electrodes, and electrode materials can be varied.

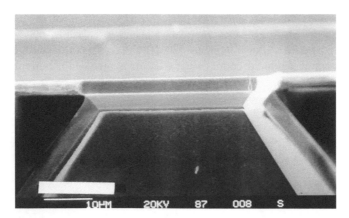

FIGURE 3.72   SEM micrograph illustrating process step D from Figure 3.71. A 30 × 30 μm Pt electrode is shown at the bottom of an electrochemical well. This Pt electrode is further contacted to the electronics from the back.

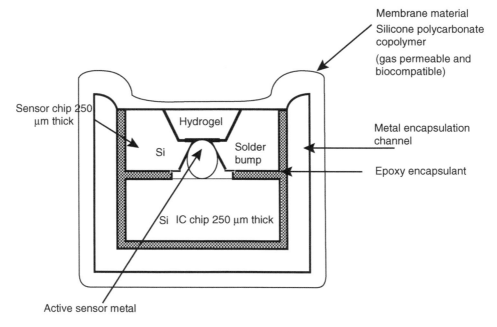

FIGURE 3.73   Schematic of the bonding scheme between sensor wafer and IC. The schematic is a cut-through of the catheter.

silicon wafer (Figure 3.73). Separating the chemistry from the electronics in this way provides extra protection for the electronics from the electrolyte as well as from the electronics, and chemical sensor manufacture can proceed independently. Depending on the type of sensor element, one or more electrodes are fabricated at the bottom of the electrochemical cells. For example, shown in Figure 3.74 is an almost completed Severinghaus CO_2 sensor with an Ag/AgCl electrode as the reference electrode (left) and an IrO_x pH-sensitive electrode, both at the bottom of one electrochemical well. The metal electrodes are electrically isolated from each other by the SiO_2 passivation layer over the surface of the silicon wafer. To complete the CO_2 sensor, we silk-screen a hydrogel containing an electrolytic medium into the silicon sensor cavity and dip-coat the sensor into a silicone-polycarbonate rubber solution to form the gas-permeable membrane. For hydrogel inside the micromachined well, we use poly(2-hydroxyethyl)methacrylate (PHEMA) or polyvinylalcohol (PVA).

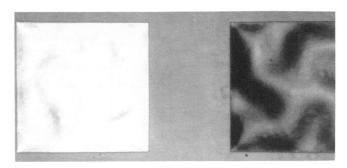

FIGURE 3.74   SEM micrograph of an Ag/AgCl (left) and IrO_x (right) electrode at the bottom of an anisotropically etched well in Si. This electrochemical cell forms the basis for a Severinghaus CO_2 sensor.

The concept of putting the sensor chemistries and the electronics on opposite sides of a substrate is a very important design feature we decided upon several years ago in view of the overwhelming problems encountered in building chemical sensors based on ISFETs or EGFETs (extended gate field effect transistors) [Madou and Morrison, 1989].

Example 3. Disposable Electrochemical Valves

One of the most difficult aspects of developing a microfluidic system is the miniaturization of valves. Most current MEMS valve technologies are still too complicated, large, power hungry, and expensive for deployment in such applications as disposables (prohibitive cost) or implants (excessive required power and size). A more appropriate technology for a valve in a disposable device, such as a diagnostic panel for blood electrolytes or an implantable drug delivery system, may be a "sacrificial", or "use-once" valve. Such a valve could ensure device sterility, isolate reagents until they are required, and even entrap a vacuum to draw sample into the device (like a vacutainer blood-collection tube). Most current MEMS valves have moving parts, such as diaphragm valves, that are micromachined in silicon and are prone to malfunction through clogging. Valves without moving parts are obviously preferred, and some such valveless systems have been incorporated in electrokinetic and centrifugal fluidic platforms. These "valves" form only a temporary barrier for liquids but not for vapors. There is a need for an even simpler valve, also without moving parts, that forms a barrier for both liquids and vapors. Creating a vapor barrier is essential if liquids are to be stored in small micromachined chambers for extended times (e.g., for a diagnostic test incorporating different on-chip reagents). In the absence of such vapor barriers, liquid would be distributed over time, driven by the gradients in vapor pressure above each individual chemical reservoir.

In researching the structures described in Example 2, we discovered that a small current applied between a counter electrode and a thin suspended metal electrode (Figure 3.75), causes anodic dissolution of the metal or local electrolysis of water depending on the nature of the valve metal. In both cases, the applied potential leads to bursting of the thin metal barriers. Thin suspended metal membranes, such as silver or Au, can thus be burst open by passing a small current from the metal via a contacting electrolyte solution to a counterelectrode. An Ag valve, for example, can be opened with an applied bias as low as 1 to 1.5 V. Although the "one-use" microvalve in Figure 3.75 involves an Ag membrane suspended over an orifice, this patented electrochemical valve technology is generic, and a wide variety of metals may be used [Madou and Tierney, 1994]. Small amounts of drugs can be drop deposited into the Si micromachined chambers, and the chamber may be closed off using a polymeric laminate. The structure shown in Figure 3.75 has also been fabricated employing a dry photoresist replacing the Si as structural material. Arrays of the element shown in Figure 3.75 can be made, and the individually addressable metal covers can then be opened electronically, releasing the drugs stored in the Si or polymer chambers. Different drugs can be released at different times and by opening more "holes" the rate of drug delivery can be set. This is illustrated in Figure 3.76 [Seetharaman et al., 2000]. One application for this type of valve is responsive drug delivery

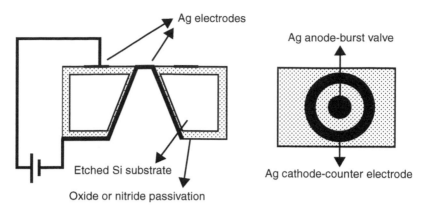

FIGURE 3.75   Disposable electrochemical valve: principle of operation.

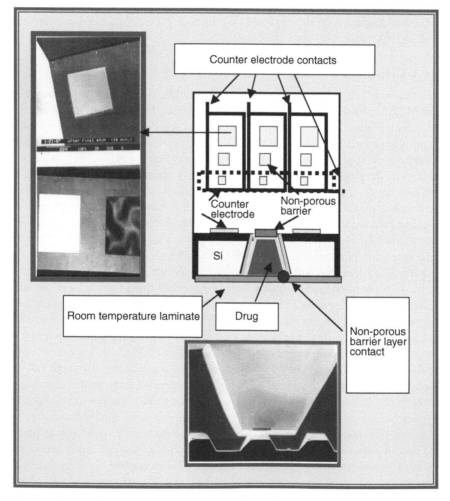

FIGURE 3.76   Metal electrodes are "blasted" open electrochemically. By making electrode arrays, the number of openings is selectable. The metal valve material may be Ag, Pt, Au, etc. In the SEM picture on the left, we show a Pt electrode (top), and a set of a Ag/AgCl electrode and an IrOx electrode (bottom). In the latter case the pH of the drug inside the reservoir can be monitored by measuring the voltage between the Ag/AgCl and IrOx electrode.

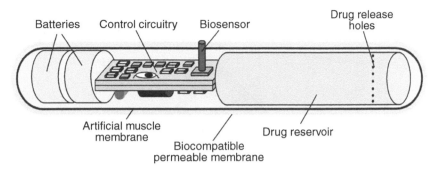

FIGURE 3.77 Model of responsive drug delivery system (pharmacy-on-a-chip). Responsive drug delivery pill type configuration. A video on the operation of the device can be found at <http://www.biomems.net/>.

in a smart pill (pharmacy-on-a-chip) as sketched in Figure 3.77. In this type of implant, openings in a drug reservoir are regulated by a biological stimulus (say, the concentration of glucose sensed by a glucose biosensor) so that a patient receives only the amount of drug that the body requires (e.g., the correct amount of insulin). A doctor may intervene via a telemetric link. This technology is currently being pursued by CHIPRx (<http://www.chiprx.com/>).

Example 4. Self-Aligned Vertical Mirrors and V-Grooves for a Magnetic Micro-Optical Matrix Switch

Hélin et al. fabricated an elegant monolithic magneto-mechanical optical switch for dynamically reconfigurable DWDM (dense wavelength division multiplexing) networks [Hélin et al., 2000]. Arrays of optical switches allow for the rapid reconfiguration of optical networks by altering the light path in a system of intersecting fibers. A wavelength division multiplexer (WDM) adds or deletes extra optical channels. A switch routing light from a single fiber into any of N output fibers is a $1 \times N$ switch. An $M \times N$ switch routes any one of M inputs to any one of N outputs. The $M \times N$ designation is the order of the optical switch. Commercially available optical switch arrays from companies such as E-Tek Dynamics (now JDS Uniphase), San Jose, California (<http://www.jdsu.com/>); and DiCon Fiberoptics, Inc., Berkeley, California (<http://www.diconfiberoptics.com/>); are limited to 1×2 or 2×2 [Maluf, 2000]. Using MEMS, the prospect is to deliver 64×64 or even larger arrays in the case of Xros (now owned by Nortel).

To make a low-cost, batch-machined switch, Hélin et al. use a simple, one-level mask process on a (100) Si wafer, which allows for the simultaneous fabrication of (100) sidewalls for high quality mirrors and (111) V-grooves for self-alignment between V-grooves and mirrors. The etching principle, a nice illustration of the wet etching processes described in this chapter, is shown in Figure 3.78. The 45° angle between the <100> and <110> directions is used to self-align the vertical mirrors and V-grooves. As the bottom and sidewall planes are all from the same {100} family, the lateral underetch rate is equal to the vertical etch rate (see also Figure 3.10). Underetching in the <100> direction creates the upstanding mirror while V-grooves (A, B, C, and D) are formed in the <110>-directions for optical fiber alignment. Contrary to DRIE-based processes, in which ripples etched in the vertical walls are inevitable, wet anisotropic etching fulfills the requirements of high surface quality mirrors, reducing optical losses. Because intersecting (111) planes stop the etching, the width of the V-groove in the mask layout fixes the position of the optical axis. The thickness of the mirror is determined by timing the etch. Cr/Au layers are deposited by vacuum evaporation to finish the mirror manufacture. The etching principle outlined in Figure 3.78 can be expanded to an $M \times N$ matrix switch as shown in Figure 3.79A for a 2×2 and 1×8 switch.

The front-side wafer etching process is preceded by a back-side etching of a cantilever beam, whose role is to support the mirror and allow it to move. The thickness of the cantilever support plate is determined by a

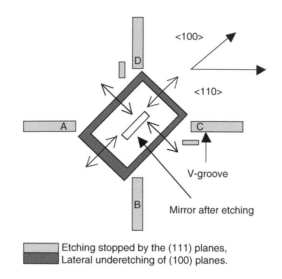

FIGURE 3.78 Principle of the self-aligned optical switch structure (arrows indicate the underetching).

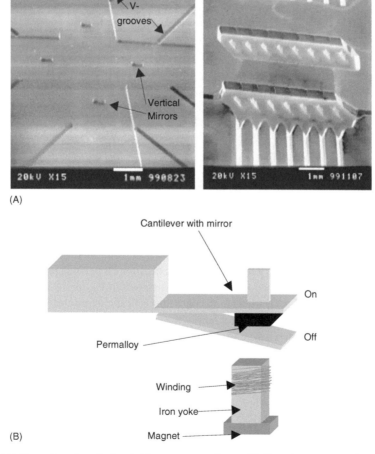

FIGURE 3.79 (A) Examples of optical switching structures for a self-aligned matrix switch; 2 × 2 bidirectional switch (left) and 1 × 8 (right). (B) Principle of switch operation with a self-latching system. (Courtesy of Dr. H. Fujita, University of Tokyo.)

timed etch, and the cantilever with mirror is actuated using electromagnetic actuation. The actuation principle is demonstrated in Figure 3.79B.

3.13 Surface Micromachining: Introduction

Bulk micromachining means that three-dimensional features are etched into the bulk of crystalline and noncrystalline materials. In contrast, surface micromachined features are built up, layer by layer, on the surface of a substrate (e.g., a single-crystal silicon wafer). Dry etching defines the surface features in the x, y plane, and wet etching releases them from the plane by undercutting. In surface micromachining, shapes in the x, y plane are unrestricted by the crystallography of the substrate. For illustration, in Figure 3.80, we compare an absolute pressure sensor based on poly-Si and made by surface micromachining with one made by bulk micromachining in single-crystal Si. Not reflected in this figure is the fact that the surface micromachined devices typically end up quite a bit smaller than their bulk micromachined counterparts.

The nature of the deposition processes involved determines the limited height of surface micromachined features (Hal Jerman, from EG&G's IC Sensors, called them 2.5 D features) [Jerman, 1994]. Specifically, low-pressure chemical vapor deposited (LPCVD) polycrystalline silicon (poly-Si) films generally are a

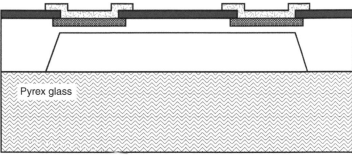

Surface micromachined absolute pressure sensor

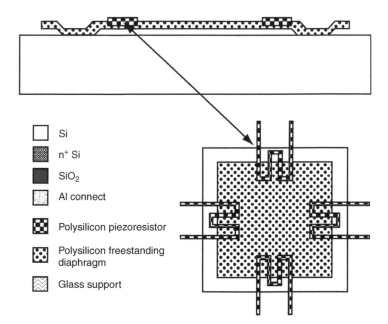

☐ Si

▓ n⁺ Si

■ SiO$_2$

Al connect

Polysilicon piezoresistor

Polysilicon freestanding diaphragm

Glass support

FIGURE 3.80 Comparison of bulk micromachined and surface micromachined absolute pressure sensors equipped with piezoresistive elements. Top: Bulk micromachining in single crystal Si. Bottom: Surface micromachining with poly-Si.

few microns high (low z), in contrast with wet bulk micromachining where only the wafer thickness limits the feature height. A low z may be a drawback for some sensors. For example, it would be difficult to fashion a large inertial mass for an accelerometer from thin poly-Si plates (a commercial surface micromachined accelerometer, the ADLX05, has an inertial mass of only 0.3 μg). Not only do many parameters in the LPCVD polysilicon process need to be controlled very precisely; subsequent high-temperature annealing (say, at temperatures of about 580°C) is needed to transform the deposited amorphous silicon into polysilicon — the main structural material in surface micromachining. Even with the best possible process control, polysilicon has some material disadvantages over single-crystal Si. For example, it generates a somewhat smaller yield strength (values between two and ten times smaller have been reported) [Greek et al., 1995] and has a lower piezoresistivity [Berre et al., 1996]. Moreover, since one grain's diameter may constitute a significant fraction of the thickness of a mechanical member, the effective Young's modulus may exhibit variability from sensor to sensor [Howe, 1995]. An important positive attribute of poly-Si is that its material properties, although somewhat inferior to those of single-crystal Si, are far superior to those of metal films, and most of all because they are isotropic, design is rendered dramatically simpler than with single-crystal material. Dimensional uncertainties may be of greater concern than material issues. Although absolute dimensional tolerances obtained with lithography techniques can be submicron, relative tolerances are poor, perhaps 1% on the length of a 100 μm long feature. The situation becomes even more critical with yet smaller feature sizes. Although the relatively coarse dimensional control in the micro-domain is not specific to surface micromachining, there is no crystallography to rely on for improved dimensional control as in the case of wet bulk micromachining. Moreover, since the mechanical members in surface micromachining tend to be smaller, more post-fabrication adjustment of the features is required to achieve reproducible characteristics. Finally, the wet process for releasing structural elements from a substrate tends to cause sticking of suspended structures to the substrate, or stiction, introducing another disadvantage associated with surface micromachining.

Some of the problems associated with surface micromachining mentioned above have been resolved by process modifications and/or alternative designs, and the technique has rapidly gained commercial interest, mainly because it is the most IC-compatible micromachining process developed to date. Moreover, in the last 10 to 15 years, processes such as silicon on insulator (SOI) [Diem et al., 1993], hinged poly-Si [Pister, 1992], Keller's molded milli-scale polysilicon [Keller and Ferrari, 1994], thick (10 μm and beyond) poly-Si [Lange et al., 1995], Sandia's Ultra-Planar Multi-level MEMS Technology (SUMMIT) (<http://mems.sandia.gov/scripts/index.asp>), and LIGA and LIGA-like processes have further enriched the surface micromachining arsenal. Some preliminary remarks on each of these surface micromachining extensions follow.

Silicon crystalline features, anywhere between fractions of a micron to 100 μm high, can be readily obtained by surface micromachining of the epi silicon or fusion-bonded silicon layer of SOI wafers [Noworolski et al., 1995]. Structural elements made from these single-crystalline Si layers result in more reproducible and reliable sensors. SOI or epi-micromachining combines the best features of surface micromachining (i.e., IC compatibility and freedom in x, y shapes) with the best features of bulk micromachining (superior single-crystal Si properties). Moreover, SOI surface micromachining frequently involves fewer process steps and offers better control over the thickness of crucial building blocks. Given the poor reproducibility of mechanical properties and generally poor electronic characteristics of polysilicon films, SOI machining may surpass the poly-Si technology for fabricating high performance devices.

The fabrication of poly-Si planar structures for subsequent vertical assembly by mechanical rotation around micromachined hinges dramatically increases the plethora of designs feasible with poly-Si [Pister, 1992]. Today, erecting these poly-Si structures with the probes of an electrical probe station or, occasionally, assembly by chance in a HF etch or DI water rinse [Chu et al., 1995] represent too complicated or unreliable post-release assembly methods for commercial acceptance, and alternative self-assembly means are urgently needed.

While at the University of California at Berkeley, Keller, now at MEMS Precision Instruments (<http://www.memspi.com>), introduced a combination of surface micromachining and LIGA-like molding processes [Keller and Ferrari, 1994] in the HEXSIL (HEXagonal honeycomb polySILicon) process, a technology enabling the fabrication of tall three-dimensional microstructures without post-release assembly. Using CVD processes, generally only thin films (~2 to 5 μm) can be deposited on flat surfaces. If, however, these surfaces are the opposing faces of deep narrow trenches, the growing films will merge to form solid beams. Releasing of such polysilicon structures and the incorporation of electroplating steps expand the surface micromachining bandwidth in terms of choice of materials and accessible feature heights. In this fashion, high-aspect-ratio structures normally associated with LIGA can now also be made of CVD polysilicon.

Applying classic LPCVD to obtain poly-Si deposition is a slow process. For example, a layer of 10 μm typically requires a deposition time of 10 hr. Consequently, most micromachined structures are based on layer thicknesses in the 2 to 5 μm range. Based on dichlorosilane (SiH_2Cl_2) chemistry, Lange et al. (1995) developed a CVD process in a vertical epitaxy batch reactor with deposition rates as high as 0.55 μm/min at 1000°C. The process yields acceptable deposition times for thicknesses in the 10 μm range. The highly columnar poly-Si films are deposited on sacrificial SiO_2 layers and exhibit low internal tensile stress making them suitable for surface micromachining.

Thick layers of polyimide and other new UV resists have also received a lot of attention as important new extensions of surface micromachining. Due to their transparency to exposing UV light, they can be transformed into tall surface structures with LIGA-like high aspect ratios. They may also be electroplated and micromolded in any plastic of choice.

In the remainder of this chapter, we first review thin film material properties in general, focusing on significant differences with bulk properties of the same material, followed by a review of the main surface micromachining processes. Next, we clarify the recent extensions of the surface micromachining technique listed above. Because of the complexity of the many parameters influencing thin film properties, we then present case studies for the most commonly used thin film materials.

In the surface micromachining examples at the end of this chapter, we first look at a lateral resonator. Resonators have found an important industrial application in accelerometers and gyros introduced by analog devices. The second example involves TI's Digital Micromirror Device, a chip now found in many projectors. The last example is the fabrication of an SOI high-sensitivity piezoresistive cantilever for label-free immunosensing.

3.14 Historical Note

The first example of a surface micromachine for an electromechanical application consisted of an under-etched metal cantilever beam for a resonant gate transistor made by Nathanson in 1967 [Nathanson et al., 1967]. By 1970, a first suggestion for a magnetically actuated metallic micromotor emerged [Dutta, 1970]. Because of fatigue problems, metals are not typically used as mechanical components. The surface micro-machining method as we know it today was first demonstrated by Howe and Muller in the early 1980s and relied on polysilicon as the structural material [Howe and Muller, 1982]. These pioneers and Guckel (1985), an early contributor to the field, produced free-standing LPCVD polysilicon structures by removing the oxide layers on which the polysilicon features were formed. Howe's first device consisted of a resonator designed to measure the change of mass upon adsorption of chemicals from the surrounding air. However, this gas sensor does not necessarily represent a good application of a surface micromachined electrostatic structure, since humidity and dust foul the thin air gap of such an unencapsulated microstructure in a minimal amount of time. Later, mechanical structures, especially hermetically sealed mechanical devices, provided proof that the IC revolution could be extended to electromechanical systems [Howe, 1995]. In these structures, the height (z-direction) typically is limited to less than 10 μm, ergo the name *surface micromachining*.

The first survey of possible applications of poly-Si surface micromachining was presented by Gabriel et al. in 1989 [Gabriel et al., 1989]. Microscale movable mechanical pin joints, springs, gears, cranks, sliders, sealed cavities, and many other mechanical and optical components have been demonstrated in the laboratory [Muller, 1987; Fan et al., 1987]. For a while in the early 1990s, it seemed that every MEMS group in the United States was trying to make surface micromachined micromotors. Micromotors may as yet lack practical use, but just as the ion-sensitive field effect transistor (ISFET) galvanized the chemical sensor community in trying out new chemical sensing approaches, micromotors energized the micromachining research community to fervently explore miniaturization of a wide variety of mechanical sensors and actuators. Micromotors also brought about the christening of the micromachining field as micro-electromechanical systems, or MEMS. In 1991, Analog Devices, in Norwood, Massachusetts, announced the first commercial product based on surface micromachining, namely the ADXL-50, a 50-g accelero-meter for activating air-bag deployment ["Analog Devices Combine Micromachining with BICMOS," 1991]. By the year 2001, Analog Devices was making 2 million surface micromachined accelerometers per month (at $4 per device in volume). A second commercial success for surface micromachining was based on Texas Instruments' Digital Micromirror Device™, or DMD. This surface micromachined movable aluminum mirror is a digital light switch that precisely controls a light source for projection displays and hard copy applications [Hornbeck, 1995]. The commercial acceptance of this second application confirmed the stay-ing power of surface micromachining.

Surface micromachining is also an established manufacturing process at Cronos, Research Triangle Park (now a JDS Uniphase Company), North Carolina, and Robert Bosch, Stuttgart, Germany.

3.15 Mechanical Properties of Thin Films

3.15.1 Introduction

Thin films in surface micromachines must satisfy a large set of rigorous chemical, structural, mechanical, and electrical requirements. Excellent adhesion, low residual stress, low pinhole density, good mechani-cal strength, and chemical resistance all may be required simultaneously. For many microelectronic thin films, the material properties depend strongly on the details of the deposition process and the growth conditions. In addition, some properties may depend on postdeposition thermal processing, referred to as annealing. Furthermore, the details of thin-film nucleation and growth may depend on the specific substrate or on the specific surface orientation of the substrate. Although the properties of a bulk mate-rial might be well characterized, its thin-film form may have properties substantially different from those of the bulk. For example, thin films generally display smaller grain size than bulk materials. An overwhelm-ing reason for the many differences stems from the properties of thin films, which exhibit a higher sur-face-to-volume ratio than large chunks of material and are strongly influenced by the surface properties.

In this chapter, we focus on the physical characteristics of the resulting thin deposits. Some terminology characterizing thin films and their deposition is introduced in Table 3.13.

TABLE 3.13   Thin Film Terms Used in Characterizing Deposition

Term	Remark
Film	Bond energy <10 kcal/mole
Chemisorbed film	Bond energy >20 kcal/mole
Nucleation	Adatoms forming stable clusters
Condensation	Initial formation of nuclei
Island formation	Nuclei grow in three dimensions, especially along the substrate surface
Coalescence	Nuclei contact each other, and larger, rounded shapes form
Secondary nucleation	Areas between islands filled in by secondary nucleation, resulting in a continuous film
Grain size of thin film	Generally smaller than for bulk materials and function of deposition and annealing conditions (higher T, larger grains)
Surface roughness	Lower at high temperatures except when crystallization starts; at low temperature the roughness is higher for thicker films, also oblique deposition and contamination increases roughness
Epitaxial and amorphous films	Very low surface roughness
Density	More porous deposits are less dense; density reveals a lot about the film structure
Crystallographic structure	Adatom mobility: amorphous, polycrystalline, single crystal or fiber texture or preferred orientation

Since thin films were originally not intended for load-bearing applications, their mechanical properties have largely been ignored. The last 18 years saw the development of a strong appreciation for understanding the mechanical properties as essential for improving the reliability and lifetime of thin films, even in nonstructural applications [Vinci and Braveman, 1991]. Surface micromachining contributes heavily to this understanding.

3.15.2 Adhesion

The importance of adhesion of various films to one another and to the substrate in overall IC performance and reliability cannot be stressed enough. As mechanical pulling forces might be involved, adhesion is even more crucial in micromachining. If films lift from the substrate under a repetitive, applied mechanical force, the device will fail. Classical adhesion tests include the Scotch® tape test, abrasion method, scratching, deceleration (ultrasonic and ultracentrifuge techniques), bending, pulling, etc. [Campbell, 1970]. Micromachined structures, because of their sensitivity to thin-film properties, enable some innovative new ways of in situ adhesion measurement. Figure 3.81 illustrates how a suspended membrane may be used for adhesion measurements. In Figure 3.81A, the membrane is suspended but still adherent to the substrate. Figure 3.81B shows the membrane after it has been peeled from its substrate by an applied load (gas pressure). Figure 3.81C illustrates the accompanying P(ressure)–V(olume) cycle, in which the membrane is inflated, peeled, and then deflated. The shaded portion of Figure 3.81C illustrates the P–V work

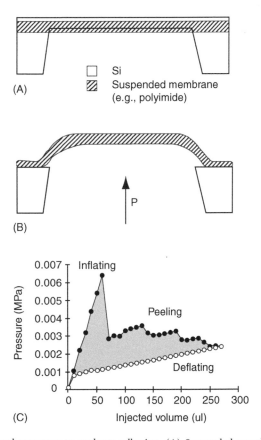

FIGURE 3.81 Micromachined structure to evaluate adhesion. (A) Suspended membrane. (B) Partially detached membrane-outward peel. (C) Pressure-volume curve during inflate-deflate cycle. (Reprinted with permission from Senturia, S. [1987] "Can We Design Microrobotic Devices without Knowing the Mechanical Properties of Materials?" in *Proceedings: IEEE Micro Robots and Teleoperators Workshop*, Hyannis, MA, pp. 3/1–5.)

creating the new surface, which equals the average work of adhesion for the film-substrate interface times the area peeled during the test [Senturia, 1987].

Cleanliness of a substrate is a *conditio sine qua non* for good film adhesion. Roughness, providing more bonding surface area and mechanical interlocking, further improves it. Adhesion also improves with increasing adsorption energy of the deposit and/or increasing number of nucleation sites in the early growth stage of the film. Sticking energies between film and substrate range from less than 10 kcal/mole in physisorption to more than 20 kcal/mole for chemisorption. The weakest form of adhesion involves Van der Waals forces only (see also Table 3.13).

It is highly advantageous to include a layer of oxide-forming elements between a metal and an oxide substrate. These adhesion layers, such as Cr, Ti, Al, etc., provide good anchors for subsequent metallization. Intermediate film formation allowing a continuous transition from one lattice to the other results in the best adhesion. Adhesion also improves when formation of intermetallic metal alloys takes place.

3.15.3 Stress in Thin Films

3.15.3.1 Stress in Thin Films — Qualitative Description

Film cracking, delamination, and void formation may all be linked to film stress. Nearly all films foster a state of residual stress due to mismatch in the thermal expansion coefficient, nonuniform plastic deformation, lattice mismatch, substitutional or interstitial impurities, and growth processes. Figure 3.82 lists stress-causing factors categorized as either intrinsic or extrinsic [Krulevitch, 1994]. The intrinsic stresses (also growth stresses) develop during the film nucleation. Extrinsic stresses are imposed by unintended

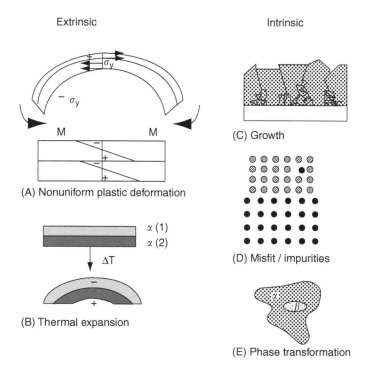

FIGURE 3.82   Examples of intrinsic and extrinsic residual stresses. (A) Nonuniform plastic deformation results in residual stresses upon unloading; M = bonding moment. (B) Thermal expansion mismatch between two materials bonded together; α (1) and α (2) are thermal expansion coefficients. (C) Growth stresses evolve during film deposition. (D) Misfit stresses due to mismatches in lattice parameters in an epitaxial film and stresses from substitutional or interstitial impurities. (E) Volume changes accompanying phase transformations cause residual stresses. (After Krulevitch, P.A., *Micromechanical Investigations of Silicon and Ni-Ti-Cu Thin Films*, Ph.D. thesis, University of California, Berkeley.)

external factors such as temperature gradients or sensor-package-induced stresses. Thermal stresses, the most common type of extrinsic stresses, are well understood and often rather easy to calculate (see below). They arise either in a structure with inhomogeneous thermal expansion coefficients subjected to a uniform temperature change or in a homogeneous material exposed to a thermal gradient [Krulevitch, 1994]. Intrinsic stresses in thin films often are larger than thermal stresses. They usually are a consequence of the nonequilibrium nature of the thin-film deposition process. For example, in chemical vapor deposition, depositing atoms (adatoms) may at first occupy positions other than the lowest-energy configuration. With too high a deposition rate or too low adatom surface mobility, these first adatoms may become pinned by newly arriving adatoms resulting in the development of intrinsic stress. Other types of intrinsic stresses illustrated in Figure 3.82 include transformation stresses occurring when part of a material undergoes a volume change during a phase transformation; misfit stresses arising in epitaxial films due to lattice mismatch between film and substrate; and impurities, either interstitial or substitutional, that cause intrinsic residual stresses due to the local expansion or contraction associated with point defects. Intrinsic stress in a thin film does not suffice to result in delamination unless the film is quite thick. For example, to overcome a low adsorption energy of 0.2 eV, a relatively high stress of about 5×10^9 dyn/cm^2 (10^7 dyn/cm^2 = 1 MPa) is required [Campbell, 1970]. High stress can result in buckling or cracking of films.

The stress developing in a film during the initial phases of a deposition may be compressive (i.e., the film tends to expand parallel to the surface), causing buckling and blistering or delamination in extreme cases (especially with thick films). Alternatively, thin films may be in tensile stress (i.e., the film tends to contract), which may lead to cracking if forces high enough to exceed the film material's fracture limit are present. Subsequent rearrangement of the atoms, either during the remainder of the deposition or with additional thermal processing, can lead to further densification or expansion, decreasing remaining tensile or compressive stresses, respectively.

The mechanical response of thin-film structures is affected by the residual stress even if the structures do not fail. For example, if the residual stress varies in the direction of film growth, the resulting built-in bending moment will warp released structures, such as cantilever beams. The presence of residual stress also alters the resonant frequency of thin-film, resonant microstructures (see Equation [3.63] below) [Pratt et al., 1992]. In addition, residual stress can lead to degradation of electrical characteristics and yield loss through defect generation [Krulevitch, 1994]. For example, it was found that the resistivity of stressed metallic films is higher than that of their annealed counterparts. Residual stress has been used advantageously in a few cases, such as in self-adjusting microstructures [Judy, et al.], and for altering the shape-set configuration in shape memory alloy films [Krulevitch, 1994].

In general, the stresses in films, by whatever means produced, are in the range of 10^8 to 5×10^{10} dyn/cm^2 and can be either tensile or compressive. For normal deposition temperatures (50 to a few hundred °C), the stress in metal films typically ranges from 10^8 to 10^{10} dyn/cm^2 and is tensile, with the refractory metals (Mo, Ta, Nb, W, and Ta) at the upper end and the soft metals (Cu, Ag, Au, Al) at the lower end. At low substrate temperatures, metal films tend to exhibit tensile stress. This often decreases in a linear fashion with increasing substrate temperature, finally going through zero or even becoming compressive. The changeover to compressive stress occurs at lower temperatures for lower melting point metals. The mobility of the adatoms is key to understanding the ranking for refractory and soft metals. A metal such as aluminum has a low melting temperature and a corresponding high diffusion rate even at room temperature, so usually it is fairly stress free. By comparison, tungsten has a relatively high melting point and a low diffusion rate, and it tends to accumulate more stress when sputter deposited. With dielectric films, stresses often are compressive and have slightly lower values than commonly noted in metals.

Tensile films result, for example, when a process by-product is present during deposition and later driven off as a gas. If the deposited atoms are not sufficiently mobile to fill in the holes left by these departing by-products, the film will contract and go in tension. Nitrides deposited by plasma CVD are usually compressive due to the presence of hydrogen atoms in the lattice. By annealing, which drives the hydrogen out, the films can become highly tensile. Annealing also has a dramatic effect on most oxides. Oxides often are porous enough to absorb or give off a large amount of water. Full of water, they are compressive; devoid of water, they are tensile. Thermal SiO$_2$ is compressive even when dry. If atoms are jammed in place (such as with sputtering), the film tends to act compressively. The stress in a thin film also varies with depth. The RF power of a plasma-enhanced CVD (PECVD) deposition influences stress; for example, a thin film may start out tensile, decrease as the power increases, and finally become compressive with further RF power increase. CVD equipment manufacturers concentrate on building stress-control capabilities into new equipment by controlling plasma frequency.

3.15.3.2   Stress in Thin Films on Thick Substrates—Quantitative Analysis

The total stress in a thin film typically is given by:

$$\sigma_{tot} = \sigma_{th} + \sigma_{int} + \sigma_{ext} \tag{3.48}$$

That is, the sum of any intentional external applied stress (σ_{ext}), the thermal stress (σ_{th}, an unintended external stress), and different intrinsic components (σ_{int}). With constant stress through the film thickness, the stress components retain the form of:

$$\sigma_x = \sigma_x(x, y)$$

$$\sigma_y = \sigma_y(x, y)$$

$$\tau_{xy} = \tau_{xy}(x, y) \tag{3.49}$$

$$\tau_{xz} = \tau_{yz} = \sigma_z = 0$$

In other words, the three nonvanishing stress components are functions of x and y alone. No stress occurs in the direction normal to the substrate (z). With x, y as principal axes, the shear stress τ_{xy} also vanishes [Chou and Pagano, 1967] and Equation (3.49) reduces to the following strain–stress relationships:

$$\varepsilon_x = \frac{\sigma_x}{E} - \frac{\nu \sigma_y}{E}$$

$$\varepsilon_y = \frac{\sigma_y}{E} - \frac{\nu \sigma_x}{E} \tag{3.50}$$

$$\sigma_z = 0$$

In the isotropic case, $\varepsilon = \varepsilon_x = \varepsilon_y$ so that $\sigma_x = \sigma_y = \sigma$, or:

$$\sigma = \left(\frac{E}{1 - \nu} \right) \varepsilon \tag{3.51}$$

where the Young's modulus of the film and the Poisson ratio of the film act independently of orientation. The quantity $E/(1 - \nu)$ often is called the biaxial modulus. Uniaxial testing of thin films is difficult, prompting the use of the biaxial modulus rather than Young's uniaxial modulus. Plane stress, as described here, is a good approximation only when several thicknesses (say three) away from the edge of the film.

   Thermal Stress. Thermal stresses develop in thin films when high-temperature deposition or annealing are involved, and they usually are unavoidable due to mismatch of thermal expansion coefficients between film and substrate. The problem of a thin film under residual thermal stress can be modeled by considering a thought experiment involving a stress-free film at high temperature on a thick substrate. Imagine detaching the film from the high-temperature substrate and cooling the system to room temperature. Usually, the substrate dimensions undergo minor shrinkage in the plane while the film's dimensions may reduce significantly. To reapply the film to the substrate and achieve complete coverage, the film needs stretching with a biaxial tensile load to a uniform radial strain ε, followed by perfect bondage to the rigid substrate and load removal. The film stress is assumed to be the same in the stretched and free-standing film as in the film bonded to the substrate; that is, no relaxation occurs in the bonding process.

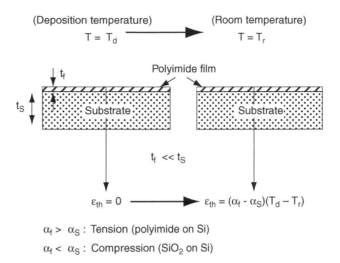

$\alpha_f > \alpha_S$: Tension (polyimide on Si)

$\alpha_f < \alpha_S$: Compression (SiO$_2$ on Si)

FIGURE 3.83 Thermal stress. Tension and compression are determined by the relative size of thermal expansion coefficients of film and substrate. Suppose a strain-free film at deposition temperature T_d is cooled to room temperature T_r on a substrate with a different coefficient of thermal expansion.

To calculate the thermal residual stress from Equation (3.51), the elastic moduli of the film must be known, as well as the volume change associated with the residual stress, i.e., the thermal strain ε_{th} resulting from the difference in the coefficients of thermal expansion between the film and the substrate.

Let us now consider whether qualitatively the above assumptions apply to the measurement of thin films on Si wafers. Such films typically measure 1 μm thick and are deposited on 4 in. wafers nominally 550 μm thick. In this case, the substrate measures nearly three orders of magnitude thicker than the film, and because the bending stiffness is proportional to the thickness cubed, the substrate essentially is rigid relative to the film. The earlier assumptions clearly apply.

Figure 3.83 portrays a quantitative example in which a polyimide film, strain-free at the deposition temperature (T_d) of 400°C, is cooled to room temperature T_r (25°C) on a Si substrate with a different coefficient of thermal expansion. The resulting strain is given by:

$$\varepsilon_{th} = \int [\alpha_f(T) - \alpha_s(T)] dT \tag{3.52}$$

where α_f and α_s represent the coefficients of thermal expansion for the polyimide film and the Si substrate respectively. The thermal strain can be of either sign, based on the relative values of α_f and α_s: positive is tensile, negative is compressive. Polyimide features a thermal expansion ($\alpha_f = 70 \times 10^{-6}$°C^{-1}) larger than the thermal expansion coefficient of Si ($\alpha_s = 2.6 \times 10^{-6}$°C^{-1}); hence, a tensile stress is expected. With SiO$_2$ grown or deposited on silicon at elevated temperatures, a compressive component [α_f (0.35 × 10^{-6}°C^{-1}) < α_s (2.6 × 10^{-6}°C^{-1})] is expected. Assuming that the coefficients of thermal expansion are temperature independent, Equation (3.52) simplifies to:

$$\varepsilon_{th} = (\alpha_f - \alpha_s)(T_d - T_r) \tag{3.53}$$

The calculated thermal strain ε_{th} for polyimide on Si then measures 25 × 10^{-3} at room temperature. The biaxial modulus ($E/(1 - v)$), with $E = 3$ GPa and $v = 0.4$, equals 5 GPa, and the residual stress σ, from Equation (3.51), is 125 MPa and tensile.

Intrinsic Stress. The intrinsic stress σ_i reflects the internal structure of a material and is less clearly understood than the thermal stress, which it often dominates [Guckel]. Several phenomena may contribute to σ_i, making its analysis very complex. Intrinsic stress depends on thickness, deposition rate (locking in defects), deposition temperature, ambient pressure, method of film preparation, type of substrate used (lattice mismatch), incorporation of impurities during growth, and so on. Some semiquantitative descriptions of various intrinsic stress-causing factors follow:

- Doping ($\sigma_{int} > 0$ or $\sigma_{int} < 0$). When doping Si, the atomic or ionic radius of the dopant and the substitutional site determine the positive or negative intrinsic stress ($\sigma_{int} > 0$ or tensile, and $\sigma_{int} < 0$ or compressive). With boron-doped poly-Si, a small atom compared to Si, the film is expected to be tensile ($\sigma_{int} > 0$); with phosphorous doping, a large atom compared to Si, the film is expected to be compressive ($\sigma_{int} < 0$).
- Atomic peening ($\sigma_{int} < 0$). Ion bombardment by sputtered atoms and working gas densifies thin films, rendering them more compressive. Magnetron sputtered films at low working pressure (<1 Pa) and low temperature often exhibit compressive stress.
- Microvoids (int > 0). Microvoids may arise when by-products during deposition escape as gases and the lateral diffusion of atoms evolves too slowly to fill all the gaps, resulting in a tensile film.
- Gas entrapment ($\sigma_{int} < 0$). As an example, we can cite the hydrogen trapped in Si_3N_4. Annealing removes the hydrogen, and a nitride film, compressive at first, may become tensile if the hydrogen content is sufficiently low.
- Shrinkage of polymers during cure ($\sigma_{int} > 0$). The shrinkage of polymers during curing may lead to severe tensile stress, as becomes clear in the case of polyimides. Special problems are associated with measuring the mechanical properties of polymers, as they exhibit a time-dependent mechanical response (viscoelasticity), a potentially significant factor in the design of mechanical structures in which polymers are subjected to sustained loads [Maseeh and Senturia, 1990].
- Grain boundaries ($\sigma_{int} = ?$). Based on intuition, it is expected that the interatomic spacing in grain boundaries differs depending on the amount of strain, thus contributing to the intrinsic stress. But the origin of, for example, the compressive stress in polysilicon and how it relates to the grain structure and interatomic spacing are not yet completely clear (see also below on coarse- and fine-grated Si).

For further reading on thin-film stress, refer to Hoffman (1976, 1975). For a short tutorial, visit <http://www.uccs.edu/~tchriste/courses/PHYS549/549-lectures/mechchar.html>.

3.15.4 Stress-Measuring Techniques

3.15.4.1 Introduction

A stressed thin film will bend a thin substrate by a measurable degree. A tensile stress will bend the surface and render it concave; a compressive stress renders the surface convex. The most common methods for measuring the stress in a thin film are based on this substrate bending principle. The deformation of a thin substrate due to stress is measured either by observing the displacement of the center of a circular disk or by using a thin cantilevered beam as a substrate and calculating the radius of curvature of the beam and hence the stress from the deflection of the free end. More sophisticated local stress measurements use analytical tools such as X-ray [Wong, 1978] acoustics, Raman spectroscopy [Nishioka et al., 1985], infrared spectroscopy [Marco et al., 1991], and electron-diffraction techniques. Local stress does not necessarily mean the same as the stress measured by substrate bending techniques because stress is defined microscopically, while deformations are induced mostly macroscopically. The relation between macroscopic forces and displacements and internal differential deformation, therefore, must be modeled carefully. Local stress measurements may also be made using in situ surface micromachined structures such as strain gauges made directly out of the film to be tested [Mehregany et al., 1987]. The deflections of thin suspended and pressurized micromachined membranes may be measured by mechanical probe [Jaccodine and Schlegel, 1966], laser [Bromley et al., 1983], or microscope [Allen et al., 1987]. Intrinsic stress influences the frequency response of microstructures (see Equation [3.63]), which can be measured by laser [Zhang et al., 1991], spectrum analyzer [Pratt et al., 1991], or stroboscope. Whereas residual stress can be determined from wafer curvature and microstructure deflection data, material structure of the film can be studied by X-ray diffraction and transmission electron microscopy (TEM). Krulevitch, among others, attempted to link the material structure of poly-Si and its residual strain

[Krulevitch, 1994]. Below, we will review stress-measuring techniques, starting with the more traditional ones and subsequently clarifying the problems and opportunities in stress measuring with surface micromachined devices.

3.15.4.2 Disk Method

For all practical purposes, only stresses in the x and y directions are of interest in determining overall thin-film stress, as a film under high stress can only expand or contract by bending the substrate and deforming it in a vertical direction. Vertical deformations will not induce stresses in a substrate because it freely moves in that direction. The latter condition enables us to obtain quite accurate stress values by measuring changes in bow or radius of curvature of a substrate. The residual stresses in thin films are large, and sensitive optical or capacitive gauges may measure the associated substrate deflections.

The disk method, which is most commonly used, is based on a measurement of the deflection in the center of the disk substrate (say, a silicon wafer) before and after processing. Because any change in wafer shape is directly attributable to the stress in the deposited film, it is relatively straightforward to calculate stress by measuring these changes. Stress in films using this method is found through the Stoney equation [Hoffman, 1976], relating film stress to substrate curvature, as follows:

$$\sigma = \frac{1}{R} \frac{E}{6(1-v)} \frac{T^2}{t} \tag{3.54}$$

where R = measured radius of curvature of the bent substrate, $E/(1-v)$ = biaxial modulus of the substrate, T = thickness of the substrate and t = thickness of the applied film [Singer 1992] The underlying assumptions include the following:

- The disc substrate is thin and has transversely isotropic elastic properties with respect to the film normal.
- The applied film thickness is much less than the substrate thickness.
- The film thickness is uniform.
- Temperature of the disk substrate/film system is uniform.
- The disc substrate/film system is mechanically free.
- The disc substrate without film has no bow.
- Stress is equi-biaxial and homogeneous over the entire substrate.
- Film stress is constant through the film thickness.

For most films on Si, we assume that $t = T$; for example, t/T measures $\sim 10^{-3}$ for thin films on Si. The legitimacy of the uniform thickness, homogeneous, and equi-biaxial stress assumptions depends on the deposition process. Chemical vapor deposition (CVD) is a widely used process, as it produces relatively uniform films; however, sputter-deposited films can vary considerably over the substrate. In regard to the assumption of stress uniformity with film thickness, residual stress can vary considerably through the thickness of the film. Equation (3.54) gives only an average film stress in such cases. In cases where thin films are deposited onto anisotropic single-crystal substrates, the underlying assumption of a substrate with transversely isotropic elastic properties with respect to the film normal is not completely justified. Using single-crystal silicon substrates possessing moderately anisotropic properties such as <100> or <111> oriented wafers (Equation [3.20]) satisfies the transverse isotropy argument. Any curvature inherent in the substrate must be measured before film deposition and algebraically added to the final measured radius of curvature. To give an idea of the degree of curvature, 1 µm of thermal oxide may cause a 30 µm warp of a 4 in silicon wafer, corresponding to a radius of curvature of 41.7 m.

The following companies offer practical disk method-based instruments to measure stress on wafers: ADE Corp. (Newton, MA, <http://www.adesemiconductor.com/applications.shtml#thinstress>), Tropel (acquired by Corning in 2001) (Fairport, NY, <http://www.corning.com/semiconductoroptics/inside_semiconductor_optics/tropel.asp>), and KLA Tencor Instruments (Mountain View, CA, <http://www.kla-tencor.com/HomePage.asp?version=flash> [Singer]). Figure 3.84A illustrates the sample output from Tencor's optical stress analysis system. Figure 3.84B represents the measuring principle of Ionic

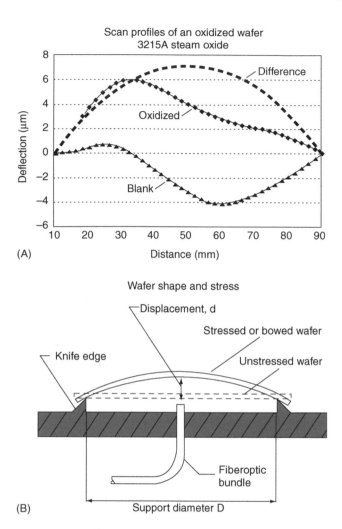

FIGURE 3.84   Curvature measurement for stress analysis. (A) Sample output from Tencor's FLX 2908 stress analysis instrument, showing how stress is derived from changes in wafer curvature. (B) The reflected light technique, used by Ionic Systems to measure wafer curvature. (Reprinted with permission from Singer, P. [1992] "Film Stress and How to Measure It," *Semicond. Int.*, **15**, 54–58, 1992.)

Systems' optical stress analyzer. None of the above techniques satisfies the need for measuring stress in low modulus materials such as polyimide. For the latter applications, the suspended membrane approach (see below) is better suited.

3.15.4.3   Uniaxial Measurements of Mechanical Properties of Thin Films

Many problems associated with handling thin films in stress test equipment may be bypassed by applying micromachining techniques. One simple example of problems encountered with thin films is the measurement of uniaxial tension to establish the Young's modulus. This method, effective for macroscopic samples, proves problematic for small samples. The test formula is illustrated in Figure 3.85A. The gauge length L in this figure represents the region we allow to elongate and the area A ($= W \times H$) is the cross section of the specimen. A stress F/A is applied and measured with a load cell; the strain $\delta L/L$ is measured with an LVDT or another displacement transducer (a typical instrument used is the Instron 1123). The Young's modulus is then deduced from:

$$E = \left(\frac{F}{A} \right) \left(\frac{L}{\delta L} \right) \tag{3.55}$$

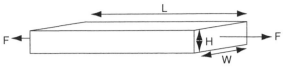

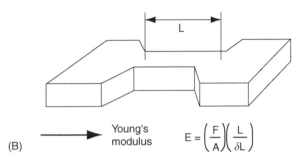

Stress : F/A (area A = W × H)

Strain : δ/L/L

L ="gauge length" (region we allow to elongate)

A, under elongation, will contract by (W+H) δ v L
unless a dog-bone structure is used

(A)

Young's
modulus $E = \left(\dfrac{F}{A}\right)\left(\dfrac{L}{\delta L}\right)$

(B)

FIGURE 3.85   Measuring Young's modulus. (A) With a bar-shaped structure. (B) With a dog-bone, Instron specimen. (After Senturia, S. in Maseeh, F., and Senturia, S.D., [1990] "Viscoelasticity and Creep Recovery of Polyimide Thin Films," in *Technical Digest: 1990 Solid State Sensor and Actuator Workshop*, Hilton Head Island, SC, pp. 55–60.)

The obvious problem for either small or large samples is how to grip the sample without changing *A*. Under elongation, *A* will indeed contract by $(w + H)\delta vL$. In general, making a dog bone–shaped structure (Instron specimen) solves that problem, as shown in Figure 3.85B. Still, the grips introduced in an Instron sample can produce end effects and uncertainties in determining *L*. Making Instron specimens in thin films is even more of a challenge, since the thin film needs to be removed from the surface, possibly changing the stress state, while the removal itself may modify the film.

As in the case of adhesion, some new techniques for testing stress in thin films based on micromachining are being explored. These microtechniques prove more advantageous than the whole wafer disc technique in that they are able to make local measurements.

The fabrication of micro-Instron specimens of thin polyimide samples is illustrated in Figure 3.86A. Polyimide is deposited on a p^+ Si membrane in multiple coats. Each coat is prebaked at 130°C for 15 min. After reaching the desired thickness, the film is cured at 400°C in nitrogen for 1 hr (A). The polyimide is then covered with a 3000 Å layer of evaporated aluminum (B). The aluminum layer is patterned by wet etching (in phosphoric-acetic-nitric solution, referred to as *PAN etch*) to the Instron specimen shape (C). Dry etching transfers the pattern to the polyimide (D). After removing the Al mask by wet etching, the p^+ support is removed by a wet isotropic etch (HNA) or a SF_6 plasma etch [Bressers et al., 1996], and finally the side silicon is removed along four pre-etched scribe lines, releasing the residual stress (F). The remaining silicon acts as supports for the grips of the Instron [Maseeh-Tehrani, 1990]. The resulting structure can be manipulated like any other macrosample without the need for removal of the film from its substrate. This technique enables the gathering of stress/strain data for a variety of commercially available polyimides [Maseeh and Senturiah, 1989]. A typical measurement result for Dupont's polyimide 2525, illustrated in Figure 3.86B, gives a break stress and strain of 77 MPa (σ_b) and 2.7% (ε_b), respectively, and 3350 MPa for the Young's modulus.

Figure 3.87A illustrates a micromachined test structure able to establish the strain and the ultimate stress of a thin film [Senturia, 1987]. A suspended rectangular polymer membrane is patterned into an asymmetric structure before removing the thin supporting Si. Once released, the wide suspended strip

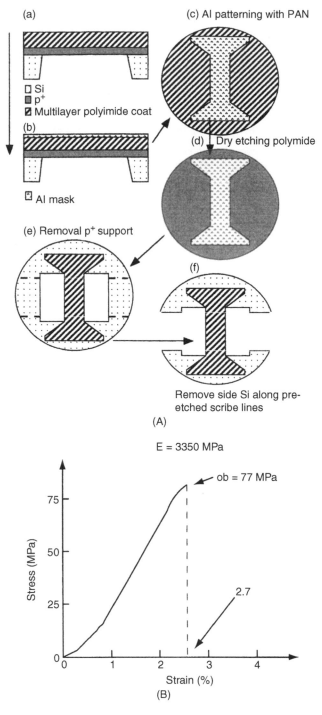

FIGURE 3.86 Uniaxial stress measurement. (A) Fabrication process of a dog-bone sample for measurement of uniaxial strain. (B) Stress versus strain for Du Pont's 2525 polyimide. (Courtesy of Dr. F. Maseeh, IntelliSense.)

(width W_1) pulls on the thinner necks (total width W_2), resulting in a deflection δ from its original mask position toward the right to its final position after release. The residual tensile stress in the film drives the deformation d as shown in Figure 3.87A. By varying the geometry, it is possible to create structures exhibiting small strain in the thinner sections as opposed to others that exceed the ultimate strain of the film.

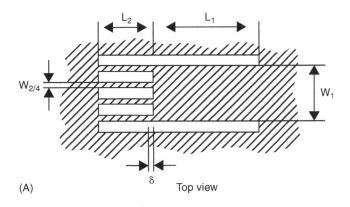

(A) Top view

(B)

FIGURE 3.87 Ultimate strain. (A) Test structures for stress-to-modulus (strain) and ultimate stress measurements. (B) Two released structures, one of which has exceeded the ultimate strain of the film, resulting in fracture of the necks. (Reprinted with permission from Senturia, S. [1987] "Can We Design Microrobotic Devices without Knowing the Mechanical Properties of the Materials?" in *Proceedings: IEEE Micro Robots and Teleoperators Workshop*, Hyannis, MA, pp. 3/1–5.)

For structures where the strain is small enough to be modeled with linear elastic behavior, the deflection δ can be related to the strain as follows:

$$\varepsilon = \frac{\sigma}{E} = \frac{\delta \left(\dfrac{W_1}{L_1} + \dfrac{W_2}{L_2} \right)}{W_1 - W_2} \tag{3.56}$$

where the geometries are defined as illustrated in Figure 3.87A. Figure 3.87B displays a photograph of two released structures, one with thicker necks and the other with necks so thin that they fractured upon release of the film. Based on the residual tensile strain of the film and the geometry of the structures that failed, the ultimate strain of the particular polyimide used was determined to be 4.5%.

Using similar micromachined tensile test structures, Biebl et al. (1993) measured the fracture strength of undoped and doped polysilicon and found 2.84 ± 0.09 GPa for undoped material and 2.11 ± 0.10 GPa in the case of phosphorous doping, 2.77 ± 0.08 GPa for boron doping, and 2.70 ± 0.09 GPa for arsenic doping. No statistically significant differences were observed between samples released using concentrated HF or buffered HF. However, a 17% decrease of the fracture stress was observed for a 100% increase in etching time. These data contrast with Greek et al.'s (1995) in situ tensile strength test result of 768 MPa for an undoped poly-Si film, a mean tensile strength almost ten times less than that of single-crystal Si

(6 GPa) [Ericson and Schweitz, 1990]. We normally expect polycrystalline films to be stronger than single-crystal films (see below under Strength of Thin Films). Greek et al. (1995) explain this discrepancy for poly-Si by pointing out that their polysilicon films have a very rough surface as compared with single-crystal material, containing many locations of stress concentration where a fracture crack can be initiated.

3.15.4.4   Biaxial Measurements of Mechanical Properties of Thin Films: Suspended Membrane Methods

We noted earlier that none of the disk stress-measuring techniques are suitable for measuring stress in low-modulus tensile materials such as polyimides. Suspended membranes are very convenient for this purpose. The same micromachined test structure used for adhesion testing, sometimes called the blister test, as shown in Figure 3.81, can measure the tensile stress in low-modulus materials. This type of test structure ensues from shaping a silicon diaphragm by conventional anisotropic etching, followed by applying the coating, and finally, removing the supporting silicon from the back with an SF_6 plasma [Senturia, 1987]. By pressurizing one side of the membrane and measuring the deflection, one can extract both the residual stress and the biaxial modulus of the membrane. Pressure to the suspended film can be applied by a gas or by a point-load applicator [Vinci and Braveman, 1991]. The load-deflection curve at moderate deflections (strains less than 5%) answers to:

$$p = C_1 \frac{\sigma t d}{a^2} + C_2 \left(\frac{E}{1 - v} \right) \frac{t d^3}{a^4} \qquad (3.57)$$

where p = pressure differential across the film, d = center deflection, a = initial radius, t = membrane thickness, and σ = initial film stress.

In the simplified Cabrera model for circular membranes, the constants C_1 and C_2 equal 4 and 8/3, respectively. For more rigorous solutions for both circular and rectangular membranes and references to other proposed models, refer to Maseeh-Tehrani (1990). The relation in Equation (3.57) can simultaneously determine s and the biaxial modulus $E/(1 - v)$; plotting pa^2/dt vs. $(d/a)^2$ should yield a straight line. The residual stress can be extracted from the intercept and the biaxial modulus from the slope of the least-squares-best-fit line [Senturia, 1987]. A typical result obtained via such measurements is represented in Figure 3.88. For the same Dupont polyimide 2525, measuring a Young's modulus of 3350 MPa in the uniaxial test (Figure 3.86B), the measurements give 5540 MPa for the biaxial modulus and 32 MPa for the residual stress. The residual stress-to-biaxial modulus ratio, also referred to as the residual biaxial strain, thus reaches 0.6%. The latter quantity must be compared to the ultimate strain when evaluating potential reliability problems associated with cracking of films. By loading the membranes to the elastic limit point, yield stress and strain can be determined as with the uniaxial test.

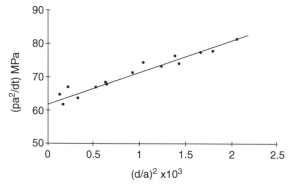

FIGURE 3.88   Load deflection data of a polyimide membrane (Du Pont 2525). From the intercept, a residual stress of 32 MPa was calculated and from the slope, a biaxial modulus of 5540 MPa. (Reprinted with permission from Senturia, S. [1987] "Can We Design Microrobotic Devices without Knowing the Mechanical Properties of the Materials?" in *Proceedings: IEEE Micro Robots and Teleoperators Workshop*, Hyannis, MA, pp. 3/1–5.)

3.15.4.5 Poisson Ratio for Thin Films

The Poisson ratio for thin films presents more difficulties to measure than the Young's modulus, as thin films tend to bend out of plane in response to in-plane shear. Maseeh and Senturia (1989) combine uni-axial and biaxial measurements to calculate the in-plane Poisson ratio of polyimides. For example, for the Dupont polyimide 2525, they determined 3350 MPa for E and 5540 MPa for the biaxial modulus ($E/(1 - v)$) leading to 0.41 ± 0.1 for the Poisson's ratio (v). The errors on both the biaxial and uniaxial measurements need to be reduced to develop more confidence in the extracted value of the Poisson's ratio. At present, the precision on the Poisson's ratio is limited to about 20%.

3.15.4.6 Other Surface Micromachined Structures to Gauge Intrinsic Stress

Various other surface micromachined structures have been used to measure mechanical properties of thin films. We will give a short review here, but the interested reader might want to consult the original references for more details.

Clamped-Clamped Beams. Several groups have used rows of clamped-clamped beams (bridges) with incrementally increasing lengths to determine the critical buckling load and hence deduce the residual compressive stress in polysilicon films (Figure 3.89A [Sekimoto et al., 1982; Guckel et al., 1985]). The

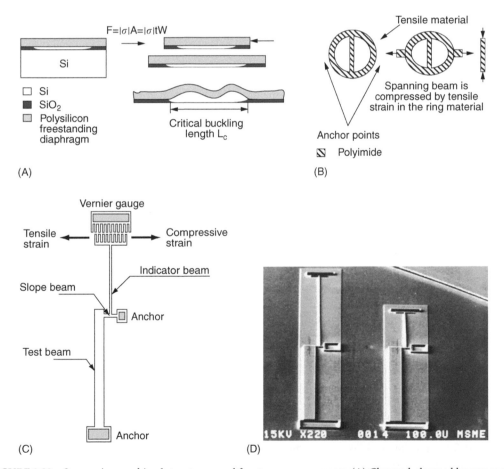

FIGURE 3.89 Some micromachined structures used for stress measurements. (A) Clamped-clamped beams: measuring the critical buckling length of clamped-clamped beams enables measurement of residual stress. (B) Crossbar rings: tensile stress can be measured by buckling induced in the crossbar of a ring structure. (C) A schematic of a strain gauge capable of measuring tensile or compressive stress. (D) SEM microphotograph of two strain gauges. (Figures c and d reprinted with permission from Lin, L. [1993] Selective Encapsulations of MEMS: Micro Channels, Needles, Resonators, and Electromechanical Filters, Ph.D. thesis, University of California, Berkeley.)

residual strain, $\varepsilon = \sigma/E$, is obtained from the critical length, L_c, at which buckling occurs (Euler's formula for elastic instability of struts):

$$\varepsilon = \frac{4\pi^2}{A} \frac{I}{L_c^2} \tag{3.58}$$

where A is the beam cross-sectional area and I the moment of inertia. As an example, with a maximum beam length of 500 μm and a film thickness of 1.0 μm, the buckling beam method can detect compressive stress as small as 0.5 MPa. This simple Euler approach does not take into consideration additional effects such as internal moments resulting from gradients in residual stress.

Ring Crossbar Structures. Tensile strain can be measured by a series of rings (Figure 3.89B) constrained to the substrate at two points on a diameter and spanned orthogonally by a clamped-clamped beam. After removal of the sacrificial layer, tensile strain in the ring places the spanning beam in compression; the critical buckling length of the beam can be related to the average strain [Guckel et al., 1988].

Vernier Gauges. Both clamped-clamped beams and ring structures need to be implemented in entire arrays of structures. They do not allow easy integration with active microstructures due to space constraints. As opposed to proof structures, one might use vernier gauges to measure the displacement of structures induced by residual strain [Lin, 1993]. The idea was first explored by Kim (1991), whose device consisted of two cantilever beams fixed at two opposite points. The end movement of the beams caused by the residual strain was measured by a vernier gauge. This method requires only one structure, but the best resolution for strain measurement reported is only 0.02% for 500-μm beams. Moreover, the vernier gauge device may indicate an erroneous strain when an out-of-plane strain gradient occurs [Lin, 1993]. Other types of direct strain measurement devices are the T- and H-shaped structures from Allen et al. (1987) and Mehregany et al. (1987). Optical measurement of the movement at the top of the T- or H-shaped structures becomes possible only with very long beams (greater than 2.5 μm). They occupy large areas, and their complexity requires finite element methods to analyze their output. The same is true for the strain magnification structure by Goosen et al. (1993) This structure measures strain by interconnecting two opposed beams such that the residual strain in the beams causes a third beam to rotate as a gauge needle. The rotation of the gauge needle quantifies the residual strain. A schematic of a micromachined strain gauge capable of measuring tensile or compressive residual stress, as shown in Figure 3.89C, was developed by Lin at the University of California, Berkeley [Lin, 1993]. Figure 3.89D represents a scanning electron microscope (SEM) photograph of Lin's strain gauge. This gauge by far outranks the various in situ gauges explored. The strain gauge uses only one structure, can be fabricated in situ with active devices, determines tensile or compressive strain under optical microscopes, has a fine resolution of 0.001%, and resists to the out-of-plane strain gradient. When the device is released in the sacrificial etch step, the test beam (length L_t) expands or contracts depending on the sign of the residual stress in the film, causing the compliant slope beam (length L_s) to deflect into an "S" shape. The indicator beam (length L_i), attached to the deforming beam at its point of inflection, rotates through an angle θ, and the deflection δ is read on the vernier scale. The residual strain is calculated as [Lin, 1993]:

$$\varepsilon_f = \frac{2L_s\delta}{3L_iL_tC} \tag{3.59}$$

where C is a correction factor required by the presence of the indicator beam [Lin, 1993]. This equation was derived from simple beam theory relations and assumes that no out-of-plane motion will occur. The accuracy of the strain gauge is greatly improved because its output is independent of both the thickness of the deposited film and the cross-section of the microstructure. Krulevitch used these devices to measure residual stress in in situ phosphorous-doped poly-Si films [Krulevitch, 1994], while Lin tested LPCVD silicon-rich silicon nitride films with it [Lin, 1993].

An improved micromachined indicator structure inspired by Lin's work was built by Ericson et al. [Ericson et al., 1995]. By reading an integrated nonius scale in an SEM or an optical microscope, internal stress was measured with a resolution better than 0.5 Mpa [Ericson et al., 1995; Benitez et al, 1995]. Both thick (10 μm) and thin (2 μm) poly-Si films were characterized this way.

Lateral Resonators. Biebl et al. (1995) extracted the Young's modulus of in situ phosphorus-doped polysilicon by measuring the mechanical response of poly-Si linear lateral comb-drive resonators (see Figure 3.97). The results reveal a value of $130 \pm 5\,GPa$ for the Young's modulus of highly phosphorus-doped films deposited at 610°C with a phosphine-to-silane mole ratio of 1.0×10^{-2} and annealing at 1050°C. For a deposition at 560°C with a phosphine to silane ratio of 1.6×10^{-3}, a Young's modulus of $147 \pm 6\,GPa$ was extracted.

3.15.4.7 Stress Nonuniformity Measurement by Cantilever Beams and Cantilever Spirals

The uniformity of stress through the depth of a film introduces a property that is extremely important to control. Variations in the magnitude and direction of the stress in the vertical direction can cause cantilevered structures to curl toward or away from the substrate. Stress gradients in the polysilicon film must thus be controlled to ensure predictable behavior of designed structures when released from the substrate. To determine the thickness variation in residual stress, noncontact surface profilometer measurements on an array of simple cantilever beams [Core et al., 1993; Chu et al., 1992] or cantilever spirals can be used [Fan et al., 1990].

Cantilever Beams. The deflections resulting from stress variation through the thickness of simple cantilever beams after their release from the substrate are shown in Figure 3.90A. The bending moment causing deflection of a cantilever beam follows out of pre-release residual stress and is given by:

$$M = \int_{-t/2}^{t/2} zb\sigma(z)\,dz \tag{3.60}$$

where $\sigma(z)$ represents the residual stress in the film as a function of thickness, and b stands for cantilever width. Assuming a linear strain gradient Γ (physical dimensions 1/length) such that $\sigma(z) = E\Gamma z$, Equation (3.60) converts to:

$$\Gamma = \frac{M(12)}{Ebt^3} = \frac{M}{EI} \tag{3.61}$$

where the moment of inertia I for a rectangular cross section is given by $I = bt^3/12$. The measured deflection z (i.e., the vertical deflection of the cantilever's endpoint) from beam theory for a cantilever with an applied end moment is given as:

$$z = \frac{ML^2}{2EI} = \frac{\Gamma L^2}{2} \tag{3.62}$$

Figure 3.90B represents a topographical contour map of an array of polysilicon cantilevers. The cantilevers vary in length from 25 to 300 μm by 25 μm increments. Notice that the tip of the longest cantilever resides at a lower height (approximately 0.9 μm closer to the substrate) than the anchored support, indicating a downward bending moment [Core et al., 1993]. The gradients can be reduced or eliminated with a high-temperature anneal. With integrated electronics on the same chip, long high-temperature processing must be avoided. Therefore, stress gradients can limit the length of cantilevered structures used in surface-micromachined designs.

The use of SOI-based micromachined cantilevers for measuring surface stress induced during adsorption of biomolecules is illustrated in Example 3 of "Polysilicon Surface Micromachining Examples".

Cantilever Spirals. Residual stress gradients can also be measured by Fan's cantilever spiral as shown in Figure 3.91A [Fan et al., 1988]. Spirals anchored at the inside spring upward, rotate, and contract with positive strain gradient (tending to curl a cantilever upward), while spirals anchored at the outside deflect in a similar manner in response to a negative gradient. Theoretically, positive and negative gradients produce spirals with mirror symmetry [Fan et al., 1990]. The strain gradient can be determined from spiral structures by measuring the amount of lateral contraction, the change in height, or the amount of rotation. Krulevitch presented the computer code for the spiral simulation in his doctoral thesis [Krulevitch, 1994]. Figure 3.91B shows a simulated spiral with a bending moment of $\Gamma = \pm 3.0\,\mu m^{-1}$ after release.

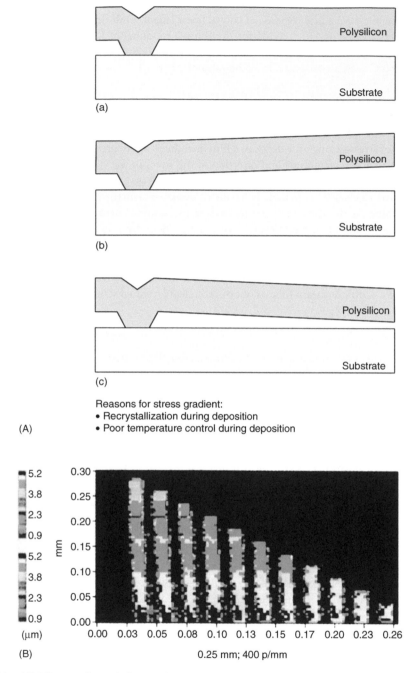

FIGURE 3.90 (A) Microcantilever deflection for measuring stress nonuniformity: a. no gradient; b. higher tensile stress near the surface; and c. lower tensile stress near the surface. (B) Topographical contour map of polysilicon cantilever array. (Reprinted with permission from Core, T.A., et al. [1993] "Fabrication Technology for an Integrated Surface-Micromachined Sensor," *Solid State Technol.* **36**, pp. 39–47.)

Krulevitch compared all the above surface micromachined structures for stress and stress gradient measurements on poly-Si films. His comments are summarized in Table 3.14 [Krulevitch, 1994]. Krulevitch found that the fixed-fixed beam structures for determining compressive stress from the buckling criterion produced remarkably self-consistent and repeatable results. Wafer curvature stress profiling proved reliable for determining average stress and the true stress gradient as compared with micromachined spirals.

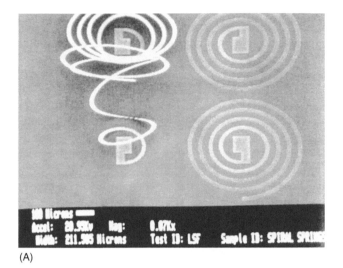

(A)

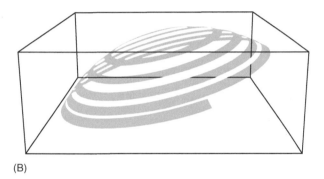

(B)

FIGURE 3.91   Cantilever spirals for stress gradient measurement. (A) SEM of micrographs of spirals from an as-deposited poly-Si. (Courtesy of Dr. L.-S. Fan, IBM, Alamden Research Center.) (B) Simulation of a thin-film micro-machined spiral with G = 3.0 mm. (Reprinted with permission from Krulevitch, P.A. [1994] Micromechanical Investigations of Silicon and Ni-Ti-Cu Thin Films, Ph.D. thesis, University of California, Berkeley.)

TABLE 3.14   Summary of Various Techniques for Measuring Residual Film Stress

Measurement Technique	Measurable Stress State	Remarks
Wafer curvature	Stress gradient, average stress	Average stress over entire wafer, provides true stress gradient, approx. 5 MPa resolution
Vernier strain gauges	Average stress	Local stress, small dynamic range, resolution = 2 MPa
Spiral cantilevers	Stress gradient	Local stress, provides equivalent linear gradient
Curling beam cantilevers	Large positive stress gradient	Local stress, provides equivalent linear gradient
Fixed-fixed beams	Average compressive stress	Local stress measurement

Source: Reprinted with permission from Krulevitch, P.A. (1994) Micromechanical Investigations of Silicon and Ni-Ti-Cu Thin Films, Ph.D. thesis, University of California, Berkeley.

Measurements of curled cantilevers could not be used much, as the strain gradients mainly were negative for poly-Si, leading to cantilevers contacting the substrate. The strain gauge dial structures were useful over a rather limited strain-gradient range. With too large a strain gradient, curling of the long beams overshadows expansion effects and makes the vernier indicator unreadable.

3.15.5   Strength of Thin Films

Due to the high activation energy for dislocation motion in silicon (2.2 eV), hardly any plastic flow occurs in single-crystalline silicon for temperatures lower than 673°C. Grain boundaries in poly-Si block dislocation motion; hence, polysilicon films can be treated as an ideal brittle material at room temperature [Biebl et al., 1995]. High yield strengths often are obtained in thin films with values up to 200 times as large as those found in the corresponding bulk material [Campbell, 1970]. In this light, the earlier quoted fracture stresses of poly-Si, between two and ten times smaller than that of bulk single-crystal, are surprising. Greek et al. [Greek et al, 1995] explain this deviation by pointing at the high surface roughness of poly-Si films compared to single-crystal Si. They believe that a reduction in surface roughness would improve the tensile fracture strength considerably.

Indentation (hardness) testing is very common for bulk materials in which the direct relationship between bulk hardness and yield strength is well known. It can be measured by pressing a hard, specially shaped point into the surface and observing indentation. This type of measurement is of little use for measuring thin films below 5×10^4 Å thick. Consequently, very little is known about the hardness of thin films. Recently, specialized instruments have been constructed (e.g., the Nanoindenter) in which load and displacement data are collected while the indentation is being introduced in a thin film. This eliminates the errors associated with later measurement of indentation size and provides continuous monitoring of load/displacement data similar to a standard tensile test. Load resolution may be 0.25 μN, displacement resolution 0.2 to 0.4 nm, and x–y sample position accuracy 0.5 μm. Empirical relations have correlated hardness with Young's modulus and with uniaxial strength of thin films. Hardness calculations must include both plastic and long-distance elastic deformation. If the indentation is deeper than 10% of the film, corrections for elastic hardness contribution of the substrate must also be included [Vinci and Braveman, 1991]. Mechanical properties such as hardness and modulus of elasticity can be determined on the micro- to picoscales using AFM [Bushan, 1996]. Bushan provides an excellent introduction to this field in the *Handbook of Micro/Nanotribology* [Bushan, 1995].

3.16   Surface Micromachining Processes

3.16.1   Basic Process Sequence

A surface micromachining process sequence for the creation of a simple freestanding poly-Si bridge is illustrated in Figure 3.92 [Howe and Muller, 1983; Howe, 1985]. A sacrificial layer, also called a spacer layer or base, is deposited on a silicon substrate coated with a dielectric layer as the buffer/isolation layer (Figure 3.92A). Phosphosilicate glass (PSG) deposited by LPCVD stands out as the best material for the sacrificial layer, because it etches even more rapidly in HF than SiO_2. To obtain a uniform etch rate, the PSG film must be densified by heating the wafer to 950 to 1100°C in a furnace or a rapid thermal annealer (RTA) [Yun, 1992]. With a first mask, the base is patterned as shown in Figure 3.92B. Windows are opened in the sacrificial layer, and a microstructural thin film (whether consisting of polysilicon, metal, alloy, or a dielectric material) is conformally deposited over the patterned sacrificial layer (Figure 3.92C). Furnace annealing, in the case of polysilicon at 1050°C in nitrogen for one hour, reduces stress stemming from thermal expansion coefficient mismatch and nucleation and growth of the film. Rapid thermal annealing has been found effective for reducing stress in polysilicon as well [Yun, 1992]. With a second mask, the microstructure layer is patterned, usually by dry etching in a $CF_4 + O_2$ or a $CF_3Cl + Cl_2$ plasma (Figure 3.92D) [Adams, 1988]. Finally, selective wet etching of the sacrificial layer, say in 49% HF, leaves a freestanding micromechanical structure (Figure 3.92E). The surface micromachining technique is applicable to combinations of thin films and lateral dimensions where the sacrificial layer can be etched without significant etching or attack of the microstructure, the dielectric, or the substrate. Typically, a surface micromachining stack may contain a total of four to five structural and sacrificial layers, but more are possible; the poly-Si surface machining process at Sandia's SUMMIT, for example, stacks up to five polysilicon and five oxide layers.

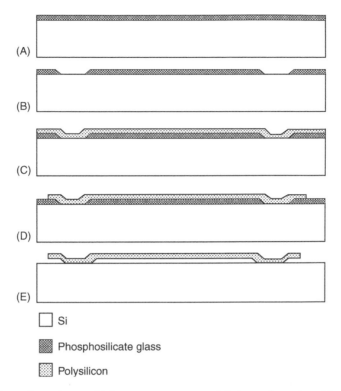

(A)

(B)

(C)

(D)

(E)

☐ Si

▓ Phosphosilicate glass

▒ Polysilicon

FIGURE 3.92 Basic surface micromachining process sequence. (A) Spacer layer deposition (the thin dielectric insulator layer is not shown). (B) Base patterning with mask 1. (C) Microstructure layer deposition. (D) Pattern microstructure with mask 2. (E) Selective etching of spacer layer.

3.16.2 Fabrication Step Details

3.16.2.1 Pattern Transfer to SiO$_2$ Buffer/Isolation Layer

A blanket n$^+$ diffusion of the Si substrate, defining a ground plane, often outlines the very first step in surface micromachining, followed by a passivation step of the substrate, for example with 0.15 µm thick LPCVD nitride on a 0.5 µm thermal oxide. Suppose the buffer/isolation passivation layer as illustrated in Figure 3.93 itself needs to be patterned, perhaps to make a metal contact pad onto the Si substrate. Then the appropriate fabrication step is a pattern transfer to the thin isolation film as shown in Figure 3.93, illustrating a wet pattern transfer to a 1 µm thick thermal SiO$_2$ film with a 1 µm resist layer. Typically, an isotropic etch such as buffered HF (e.g., BHF [5:1], which is five parts NH$_4$F and one part concentrated HF) is used (unbuffered HF attacks the photoresist). This solution etches SiO$_2$ at a rate of 100 nm/min, and the creation of the opening to the underlying substrate takes about 10 min. The etch progress may be monitored optically (color change) or by observing the hydrophobic/hydrophilic behavior [Hermansson et al., 1991] of the etched layer. With a resist opening L_m the undercut typically measures the same thickness as the oxide thickness t_{SiO_2}. In other words, the contact pad will have a size of $L_m + 2t_{SiO_2}$. The undercut worsens with loss of photoresist adhesion during etching. Using an adhesion promoter such as HMDS (hexamethyldisilazane) proves useful in such cases. A new bake after 5 min of etching is a good procedure to maintain the resist integrity. After the isotropic etch, the resist is stripped in a piranha etch bath. This strong oxidizer grows about 3 nm of oxide back in the cleared window. To remove the oxide resulting from the piranha, a dip in diluted BHF suffices. After cleaning and drying, the substrate is ready for contact metal and base material deposition. Applying a dry etch (say, a CF$_4$-H$_2$ plasma) to open the window in the oxide would eliminate undercutting of the resist but requires a longer setup time. With an LPCVD low stress

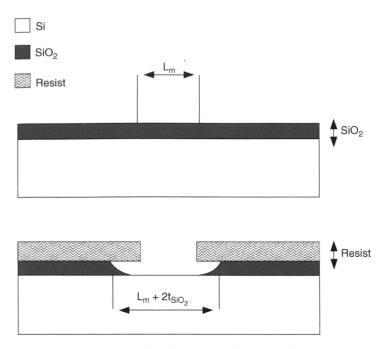

FIGURE 3.93 Wet etch pattern transfer to a thin thermal SiO$_2$ film for the fabrication of a contact pad to the Si substrate.

nitride on top of a thermal SiO$_2$, an often-used combination for etching the buffer/isolation layer is a dry etch (say, a SF$_6$ plasma) followed by a 5:1 BHF [Tang, 1990].

3.16.2.2 Base Layer (Also Spacer or Sacrificial Layer) Deposition and Etching

A thin LPCVD phosphosilicate glass (PSG) layer (say, 2 μm thick) is a preferred base, spacer, or sacrificial layer material. Adding phosphorous to SiO$_2$ to produce PSG enhances the etch rate in HF [Monk et al., 1992; Tenney and Ghezzo, 1973]. Other advantages to using doped SiO$_2$ include its utility as a solid state diffusion dopant source to make subsequent polysilicon layers electrically conductive and helping to control window taper (see below). As deposited phosphosilicate displays a nonuniform etch rate in HF and must be densified, typically carried out in a furnace at 950°C for 30 min to 1 hr in a wet oxygen ambient. The etch rate in BHF can be used as the measure of the densification quality. The base window etching stops at the buffer isolation layer, often a Si$_3$N$_4$/SiO$_2$ layer that also forms the permanent passivation of the device. Windows in the base layer are used to make anchors onto the buffer/isolation layer for mechanical structures.

The edges of the etched windows in the base may need to be tapered to minimize coverage problems with subsequent structural layers, especially if these layers are deposited with a line-of-sight deposition technique. An edge taper is introduced through an optimization of the plasma etch conditions, through the introduction of a gradient of the etch rate, or by reflow of the etched spacer. Figure 3.94A depicts the reflow process of a PSG spacer after patterning in a dry etch. Viscous flow at higher temperature smooths the edge taper. The ability of PSG to undergo viscous deformation at a given temperature primarily is a function of the phosphorous content in the glass; reflow profiles get progressively smoother the higher the phosphorous concentration — reflecting the corresponding enhancement in viscous flow [Levy and Nassau, 1986]. In Figure 3.94B, ion implantation of PSG has created a rapidly etching, damaged PSG layer [North et al., 1978; Goetzlich and Ryssel, 1981; White, 1980]. The steady state taper is a function of the etch rate in PSG etchants (e.g., BHF).

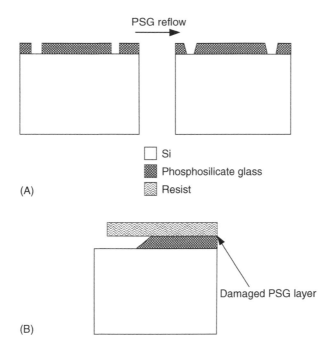

FIGURE 3.94   Edge taper of spacer layer. (A) Taper by reflow of PSG. (B) Taper by ion implantation of PSG.

3.16.2.3   Deposition of Structural Material

For the best step coverage of a structural material over the base window, chemical vapor deposition is preferred. If a physical deposition method must be used, sputtering is preferred over line-of-sight deposition techniques, which lead to the poorest step coverage. In the latter case, edge taper could be introduced advantageously.

The most widely used structural material in surface micromachining is polysilicon (poly-Si, or simply poly). Polysilicon is deposited by low-pressure (25 to 150 Pa) chemical vapor deposition (LPCVD) in a furnace (a poly chamber) at about 600°C. The undoped material is usually deposited from pure silane, which thermally decomposes according to the reaction:

$$SiH_4 \rightarrow Si + 2H_2 \qquad\qquad \text{(Reaction 3.16)}$$

Typical process conditions may consist of a temperature of 605°C, a pressure of 550 mTorr (73 Pa), and a silane flow rate of 125 sccm. Under those conditions, a normal deposition rate is 100 Å/min. To deposit a 1 μm film will take about 90 min. Sometimes, the silane is diluted by 70 to 80% nitrogen. The silicon is deposited at temperatures ranging from 570 to 580°C for fine-grained poly-Si to 620 to 650°C for coarse-grained poly-Si. The characteristics of these two types of poly-Si materials will be compared in the materials case studies below. Furnace annealing of the poly-Si film at 1050°C in nitrogen for 1 hr is commonly employed to reduce stress stemming from thermal expansion coefficient mismatch and nucleation and growth of the polysilicon film. Deposition of Si from a low temperature PECVD is also possible but results in amorphous silicon.

To make parts of the microstructure conductive, dopants can be introduced in the poly-Si film by adding dopant gases to the silane gas stream, by drive-in from a solid dopant source, or by ion implantation. When doping from the gas phase, the dopant can be readily controlled in the range of 10^{19} to 10^{21}/cm³. Polysilicon deposition rates, in the case of gas phase doping (in situ doping), may be significantly affected. For example, decreases in poly-Si deposition rate by as much as a factor of 25 have been reported in phosphine and arsine doping. The effect is associated with the poisoning of reaction sites by phosphine and arsine [Adams, 1988]. The lower deposition rate of in situ phosphine doping can be mitigated by reducing the ratio of phosphine to silane flow by one third [Howe, 1995]. With the latter flow regime, deposition at 585°C (for a 2 μm thick film), followed by 900°C rapid thermal annealing for 7 min, results in a polysilicon with low residual stress, negligible stress gradient, and low resistivity [Howe, 1995; Biebl et al., 1995]. In situ boron doping, in contrast to arsine and phosphine doping, accelerates the polysilicon deposition rate through an enhancement of silane adsorption induced by the boron presence [Fresquet et al., 1995]. Film thickness uniformity for doped films typically is less than 1%, and sheet resistance uniformity is less than 2%. Alternatively, poly-Si may be doped from PSG films sandwiching the undoped poly-Si film. By annealing such a sandwich at 1050°C in N_2 for one hour, the polysilicon is symmetrically doped by diffusion of dopant from the top and the bottom layers of PSG. Symmetric doping results in a polysilicon film with a moderate compressive stress. The resulting uniform grain texture avoids gradients in the residual stress that would cause bending moments warping microstructures upon release. Finally, ion implantation of undoped polysilicon followed by high-temperature dopant drive-in also leads to conductive polysilicon. This polysilicon has a moderate tensile stress, with a strain gradient that causes cantilevers to deflect toward the substrate [Core et al., 1993]. The poly-Si is now ready for patterning by RIE in, say, a CF_4-O_2 plasma.

Although the mechanical properties are less understood than for single-crystal Si, microstructures based on poly-Si as a mechanical member have been commercialized swiftly ["Analog Devices Combine Micromachining with BICMOS," *Semicond. Int.* 14, p. 17, 1991; Core et al., 1993].

Surface micromachining of thin single-crystalline Si layers in epitaxial layers or fusion bonded and etched back Si as well as surface micromachining involving resists such as polyimides is also very common and is discussed separately below. Other structural materials used in surface micromachining include aluminum, SiO_2, silicon nitride, silicon oxynitride, polyimide, diamond, SiC, sputtered Si, GaAs, tungsten, α-Si:H, Ni, W, Al, and others. More information on the materials properties of these materials in thin film form is provided in the sections below.

3.16.2.4 Selective Etching of Spacer Layer

Selective Etching. To create movable micromachines, the microstructures must be freed from the spacer layers. The challenge in freeing microstructures by undercutting is evident from Figure 3.95. After patterning the poly-Si by RIE in, say, an SF_6 plasma, it is immersed in an HF solution to remove the underlying sacrificial layer, releasing the structure from the substrate. Commonly, a layer of sacrificial

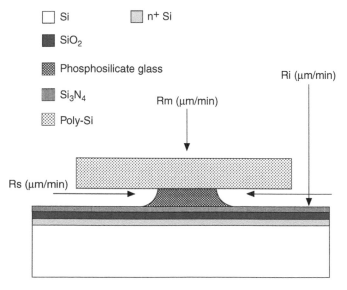

FIGURE 3.95 Selective etching of spacer layer.

phosphosilicate glass between 1 and 2000 µm long and 0.1 to 5 µm thick is etched in concentrated, dilute, or buffered HF. The spacer etch rate R_s should be faster than the attack on the microstructural element R_m and that of the insulator layer R_i. For this type of complete undercutting, only wet etchants can be used. Etching narrow gaps and undercutting wide areas with BHF can take hours. To shorten the etch time, extra apertures in the microstructures are sometimes provided for additional access to the spacer layer. The etch rate of PSG, the most common spacer material, increases monotonically with dopant concentration, and thicker sacrificial layers etch faster than thinner layers [Monk et al., 1994].

The selectivity ratios for spacer layer, microstructure, and buffer layer are not infinite [Lober and Howe, 1988], and in some instances even silicon substrate attack by BHF is observed under polysilicon/spacer regions [Fan et al., 1988; Mehregany et al., 1988]. It also has been shown that during an HF release step the mechanical properties of polysilicon, including residual stress, Young's modulus, and fracture strain, are affected [Walker et al., 1991]. Heavily phosphorous-doped polysilicon is especially prone to attack by BHF. In general, the Young's modulus and fracture strain of a thin polysilicon film decrease with increasing exposure time to HF and increasing HF concentration. Silicon nitride deposited by LPCVD etches much more slowly in HF than oxide films, making it a more desirable isolation film. When depositing this film with a silicon-rich composition, the etch rate is even slower (15 nm/min) [Tang, 1990]. Eaton et al. (1995) compared oxide and nitride etching in a 1:1 HF:H$_2$O and in a 1:1 HF:HCl solution and concluded that the HCl-based etch yielded both faster oxide etch rates (617 nm/min vs. 330 nm/min) and slower nitride etch rates (2 nm/min vs. 3.6 nm/min), providing a much greater selectivity of the oxide to silicon nitride (310 vs. 91!). The same authors also studied the optimal composition of a sacrificial oxide for the fastest possible etching in their most selective 1:1 HF:HCl etch. The faster sacrificial layer etch limited the damage to nitride structural elements. Their results are summarized in Table 3.15. A densified CVD SiO$_2$ was used as a control, and a 5%/5% borophosphosilicate glass (BPSG) was found to etch the fastest.

Watanabe et al. (1995), using low-pressure vapor HF, found high etch ratios of PSG and BPSG to thermal oxides of over 2000, with the BPSG etching slightly faster than the PSG. We will see further that low-pressure vapor HF also leads to less stiction of structural elements to the substrate. Table 3.16 presents etch rate and etch ratios for R_i and R_s in BHF (7:1) for a few selected materials.

Detailed studies on the etching mechanism of oxide spacer layers were undertaken by Monk et al. (1994a, 1994b). They found that the etching reaction shifts from being kinetic controlled to diffusion controlled as the etch channel becomes longer. This affects mainly large-area structures, as diffusion limitations were observed only after approximately 200 µm of channel etching or 15 min in concentrated HF.

TABLE 3.15 Etch Rate in 1:1 HF:HCl of a Variety of Sacrificial Oxides

Thin oxide	Lateral etch rate (Å/min)
CVD SiO$_2$ (densified at 1050°C for 30 min.)	6170
Ion implanted and densified CVD SiO$_2$ (P, 8 10^{15}/cm^2, 50 keV)	8330
Phosphosilicate (PSG)	11330
5%/5% borophosphosilicate (BPSG)	41670

Source: Adapted with permission from Eaton, W.P., and Smith, J.H. (1995) "A CMOS-Compatible, Surface-Micromachined Pressure Sensor for Aqueous Ultrasonic Application," in *Smart Structures and Materials 1995: Smart Electronics (Proceedings of the SPIE)*, San Diego, 1995, pp. 258–65.

TABLE 3.16 Etching of Spacer Layer and Buffer Layer in BHF (7:1)

Property	Material		
	LPCVD Si$_3$N$_4$	LP CVD SiO$_2$	LP CVD 7% PSG
Etch rate	7–12 Å/min (R$_i$)	700 Å/min (R$_s$)	~10,000 Å/min (R$_s$)
Selectivity ratio	1	60–100	800–1200 Å/min

TABLE 3.17   Etchants-Spacer and Microstructural Layer

Etchant	Buffer/isolation	Spacer	Microstructure
Buffered HF (5:1,NH$_4$F:conc.HF)	LPCVD Si$_3$N$_4$/thermal SiO$_2$	PSG	Poly-Si[a]
RIE using CHF$_3$ BHF (6:1)	LPCVD Si$_3$N$_4$	LP CVD SiO$_2$	CVD Tungsten[b]
KOH	LPCVD Si$_3$N$_4$/thermal SiO$_2$	Poly-Si, porous Si at room temperature)	Si$_3$N$_4$[c] SOI[d]
Ferric chloride	Thermal SiO$_2$	Cu	Polyimide[e]
HF	LPCVD Si$_3$N$_4$/thermal SiO$_2$	PSG	Polyimide[f]
Phosphoric/acetic acid/nitric acid (PAN or 5:8:1:1 water:phosphoric: acetic:nitric)	Thermal SiO$_2$	Al	PE CVD Si$_3$N$_4$ Nickel[g]
Ammonium iodide /iodine alcohol	Thermal SiO$_2$	Au	Ti[h]
Ethylene-diamine/pyrocathecol (EDP)	Thermal SiO$_2$	Poly-Si	SiO$_2$

[a] Howe and Muller (1983), Howe (1985), Guckel and Burns (1984); [b] Chen and MacDonald (1991); [c] Sugiyama et al. (1986; 1987); [d] Gennissen et al. (1995); [e] Kim and Allen (1991); [f] Suzuki et al. (1994); [g] Chang et al. (1991); Scheeper et al. (1991); [h] Yamada and Kuriyama (1991).

Eaton and Smith (1996) developed a release etch model that is an extension of the work done by Monk et al. (1994a, 1994b) and Liu et al. (1993).

Etching is followed by rinsing and drying. Extended rinsing causes a native oxide to form on the surface of the polysilicon structure. Such a passivation layer often is desirable and can be formed more easily by a short dip in 30% H$_2$O$_2$.

Etchant-Spacer-Microstructure Combinations. A wide variety of etchant, spacer, and structural material combinations have been used; a limited listing is presented in Table 3.17. One interesting case concerns poly-Si as the sacrificial layer. This was used, for example, in the fabrication of a vibration sensor at Nissan Motor Co. [Nakamura et al., 1985]. In this case, poly-Si is etched in KOH from underneath a nitride/poly-silicon/nitride sandwich cantilever. Also, a solution of HNO$_3$ and BHF can be used to etch poly-Si, but it proves difficult to control. Using aqueous solutions of NR$_4$OH, where R is an alkyl group, provides a better etching solution for poly-Si with greater selectivity with respect to silicon dioxide and phosphosilicate glass. The relatively slow etch rate enables better process control [Bassous and Liu, 1978], and the etchant does not contain alkali ions, making it more CMOS compatible. With tetramethyl ammonium hydroxide (TMAH), the etch rate of CVD poly-Si, deposited at 600°C from SiH$_4$, follows the rates of the (100) face of single-crystal Si and is dopant dependent. The selectivity of Si/SiO$_2$ and Si/PSG at temperatures below 45°C is measured to be about 1000. Hence, a layer of 500 Å PSG can be used as the etch mask for 10,000 Å of poly-Si. In addition, porous Si has been used as a sacrificial layer in micromachining. The high surface area of this material makes for rapid etching in KOH [Gennissen et al., 1995].

3.16.2.5   Stiction

Stiction during Release. The use of sacrificial layers enables the creation of very intricate movable polysilicon surface structures. An important limitation of such polysilicon shapes is that large-area structures tend to deflect through stress gradients or surface tension induced by trapped liquids and attach to the substrate/isolation layer during the final rinsing and drying step, a stiction phenomenon that may be related to hydrogen bonding or residual contamination. Recently, great strides were made toward a better understanding and prevention of stiction.

The sacrificial layer removal with a buffered oxide etch followed by a long, thorough rinse in deionized water and drying under an infrared lamp typically represent the last steps in the surface micromachining sequence. As the wafer dries, the surface tension of the rinse water pulls the delicate microstructure to the substrate where a combination of forces, probably van der Waals forces and hydrogen bonding, keeps it firmly attached (Figure 3.96) [Core et al., 1993]. Once the structure is attached to the substrate by stiction, the mechanical force needed to dislodge it usually is large enough to damage the micromechanical structure [Lober and Howe, 1988; Guckel et al., 1987; Alley et al., 1988]. The same phenomena are thought to be involved in room temperature wafer bonding. We will not further dwell upon the mechanics of the stiction

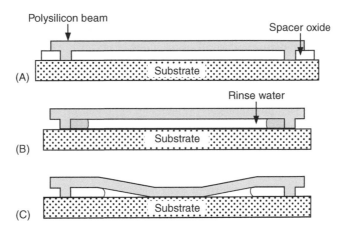

FIGURE 3.96 Stiction phenomenon in surface micromachining and the effect of surface tension on micromechanical structures. (A) Unreleased beam. (B) Released beam before drying. (C) Released beam pulled to the substrate by capillary forces as the wafer dries.

process here; for more information refer to the theoretical and experimental analysis of the mechanical stability and adhesion of microstructures under capillary forces by Mastrangelo et al. (1993a; 1993b).

Creating stand-off bumps on the underside of a poly-Si plate [Tang, 1990; Fan, 1989] and adding meniscus-shaping microstructures to the perimeter of the microstructure are mechanical means to help reduce sticking [Abe et al., 1995]. Fedder et al. (1992) used another mechanical approach to avoid stiction by temporarily stiffening the microstructures with polysilicon links. These very stiff structures are not affected by liquid surface tension forces, and the links are severed afterward with a high current pulse once the potentially destructive processing is complete. Yet another mechanical approach to avoid stiction involves the use of sacrificial supporting polymer columns. A portion of the sacrificial layer is substituted by polymer spacer material, spun-on after partial etch of the oxide glass. After completion of the oxide etch, the polymer spacer prevents stiction during evaporative drying. Finally, an isotropic oxygen plasma etches the polymer to release the structure [Mastrangelo and Saloka, 1993].

Ideally, to ensure high yields, contact between structural elements and the substrate should be avoided during processing. In a liquid environment, however, this may become impossible due to the large surface tension effects. Consequently, most solutions to the stiction problem involve reducing the surface tension of the final rinse solution by physico-chemical means. Lober et al. (1988a) for example, tried HF vapor, and Guckel et al. (1989, 1990) used freeze-drying of water/methanol mixtures. Freezing and sublimating the rinse fluid in a low-pressure environment gives improved results by circumventing the liquid phase. Takeshima et al. (1991) used t-butyl alcohol freeze-drying. Since the freezing point of this alcohol lies at 25.6°C, it is possible to perform freeze-drying without special cooling equipment. Supercritical drying results in increased microstructure yields [Mulhern et al., 1993]. With this technique, the rinse fluid is displaced with a liquid that can be driven into a supercritical phase under high pressure. This supercritical phase does not exhibit surface tension, which allows for the drying of released microstructures without sticking. Typically, CO_2 at 35°C and 1100 psi is used.

Kozlowski et al. (1995) substituted HF in successive exchange steps by the monomer divinylbenzene to fabricate very thin (500-nm) micromachined polysilicon bridges and cantilevers. The monomer was polymerized under UV light at room temperature and was removed in an oxygen plasma. Analog Devices applied a proprietary technique involving only standard IC process technology in the fabrication of a micro-accelerometer to eliminate yield losses due to stiction [Core et al., 1993].

In-Use Stiction. Stiction remains a fundamental reliability issue due to contact with adjacent surfaces after release. Stiction-free passivation that can survive the packaging temperature cycle is not known at present [Howe, 1995]. Attempted solutions are summarized below.

Adhesive energy may be minimized in a variety of ways, for example by forming bumps on surfaces (see above) or roughening of opposite surface plates [Alley et al., 1993]. Also, self-assembled monolayer coatings have been shown to reduce surface adhesion and to be effective at friction reduction in bearings at the same time [Alley et al., 1992]. Making the silicon surface very hydrophobic and coating it with diamond-like carbon are other potential solutions for preventing postrelease stiction of polysilicon microstructures. Man et al. (1996) eliminate postrelease adhesion in microstructures by using a thin conformal fluorocarbon film. The film eliminates the adhesion of polysilicon beams up to 230 μm long even after direct immersion in water. The film withstands temperatures as high as 400°C, and wear tests show that the film remains effective after 108 contact cycles. Along the same line, ammonium fluoride-treated Si surfaces are thought to be superior to the HF-treated surfaces due to a more complete hydrogen termination, which leads to a cleaner hydrophobic surface [Houston et al., 1995]. Gogoi et al. (1995) introduced electromagnetic pulses for postprocessing release of stuck microstructures.

3.16.3 Control of Film Stress

After reviewing typical surface micromachining process sequences, we are ready to investigate some of the mechanical properties of the fabricated mechanical members. Consider a straight lateral resonator as shown in Figure 3.97A. Electrostatic force is applied by a drive comb to a suspended shuttle. Its motion is detected capacitively by a sense comb. Many applications require tight control over the resonant frequency f_0. An analytical approximation for f_0 of this type of resonator can be deduced from Rayleigh's method:

$$f_0 = \frac{1}{2\pi} \sqrt{\frac{4EtW^3}{mL^3} + \frac{24\sigma_r tW}{5mL}} \tag{3.63}$$

where E represents the Young's modulus of polysilicon; L, W, and t are the length, width, and thickness of the flexures; and m stands for the mass of the suspended shuttle (of the order of 10^{-9} kg or less). For typical values ($L = 150$ μm and $W = t = 2$ μm) and a small tensile residual stress, the resonant frequency f_0 is between 10 and 100 kHz [Howe, 1995]. For typical values of L/W, the stress term in Equation (3.63) dominates the bending term. Any residual stress σ_r obviously will affect the resonant frequency. Consequently, stress and stress gradients represent critical stages for microstructural design.

One of the many challenges of any surface-micromachining process is to control the intrinsic stresses in the deposited films. Several techniques can be used to control film stress. Some we detailed before, but we list them again for completeness:

- Large-grained poly-Si films, deposited at around 625°C, have a columnar structure and are always compressive. Compressive stress can cause buckling in constrained structures. Annealing at high temperatures, between 900 and 1150°C, in nitrogen significantly reduces the compressive stress in as-deposited poly-Si [Guckel et al., 1985; Howe and Muller, 1983] and can eliminate stress gradients. No significant structural changes occur when annealing a columnar poly-Si film. The annealing process is not without danger in cases where active electronics are integrated on the same chip. Rapid thermal annealing might provide a solution (see IC Compatibility, below).
- Undoped poly-Si films are in an amorphous state when deposited at 580°C or lower. The stress and the structure of this low-temperature material depend on temperature and partial pressure of the silane. A low-temperature anneal leads to a fine-grained poly-Si with low tensile stress and very smooth surface texture [Guckel et al., 1989]. Tensile rather than compressive films are a necessity if lateral dimensions of clamped structures are not to be restricted by compressive buckling. Conducting regions are formed in this case by ion implantation. Fine-grained and large-grained poly-Si are compared in more detail further below.
- Phosphorous [Murarka and Retajczyk, 1983; Orpana and Korhonen, 1991; Lin et al., 1993], boron [Orpana and Korhonen, 1991; Choi and Hearn, 1984; Ding and Ko, 1991], arsenic [Orpana and Korhonen, 1991], and carbon [Hendriks et al., 1983] doping have all been shown to affect the state

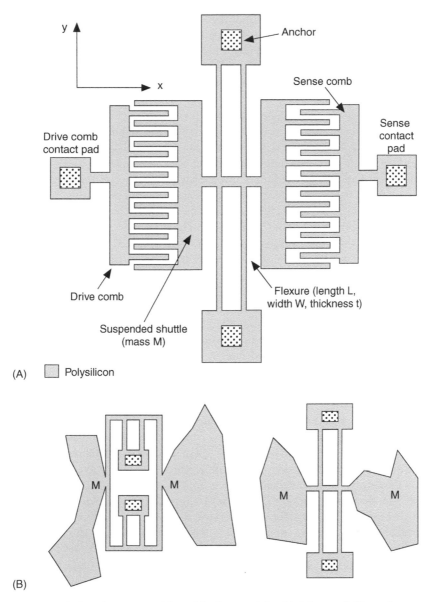

FIGURE 3.97 Layout of a lateral resonator with straight flexures. (A) Folded flexures (left) to release stress are compared with straight flexures (right). (B) M is shuttle mass.

of residual stress in poly-Si films. In the case of single-crystal Si, to compensate for strain induced by dopants one can implant with atoms with the opposite atomic radius vs. silicon. Similar approaches would most likely be effective for poly-Si as well.

- Tang et al. (1990) developed a technique that sandwiches a poly-Si structural layer between a top and bottom layer of PSG and lets the high temperature anneal drive in the phosphorous symmetrically, producing low stress poly-Si with a negligible stress gradient.

- Another stress-reduction method is to vary the materials' composition, something readily done in CVD processes. An example of this method is the Si enrichment of Si_3N_4, which reduces the tensile stress [Sekimoto et al., 1982; Guckel et al., 1986].

- During plasma-assisted film deposition processes, stress can be influenced dramatically. In a physical deposition process such as sputtering, stress control involves varying gas pressure and substrate bias. In plasma-enhanced chemical vapor deposition (PECVD), the RF power, through increased ion-bombardment, influences stress. In this way, the stress in a thin film starts out tensile, decreases as the power increases, and finally becomes compressive with further RF power increase. PECVD equipment manufacturers are also working to build stress-control capabilities into new equipment by controlling plasma frequency. In CVD, stress control involves all types of temperature treatment programs.

- A clever mechanical design might facilitate structural stress relief [Tang et al., 1989]. By folding the flexures in the lateral resonator in Figure 3.97B, and by the overall structural symmetry, the relaxation of residual polysilicon stress is possible without structural distortion. By folding the flexures, the resonant frequency, f_0, becomes independent of σ_r (see Equation [3.63]). The springs in the resonator structure provide freedom of travel along the direction of the comb-finger motions (x) while restraining the structure from moving sideways (y), thus preventing the comb fingers from shorting out the drive electrodes. In this design, the spring constant along the y direction must be higher than along the x direction (i.e., $k_y \gg k_x$). The suspension should allow for the relief of the built-in stress of the polysilicon film and the axial stress induced by large vibrational amplitudes. The folded-beam suspensions meet both criteria. They enable large deflections in the x direction (perpendicular to the length of the beams) while providing stiffness in the y direction (along the length of the beams). Furthermore, the only anchor points (see Figure 3.97B) for the whole structure reside near the center, thus allowing the parallel beams to expand or contract in the y direction, relieving most of the built-in and induced stress [Tang, 1990]. Tang also modeled and built spiral and serpentine springs supporting torsional resonant plates. An advantage of the torsional resonant structures is that they are anchored only at the center, enabling radial relaxation of the built-in stress in the polysilicon film [Tang, 1990]. For some applications, the design approach with folded flexures is an attractive way to eliminate residual stress. However, a penalty for using flexures is increased susceptibility to out-of-plane warpage from residual stress gradients through the thickness of the polysilicon microstructure [Howe, 1995].

- Corrugated structural members, invented by Jerman for bulk micromachined sensors [Jerman, 1990], also reduce stress effectively. In the case of a single-crystal Si membrane, stress may be reduced by a factor of 1000 to 10,000 [Spiering et al., 1991]. One of the applications of such corrugated structures is the decoupling of a mechanical sensor from its encapsulation. By reducing the influence of temperature changes and packaging stress [Spiering et al., 1991; Offereins et al., 1991], thermal stress alone can be reduced by a factor of 120 [Spiering et al., 1993]. Besides stress release, corrugated structures enable much larger deflections than do similar planar structures. This type of structural stress release was studied in some detail for single-crystal Si [Zhang and Wise, 1994], polyimide [Mullem et al., 1991], and LPCVD silicon nitride membranes [Scheeper et al., 1994]; however, the quantitative influence of corrugated poly-Si structures still requires investigation. Figure 3.98 illustrates a fabrication sequence for a polyimide corrugated structure. The sacrificial Al in step 4 may be etched away by a mixture of phosphoric acidic, acetic acid, and nitric acid (PAN, see Table 3.17).

3.16.4 Dimensional Uncertainties

The often-expressed concerns about run-to-run variability in material properties of polysilicon or other surface micromachined materials are somewhat misplaced, Howe points out [Howe, 1995]. He contrasts the relatively large dimensional uncertainties inherent to any lithography technique with poly-Si quality factors of up to 100,000 and long-term (>3 years) resonator frequency variation of less than 0.02 Hz. We follow his calculations here to prove the relative importance of the dimensional uncertainties. The shuttle mass m of a resonator as shown in Figure 3.97 is proportional to the thickness (t) of the polysilicon film, and neglecting the residual stress term, Equation (3.63) reduces to:

$$f_0 \propto \left(\frac{W}{L} \right)^{\frac{3}{2}} \tag{3.64}$$

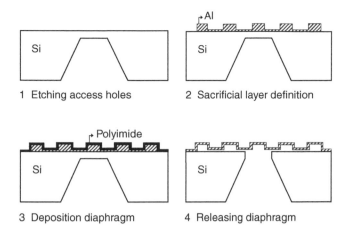

FIGURE 3.98 Schematic view of the fabrication process of a polyimide corrugated diaphragm. (Reprinted with permission from van Mullem, C.J. et al. [1991] "Large Deflection Performance of Surface Micromachined Corrugated Diaphragms," in *6th International Conference on Solid-State Sensors and Actuators [Transducers '91]*, San Francisco, pp. 1014–17.)

In case the residual stress term dominates in Equation (3.63), the resonant frequency is expressed as:

$$f_0 \propto \left(\frac{W}{L} \right)^{\frac{1}{2}} \tag{3.65}$$

The width-to-length ratio is affected by systematic and random variations in the masking and etching of the microstructural polysilicon. For 2-μm thick structural polysilicon patterned by a wafer stepper and etched with a reactive-ion etcher, a reasonable estimate for the variation in linear dimension of etched features Δ is about 0.2 μm (10% relative tolerance).

From Equation (3.64), the variation Δ in lateral dimensions will result in an uncertainty δf_0 in the lateral frequency of:

$$\frac{\delta f_0}{f_0} \approx \frac{3}{2} \left(\frac{\Delta}{W} \right) \tag{3.66}$$

for a case where the residual stress can be ignored. With a nominal flexure width of $W = 2\,\mu$m, the resulting uncertainty in resonant frequency is 15%. For the stress-dominated case, Equation (3.65) indicates that the uncertainty is:

$$\frac{\delta f_0}{f_0} \approx \frac{1}{2} \left(\frac{\Delta}{W} \right) \tag{3.67}$$

The same 2 μm wide flexure would then lead to a 5% uncertainty in resonant frequency.

Interestingly, the stress-free case exhibits the most significant variation in the resonant frequency. In either case, resonant frequencies must be set by some postfabrication frequency trimming or other adjustment.

3.16.5 Sealing Processes in Surface Micromachining

Sealing cavities to hermetically enclose sensor structures is a significant attribute of surface micromachining. Sealing cavities in surface micromachines often embodies an integral part of the overall fabrication process and presents a desirable chip-level, batch packaging technique. The resulting surface packages (microshells) are much smaller than typical bulk micromachined ones. A sealed cavity made with epi-micromachining is shown in Figure 3.107, and many more details on packaging using surface micromachining techniques are provided in Chapter 8 of Madou (2002).

3.16.6   IC Compatibility

Putting detection and signal conditioning circuits right next to the sensing element enhances the performance of the sensing system, especially when dealing with high-impedance sensors. A key benefit of surface micromachining, besides small device size and single-sided wafer processing, is its compatibility with CMOS processing. IC compatibility implies simplicity and economy of manufacturing. In the examples at the end of this chapter, we will discuss how Analog Devices used a mature 4-μm BICMOS process to integrate electronics with a surface micromachined accelerometer.

To develop an appreciation of integration issues involved in combining a CMOS line with surface micromachining, we highlight Yun's (1992) comparison of CMOS circuitry and surface micromachining processes in Table 3.18a. Surface micromachining processes are similar to IC processes in several aspects. Both processes use similar materials, lithography, and etching techniques. CMOS processes involve at least ten lithography steps in which lateral small feature size plays an important role. Some processing steps, such as gate and contact patterning, are critical to the functionality and performance of the CMOS circuits. Furthermore, each processing step is strongly correlated with other steps. A change in any one of the processing steps will lead to modifications in a number of other steps in the process. In contrast, surface micromachining is relatively simple. It usually consists of two to six masks, and the feature sizes are much larger. The critical processing steps, such as structural poly-Si, often are self-aligned, which eliminates lithographic alignment. The CMOS process is mature, quite generic, and fine tuned, while surface micromachining strongly depends on the application and still needs maturation.

Table 3.18b presents the critical temperatures associated with the LPCVD deposition of a variety of frequently used materials in surface micromachining. Polysilicon is used for structural layers, and thermal SiO$_2$, LPCVD SiO$_2$, and PSG are used as sacrificial layers; silicon nitride is used for passivation. The highest-temperature process in Table 3.18b is 1050°C and is associated with the annealing step to release stress in the polysilicon layers. Doped polysilicon films deposited by LPCVD under conventional IC conditions usually are in a state of compression that can cause mechanically constrained structures such as bridges and diaphragms to buckle. At about 1000°C, the annealing step above promotes crystallite growth and reduces the strain. If a polysilicon microstructure is built after the CMOS active electronics have been implemented (a so-called post-CMOS procedure), temperatures above 950°C must be avoided, as junction migration will take place at those temperatures. This is especially true with devices incorporating

TABLE 3.18   Surface Micromachining and CMOS

A. Comparison of CMOS and surface micromachining		
	CMOS	Surface micromachining
Common features		Silicon based processes, same materials, same etching principles
Process flow	Standard	Application specific
Vertical dimension	~1 μm	~1–5 μm
Lateral dimension	<1 μm	2–10 μm
Complexity (no. masks)	>10	2–6

B. Critical process temperatures for microstructures		
	Temperature (°C)	Material
LP CVD deposition	450	Low temperature oxide (LTO)/PSG
LP CVD deposition	610	Low stress poly-Si
LP CVD deposition	650	Doped poly-Si
LP CVD deposition	800	Nitride
Annealing	950	PSG densification
	1050	Poly-Si stress annealing

Source: Reprinted with permission from Yun, W. (1992) A Surface Micromachined Accelerometer with Integrated CMOS Detection Circuitry, Ph.D. thesis, University of California, Berkeley.

shallow junctions where migration might be a problem at temperatures as low as 800°C. The degradation of the aluminum metallization presents a bigger problem. Aluminum typically is used as the interconnect material in the conventional CMOS process. At temperatures of 400 to 450°C, the aluminum metallization will start suffering. Anneal temperatures (densification of the PSG and stress anneal of the poly-Si) only account for some of the concerns; in general, several compatibility issues must be considered: (1) deposition and anneal temperatures, (2) passivation during micromachining etching steps, and (3) surface topography.

Yun (1992) compared three possible approaches to building integrated microdynamic systems: pre-, mixed, and post-CMOS microstructural processes. He concluded that building up the microstructures after implementation of the active electronics offers the best results.

To avoid problems with microstructure topography, which commonly includes step heights of 2 to 3 μm, the CMOS module is fabricated before the microstructure module in a post-CMOS process. Although this solves topography problems, it introduces constraints on the CMOS. In a post-CMOS process, the electronic circuitry is passivated to protect it from the subsequent micromachining processes. The standard IC processing may be performed at a regular IC foundry, while the surface micromachining occurs as an add-on in a specialized sensor fabrication facility. LPCVD silicon nitride (deposited at 800°C, see Table 3.18b) is stable in HF solutions and is the preferred passivation layer for the IC during the long release etching step. PECVD nitride can be deposited at around 320°C, but it displays relatively poor step coverage, while pinholes in the film allow HF to diffuse through and react with the oxide underneath. LPCVD nitride is conformably deposited. While it shows fewer pinholes, circuitry needs to be able to survive the 800°C deposition temperature.

Aluminum metallization must be replaced by another interconnect scheme to raise the post-CMOS temperature ceiling higher than 450°C. Tungsten — which is refractory, shows low resistivity, and has a thermal expansion coefficient matching that of Si — is an obvious choice. One problem with tungsten metallization is that tungsten reacts with silicon at about 600°C to form WSi_2, implying the need for a diffusion barrier. The process sequence for the tungsten metallization developed at Berkeley is shown in Figure 3.99. A diffusion barrier consisting of $TiSi_2$ and TiN is used. The TiN film forms during a 30 s sintering step to 600°C in N_2. Rapid thermal annealing, with its reduced time at high temperatures (10 s to 2 min) and high ramp rates (~150°C/s), allows very precise process control as well as a dramatic reduction of thermal budgets, reducing duress for the active on-chip electronics. Titanium silicide is formed at the interface of titanium and silicon while titanium nitride forms simultaneously at the exposed surface of the titanium film. The $TiSi_2$/TiN forms a good diffusion barrier against the formation of WSi_2 and simultaneously provides an adhesion and contact layer for the W metallization.

To avoid the junction migration in a post-CMOS process, rapid thermal annealing is used for both the PSG densification and polysilicon stress anneal: 950°C for 30 s for the PSG densification and 1000°C for 60 s for the stress anneal of the poly-Si. Alternatively, fine-grained polysilicon can be used, as it yields a controlled tensile strain with low-temperature annealing [Guckel et al., 1988]. Despite some advantages, the post-CMOS process with tungsten metallization is not the preferred implementation. Hillock formation in the W lines during annealing and high contact resistance remain problems [Howe, 1995]. The finely tuned CMOS fabrication sequence may also be affected by the heavily doped structural and sacrificial layers. Most importantly, the use of tungsten for circuit interconnect is not consistent with mainstream IC technologies, where aluminum interconnects predominate. Given that IC manufacturers have already invested enormous resources into the development of multilevel aluminum interconnect technologies, and further given the inferior resistivity of tungsten compared to aluminum, the described tungsten-based post-CMOS process, although useful as a demonstration tool, is not likely to be adopted in industry [Nguyen, 1998].

The mixed CMOS/micromachining approach implements a processing arrangement that puts the processes in a sequence to minimize performance degradation for both electronic and mechanical components. According to Yun, this requires significant modifications to the CMOS fabrication sequence. Nevertheless, Analog Devices relied on such an interleaved process sequence to build the first commercially available integrated micro-accelerometer (see Example 1 in "Polysilicon Surface Micromachining

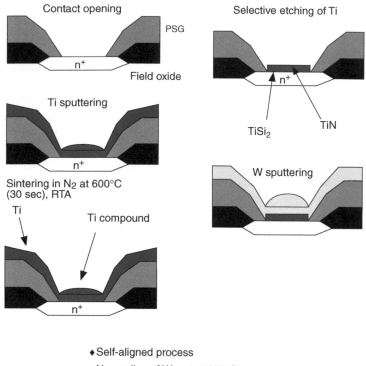

FIGURE 3.99   Tungsten metallization process in a modified CMOS process. (Adapted with permission from Yun, W. [1992] A Surface Micromachined Accelerometer with Integrated CMOS Detection Circuitry, Ph.D. thesis, University of California, Berkeley.)

Examples"). The modifications required on a standard BICMOS line are minimal: to facilitate integration of the IC and to surface micromachine the thickness of deposited microstructural films, the line is limited to 1 to 4 μm. Relatively deep junctions permit thermal processing for the sensor poly-Si anneal, and interconnections to the sensor are made only via n^+ underpasses. Therefore, no metallization is present in the sensor area. This industrial solution remains truer to the traditional IC process experience than the post-CMOS procedure. Howe recently detailed another example of such a mixed process [Howe, 1995]. In Howe's scenario, the micromachining sequence is inserted after the completion of the electronic structures but prior to contact etching or aluminum metallization. By limiting the polysilicon annealing to 7 min at 900°C, only minor dopant redistribution is expected. Contact and metallization lithography and etching are more complex now due to the severe topography of the poly-Si microstructural elements. These processes, however, have their own associated limitations: mixed processes often require longer, more expensive development periods for new product lines.

The pre-CMOS approach is to fabricate microstructural elements before any CMOS process steps. At first glance, this seems like an attractive approach, as no major modifications would be needed for process integration. However, due to the vertical dimensions of microstructures, step coverage is a problem for the interconnection between the sensor and the circuitry (the latest approach introduced by Howe faces the same dilemma). Passivation of the microstructure during the CMOS process can also become problematic. Furthermore, the fine-tuned CMOS fabrication sequences, such as gate oxidation, can be affected by the heavily doped structural layers. Consequently, this approach is used only for special applications

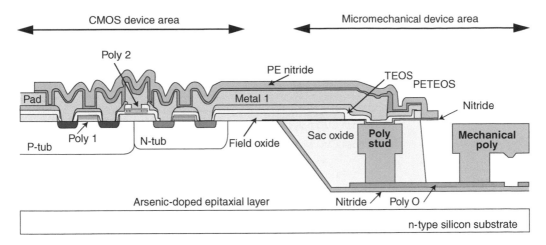

FIGURE 3.100 A schematic cross-section of the embedded micromechanics approach to Sandia's CMOS/MEMS integration.

[Yun, 1992]. Precircuit processes may place limitations on foundry-based fabrication schemes, since circuit foundries may be sensitive to contamination from MEMS foundries.

A unique pre-CMOS process (micromechanics-first) that overcomes the planarity issues of building the MEMS before the CMOS has been developed at Sandia National Labs [Smith et al., 1995]. In this approach, micromechanical devices are fabricated in a trench etched in a silicon epilayer. After the mechanical components are complete, the trench is filled with oxide, planarized using chemical-mechanical polishing (CMP), and sealed with a nitride membrane. The flat wafer with the embedded micromechanical devices is then processed by means of conventional CMOS processing. Additional steps are added at the end of the CMOS process to expose and release the embedded micromechanical devices. A cross-section of a structure made with this technology is shown in Figure 3.100.

The SPIE "Smart Structures and Materials 1996" meeting in San Diego, CA, had two complete sessions dedicated to the crucial issue of integrating electronics with polysilicon surface micromachining [Varadan, V.K., and McWhorter, P.J., 1996]. Research aimed at achieving a truly modular merged circuits + microstructures technology is still ongoing.

Since its inception in December 1992, the Multi-User MEMS Process (MUMPs®) has quickly become a well established commercial program that provides customers with cost-effective access to surface micromachining for prototyping activities. It caused many researchers to focus on design and testing of new MEMS concepts rather than building yet another clean room. MUMPs® was approaching its 40th run in December 2000 (<http://www.memscap.com/memsrus/crmumps.html>). Another source for surface micromachining prototyping is at Sandia, with SUMMIT.

Some surface micromachining tutorials on the web may be found at the following URLs:

- http://mems.cwru.edu/shortcourse/partII_2.html (from Case Western Reserve University).
- <http://www.memsrus.com/svcssurf.html (from Cronos)>
- Bill Trimmer's homepage, <http://home.earthlink.net/~trimmerw/mems/tutorials.html>.

3.17 Poly-Si Surface Micromachining Modifications

3.17.1 Porous Poly-Si

In the first part of this chapter, we discussed the transformation of single-crystal Si into a porous material with porosity and pore sizes determined by the current density, type and concentration of the dopant, and the hydrofluoric acid concentration. A transition from pore formation to electropolishing is reached

1. Deposit CVD silicon nitride on silicon wafer.

2. Pattern and plasma-etch silicon nitride.

3. Deposit CVD polysilicon.

4. Deposit CVD silicon nitride.

5. Pattern and plasma-etch nitride and polysilicon.

FIGURE 3.101 Masking and process sequence preparing a wafer for laterally grown porous polysilicon. (Reprinted with permission from Field, L. [1987] "Low-Stress Nitrides for Use in Electronic Devices," in *EECS/ERL Res. Summary,* University of California, Berkeley pp. 42–43.)

by raising the current density and/or by lowering the hydrofluoric acid concentration [Memming and Schwandt, 1966; Zhang et al., 1989]. Porous silicon can also be formed under similar conditions from LPCVD poly-Si [Anderson et al., 1994]. In this case, pores roughly follow the grain boundaries of the polysilicon. Figure 3.101 illustrates the masking and process sequence to prepare a wafer to make thin layers of porous Si between two insulating layers of low stress silicon nitride [Field, 1987]. After the process steps outlined in Figure 3.101, the wafer is put in a Teflon® test fixture, protecting the back from HF attack. An electrical contact is established on the back of the wafer, and a potential is applied with respect to a Pt-wire counterelectrode immersed in the same HF solution. Electrolytes consisting of 5 to 49% wt HF and current densities from 0.1 to 50 A cm^{-2} are used. The advance of a pore-etching front, growing parallel to the wafer surface, may be monitored using a line-width measurement tool. The highest observed rate of porous-silicon formation is 15 μm min^{-1} (in 25%wt HF). In the electropolishing regime, at the highest currents, the etch rate is diffusion limited but the reaction is controlled by surface reaction kinetics in the porous Si growth regime.

By changing the conditions from pore formation to electropolishing and back to porous Si, an enclosed chamber may be formed with porous poly-Si walls (plugs) and "floor and ceiling" silicon nitride layers. Sealing of the cavities by clogging the microporous poly-Si was attempted by room temperature oxidation in air and in a H_2O_2 solution. Leakage through the porous plug persisted after those room temperature oxidation treatments, but with a Ag deposition from a 400 mM $AgNO_3$ solution and subsequent atmospheric tarnishing (48 hr, Ag_2S), the chambers appeared to be sealed, as determined from the lack of penetration by methanol. This technology might open possibilities for filling cavities with liquids and gases under low-temperature conditions. The chamber provided with a porous plug might also make a suitable on-chip electrochemical reference electrolyte reservoir.

Hydrofluoric acid can penetrate thin layers of poly-Si either at foreign particle inclusion sites or at other critical film defects such as grain boundaries (see above). This way, the HF can etch underlying oxide layers creating, for example, circular regions of freestanding poly-Si, so-called blisters. The poly-Si permeability associated with blistering of poly-Si films has been applied successfully by Judy et al. (1993a, 1993b) to produce thin-shelled hollow beam electrostatic resonators from thin poly-Si films deposited onto PSG. The possible advantage of using these hollowed structures is to obtain a yet higher resonator quality factor Q. The devices were made in such a way that the 0.3 μm thick undoped poly-Si completely encased a PSG core. After annealing, the structures were placed in HF, which penetrated the poly-Si shell and dissolved away the PSG, eliminating the need for etch windows. It was not possible to discern the actual pathways through the poly-Si using TEM.

Lebouitz et al. (1995) applied permeable polysilicon etch-access windows to increase the speed of creating microshells for packaging surface micromachined components. After etching the PSG through the many permeable Si windows, the shell is sealed with 0.8 μm of low-stress LPCVD nitride.

Porous Si can be used very effectively as a sacrificial layer both in poly-Si surface micromachining and in SOI micromachining (see Table 3.17) [Gennissen et al., 1995]. The high surface area of the porous Si results in rapid etching in KOH at room temperature. The porous layer can be formed directly under a deposited layer or under the epilayer in SOI surface micromachining.

Silicon is not the only material that can be converted into a porous sponge-like material. Most semiconductor materials can be modified this way. For example, high-aspect-ratio structures were successfully made in porous anodic aluminum oxide [Routkevitch et al., 2000]. The latter enables a class of MEMS applications such as high-temperature gas microsensors, vacuum microelectronics, RF-MEMS, nozzles, filters, membranes, and so forth [Mardilovich et al., 2000].

3.17.2 Hinged Polysilicon

One way to achieve high vertical structures with surface micromachining is to build large, flat structures horizontally and then rotate them on a hinge to an upright position. Pister et al. [Burgett et al., 1992; Ross and Pister, 1994; Lin et al., 1995; Yeh et al., 1994; Lin et al., 1996] developed the poly-Si hinges shown in Figure 3.102A. On these hinges, long structural poly-Si features (1 mm and beyond) can be rotated out of the plane of a substrate. To make the hinged structures, a 2 μm thick PSG layer (PSG-1) is deposited on the Si substrate as the sacrificial material, followed by the deposition of the first polysilicon layer (2-μm thick poly-1). This structural layer of polysilicon is patterned by photolithography and dry etching to form the desired structural elements, including hinge pins to rotate them. Following the deposition and patterning of poly-1, another layer of sacrificial material (PSG-2) of 0.5-μm thickness is deposited. Contacts are made through both PSG layers to the Si substrate, and a second layer of polysilicon is deposited and patterned (poly-2), forming a staple to hold the first polysilicon layer hinge to the surface. The first and second layers of poly are separated everywhere by PSG-2 to allow the first polysilicon layer to freely rotate off the wafer surface when the PSG is removed in a sacrificial etch. After the sacrificial etch, the structures are rotated in their respective positions. This is accomplished in an electrical probe station by skillfully manipulating the movable parts with the probe needles. Once the components are in position, high friction in the hinges tends to keep them in the same position. To obtain more precise and stable control of position, additional hinges and supports are incorporated. To provide electrical contact to the vertical poly-Si structures, one can rely on the mechanical contact in the hinges, or poly-Si beams (cables) can be attached from the vertical structure to the substrate.

Pister's research team made a wide variety of hinged microstructures, including hot wire anemometers, a box dynamometer to measure forces exerted by embryonic tissue, a parallel plate gripper [Burgett et al., 1992], a micro-windmill [Ross and Pister, 1994], a micro optical bench (MOB) for free-space integrated optics [Lin et al., 1995], and a standard CMOS single piezoresistive sensor to quantify the single heart cell contractile forces of a rat [Lin et al., 1996]. One example from this group's efforts is illustrated in Figure 3.102B, showing an SEM photograph of an edge-emitting laser diode shining light onto a collimating micro-Fresnel lens [Lin et al., 1995]. The micro-Fresnel lens in the SEM photo is surface micromachined in the plane and erected on a polysilicon hinge. The lens has a diameter of 280 μm. Alignment plates at the front and the back sides of the laser are used for height adjustment of the laser spot so that the emitting spot falls exactly onto the optical axis of the micro-Fresnel lens. After assembly, the laser is electrically contacted by silver epoxy. Although this hardly outlines standard IC manufacturing practices, excellent collimating ability for the Fresnel lenses has been achieved. The eventual goal of this work is an optical bench in which micro lenses, mirrors, gratings, and other optical components are prealigned in the mask layout stage using

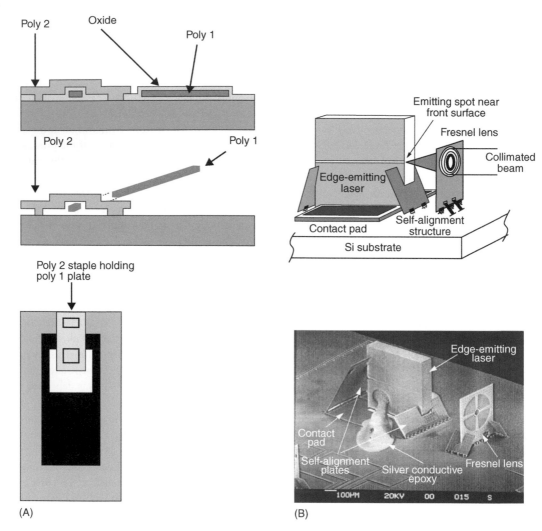

FIGURE 3.102   Microfabricated hinges. (A) Cross-section, side view, and top view of a single-hinged plate before and after the sacrificial etch. (B) Schematic (top) and SEM micrograph of the self-aligned hybrid integration of an edge-emitting laser with a micro-Fresnel lens. (Reprinted with permission from Lin, L.Y. et al. [1995] "Micromachined Integrated Optics for Free-Space Interconnections," in *Proceedings: IEEE Micro Electro Mechanical Systems [MEMS '95]*, Amsterdam, Netherlands, pp. 77–82.)

computer-aided design. Additional fine adjustment would be achieved by on-chip microactuators and micropositioners such as rotational and translational stages. Erecting these poly-Si structures with the probes of an electrical probe station or, occasionally, assembly by chance in the HF etch or deionized water rinse represent too complicated or too unreliable postrelease assembly methods for commercial acceptance. At the University of Illinois (Urbana) the Micro Actuator, Sensors, and Systems Group, headed by Dr. Chang Liu, uses magnetic actuation to position a large array of hinged microstructures in parallel. A magnetic material, permalloy, is electroplated and integrated into the hinged microstructures. The result may be improved manufacturability and stability of these delicate structures [Yi and Liu, 1999].

   Friction in poly-Si joints, as made by Pister, is high because friction is proportional to the surface area (s^2) and becomes dominant over inertial forces (s^3) in the microdomain. Such joints are not suitable for microrobotic applications. Although attempts have been made to incorporate poly-Si hinges in such applications [Yeh et al., 1994], plastically deformable hinges make more sense for microrobot machinery

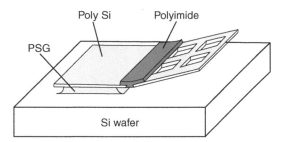

FIGURE 3.103   Flexible polyimide hinge and poly-Si plate (butterfly wing). (Reprinted with permission from Suzuki, K. et al. [1994] "Insect-Model Based Microrobot with Elastic Hinges," *J. Microelectromech. Syst.* 3, pp. 4–9.)

involving rotation of rigid components. Noting that the external skeleton of insects incorporates hard cuticles connected by elastic hinges, Suzuki et al. (1994) fabricated rigid poly-Si plates ($E = 140\,GPa$) connected by elastic polyimide hinges ($E = 3\,GPa$) as shown in Figure 3.103 (see also the section below on polyimide surface structures). Holes in the poly-Si plates shorten the PSG etch time compared with plates without holes. The plates without holes remain attached to the substrate, while the ones with holes are completely freed. Using electrostatic actuators, the structure shown in Figure 3.103 can be made to flap like the wing of a butterfly. By applying an AC voltage of 10 kHz, resonant vibration of such a flapping wing was observed [Suzuki et al., 1994].

Hoffman et al. (1995) demonstrated aluminum plastically deformable hinges on oxide movable thin plates. Oxide plates and Al hinges were etched free from a Si substrate by using XeF_2, a vapor phase etchant exhibiting excellent selectivity of Si over Al and oxide. According to the authors, this process, due to its excellent CMOS compatibility, might open the way to designing and fabricating sophisticated integrated CMOS-based sensors with rapid turnaround time.

3.17.3   Thick Polysilicon

Applying classical LPCVD to obtain poly-Si deposition is a slow process. For example, a layer of 10 μm typically requires a deposition time of 10 hr. Consequently, most micromachined structures are based on layer thicknesses in the 2 to 5 μm range. Basing their process on dichlorosilane (SiH_2Cl_2) chemistry, Lange et al. (1995) developed a CVD process in a vertical epitaxy batch reactor with deposition rates as high as 0.55 μm/min at 1000°C. The process yields acceptable deposition times for thicknesses in the 10 μm range (20 min). The highly columnar poly-Si films are deposited on sacrificial SiO_2 layers and exhibit low internal tensile stress, making them suitable for surface micromachining. The surface roughness comprises about 3% of the thickness, which might preclude some applications.

Kahn et al. (1996) made mechanical property test structures from thick undoped and in situ B-doped polysilicon films. The elastic modulus of the B-doped polysilicon films was determined as $150 \pm 30\,GPa$. The residual stress of as-deposited undoped thick polysilicon was determined as $200 \pm 10\,MPa$.

This "thick epi-polysilicon surface micromachining" has been used to fabricated accelerometers with a seismic mass thickness well beyond what regular micromachining can achieve (<http://www.sensorscan.com/sensoren/sensoren/accelerometer_1.htm>).

3.17.4   Milli-Scale Molded Polysilicon Structures

The assembly of tall three-dimensional features in the described hinged polysilicon approach is complicated by the manual assembly of fabricated microparts, which is similar to building a miniature boat in a bottle. Keller, while at the University of California, Berkeley, came up with an elegant alternative for building tall, high-aspect-ratio microstructures in a process that does not require postrelease assembly steps [Keller and Ferrari, 1994]. The technique involves deep dry etching of trenches in a Si substrate, deposition of sacrificial and structural materials in those trenches, and demolding of the deposited structural

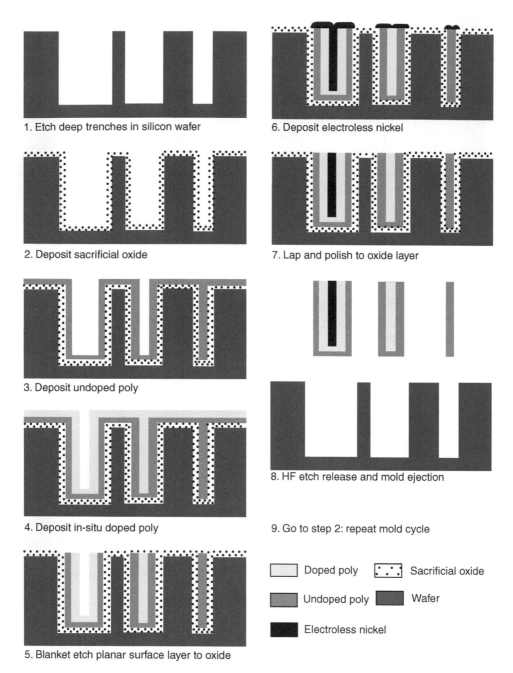

1. Etch deep trenches in silicon wafer

2. Deposit sacrificial oxide

3. Deposit undoped poly

4. Deposit in-situ doped poly

5. Blanket etch planar surface layer to oxide

6. Deposit electroless nickel

7. Lap and polish to oxide layer

8. HF etch release and mold ejection

9. Go to step 2: repeat mold cycle

Doped poly   Sacrificial oxide

Undoped poly   Wafer

Electroless nickel

FIGURE 3.104 Schematic illustration of HEXSIL process. The mold wafer may be part of an infinite loop. (Courtesy of Dr. C. Keller, MEMS Precision Instruments.)

materials by etching away the sacrificial materials. CVD processes can typically only deposit thin films (~1 to 2 µm) on flat surfaces. If, however, these surfaces are the opposing faces of deep narrow trenches, the growing films will merge to form solid beams. In this fashion, high-aspect-ratio structures that would normally be associated with LIGA now also can be made of CVD polysilicon. The procedure is illustrated in Figure 3.104. [Keller and Ferrari, 1994; Keller and Howe, 1995]. The first step is to etch deep trenches into a silicon wafer. The depth of the trenches equals the height of the desired beams and is limited to about 100 μm with aspect ratios of about 10 (say a 10 μm diameter hole with a depth of 100 μm). For trench etching, Keller uses a Cl_2 plasma etch with the following approximate etching conditions: flow rates of 200 sccm for He and 180 sccm for Cl_2, a working pressure of 425 mtorr, a power setting of 400 W, and an electrode gap of 0.8 cm. The etch rate for Si in this mode equals 1 μm/min. Thermal oxide and CVD oxide act as masks with 1 μm of oxide needed for each 20 μm of etch depth. Before the Cl_2 etch, a short 7 s SF_6 pre-etch removes any remaining native oxide in the mask openings. During the chlorine etch, a white sidewall passivating layer must be controlled to maintain perfect vertical sidewalls. After every 30 min of plasma etching, the wafers are submerged in a silicon isotropic etch long enough to remove the residue [Keller and Howe, 1995]. Beyond 100 μm, severe undercutting occurs, and the trench cross-section becomes sufficiently ellipsoidal to prevent molded parts from being pulled out. Advances in dry cryogenic etching are continually improving attainable etch depths, trench profiles, and minimum trench diameter. We can expect continuous improvements in the tolerances of this novel technique.

After plasma etching, an additional 1 μm of silicon is removed by an isotropic wet etch to obtain a smoother trench wall surface. Alternatively, to smooth sidewalls and bottoms of the trenches, a thermal wet oxide is grown and etched away. The sacrificial oxide in step 2 is made by CVD phosphosilicate glass (PSG at 450°C, 140 Å/min), CVD low-temperature oxide (LTO at 450°C), or CVD polysilicon (580°C, 65 Å/min). The latter is completely converted to SiO_2 by wet thermal oxidation at 1100°C. The PSG needs an additional reflowing and densifying anneal at 1000°C in nitrogen for 1 hr. This results in an etching rate of the sacrificial layer of ∼20 μm/min in 49% HF. The mold shown in Figure 3.104 displays three different trench widths and can be used to build integrated micromachines incorporating doped and undoped poly-Si parts as well as metal parts. The remaining volume of the narrowest trench after oxide deposition is filled completely with the first deposition of undoped polysilicon (poly 1) in step 3. The undoped poly will constitute the insulating regions in the micromachine. Undoped CVD polysilicon was formed in this case at 580°C, with a 100 sccm silane flow rate and a 300 mtorr reactor pressure, resulting in a deposition rate of 0.39 μm/hr. The deposited film under these conditions is amorphous or very fine grained. Generally speaking, CVD polysilicon films conform well to the underlying topography on the wafer and show good step coverage. With trenches of an aspect ratio in excess of 10, some thinning of the film on the sidewalls occurs. Since the narrowest trenches are completely filled in by the first deposition, they cannot accept material from later depositions. The trenches of intermediate width are lined with the first material and then completely filled in by the second deposition. In the case illustrated, the second deposition (step 4) consists of in situ doped poly-Si and forms the resistive region in the micromachine under construction. To prevent diffusion of P from the doped poly deposited on top of the narrow undoped beams, a blanket etch in step 5 is used to remove the doped surface layer prior to the anneal of the doped poly.

The third deposition, in step 6 of the example case, consists of electroless nickel plating on poly-Si surfaces but not on oxide surfaces and results in the conducting parts of the micromachine. By depositing structural layers in order of increasing conductivity, as done here, regions of different conductivity can be separated by regions of narrow trenches containing only nonconducting material. Lapping and polishing in step 7 with a 1-μm diamond abrasive in oil planarizes the top surface, readying it for HF etch release and mold ejection in step 8. Annealing of the polysilicon is required to relieve the stress before removing the parts from the wafer so they remain straight and flat. In step 8, the sacrificial oxide is dissolved in 49% HF. A surfactant such as Triton X100 is added to the etch solution to facilitate part ejection by reducing surface adhesion between the part and the mold. The parts are removed from the wafer, and the wafer may be returned to step 2 for another mold cycle.

An example micromachine, resulting from the described process, is the thermally actuated tweezers shown in Figure 3.105. These HEXSIL tweezers measure 4 mm long, 2 mm wide, and 80 μm tall. The thermal expansion beam to actuate the tweezers consists of the in situ doped poly-Si; the insulating parts are made from the undoped poly-Si material. Ni-filled poly-Si beams are used for the current supply leads. It is possible to combine the HEXSIL process with classical poly-Si micromachining, as illustrated in

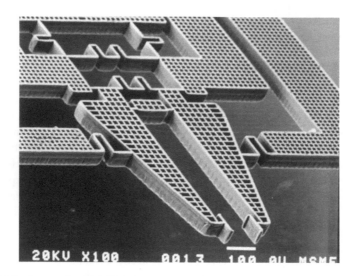

FIGURE 3.105   SEM micrograph of HEXSIL tweezers: 4 mm long, 2 mm wide, and 80 mm tall. Lead wires for current supply are made from Ni-filled poly-Si beams; in situ phosphorus-doped polysilicon provides the resistor part for actuation. The width of the beam is 8 μm: 2 μm poly-Si, 4 μm Ni, and 2 μm poly-Si. (Courtesy of Dr. C. Keller, MEMS Precision Instruments.)

Figure 3.106A, where HEXSIL forms a stiffening rib for a membrane filter fashioned by surface micromachining of a surface poly-Si layer. The surface poly-Si is deposited after HEXSIL.

A critical need in HEXSIL technology is controlled mold ejection. Keller et al. (1995) have experimented with HEXSIL-produced bimorphs, making the structure spring up after release. HEXSIL technology is now exploited commercially by Keller at MEMS Precision Instruments (<http://www.memspi.com>). Keller did his early HEXSIL work at UCB and later incorporated the Bosch deep RIE process to create tweezers structures similar to the one shown in Figure 3.105, which can now be fabricated in single-crystal Si (Figure 3.106B). MEMS Precision Instruments today makes microtweezers based on one of three processes: HEXSIL, deep RIE Bosch process, or Sandia's SUMMIT process.

3.18   Surface Micromachining Modifications Not Involving Polysilicon

3.18.1   SOI Surface Micromachining

3.18.1.1   Introduction

SOI is an exciting new approach to both IC chip making and MEMS. In SOI, bulk silicon wafers are replaced with wafers that have three layers: a thin surface layer of silicon (from a few hundred angstroms to several microns thick), an underlying layer of insulating material, and a support or "handle" silicon wafer. The insulating layer, usually made of silicon dioxide is called the "buried oxide," or "BOX," and is typically a few thousand angstroms thick. As we will see below, this thin buried oxide layer sandwiched between two layers of Si can be achieved in several different ways. When transistors are built within the thin top silicon layer, they switch signals faster (up to 10 GHz), run at lower voltages, and are much less vulnerable to signal noise from background cosmic ray particles. Furthermore, on an SOI wafer, each transistor is isolated from its neighbor by a complete layer of silicon dioxide, which makes them immune to "latch-up" problems, and they can be spaced closer together than transistors built on bulk silicon wafers. Building circuits on SOI thus allows for more compact chip designs, resulting in smaller IC devices (with higher production yield) and more chips per wafer (increasing fabrication productivity). Below, we present a short review of the three most popular methods for producing SOI wafers and then introduce some uses of SOI in MEMS.

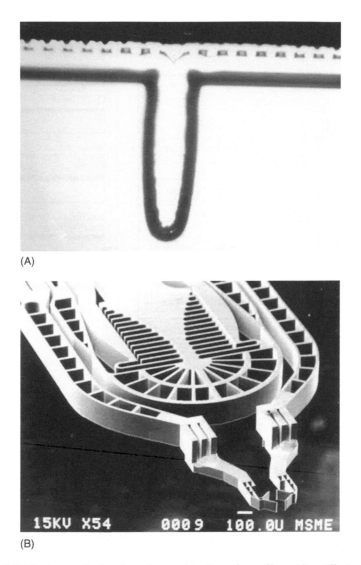

(A)

(B)

FIGURE 3.106 (A) SEM micrograph of surface micromachined membrane filter with a stiffening rib (50 mm high). Original magnification 1000×. (B) Single-crystal microtweezers made by the Bosch deep RIE process. (Courtesy of Dr. C. Keller, MEMS Precision Instruments.)

3.18.1.2   SOI Wafer Fabrication Techniques

Three major techniques currently are applied to produce SOI wafers (Inset 3.10): SIMOX (Separated by IMplanted OXygen), the Si fusion bonded (SFB) wafer technique, and zone-melt recrystallized (ZMR) polysilicon. With SIMOX, standard Si wafers are implanted with oxygen ions and then annealed at high temperatures (1300°C). The oxygen and silicon combine to form a silicon oxide layer beneath the silicon surface. The oxide layer's thickness and depth are controlled by varying the energy and dose of the implant and the anneal temperature. In some cases, a CVD process deposits additional epitaxial silicon on the top silicon layer. Attempts have also been made to implant nitrogen in Si to create abrupt etch stops. At high enough energies, the implanted nitrogen is buried 1/2 to 1 μm deep. At a high enough dose, the etching in that region stops. It is not necessary to implant the stoichiometric amount of nitrogen concentration; a dose lower by a factor of 2 to 3 suffices. After implantation, it is necessary to anneal the wafer because the implantation destroys the crystal structure at the surface of the wafer.

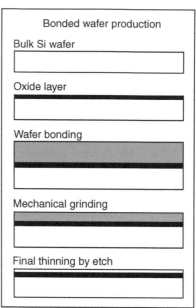

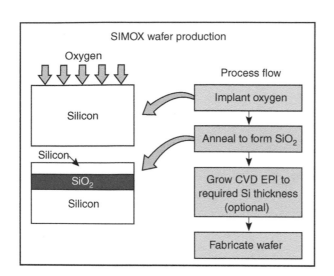

INSET 3.10   How SOI wafers are made.

The Si fusion bonded wafer process starts with an oxide layer (typically about 1 μm) grown on a standard Si wafer. That wafer is then bonded to another wafer, with the oxide sandwiched between. For the bonding, no mechanical pressure or other forces are applied. The sandwich is annealed at 1100°C for 2 hr in a nitrogen ambient, which creates a strong bond between the two wafers. One of the wafers is then ground to a thickness of a few microns using mechanical polishing and CMP.

A third process for making SOI structures is to recrystallize polysilicon (e.g., with a laser, an electron-beam, or a narrow strip heater) that has been deposited on an oxidized silicon wafer. This process is called zone melting recrystallization (ZMR) and is used primarily for local recrystallization but has not yet been explored much for use in micromachining applications.

The crystalline perfection of conventional silicon wafers in SFB and ZMR is completely maintained in the SOI layer, as the wafers do not suffer from implant-induced defects. By using plasma etching, wafers with a top Si layer thickness of as little as 1000 to 3000 Å with total thickness variations of less than 200 Å can be made [Dunn, 1993]. SOI top Si layers of 2 μm thick are more standard [Abe and Matlock, 1990]. A tremendous amount of effort is spent in the IC industry on controlling the SOI thin Si layer thickness, which benefits any narrow tolerance IC and micromachine.

Etched-back fusion-bonded Si wafers and SIMOX are employed extensively to build both ICs and micromachines, and both are commercially available. Two vendors are Silicon Genesis (SiGen) <http://www.sigen.com> and SOITEC <http://www.soitec.com/en/products/p_1.htm>.

3.18.1.3   SOI Use in MEMS

Overview. Kanda (1991) reviews different types of SOI wafers in terms of their micromachining and IC applications. Working with SOI wafers, he points out, offers several advantages over bulk Si wafers: fewer process steps are needed for feature isolation; parasitic capacitance is reduced; and power consumption is lowered. In the IC industry, SOI wafers are principally used for high-speed CMOS ICs (<10 GHz), smart power ICs (voltages up to 100 V), three-dimensional ICs, and radiation-hardened devices [Kuhn and Rhee, 1973].

An important use of silicon on insulator wafers is in the production of high-temperature sensors. Compared with p–n junction isolation, which is limited to about 125°C, much higher-temperature (up to 300°C) devices are possible based on the dielectric insulation of SOI. A wide variety of SOI surface micromachined structures have been explored including pressure sensors, accelerometers, torsional micromirrors,

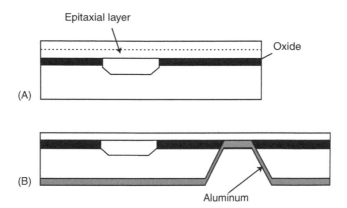

FIGURE 3.107 (A) A wafer sandwich after grind-and-polish step. (B) A wafer after electrochemical etch-back in KOH, buried oxide removal, and aluminum deposition. (From Noworolski, J. M. et al. [1995] "Fabrication of SOI Wafers with Buried Cavities Using Silicon Fusion Bending and Electrochemical Etchback," in *8th International Conference on Solid-State Sensors and Actuators [Transducers '95]*, Stockholm, pp. 71–74.)

light sources, and optical choppers [Diem et al., 1993; Noworolski et al., 1995]. Often, in those micromachining applications, SOI wafers are employed to produce an etch stop. The silicon fusion bonded (SFB) method offers the more versatile MEMS approach due to the associated potential for thicker single-crystal layers and the option of incorporating buried cavities, facilitating micromachine packaging. Sensors manufactured by means of SFB now are commercially available [Pourahmadi et al., 1992]. The SIMOX approach is less labor intensive and holds better membrane thickness control. An important expansion of the SOI technique is selective epitaxy. The latter enables a wide range of new mechanical structures and enables novel etch-stop methods [Gennissen et al., 1995] as well as electrical and/or thermal separation and independent optimization of active sensor and readout electronics [Bartek et al., 1995]. In most cases, SOI machining involves dry anisotropic etching to etch a pattern into the Si layer on top of the insulator. These structures then are released by etching the sacrificial buried SiO_2 insulator layer, which displays a thickness with very high reproducibility (400 ± 5 nm) and uniformity ($< \pm 5$ nm), especially in the case of SIMOX. Etched free cantilevers and membranes consist of single-crystalline silicon with thicknesses ranging from microns and submicrons (SIMOX) up to hundreds of microns (SFB).

The increasing use of SOI technology for fabricating MEMS devices has brought its own set of unique problems that must be overcome for applications where CD control is important. For instance, the etching of deep trenches in SOI wafers necessitates a fast etch process followed by a slow etch process as soon as the buried dioxide layer is reached [Maluf, 2000]. The main problem in plasma processing of SOI is the notching at the oxide interface, which results from the deflection of the incident ions by charging up the insulating oxide surface. The notch width increases with time, and it is here that etch uniformity begins to play an important role as overetching times to clear the wafer can become significant. For example, a $400\,\mu m$ deep etch with $+5\%$ uniformity requires a $>10\%$ overetch to ensure that all features etch to the required depth. Some areas of the wafer will thus be subject to a 40-μm overetch, and precise control of the profile during overetching becomes essential (<http://www.stsystems.com/soi.html>).

Creating Buried Cavities in Silicon by Silicon Fusion Bonded Micromachining. Silicon fusion bonding enables the formation of thick single-crystal layers with cavities built in. An example is shown in Figure 3.107 [Noworolski et al., 1995]. The device pictured involves two 4-in <100> wafers: a handle wafer and a wafer used for the SOI surface. The p-type (3 to 7 W-cm) handle wafer is thermally oxidized at 1100°C to obtain a 1-μm thick oxide. Thermal oxidation enables thicker oxides than the ones formed in SIMOX by ion implantation and avoids the potential implantation damage in the working material. To make a buried cavity, the oxide is patterned and etched. To produce yet deeper cavities, the Si handle wafer may be etched as well (as in Figure 3.107). In the case shown, the top wafer consists of the same p-type substrate material as the handle wafer with a 2 to 30 μm thick n-type epitaxial layer. The epitaxial layer determines

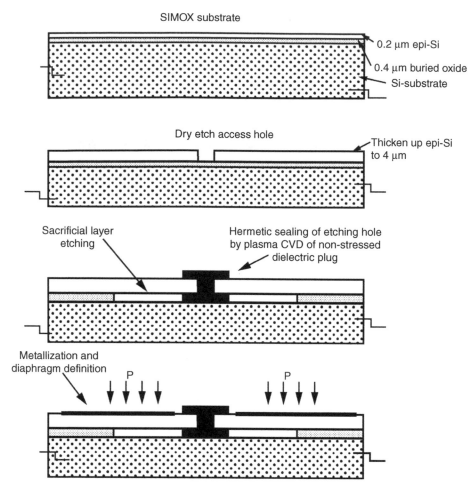

FIGURE 3.108 Process sequence of a SIMOX absolute capacitive pressure sensor by Diem et al. (Reprinted with permission from Diem, B. et al. [1993] "SOI (SIMOX) as a Substrate for Surface Micromachining of Single Crystalline Silicon Sensors and Actuators," in *7th International Conference on Solid-State Sensors and Actuators [Transducers '93],* Yokohama, pp. 233–36.)

the thickness of the final mechanical material. The epitaxial layer is fusion bonded to the cavity side of the handle wafer (2 hr at 1100°C). The top wafer is then partially thinned by grinding and polishing (Figure 3.107A). An insulator is deposited and patterned on the back side of the handle wafer to etch access holes to the insulator. After the insulator at the bottom of the etch hole is removed by a buffered oxide etch (BOE), aluminum is sputtered and sintered to make contact to the n-type epilayer for the electrochemical etch back of the remaining p-type material (Figure 3.107B). The final single-crystal silicon thickness is uniform to within $\pm 0.05\,\mu m$ (standard deviation) and does not require a costly high-accuracy polish step.

Draper Laboratory is using SOI processes in the development of inertial sensors, gyros, and accelerometers as an alternative to devices fabricated by the dissolved wafer process (see Example 1 in Wet Bulk Micromachining Examples). The main advantage is that the former consists of an all Si process rather than a Si/Pyrex sandwich [Greiff, 1995].

Fabricating Pressure Sensors in SIMOX Surface Micromachining. Both capacitive and piezoresistive pressure sensors were microfabricated from SIMOX wafers [Diem et al., 1993]. Figure 3.108 illustrates the process sequence by Diem et al. (1993) for fabricating an absolute capacitive pressure sensor. The $0.2\,\mu m$ silicon surface layer of the SIMOX wafer is thickened with doped epi-Si to $4\,\mu m$. An access hole is RIE etched in the Si layer, and vacuum cavity and electrode gap are obtained by etching the SiO_2 buried layer.

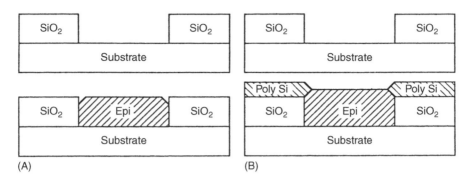

FIGURE 3.109 Selective deposition of epitaxial silicon. (A) Selective deposition of epi Si on Si in SiO$_2$ windows. (B) Simultaneous deposition of epi Si on Si, and poly Si on SiO$_2$. (Reprinted with permission from Wolf, S., and Tauber, R. N. [1987] *Silicon Processing for the VLSI Era*, Sunset Beach, Lattice Press.)

Since the buried thick oxide layer exhibits a very high reproducibility and homogeneity over the whole wafer (0.4 μm ± 5 nm), the resulting vacuum cavity and electrode gap after etching also are very well controlled. The small gap results in relatively high capacitance values between the free membrane and bulk substrate (20 pF/mm^2). Diaphragm diameter, controlled by the SiO$_2$ etching, is up to several hundred micrometers (±2 μm). The etching hole is hermetically sealed under vacuum by plasma CVD deposition of nonstressed dielectric layer plugs.

With the above scheme, Diem et al. realized an absolute pressure sensor with a size of less than 1.5 mm^2. The temperature dependence of a capacitive sensor is due mainly to the temperature coefficient of the offset capacitance. Therefore, a temperature compensation is needed for high-accuracy sensors. A drastic reduction of the temperature dependence is obtained by a differential measurement, especially if the reference capacitor resembles the sensing capacitor. A reference capacitor is designed with the membrane blocked by several plugs for pressure insensitivity. The localization and the number of plugs are modeled by finite element analysis (FEA) (ANSYS software was used) to get a deformation lower than 1% of the active sensor's deformation. Even without temperature calibration, the high output of the differential signal results in an overall output error better than ±2% over the whole temperature range (–40 to +125°C) compared with 10% for nondifferential measurements. The temperature coefficient of the sensitivity is about 100 ppm/°C, which agrees with the theoretical variation of the Young's modulus of silicon. A piezoresistive sensor can be achieved by implanting strain gauges in the membrane. Although SIMOX wafers are more expensive than regular wafers, they come with several process steps embedded, and they make packaging easier.

Selective Epitaxy Surface Micromachining. Selective epitaxy can be a promising approach for creating novel microstructures. Figure 3.109 illustrates the selective deposition of epi-Si on a Si substrate through a SiO$_2$ window [Wolf and Tauber, 1987]. The same figure also demonstrates the simultaneous deposition of poly-Si on SiO$_2$ and crystalline epi-Si on Si, creating the basis for a structure featuring an epi-Si anchor with poly-Si side arms.

Neudeck et al. [Neudeck, 1990; Schubert and Neudeck, 1990] at Purdue, and Gennissen et al. [Gennissen et al., 1995; Bartek et al., 1995] at Twente, proved that selective epitaxy can also be applied for automatic etch stop on buried oxide islands. Figure 3.110A demonstrates how epitaxial lateral overgrowth (ELO) can bury oxide islands. After removal of the native oxides from the seed windows, epi is grown for 20 min at 950°C and at 60 torr using a Si$_2$H$_2$Cl$_2$-HCl-H$_2$ gas system. The epi growth front moves parallel to the wafer surface while growing in the lateral direction, leaving a smooth planar surface. During epi growth, the HCl prevents poly nucleation on the nonsilicon areas. The epi quality is strongly dependent on the orientation of the seed holes in the oxide. Seed holes oriented in the <100> direction lead to the best epi material and surface quality. Selective epi's other big problem for fabrication remains sidewall defects [Bashir et al., 1995]. The buried oxide islands stop the KOH etch of the substrate, enabling formation of beams and membranes as shown in Figure 3.110B. This technique might form the basis of many high-performance microstructures. The Purdue and Twente groups also work on confined selective epitaxial

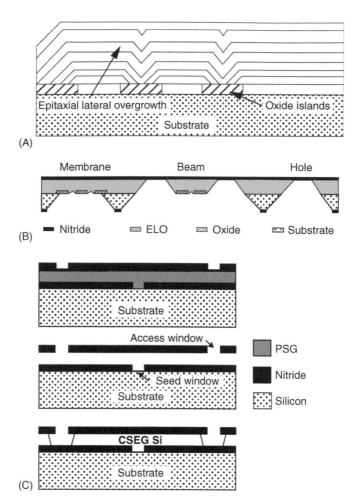

FIGURE 3.110 Micromachining with epi-Si. (A) Lateral overgrowth process of epi-Si (ELO, epitaxial lateral overgrowth) out of <100>-oriented holes in an oxide mask. (B) KOH etch stop on buried oxide islands or front side nitride. (C) Principle of confined selective epitaxial growth. (Reprinted with permission from papers presented at Transducers '95, Stockholm, Sweden, 1995.)

growth (CSEG), a process pioneered by Neudeck et al. [Schubert and Neudeck, 1990]. In this process, a micromachined cavity is formed above a silicon substrate with a seed contact window to the silicon substrate and access windows for epi-Si (Figure 3.110C) [Bartek et al., 1995]. Low-stress silicon-rich nitride layers act as structural layers to confine epitaxial growth; PSG is used as sacrificial material. This confined selective epitaxial growth technique allows electrical and/or thermal isolation separation as well as independent optimization of active sensor and readout electronic areas.

3.18.1.4 SOI vs. Poly-Si Surface Micromachining

The power of poly-Si surface micromachining lies mainly in its CMOS compatibility. When deposited on an insulator, both poly-Si and single-crystal layers enable higher operating temperatures (>200°C) than bulk micromachined sensors featuring p-n junction isolation only (130°C max) [Luder, 1986]. An additional benefit for SOI-based micromachining is IC compatibility combined with single-crystal Si performance excellence. The maximum gauge factor (see Equation [3.28]) of a poly-Si piezoresistor is about 30, roughly 15 times larger than that of a metal strain gauge but only one third that of an indiffused resistor in single-crystal Si [Obermeier and Kopystynski, 1992]. Higher piezoresistivity and fracture stress would seem to favor SOI for sensor manufacture. However, there is an important counter argument: the piezoresistivity and fracture stress in poly-Si are isotropic, a major design simplification. Moreover, by laser recrystallization, the gauge factor of poly-Si might increase to above 50 [Voronin et al., 1992], and by appropriate boron doping the temperature coefficient of resistance (TCR) can actually reach 0 vs. a TCR of, say, $1.7 \times 10^{-3} \mathrm{K}^{-1}$ for single-crystal p-type Si. Neither technical nor cost issues will be the deciding factor in determining which technology will become dominant in the next few years. Micromachining is very much a hostage to trends in the IC industry; promising technologies such as GaAs and micromachining do not necessarily take off, in no small part because of the invested capital in some limited sets of standard silicon technologies. On this basis, SOI surface micromachining is the favored candidate; SOI extends silicon's technological relevance and experiences, increasing investment from the IC industry and benefiting SOI micromachining [Kanda, 1991].

Based on the above, we believe that SOI micromachining not only introduces an improved method of making many simple micromachines, but it also will probably become the favored approach of the IC industry. A summary of SOI advantages is listed below:

- IC industry employment in all type of applications, such as MOS, bipolar digital, bipolar linear, power devices, BICMOS, CCDs, heterojunction bipolar [Burggraaf, 1991], etc.
- Batch packaging through embedded cavities
- CMOS compatibility
- Substrate industrially available at lower and lower cost (about \$200 today)
- Excellent mechanical properties of the single-crystalline surface layer
- Freedom of shapes in the *x–y* dimensions and continually improving dry etching techniques resulting in larger aspect ratios and higher features
- Freedom to choose a very well controlled range of thicknesses of epi surface layers
- SiO_2 buried layer as sacrificial and insulating layer and excellent etch stop
- Dramatic reduction of process steps as the SOI wafer comes with several "embedded" process steps
- High-temperature operation

A tutorial on SOI can be found at <http://www.sigen.com/whatissoi.html>.

3.18.2 Resists as Structural Elements and Molds in Surface Micromachining

3.18.2.1 Introduction

Deep UV photoresists enable the molding of high-aspect-ratio microstructures in a wide variety of moldable materials, or they are used directly as structural elements in the case of permanent photoresists.

3.18.2.2 UV Depth Lithography

Polyimide Surface Structures. Polyimide surface structures, due to their transparency to exposing UV light, can be made very high and exhibit LIGA-like high aspect ratios. By using multiple coats of spun-on polyimide, thick suspended plates are possible. Moreover, composite polyimide plates can be made, depositing and patterning a metal film between polyimide coats. Polyimide surface microstructures are typically released from the substrate by selectively etching an aluminum sacrificial layer (see Figure 3.98), although Cu and PSG (e.g., in the butterfly wing in Figure 3.103) have been used as well (see Table 3.17).

Because polyimides are easily deformed, they usually do not qualify as mechanical members but have been used, for example, in plastically deformable hinges (see Figure 3.103) [Suzuki et al., 1994; Hoffman et al., 1995]. An early result in polyimide surface micromachining was obtained at SRI International, where polyimide pillars (spacers) about 100 µm in height were used to separate a Si wafer, which was equipped with a field emitter array, from the display glass plate of a flat panel display [Brodie et al., 1990]. The flat panel display and an SEM picture of the pillars are shown in Figure 3.111. The Probimide 348 FC formulation of Ciba-Geigy was used. This viscous precursor formulation (48% by weight of a polyamic ester, a surfactant for wetting, and a sensitizer) with a 3500 cs viscosity was applied to the Si substrate and

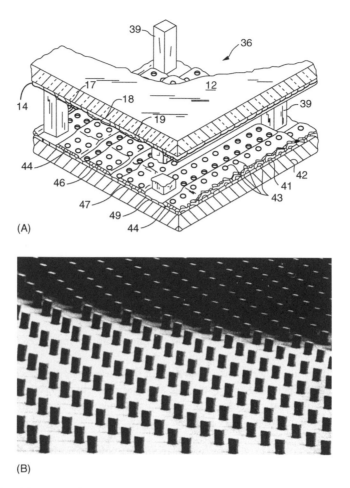

FIGURE 3.111   Polyimide structural elements. (A) Micromachined flat panel display. Number 39 represents one of the spacer pillars in the matrix of polyimide pillars (100 µm high). The spacer array separates the emitter plate from the front display plate. (B) SEM of the spacer matrix. (Courtesy of Dr. I. Brodie.) The height of the pillars is similar to what can be accomplished with LIGA. Since only a simple UV exposure was used, this polyimide is referred to as poor man's LIGA process, or pseudo-LIGA. (From Brodie, I. et al., U.S. Patent 4,923,421, 1990.)

formed into a film of a 125 µm thickness by spinning. A 30 to 40 min prebake at about 100°C removed the organic solvents from the precursor. The mask with the pillar pattern was then aligned to the wafer coated with the precursor and subjected to about 20 min of UV radiation. After driving off moisture by another baking operation, the coating, still warm, was spray developed (QZ 3301 from Ciba-Geigy), revealing the desired spacer matrix. By baking the polyimide at 100°C in a high vacuum (10^{-9} torr), the pillars shrank to about 100 µm, while the polyimide became more dense and exhibited greater structural integrity.

Frazier et al. (1993) obtained a height-to-width aspect ratio of about 7 with polyimide structures. Ultraviolet was used to produce structures with heights in the range of 30 to 50 µm. At greater heights, the verticality of the sidewalls became relatively poor. Spun-on thickness in excess of 60 µm in a single coat was obtained for both Ciba-Geigy and Du Pont commercial UV-exposable, negative-tone polyimides. Using a G-line mask aligner, an exposure energy of 350 mJ/cm² was sufficient to develop a pattern with the Ciba-Geigy QZ 3301 developer. Allen and his team combined polyimide insert molds with electrode-position to make a wide variety of metal structures [Ahn et al., 1993].

Other Resists Used for UV Depth Lithography Surface Micromachining. Besides polyimides, research on novolak-type resists also is leading to higher three-dimensional features. Lochel et al. (1994, 1996a, 1996b) use novolak, positive tone resists of high viscosity (e.g., AZ 4000 series, Hoechst). They deposit, in a multiple-coating process, layers up to 200 μm thick in a specially designed spin coater that incorporates a co-rotating cover. The subsequent UV lithography yields patterns with aspect ratios up to 10, steep edges (more than 88°), and a minimum feature size down to 3 μm. By combining this resist technology with sacrificial layers and electroplating, a wide variety of three-dimensional microstructures result.

Along the same line, researchers at IBM experimented with Epon SU-8 (Shell Chemical), an epoxy-based, onium-sensitized, UV transparent negative photoresist used to produce high-aspect-ratio (>10:1) features as well as straight sidewalled images in thick film (>200 μm) using standard lithography [Acosta et al., 1995; LaBianca et al., 1995]. SU-8 imaged films were used as stencils to plate permalloy for magnetic motors [Acosta et al., 1995].

Patterns generated with these thick resist technologies should be compared with LIGA-generated patterns, not only in terms of aspect ratio but also in terms of sidewall roughness and sidewall run-out. Such a comparison will determine which surface machining technique to employ for the job at hand.

3.19 Comparison of Bulk Micromachining with Surface Micromachining

Surface and bulk micromachining have many processes in common. Both techniques rely heavily on photolithography; oxidation; diffusion and ion implantation; LPCVD and PECVD for oxide, nitride, and oxynitride; plasma etching; use of polysilicon; and metallizations with sputtered, evaporated, and plated Al, Au, Ti, Pt, Cr, and Ni. Where the techniques differ is in the use of anisotropic etchants, anodic and fusion bonding, (100) vs. (110) starting material, p^+ etch stops, double-sided processing and electrochemical etching in bulk micromachining, and the use of dry etching in patterning and isotropic etchants in release steps for surface micromachines. Combinations of substrate and surface micromachining also frequently appear. The use of polysilicon avoids many challenging processing difficulties associated with bulk micromachining and offers new degrees of freedom for the design of integrated sensors and actuators. Design freedom includes many more possible shapes in the x-y plane and the ease of integration of several sensors on one die (e.g., a two-axis accelerometer). The technology combined with sacrificial layers also allows the nearly indispensable further advantage of in situ assembly of the tiny mechanical structures because the structures are preassembled as a consequence of the fabrication sequence. Another advantage focuses on thermal and electrical isolation of polysilicon elements. Polycrystalline piezoresistors can be deposited and patterned on membranes of other materials, for example on a SiO_2 dielectric. This configuration is particularly useful for high-temperature applications. The p–n junctions act as the only electrical insulation in the single-crystal sensors, resulting in high leakage currents at high temperatures, whereas current leakage for the poly-Si/SiO$_2$ structure virtually does not exist. The limits of surface micromachining are quite striking. CVD silicon usually caps at layers no thicker than 1 to 2 μm because of residual stress in the films and the slow deposition process (thick poly-Si needs further investigation). A combination of a large variety of layers may produce complicated structures, but each layer is still limited in thickness. Also, the wet chemistry needed to remove the interleaved layers may require many hours of etching (except when using the porous Si option discussed above), and even then stiction often results.

The structures made from polycrystal silicon exhibit inferior electronic and slightly inferior mechanical properties compared with single-crystal silicon. For example, poly-Si has a lower piezoresistive coefficient (resulting in a gauge factor of 30 vs. 90 for single-crystal Si), and it has a somewhat lower mechanical fracture strength. Poly-Si also warps due to the difference of thermal expansion coefficient between polysilicon and single-crystal silicon. Its mechanical properties strongly depend on processing procedures and parameters.

Table 3.19 introduces a comparison of surface micromachining with wet bulk micromachining. The status depicted reflects the mid-1990s and only includes poly-Si surface micromachining. As discussed, SOI micromachining, thick poly-Si, hinged poly-Si, polyimide, and millimeter-molded poly-Si structures have dramatically expanded the application bandwidth of surface micromachining. In the chapter on

TABLE 3.19   Comparison of Bulk Micromachining with Surface Micromachining

Bulk micromachining	Surface micromachining
Large features with substantial mass and thickness	Small features with low thickness and mass
Utilizes both sides of the wafer	Multiple deposition and etching required to build up structures
Vertical dimensions: one or more wafer thicknesses	Vertical dimensions are limited to the thickness of the deposited layers ($\sim 2\,\mu m$) leading to compliant suspended structures with the tendency to stick to the support
Generally involves laminating Si wafer to Si or glass	Surface micromachined device has its built-in support, and is more cost effective
Piezoresistive or capacitive sensing	Capacitive and resonant sensing mechanisms
Wafers may be fragile near the end of the production	Cleanliness critical near end of process
Sawing, packaging, testing is difficult	Sawing, packaging, testing is difficult
Some mature products and producers	No mature products or producers
Not very compatible with IC technology	Natural but complicated integration with circuitry. Integration is often required due to the tiny capacitive signals

Source: Adapted with permission from Jerman, H. (1994) "Bulk Silicon Micromachining," hard copies of viewgraphs presented in Banff, Canada.

TABLE 3.20   Comparison of Materials Properties of Si Single Crystal with Crystalline Polysilicon

Material property	Single Crystal Si	Poly-Si
Thermal conductivity (W/cm °K)	1.57	Strong function of the grain structure of the film ; 0.30–0.35 (for fine grains and double that for larger grains)
Thermal expansion (10^{-6}/°K)	2.33	2–2.8
Specific heat (cal/g °K)	0.169	0.169
Piezoresistive coefficients	n-Si ($\pi_{11} = -102.2$) p-Si ($\pi_{44} = +138.1$), e.g. gauge factor of 90	E.g., gauge factor of 30 (>50 with laser recrystallization)
Refractive index		4.1 at 600 nm
Mobility (cm²/V/sec)	Holes: 600 Electrons: 1500	Maximum for electrons: 30
Density (cm^{-3})	2.32	2.32
Fracture strength (GPa)	6	0.8 to 2.84 (undoped poly-Si)
Dielectric constant	11.9	Sharp maxima of 4.2 and 3.4 eV at 295 and 365 nm, respectively
Residual stress	none	Depends on structure. As deposited films are compressive
Temperature resistivity coefficient (°C^{-1})TCR	0.0017 (p-type)	0.0012 Nonlinear, + or − through selective doping, increases with decreasing doping level can be made 0!
Poisson ratio	0.262 max for (111)	0.23
Young's modulus (10^{11} N/m²)	1.90 (111)	1.61
Resistivity at room temperature (Ω.cm)	Depends on doping	Strong function of the grain structure of the film; plateaus at 4×10^{-4} above 1 10 $\times$ 21 cm^{-3} P (always higher than for SCS)

Source: Based on Lin (1993), Adams (1988), Kamins, (1988), and Heuberger (1989).

LIGA, we will see how X-ray lithography can further expand the z direction for new surface micromachined devices with unprecedented aspect ratios and extremely low surface roughness. In Table 3.20, we compare physical properties of single-crystal Si with those of poly-Si.

Although polysilicon can be an excellent mechanical material, it remains a poor electronic material. Reproducible mechanical characteristics are difficult and complex to realize consistently. Fortunately, SOI surface micromachining and other newly emerging surface micromachining techniques can alleviate many of the problems [Petersen et al., 1991].

TABLE 3.21   Comparison of Different Deposition Techniques

	Atmospheric CVD (AP CVD)	Low T LP CVD	Medium T LP CVD	Plasma assisted CVD (PE CVD)
Temp (°C)	300–500	300–500	500–900	100–350
Materials	SiO_2, P-glass	SiO_2, P-glass, BP-glass	Poly-Si, SiO_2, P-glass, BP-glass Si_3N_4, SiON	SiN, SiO_2, SiON
Uses	Passivation, insulation, spacer	Passivation, insulation, spacer	Passivation, gate metal, structural element, spacer	Passivation, insulation, structural elements
Throughput	High	High	High	Low
Step coverage	Poor	Poor	Conformal	Poor
Particles	Many	Few	Few	Many
Film properties	Good	Good	Excellent	Poor

Note: P-glass = phosphorus-doped glass; BP-glass = borophosphosilicate glass.
Source: Adapted with permission from Adams, A.C. (1988) "Dielectric and Polysilicon Film Deposition," in *VLSI Technology*, Sze, S.M., ed., McGraw-Hill, New York.

3.20   Materials Case Studies

3.20.1   Introduction

Thin-film properties prove not only difficult to measure but also to reproduce, given the many influencing parameters. Dielectric and polysilicon films can be deposited by evaporation, sputtering, and molecular beam techniques. In VLSI and surface micromachining, none of these techniques is as widely used as CVD. The major problems associated with the former methods are defects caused by excessive wafer handling, low throughput, poor step coverage, and nonuniform depositions. From the comparison of CVD techniques in Table 3.21, we can conclude that LPCVD at medium temperatures prevails above all others. VLSI devices and integrated surface micromachines require low processing temperatures to prevent movement of shallow junctions, uniform step coverage, few process-induced defects (mainly from particles generated during wafer handling and loading), and high wafer throughput to reduce cost. These requirements are best met by hot-wall, low-pressure depositions [Iscoff, 1991]. While depositing a material with LPCVD, the following process parameters can be varied: deposition temperature, gas pressure, flow rate, and deposition time.

Table 3.22 cites some approximate mechanical properties of microelectronic materials. The numbers for thin film materials must be approached as approximations; the various parameters affecting mechanical properties of thin films will become clear in the case studies below.

3.20.2   Polysilicon Deposition and Material Structure

3.20.2.1   Introduction

The IC industry applies LPCVD polysilicon in applications ranging from simple resistors, gates for MOS transistors, thin-film transistors (TFT) based on amorphous hydrogenated silicon (α-Si:H), DRAM cell plates, and trench fills, as well as in emitters in bipolar transistors and conductors for interconnects. For the last application, highly doped polysilicon is especially suited; it is easy to establish ohmic contact; it is light insensitive and corrosion resistant; and its rough surface promotes adhesion of subsequent layers. Doping elements such as arsenic, phosphorous, or boron reduce the resistivity of the polysilicon. Commercial LPCVD equipment is available to deposit polysilicon on wafers up to eight inches in diameter. A typical batch is 100 to 200 wafers. Polysilicon also has emerged as the central structural/mechanical material in surface micromachining, and a closer look at the influence of deposition methodology on its materials characteristics is warranted.

TABLE 3.22   Approximate Mechanical Properties of Microelectronic Materials

	E (GPa)	μ	α (1/°C)	σ_0
Substrates				
Silicon	190	0.23	2.6×10^{-6}	–
Alumina	~415	–	8.7×10^{-6}	–
Silica	73	0.17	0.4×10^{-6}	–
Films				
Polysilicon	160	0.23	2.8×10^{-6}	Varies
Thermal SiO$_2$	70	0.20	0.35×10^{-6}	Compressive, e.g. 350 MPa
PE CVD SiO$_2$			2.3×10^{-6}	
LP CVD Si$_3$N$_4$	270	0.27	1.6×10^{-6}	Tensile
Aluminum	70	0.35	25×10^{-6} (high!)	Varies
Tungsten (W)	410 (stiff!)	0.28	4.3×10^{-6}	Varies
Polyimide	3.2	0.42	$20–70 \times 10^{-6}$ (very high)	Tensile

Note: E = Young's modulus, μ = Poisson ration, α = coefficient of thermal expansion, σ_0 = residual stress.
Source: Based with permission on lecture notes from Senturia, S.D., and Howe, R.T. (1990) "Mechanical Properties and CAD," Lecture Notes, MIT, Boston.

3.20.2.2   Undoped Poly-Si

The properties of low-pressure chemical vapor deposited (LPCVD) undoped polysilicon films are determined by the nucleation and growth of the silicon grains. In most cases, the Si wafers upon which we deposit polysilicon are placed vertically and closely spaced together in quartz boats. Because of the close spacing, the deposition process is operated in a reaction-limited regime to obtain uniform deposition across the wafers. LPCVD Si films, grown slightly below the crystallization temperature (about 600°C for LPCVD), initially form an amorphous solid that subsequently may crystallize during the deposition process [Krulevitch, 1994; Guckel et al., 1990]. The CVD method results in amorphous films when the deposition temperatures are well below the melting temperature of Si (1410°C). The subsequent transition from amorphous to crystalline depends on atomic surface mobility and deposition rate. At low temperatures, surface mobility is low, and nucleation and growth are limited. Newly deposited atoms become trapped in random positions and, once buried, require a substantial amount of time to crystallize because solid state diffusion is significantly lower than surface mobility. That is why, for low-temperature deposition, amorphous layers start to crystallize only after sufficient time at temperature in the reactor. Working at temperatures between 580 and 591.5°C, Guckel et al. (1990) produced mostly amorphous films. However, Krulevitch, working at only slightly higher temperatures (605°C) and probably leaving the films longer in the LPCVD setup, produced crystallized films.

In Figure 3.112 we show the transition from polysilicon to monocrystalline Si during CVD deposition at different temperatures. Upon crossing the transition temperature between amorphous and crystalline growth, crystalline growth immediately initiates at the substrate due to the increased surface mobility, which allows adatoms to find low-energy, crystalline positions from the start of the deposition process. The deposition temperature at which the transition from amorphous to a crystalline structure occurs depends on many parameters, such as deposition rate, partial pressure of hydrogen, total pressure, presence of dopants, and presence of impurities (O, N, or C) [Adams, 1988]. In the crystalline regime, numerous nucleation sites form, resulting in a transition zone of a multitude of small grains at the film–substrate interface to columnar crystallites on top, as shown in the schematic of a 620–650°C columnar film in Figure 3.113. In this figure, a transition zone of small, randomly oriented grains is sketched near the SiO$_2$ layer. The rate of crystallization is faster here than the deposition rate. Columnar grains ranging between 0.03 and 0.3 µm in diameter form on top of the small grains [Adams, 1988]. The columnar coarse grain structure arises from a process of growth competition among the small grains, during which those grains preferentially oriented for fast vertical growth survive at the expense of misoriented slowly growing grains [Drift, 1967; Matson and Polysakov, 1977]. The lower the deposition temperature, the smaller the initial grain size will be. At 700°C, films also are columnar; however, the grains are

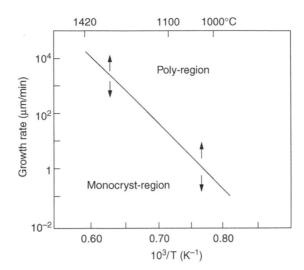

FIGURE 3.112 Growth rate versus temperature. Epitaxy to CVD transition in CVD polysilicon deposition and epitaxial monocrystalline Si deposition. (Reprinted with permission from Wolf, S., and Tauber, R.N. [1987] *Silicon Processing for the VLSI Era*, Sunset Beach, Lattice Press.)

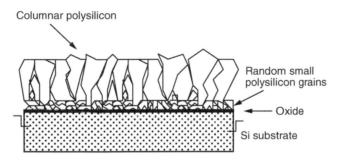

FIGURE 3.113 Schematic of compressive poly-Si formed at 620–650°C. The columnar coarse-grain structure arises from a process of grain growth competition among the small grains, during which those grains preferentially oriented for fast vertical growth survive at the expense of misoriented slowly growing grains. (After Krulevitch, P. A. [1994] *Micromechanical Investigations of Silicon and Ni-Ti-Cu Thin Films*, Ph.D. thesis, University of California, Berkeley.)

cylindrical, extending through the thickness of the entire film, and there is no transition zone near the SiO_2 interface [Krulevitch, 1994].

Stress in poly-Si films was found to vary significantly with deposition temperature and silane pressure. Guckel et al. (1990) found that their mainly amorphous films deposited at temperatures below 600°C proved highly compressive with strain levels as high as –0.67%. At temperatures barely above 600°C, Krulevitch reports tensile films, whereas for yet higher temperatures ($\geq$620°C) the stress again turns compressive. While films deposited at temperatures greater than 630°C all turned out compressive, the magnitude of the compression decreased with increasing temperature. The stress gradient in the poly-Si films explains why compressive undoped and unannealed poly-Si beams tend to curl upward (positive stress gradient) when released from the substrate [Lober et al., 1988].

Using high-resolution transmission electron microscopy, Guckel et al. (1990) and Krulevitch (1994) found a strong correlation between the material's microstructure and the exhibited stress. Guckel et al. found that in their mainly amorphous films deposited at temperatures below 600°C a region near the substrate interface crystallized during growth with grains between 100 and 4000 Å. Krulevitch found that

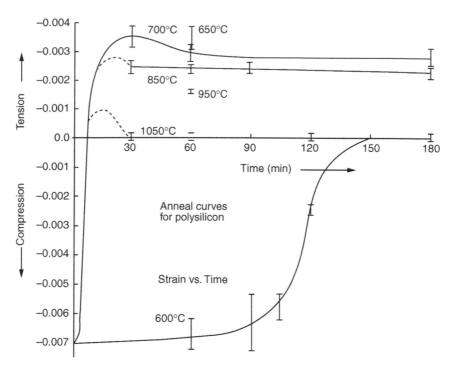

FIGURE 3.114   Anneal curves for poly-Si. Strain versus anneal time. Upper curves: low-temperature film. Lower curve: high-temperature film. (Reprinted with permission from Guckel, H. et al. [1988] "Fine-Grained Polysilicon Films with Built-in Tensile Strain," *IEEE Trans. Electron. Dev.* 35, pp. 800–1.)

tensile, low-temperature films (605°C) have Si grains dispersed throughout the film thickness. Krulevitch suggests that the compressive stress in the higher temperature compressive films ($\geqslant$620°C) relates to the competitive growth mechanism of the columnar grains. The same author concluded that thermal sources of stress are insignificant.

Importantly, Guckel et al. discovered that annealing in nitrogen or under vacuum converts the low-temperature films with compressive built-in strain (−0.007) to a tensile strain with controllable strain levels between 0 and +0.003 (Figure 3.114). During anneal, no grain size increase was noticed (100 to 4000 Å), but a slight increase in surface roughness was measured. This type of poly-Si is referred to as fine-grain poly-Si (also Wisconsin poly-Si). Guckel et al. explain this strain field reversal as follows: as the amorphous region of the film crystallizes, it attempts to contract, but due to the substrate constrained newly crystallized region, a tensile stress results. Higher-temperature films during an anneal also become less strained, but the strain remains compressive (see lower curve in Figure 3.114). Moreover, in this case, grain size does increase, and the surface turns considerably rougher. The latter is called coarse-grain poly-Si. Fine-grain poly-Si, with its tensile strain, is preferable; however, it cannot be doped to as low a resistivity as coarse-grain polysilicon. Hence, fine-grain polysilicon should be considered as a structural material rather than an electronic material.

Summarizing, stress in poly-Si depends on the material's microstructure, with tension arising from the amorphous to crystalline transformation during deposition and compression from the competitive grain growth mechanism.

Polysilicon deposited at 600 to 650°C has a {110}-preferred orientation. At higher temperatures, the {100} orientation dominates. Dopants, impurities, and temperature influence this preferred orientation [Adams, 1988]. Drosd and Washburn (1982) introduced a model explaining the experimental observation that regrowth of amorphous Si is faster for {100} surfaces, followed by {110} and {111}, which are 2.3 and 20 times slower respectively. Interestingly, the latter also pinpoints the order of fastest to slowest etching of the crystallographic planes in alkaline etchants. As discussed earlier in this chapter, Elwenspoek

TABLE 3.23 Comparison of coarse-grain and fine-grain poly-Si

	Coarse-grain poly-Si	Fine-grain poly-Si
Temperature of deposition (°C)	620–650	570–591.5
Surface roughness	Rough >50 Å	Smooth <15 Å
Grain size	Undoped: 160–320 Å as deposited In-situ P doped: 240–400 Å	Very small grains
As deposited strain	0.002 (compressive)	−0.007 (compressive)
Effect of high temperature anneal	• Grains size increases; • residual strain decreases but remains compressive; • reduced bending moment	Grain size increases to 100 Å [Guckel et al., 1988] (others have found 700–900 Å); large variation in strain (see Figure 3.113): from compressive to tensile
Dry and wet etch rate	Higher for doped material, depends on dopant concentration	Higher for doped material, depends on dopant concentration
Texture	<110> as deposited; <311> in-situ P doped	No texture as deposited, depends on dopant concentration, <111> after 900–1000°C anneal [Harbeke et al., 1983]

et al. (1994) used this observation of symmetry between etching and growing of Si planes as an important insight to develop a new theory explaining anisotropy in etching. In Table 3.23, we compare the discussed coarse- and fine-grain poly-Si forms.

The above picture is further be complicated by the controlling nature of the substrate. For example, depositing amorphous Si (α-Si) at even lower temperatures of 480°C from disilane (Si_2H_6), and crystallizing it by subsequent annealing at 600°C demonstrated a large dependency of crystallite size on the underlying SiO_2 surface condition. Treating the surface with $HF:H_2O$ or $NH_4OH:H_2O_2:H_2O$ leads to poly-Si films with a large grain size, two or three times as large as without SiO_2 treatment, believed to be the consequence of nucleation rate suppression [Shimizu and Ishihara, 1995].

Abe and Reed made low-strain polysilicon thin film by DC-magnetics sputtering and postannealing. The films showed very small regional stress and very smooth texture. The deposition rate was 193 Å/min, and the substrate was neither cooled nor heated. The average roughness was found to be comparable to the surface roughness of polished bare silicon substrates [Abe and Reed, 1996].

3.20.2.3 Doped Poly-Si

To produce micromachines, doped poly-Si is used far more frequently than undoped poly-Si. Dopants decrease the resistivity to produce conductors and control stress. Polysilicon can be doped by diffusion, implantation, or the addition of dopant gases during deposition (in situ doping).

Doping poly-Si films in situ reduces the number of processing steps required for producing doped micro devices by eliminating the need for a subsequent high-temperature step associated with a diffusion or ion-implantation anneal. It also provides the potential for more uniform doping throughout the film thickness. In situ doping of poly-Si is accomplished by maintaining a constant PH_3 to SiH_4 gas flow ratio of about 1 vol% in a hot-wall LPCVD setup. At this ratio, the phosphorous content in the film appears above the saturation limit, and the excess dopant segregates at the grain boundaries [Adams, 1988]. Phosphorus diffuses significantly faster in polysilicon than in single-crystalline silicon. The diffusion takes place primarily along the grain boundaries. The diffusivity in thin films of polysilicon (i.e., small equi-axed grains) is about 1×10^{-12} cm^2/s. The dopant concentration of in situ doped films is normally very high ($\sim 10^{20}$ cm^{-3}). Above about $1 \times 10^{+21}$ cm^{-3}, the film resistivity reaches a plateau of 4×10^{-4} W-cm because of the low mobility of electrons or holes. The maximum mobility for the highest phosphorus-doped polysilicon is about 30 cm^2/Vs [Kamins, 1988] (see Table 3.20).

In situ phosphorus-doped poly-Si undergoes the same amorphous to crystalline growth transformation observed in the undoped film, with the material's microstructure depending on deposition temperature as well as deposition pressure. The temperature of transformation is lower for the doped films than for the undoped poly-Si and occurs between 580 and 620°C [Mulder et al., 1990; Kinsbron et al., 1983]. Phosphorus doping thus enhances crystallization in amorphous silicon [Lietoila, et al., 1982] and, due to passivation of the poly-Si surface by the phosphine gas, reduces the poly-Si deposition rate [Mulder et al., 1990]. Decreases in deposition rate by as much as a factor of 25 have been reported [Meyerson and Olbricht, 1984]. Slower deposition rates allow more time for adatoms to find crystalline sites, resulting in crystalline growth at lower temperatures. From Table 3.23, we read that the grain size of phosphorus-doped poly-Si tends to be larger (240 to 400 Å) than for the undoped material and that {311} planes show up as a texture facet in the doped material. In contrast to in situ phosphine and arsine doping, which both decrease the deposition rate, diborane doping of poly-Si to make it p$^+$ accelerates the deposition rate [Adams, 1988].

At lower deposition temperatures and higher pressures, the microstructure again consists of amorphous and crystalline regions, while at higher temperatures and lower pressures columnar films result and as deposited films exhibit compressive residual stress. The columnar films have a stress gradient that increases toward the film surface, as opposed to the gradient found in undoped columnar poly-Si. This gradient in stress most likely is due to nonuniform distribution of phosphorus throughout the film. Annealing at 950°C for 1 hr results in the same stress and stress gradient for initially columnar and initially amorphous/crystalline films (i.e., $\sigma_f = -45$ MPa and $\Gamma = +0.2$ mm^{-1}, respectively) [Krulevitch, 1994].

As with undoped poly-Si, phosphorus-doped poly-Si films with smooth surfaces (fine-grain) can be obtained by depositing in situ doped films in the amorphous state and then annealing [Harbeke et al., 1983; Hendriks and Mavero, 1991]. Phosphorus-doped poly-Si oxidizes faster than undoped poly-Si. The rate of oxidation is determined by the dopant concentration at the poly-Si surface [Adams, 1988]. The addition of oxygen to poly-Si increases the film's resistivity, and the resulting coating, semi-insulating poly-Si (SIPOS), acts as a passivating coating for high-voltage devices in the IC industry. SISPOS has not emerged in surface micromachining yet.

Four drawbacks of in situ phosphorus doping are the complexity of the deposition process, slower deposition rates [Kurokawa, 1982], reduced film thickness uniformity [Meyerson and Olbricht, 1984], and the cleaning of the reactor, which is more demanding for the doped process. Doping uniformity can be improved by modifying the reactor geometry [Mulder et al., 1990], and as discussed before (under "Deposition of Structural Material"), the lower deposition rate of in situ phosphine doping can also be mitigated by reducing the flow of phosphine/silane ratio by one third [Howe, 1995].

Diffusion is often a more effective method for doping of polysilicon than in situ doping, especially for very heavy dopings (e.g., polysilicon resistivities down to the 10^{-4} Ω-cm) of thick (e.g., 2 μm) films. However, diffusion is a high-temperature (e.g., 900 to 1000°C) process. If the diffusion step is performed for long durations (a few hours) to achieve uniform doping throughout the thickness of a-few-microns-thick films, it could destroy electronics that are fabricated on the wafer prior to polysilicon surface micromachining. If performed for a short time, dopant distribution through the film thickness will not be uniform enough, resulting in difficulties with mechanical property variations through the film thickness. As discussed earlier, the use of doped oxide sacrificial layers relaxes some of the concerns associated with doping the film uniformly by diffusion; the sacrificial doped oxide acts as a diffusion source into polysilicon.

Ion implantation can also be used for doping polysilicon. The implantation energy is typically adjusted so that the peak of the concentration profile is at the center of the film thickness. However, the resistivity of implanted polysilicon films is not as low as that possible by diffusion.

As with undoped poly-Si, the intrinsic stresses in as-deposited doped polysilicon films are large (>500 MPa) and can result in warping or curling of released micromechanical structures. The values for the fracture stress of boron-, arsenic-, and phosphorus-doped polysilicon are 2.77 ± 0.08 GPa, 2.70 ± 0.09 GPa, and 2.11 ± 0.1 GPa respectively, compared with 2.84 ± 0.09 GPa for undoped polysilicon. The lower value for phosphorus-doped material has been attributed to high surface roughness and to the large number of defects associated with extensive grain growth in highly phosphorus-doped films [Biebl and Philipsborn, 1993]. Maluf reports that the operation of poly-Si at elevated temperatures is subject to long-term instabilities, drift, and hysteresis due to slow-stress-annealing effects [Maluf, 2000]. Several sensors using polysilicon piezoresistive sense elements have been demonstrated. The piezoresistive coefficient is an average over all the orientations in polycrystalline silicon. The gauge factor ranges between 20 and 40, which is about a factor of 5 smaller than in single-crystal Si, and quickly decreases as the doping exceeds 10^{19} cm^{-3}. At doping levels of 10^{20} cm^{-3}, the temperature coefficient of resistance (TCR) of poly-Si is approximately 0.04% per °C, compared with 0.14% per °C for crystalline Si. This is an advantage for the use of poly-Si, even though the gauge factor is quite low at these high doping levels.

3.20.2.4 PECVD and Sputtered Polysilicon

The quest for low-temperature poly-Si deposition processes is driven by the need for compatibility with prefabricated aluminum-metallized CMOS circuitry. This makes the 320°C PECVD and ~350°C sputter deposition methods of polysilicon especially interesting.

PECVD films, deposited in a 50 kHz parallel-plate diode reactor, can be doped in situ and crystallized by rapid thermal annealing (RTA: 1100°C, 100 s). It was shown that small-grained PECVD films annealed by RTA have good electrical properties and gauge factors between 20 and 30, similar to those reported for other alternative types of polycrystalline silicon [Compton, 1992].

Honer et al. introduced sputtered polysilicon to the MEMS community [Honer and Kovacs, 2000]. Sputtered silicon can be used to fabricate polysilicon microstructures atop standard, aluminum metallized CMOS at temperatures below 350°C. Films with stress values lower than 100 MPa are routinely obtained.

An important benefit of both these low-temperature processes is that they are compatible not only with conventional oxide sacrificial layers but also with certain organic sacrificial layers (e.g., polyimide). Organic layers can be removed in a dry oxygen plasma etch, thus avoiding the stiction and selectivity problems associated with wet etch releases and at the same time alleviating the concerns over chemical attack on structural elements by HF.

A drawback for the sputtered boron doped Si films is that, to increase their conductivity (to 25 Ω.m^{-1}), they need to be clad in symmetric, 50 nm thick layers of titanium-tungsten.

Kamins provides a detailed study of polysilicon physical properties [Kamins, 1988]. Another good resource on poly-Si morphology and doping is at <http://mems.cwru.edu/shortcourse/ partII_2.html> (a short course) and at <http://www.iue.tuwien.ac.at/phd/puchner/> (a Ph.D. thesis). Sharpe and Edwards at John Hopkins have perhaps provided the most detailed recent polysilicon mechanical property studies [Sharpe et al., 1997]; also visit <http://www.cnde.com/people/sharpe.htm>.

3.20.3 Amorphous and Hydrogenated Amorphous Silicon

Amorphous silicon behaves quite differently from either fine- or coarse-grain poly-Si. Amorphous Si can be stress annealed at temperatures as low as 400°C [Chang et al., 1991]. This low-temperature anneal makes the material compatible with almost any active electronic component. Unfortunately, very little is known about the mechanical properties of amorphous Si.

Hydrogenated amorphous Si (α-Si:H), with its interesting electronic properties, is even less understood in terms of its mechanical properties. If the mechanical properties of hydrogenated amorphous Si were found to be as good as those of poly-Si, the material might make a better choice than poly-Si as a MEMS material, given its better electronic characteristics.

The amorphous polysilicon material produces a high breakdown strength (7 to 9 MV/cm) oxide with low leakage currents (vs. a low breakdown voltage and large leakage currents for polycrystalline Si oxides). Amorphous polysilicon also attains a broad maximum in its dielectric function without the characteristic sharp structures near 295 and 365 nm (4.2 and 3.4 eV) of crystalline poly-Si. Approximate refractive index values at a wavelength of 600 nm are 4.1 for crystalline polysilicon and 4.5 for amorphous material [Adams, 1988]. As deposited, the material is under compression, but an anneal at temperatures as low as 400°C reduces the stress significantly, even leading to tensile behavior [Chang et al., 1991].

Like poly-Si, amorphous silicon exhibits a strong piezoresistive effect. The gauge factor is about five times smaller than that of single-crystal Si, but the TCR is lower than that of Si.

Hydrogenated amorphous silicon enables the fabrication of active semiconductor devices on foreign substrates at temperatures between 200 and 300°C. The technology, first applied primarily to the manufacturing of photovoltaic panels, now is quickly expanding into the field of large-area microelectronics such as active matrix liquid crystal displays (AMLCD). It is somewhat surprising that micromachinists have not taken more advantage of this material either to power surface micromachines or implement electronics cheaply on non-Si substrates.

Spear and Le Comber (1975) showed that, in contrast to α-Si, α-Si:H could be doped both n- and p-type. Singly bonded hydrogen, incorporated at the Si dangling bonds, reduces the electronic defect density from $\sim 10^{19}$/cm^3 to $\sim 10^{16}$/cm^3 (typical H concentrations are 5 to 10 atomic percent — several orders of magnitude higher than needed to passivate all the Si dangling bonds). The lower defect density results in a Fermi level that is free to move, unlike in ordinary amorphous Si, where it is pinned. Other interesting electronic properties are associated with α-Si:H — exposure of α-Si:H to light increases photoconductivity by four to six orders of magnitude, and its relatively high electron mobility (~ 1 cm^2/V s^{-1}) enables fabrication of useful thin-film transistors. Lee et al. (1995) noted that hydrogenated amorphous silicon solar cells are an attractive means to realize an onboard power supply for integrated micromechanical systems. They point out that the absorption coefficient of α-Si:H is more than an order of magnitude larger than that of single-crystal Si near the maximum solar photon energy region of 500 nm. Accordingly, the optimum thickness of the active layer in an α-Si:H solar cell can measure 1 μm, much smaller than that of single-crystal Si solar cells. By interconnecting 100 individual solar cells in series, the measured open circuit potential reaches as high as 150 V under AM 1.5 conditions, a voltage high enough to drive onboard electrostatic actuators.

Hydrogenated amorphous silicon is manufactured by plasma-enhanced chemical vapor deposition from silane. Usually, planar RF-driven diode sources using SiH$_4$ or SiH$_4$/H$_2$ mixtures are used. Typical pressures of 75 mtorr and temperatures between 200 and 300°C allow silane decomposition with Si deposition as the dominant reaction. Decomposition occurs by electron impact ionization, producing many different neutral and ionic species [Crowley, 1992]. Deposition rates for usable device quality α-Si:H generally do not exceed ~ 2 to 5 Å/s, due to the effects of temperature, pressure, and discharge power. Table 3.24 gives state-of-the-art parameters for α-Si:H prepared by PECVD. Although its semiconducting properties are inferior to single-crystal Si, the material is finding more and more applications. Some examples are TFT switches for picture elements in AMLCDs [Holbrook and McKibben, 1992], page-wide TFT-addressed document scanners, and high-voltage TFTs capable of switching up to 500 V [Bohm, 1988]. An excellent source for further information on amorphous silicon is the book *Plasma Deposition of Amorphous Silicon-Based Materials* [Bruno et al., 1995].

3.20.4 Silicon Nitride

3.20.4.1 Introduction

Silicon nitride (SixNy) is a commonly used material in microcircuit and microsensor fabrication due to its many superior chemical, electrical, optical, and mechanical properties. The material provides an efficient passivation barrier to the diffusion of water and to mobile ions, particularly of Na$^+$. It also oxidizes slowly (about 30 times slower than silicon) and has highly selective etch rates over SiO$_2$ and Si in many etchants. Some applications of silicon nitride are optical waveguides (nitride/oxide), encapsulant (diffusion barrier to water and ions), insulators (10^{16} Ω-cm, high dielectric strength, field breakdown limit of 10^7 V/cm), mechanical protection layer, etch mask, oxidation barrier, and ion implant mask (density is 1.4 times that of SiO$_2$). Silicon nitride is also hard, with a Young's modulus higher than that of Si, and it can be used, for example, as a bearing material in micromotors [Pool, 1988].

Silicon nitride can be deposited by a wide variety of CVD techniques (APCVD, LPCVD, and PECVD), and its intrinsic stresses can be controlled by the specifics of the deposition process. Silicon nitride and silicon oxide deposited with these techniques usually exhibit too much residual stress, which hampers their use as mechanical components. However, CVD of mixed silicon oxynitride can produce substantially stress-free components.

Nitride often is deposited from SiH$_4$ or other Si containing gases and NH$_3$ in a reaction such as:

$$3SiCl_2H_2 + 4NH_3 \rightarrow Si_3N_4 + 6HCl + 6H_2 \qquad \text{(Reaction 3.17)}$$

TABLE 3.24   Typical Optoelectronic Parameters Obtained for PECVD α-Si:H

	Symbol	Parameter
Undoped		
Hydrogen content		~10%
Dark conductivity at 300 K	σ_D	~10^{-10} $(\Omega\text{-cm})^{-1}$
Activation energy	E_σ	0.8–0.9 eV
Pre-exponent conductivity factor	σ_0	>10^3 $(\Omega\text{-cm})^{-1}$
Optical band gap at 300 K	E_g	1.7–1.8 eV
Temperature variation of band gap	$E_g(T)$	2–4 × 10^{-4} eV/K
Density of states at the minimum	g_{min}	>10^{15}–10^{17} cm^3/eV
Density of states at the conduction band edge		~10^{15}/cm^3
ESR spin density	N_s	~10^{21}/cm^3-eV
Infrared spectra		2000/640 cm^{-1}
Photoluminescence peak at 77 K		~1.25 eV
Extended state mobility		
Electrons	μ_n or μ_e	>10 cm^2/V-s
Holes	μ_p or μ_h	~1 cm^2/V-s
Drift mobility		
Electrons	μ_n or μ_e	~1 cm^2/V-s
Holes	μ_p or μ_h	~10^{-2} cm^2/V-s
Conduction band tail slope		25 meV
Valence band tail slope		40 meV
Hole diffusion length		~1 μm
Doped amorphous		
n-type[a]	σ_D	10^{-2} $(\Omega\text{-cm})^{-1}$
	E_g	~0.2 eV
p-type[b]	σ_D	10^{-3} $(\Omega\text{-cm})^{-1}$
	E_g	~0.3 eV
Doped microcrystalline		
n-type[c]	σ_D	≥1 $(\Omega\text{-cm})^{-1}$
	E_g	≤0.05 eV
p-type[d]	σ_D	≥1 $(\Omega\text{-cm})^{-1}$
	E_g	≤0.05 eV

[a] –1% PH3 added to gas phase.
[b] –1% B2H6 added to gas phase.
[c] –1% PH3 added to dilute SiH_4/H_2, or 500 vppm PH_3 added to SiF_4/H_2 (8:1) gas mixtures. Relatively high powers are involved.
[d] –1% B2H6 added to dilute SiH_4/H_2.
Source: Reprinted with permission from Crowley, J.L. (1992) "Plasma Enhanced CVD for Flat Panel Displays," *Solid State Technol*, February, pp. 94–98.

In this CVD process, the stoichiometry of the resulting nitride can be moved toward a silicon-rich composition by providing excess silane or dichlorosilane compared to ammonia.

3.20.4.2   PECVD Nitride

Plasma-deposited silicon nitride, also called plasma nitride or SiN, is used as the encapsulating material for the final passivation of devices. The plasma-deposited nitride provides excellent scratch protection, serves as a moisture barrier, and prevents sodium diffusion. Because of the low deposition temperature, 300 to 350°C, the nitride can be deposited over the final device metallization. Plasma-deposited nitride and oxide both act as insulators between metallization levels, which is particularly useful when the bottom metal level is aluminum or gold. The silicon nitride that results from PECVD in the gas mixture of Reaction 3.17 has two shortcomings: high hydrogen content (in the range of 20 to 30 atomic percent) and high stress. The high compressive stress (up to 5 × 10^9 dyn/cm^2) can cause wafer warping and voiding and cracking of underlying aluminum lines [Rosler, 1991]. The hydrogen in the nitride also leads to degraded MOSFET lifetimes. To avoid hydrogen incorporation, low-hydrogen or no-hydrogen source gases

TABLE 3.25   Properties of Silicon Nitride

Deposition	LP CVD	Plasma Enhanced CVD
Temperature (°C)	700–800	250–350
Density (g/cm^3)	2.9–3.2	2.4–2.8
Pinholes	No	Yes
Throughput	High	Low
Step coverage	Conformable	Poor
Particles	Few	Many
Film quality	Excellent	Poor
Dielectric constant	6–7	6–9
Resistivity (Ω-cm)	10^{16}	10^6–10^{15}
Refractive index	2.01	1.8–2.5
Atom % H	4–8	20–25
Energy gap	5	4–5
Dielectric strength (10^6 V/cm)	10	5
Etch rate in conc. HF	200 Å/min	
Etch rate in BHF	5–10 Å/min	
Residual stress (10^9 dyne/cm^2)	1T	2 C-5T
Poisson ratio	0.27	
Young's modulus	270 GPa	
TCE	1.6×10^{-6}/°C	

Note: C = compressive; T = tensile.
Source: Based on Adams (1988), Sinha (1978), and Retajczyk et al. (1980).

such as nitrogen may be employed instead of ammonia as the nitrogen source. Also, a reduced flow of SiH_4 results in less Si-H in the film. The hydrogen content and the amount of stress in the film are closely linked. Compressive stress, for example, changes toward tensile stress upon annealing to 490°C in proportion to the Si-H bond concentration. By adding N_2O to the nitride deposition chemistry, an oxynitride forms with lower stress characteristics; however, oxynitrides are somewhat less effective as moisture and ion barriers than nitrides.

One key advantage of PECVD nitride, besides the low-temperature aspect, is the ability to control stress during deposition. Silicon nitride deposited at the typical 13.56 MHz plasma excitation frequency exhibits tensile stress of about 400 MPa. When depositing the film at 50 kHz, a compressive stress of 200 MPa is typical. At such low-frequency (high-energy) the ion bombardment results in films with low compressive stress, lower etch rates, and higher density. Therefore, the effect of ion bombardment is more pronounced at lower pressures, and better quality CVD films ensue, characterized as films with a low wet etch rate (high film density) and low compressive stress. Experimental results show that, as the reactor pressure is lowered, film stress goes from tensile to compressive, and wet etch rates decrease. Often, stress dynamically changes when the film is exposed to the atmosphere and subsequent heating. Stress also is affected by moisture exposure and temperature cycling (for a good review, see Wu and Rosler [1992]).

We compare properties of silicon nitride formed by LPCVD and PECVD in Table 3.25, and Table 3.26 highlights typical PECVD process parameters. The refractive index of the deposited silicon nitride films is a measure of impurity content and overall quality. Its value ranges from 1.8 to 2.5 for PECVD films as compared to 2.01 for stoichiometric LPCVD (see next section). The higher numbers indicate excess silicon and a low number represents an excess of oxygen.

3.20.4.3 LPCVD Nitride

In the IC industry, stoichiometric silicon nitride (Si_3N_4) is LPCVD deposited at 700 to 900°C and at 200 to 500 mtorr and functions as an oxidation mask and as a gate dielectric in combination with thermally grown SiO_2. In micromachining, LPCVD nitride serves as an important mechanical membrane material and isolation/buffer layer. The standard source gases are SiH_2Cl_2 and ammonia, and at the above specified temperatures and pressures, this leads to a deposition rate of about 30 Å/min. The deposition of the

TABLE 3.26   Silicon Nitride PECVD Process Conditions

Flow (sccm)	SiH_4	190–270
	NH_3	1900
	N_2	1000
Temperature (°C)	T.C.	350 or 400
	Wafer	~330 or 380
Pressure (torr)	1100	2.9
RF power (watt)		
Deposition rate (Å/min)		1200–1700
Refractive index		2.0

Source: Reprinted with permission from Wu, T.H.T., and Rosier, R.S. (1992) "Stress in PSG and Nitride Films as Related to Film Properties and Annealing," *Solid State Technol.* vol. **35**, pp. 65–71. Copyright 1992 PennWell Publishing Company.

TABLE 3.27   Etching Behavior of LPCVD Si_3N_4

Etchant	Temperature (°C)	Etch Rate	Selectivity of Si_3N_4:SiO_2: Si
H_3PO_4	180	100 Å/min	10:1:0.3
CF_4-4% O_2 Plasma	—	250 Å/min	3:2.5:17
BHF	25	5–10 Å/min	1:200:±0
HF (40%)	25	200 Å/min	1:>100: 0.1

amorphous silicon nitride is reaction-limited, so it deposits on both sides of the wafers with equal thickness. Dichlorosilane is used instead of silane because it results in more uniform film thickness and the wafers can be spaced closer together for larger loads. For stoichiometric silicon nitride, typical gas flows are in a 10:1 ammonia to dichlorosilane ratio. Stoichiometric silicon nitride has a large residual tensile stress of about 10^{10} dynes/cm^2 and, as a consequence, film thickness is often limited to a few thousand angstroms. By increasing the Si content in silicon nitride, the tensile film stress reduces (even to compressive), the film turns more transparent, and the HF etch rate drops. Such films result by increasing the dichlorosilane:ammonia ratio [Sakimoto et al., 1982]. At an ammonia-to-dichlorosilane ratio of approximately 1 to 6, the films are nearly stress free for deposition temperatures of 850°C and pressures of 500 mtorr [Sekimoto et al., 1982]. Figure 3.115 illustrates the effect of gas flow ratio and deposition temperature on stress and the corresponding refractive index and HF etch rate [Sakimoto et al., 1982].

Silicon-rich or low-stress nitride emerges as an important micromechanical material. Low residual stress means that relatively thick films can be deposited and patterned without fracture. Low etch rate in HF means that films of silicon-rich nitride survive release etches better than stoichiometric silicon nitride. The etch characteristics of LPCVD SixNy are summarized in Table 3.27. The properties of a LPCVD SixNy film deposited by reaction of $SiCl_2H_2$ and NH3 (5:1 by volume) at 850°C were already summarized in Table 3.25.

3.20.5   CVD Silicon Dioxides

3.20.5.1   Introduction

To a large degree, silicon owes its success to its stable oxide. By contrast, germanium's oxide is soluble in water, and GaAs produces only leaky oxides. Silicon oxides, like other dielectrics, function as insulation between conducting layers, for diffusion and ion implementation masks, diffusion from doped oxides, capping doped oxides to prevent the loss of dopants, for gettering impurities, and for passivation to protect devices from impurities, moisture, and scratches. In micromachining, silicon oxides serve the same purposes but also act as sacrificial material. Phosphorus-doped glass (PSG), also called P-glass or phosphosilicate glass, and borophosphosilicate glasses (BPSG) soften and flow at lower temperatures, enabling

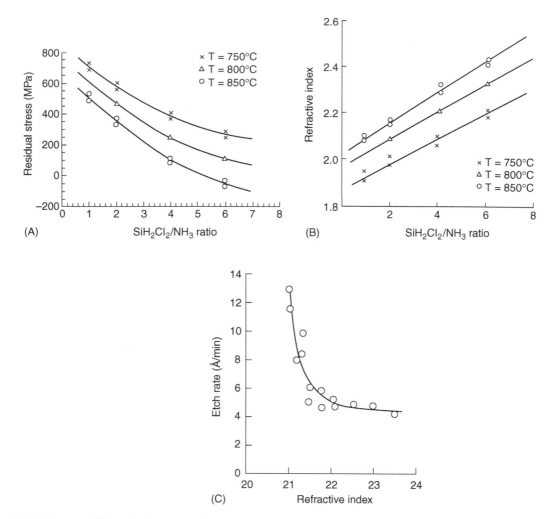

FIGURE 3.115 Silicon nitride LPCVD deposition parameters. (A) Effect of gas-flow ratio and deposition tempera-ture on stress in nitride films. (B) The corresponding index of refraction. (C) The corresponding HF etch rate. (Reprinted with permission from Sakimoto, M. et al. [1982] "Silicon Nitride Single-Layer X-ray Mask," *J. Vac. Sci. Technol.* 21, pp. 1017–21.)

the smoothing of topography. They etch much faster than SiO_2, which benefits their application as sacri-ficial material. Thermal oxidation of polysilicon is carried out similarly to that of single-crystal silicon, and the oxidation rate of undoped polysilicon falls between that of (100)– and (111)–oriented single-crystal silicon. As we saw before, phosphorous-doped polysilicon oxidizes much faster than undoped polysilicon, although this impurity enhancement of the oxidation rate is smaller in polysilicon than in single-crystal silicon. Now we consider CVD techniques to deposit doped and undoped SiO_2.

3.20.5.2 CVD Undoped SiO_2

Thermal oxidation is most commonly used for growing oxide films below 1 μm thick. LPCVD processes such as low-temperature oxidation (LTO), oxidation with tetraethylorthosilicate (TEOS), high-temperature oxidation (HTO), and plasma enhanced CVD (PECVD) enable the growth of thicker oxides (e.g., a few microns) at lower temperatures. These different CVD methods are compared with thermal SiO_2 growth in Table 3.28. In surface micromachining, LTO is most commonly used. In LTO, silicon dioxide is deposited on the wafer out of the vapor phase from the reaction of silane (SiH_4) with oxygen at relative low temperatures

TABLE 3.28 Comparison of Different Silicon Dioxide Growth/Deposition Processes

Deposition	Thermal	LTO	TEOS	HTO	PECVD
Source	O_2	$SiH_4 + O_2$	$TEOS + O_2$	$SiCl_2H_2 + N_2O$	$SiH_4 + N_2O$
Temperature (°C)	900–1100	400–450	700	900	200
Composition (typical inclusions in the films are shown in the parentheses)	SiO_2	$SiO_2(H)$	SiO_2	$SiO_2(Cl)$	$SiO_{1.9}(H)$
Step coverage	Conformal	Non-conformal	Conformal	Conformal	Non-conformal
Thermal stability	Stable	Densifies	Stable	Loses Cl	Loses H
Density (g/cm³)	2.2	2.1	2.2	2.2	2.3
Stress(10^9 dyne/cm²)	3 Compressive	3 Tensile	1 Compressive	3 Compressive	3 Compressive to 3 tensile

Source: Reprinted with permission from Adams, A.C. (1988) "Dielectric and Polysilicon Film Deposition," in *VLSI Technology*, Sze, S.M., ed., McGraw-Hill, New York.

(400 to 450°C) and low pressure (200 to 400 mtorr). Silane spontaneously combusts in the presence of oxygen (in other words, silane is pyrophoric), and to avoid a large depletion of the gases at the entrance of the furnace tube, the gases are introduced in the furnace through injectors along the length of the tube. The silane/oxygen gas mixture has also been applied in atmospheric pressure (AP) and plasma-enhanced conditions. The main advantage of silane/oxygen is the low deposition temperature; the main disadvantage is the nonconformal step coverage.

In general, TEOS brings about a better starting chemistry than the traditional silane-based CVD technologies. TEOS-based depositions lead to superior film quality in terms of step coverage and reflow properties. With an LPCVD reactor, the deposition temperature for TEOS is as high as 650 to 750°C, precluding its use over aluminum lines. In contrast, silicon dioxide is deposited by PECVD at 300 to 400°C from TEOS. CVD oxides in general feature porosity and low density. Low-frequency (high-energy) ion bombardment results in more compressive films, higher density, and lower etch rates as well as better moisture resistance [Rosler, 1991]. Use of TEOS oxide to replace SiH_4-based oxide in spin-on-glass (SOG) and photoresist planarization schemes now has become commonplace for devices with small features.

Another promising silicon dioxide deposition technology is the subatmospheric pressure CVD (SACVD). Both undoped and borophosphosilicate glass have been deposited in this fashion. The process involves an ozone and TEOS reaction at a pressure of 600 torr and at a temperature below 400°C. While offering the same good step coverage over submicron gaps, the films from the thermal reaction of ozone and TEOS exhibit relatively neutral stress and have a higher film density compared with low-pressure processes. This increased density gives the oxide greater moisture resistance, lower wet etch rate, and smaller thermal shrinkage. Compared with a 60 torr process, the film density has increased from 2.09 to 2.15 g/cm³, the wet etch rate decreased by more than 40%, and the thickness shrinkage changed from 12 to 4% after a 30 min anneal in dry N_2 at 1000°C [Lee et al., 1992].

3.20.5.3 CVD Phosphosilicate Glass Films

Adding a small amount of phosphine to the gas stream during deposition to obtain a lower melting point for the oxide (PH_3) results in a phosphosilicate glass (PSG). PSG is the same deposition process as LTO, with phosphine added to dope the glass with phosphorus from 2 to 8 wt%. Typical deposition rates are about 100 Å/min. Numerous applications of this material exist, such as

- Interlevel dielectric to insulate metallization levels
- Gettering and flow capabilities
- Passivation overcoat to provide mechanical protection for the chip from its environment
- Solid diffusion source to dope silicon with phosphorus
- Fast etching sacrificial material in surface micromachining

Both wet and dry etching rates of PSG are faster than the undoped material and depend on the dopant concentration. Profiles over steps get progressively smoother with higher phosphorous concentrations,

TABLE 3.29   Typical PSG Process Conditions

Flow (sccm)	SiH_4	150–230
	N_2O	4500
	N_2	1500
	PH_3	150[a]
Temperature (°C)	T.C.	400
	Wafer	~380–390
Pressure (torr)		2.2
RF power (watt)		1200
Deposition rate (Å/min)		4000–5000
Refractive index		1.46

[a]10% PH3 in N2.
Source: Reprinted with permission from Wu, T.H.T., and Rosler, R.S. (1992) "Stress in PSG and Nitride Films as Related to Film Properties and Annealing," *Solid State Technol.* vol. 35, pp. 65–71. Copyright 1992 PennWell Publishing Company.

TABLE 3.30   Typical BPSG Process Conditions

Parameter	Dimension	NSG	BPSG
Deposition temp.	°C	400	400
TEOS flow	g/min	0.33	0.66
O_2 flow	SLM	7.5	7.5
O_3/O_2	volume %	1	4.5
Carrier N_2 flow	SLM	18.0	18.0
B conc.	atomic %		4
P conc.	atomic %		6
Exhaust	mmH_2O	2.0	2.0
Growth rate	Å/min	1200	1800
Thickness	Å	1000	5800

Note: NSG = nondoped silicate glass; BPSG = borophosphosilicate glass.
Source: Reprinted with permission from Bonifield, T. et al. [1993] "Extended Run Evaluation of TEOS/Ozone BPSG Deposition," *Semicond. Int.*, July 1993, pp. 200–4.

reflecting the corresponding enhancement in viscous flow [Levy and Nassau, 1986]. The addition of phosphorous to LTO decreases the tensile residual stress, and at about 8% phosphorous, the stress level is reduced to about 108 dynes/cm^2 [Pliskin, 1977]. Increased ion bombardment (energy or density) in a PECVD PSG results in more stable phosphosilicate glass film with compressive as-deposited stress. Table 3.29 summarizes typical PSG process conditions [Wu and Rosler, 1992].

The addition of boron to P-glass further lowers its softening temperature. Flow occurs at temperatures between 850 and 950°C, even with phosphorous concentration as low as 4 wt%. A BPSG doped at 4% boron and 6% phosphorus is normal. BPSG deposition conditions are shown in Table 3.30 [Bonifield et al., 1993].

3.20.6   Metals in Surface Micromachining

Thin films of metals, with and without a sacrificial layer, have been used instead of polysilicon for many surface micromachining applications. These thin film structures are used to achieve specific properties, including high reflectivity (e.g., Al, Au), high mass density (e.g., W, Au, Pt), specific adsorption and adhesion characteristics (e.g., Pd, Ir, Au, Pt), and stable damping characteristics. We consider only two metals here, tungsten and aluminum. Tungsten CVD deposition is IC compatible. The material has some unique mechanical properties (see Tables 3.22 and 3.6) and can be applied selectively. Selective CVD tungsten has the unique property that tungsten will nucleate only on silicon or metal surfaces but does not deposit on dielectrics such as oxides and nitrides [Chen and MacDonald, 1991]. This enables the MEMS engineer to make multiple-level MEMS microstructures on fixed or moving substrates [MacDonald et al., 1989]. The tungsten films can be mechanically polished before the release step to obtain optically flat surfaces. Such multiple-layer tungsten structures are used to make mirrors, to add mass to structures, to make compact microactuators, and to make electrodes and electrostatic lenses to focus charged particles. In addition, the tungsten process and other similar metal processes can be used to make microelectron/ion optics for micro instruments with integrated tips.

Aluminum is employed in Texas Instruments' surface micromachined Digital Micromirror Device, and an organic polymer is used as the sacrificial layer. The aluminum metal is used both for the L-shaped flexure hinges and the mirror itself [Lin, 1996]. Details can be found in Example 2 of "Polysilicon Surface Micromachining Examples."

3.20.7 Polycrystalline Diamond and SiC Films

3.20.7.1 Diamond

If it could be micromachined, the exceptional physical properties of diamond (hardness, wear resistance, low coefficient of friction comparable with that of Teflon®, and thermal and chemical stability) would promise to expand the range of applications for microdevices. Diamond components may, for example, be as much as 10,000 times more wear-resistant than those made from silicon. Diamond is also a biocompatible material, so it could be used in the body as a drug-dispensing unit without initiating an allergic reaction. This is because carbon (the elemental ingredient of diamond) is chemically benign. In addition, diamond reduces stiction as compared with silicon, perhaps making diamond surface micromachined devices easier to manufacture. Diamond can be doped to change it from an insulator to a semiconductor. P-doping works well, but it remains difficult to make an n-type diamond.

Polycrystalline diamond film, deposited by CVD, is one potential approach to high-temperature harsh-environment MEMS [Herb et al., 1990] (see also <http://www.nasatech.com/Briefs/Feb99/NPO20529. html>). The most common polycrystalline diamond CVD process involves a flowing mixture of methane and hydrogen, typically at a total pressure of 45 torr (6 kPa) and a substrate temperature of 950°C. Problems with films formed this way include difficulty growing good quality diamond films, oxidation above 500°C for nonpassivated films, difficulty making reliable ohmic contacts to the material, reproducibility, and surface roughness of the films [Obermeier, 1995].

At Argonne National Laboratories, "ultrananocrystalline" diamond films have been deposited by a novel CVD process that may overcome some of the problems mentioned above (<http://www.techtransfer. anl.gov/techtour/diamondmems.html>). Argonne's patented CVD method was developed at first using fullerenes as the carbon source. Fullerene powder is vaporized and introduced into an argon plasma, causing the fullerenes to fragment into two-atom carbon molecules (dimers). Silicon or other substrate materials are "primed" with fine diamond powder, and when the carbon dimers settle out of the plasma onto the substrate, they arrange into a film of small diamond crystals about 3 to 5 nm in diameter. Subsequent work showed that the same result can be achieved by introducing methane into an argon plasma as long as little or no additional hydrogen is present. The Argonne method differs in two major respects from other CVD methods for making diamond film. First, the molecular building block is different (carbon dimers rather than methyl radicals), and second, little or no hydrogen is present in the plasma ($\sim$1%), while in other methods the plasma contains 97–99.5% hydrogen along with methane. The ultrananocrystalline films produced by Argonne's method are completely free of intergranular flaws and nondiamond secondary phases that degrade the properties of conventionally produced diamond films; they have crystals about 50 to 200 times smaller than those in other films and are 10 to 20 times smoother than conventional high-purity CVD diamond films.

For surface micromachining, a "starter" (nucleation) layer of nanocrystalline diamond powder on a silicon substrate is patterned, and the diamond film forms only in the areas where the nucleation layer remains. Flat, free-standing diamond films can also be grown on silicon dioxide. The diamond layer may be released from the substrate by etching away the silicon dioxide. The result is free-standing diamond structures as little as 300 nm thick with features as small as 100 nm and friction coefficients as low as 0.01.

Argonne is also developing deposition methods that permit integrated fabrication of complete devices, such as micron-sized pinwheels, gears, turbines, and micromotors, without manual assembly. Also, hollow 3-D structures can be fabricated in diamond surface micromachining. In this case, silicon is used as the sacrificial material (e.g., a disk or pin) for the desired diamond component. The diamond film is grown on the Si structure and conforms to its shape. The silicon is etched away, leaving a diamond cylinder or tube. It has been shown that this method is capable of making convex hollow structures (e.g., a hollow pyramid) — shapes that cannot be produced by conventional silicon lithographic methods.

Sandia researchers have explored amorphous diamond for micromachining. Amorphous diamond is the hardest substance known, after crystalline diamond (<http://www.sandia.gov/media/NewsRel/ NR2000/ diamond.htm>). They use pulsed laser deposition, and a simple proof-of-principle device — a diamond interdigitated comb — takes about three hours to fabricate. Subsequent annealing of the diamond film device so that it has zero stress (to prevent warpage) takes a few minutes. Before this work, amorphous diamond itself had been impractical because its tremendous internal stresses — hundreds of atmospheres — made it impossible for the material to stand alone or to coat thickly on any but the strongest substrates.

3.20.7.2   SiC

Like diamond, silicon carbide (SiC) is well known for its mechanical hardness, chemical inertness, high thermal conductivity, and electrical stability at temperatures well above 300°C, making it another excellent candidate for high-temperature MEMS. Both also exhibit piezoresistive properties. In comparison to diamond, other very attractive SiC features are that it can be doped both p- and n-type fairly easily, and it allows a natural oxide to be grown on its surface. SiC is thus an attractive candidate as an alternative semiconductor MEMS material. Because of its large band gap and good carrier mobility, it can be employed in high-temperature and high-power applications. In addition, the fabrication technology for SiC active electronic devices is based mostly on processes established in the Si microelectronics industry. One property that makes SiC films particularly attractive for micromachining is that these films can easily be patterned by dry etching using Al masks. Patterned SiC films can actually be used as passivation layers in the micromachining of the underlying Si substrate (SiC can withstand both KOH and HF etching!) [Krotz et al., 1995].

SiC crystallizes in many different polytypes, which differ from one another in the stacking sequence of a repeat unit consisting of two planes of close-packed Si and C atoms. The two most common SiC polytypes are 3C-SiC and 6H-SiC. The 3C polytype, also known as beta-SiC (or β-SiC), is the only polytype with a cubic structure. 3C-SiC crystallizes in a ZnS-type structure, hence it can be deposited on Si. Historically, most research has focused on developing 6H-SiC as a semiconductor material for high-temperature and high-power electronics. However, the small wafer size (<2 in) and the inability to grow epitaxial layers on substrates other than 6H-SiC has hindered development of the 6H polytype. Over the last ten years, there has been a growing interest in 3C-SiC as a MEMS material. As we saw above, unlike 6H-SiC, the 3C SiC polytype can be epitaxially grown on single-crystal silicon substrates. Large-area substrates enable low-cost batch processing, essential in making SiC MEMS devices viable for mass production applications (e.g., automotive). It was at Case Western Reserve University (CWRU) that the first successful depositions of spatially uniform single-crystal, 3C-SiC films on 4 in (100) silicon wafers were made. These epitaxial 3C-SiC films on Si were used to craft structures like diaphragms and cantilever beams. Since SiC films are highly resistant to KOH and EDP etching, dry etching and Al masks are employed. Because these SiC films must be grown directly on single-crystal silicon, surface micromachining with a sacrificial layer was not possible. The solution to making SiC surface micromachining possible was again invented at CWRU and involves growing polycrystalline cubic silicon carbide (poly-SiC) films of about 2 μm thickness by an APCVD process on 4 in, polysilicon-coated, (100) silicon wafers [Fleischman et al., 1996; Roy et al., 2000]. Surface micromachining is achieved by using the underlying polysilicon film as the sacrificial layer. The poly-SiC deposition process is carried out at 1280°C in a cold-wall, RF-induction heated, vertical APCVD reactor. After deposition, the SiC film is polished to reduce the surface roughness (R_a) from ~400 Å in the as-grown film to <40 Å. The poly-SiC lateral resonant device shown in Figure 3.116 was made by this surface micromachining approach, and the finished device exhibits Qs as high as 215,000 at pressures below 10^{-5} torr. Resonant frequency drifts of

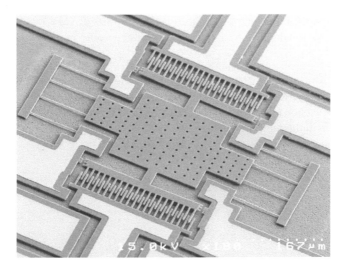

FIGURE 3.116 SEM micrograph of a released poly-SiC lateral resonant device. The suspension beam length and width are nominally 100 μm and 2.5 μm, respectively. Exposed poly-SiC shows up as dark gray, while Ni metallization appears light gray. (Courtesy of Dr. Mehregany, M., Case Western Reserve University.)

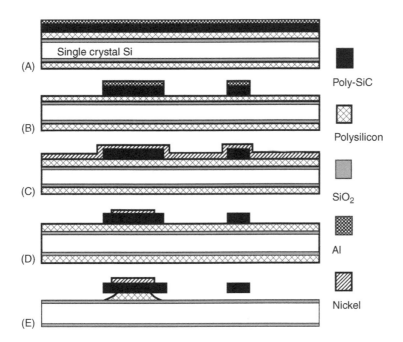

FIGURE 3.117 Schematic description of the poly-SiC resonator fabrication process showing cross-sections after: (A) a 5000 Å thick Al layer is sputter deposited over the polished poly-SiC; (B) the Al layer is patterned using lithography and anisotropic $CHF_3/O_2/He$ plasma etching to transfer the Al pattern into the poly-SiC; (C) the Al layer is stripped and a 7500 Å thick layer of Ni is deposited by sputtering; (D) patterning of the Ni film for contacts using photolithography and wet etching; and (E) release of poly-SiC in 40 wt% KOH at 40°C.

less than 18 ppm/hr have been observed, and device operation at elevated temperatures as high as 950°C has been achieved. In Figure 3.117, the poly-SiC surface micromachining steps are outlined.

Further advances in SiC deposition and patterning techniques have led to the development of single and multilevel implementations similar to the poly-Si MUMPs process. The Multi-User Silicon Carbide (MUSiC) is an eight-mask, four polycrystalline SiC (poly-SiC) layer surface micromachining process.

The state of Ohio has funded high-temperature MEMS efforts at Case Western Reserve University (<http://mems.EECS.cwru.edu/SiC/ >) and made the center a leader in the field.

3.20.7.3 GaAs

Table 3.31 compares some of the material properties for single-crystal SiC, Si, GaAs, and diamond with relevance to electronics and MEMS. Not enough good data are available yet to assemble a similar table for the polycrystalline version of the listed semiconductors. From a comparison of Si and GaAs for micro-machining applications in this table, we conclude that GaAs is a better material for thermal isolation and for higher-temperature operation. Single-crystal GaAs lends itself as a material for miniaturization applications

TABLE 3.31   Material Properties of Four Important MEMS Materials at 300 K

Property	3C-SiC	GaAs	Si	Diamond
Melting point (°C)	2830 (pressure is 35 bar; decomposes)	1238	1415	4000 (Phase change occurs)
Max. operating temp. (°C)	873	460	300	1100
Thermal conductivity (W/cm °C)	4.9	.5	1.57 Comparable to metals such as carbon steel (0.97) and Al (2.36)	20
Linear thermal expansion coeff. ($\times 10^{-6}$ °C^{-1})	4.7	5.9	2.35 The low expansion coefficient of Si is closer to quartz (7.1) than to a metal (e.g., 25 for Al) making it insensitive to thermal shock	0.08
Young's modulus (GPa)	448	75	190 (111) The elastic modulus is similar to that of steel[a]	1035
Physical stability	Excellent	Fair. Sublimation of As is a problem	Good	Excellent
Energy gap (eV)	2.39	1.42 Direct	1.12 Indirect	5.5
Chemical resistance	Very good	Poor	Good	Excellent
Electron mobility (cm^2/V s)	1000	8500	1500	2200
Hole mobility (cm^2/V s)	50	400	600	1600
Density	3.2	5.3	2.32 Lower density than Al (2.7), thus it has a high stiffness to weight ratio	3.5
Yield strength (Gpa)	21 (for 6H-SiC)	2.0	7 (steel is 2.1) IC grade Si is stronger than steel	53
Breakdown voltage ($\times 10^6$ V/cm)	2	0.4	0.3	10
Dielectric constant	9.7	13.1	11.9	5.5
Lattice constant (Å)	4.36	5.65	5.43	3.57
Knoop hardness (kg/mm^2)	3980	600	1000 (stainless steel is 660)	10,000
Sat. Electron Drift Velocity ($\times 10^7$ cm/s)	2.2	2	1.1	2.7

Sources: Data from Wu and Rosler, 1992; Sinha and Smith, 1978; Retajczyk and Sinha, 1980; Sakimoto et al., 1982; Lee et al., 1992; Pliskin, 1977; Bonifield et al., 1993; MacDonald et al., 1989; Lin, 1996; Herb et al., 1990; Obermeier, 1995; Krotz et al., 1995; Fleischman et al., 1996; Roy et al., 2000; Ericson et al., 1988; Takebe et al., 1993; Zhang and MacDonald, 1993; Hjort et al., 1990; Hjort et al., 1992; Karam et al., 1996; Morkoc et al., 1988; Yeh and Smith, 1994; 1994; Oden et al., 1996; Thundat et al., 1997; Thundat et al., 1995; Hansen et al., 2001; Moulin, 2000; Thundat et al., 2001; Harley and Kenny, 1999.

Note: Based on <http://books.nap.edu/books/0309053358/html/12.html > and < http://mems.EECS.cwru.edu/SiC/>.

because of its attractive thermal and optoelectronic properties. The material is less attractive for mechanical devices, with a yield load smaller by a factor of two compared with silicon [Ericson et al., 1988]. The many heterostructures possible with GaAs make a wide variety of optical components such as lasers and optical waveguides feasible. The piezoelectric effect enables piezoelectric transducers. Based on the high electron mobility, the material is also ideal for the measurement of magnetic fields through the Hall effect. "Macro" pressure, temperature, and vibration sensors making use of the influence of external pressure and temperature on the band gap have been built.

Besides wet [Ericson et al., 1988] and dry [Takebe et al., 1993] bulk micromachining of GaAs, some surface micromachining work has been reported [Zhang and MacDonald, 1993]. Surface micromachined sensor and actuator structures of metal on a GaAs substrate have been demonstrated in a scheme compatible with normal IC processing of GaAs. It was also shown that surface micromachined structures in epitaxial GaAs using $Al_xGa_{1-x}As$ ($x = 0.5$) as sacrificial layers are possible. By using several steps of MOCVD epitaxial layer regrowth, structures of polysilicon-like complexity were built [Hjort et al., 1990; Hjort et al., 1992].

Although GaAs enables the production of faster devices in the IC industry, its use is prohibitive due to its high cost. In the micromachining world, possible applications are optical shutters or choppers, actuators in monolithic microwave integrated circuits (MMICs), sensors using piezo- or optoelectrical properties of GaAs, or applications favoring integration of micromechanical devices with electronic circuitry for fast signal processing, high operating temperature, or high radiation tolerance. The latter applications so far have mainly lured research and development money from military agencies.

As the micromachining industry largely evolves with the IC industry, we cannot expect GaAs micromachines to succeed unless GaAs use penetrates the IC industry. In this context, Karam et al. (1996) are investigating gallium arsenide micromachining techniques using high electron mobility transistor (HEMT) and the metal Schottky field effect transistor (MESFET) foundry processes. Perhaps work on automated processes to deposit GaAs on Si could benefit both the IC and micromachining industries. For example, GaAs/Si wafers will be larger and stronger, have a better thermal conductivity, and be lighter than GaAs wafers [Morkoc et al., 1988]. Along this line, Yeh et al. trapped GaAs laser diodes in micromachined wells in a Si substrate [Yeh and Smith, 1994a, 1994b]. For a listing of dry and wet GaAs etchants, see Karam et al. (1996).

Micromachining is proving to be a good method to incorporate micromirrors in resonant optical cavities for tunable lasers. Perhaps micromachining using gallium arsenide and group III–V compound semiconductors will become a practical way to integrate RF switches, antennas, and other custom high-frequency components with ultra-high-speed electronic devices for wireless communications.

3.21 Polysilicon Surface Micromachining Examples

Example 1. Analog Devices Accelerometer

Both Robert Bosch GmbH (Stuttgart, Germany) and Analog Devices, Inc. (Norwood, Massachusetts) offer surface micromachined accelerometers based on lateral resonators. We will review only the Analog Devices ADXL accelerometer product family here. The ADXL-50 constituted the first commercially available surface micromachined MEMS structure. Today, surface micromachined accelerometers are incorporated in Ford and General Motors cars, among others, as well as inside computer game joysticks, robots, watches, shoes, etc.

To facilitate integration of their surface micromachined accelerometers with onboard electronics, Analog Devices Inc. opted for a mature 4-μm BICMOS process [Core et al., 1993]. BICMOS is a manufacturing process for semiconductor devices that combines bipolar and CMOS to give the best balance between available output current and power consumption. Figure 3.118 presents a photograph of the finished ADXL-50 accelerometer with on-chip excitation, self-test, and signal conditioning circuitry.

The suspended comb-like polysilicon structure in the center of the die is the sensitive element, and its primary axis of sensitivity lies in the plane of the die (x–y plane). In bulk micromachining, the sense axis is more often orthogonal to the plane of the die (z-axis). The polysilicon-sensing element of the ADXL family of sensors occupies only 5% of the total die area and consists of three sets of 2-μm thick polysilicon finger-like electrodes (Inset 3.11).

FIGURE 3.118   Analog Devices' ADXL-50 accelerometer with a surface micromachined capacitive sensor (center), on-chip excitation, self-test, and signal-conditioning circuitry. (Reprinted with permission from Core, T.A. et al. [1993] "Fabrication Technology for an Integrated Surface-Micromachined Sensor," *Solid State Technol.* **36**, pp. 39–47.)

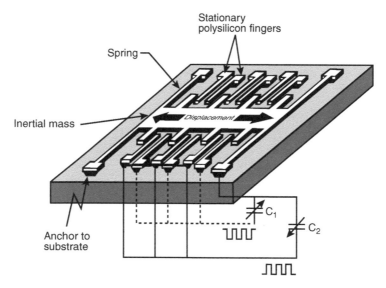

INSET 3.11   Illustration of the basic mechanical structure of Analog Devices' ADXL family of surface-micromachined accelerometers. A comb-like plate suspended from springs forms the inertial mass. Displacements of the mass are measured capacitively with respect to two sets of stationary finger-like electrodes. Based on Maluf [2000].

Two sets are anchored to the substrate, and a third set is suspended about 1 μm above the surface by means of two polysilicon beams acting as suspension springs. The fingers of the movable shuttle mass are interlaced with the fingers of the two fixed sets. In the core of the ADXL-50, the whole chip measures 500 × 625 μm and operates as an automotive airbag deployment sensor. The measurement accuracy is

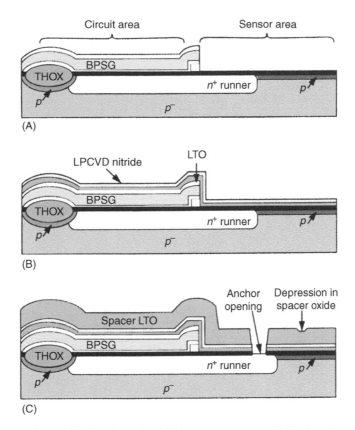

FIGURE 3.119 Preparation of IC chip for poly-Si. (A) Sensor area post-BPSG planarization and moat mask. (B) Blanket deposition of thin oxide and thin nitride layer. (C) Bumps and anchors made in LTO spacer layer. (Reprinted with permission from Core, T.A. et al. [1993] "Fabrication Technology for an Integrated Surface-Micromachined Sensor," *Solid State Technol.* **36**, pp. 39–47.)

5% over the ± 50 g range. Deceleration in the axis of sensitivity exerts a force on the central mass that in turn displaces the interleaved capacitor plates, causing a fractional change in capacitance. The overall capacitance is small, typically on the order of 100 fF and, for the ADXL05 (rated at ± 5 g), with an inertial mass of 0.3 µg only, the capacitance change is as small as 100 aF [Maluf, 2000]. These small capacitance changes necessitate on-chip integrated electronics to reduce the impact of parasitic sources. In operation, the ADXL family has a force-balance electronic control loop to prevent the mass from actual macroscopic movements, greatly improving output linearity because the center element never moves by more than a few nanometers. Applying a large-amplitude low-frequency voltage, below the natural frequency of the sensor, allows one to compensate for accelerometer plate movement by external acceleration. At the same time, the sensing excitation frequency (1 MHz) is much higher than the resonant frequency, so it does not produce an actuation force on the capacitor plates. As long as sense and actuation signals do not interfere, the sense and actuation plate may be the same.

In the sensor design, n^+ underpasses connect the sensor area to the electronic circuitry, replacing the usual heat-sensitive aluminum connect lines. Most of the sensor processing is inserted into the BICMOS process right after the borophosphosilicate glass planarization. After planarization, a designated sensor region, or moat, is cleared in the center of the die (Figure 3.119A). A thin oxide is then deposited to passivate the n^+ underpass connects, followed by a thin, low-pressure, vapor-deposited nitride to act as an etch stop (buffer layer) for the final poly-Si release etch (Figure 3.119B). The spacer or sacrificial oxide used is a 1.6-µm densified low-temperature oxide (LTO) deposited over the whole die (Figure 3.119C).

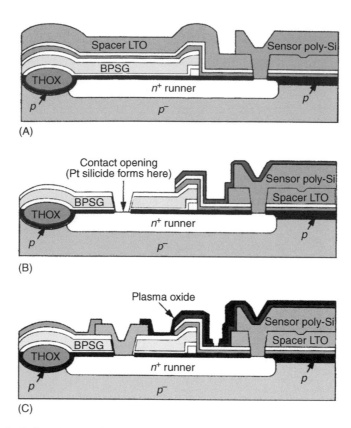

FIGURE 3.120 Poly-Si deposition and IC metallization. (A) Cross-sectional view after polysilicon deposition, implant, anneal, and patterning. (B) Sensor area after removal of dielectrics from circuit area, contact mask, and Pt silicide. (C) Metallization scheme and plasma oxide passivation and patterning. (Reprinted with permission from Core, T.A. et al. [1993] "Fabrication Technology for an Integrated Surface-Micromachined Sensor," *Solid State Technol.* 36, pp. 39–47.)

In a first timed etch, small depressions that will form bumps or dimples on the underside of the polysilicon sensor are created in the LTO layer. These will limit stiction in case the sensor comes in contact with the substrate. A subsequent etch cuts anchors into the spacer layer to provide regions of electrical and mechanical contact (Figure 3.119). The 2-μm thick sensor poly-Si is then deposited, implanted, annealed, and patterned (Figure 3.120A). The relatively deep junctions of the BICMOS process permit the polysilicon thermal anneal as well as brief dielectric densifications without resulting in degradation of the electronic functions. Next is the IC metallization, which starts with the removal of the sacrificial spacer oxide from the circuit area along with the LPCVD nitride and LTO layer. A low-temperature oxide is deposited on the poly-Si sensor part, and contact openings appear in the IC part of the die where platinum is deposited to form a platinum silicide (Figure 3.120B). The trimmable thin-film material, TiW barrier metal, and Al/Cu interconnect metal are sputtered on and patterned in the IC area. The circuit area is then passivated in two separate deposition steps. First, plasma oxide is deposited and patterned (Figure 3.120C), followed by a plasma nitride (Figure 3.119A) to form a seal with the earlier deposited LPCVD nitride. The nitride acts as an HF barrier in the subsequent long etch release. The plasma oxide left on the sensor acts as an etch stop for the removal of the plasma nitride (Figure 3.121A). Subsequently, the sensor area is prepared for the final release etch. The undensified dielectrics are removed from the sensor, and the final protective resist mask is applied. The photoresist protects the circuit area from the long-term buffered oxide etch (Figure 3.121B). The final device cross-section is shown in Figure 3.121C.

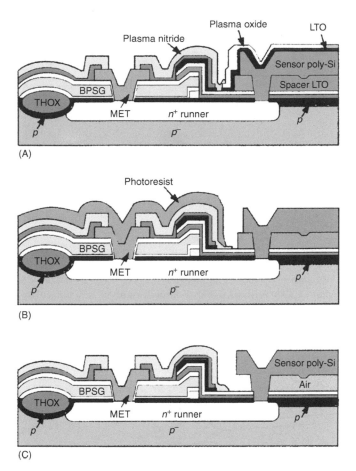

FIGURE 3.121 Prerelease preparation and release. (A) Post-plasma nitride passivation and patterning. (B) Photoresist protection of the IC. (C) Freestanding released poly-Si beam. (Reprinted with permission from Core, T.A. et al. [1993] "Fabrication Technology for an Integrated Surface-Micromachined Sensor," *Solid State Technol.* **36**, pp. 39–47.)

Example 2. TI Micromirrors

In 1987, the first Digital Micromirror Device (DMD) was developed at Texas Instruments (U.S. Patent 4,615,595, October 7, 1986). A typical DMD consists of a two-dimensional array of optical switching elements (pixels) on a silicon substrate as shown in Figure 3.122. Two pixels are schematically illustrated in Figure 3.123 along with the underlying Si chip and circuitry. Each pixel is made up of a reflective aluminum micromirror supported from a central post. The central mirror post at the back of the mirror is mounted on a lower aluminum metal platform — the yoke. The yoke is suspended above the silicon substrate by thin compliant L-shaped hinges (made from a proprietary Al alloy) anchored to the underlying substrate by two stationary posts. The different components of an individual micromirror are illustrated in Figure 3.124. Two bias electrodes tilt the mirror either $+10°$ or $-10°$ by applying 24 V between one electrode or the other and the yoke. Off-axis illumination of the Al mirror reflects into the projection lens only when the micromirror is in its $+10°$ state, producing a bright appearance, or ON state. In the flat position and in the $-10°$ state, the pixel appears dark. In the fully deflected position, the yoke touches a landing site with its landing tips, and because the landing site is biased at the same voltage, electrical shorting is prevented. Once the applied voltage is removed, the springy Al alloy hinges restore the micromirror to its initial position. A standard DMD microchip contains more than 442,000 switchable mirrors on a 5/8-in wide surface. Mirrors are switched according to memory impulses stored in static random-access memory (SRAM) cells beneath the tiny array. The mirrors can be independently cycled at 100,000 flips

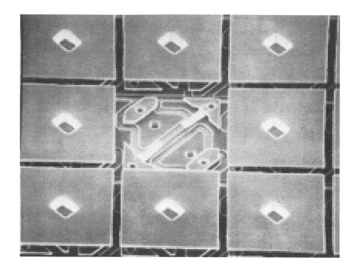

FIGURE 3.122 Texas Instruments DMD pixel array. One pixel has been removed to show the Si chip below the reflective aluminum mirrors. Reprinted with permission from <http://www.semiconductor.net/semiconductor/issues/issues/2000/200002/six0002et.asp>

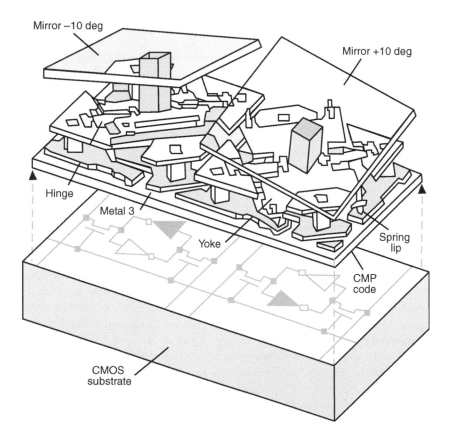

FIGURE 3.123 Schematic of two pixels in a Texas Instruments of a DMD. The mirrors are made transparent for clarity of the drawing. The +10° mirror is in the ON position and the −10° is in the OFF position. (Based on <http://www.spie.org/web/oer/october/oct98/tv.html>

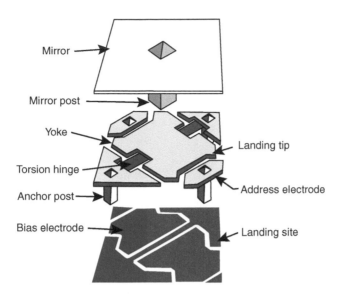

FIGURE 3.124 Illustration of the various components of a single DMD pixel. The basic structure consists of a bottom aluminum layer containing two electrodes, a middle aluminum layer containing a yoke suspended by two torsional hinges, and a top reflective aluminum mirror. An applied electrostatic voltage on a bias electrode deflects the yoke and the mirror toward that electrode. A pixel measures approximately 17 μm on a side. (Adapted with permission from Maluf, N. [2000] *An Introduction to Microelectromechanical Systems Engineering*, Artech House, Boston.).

per second. Greys can be achieved by multiple mirrors or by flipping the mirror quickly between black and white states. Colors can be generated by having the incident light shine through a rotating disk divided into three colored segments, red, blue, and green. A separate image is generated for each color, timed to appear as the appropriate segment covers the light source. Printing and display technology can now enjoy the advantage of digital fidelity and digital stability.

The surface micromachining process to fabricate DMDs on wafers incorporating CMOS electronic address and control circuitry is illustrated in Figure 3.125. Because of the underlying active circuitry and the presence of Al metal for connectors and MEMS structures, all micromachining process steps are carried out at temperatures below 400°C. After completing the CMOS circuitry, a thick oxide is deposited over the whole Si wafer and is chemomechanically polished (CMP) to provide a flat surface to start building the mirror array. A sputter-deposited Al layer is patterned to provide bias and address electrodes, landing pads, and electrical interconnects to the underlying electronics. Hardened photoresist is used as the sacrificial material (Figure 3.125A). A proprietary Al alloy is sputter deposited to form the hinges for the mirror. It is the nature of this Al alloy that secures the mechanical integrity of the mirror actuation. Subsequently, the torsion hinge regions are protected by a patterned thin PECVD deposited silicon dioxide (Figure 3.125B). In the next step, a thicker coat of another proprietary aluminum alloy is deposited to form the yoke structure; this new coat of Al buries the thin oxide hinge mask. A second PECVD oxide mask is deposited over this second level of Al metal and patterned in the shape of the yoke and anchor posts (Figure 3.125C). In a dry etch step, the exposed aluminum areas are removed down to the organic sacrificial resist except where the oxide hinge mask remains. In those regions, only the thick yoke metal is removed, stopping on the SiO_2 mask and so preserving the underlying hinge structure (Figure 3.125D). Both thin layers of PECVD oxide mask are stripped before a second layer of sacrificial resists is deposited, UV-hardened, and patterned. A third aluminum alloy is sputter deposited and defines the mirror and the central mirror post. Again, a thin layer of PECVD silicon dioxide is deposited and patterned to define the mirror (Figure 3.125E). An oxygen plasma etch removes both sacrificial layers and releases the micromirrors (Figure 3.125F). Finally, after release, a special passivation step deposits a thin antistiction layer to prevent adhesion between the yoke tips and the landing pads. Because the weight of the micromirrors is

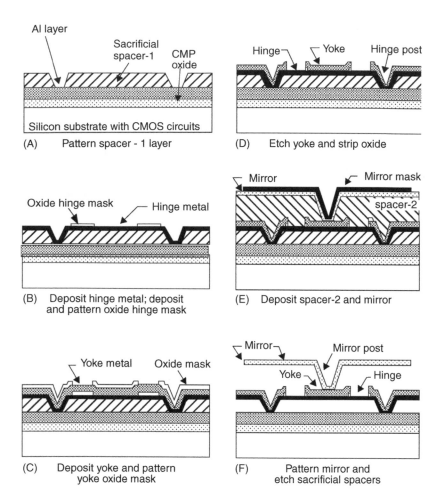

FIGURE 3.125   Fabrication steps of the Texas Instruments DMD. Steps A to F are explained in the text. (Based with permission on Maluf, N. [2000] *An Introduction to Microelectromechanical Systems Engineering*, Artech House, Boston.)

insignificant, the DMD micromirrors can withstand 1500 g mechanical shocks. Optimization of the hinge metal alloy and fabrication processes has resulted in a mean time between failure (MTBF) of more than 100,000 hr. Invented by Larry J. Hornbeck at TI, DMD is the key component in more than 17 projector brands and has brought digital cinema to *Star Wars* and *Toy Story II*. For this amazing piece of engineering, TI's Hornbeck received the first Emmy award ever bestowed for a projection display technology (<http://www.spie.org/web/oer/october/oct98/tv.html>).

A competing technology, the Actuated Mirror Array (AMA) (now Daewoo's TMA), invented by Gregory Um, is also a MEMS technology. Unlike DMD, TMA is built with piezo materials and, for high resolution, its large array size and consequent cost still pose problems. But it achieves brightness 15% higher than any other projector. At the University of Wisconsin, arrays for "gene expression analysis," usually produced with lithographic masks, are being replaced with the same DMD used for projection display to make "virtual masks" containing nearly half a million features.

Example 3. Design of SOI-Based High-Sensitivity Piezoresistive Cantilevers for Label-Less Sensing

Micromachined single-crystal Si cantilevers of the type shown in Figure 3.126 can be used for real-time, in situ measurements of physical parameters such as viscosity [Oden et al., 1996], pressure, density, flow rate, and temperature. The latter is achieved simply by coating a cantilever with metal on one side to

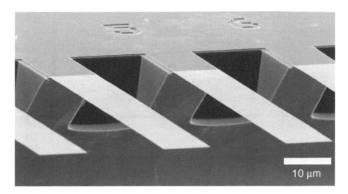

FIGURE 3.126 Scanning electron micrograph of a section of a microfabricated silicon cantilever array (eight cantilevers, each 1 μm thick, 500 μm long, and 100 μm wide, with a pitch of 250 μm, spring constant 0.02 Nm^{-1}. (Reprinted with permission from Fritz et al. [2000] "Translating Biomolecular Recognition into Nanomechanics," *Science* **288**, pp. 316–18. Copyright 2000 American Association for the Advancement of Science. [Courtesy of Dr. J. Fritz, MIT Media Lab.])

form a bimetal [Thundat et al., 1997]. Mercury vapor [Thundat et al., 1997], moisture, volatile mercaptans, DNA hybrization [Thundat et al., 1995], discrimination of single-nucleotide (SNP) mismatches in DNA [Hansen et al., 2001], protein conformational changes [Moulin, 2000], and antibody–antigen binding may also be monitored using cantilevers [Thundat et al., 2001]. For chemical and biochemical sensing, the microcantilever surface must be derivatized with chemically selective coatings, for example, through self-assembled alkanethiols or organosilane films, direct covalent attachment of molecular receptors, or dip coating. The cantilever measuring principle may be based on the detection of changes in resonance response comprising frequency, phase, amplitude, and Q-factor. An alternative approach involves the measurement of bending induced by the adsorption of molecules onto a thin microcantilever that has two chemically different opposing surfaces. The resonance variations are due to mass loading, surface stress, damping, or a combination of these variables, while bending is due to differential surface stresses on opposing sides of the cantilever.

In the current example, we consider the design of a cantilever optimized for maximum mass and bending stress sensitivity — sensitive to discriminate between DNA fragments differing in length by one base pair only. Most importantly, the sensor and accompanying instrumentation should not require a fluorescence or radioactive label and detection thereof to ascertain the binding of any type of biological affinity pair, and it should be inexpensive and compact. Bending and resonance frequency shifts of a cantilever can be measured with high precision using optical reflection with a diode laser and a linear position sensitive detector (PSD), but piezoresistive, capacitive, and piezoelectric measurements are also in use. All these various measuring options are commonly employed in atomic force microscopy (AFM). In this example, we want to implement an integrated piezoresistive resistor for position sensing, as this approach affords lower cost and portability [Harley and Kenny, 1999].

Consider a uniform beam with a rectangular cross-section that is fixed at one end and deflected at the other (Inset 3.12). Such a beam experiences lengthwise stress that is compressive below the centerline and tensile above it. The effective spring constant k of this beam is given by:

$$k = \frac{EWt^3}{4L^3} \tag{3.68}$$

with E the Young's modulus and W, t, and L the width, thickness, and length of the beam respectively. Representative values for W, t, and L are 20 μm, 0.6 μm, and 100 μm, with a resulting value for k of 0.1 N/m.

The natural frequency f_0 of a simple spring-mass system is given by:

$$f_0 = \frac{1}{2\pi}\sqrt{\frac{k}{m}} \tag{3.69}$$

with k the spring constant and m the mass.

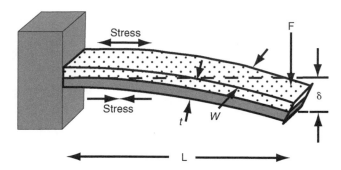

INSET 3.12   Stresses in a uniform cantilever beam when deflected.

TABLE 3.32   Gravimetric sensitivity of acoustic devices

Device Type	Resonance Frequency f_0 (MHz)	S_m Relation	S_m (cm²/g)	MDMD (ng/cm²)
Microcantilever with distributed load	5-0.02	$S_m\sim 1/\rho t$	10,000	0.02
Microcantilever with end load	5-00.2	$S_m\sim 1/2\rho t$	5,000	0.04
SAW	112	$S_m\sim 1/\rho\lambda$	151	1.2
Quartz microbalance	6	$S_m\sim 1/\rho t$	14	10
Acoustic plate mode device	104	$S_m\sim 1/\rho t$	65	1.0
Flexural plate wave (FPW)	2.6	$S_m\sim 1/\rho t$	951	0.4

The effective mass of the beam is related to the mass of the beam m_b through $m = nm_b$, where n is a geometric parameter, which for a rectangular bar is 0.24 [Chen et al., 1995]. Under the assumption of negligible variation in spring constant and uniformly distributed mass loading of the beam, Equation (3.69) also describes the mass dependence of the cantilever, in which case the effective mass is that of the cantilever plus adsorbate. Adding a discrete mass m_d at the very end of the cantilever defines a total effective mass m_T of the cantilever-adsorbate system as $m_T = m + m_d$. This results in an expression for the resonant frequency given by Equation (3.69) by substituting Equation (3.68) into Equation (3.69):

$$f_0 = \frac{1}{2\pi}\sqrt{\frac{EWt^3}{4m_T L^3}} = \frac{1}{2\pi}\sqrt{\frac{EWt^3}{4L^3(m_d + 0.24WtL\rho)}} \tag{3.70}$$

where ρ is the density of the cantilever. To compare the mass sensitivity S_m (which is defined as $S_m = \lim 1/f\,\Delta f/\Delta m = 1/f\,df/dm$, with Δm and dm normalized to the active area of the device) of a cantilever sensor to that of other gravimetric devices (as tabulated in Table 3.32]), we take the mass derivative of Equation (3.70) and divide by the operational frequency. For an end-loaded cantilever (mass only allowed to accumulate in a small area at the free end of the beam), this results in:

$$S_m = \frac{-\xi}{2\rho(\xi t_d + 0.24t)} \tag{3.71}$$

where ξ and t_d are the fractional area coverage and thickness of the deposited mass. The negative sign denotes that the frequency is decreasing with increasing mass piling up. In the case in which mass loading is distributed evenly over the cantilever surface, S_m is given by:

$$S_m = \frac{1}{\rho t} \tag{3.72}$$

The positive sign indicates that as mass is added the resonant frequency increases, corresponding to the increase of the cross-sectional thickness of the beam. Table 3.32 also lists the minimum detectable mass density (MDMD) estimated from Equation (3.72) by assuming a ratio of minimum detectable frequency shift to operation frequency $\Delta f/f$ of 2×10^{-7}:

$$\Delta m_{\min} = \frac{1}{s_m} \frac{\Delta f_{\min}}{f} \tag{3.73}$$

where $\Delta m_{\min}$ and $\Delta f_{\min}$ are the minimum detectable mass density and minimum detectable frequency change, respectively.

For end-loaded devices, an MDMD of 0.04 ng/cm^2 is calculated, and for a distributed device 0.02 ng/cm^2. From the numbers in Table 3.32, cantilevers in principle are the most sensitive and have the lowest detection limit. Cantilevers are projected to be about an order of magnitude better than flexural plate wave devices (FPW), the next best gravimetric sensor type. From this table, for mass sensing with a cantilever a uniform mass loading is preferred. From Equations (3.71) and (3.72), it is further advantageous to make the cantilever as low in density (ρ) and as thin as possible (t). In practice, using an optical detection method, a mass sensitivity as small as 0.7 picograms has been reported [Thundat et al., 1995].

The above design considerations involve a beam in resonance mode. Now let us consider the optimal design rules for measuring bending induced by differential surface stresses. This is especially important for measurements in liquids; immersion of a cantilever in water damps the resonance response to a value approximately an order of magnitude less than in air, while the bending response remains unaffected by the presence of water. Moreover, Thundat et al. claim that adsorption-induced stress sensors have a sensitivity three orders of magnitude higher than frequency-variation-based sensors [Thundat et al., 2001]. In practice, using optical techniques to measure deflection, surface stresses as small as $\sim 10^{-4}$ Nm^{-1} have been measured [O'Shea et al., 1996]. Here, we want to implement a simpler piezoresistive measurement. Unfortunately, this usually comes at the expense of resolution. The aim then is to design the cantilever and the piezoresistor such that the sensitivity obtained with the optical methods can be maintained.

For the analysis of the degree of bending caused by differential stress, we examine Stoney's formula. In 1905, Stoney derived a relation between adsorption-induced surface stress and the radius of curvature of a thin substrate. In the case in which the substrate is a thin cantilever, the Stoney equation is given by:

$$\frac{1}{R} = \frac{6(1 - v)}{Et^2} \sigma \tag{3.74}$$

where R corresponds to the radius of curvature of the cantilever, t its thickness, σ the differential surface stress ($\sigma_1 - \sigma_2$, i.e., the difference in surface stress between the top and bottom surfaces, in units of N/m), and v and E are Poisson's ratio and Young's modulus for the substrate respectively (see also Equation [3.54] for bending of a thin disk).

From geometric considerations, the radius of curvature is related to the displacement of the free end of the cantilever, δ, and its length L as $1/R = 2\delta/L^2$. Combining this last expression with Equation (3.74) we derive the cantilever displacement as a function of the differential surface stress as:

$$\delta = \frac{3L^2(1 - v)}{Et^2}(\sigma_1 - \sigma_2) \tag{3.75}$$

From this equation, at a constant differential surface stress, to maximize the deflection we want to decrease t and E, increase L, and decrease v. To maximize sensitivity, the piezoresistor should be placed in the zone of maximal stress. The stress in the beam is zero at the centerline and increases linearly with the distance away from the centerline. The stress also increases toward the base of the cantilever so that the highest sensitivity will be achieved with a piezoresistor placed at the surface of the cantilever beam near the base. The stress at that point can be calculated to be:

$$\sigma_{max} = \frac{6L}{Wt^2} F = \frac{3Et}{2L^2} \delta \tag{3.76}$$

where F is the applied force (i.e., $k \times \delta$). (See also Equation [3.32] for plate bending.) The resulting fractional resistance change is given by the piezoresistive relation:

$$\frac{\Delta R}{R} = \pi_l \sigma_{max} = \beta \frac{6L\pi_l}{Wt^2} F = \beta \frac{3Et\pi_l}{2L^2} \delta \tag{3.77}$$

where π_l is the longitudinal piezoresistive coefficient of silicon at the operating temperature and at a given doping. The cantilever is oriented along the <110> crystallographic axis of silicon where the piezoresistive coefficient π_l is maximum (see Table 3.5) [Tortonese et al., 1991; Tortonese et al., 1993]. The coefficient β is a correction factor between 0 and 1 and accounts for the fact that the resistor is not limited to the surface of the cantilever but has a certain depth. Its value depends on the silicon doping depth profile and the thickness of the beam. The doping depth is determined by the implantation parameters and subsequent thermal processes. It is important that the resistor be made shallow so that the current flows as close as possible to the surface of the cantilever. In an extreme case, a uniformly doped cantilever would exhibit no net piezoresistive response because opposite sign stresses on the top and the bottom of the cantilever would give an equal but opposite contribution to the change in resistance. Substituting Equation (3.75) in Equation (3.77), we obtain the expression for fractional resistance change in terms of surface stress:

$$\frac{\Delta R}{R} = \beta \frac{3\pi_L(1-v)}{t} (\sigma_1 - \sigma_2) \tag{3.78}$$

Thinning the cantilever is thus the principal means to obtain the largest resistivity change. Making the overall device smaller (W, L, and t) is helpful for making denser arrays and is simpler to achieve with piezoresistive cantilevers, as they are easier to make smaller than the ones that are used in conjunction with an optical detection technique, principally because their surface area is not limited by the laser spot size (typically about $30\,\mu m$).

We now address the need to design in a reference cantilever. Immersion of cantilevers in a liquid results in two reported long-term drift phenomena. First, thermal effects (especially where a gold layer on one surface creates a sensitive bimetallic) occur slowly over a period of hours. Second, another slow effect, which takes up to 10 hr to stabilize, seems associated with slow rearrangement of surface adsorbates [Moulin, 2000]. Both drift phenomena imply that a differential-type measurement will be beneficial to extract the data from adjoining cantilevers and to exclude other common-mode drift phenomena such as low-frequency vibrations. The latter approach was implemented by Fritz et al., who demonstrated that, in doing so, a single mismatch between two DNA sequences could be detected [Fritz et al., 2000]. From the above, it also would be preferable to derivatize the cantilever pairs with the required chemical coatings without using a metal deposit. This can reduce the temperature effect and lessen the mass of the sensor.

The fabrication challenge is thus to make an array with the thinnest possible cantilevers and with an extremely shallow doped layer for the piezoresistor at its base so as to keep β in Equation (3.77) as close to 1 as possible. Moreover, a symmetrical Wheatstone bridge design with a built-in reference cantilever is required to reduce the described background drift and common-mode vibrations. For the fabrication of large arrays of cantilever pairs, one would like to have them all as identical as possible; also if possible the chemical coatings should be deposited directly on the beam rather than on a deposited metal layer. Our proposed design incorporates features from various MEMS research groups around the world. The first concerns the depth of the doped region in the piezoresistors. For very thin beams, it is difficult to confine the doped region to the surface of the beam. Since activation of dopants is achieved by annealing, dopant diffusion is unavoidable.

To maximize β, we rely on Harley et al., who made cantilevers under 1000 Å thick (870–900 Å) with lengths ranging from 10 to 350 μm and widths from 2 to 44 μm [Harley and Kenny, 1999]. To reduce the depth beyond the capabilities of conventional implantation, this team used vapor-phase epitaxial growth to deposit the boron-doped layer. The boron atoms are incorporated into the lattice during the epitaxy, so an activating anneal is not required. Furthermore, since there is no damage-enhanced mobility, some high-temperature

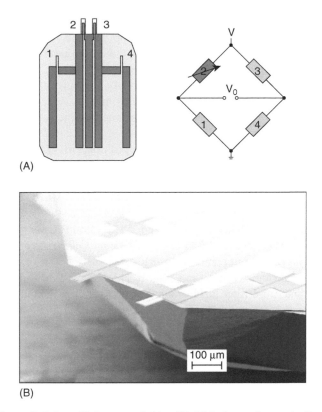

FIGURE 3.127 (A) Thermally balanced Wheatstone bridge. (B) SEM of a cantilever pair. (Sensor and reference courtesy of Dr. Anja Boisen, Technical University Denmark.)

steps can be tolerated. Their fabrication starts by growing and removing a 2000 Å thermal oxide from the epi Si layer of a 10 W-cm p-type SIMOX SOI wafer, thinning the Si layer to 800 Å. A 30-s HCl clean in the epichamber removes another 100 Å, before a 300 Å of $4 \times 10^{19}\,cm^{-3}$ boron-doped Si is grown over the entire wafer. The boron-doped layer defines the thickness of the resistors. The intermediate oxide layer is used as an etch stop during the process of etching the bulk silicon substrate in one of the last fabrication steps. In a first step, the cantilevers are patterned and plasma etched. Boron contacts are implanted at $1 \times 10^{15}\,cm^{-2}$, 30 keV, followed by a 200 Å growth of passivating thermal oxide during a 3 hr anneal at 700°C. For contacting the piezoresistors, aluminum is deposited and annealed in a forming gas at 400°C for 1 hr. Subsequently, a release mask is sputtered on the back of the wafer, and a Bosch DRIE is used to release the cantilevers. The DRIE stops at the buried oxide, which is removed with a 6:1 buffered oxide etch (BOE). For implementing a reference probe, one could take a cue from Thaysen et al. (2000), who fabricated a thermally symmetrical Wheatstone bridge as shown in Figure 3.127.

References

Abe, T., and Matlock, J.H. (1990) "Wafer Bonding Technique for Silicon-on-Insulator Technology," *Solid State Technol.* November, pp. 39–40.

Abe, T., Messner, W.C., and Reed, M.L. (1995) "Effective Methods to Prevent Stiction during Post-Release-Etch Processing," in *Proceedings: IEEE Micro Electro Mechanical Systems (MEMS '95)*, Amsterdam, Netherlands, pp. 94–99.

Abe, T., and Reed, M.L. (1996) "Low Strain Sputtered Polysilicon for Micromechanical Structures," in *Ninth Annual International Workshop on Micro Electro Mechanical Systems*, San Diego, pp. 258–62.

Abu-Zeid, M.M., Kendall, D.L., Guel, G.R., and Galeazzi, R. (1984) "Corner Undercutting in Anisotropically Etched Isolation Contours," *J. Electrochem. Soc.* 131, pp. 2138–42.

Abu-Zeid, M.M., Kendall, D.L., Guel, G.R., and Galeazzi, R. (1985) "Abstract 275," *JECS*, Toronto, Canada, p. 400.

Acosta, R.E., Ahn, C., Babich, I.V., Cooper, E.I., Cotte, J.M., Horkans, W.J., Jahnes, C., Krongelb, S., Kwietniak, K.T., Labianca, N.C., O'Sullivan, E.J.M., Pomerene, A.T., Rath, D.L., Romankiw, L.T., and Tornello, J.A. (1995) "Integrated Variable Reluctance Magnetic Mini-Motor," *J. Electrochem. Soc.* 95-2, pp. 494–95.

Adams, A.C. (1988) "Dielectric and Polysilicon Film Deposition," in *VLSI Technology*, Sze, S.M. ed., McGraw-Hill, New York.

Ahn, C.H., Kim, Y.J., and Allen, M.G. (1993) "A Planar Variable Reluctance Magnetic Micromotor with a Fully Integrated Stator and Wrapped Coils," in *Proceedings: IEEE Micro Electro Mechanical Systems (MEMS '93)*, Fort Lauderdale, pp. 1–6.

Alavi, M., Buttgenbach, S., Schumacher, A., and Wagner, H.J. (1991) "Laser Machining of Silicon for Fabrication of New Microstructures," in *6th International Conference on Solid-State Sensors and Actuators (Transducers '91)*, San Francisco, pp. 512–15.

Alavi, M., Buttgenbach, S., Schumacher, A., and Wagner, H.J. (1992) "Fabrication of Microchannels by Laser Machining and Anisotropic Etching of Silicon," *Sensor. Actuator. A,* A32, pp. 299–302.

Allen, D.M. (1986) *The Principles and Practice of Photochemical Machining and Photoetching*, Adam Hilger, Bristol and Boston.

Allen, M.G., Mehregany, M., Howe, R.T., and Senturia, S.D. (1987) "Microfabricated Structures for the In Situ Measurement of Residual Stress, Young's Modulus and Ultimate Strain of Thin Films," *Appl. Phys. Lett.* 51, pp. 241–43.

Alley, R.L., Cuan, G.J., Howe, R.T., and Komvopoulos, K. (1988) "The Effect of Release Etch Processing on Surface Microstructure Stiction," in *Technical Digest: 1988 Solid State Sensor and Actuator Workshop*, Hilton Head Island, SC, pp. 202–7.

Alley, R.L., Howe, R.T., and Komvopoulos, K. (1992) "The Effect of Release-Etch Processing on Surface Microstructure Stiction," in *Technical Digest: 1992 Solid State Sensor and Actuator Workshop*, Hilton Head Island, SC, pp. 202–7.

Alley, R.L., Mai, P., Komvopoulos, K., and Howe, R.T. (1993) "Surface Roughness Modifications of Interfacial Contacts in Polysilicon Microstructures," in *7th International Conference on Solid-State Sensors and Actuators (Transducers '93)*, Yokohama, pp. 288–91.

Allongue, P., Costa-Kieling, V., and Gerischer, H. (1993) "Etching of Silicon in NaOH Solutions," parts 1 and 2," *J. Electrochem. Soc* 140, pp. 1009–18 and 1018–26.

Ammar, E.S., and Rodgers, T.J. (1980) "UMOS Transistors on (110) Silicon," *IEEE Trans. Electron Devices* ED-27, pp. 907–14.

"Analog Devices Combine Micromachining with BICMOS," (1991) *Semicond. Int.* 14, p. 17.

Anderson, R.C., Muller, R.S., and Tobias, C.W. (1994) "Porous Polycrystalline Silicon: A New Material for MEMS," *J. Microelectromech. Syst.* 3, pp. 10–18.

Archer, V.D. (1982) "Methods for Defect Evaluation of Thin <100> Oriented Silicon in Epitaxial Layers Using a Wet Chemical Etch," *J. Electrochem. Soc.* 129, pp. 2074–76.

Arita, Y., and Sunohara, Y. (1977) "Formation and Properties of Porous Silicon Film," *J. Electrochem. Soc.* 124, pp. 285–95.

Asano, M., Cho, T., and Muraoko, H. (1976) "Applications of Choline in Semiconductor Technology," *Electrochem. Soc. Extend. Abstr.* 354, pp. 911–13.

Barret, S., Gaspard, F., Herino, R., Ligeon, M., Muller, F., and Rong, I. (1992) "Porous Silicon as a Material in Microsensor Technology," *Sensor. Actuator. A* 33, pp. 19–24.

Bartek, M., Gennissen, P.T.J., French, P.J., and Wolffenbuttel, R.F. (1995) "Confined Selective Epitaxial Growth: Potential for Smart Silicon Sensor Fabrication," in *8th International Conference on Solid-State Sensors and Actuators (Transducers '95)*, Stockholm, June, pp. 91–94.

Barth, P. (1984) "Si in Biomedical Applications," *Micro-Electron. Photonic. Mater. Sensor. Technol.*

Barth, P.W. (1990) "Silicon Fusion Bonding for Fabrication of Sensors, Actuators and Microstructures," *Sensor. Actuator. A* **A23**, pp. 919–26.

Barth, P.W., Shlichta, P.J., and Angel, J.B. (1985) "Deep Narrow Vertical-Walled Shafts in <110> Silicon," in *3rd International Conference on Solid-State Sensors and Actuators*, Philadelphia, pp. 371–73.

Bashir, R., Neudeck, G.W., Yen, H., Kvam, E.P., and Denton, J.P. (1995) "Characterization of Sidewall Defects in Selective Epitaxial Growth of Silicon," *J. Vac. Sci. Technol.* **11**, pp. 923–28.

Bassous, E., and Liu, C.-Y. (1978) "Polycrystalline Silicon Etching with Tetramethylammonium Hydroxide," U.S. Patent 4,113,551.

Bassous, E., Taub, H.H., and Kuhn, L. (1977) "Ink Jet Printing Nozzle Arrays Etched in Silicon," *Appl. Phys. Lett.* **31**, pp. 135–37.

Baum, T., and Schiffrin, D.J. (1997) "AFM Study of Surface Finish Improvement by Ultrasound in the Anisotropic Etching of Si <100> in KOH for Micromachining Applications," *J. Micromech. Microeng.* **7**, pp. 338–42.

Bean, K. (1978) "Anisotropic Etching of Silicon," *IEEE Trans. Electron Devices* **ED-25**, pp. 1185–93.

Bean, K.E., and Runyan, W.R. (1977) "Dielectric Isolation: Comprehensive, Current and Future," *J. Electrochem. Soc.* **124**, pp. 5C–12C.

Bean, K.E., Yeakley, R.L., and Powell, T.K. (1974) "Orientation Dependent Etching and Deposition of Silicon," *J. Electrochem. Soc.* **121**, p. 87C.

Benitez, M.A., Esteve, J., Benrakkad, M.S., Morante, J.R., Samitier, J., and Schweitz, J.Å. (1995) "Stress Profile Characterization and Test Structures Analysis of Single and Double Ion Implanted LPCVD Polycrystalline Silicon," in *8th International Conference on Solid-State Sensors and Actuators (Transducers '95)*, Stockholm, pp. 88–91.

Bennema, P. (1984) "Spiral Growth and Surface Roughening: Developments since Burton, Cabrera, and Frank," *J. Cryst. Growth* **69**, pp. 182–97.

Berre, M.L., Kleinmann, P., Semmache, B., Barbier, D., and Pinard, P. (1996) "Electrical and Piezoresistive Characterization of Boron-Doped LPCVD Polycrystalline Silicon under Rapid Thermal Annealing." Sensors and Actuators A-Physical 54, pp. 700–703. In *Smart Electronics and MEMS: Proceedings of the Smart Structures and Materials 1996 Meeting*, Varadan, V.K., and McWhorter, P.J., eds., San Diego, vol. 2722.

Biebl, M., Mulhern, G.T., and Howe, R.T. (1995a) "In Situ Phosphorous-Doped Polysilicon for Integrated MEMS," in *International Conference on Solid-State Sensors and Actuators (Transducers '95)*, Stockholm pp. 198–201.

Biebl, M., Brandl, G., and Howe, R.T. (1995b) "Young's Modulus of In Situ Phosphorous-Doped Polysilicon," in *8th International Conference on Solid-State Sensors and Actuators (Transducers '95)*, Stockholm, pp. 80-3.

Bogenschutz, A.F., Locherer, K.-H., Mussinger, W., and Krusemark, W. (1967) "Chemical Etching of Semiconductors in HNO3-HF-CH3COOH," *J. Electrochem. Soc.* **114**, pp. 970–73.

Bogh, A. (1971) "Ethylene Diamine-Pyrocatechol-Water Mixture Shows Etching Anomaly in Boron-Doped Silicon," *J. Electrochem. Soc.* **118**, pp. 401–2.

Bohm, M. (1988) "Advances in Amorphous Silicon Based Thin Film Microelectronics," *Solid State Technol.* **31**, September 1988, pp. 125–31.

Boivin, L.P. (1974) "Thin Film Laser-to Fiber Coupler," *Appl. Opt.* **13**, pp. 391–95.

Bomchil, G., Herino, R., and Barla, K. (1986) "Formation and Oxidation of Porous Silicon for Silicon on Insulator Technologies," in *Energy Beam-Solid Interactions and Transient Thermal Processes*, Nguyen, V.T., and Cullis, A., eds., Les Ulis: Les Editions de Physique, p. 463.

Bonifield, T., Hewes, K., Merritt, B., Robinson, R., Fisher, S., and Maisch, D. (1993) "Extended Run Evaluation of TEOS/Ozone BPSG Deposition," *Semicond. Int.* July, pp. 200–4.

Brantley, W.A. (1973) "Calculated Elastic Constants for Stress Problems Associated with Semiconductor Devices," *J. Appl. Phys.* **44**, pp. 534–35.

Bressers, P.M.M.C., Kelly, J.J., Gardeniers, E.J.G., and Elwenspoek, M. (1996) "Surface Morphology of p-Type (100) Silicon Etched in Aqueous Alkaline Solution," *J. Electrochem. Soc.* **143**, pp. 1744–50.

Britland, S., Moores, G.R., Clark, P., and Connolly, P. (1990) "Patterning and Cell Adhesion and Movement on Artificial Substrate: A Simple Method," *J. Anat.* **170**, pp. 235–36.

Brodie, I., Gurnick, H.R., Holland, C.E., and Moessner, H.A. (1990) "Method for Providing Polyimide Spacers in a Field Emission Panel Display," U.S. Patent 4,923,421.

Brodie, I., and Muray, J.J. (1982) *The Physics of Microfabrication*, Plenum Press, New York.

Bromley, E.I., Randall, J.N., Flanders, D.C., and Mountain, R.W. (1983) "A Technique for the Determination of Stress in Thin Films," *J. Vac. Sci. Technol.* **B1**, pp. 1364–66.

Bruno, G., Capezzuto, P., and Madan, A. eds. (1995) *Plasma Deposition of Amorphous Silicon-Based Materials*, Academic Press, Boston.

Bryzek, J., Petersen, K., Mallon, J.R., Christel, L., and Pourahmadi, F. (1990) *Silicon Sensors and Microstructures*, Novasensor, Fremont, CA.

Burgett, S.R., Pister, K.S., and Fearing, R.S. (1992) "Three-Dimensional Structures Made with Microfabricated Hinges," in *ASME 1992, Micromechanical Sensors, Actuators, and Systems*, Anaheim, CA, pp. 1–11.

Burggraaf, P. (1991) "Epi's Leading Edge," *Semicond. Int.*, June pp. 67–71.

Buser, R.A., and de Rooij, N.F. (1988) "Monolithishes Kraftsensorfeld," *VDI-Berichte* Nr. 677.

Bushan, B. ed. (1995) *Handbook of Micro/Nanotribology*, CRC Press, Boca Raton.

Bushan, B. (1996) "Nanotribology and Nanomechanics of MEMS Devices," in *Proceedings: IEEE Ninth Annual International Workshop on Micro Electro Mechanical Systems*, San Diego, pp. 91–98.

Buttgenbach, S. (1991) *Mikromechanik*, Teubner Studienbucher, Stuttgart.

Cahill, S.S., Chu, W., and Ikeda, K. (1993) "High Aspect Ratio Isolated Structures in Single Crystal Silicon," in *7th International Conference on Solid-State Sensors and Actuators (Transducers '93)*, Yokohama, pp. 250–53.

Callister, D.W. (1985) *Materials Science and Engineering*, Wiley, New York.

Campbell, D.S. (1970) "Mechanical Properties of Thin Films," in *Handbook of Thin Film Technology*, Maissel, L.I., and Glang, R., eds., McGraw-Hill, New York.

Canham, L.T. (1990) "Silicon Quantum Wire Array Fabrication by Electrochemical and Chemical Dissolution of Wafers," *Appl. Phys. Lett.* **57**, pp. 1046–50.

Chang, S., Eaton, W., Fulmer, J., Gonzalez, C., Underwood, B., Wong, J., and Smith, R.L. (1991) "Micromechanical Structures in Amorphous Silicon," in *6th International Conference on Solid-State Sensors and Actuators (Transducers '91)*, San Francisco, pp. 751–54.

Chen, G.Y., Thundat, T., Wachter, E.A., and Warmack, R.J. (1995) "Adsorption-Induced Surface Stress and Its Effects on Resonance Frequency of Cantilevers," *J. Appl. Phys.* **77**, pp. 3618–22.

Chen, L.-Y., and MacDonald, N.C. (1991) "A Selective CVD Tungsten Process for Micromotors," in *6th International Conference on Solid-State Sensors and Actuators (Transducers '91)*, San Francisco, pp. 739–42.

Cho, S.T. (1995) "A Batch Dissolved Wafer Process for Low Cost Sensor Applications," in *Micromachining and Microfabrication Process Technology 2 (Proceedings of the SPIE)*, Austin, pp. 10–17.

Choi, M.S., and Hearn, E.W. (1984) "Stress Effects in Boron-Implanted Polysilicon Films," *J. Electrochem. Soc.* **131**, pp. 2443–46.

Chou, P.C., and Pagano, N.J. (1967) *Elasticity: Tensor, Dyadic, and Engineering Approaches*, Dover Publications, New York.

Chu, P.B., Nelson, P.R., Tachiki, M.L., and Pister, K.S.J. (1995) "Dynamics of Polysilicon Parallel-Plate Electrostatic Actuators," in *8th International Conference on Solid-State Sensors and Actuators (Transducers '95)*, Stockholm, pp. 356–59.

Chu, T.L., and Gavaler, J.R. (1965) "Dissolution of Silicon and Junction Delineation in Silicon by the CrO_3-HF-H_2O System," *Electrochim. Acta.* **10**, pp. 1141–48.

Chu, W.H., Mehregany, M., Ning, X., and Pirouz, P. (1992) "Measurement of Residual Stress-Induced Bending Moment of p+ Silicon Films," *Mat. Res. Soc. Symp. Proc.* **239**, p. 169.

Chuang, S.F., and Smith, R.L. (1988) "Preferred Crystallographic Directions of Pore Propagation in Porous Silicon Layers," in *Technical Digest: 1988 Solid State Sensor and Actuator Workshop*, Hilton Head Island, SC, pp. 151–53.

Ciarlo, D.R. (1987) "Corner Compensation Structures for (110) Oriented Silicon," in *Proceedings: IEEE Micro Robots and Teleoperators Workshop*, Hyannis, MA, pp. 6/1–4.

Clark, L.D., and Edell, D.J. (1987) "KOH:H_2O Etching of (110) Si, (111) Si, SiO2, and Ta: An Experimental Study," in *Proceedings: IEEE Micro Robots and Teleoperators Workshop*, Hyannis, MA, pp. 5/1–6.

Clark, L.D., Edell, D.J., McNeil, V.M., and Toi, V.V. (1992) "Factors Influencing the Biocompatibility in Insertable Silicon Microshafts in Cerebral Cortex," *IEEE Trans. Biomed. Eng.* **39**, pp. 635–43.

Clemens, D.P. (1973) "Anisotropic Etching of Silicon on Sapphire," vol. 407 *J. Electrochem. Soc.* **120**, C240–C240.

Compton, D.R., "PECVD: A Versatile Technology," *Semicond. Int.* July, pp. 60–65.

Connolly, P., Moores, G.R., Monaghan, W., Shen, J., Britland, S., and Clark, P. (1992) "Microelectronic and Nanoelectronic Interfacing Techniques for Biological Systems," *Sensor. Actuator. B* **B6**, pp. 13–21.

Cooper, J.M., Barker, J.R., Magill, J.V., Monaghan, W., Robertson, M., Wilkinson, C.D.W., Curties, A.S.G., and Moores, G.R. (1993) "A Review of Research in Bioelectronics at Glasgow University," *Biosens. Bioelectron.* **8**, pp. R22–R30.

Core, T.A., Tsang, W.K., and Sherman, S.J. (1993) "Fabrication Technology for an Integrated Surface-Micromachined Sensor," *Solid State Technol.* **36**, pp. 39–47.

Craven, D., Yu, K., and Pandhumsoporn, T. (1995) "Etching Technology for 'Through-the-Wafer' Silicon Etching," in *Micromachining and Microfabrication Process Technology (Proceedings of the SPIE)*, Austin, pp. 259–63.

Crowley, J.L. (1992) "Plasma Enhanced CVD for Flat Panel Displays," *Solid State Technol.* February, pp. 94–8.

d'Aragona, F.S. (1972) "Dislocation Etch for (100) Planes in Silicon," *J. Electrochem. Soc.* **119**, pp. 948–51.

Dash, W.C. (1956) "Copper Precipitation on Dislocation in Silicon", *J. Appl. Phys.* **27**, p. 1193.

Declercq, M.J., DeMoor, J.P., and Lambert, J.P. (1975) "A Comparative Study of Three Anisotropic Etchants for Silicon," *Electrochem. Soc. Abstr.* 75-2, p. 446.

Declercq, M.J., Gerzberg, L., and Meindl, J.D. (1975) "Optimization of the Hydrazine-Water Solution for Anisotropic Etching of Silicon in Integrated Circuit Technology," *J. Electrochem. Soc.* **122**, 545–52.

Diem, B., Delaye, M.T., Michel, F., Renard, S., and Delapierre, G. (1993) "SOI(SIMOX) as a Substrate for Surface Micromachining of Single Crystalline Silicon Sensors and Actuators," in *7th International Conference on Solid-State Sensors and Actuators (Transducers '93)*, Yokohama, pp. 233–36.

Ding, X., and Ko, W. (1991) "Buckling Behavior of Boron-Doped P$^+$ Silicon Diaphragms," in *6th International Conference on Solid-State Sensors and Actuators (Transducers '91)*, San Francisco, pp. 201–4.

Doan, V.V., and Sailor, M.J. (1992) "Luminescent Color Image Generation on Porous Si," *Science* **256**, pp. 1791–92.

Drosd, R., and Washburn, J. (1982) "Some Observations on the Amorphous to Crystalline Transformation in Silicon," *J. Appl. Phys.* **53**, pp. 397–403.

Dunn, P.N. (1993) "SOI: Ready to Meet CMOS Challenge," *Solid State Technol.* October, pp. 32–35.

Durant, W. (1957) *The Reformation: A History of European Civilzation from Wyclif to Calvin*, MJF Books, New York, pp. 1300–564.

Dutta, B. (1970) "Integrated Micromotor Concepts," in *Int. Conf. on Microelectronic Circuits and Systems Theory*, Sydney, pp. 36–37.

Eaton, W.P., and Smith, J.H. (1995) "A CMOS-Compatible, Surface-Micromachined Pressure Sensor for Aqueous Ultrasonic Application," in *Smart Structures and Materials 1995: Smart Electronics (Proceedings of the SPIE)*, San Diego, pp. 258–65.

Eaton, W.P., and Smith, J.H. (1996) "Release-Etch Modeling for Complex Surface Micromachined Structures," in *SPIE Proceedings: Micromachining and Microfabrication Process Technology 2*, Austin, pp. 80–93.

Elwenspoek, M., Gardeniers, H., de Boer, M., and Prak, A. (1993) "On the Mechanism of Anisotropic Etching of Silicon," *J. Electrochem. Soc.* **140**, pp. 2075–80.

Elwenspoek, M., Gardeniers, H., de Boer, M., and Prak, A. (1994) "Micromechanics," Report No. 122830, University of Twente.

Elwenspoek, M., Lindberg, U., Kok, H., and Smith, L. (1994) "Wet Chemical Etching Mechanism of Silicon," in *IEEE International Workshop on Micro Electro Mechanical Systems (MEMS '94)*, Oiso, Japan, pp. 223–28.

Elwenspoek, M., and van der Weerden, J.P. (1987) "Kinetic Roughening and Step Free Energy in the Solid-on-Solid Model and on Naphtalene Crystals," *J. Phys. A: Math. Gen.* **20**, pp. 669–78.

Enoksson, P. (1997) "New Structure for Corner Compensation in Anisotropic KOH Etching," *J. Micromech. Microeng.* **7**, p. 141.

Eremenko, V.G., and Nikitenko, V.I. (1972) "Electron Microscope Investigation of the Microplastic Deformation Mechanisms by Indentation," *Phys. Stat. Sol. A* **14**, pp. 317–30.

Ericson, F., Greek, S., Soderkvist, J., and Schweitz, J.-Å. (1995) "High Sensitive Internal Film Stress Measurement by an Improved Micromachined Indicator Structure," in *8th International Conference on Solid-State Sensors and Actuators (Transducers '95)*, Stockholm, pp. 84–7.

Ericson, F., Johansson, S., and Schweitz, J.-Å. (1988) "Hardness and Fracture Toughness of Semi-conducting Materials Studied by Indentation and Erosion Techniques," *J. Mater. Sci. Eng.* **A105/106**, pp. 131–41.

Ericson, F., and Schweitz, J.-Å. (1990) "Micromechanical Fracture Strength of Silicon," *J. Appl. Phys.* **68**, pp. 5840–44.

Fan, L.-S. (1989) Integrated Micromachinery: Moving Structures on Silicon Chips, Ph.D. thesis, University of California, Berkeley.

Fan, L.-S., Muller, R.S., Yun, W., Huang, J., and Howe, R.T. (1990) "Spiral Microstructures for the Measurement of Average Strain Gradients in Thin Films," in *Proceedings: IEEE Micro Electro Mechanical Systems (MEMS '90)*, Napa Valley, CA, pp. 177–81.

Fan, L.-S., Tai, Y.C., and Muller, R.S. (1987) "Pin Joints, Gears, Springs, Cranks, and Other Novel Micromechanical Structures," in *4th International Conference on Solid-State Sensors and Actuators (Transducers '87)*, Tokyo, pp. 849–52.

Fan, L.-S., Tai, Y.C., and Muller, R.S. (1988) "Integrated Movable Micromechanical Structures for Sensors and Actuators," *IEEE Trans. Electron Devices* **35**, pp. 724–30.

Faust, J.W., and Palik, E.D. (1983) "Study of the Orientation Dependent Etching and Initial Anodization of Si in Aqueous KOH," *J. Electrochem. Soc.* **130**, pp. 1413–20.

Fedder, G.K., Chang, J.C., and Howe, R.T. (1992) "Thermal Assembly of Polysilicon Microactuators with Narrow-Gap Electrostatic Comb Drive," in *Technical Digest: 1992 Solid State Sensor and Actuator Workshop*, Hilton Head Island, SC, pp. 63–68.

Field, L. (1987) "Low-Stress Nitrides for Use in Electronic Devices," in *EECS/ERL Res. Summary*, University of California, Berkeley, pp. 42–43.

Finne, R.M., and Klein, D.L. (1967) "A Water-Amine-Complexing Agent System for Etching Silicon," *J. Electrochem. Soc.* **114**, pp. 965–70.

Fleischman, A.J., Roy, S., Zorman, C.A., Mehregany, M., and Matus, L.G. (1996) "Polycrystaline Silicon Carbide for Surface Micromachining," in *Ninth Annual International Workshop on Micro Electro Mechanical Systems*, San Diego, pp. 234–38.

Forster, J.H., and Singleton, J.B., "Beam-Lead Sealed Junction Integrated Circuits," *Bell Lab. Rec.* **44**, pp. 313–17, 1966.

Frazier, A.B., and Allen, M.G. (1993) "Metallic Microstructures Fabricated Using Photosensitive Polyimide Electroplating Molds," *J. Microelectromech. Syst.* **2**, pp. 87–94.

Fresquet, G., Azzaro, C., and Couderc, J.-P. (1995) "Analysis and Modeling of In Situ Boron-Doped Polysilicon Deposition by LP CVD," *J. Electrochem. Soc.* **142**, pp. 538–47.

Fritz, J., Baller, M.K., Lang, H.P., Rothuizen, H., Vettinger, P., Meyer, E., Güntherodt, H.-J., Gerber, C., and Gimzewski, J.K. (2000) "Translating Biomolecular Recognition into Nanomechanics," *Science* **288**, pp. 316–18.

Gabriel, K., Jarvis, J., and Trimmer, W. (1989) "Small Machines, Large Opportunities: A Report on the Emerging Field of Microdynamics," National Science Foundation.

Gennissen, P.T.J., Bartke, M., French, P.J., Sarro, P.M., and Wolffenbuttel, R.F. (1995) "Automatic Etch Stop on Buried Oxide Using Epitaxial Lateral Overgrowth," in *8th International Conference on Solid-State Sensors and Actuators (Transducers '95)*, Stockholm, June, pp. 75–78.

Gennissen, P.T.J., French, P.J., Munter, D.P.A.D., Bell, T.E., Kaneko, H., and Sarro, P.M. (1995) "Porous Silicon Micromachining Techniques for Accelerometer Fabrication," in *Proceedings ESSDERC '95*, The Hague, pp. 593–96.

Ghandi, S.K. (1968) *The Theory and Practice of Micro-Electronics*, Wiley, New York.

Glembocki, O.J., Stahlbush, R.E., and Tomkiewicz, M. (1985) "Bias-Dependent Etching of Silicon in Aqueous KOH," *J. Electrochem. Soc.* **132**, pp. 145–51.

Goetzlich, J., and Ryssel, H. (1981) "Tapered Windows in Silicon Dioxide, Silicon Nitride, and Polysilicon Layers by Ion Implantation," *J. Electrochem. Soc.* **128**, pp. 617–19.

Gogoi, B.P., and Mastrangelo, C.H. (1995) "Post-Processing Release of Microstructures by Electromagnetic Pulses," in *8th International Conference on Solid-State Sensors and Actuators (Transducers '95)*, Stockholm, June, pp. 214–17.

Goosen, J.F.L., van Drieenhuizen, B.P., French, P.J., and Wolfenbuttel, R.F. (1993) "Stress Measurement Structures for Micromachined Sensors," in *7th International Conference on Solid-State Sensors and Actuators (Transducers '93)*, Yokohama, pp. 783–86.

Gravesen, P. (1986) Silicon Sensors, status report for the industrial engineering thesis, DTH Lyngby, Denmark.

Greek, S., Ericson, F., Johansson, S., and Schweitz, J.-Å. (1995) "In Situ Tensile Strength Measurement of Thick-Film and Thin-Film Micromachined Structures," in *8th International Conference on Solid-State Sensors and Actuators (Transducers '95)*, Stockholm, pp. 56–59.

Greenwood, J.C. (1969) "Ethylene Diamine-Cathechol-Water Mixture Shows Preferential Etching of p-n Junction," *J. Electrochem. Soc.* **116**, pp. 1325–26.

Greenwood, J.C. (1984) "Etched Silicon Vibrating Sensor," *J. Phys. E, Sci. Instrum.* **17**, pp. 650–52.

Greenwood, J.C. (1988) "Silicon in Mechanical Sensors," *J. Phys. E, Sci. Instrum.* **21**, 1114–28.

Greiff, P. (1995) "SOI-Based Micromechanical Process," in *Micromachining and Microfabrication Process Technology (Proceedings of the SPIE)*, Austin, pp. 74–81.

Guckel, H., and Burns, D.W. (1984) "Planar Processed Polysilicon Sealed Cavities for Pressure Transducer Arrays," in *Technical Digest: IEEE International Electron Devices Meeting (IEDM '84)*, San Francisco, pp. 223–25.

Guckel, H., and Burns, D.W. (1985) "A Technology for Integrated Transducers," in *International Conference on Solid-State Sensors and Actuators*, Philadelphia, pp. 90–92.

Guckel, H., Burns, D.W., Nesler, C.K., and Rutigliano, C.R. (1987) "Fine Grained Polysilicon and Its Application to Planar Pressure Transducers," in *4th International Conference on Solid-State Sensors and Actuators (Transducers '87)*, Tokyo, pp. 277–82.

Guckel, H., Burns, D.W., Tilmans, H.A.C., DeRoo, D.W., and Rutigliano, C.R. (1988) "Mechanical Properties of Fine Grained Polysilicon: The Repeatability Issue," in *Technical Digest: 1988 Solid State Sensor and Actuator Workshop*, Hilton Head Island, SC, pp. 96–99.

Guckel, H., Burns, D.W., Tilmans, H.A.C., Visser, C.C.G., DeRoo, D.W., Christenson, T.R., Klomberg, P.J., Sniegowski, J.J., and Jones, D.H. (1988) "Processing Conditions for Polysilicon Films with Tensile Strain for Large Aspect Ratio Microstructures," in *Technical Digest: 1988 Solid State Sensor and Actuator Workshop*, Hilton Head Island, SC, pp. 51–56.

Guckel, H., Burns, D.W., Visser, C.C.G., Tilmans, H.A.C., and Deroo, D. (1988) "Fine-Grained Polysilicon Films with Built-in Tensile Strain," *IEEE Trans. Electron Devices* **35**, pp. 800–1.

Guckel, H., Randazzo, T., and Burns, D.W. (1985) "A Simple Technique for the Determination of Mechanical Strain in Thin Films with Applications to Polysilicon," *J. Appl. Phys.* **57**, pp. 1671–75.

Guckel, H., Showers, D.K., Burns, D.W., Rutigliano, C.R., and Nesler, C.G. (1986) "Deposition Techniques and Properties of Strain Compensated LP CVD Silicon Nitride," in *Technical Digest: 1986 Solid State Sensor and Actuator Workshop*, Hilton Head Island, SC.

Guckel, H., Sniegowski, J.J., Christenson, T.R., Mohney, S., and Kelly, T.F. (1989) "Fabrication of Micromechanical Devices from Polysilicon Films with Smooth Surfaces," *Sensor. Actuator.* **20**, pp. 117–21.

Guckel, H., Sniegowski, J.J., Christenson, T.R., and Raissi, F. (1990) "The Application of Fine-Grained, Tensile Polysilicon to Mechanically Resonant Transducers," *Sensor. Actuator. A* **A21**, pp. 346–51.

Hallas, C.E., "Electropolishing Silicon," *Solid State Technol.* **14**, pp. 30–32.

Hansen, K.M., Ji, H.-F., Wu, G., Datar, R., Cote, R., Majumdar, A., and Thundat, T. (2001) "Cantilever-Based Optical Deflection Assay for Discrimination of DNA Single-Nucleotide Mismatches," *Anal. Chem.*

Harbeke, G., Krausbauer, L., Steigmeier, E.F., and Widmer, A.E. (1983) "LPCVD Polycrystalline Silicon: Growth and Physical Properties of In-Situ Phosphorous Doped and Undoped Films," *RCA Rev.* **44**, pp. 287–313.

Harley, J.A., and Kenny, T.W. (1999) "High-Sensitivity Cantilevers Under 1000 Å Thick," *Appl. Phys. Lett.* **75**, pp. 289–91.

Harris, T.W. (1976) *Chemical Milling*, Clarendon Press, Oxford.

Hélin, P., Bourouina, T., Fujita, H., Maekoba, H., Cugat, O., and Reyne, G. (2000) "Self-Aligned Vertical Mirrors and V-Grooves for Magnetic Micro Optical Matrix Switch," in *Nano et Micro Technologies*, D. Hauden, ed., Hermes Science Publications, pp. 55–87.

Heller, M.J., Forster, A.H., and Tu, E. (2000) "Active Microelectronic Chip Devices which Utilize Controlled Electrophoretic Fields for Multiplex DNA Hybridization and Other Genomic Applications," *Electrophoresis* **21**, pp. 157–64.

Hendriks, M., Delhez, R., and Radelaar, S. (1983) "Carbon Doped Polycrystalline Silicon Layers," *Studies in Inorganic Chemistry*, Elsevier, Amsterdam, p. 193.

Hendriks, M., and Mavero, C. (1991) "Phosphorous Doped Polysilicon for Double Poly Structures: Part 1. Morphology and Microstructure," *J. Electrochem. Soc.* **138**, pp. 1466–70.

Herb, J.A., Peters, M.G., Terry, S.C., and Jerman, J.H. (1990) "PECVD Diamond Films for Use in Silicon Microstructures," *Sensor. Actuator. A* **A23**, pp. 982–87.

Hermansson, K., Lindberg, U., Hok, B., and Palmskog, G. (1991) "Wetting Properties of Silicon Surfaces," in *6th International Conference on Solid-State Sensors and Actuators (Transducers '91)*, San Francisco pp. 193–96.

Herr, E., and Baltes, H. (1991) "KOH Etch Rates of High-Index Planes from Mechanically Prepared Silicon Crystals," in *6th International Conference on Solid-State Sensors and Actuators (Transducers '91)*, San Francisco, pp. 807–10.

Hesketh, P.J., Ju, C., Gowda, S., Zanoria, E., and Danyluk, S. (1993) "Surface Free Energy Model of Silicon Anisotropic Etching," *J. Electrochem. Soc.* **140**, pp. 1080–85.

Heuberger, A. (1989) *Mikromechanik*, Springer Verlag, Heidelberg.

Hirata, M., Suzuki, K., and Tanigawa, H. (1988) "Silicon Diaphragm Pressure Sensors Fabricated by Anodic Oxidation Etch-Stop," *Sensor. Actuator.* **13**, pp. 63–70.

Hjort, K., Schweitz, J.-Å., Andersson, S., Kordina, O., and Janzen, E. (1992) "Epitaxial Regrowth in Surface Micromachining of GaAs," in *Proceedings: IEEE Micro Electro Mechanical Systems (MEMS '92)*, Travemunde, Germany, pp. 83–86.

Hjort, K., Schweitz, J.-Å., and Hok, B. (1990) "Bulk and Surface Micromachining of GaAs Structures," in *Proceedings: IEEE Micro Electro Mechanical Systems (MEMS '90)*, Napa Valley, CA, pp. 73–76.

Hoffman, E., Warneke, B., Kruglick, E., Weigold, J., and Pister, K.S.J. (1995) "3D Structures with Piezoresistive Sensors in Standard CMOS," in *Proceedings: IEEE Micro Electro Mechanical Systems (MEMS '95)*, Amsterdam, pp. 288–93.

Hoffman, R.W. (1975) "Stresses in Thin Films: The Relevance of Grain Boundaries and Impurities," *Thin Solid Films* **34**, pp. 185–90.

Hoffman, R.W. (1976) "Mechanical Properties of Non-Metallic Thin Films," in *Physics of Nonmetallic Thin Films (NATO Advanced Study Institutes Series: Series B, Physics)*, Dupuy, C.H.S., and Cachard, A.A., eds., Plenum Press, New York, pp. 273–353.

Hoffmeister, W. (1969) Determination of the Etch Rate of Silicon in Buffered HF Using a 31Si Tracer Method," *Int. J. Appl. Radiat. Isotope.* **2**, p. 139.

Holbrook, D.S., and McKibben, J.D. (1992) "Microlithography for Large Area Flat Panel Display Substrates," *Solid State Technol.* May, pp. 166–72.

Holland, C.E., Westerberg, E.R., Madou, M.J., and Otagawa, T. (1988) "Etching Method for Producing an Electrochemical Cell in a Crystalline Substrate," U.S. Patent 4,764,864.

Honer, K.A., and Kovacs, G.T.A. (2000) "Sputtered Silicon for Integrated MEMS Applications," in *Solid-State Sensor and Actuator Workshop*, Transducers Research Foundation, Hilton Head Island, SC, pp. 308–11.

Horch, K., Normann, R.A., and Schmidt, S. (1993) "Biocompatibility of Silicon-Based Electrode Arrays Implanted in Feline Cortical Tissue," *J. Biomed. Mater. Res.* **27**, pp. 1393–99.

Hornbeck, L.J. (1995) "Projection Displays and MEMS: Timely Convergence for a Bright Future," in *Micromachining and Microfabrication Process Technology (Proceedings of the SPIE)*, Austin, p. 2.

Houston, M.R., Maboudian, R., and Howe, R. (1995) "Ammonium Fluoride Anti-Stiction Treatment for Polysilicon Microstructures," in *8th International Conference on Solid-State Sensors and Actuators (Transducers '95)*, Stockholm, June, pp. 210–13.

Howe, R.T. (1985) "Polycrystalline Silicon Microstructures," in *Micromachining and Micropackaging of Transducers*, Fung, C.D., Cheung, P.W., Ko, W.H., and Fleming, D.G., eds., Elsevier, New York, pp. 169–87.

Howe, R.T. (1995) "Recent Advances in Surface Micromachining," in *Technical Digest: 13th Sensor Symposium*, Tokyo, pp. 1–8.

Howe, R.T. (1995) "Polysilicon Integrated Microsystems: Technologies and Applications," in *8th International Conference on Solid-State Sensors and Actuators (Transducers '95)*, Stockholm, June, pp. 43–46.

Howe, R.T., and Muller, R.S. (1982) "Polycrystalline Silicon Micromechanical Beams," in *Spring Meeting of the Electrochemical Society*, Montreal, pp. 184–85.

Howe, R.T., and Muller, R.S. (1983) "Polycrystalline Silicon Micromechanical Beams," *J. Electrochem. Soc.* **130**, 1420–23.

Howe, R.T., and Muller, R.S. (1983) "Stress in Polycrystalline and Amorphous Silicon Thin Films," *J. Appl. Phys.* **54**, pp. 4674–75.

Hu, J.Z., Merkle, L.D., Menoni, C.S., and Spain, I.L. (1986) "Crystal Data for High-Pressure Phases of Silicon," *Phys. Rev. B* **34**, pp. 4679–84.

Hu, S.M., and Kerr, D.R. (1967) "Observation of Etching of n-Type Silicon in Aqueous HF Solutions," *J. Electrochem. Soc.* **114**, p. 414.

Huff, M.A., and Schmidt, M.A. (1992) "Fabrication, Packaging, and Testing of a Wafer Bonded Microvalve," in *Technical Digest: 1992 Solid State Sensor and Actuator Workshop*, Hilton Head Island, SC., pp. 194–197.

IMEC, "Silicon Detectors," IMEC Brochure, 1994.

Iscoff, R. (1991) "Hotwall LP CVD Reactors: Considering the Choices," *Semicond. Int.* June, pp. 60–64.

Jaccodine, R.J., and Schlegel, W.A. (1966) "Measurements of Strains at Si-SiO$_2$ Interface," *J. Appl. Phys.* **37**, pp. 2429–34.

Jackson, K.A. (1966) "A Review of the Fundamental Aspects of Crystal Growth," Proceedings of an International Conference on Crystal Growth, Boston, June 20–24, pp. 17–24.

Jackson, T.N., Tischler, M.A., and Wise, K.D. (1981) "An Electrochemical P-N Junction Etch-Stop for the Formation of Silicon Microstructures," *IEEE Electron Device Lett.* **EDL-2**, pp. 44–45.

Jenkins, M.W. (1977) "New Preferential Etch for Defects in Silicon Crystals", *J. Electrochem. Soc.* **124**, p. 757.

Jerman, H. (1994) "Bulk Silicon Micromachining," hard copies of viewgraphs presented in Banff, Canada.

Jerman, J.H. (1990) "The Fabrication and Use of Micromachined Corrugated Silicon Diaphragms," *Sensor. Actuator. A* **A23**, pp. 988–92.

Joseph, J., Madou, M., Otagawa, T., Hesketh, P., and Saaman, A. (1989) "Catheter-Based Micromachined Electrochemical Sensors," in *Catheter-Based Sensing and Imaging Technology (Proceedings of the SPIE)*, Los Angeles, pp. 18–22.

Judy, M.W., Cho, Y.H., Howe, R.T., and Pisano, A.P. (1991) "Self-Adjusting Microstructures (SAMS)," in *Proceedings: IEEE Micro Electro Mechanical Systems (MEMS '91)*, Nara, Japan, pp. 51–56.

Judy, M.W., and Howe, R.T. (1993a) "Hollow Beam Polysilicon Lateral Resonators," in *Proceedings: IEEE Micro Electro Mechanical Systems (MEMS '93)*, Fort Lauderdale, pp. 265–71.

Judy, M.W., and Howe, R.T. (1993b) "Highly Compliant Lateral Suspensions Using Sidwall Beams," in *7th International Conference on Solid-State Sensors and Actuators (Transducers '93)*, Yokohama, pp. 54–57.

Jung, K.H., Shih, S., and Kwong, D.L. (1993) "Developments in Luminescent Porous Si," *J. Electrochem. Soc.* **140**, pp. 3046–64.

Kahn, H., Stemmer, S., Nandakumar, K., Heuer, A.H., Mullen, R.L., Ballarini, R., and Huff, M.A. (1996) "Mechanical Properties of Thick, Surface Micromachined Polysilicon Films," *Ninth Annual International Workshop on Micro Electro Mechanical Systems (MEMS '96)*, San Diego, pp. 343–48.

Kamins, T. (1988) *Polycrystalline Silicon for Integrated Circuits*, Kluwer, Boston.

Kaminsky, G. "Micromachining of Silicon Mechanical Structures," *J. Vac. Sci. Technol.* B3, pp. 1015–24.

Kanda, Y. (1982) "A Graphical Representation of the Piezoresistance Coefficients in Silicon," *IEEE Trans. Electron Devices* **ED-29**, pp. 64–70.

Kanda, Y. (1991) "What Kind of SOI Wafers Are Suitable for What Type of Micromachining Purposes?" in *6th International Conference on Solid-State Sensors and Actuators (Transducers '91)*, San Francisco, pp. 452–55.

Karam, J.M., Courtois, B., Holjo, M., Leclercq, J.L., and Viktorotovitch, P. (1996) "Collective Fabrication of Gallium Arsenide Based Microsystems," in *SPIE: Micromachining and Microfabrication Process Technology 2*, Austin, pp. 315–24.

Keller, C., and Ferrari, M. (1994) "Milli-Scale Polysilicon Structures," in *Technical Digest: 1994 Solid State Sensor and Actuator Workshop*, Hilton Head Island, SC, pp. 132–37.

Keller, C.G., and Howe, R.T. (1995) "Hexsil Bimorphs for Vertical Actuation," in *8th International Conference on Solid-State Sensors and Actuators (Transducers '95)*, Stockholm, pp. 99–102.

Keller, C.G., and Howe, R.T. (1995) "Nickel-Filled Hexsil Thermally Actuated Tweezers," in *8th International Conference on Solid-State Sensors and Actuators (Transducers '95)*, Stockholm, pp. 376–79.

Kendall, D.L. (1975) "On Etching Very Narrow Grooves in Silicon," *Appl. Phys. Lett.* **26**, pp. 195–98.

Kendall, D.L. (1979) "Vertical Etching of Silicon at Very High Aspect Ratios," *Ann. Rev. Mater. Sci.* **9**, pp. 373–403.

Kendall, D.L., and Guel, G.R. (1985) "Orientation of the Third Kind: The Coming of Age of (110) Silicon," in *Micromachining and Micropackaging of Transducers*, Fung, C.D. ed., Elsevier, New York, pp. 107–24.

Kenney, D.M. (1967) "Methods of Isolating Chips of a Wafer of Semiconductor Material," U.S. Patent 3,332,137.

Kern, W. (1978) "Chemical Etching of Silicon, Germanium, Gallium Arsenide, and Gallium Phosphide," *RCA Rev.* **39**, pp. 278–308.

Kern, W., and Deckert, C.A. (1978) "Chemical Etching," in *Thin Film Processes*, Vossen, J.L., and Kern, W., eds., Academic Press, Orlando.

Khazan, A.D. (1994) *Transducers and Their Elements*, PTR Prentice Hall, Englewood Cliffs, NJ.

Kim, B., and Dong-II, D.C. (1998) "Aqueous KOH Etching of Silicon (110): Etch Characteristics and Compensation Methods for Convex Corners," *J. Electrochem. Soc.* **145**, p. 2499.

Kim, C.J. (1991) Silicon Electromechanical Microgrippers: Design, Fabrication, and Testing, Ph.D. thesis, University of California, Berkeley.

Kim, S.C., and Wise, K. (1983) "Temperature Sensitivity in Silicon Piezoresistive Pressure Transducers," *IEEE Trans. Electron Devices* **ED-30**, pp. 802–10.

Kim, Y.W., and Allen, M.G. (1991) "Surface Micromachined Platforms Using Electroplated Sacrificial Layers," in *6th International Conference on Solid-State Sensors and Actuators (Transducers '91)*, San Francisco, pp. 651–54.

Kinsbron, E., Sternheim, M., and Knoell, R. (1983) "Crystallization of Amorphous Silicon Films during Low Pressure Chemical Vapor Deposition," *Appl. Phys. Lett.* **42**, pp. 835–37.

Kittel, C. (1976) *Introduction to Solid State Physics*, Wiley, New York.

Klaassen, E.H., Reay, R.J., Storment, C., Audy, J., Henry, P., Brokaw, P.A.P., and Kovacs, G.T.A. (1996) "Micromachined Thermally Isolated Circuits," in *Proceedings: Solid-State Sensors and Actuators Workshop*, Hilton Head Island, SC, pp. 127–31.

Klein, D.L., and D'Stefan, D.J. (1962) "Controlled Etching of Silicon in the HF-HNO$_3$ System," *J. Electrochem. Soc.* **109**, pp. 37–42.

Kloeck, B., Collins, S.D., de Rooij, N.F., and Smith, R.L. (1989) "Study of Electrochemical Etch-Stop for High-Precision Thickness Control of Silicon Membranes," *IEEE Trans. Electron Devices* **36**, pp. 663–69.

Kovacs, G.T.A., Maluf, N.I., and Petersen, K.E., "Bulk Micromachining of Silicon," in *Proceedings of the IEEE* 86, pp. 1536–51.

Kozlowski, F., Lindmair, N., Scheiter, T., Hierold, C., and Lang, W. (1995) "A Novel Method to Avoid Sticking of Surface Micromachined Structures," in *8th International Conference on Solid-State Sensors and Actuators (Transducers '95)*, Stockholm, June, pp. 220–23.

Kragness, R.C., and Waggener, H.A. (1973) "Precision Etching of Semiconductors," U.S. Patent 3,765,969.

Krotz, G., Legner, W., Wagner, C., Moller, H., Sonntag, H., and Muller, G. (1995) "Silicon Carbide as a Mechanical Material," in *8th International Conference on Solid-State Sensors and Actuators (Transducers '95)*, Stockholm, pp. 186–89.

Krulevitch, P.A. (1994) Micromechanical Investigations of Silicon and Ni-Ti-Cu Thin Films, Ph.D. thesis, University of California, Berkeley.

Kuhn, G.L., and Rhee, C.J. (1973) "Thin Silicon Film on Insulating Substrate," *J. Electrochem. Soc.* **120**, pp. 1563–66.

Kurokawa, H. (1982) "P-Doped Polysilicon Film Growth Technology," *J. Electrochem. Soc.* **129**, pp. 2620–24.

LaBianca, N.C., Gelorme, J.D., Cooper, E., O'Sullivan, E., and Shaw, J. (1995) "High Aspect Ratio Optical Resist Chemistry for MEMS Applications," in *JECS 188th Meeting*, Chicago, pp. 500–1.

Lambrechts, M., and Sansen, W. (1992) *Biosensors: Microelectrochemical Devices*, Institute of Physics Publishing, Philadelphia.

Lammel, G., and Renaud, P. (2000) "Two Mask Tunable Optical Filter of Porous Silicon as Microspectrometer," in *Eurosensors 14*, de Reus R., and Bouwstra S., eds., Copenhagen, pp. 183–84.

Lange, P., Kirsten, M., Riethmuller, W., Wenk, B., Zwicker, G., Morante, J.R., Ericson, F., and Schweitz, J.-Å. (1995) "Thick Polycrystalline Silicon for Surface Micromechanical Applications: Deposition, Structuring, and Mechanical Characterization," in *8th International Conference on Solid-State Sensors and Actuators (Transducers '95)*, Stockholm, 1995, pp. 202–5.

Lebouitz, K.S., Howe, R.T., and Pisano, A.P. (1995) "Permeable Polysilicon Etch-Access Windows for Microshell Fabrication," in *8th International Conference on Solid-State Sensors and Actuators (Transducers '95)*, Stockholm, 1995, pp. 224–27.

Lee, D.B. (1969) "Anisotropic Etching of Silicon," *J. Appl. Phys.* **40**, pp. 4569–74.

Lee, J.B., Chen, Z., Allen, M.G., Rohatgi, A., and Arya, R. (1995) "A Miniaturized High-Voltage Solar Cell Array as an Electrostatic MEMS Power Supply," *J. Microelectromech. Syst.* **4**, pp. 102–8.

Lee, J.G., Choi, S.H., Ahn, T.C., Hong, C.G., Lee, P., Law, K., Galiano, M., Keswick, P., and Shin, B. (1992) "SA CVD: A New Approach for 16 Mb Dielectrics," in *Semicond. Int.* May, pp. 115–20.

Lehmann, V. (1993) "The Physics of Macropore Formation in Low Doped n-Type Silicon," *J. Electrochem. Soc.* **140**, pp. 2836–43.

Lehmann, V. (1996) "Porous Silicon: A New Material for MEMS," in *Ninth Annual International Workshop on Micro Electro Mechanical Systems (MEMS '96)*, San Diego, pp. 1–6.

Lehmann, V., and Foll, H. (1990) "Formation Mechanism and Properties of Electrochemically Etched Trenches in n-Type Silicon," *J. Electrochem. Soc.* **137**, pp. 653–59.

Lehmann, V., and Gosele, U. (1991) "Porous Silicon Formation: A Quantum Wire Effect," *Appl. Phys. Lett.* **58**, pp. 856–58.

Lepselter, M.P. (1966) "Beam Lead Technology," *Bell. Sys. Tech. J.* **45**, pp. 233–54.

Lepselter, M.P. (1967) "Integrated Circuit Device and Method," U.S. Patent 3,335,338.

Levy, R.A., and Nassau, K. (1986) "Viscous Behavior of Phosphosilicate and Borophosphosilicate Glasses in VLSI Processing," *Solid State Technol.* October pp. 123–30.

Levy-Clement, C., Lagoubi, A., Tenne, R., and Neumann-Spallart, M. (1992) "Photoelectrochemical Etching of Silicon," *Electrochim. Acta* **37**, pp. 877–88.

Lewerenz, H.J., Stumper, J., and Peter, L.M. (1988) "Deconvolution of Charge Injection Steps in Quantum Yield Multiplication on Silicon," *Phys. Rev. Lett.* **61**, pp. 1989–92.

Li, G., Hubbard, T., and Antonsson, E.K. (1998) "EGS: On-line WWW Wet Etch Simulator," presented at *IEEE MSM '98*, Santa Clara, CA..

Lietoila, A., Wakita, A., Sigmon, T.W., and Gibbons, J.F. (1982) "Epitaxial Regrowth of Intrinsic, 31P-Doped and Compensated (31P + 11B-Doped) Amorphous Si," *J. Appl. Phys.* **53**, pp. 4399–405.

Lin, G., Pister, K.S.J., and Roos, K.P. (1996) "Standard CMOS Piezoresistive Sensor to Quantify Heart Cell Contractile Forces," in *Ninth Annual International Workshop on Micro Electro Mechanical Systems (MEMS '96)*, San Diego, pp. 150–55.

Lin, L. (1993) Selective Encapsulations of MEMS: Micro Channels, Needles, Resonators, and Electromechanical Filters, Ph.D. thesis, University of California, Berkeley.

Lin, L., Howe, R.T., and Pisano, A.P. (1993) "A Novel In Situ Micro Strain Gauge," in *Proceedings: IEEE Micro Electro Mechanical Systems (MEMS '93)*, Fort Lauderdale, pp. 201–6.

Lin, L.Y., Lee, S.S., Wu, M.C., and Pister, K.S.J. (1995) "Micromachined Integrated Optics for Free-Space Interconnections," in *Proceedings: IEEE Micro Electro Mechanical Systems (MEMS '95)*, Amsterdam, pp. 77–82.

Lin, T.-H. (1996) "Flexure-Beam Micromirror Devices and Potential Expansion for Smart Micromachining," in *Smart Electronics and MEMS (Proceedings of the SPIE)*, San Diego, pp. 20–29.

Lin, V.S., Motesharei, K., Dancil, K.S., Sailor, M.J., and Ghadiri, M.R. (1997) "A Porous Silicon-Based Optical Interferometric Biosensor," *Science* **278**, 840–43.

Linde, H., and Austin, L. (1992) "Wet Silicon Etching with Aqueous Amine Gallates," *J. Electrochem. Soc.* **139**, pp. 1170–74.

Liu, J., Tai, Y.-C., Lee, J., Pong, K.-C., Zohar, Y., and Ho, C.-H. (1993) "In Situ Monitoring and Universal Modeling of Sacrificial PSG Etching Using Hydrofluoric Acid," in *Proceedings: Micro Electro Mechanical Systems (MEMS '93)*, Fort Lauderdale, pp. 71–6.

Lober, T.A., and Howe, R.T. (1988a) "Surface Micromachining for Electrostatic Microactuator Fabrication," in *Technical Digest: 1988 Solid State Sensor and Actuator Workshop*, Hilton Head Island, SC.

Lober, T.A., Huang, J., Schmidt, M.A., and Senturia, S.D. (1988b) "Characterization of the Mechanisms Producing Bending Moments in Polysilicon Micro-Cantilever Beams by Interferometric Deflection Measurements," in *Technical Digest: 1988 Solid State Sensor and Actuator Workshop*, Hilton Head Island, SC, pp. 92–95.

Lochel, B., Maciossek, A., Quenzer, H.J., and Wagner, B. (1994) "UV Depth Lithography and Galvanoforming for Micromachining," in *Electrochemical Microfabrication 2*, Miami Beach, pp. 100–11.

Lochel, B., Maciossek, A., Quenzer, H.J., and Wagner, B. (1996a) "Ultraviolet Depth Lithography and Galvanoforming for Micromachining," *J. Electrochem. Soc.* **143**, pp. 237–44.

Lochel, B., Rothe, M., Fehlberg, S., Gruetzner, G., and Bleidiessel, G. (1996b) "Influence of Resist Baking on the Pattern Quality of Thick Photoresists," in *SPIE-Micromachining and Microfabrication Process Technology 2*, Austin, pp. 174–81.

Long, M.K., Burdick, J.W., and Antonsson, E.K. (1999) "Design of Compensation Structures for Anisotropic Etching," presented at *MSM '99: Modeling and Simulation of Microsystems, Semiconductors, Sensors and Actuators*, San Juan, Puerto Rico.

Luder, E. (1986) "Polcrystalline Silicon-Based Sensors," *Sensor. Actuator.* **10**, pp. 9–23.

MacDonald, N.C., Chen, L.Y., Yao, J.J., Zhang, Z.L., McMillan, J.A., Thomas, D.C., and Haselton, K.R. (1989) "Selective Chemical Vapor Deposition of Tungsten for Microelectromechanical Structures," *Sensor. Actuator.* **20**, pp. 123–33.

Madou, M.J. (1994) "Compatibility and Incompatibility of Chemical Sensors and Analytical Equipment with Micromachining," in *Technical Digest: 1994 Solid State Sensor and Actuator Workshop*, Hilton Head Island, SC, pp. 164–71.

Madou, M.J. (2002) *Fundamentals of Microfabrication*, 2nd ed., CRC Press, Boca Raton.

Madou, M.J., Frese, K.W., and Morrison, S.R. (1980) "Photoelectrochemical Corrosion of Semiconductors for Solar Cells," *SPIE* **248**, pp. 88–95.

Madou, M.J., Loo, B.H., Frese, K.W., and Morrison, S.R. (1981) "Bulk and Surface Characterization of the Silicon Electrode," *Surf. Sci.* **108**, pp. 135–52.

Madou, M.J., and Morrison, S.R. (1989) *Chemical Sensing with Solid State Devices*, Academic Press, New York.

Madou, M.J., and Otagawa, T. (1989) "Microelectrochemical Sensor and Sensor Array," U.S. Patent 4,874,500.

Madou, M.J., and Tierney, M. (1994) "Micro-Electrochemical Valves and Methods," U.S. Patent 5,368,704.

Maluf, N. (2000) *An Introduction to Microelectromechanical Systems Engineering*, Artech House, Boston.

Man, P.F., Gogoi, B.P., and Mastrangelo, C.H. (1996) "Elimination of Post-Release Adhesion in Microstructures Using Thin Conformal Fluorocarbon Films," in *Ninth Annual International Workshop on Micro Electro Mechanical Systems (MEMS '96)*, San Diego, pp. 55–60.

Marco, S., Samitier, J., Ruiz, O., Morante, J.R., Esteve-Tinto, J., and Bausells, J. (1991) "Stress Measurements of SiO$_2$-Polycrystalline Silicon Structures for Micromechanical Devices by Means of Infrared Spectroscopy Technique," in *6th International Conference on Solid-State Sensors and Actuators (Transducers '91)*, San Francisco, pp. 209–12.

Mardilovich, P., Routkevitch, D., and Govyadinov, A. (2000) "Hybrid Micromachining and Surface Microstructuring of Alumina Ceramic," in Hesketh, P.J., Hughes, H.G., Bailey, W.E., Misra, D., Ang, S.S., and Davidson, J.L., eds., The Electrochemical Society, Inc. in Proceeding Volumes 2000–19.

Maseeh, F., and Senturia, S.D. (1989) "Elastic Properties of Thin Polyimide Films," in *Polyimides: Materials, Chemistry, and Characterization*, Feger, C., Khojasteh, M.M., and McGrath, J.E., eds., Elsevier Science Publishers B.V., Amsterdam, pp. 575–84.

Maseeh, F., and Senturia, S.D. (1990) "Plastic Deformation of Highly Doped Silicon," *Sensor. Actuator. A* **A23**, pp. 861–65.

Maseeh, F., and Senturia, S.D. (1990) "Viscoelasticity and Creep Recovery of Polyimide Thin Films," in *Technical Digest: 1990 Solid State Sensor and Actuator Workshop*, Hilton Head Island, SC, pp. 55–60.

Maseeh-Tehrani, F. (1990) Characterization of Mechanical Properties of Microelectronic Thin Films, Ph.D. thesis, Massachusetts Institute of Technology.

Mastrangelo, C.H., and Hsu, C.H. (1993a) "Mechanical Stability and Adhesion of Microstructures under Capillary Forces: Part 1. Basic Theory," *J. Microelectromech. Syst.* **2**, pp. 33–43.

Mastrangelo, C.H., and Hsu, C.H. (1993b) "Mechanical Stability and Adhesion of Microstructures under Capillary Forces: Part 2. Experiments," *J. Microelectromech. Syst.* **2**, pp. 44–55.

Mastrangelo, C.H., and Saloka, G.S. (1993) "A Dry-Release Method Based on Polymer Columns for Microstructure Fabrication," in *Proceedings: IEEE Micro Electro Mechanical Systems (MEMS '93)*, Fort Lauderdale, pp. 77–81.

Matson, E.A., and Polysakov, S.A. (1977) "On the Evolutionary Selection Principle in Relation to the Growth of Polycrystalline Silicon Films," *Phys. Sta. Sol. (a)* **41**, pp. K93–K95.

Matsumura, M., and Morrison, S.R. (1983) "Photoanodic Properties of an n-Type Silicon Electrode in Aqueous Solution Containing Fluorides," *J. Electroanal. Chem.* **144**, pp. 113–20.

Mayer, G.K., Offereins, H.L., Sandmeier, H., and Kuhl, K. (1990) "Fabrication of Non-Underetched Convex Corners in Ansisotropic Etching of (100)-Silicon in Aqueous KOH with Respect to Novel Micromechanic Elements," *J. Electrochem. Soc.* **137**, pp. 3947–51.

McNeil, V.M., Wang, S.S., Ng, K.-Y., and Schmidt, M.A. (1990) "An Investigation of the Electrochemical Etching of (100) Silicon in CsOH and KOH," *in Technical Digest: 1990 Solid State Sensor and Actuator Workshop, Hilton Head Island, SC*, pp. 92–97.

Meek, R.L. "Electrochemically Thinned N/N+ Epitaxial Silicon: Method and Applications," *J. Electrochem. Soc.* **118**, pp. 1240–46.

Mehregany, M., Gabriel, K.J., and Trimmer, W.S.N. (1988) "Integrated Fabrication of Polysilicon Mechanisms," *IEEE Trans. Electron Devices* **35**, pp. 719–23.

Mehregany, M., Howe, R.T., and Senturia, S.D. (1987) "Novel Microstructures for the In Situ Measurement of Mechanical Properties of Thin Films," *J. Appl. Phys.* **62**, pp. 3579–84.

Mehregany, M., and Senturia, S.D. (1988) "Anisotropic Etching of Silicon in Hydrazine," *Sensor. Actuator.* **13**, pp. 375–90.

Memming, R., and Schwandt, G. (1966) "Anodic Dissolution of Silicon in Hydrofluoric Acid Solutions," *Surf. Sci.* **4**, pp. 109–24.

Metzger, H., and Kessler, F.R. (1970) "Der Debye-Sears Effect zur Bestimmung der Elastischen Konstanten von Silicium," *Z. Naturf.* **A25**, pp. 904–6.

Meyerson, B.S., and Olbricht, W. (1984) "Phosphorous-Doped Polycrystalline Silicon via LPCVD: 1. Process Characterization," *J. Electrochem. Soc.* **131**, pp. 2361–65.

Middlehoek, S., and Audet, S.A. (1989) *Silicon Sensors*, Academic Press, San Diego.

Middlehoek, S., and Dauderstadt, U. (1994) "Haben Mikrosensoren aus Silizium eine Zukunft?" *Technische Rundschau* July, pp. 102–5.

Mlcak, R., Tuller, H.L., Greiff, P., and Sohn, J. (1993) "Photo Assisted Electromechanical Machining of Micromechanical Structures," in *Proceedings: IEEE Micro Electro Mechanical Systems (MEMS '93)*, Fort Lauderdale, pp. 225–29.

Monk, D.J., Soane, D.S., and Howe, R.T. (1992) "LPCVD Silicon Dioxide Sacrificial Layer Etching for Surface Micromachining," in *Smart Materials Fabrication and Materials for Micro-Electro-Mechanical Systems*, San Francisco, pp. 303–10.

Monk, D.J., Soane, D.S., and Howe, R.T. (1994a) "Hydrofluoric Acid Etching of Silicon Dioxide Sacrifical Layers: Part 1. Experimental Observations," *J. Electrochem. Soc.* **141**, pp. 264–69.

Monk, D.J., Soane, D.S., and Howe, R.T. (1994b) "Hydrofluoric Acid Etching of Silicon Dioxide Sacrificial Layers: Part 2. Modeling," *J. Electrochem. Soc.* **141**, pp. 270–74.

Morkoc, H., Unlu, H., Zabel, H., and Otsuka, N. (1988) "Gallium Arsenide on Silicon: A Review," *Solid State Technol.* March, pp. 71–6.

Morrison, S.R. (1980) *Electrochemistry on Semiconductors and Oxidized Metal Electrodes*, Plenum Press, New York.

Moulin, A.M. (2000) "Microcantilever-Based Biosensors," *Ultramicroscopy* **82**, pp. 23–31.

Mulder, J.G.M., Eppenga, P., Hendriks, M., and Tong, J.E. (1990) "An Industrial LP CVD Process for In Situ Phosphorus-Doped Polysilicon," *J. Electrochem. Soc.* **137**, pp. 273–79.

Mulhern, G.T., Soane, D.S., and Howe, R.T. (1993) "Supercritical Carbon Dioxide Drying of Microstructures," in *7th International Conference on Solid-State Sensors and Actuators (Transducers '93)*, Yokohama, pp. 296–99.

Muller, R.S. (1987) "From ICs to Microstructures: Materials and Technologies," in *Proceedings: IEEE Micro Robots and Teleoperators Workshop*, Hyannis, MA, pp. 2/1–5.

Muraoka, H., Ohashi, T., and Sumitomo, T. (1973) "Controlled Preferential Etching Technology," in *Semiconductor Silicon 1973*, Chicago pp. 327–38.

Murarka, S.P., and Retajczyk, T.F.J. (1983) "Effect of Phosphorous Doping on Stress in Silicon and Polycrystalline Silicon," *J. Appl. Phys.* **54**, pp. 2069–72.

Nakamura, M., Hoshino, S., and Muro, H. (1985) "Monolithic Sensor Device for Detecting Mechanical Vibration," in *Densi Tokyo, 24 (IEE Tokyo Section*, pp. 87–88.

Nathanson, H.C., Newell, W.E., Wickstrom, R.A., and Davis, J.R. (1967) "The Resonant Gate Transistor," *IEEE Trans. Electron Devices* **ED-14**, pp. 117–33.

National Semiconductor (1974) "Transducers, Pressure, and Temperature," catalog, Sunnyvale, CA.

Neudeck, G.W., et al. (1990) "Three Dimensional Devices Fabricated by Silicon Epitaxial Lateral Overgrowth," *J. Electron. Mater.* **19**, pp. 1111–17.

Nguyen, C.T.-C. (1988) "Frequency-Selective MEMS for Miniaturized Communication Devices," *in Aerospace Conference, Snowmass, Colo.: IEEE* 1, pp. 445–60.

Nikanorov, S.P., Burenkov, Y.A., and Stepanov, A.V. (1972) "Elastic Properties of Silicon," *Sov. Phys.Solid State* 13, pp. 2516–18.

Nikpour, B., Landsberger, L.M., Hubbard, T.J., Kahrizi, M., and Iftimie, A. (1998) "Concave Corner Compensation between Vertical (010)–(001) Planes Anisotropically Etched in Si(100)," *Sensor. Actuator. A* 66, p. 299.

Nishioka, T., Shinoda, Y., and Ohmachi, Y. (1985) "Raman Microprobe Analysis of Stress in Ge and GaAs/Ge on Silicon Dioxide-Coated Silicon Substrates," *J. Appl. Phys.* 57, pp. 276–81.

North, J.C., McGahan, T.E., Rice, D.W., and Adams, A.C. (1978) "Tapered Windows in Phosphorous-Doped Silicon Dioxide by Ion Implantation," *IEEE Trans. Electron Devices* ED-25, pp. 809–12.

Noworolski, J.M., Klaassen, E., Logan, J., Petersen, K., and Maluf, N. (1995) "Fabrication of SOI Wafers with Buried Cavities Using Silicon Fusion Bonding and Electrochemical Etchback," in *8th International Conference on Solid-State Sensors and Actuators (Transducers '95)*, Stockholm, pp. 71–74.

Nunn, T., and Angell, J. (1975) "An IC Absolute Pressure Transducer with Built-in Reference Chamber," in *Workshop on Indwelling Pressure Transducers and Systems*, Cleveland, pp. 133–36.

Obermeier, E. (1995) "High Temperature Microsensors Based on Polycrystalline Diamond Thin Films," in *8th International Conference on Solid-State Sensors and Actuators (Transducers '95)*, Stockholm, pp. 178–81.

Obermeier, E., and Kopystynski, P. (1992) "Polysilion as a Material for Microsensor Applications," *Sensor. Actuator. A* A30, pp. 149–55.

Oden, P.I., Chen, G.Y., Steele, R.A., Warmack, R.J., and Thundat, T. (1996) "Viscous Drag Measurements Utilizing Microfabricated Cantilevers," *Appl. Phys. Lett.* 68, pp. 1465–69.

Offereins, H.L., Sandmaier, H., Folkmer, B., Steger, U., and Lang, W. (1991) "Stress Free Assembly Technique for a Silicon Based Pressure Sensor," in *6th International Conference on Solid-State Sensors and Actuators (Transducers '91)*, San Francisco, pp. 986–89.

Ohwada, K., Negoro, Y., Konaka, U., and Oguchi, T. (1995) "Groove Depth Uniformization in (110) Si Anisotropic Etching by Ultrasonic Wave and Application to Accelerometer Fabrication," in *Proceedings: IEEE Micro Electro Mechanical Systems Conference*, Amsterdam, pp. 100–5.

O'Neill, P. (1980) "A Monolithic Thermal Converter," *Hewlett-Packard J.* 31, pp. 12–13.

Orpana, M., and Korhonen, A.O. (1991) "Control of Residual Stress in Polysilicon Thin Films by Heavy Doping in Surface Micromachining," in *6th International Conference on Solid-State Sensors and Actuators (Transducers '91)*, San Francisco, pp. 957–60.

O'Shea, S.J., Welland, M.E., Brunt, T.A., Ramadan, A.R., and Rayment, T. (1996) *J. Vac. Sci. Technol.* B14, p. 1383.

Palik, E.D., Bermudez, V.M., and Glembocki, O.J. (1985) "Ellipsometric Study of the Etch-Stop Mechanism in Heavily Doped Silicon," *J. Electrochem. Soc.* 132, pp. 135–41.

Palik, E.D., Faust, J.W., Gray, H.F., and Green, R.F. (1982) "Study of the Etch-Stop Mechanism in Silicon," *J. Electrochem. Soc.* 129, pp. 2051–59.

Palik, E.D., Glembocki, O.J., and Heard, J.I. (1987) "Study of Bias-Dependent Etching of Si in Aqueous KOH," *J. Electrochem. Soc.* 134, pp. 404–9.

Palik, E.D., Glembocki, O.J., and Stahlbush, R.E. (1988) "Fabrication and Characterization of Si Membranes," *J. Electrochem. Soc.* 135, pp. 3126–34.

Palik, E.D., Gray, H.F., and Klein, P.B. (1983) "A Raman Study of Etching Silicon in Aqueous KOH," *J. Electrochem. Soc.* 130, pp. 956–59.

Peeters, E. (1994) *Process Development for 3D Silicon Microstructures, with Application to Mechanical Sensor Design*, Ph.D. thesis, Catholic University of Louvain, Belgium.

Peeters, E., Lapadatu, D., Sansen, W., and Puers, B. (1993a) "PHET: An Electrodeless Photovoltaic Electrochemical Etch-Stop Technique," in *7th International Conference on Solid-State Sensors and Actuators (Transducers '93)*, Yokohama, pp. 254–57.

Peeters, E., Lapadatu, D., Sansen, W., and Puers, B. (1993b) "Developments in Etch-Stop Techniques," presented at *4th European Workshop on Micromechanics (MME '93)*, Neuchatel, Switzerland.

Petersen, K., Gee, D., Pourahmadi, F., Craddock, R., Brown, J., and Christel, L. (1991) "Surface Micromachined Structures Fabricated with Silicon Fusion Bonding," in *6th International Conference on Solid-State Sensors and Actuators (Transducers '91)*, San Francisco, pp. 397–99.

Petersen, K.E. (1982) "Silicon as a Mechanical Material," *Proceedings of the IEEE* 70, pp. 420–57.

Pfann, W.G. (1961) "Improvement of Semiconducting Devices by Elastic Strain," *Solid State Electron.* 3, pp. 261–67.

Pister, K.S.J. (1992) "Hinged Polysilicon Structures with Integrated CMOS TFTs," in *Technical Digest: 1992 Solid State Sensor and Actuator Workshop*, Hilton Head Island, SC, pp. 136–39.

Pliskin, W.A. (1977) "Comparison of Properties of Dielectric Films Deposited by Various Methods," *J. Vac. Sci. Technol.* 14, pp. 1064–81.

Pool, R. (1988) "Microscopic Motor Is a First Step," *Res. News* October, pp. 379–80.

Pourahmadi, F., Christel, L., and Petersen, K. (1992) "Silicon Accelerometer with New Thermal Self-Test Mechanism," in *Technical Digest: 1992 Solid State Sensor and Actuator Workshop*, Hilton Head Island, SC, pp. 122–25.

Pratt, R.I., Johnson, G.C., Howe, R.T., and Chang, J.C. (1991) "Micromechanical Structures for Thin Film Characterization," in *6th International Conference on Solid-State Sensors and Actuators (Transducers '91)*, San Francisco, pp. 205–8.

Pratt, R.I., Johnson, G.C., Howe, R.T., and Nikkel, D.J.J. (1992) "Characterization of Thin Films Using Micromechanical Structures," *Mat. Res. Soc. Symp. Proc.* 276, pp. 197–202.

Price, J.B. (1973) "Anisotropic Etching of Silicon with KOH-H_2O-Isopropyl Alcohol," in *Semiconductor Silicon 1973*, Chicago, pp. 339–53.

Puers, B., and Sansen, W. (1990) "Compensation Structures for Convex Corner Micromachining in Silicon," *Sensor. Actuator. A* A23, pp. 1036–41.

Puers, R. (1991) "Mechanical Silicon Sensors at K.U. Leuven," in *Proceedings: Themadag Sensoren*, Rotterdam, pp. 1–8.

Pugacz-Muraszkiewicz, I.J., and Hammond, B.R. (1977) "Application of Silicates to the Detection of Flaws in Glassy Passivation Films Deposited on Silicon Substrates," *J. Vac. Sci Technol.* 14, pp. 49–53.

Raley, N.F., Sugiyama, F., and Duzer, T.V. (1984) "(100) Silicon Etch-Rate Dependence on Boron Concentration in Ethylenediamine-Pyrocatechol-Water Solutions," *J. Electrochem. Soc.* 131, pp. 161–71.

Reay, R.J., Klaassen, E.H., and Kovacs, G.T.A. (1994) "Thermally and Electrically Isolated Single-Crystal Silicon Structures in CMOS Technology," *IEEE Electron Device Lett.* 15, pp. 309–401.

Rehrig, D.L. (1990) "In Search of Precise Epi Thickness Measurements," *Semicond. Int.* 13, pp. 90–95.

Reisman, A., Berkenbilt, M., Chan, S.A., Kaufman, F.B., and Green, D.C. (1979) "The Controlled Etching of Silicon in Catalyzed Ethylene-Diamine-Pyrocathechol-Water Solutions," *Electrochem. Soc.* 126, pp. 1406–14.

Retajczyk, T.F.J., and Sinha, A.K. (1980) "Elastic Stiffness and Thermal Expansion Coefficients of Various Refractory Silicides and Silicon Nitride Films," *Thin Solid Films* 70, pp. 241–47.

Robbins, H., and Schwartz, B. (1959) "Chemical Etching of Silicon: 1. The System, HF, HNO_3, and H_2O," *J. Electrochem. Soc.* 106, pp. 505–8.

Robbins, H., and Schwartz, B. (1960) "Chemical Etching of Silicon: 2. The System HF, HNO_3, H_2O, and $HC_2C_3O_2$," *J. Electrochem. Soc.* 107, pp. 108–11.

Rodgers, T.J., Hiltpold, W.R., Frederick, B., Barnes, J.J., Jenné, F.B., and Trotter, J.D. (1977) "VMOS Memory Technology," *IEEE J. Solid-State Circuits* SC-12, pp. 515–23.

Rodgers, T.J., Hiltpold, W.R., Zimmer, J.W., Marr, G., and Trotter, J.D. (1976) "VMOS ROM," *IEEE J. Solid-State Circuits* SC-11, pp. 614–22.

Rosler, R.S., "The Evolution of Commercial Plasma Enhanced CVD Systems," *Solid State Technol.* June, pp. 67–71.

Ross, M., and Pister, K. (1994) "Micro-Windmill for Optical Scanning and Flow Measurement," in *Eurosensors 8 (Sens. Actuators A)*, Toulouse, France, pp. 576–79.

Routkevitch, D., Govyadinov, A., and Mardilovich, P. (2000) "High Aspect Ratio, High Resolution Ceramic Mems," in *2000 International Mechanical Engineering Congress and Exposition*, pp. 1–6.

Roy, S., McIlwain, A.K., DeAnna, R.G., Fleischman, A.J., Burla, R.K., Zorman, C.A., and Mehregany, M. (2000) "SiC Resonator Devices for High Q and High Temperature Applications," in *Solid-State Sensor and Actuator Workshop*, Transducers Research Foundation, Hilton Head Island, SC, pp. 22–25.

Sakimoto, M., Yoshihara, H., and Ohkubo, T. (1982) "Silicon Nitride Single-Layer X-ray Mask," *J. Vac. Sci. Technol.* 21, pp. 1017–21.

Samaun, S., Wise, K.D., and Angell, J.B. (1973) "An IC Piezoresistive Pressure Sensor for Biomedical Instrumentation," *IEEE Trans. Biomed. Eng.* 20, pp. 101–9.

Sandifer, J.R., and Voycheck, J.J. (1999) "A Review of Biosensor and Industrial Applications of pH-ISFETs and an Evaluation of Honeywell's 'DuraFET,'" *Mikrochim. Acta* 131, pp. 91–98.

Sandmaier, H., Offereins, H.L., Kuhl, K., and Lang, W. (1991) "Corner Compensation Techniques in Anisotropic Etching of (100)-Silicon Using Aqueous KOH," in *6th International Conference on Solid-State Sensors and Actuators (Transducers '91)*, San Francisco, pp. 456–59.

Saro, P.M., and van Herwaarden, A.W. (1986) "Silicon Cantilever Beams Fabricated by Electrochemically Controlled Etching for Sensor Applications," *J. Electrochem. Soc.* 133, pp. 1722–29.

Scheeper, P.R., Olthuis, W., and Bergveld, P. (1991) "Fabrication of a Subminiature Silicon Condenser Microphone Using the Sacrificial Layer Technique," in *6th International Conference on Solid-State Sensors and Actuators (Transducers '91)*, San Francisco, pp. 408–11.

Scheeper, P.R., Olthuis, W., and Bergveld, P. (1994) "The Design, Fabrication, and Testing of Corrugated Silicon Nitride Diaphragms," *J. Microelectromech. Syst.* 3, pp. 36–42.

Schimmel, D.G. (1979) "Defect Etch for (100) Silicon Evaluation", *J. Electrochem. Soc.* 126, p. 479.

Schimmel, D.G., and Elkind, M.J. (1973) "An Examination of the Chemical Staining of Silicon," *J. Electrochem. Soc.* 125, pp. 152–55.

Schnable, G.L., and Schmidt, P.F. (1976) "Applications of Electrochemistry to Fabrication of Semiconductor Devices," *J. Electrochem. Soc.* 123, pp. 310C–15C.

Schnakenberg, U., Benecke, W., and Lochel, B. (1990) "NH$_4$OH-Based Etchants for Silicon Micromachining," *Sensor. Actuator. A* A23, pp. 1031–35.

Schubert, P.J., and Neudeck, G.W. (1990) "Confined Lateral Selective Epitaxial Growth of Silicon for Device Fabrication," *IEEE Electron Device Lett.* 11, pp. 181–83.

Schumacher, A., Wagner, H.-J., and Alavi, M. (1994) "Mit Laser und Kalilauge," *Technische Rundschau* 86, pp. 20–23.

Schwartz, B., and Robbins, H. (1961) "Chemical Etching of Silicon-III: A Temperature Study in the Acid System," *J. Electrochem. Soc.* 108, pp. 365–72.

Schwartz, B., and Robbins, H. (1976) "Chemical Etching of Silicon-IV: Etching Technology," *J. Electrochem. Soc.* 123, pp. 1903–9.

Searson, P.C., Prokes, S.M., and Glembocki, O.J. (1993) "Luminescence at the Porous Silicon/Electrolyte Interface," *J. Electrochem. Soc.* 140, pp. 3327–31.

Seetharaman, S., Ke-Qin, H., and Madou, M. (2000) "Microactuators toward Microvalves for Responsive Controlled Drug Delivery," *Sensor. Actuator. B* 67, pp. 149–60.

Seidel, H. (1986) "Der Mechanismus des Siliziumätzens in alkalischen Lösungen", Ph.D. thesis, FU Berlin, Germany.

Seidel, H. (1987) "The Mechanism of Anisotropic Silicon Etching and Its Relevance for Micromachining," in *Digest of Technical Papers, Transducers '87, 4th Intl. Conf. Solid-State Sensors and Actuators*, pp. 120–25.

Seidel, H. (1989) "Nasschemische Tiefenatztechnik," in *Mikromechanik*, Heuberger, A., ed., Springer Verlag, Heidelberg.

Seidel, H. (1990) "The Mechanism of Anisotropic Electrochemical Silicon Etching in Alkaline Solutions," in *Technical Digest: 1990 Solid State Sensor and Actuator Workshop*, Hilton Head Island, SC, pp. 86–91.

Seidel, H., and Csepregi, L. (1988) "Advanced Methods for the Micromachining of Silicon," in *Technical Digest: 7th Sensor Symposium*, Tokyo, pp. 1–6.

Seidel, H., Csepregi, L., Heuberger, A., and Baumgartel, H. (1990a) "Anisotropic Etching of Crystalline Silicon in Alkaline Solutions: 2. Influence of Dopants," *J. Electrochem. Soc.* **137**, pp. 3626–32.

Seidel, H., Csepregi, L., Heuberger, A., and Baumgartel, H. (1990b) "Anisotropic Etching of Crystalline Silicon in Alkaline Solutions: 1. Orientation Dependence and Behavior of Passivation Layers," *J. Electrochem. Soc.,* **137**, pp. 3612–26.

Sekimoto, M., Yoshihara, H., and Ohkubo, T. (1982) "Silicon Nitride Single-Layer X-ray Mask," *J. Vac. Sci. Technol.* **21**, pp. 1017–21.

Senturia, S. (1987) "Can We Design Microrobotic Devices Without Knowing the Mechanical Properties of Materials?" in *Proceedings: IEEE Micro Robots and Teleoperators Workshop*, Hyannis, MA, pp. 3/1–5.

Senturia, S.D., and Howe, R.T. (1990) Mechanical Properties and CAD, lecture notes, Massachusetts Institute of Technology.

Sharpe, W.N., Yuan, B., Vaidyanathan, R., and Edwards, R.L. (1997) "Measurements of Young's Modulus, Poisson's Ratio, and Tensile Strength of Polysilicon," in *Tenth IEEE International Workshop on Microelectromechanical Systems*, Nagoya, Japan, pp. 424–29.

Shengliang, Z., Zongmin, Z., and Enke, L. (1987) "The NH_4F Electrochemical Etching Method of Silicon Diaphragm for Miniature Solid-State Pressure Transducer," in *4th International Conference on Solid-State Sensors and Actuators (Transducers '87)*, Tokyo, pp. 130–33.

Shimbo, M., Furukawa, K., and Tanzawa, K. (1986) "Silicon-to-Silicon Direct Bonding Method," *J. Appl. Phys.* **60**, pp. 2987–89.

Shimizu, T., and Ishihara, S. (1995) "Effect of SiO_2 Surface Treatment on the Solid-Phase Crystallization of Amorphous Silicon Films," *J. Electrochem. Soc.* **142**, pp. 298–302.

Shockley, W. (1963) "Method of Making Thin Slices of Semiconductive Material," U.S. Patent 3,096,262.

Singer, P. (1992) "Film Stress and How to Measure It," *Semicond. Int.* **15**, pp. 54–58.

Sinha, A.K., and Smith, T.E. (1978) "Thermal Stresses and Cracking Resistance of Dielectric Films," *J. Appl. Phys.* **49**, pp. 2423–26.

Sirtl, E., and Adler, A. (1961) "Chromsaure-Flusssaure Als Spezifisches System Zur Atzgrubenentwicklung Auf Silizium", *Z. Metallkd.* **52**, p. 529.

Smith, C.S. (1954) "Piezoresistance Effect in Germanium and Silicon," *Phys. Rev.* **94**, pp. 42–49.

Smith, J., Montague, S., Sniegowski, J., and McWhorter, P. (1995) "Embedded Micromechanical Devices for Monolithic Integration of MEMs with CMOS," in *Technical Digest: IEEE International Electron Devices Meeting (IEDM '95)*, Washington, DC, pp. 609–12.

Smith, R.L., and Collins, S.D. (1990) "Thick Films of Silicon Nitride," *Sensor. Actuator. A* **A23**, pp. 830–34.

Spear, W.E., and Comber, P.G.L. (1975) "Substitutional Doping of Amorphous Silicon," *Solid State Commun.* **17**, pp. 1193–96.

Spiering, V.L., Bouwstra, S., Burger, J., and Elwenspoek, M. (1993) "Membranes Fabricated with a Deep Single Corrugation for Package Stress Reduction and Residual Stress Relief," in *4th European Workshop on Micromechanics (MME '93)*, Neuchatel, Switzerland, pp. 223–27.

Spiering, V.L., Bouwstra, S., Spiering, R.M.E.J., and Elwenspoek, M. (1991) "On-Chip Decoupling Zone for Package-Stress Reduction," in *6th International Conference on Solid-State Sensors and Actuators (Transducers '91)*, San Francisco, pp. 982–85.

Steinsland, E., Nese, M., Hanneborg, A., Bernstein, R.W., Sandmo, H., and Kittilsland, G. (1995) "Boron-Etch Stop in TMAH Solutions," in *Proc. Transducers '95, 8th Int. Conf. Solid-State Sensors and Actuators*, Stockholm, pp. 190–93.

Stoller, A.I. (1970) "The Etching of Deep Vertical-Walled Patterns in Silicon," *RCA Rev.* **31**, pp. 271–75.

Stoller, A.I., Speers, R.F., and Opresko, S. (1970) "A New Technique for Etch Thinning Silicon Wafers," *RCA Rev.* **31**, pp. 265–70.

Stoller, A.I., and Wolff, N.E. (1966) "Isolation Techniques for Integrated Circuits," in *Proceedings: Second International Symposium on Microelectronics*, Munich, 1966.

Sugiyama, S., Kawakata, K., Abe, M., Funabashi, H., and Igarashi, I. (1987) "High-Resolution Silicon Pressure Imager with CMOS Processing Circuits," in *4th International Conference on Solid-State Sensors and Actuators (Transducers '87)*, Tokyo, pp. 444–47.

Sugiyama, S., Suzuki, T., Kawahata, K., Shimaoka, K., Takigawa, M., and Igarashi, I. (1986) "Micro-Diaphragm Pressure Sensor," in *Technical Digest: IEEE International Electron Devices Meeting (IEDM '86)*, pp. 184–87.

Sundaram, K.B., and Chang, H.-W. (1993) "Electrochemical Etching of Silicon by Hydrazine," *J. Electrochem. Soc.* **140**, pp. 1592–97.

Suzuki, K., Shimoyama, I., and Miura, H. (1994) "Insect-Model Based Microrobot with Elastic Hinges," *J. Microelectromech. Syst.* **3**, pp. 4–9.

Tabata, O., Asahi, R., Funabashi, H., Shimaoka, K., and Sugiyama, S. (1992) "Anisotropic Etching of Silicon in TMAH Solutions," *Sensor. Actuator. A* **A34**, pp. 51–57.

Tabata, O., Asahi, R., and Sugiyama, S. (1990) "Anisotropic Etching with Quarternary Ammonium Hydroxide Solutions," in *Technical Digest: 9th Sensor Symposium*, Tokyo, pp. 15–18.

Tabata, O., Asahi, R., and Sugiyama, S. (1995) "pH-Controlled TMAH Etchants for Silicon Micromachining," in *8th International Conference on Solid-State Sensors and Actuators (Transducers '95)*, Stockholm, pp. 83–86.

Takebe, T., Yamamoto, T., Fujii, M., and Kobayashi, K. (1993) "Fundamental Selective Etching Characteristics of HF + H_2O_2 + H_2O Mixtures for GaAs," *J. Electrochem. Soc.* **140**, pp. 1169–80.

Takeshima, N., Gabriel, K.J., Ozaki, M., Takahashi, J., Horiguchi, H., and Fujita, H. (1991) "Electrostatic Parallelogram Actuators," in *6th International Conference on Solid-State Sensors and Actuators (Transducers '91)*, San Francisco, pp. 63–66.

Tan, T.Y., Foll, H., and Hu, S.M. (1981) "On the Diamond-Cubic to Hexagonal Phase Transformation in Silicon," *Phil. Mag. A* **44**, pp. 127–40.

Tang, W.C., Nguyen, T.-C.H., and Howe, R.T. (1989a) "Laterally Driven Polysilicon Resonant Microstructures," in *Proceedings: IEEE Micro Electro Mechanical Systems (MEMS '89)*, Salt Lake City, pp. 53–59.

Tang, W.C., Nguyen, T.H., and Howe, R.T. (1989b) "Laterally Driven Polysilicon Resonant Microstructures," *Sensor. Actuator.* **20**, pp. 25–32.

Tang, W.C.K. (1990) Electrostatic Comb Drive for Resonant Sensor and Actuator Applications, Ph.D. thesis, University of California, Berkeley.

Tenney, A.S., and Ghezzo, M. (1973) "Etch Rates of Doped Oxides in Solutions of Buffered HF," *J. Electrochem. Soc.* **120**, pp. 1091–95.

Texas Instruments, "Thermal Character Print Head," Texas Instruments, Austin, TX.

Thaysen, J., Boisen, A., Hansen, O., and Bouwstra, S. (2000) "Atomic Force Microscopy Probe with Piezoresistive Read-Out and a Highly Symmetrical Wheatstone Bridge Arrangement," *Sensor. Actuator., A* **83**, pp. 47–53.

Theunissen, M.J., Apples, J.A., and Verkuylen, W.H.C.G. (1970) "Applications of Preferential Electrochemical Etching of Silicon to Semiconductor Device Technology," *J. Electrochem. Soc.* **117**, pp. 959–65.

Theunissen, M.J., Apples, J.A., and Verkuylen, W.H.C.G. (1972) "Etch Channel Formation during Anodic Dissolution of n-Type Silicon in Aqueous Hydrofluoric Acid," *J. Electrochem. Soc.* **119**, pp. 351–60.

Thundat, T., Bottomley, L.A., Meller, S., Velander, W.H., and Tassell, R.V. (2001) "Microcantilever Immunosensors," in *Immunoassays: Methods and Protocols*, Ghindilis, A.L., Pavlov, A.R., and Atanajov, P.B., eds., Humana Press, Totowa, NJ.

Thundat, T., Chen, G.Y., Warmack, R.J., Allison, D.P., and Wachter, E.A., "Vapor Detection Using Resonating Microcantilevers," *Anal. Chem.* **67**, pp. 519–21.

Thundat, T., Oden, P.I., and Warmack, R.J. (1997) "Microcantilever Sensors," *Microscale Thermophys. Eng.* **1089–3954/97**, pp. 1:185–99.

Thundat, T., Wachter, E.A., Sharp, S.L., and Warmack, R.J. (1995) "Detection of Mercury Vapor Using Resonating Cantilevers," *Appl. Phys. Lett.* **66**, pp. 1695–97.

Timoshenko, S.P., and Woinowsky-Krieger, S. (1959) *Theory of Plates and Shells*, McGraw-Hill, New York.

Tong, Q.-Y., Cha, G., Gafiteanu, R., and Gosele, U. (1994) "Low Temperature Wafer Direct Bonding," *J. Microelectromech. Syst.* **3**, pp. 29–35.

Tortonese, M., Barrett, R.C., and Quate, C.F. (1993) "Atomic Resolution with an Atomic Force Microscope Using Piezoresistive Detection," *Appl. Phys. Lett.* **62**, pp. 834–36.

Tortonese, M., Yamada, H., Barrett, R.C., and Quate, C.F. (1991) "Atomic Force Microscopy Using a Piezoresistive Cantilever," in *Proceedings: Transducers '91.*, IEEE, Piscataway, NJ, pp. 448–51.

Tuck, B. (1975) "Review: The Chemical Polishing of Semiconductors," *J. Mater. Sci.* **10**, pp. 321–39.

Tuckerman, D.B., and Pease, R.F.W. (1981) "High-Performance Heat Sinking for VLSI," *IEEE Electron Device Lett.* **EDL-2**, pp. 126–29.

Tufte, O.N., Chapman, P.W., and Long, D. (1962) "Silicon Diffused-Element Piezoresistive Diaphragms," *J. Appl. Phys.* **33**, p. 3322.

Turner, D.R. (1958) "Electropolishing Silicon in Hydrofluoric Acid Solutions," *J. Electrochem. Soc.* **105**, pp. 402–8.

Uhlir, A. (1956) "Electrolytic Shaping of Germanium and Silicon," *Bell Syst. Tech. J.* **35**, pp. 333–47.

Unagami, T. (1980) "Formation Mechanism of Porous Silicon Layer by Anodization in HF Solution," *J. Electrochem. Soc.* **127**, pp. 476–83.

van Dijk, H.J.A. (1972) "Method of Manufacturing a Semiconductor Device and Semiconductor Device Manufactured by Said Method," U.S. Patent 3,640,807.

van Dijk, H.J.A., and de Jonge, J. (1970) "Preparation of Thin Silicon Crystals by Electrochemical Thinning of Epitaxially Grown Structures," *J. Electrochem. Soc.* **117**, pp. 553–54.

van der Drift, A. (1967) "Evolutionary Selection: A Principle Governing Growth Orientation in Vapour-Deposited Layers," *Philips Res. Rep.* **22**, pp. 267–88.

van Mullem, C.J., Gabriel, K.J., and Fujita, H. (1991) "Large Deflection Performance of Surface Micromachined Corrugated Diaphragms," in *6th International Conference on Solid-State Sensors and Actuators (Transducers '91)*, San Francisco, pp. 1014–17.

Varadan, V.K., and P.J. McWhorter, eds., "Smart Electronics and MEMS," *Proceedings of the Smart Structures and Materials 1996 Meeting,* San Diego: SPIE 2722, pp. 46–54.

Vinci, R.P., and Braveman, J.C. (1991) "Mechanical Testing of Thin Films," in *6th International Conference on Solid-State Sensors and Actuators (Transducers '91)*, San Francisco, pp. 943–48.

von Recum, Andreas, F., Cooke and Francis, W. "Soft Tissue Implants with Micron Scale Surface Texture," U.S. Patents 4,871,366 and 4,846,834.

Voronin, V.A., Druzhinin, A.A., Marjamora, I.I., Kostur, V.G., and Pankov, J.M. (1992) "Laser-Recrystallized Polysilicon Layers in Sensors," *Sensor. Actuator. A* **A30**, pp. 143–47.

Waggener, H.A. (1970) "Electrochemically Controlled Thinning of Silicon," *Bell. Sys. Tech. J.* **49**, pp. 473–75.

Waggener, H.A., and Dalton, J.V. (1972) "Control of Silicon Etch Rates in Hot Alkaline Solutions by Externally Applied Potentials," *J. Electrochem. Soc.* **119**, p. 236C.

Waggener, H.A., Kragness, R.C., and Tyler, A.L. (1976a) "Anisotropic Etching for Forming Isolation Slots in Silicon Beam Leaded Integrated Circuits," in *Technical Digest: IEEE International Electron Devices Meeting*, Washington, DC, p. 68.

Waggener, H.A., Krageness, R.C., and Tyler, A.L. (1967b) "Two-Way Etch," *Electronics* **40**, p. 274.

Walker, J.A., Gabriel, K.J., and Mehregany, M. (1991) "Mechanical Integrity of Polysilicon Films Exposed to Hydrofluoric Acid Solutions," *J. Electron. Mater.* **20**, pp. 665–70.

Watanabe, H., Ohnishi, S., Honma, I., Kitajima, H., Ono, H., Wilhelm, R.J., and Sophie, A.J.L. (1995) "Selective Etching of Phosphosilicate Glass with Low Pressure Vapor HF," *J. Electrochem. Soc.* **142**, pp. 237–43.

Watanabe, Y., Arita, Y., Yokoyama, T., and Igarashi, Y. (1975) "Formation and Properties of Porous Silicon and Its Applications," *J. Electrochem. Soc.* **122**, pp. 1351–55.

Weinberg, M., Bernstein, J., Borenstein, J., Campbell, J., Cousens, J., Cunningham, B., Fields, R., Greiff, P., Hugh, B., Niles, L., and Sohn, J. (1996) "Micromachining Inertial Instruments," in *Micromachining and Microfabrication Process Technology 2*, Austin, pp. 26–36.

Weirauch, D.F. (1975) "Correlation of the Anisotropic Etching of Single-Crystal Silicon Spheres and Wafers," *J. Appl. Phys.* **46**, pp. 1478–83.

Wen, C.P., and Weller, K.P. (1972) "Preferential Electro-Chemical Etching of p^+ Silicon in an Aqueous HF-H_2SO_4 Electrolyte," *J. Electrochem. Soc.* **119**, pp. 547–48.

White, L.K. (1980) "Bilayer Taper Etching of Field Oxides and Passivation Layers," *J. Electrochem. Soc.* **127**, pp. 2687–93.

Williams, K.R., and Muller, R.S. (1996) "Etch Rates for Micromachining Processes," *J. Electrochem. Soc.* **137**, pp. 3612–32.

Wise, K.D., Robinson, M.G., and Hillegas, W.J. (1981) "Solid State Processes to Produce Hemispherical Components for Inertial Fusion Targets," *J. Vac. Sci. Technol.* **18**, pp. 1179–82.

Wise, K.D., Robinson, M.G., and Hillegas, W.J. (1985) "Silicon Micromachining and Its Applications to High Performance Integrated Sensors," in *Micromachining and Micropackaging of Transducers*, Fung, C.D., Cheung, P.W., Ko, W.H., and Fleming, D.G. eds., Elsevier, New York, pp. 3–18.

Wolf, S., and Tauber, R.N. (1987) *Silicon Processing for the VLSI Era*, Lattice Press, Sunset Beach.

Wong, A. (1990) "Silicon Micromachining," viewgraphs presented in Chicago.

Wong, S.M. (1978) "Residual Stress Measurements on Chromium Films by X-ray Diffraction Using the sin2 Y Method," *Thin Solid Films* **53**, pp. 65–71.

Worthman, J.J., and Evans, R.A. (1965) "Young's Modulus, Shear Modulus, and Poisson's Ratio in Silicon and Germanium," *J. Appl. Phys.* **36**, pp. 153–56.

Wu, T.H.T., and Rosler, R.S. (1992) "Stress in PSG and Nitride Films as Related to Film Properties and Annealing," *Solid State Technol.* May, pp. 65–71.

Wu, X., and Ko, W.H. (1987) "A Study on Compensating Corner Undercutting in Anisotropic Etching of (100) Silicon," in *4th International Conference on Solid-State Sensors and Actuators (Transducers '87)*, Tokyo, pp. 126–29.

Wu, X.P., Wu, Q.H., and Ko, W.H. (1985) "A Study on Deep Etching of Silicon Using EPW," in *3rd International Conference on Solid-State Sensors and Actuators (Transducers '85)*, Philadelphia, pp. 291–94.

Wu, X.P., Wu, Q.H., and Ko, W.H. (1986) "A Study on Deep Etching of Silicon Using Ethylene-Diamine-Pyrocathechol-Water," *Sensor. Actuator.* **9**, pp. 333–43.

Wu, X.-P., and Ko, W.H. (1989) "Compensating Corner Undercutting in Anisotropic Etching of (100) Silicon," *Sensor. Actuator.* **18**, pp. 207–15.

Yamada, K., and Kuriyama, T. (1991) "A New Modal Mode Controlling Method for a Surface Format Surrounding Mass Acceleromenter," in *6th International Conference on Solid-State Sensors and Actuators (Transducers '91)*, San Francisco, pp. 655–58.

Yamana, M., Kashiwazaki, N., Kinoshita, A., Nakano, T., Yamamoto, M., and Walton, W.C. (1990) "Porous Silicon Oxide Layer Formation by the Electrochemical Treatment of a Porous Silicon Layer," *J. Electrochem. Soc.* **137**, pp. 2925–27.

Yang, K.H. (1984) "An Etch for Delineation of Defects in Silicon," *J. Electrochem. Soc.* **131**, pp. 1140–45.

Yeh, H.J., and Smith, J.S. (1994a) "Fluidic Self-Assembly of Microstructures and Its Application to the Integration of GaAs on Si," in *IEEE International Workshop on Micro Electro Mechanical Systems (MEMS '94)*, Oiso, Japan, pp. 279–84.

Yeh, H.J., and Smith, J.S. (1994b) "Integration of GaAs Vertical-Cavity Surface-Emitting Laser on Si by Substrate Removal," *Appl. Phys. Lett.* **64**, pp. 1466–68.

Yeh, R., Kruglick, E.J., and Pister, K.S.J. (1994) "Towards an Articulated Silicon Microrobot," in *ASME 1994, Micromechanical Sensors, Actuators, and Systems*, Chicago, pp. 747–54.

Yi, Y.W., and Liu, C. (1999) "Magnetic Actuation of Hinged Microstructures," *J. Microelectromech. Syst.* **8**, pp. 10–17.

Yoshida, T., Kudo, T., and Ikeda, K. (1992) "Photo-Induced Preferential Anodization for Fabrication of Monocrystalline Micromechanical Structures," in *Proceedings: IEEE Micro Electro Mechanical Systems (MEMS '92)*, Travemunde, Germany, pp. 56–61.

Yun, W. (1992) A Surface Micromachined Accelerometer with Integrated CMOS Detection Circuitry, Ph.D. thesis, University of California, Berkeley.

Zhang, L.M., Uttamchandani, D., and Culshaw, B. (1991) "Measurement of the Mechanical Properties of Silicon Microresonators," *Sensor. Actuator. A* **A29**, pp. 79–84, 1991.

Zhang, M., Desai, T., and Ferrari, M. (1998) "Proteins and Cells on PEG Immobilized Silicon Surfaces Biomaterials," *Biomaterials* **19**, 953–60.

Zhang, Q., Liu, L.L., and Li, Z. (1996) "A New Approach to Convex Corner Compensation for Anisotropic Etching of (100) Si in KOH," *Sensor. Actuator. A* **56**, p. 251.

Zhang, X.G. (1991) "Mechanism of Pore Formation on n-Type Silicon," *J. Electrochem. Soc.* **138**, pp. 3750–56.

Zhang, X.G., Collins, S.D., and Smith, R.L. (1989) "Porous Silicon Formation and Electropolishing of Silicon by Anodic Polarization in HF Solution," *J. Electrochem. Soc.* **136**, pp. 1561–65.

Zhang, Y., and Wise, K.D. (1994) "Performance of Non-Planar Silicon Diaphragms under Large Deflections," *J. Microelectromech. Syst.* **3**, pp. 59–68.

Zhang, Z.L., and MacDonald, N.C. (1993) "Fabrication of Submicron High-Aspect-Ratio GaAs Actuators," *J. Microelectromech. Syst.* **2**, pp. 66–73.

4

LIGA and Micromolding

Guangyao Jia and
Marc J. Madou
University of California, Irvine

4.1 Introduction

LIGA is the German acronym for X-ray lithography (*X-ray lithographie*), electrodeposition (*galvanoformung*), and molding (*abformtechnik*). The process involves a thick layer of X-ray resist (from microns to centimeters), high-energy X-ray radiation exposure, and development to arrive at a three-dimensional resist structure. Subsequent metal deposition fills the resist mold with a metal, and after resist removal, a freestanding metal structure results [IMM, 1995]. The metal shape may be a final product or serve as a mold insert for precision plastic molding. Molded plastic parts may in turn be final products or lost molds (see Figure 4.1). The plastic mold retains the same shape, size, and form as the original resist structure but is produced quickly and inexpensively as part of an infinite loop. The plastic lost mold may generate metal parts in a second electroforming process or generate ceramic parts in a slip casting process.

The bandwidth of possible sizes in all three dimensions renders LIGA useful for manufacture of microstructures (micron and submicron dimensions) and packages for these microstructures (millimeter and centimeter dimensions), and even for the connectors from those packages to the "macro world" (electrical, e.g., through-vias; or physical, e.g., gas in- and outlets).

Once LIGA was established in the research community, interest in other micro- and nano-replication methods became more pronounced. Given the cost of the LIGA equipment, various LIGA-like processes took center stage. These pseudo-LIGA methods involve the replication of masters created by alternate means such as deep reactive ion etching (DRIE) and novel ultraviolet thick photoresists. This more generalized lithography and replication procedure is illustrated in Figure 4.2.

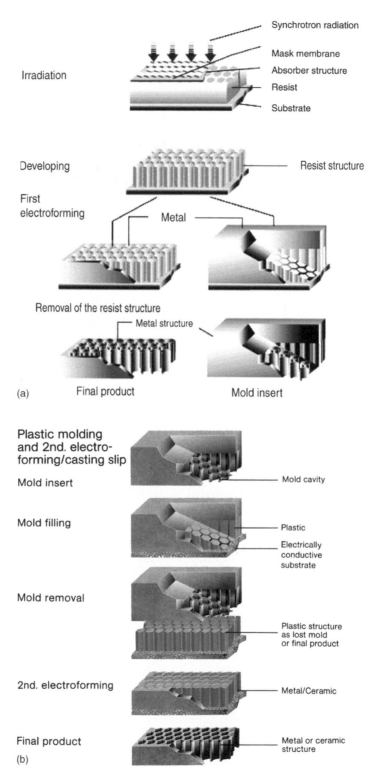

FIGURE 4.1   (a) Basic LIGA process steps X-ray deep-etch lithography and first electroforming. (b) Plastic molding and second electroforming/slip casting. (Reprinted with permission from Lehr, H., and Schmidt, M. [1995] "The LIGA Technique," commercial brochure, IMM GmbH, Mainz-Hechtsheim.)

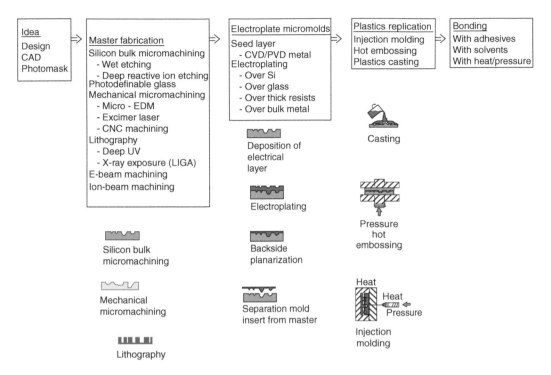

FIGURE 4.2   Process flow for plastic microfabrication.

Micromachining techniques are reshaping manufacturing approaches for a wide variety of small parts. Frequently, IC-based batch microfabrication methods are considered along with more traditional serial machining methods. In this evolution, LIGA and pseudo-LIGA processes constitute "handshake-technologies" bridging IC and classical manufacturing technologies. The capacity of LIGA and pseudo-LIGA for creating a wide variety of shapes from different materials makes these methods akin to classical machining, with the added benefit of unprecedented aspect ratios and absolute tolerances rendered possible by lithography and other high-precision mold fabrication techniques.

In this chapter, after a historical introduction to LIGA, we will analyze the process steps depicted in Figures 4.1 and 4.2. We start with a description of the different applications and technical characteristics of synchrotron radiation and then present an introduction to the crucial issues involved in making X-ray masks optimized for LIGA. Emphasis is on the two most important additive steps in LIGA and LIGA-like methods: electro- and electroless deposition of metals and plastic micromolding. Micromolding with an elastomeric mold, one of the many types of soft lithography, is widely used in research due to its low cost and easy implementation. Bonding of plastic molded micro parts, including plastic welding, organic solvent bonding, and bonding with thermoset adhesives, is covered in Madou (2002).

4.2   LIGA — Background

4.2.1   History

LIGA combines the sacrificial wax molding method, known since the time of the Egyptians, with X-ray lithography and electrodeposition. Combining electrodeposition and X-ray lithography was first carried out by Romankiw and co-workers at IBM as early as 1975 [Spiller et al., 1976]. These authors made high-aspect-ratio metal structures by plating gold in X-ray-defined resist patterns of up to 20 μm thick. They had, in other words, already invented "LIG"; that is, LIGA without the *abformung* (molding) [Spiller et al.,

1976]. This IBM work was an extension of through-mask plating, also pioneered by Romankiw et al. in 1969, and was geared toward the fabrication of thin film magnetic recording heads [Romankiw et al., 1970] (see Figure 4.30). The addition of plastic molding to the lithography and plating process was realized by Ehrfeld et al. (1982) at the Karlsruhe Nuclear Research Center (the Kernforschungszentrum Karlsruhe, or KfK), in 1982. By adding molding, these pioneers recognized the broader implications of LIGA as a new means of low-cost manufacturing of a wide variety of micro parts with unprecedented accuracies from various materials previously impossible to batch fabricate [Becker et al., 1982]. In Germany, LIGA originally developed almost completely outside of the semiconductor industry. In the United States, it was the late Henry Guckel who, starting in 1988, repositioned the field in light of semiconductor process capabilities and brought it closer to standard manufacturing processes.

The development of the LIGA process initiated by KfK was intended for the mass production of micron-sized nozzles for uranium-235 enrichment (see Figure 4.3) [Becker et al., 1982]. The German group used synchrotron radiation from a 2.5 GeV storage ring for the exposure of the poly(methyl–methacrylate) (PMMA) resist.

Today, LIGA and LIGA-like processes are researched in many laboratories around the world, and developing the ideal means of fabricating micromolds for the large-scale production of precise micromachines

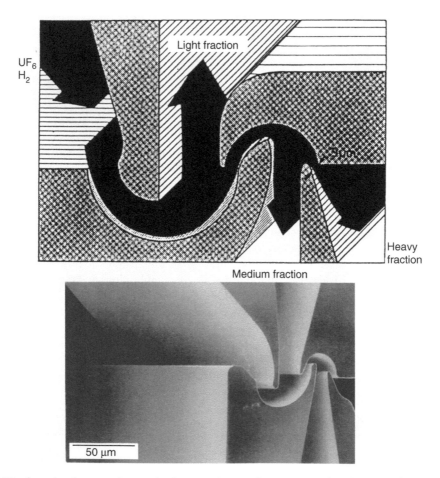

FIGURE 4.3 Scanning electron micrograph of a separation nozzle structure produced by electroforming with nickel using a micromolded PMMA template. This nozzle represents the first actual product ever made by LIGA. (Reprinted with permission from Hagmann, P. et al. [1987] "Fabrication of Microstructures with Extreme Structural Heights by Reaction Injection Molding," presented at *First Meeting of the European Polymer Federation*, European Symposium on Polymeric Materials, Lyon, France.)

remains an elusive goal. In LIGA, mold inserts are made via X-ray lithography, but depending on the dimensions of the micro parts, the accuracy requirements, and the fabrication costs, mold inserts may also be realized by e-beam writing, computer numerically controlled (CNC) machining, wet Si bulk micromachining, deep UV resists, deep reactive ion etching (DRIE), ultrasonic cutting, excimer laser ablation, electrodischarge machining (EDM), and laser cutting (see Figure 4.2).

4.2.2 Synchrotron Orbital Radiation (SOR)

4.2.2.1 Introduction

Lithography based on synchrotron radiation, also called synchrotron orbital radiation (SOR), is primarily pursued with the aim of adopting the technology as an industrial tool for the large-scale manufacture of microelectronic circuits with characteristic dimensions in the submicron range [Waldo and Yanof, 1991; Hill, 1991]. Synchrotron radiation sources "outshine" electron impact and plasma sources for generating X-rays. They emit a much higher flux of usable collimated X-rays, thereby allowing shorter exposure times and larger throughputs. The pros and cons of X-ray radiation for lithography in IC manufacture are summarized in Table 4.1.

Despite the many promising features of X-ray lithography, the technique still lacks mainstream acceptance in the IC industry. In 1991, experts projected that X-ray lithography would be in use by 1995 for 64-Mb DRAM manufacture, with critical dimensions (CDs) around 0.3 to 0.4 μm. With more certainty, they projected that the transition to X-rays would occur with the 0.2 to 0.3 μm CDs of 256-Mb DRAMs by 1998 [Waldo and Yanof, 1991]. Both dates passed without the emergence of the industrial use of X-rays. Continued improvements in optical lithography outpace the industrial use of X-ray lithography for IC applications. However, its use for prototype development on a small scale will no doubt continue.

In addition to being an option for next-generation IC lithography, X-rays are also used in the fabrication of 3-D microstructures. In LIGA, synchrotron radiation is used only in the lithography step. Other micromachining applications for SOR do exist. Urisu and his colleagues, for example, explored the use of synchrotron radiation for radiation-excited chemical vapor deposition and etching [Urisu and Kyuragi, 1987]. Micromachinists are hoping to piggyback X-ray lithography research and development efforts for the fabrication of micromachines onto major IC projects. The use of X-ray lithography for fabricating micro devices other than integrated circuits does not yet present a large business opportunity by itself. Not having a major IC product line associated with X-ray lithography makes it difficult to justify the use of X-ray lithography for micromachining, especially since other, less expensive micromachining technologies have not yet opened up the type of mass markets expected in the IC world. The fact that the X-rays used in LIGA are shorter in wavelength than in the IC application (2 to 10 Å vs. 20 to 50 Å) also puts micromachinists at a disadvantage. For example, the soft X-rays in the IC industry may eventually be generated from a much less expensive source, such as a transition radiation source [Goedtkindt et al., 1991]. Also, nontraditional IC materials are frequently employed in LIGA. The fabrication of X-ray masks poses more difficulties than masks for IC applications. Rotation and slanting of the X-ray masks to craft nonvertical

TABLE 4.1   Pros and Cons of SOR X-Ray Lithography for IC Manufacture

Pros	Cons
Lithography process insensitive to resist thickness, exposure time, and development time (large DOF)	Resist not very sensitive (not too important because of the intense light source)
Absence of backscattering results in insensitivity to substrate type, reflectivity and topography, pattern geometry and proximity, dust and contamination	Masks very difficult and expensive to make
High resolution <0.2 μm	Very high start-up investment
Some have suggested high throughput.	It is not proven as a system yet.
	Radiation effects on SiO_2 can be involved.

TABLE 4.2   SOR Applications

Application area	Instruments/technologies needed
Structural analysis	
Atoms	Photoelectron spectrometers
Molecules	Absorption spectrometers
Very large molecules	Fluorescent spectrometers
Proteins	Diffraction cameras
Cells	Scanning electron microscope (to view topographical radiographs)
Crystals	Time resolved X-ray diffractometers
Polycrystals	
Chemical analysis	
Trace	Photoelectron spectrometers
Surface	(Secondary ion) mass spectrometer
Bulk	Absorption/fluorescence spectrometers
	Vacuum systems
Microscopy	
Photoelectron	Photoemission microscopes
X-ray	X-ray microscopes SEM (for viewing)
	Vacuum systems
Micro/nanofabrication	
X-ray lithography	Steppers, mask making
Photochemical deposition of thin films	Vacuum systems
Etching	LIGA process
Medical diagnostics	
Radiography	X-ray cameras and equipment
Angiography and tomography	Computer aided display
Photochemical reactions	
Preparation of novel materials	Vacuum systems
	Gas-handling equipment

Source: Adapted with permission from Nippon Telegraph and Telephone Corporation (NNT), 1991.

walls further differentiates LIGA exposure stations. All these factors make exploring LIGA a challenge. However, given sufficient research and development money, large markets are likely to emerge over the next five to ten years. These markets could be in the manufacture of devices with stringent requirements imposed on resolution, aspect ratio, structural height, and parallelism of structural walls. Optical applications for the information technology (IT) field seem particularly attractive early product targets.

So far, it is the research community that has primarily benefited from the availability of SOR photon sources. With its continuously tunable radiation across a very wide photon range, highly polarized and directed into a narrow beam, SOR provides a powerful probe of atomic and molecular resonances. Other types of photon sources prove unsatisfactory for these applications in terms of intensity or energy spread. As can be concluded from Table 4.2, applications of SOR beyond lithography range from structural and chemical analysis to microscopy, angiography, and even to the preparation of new materials.

4.2.2.2   Technical Aspects

Some important concepts associated with synchrotron radiation (such as the bending radius of the synchrotron magnet, magnetic field strength, beam current, critical wavelength, and total radiated power) require introduction. Figure 4.4 presents a schematic of an X-ray exposure station. Electrons are injected into the ring, where they are maintained at energies anywhere from 10^6 to 10^9 eV. The cone of radiation shown in this figure is the electromagnetic radiation emitted by the circling electrons due to the radial acceleration that keeps them in the orbit of the electron synchrotron or storage ring. For high-energy particle studies, this radiation, emitted tangential to the circular electron path (Bremsstrahlung), limits the maximum energy the electrons can attain. Bremsstrahlung is a nuisance for studies of the composition of the atomic nucleus in which high-energy particles are smashed into the nucleus. To minimize the

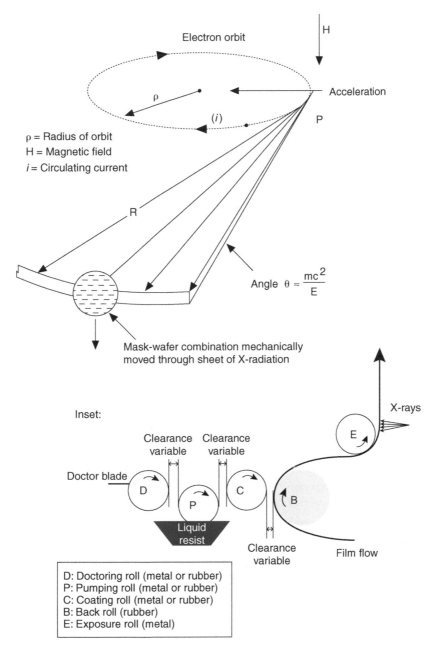

FIGURE 4.4 Schematic of an X-ray exposure station with a synchrotron radiation source. The X-ray radiation cone (opening θ) is tangential to the electron's path, describing a line on an intersecting substrate. Inset: A vision for the future, continuous micromanufacturing.

Bremsstrahlung, physicists desire ever bigger synchrotrons. The energy lost in the emission process is made up in a radio frequency (RF) cavity, where electrons are accelerated back up to the storage ring energy. Injection of electrons must be repeated a few times a day, since the electron current slowly decays due to leakage. For X-ray lithography applications, electrical engineers want to maximize the X-ray emission and build small synchrotrons instead (the radius of curvature for a compact, superconducting synchrotron, for example, is 2 m). The operating cost of these magnets is primarily that of the liquid helium refrigeration. Once high-T_c (critical temperature, the temperature at which the resistance falls to zero) superconducting materials can be made in bulk, compact storage rings will become extremely attractive. The angular opening of the radiation cone in Figure 4.4 is determined by the electron energy, E, and is given by:

$$\theta \approx \frac{mc^2}{E} = \frac{0.5}{E(\text{GeV})} \ (\text{mrad}) \tag{4.1}$$

The X-ray light bundle with the cone opening, θ, describes a horizontal line on an intersecting substrate as the X-ray bundle is tangent to the circular electron path. In the vertical direction, the intensity of the beam exhibits a Gaussian distribution, and the vertical exposed height on the intersecting substrate can be calculated knowing θ and R, the distance from the radiation point P to the substrate. With $E = 1$ GeV, $\theta = 1$ mrad, and $R = 10$ m, the exposed area in the vertical direction measures about 0.5 cm. To expose a substrate homogeneously over a wider vertical range, the sample is moved vertically through the irradiation band with a precision scanner, for example, at a speed of 10 mm/s over a 100 mm scanning distance. Usually, the substrate is stepped up and down repeatedly until the desired X-ray dose is obtained. It is interesting to note that an SOR setup affords continuous lithography, a prospect that may make disposable ICs and MEMS a possibility. Rolls of dry X-ray photoresist or X-ray photoresist-covered foils could pass through the exposure beam, resulting in a continuous lithography process — that is, a "beyond batch" type of approach (see inset in Figure 4.4).

The electron energy E in Equation (4.1) is given by:

$$E(\text{GeV}) = 0.29979 \, B \, (\text{Tesla}) \, \rho \, (\text{meters}) \tag{4.2}$$

where B is the magnetic field and ρ is the radius of the circular path of the electrons in the synchrotron.

The total radiated power can be calculated from the energy loss of the electrons per turn and is given by:

$$P(\text{kW}) = \frac{88.47 E^4 i}{\rho} \tag{4.3}$$

where i is the beam current.

The emission of the synchrotron electrons is a broad spectrum without characteristic peaks or line enhancements, and its distribution extends from the microwave region through the infrared, visible, ultraviolet, and into the X-ray region. The critical wavelength, λ_c, is defined so that the total radiated power at lower wavelengths equals the radiated power at higher wavelengths, and is given by:

$$\lambda_c(\text{Å}) = \frac{5.59 \rho \, (\text{m})}{E^3 (\text{GeV})} \tag{4.4}$$

Equation (4.3) shows that the total radiated power increases with the fourth power of the electron energy. From Equation (4.4), we appreciate that the spectrum shifts toward shorter wavelengths with the third power of the electron energy.

The dose variation absorbed in the top vs. the bottom of an X-ray resist should be kept small so that the top layer does not deteriorate before the bottom layers are sufficiently exposed. Since the depth of penetration increases with decreasing wavelength, synchrotron radiation of very short wavelengths is needed to pattern thick resist layers. To obtain good aspect ratios in LIGA structures, the critical wavelength ideally should be 2 Å. Bley et al. (1992) at KfK designed a new synchrotron optimized for LIGA. They proposed a magnetic flux density, B, of 1.6285 T; a nominal energy, E, of 2.3923 GeV; and a bending radius, ρ, of 4.9 m. With those parameters, Equation (4.4) results in the desired λ_c of 2 Å and an opening angle of radiation, based on Equation (4.1), of 0.2 mrad (in practice, this angle will be closer to 0.3 mrad due to electron beam emittance).

The X-rays traveling from the ring to the sample site are held in a high vacuum. The sample itself is kept either in air or in a helium (He) atmosphere. The inert atmosphere prevents corrosion of the exposure chamber, mask, and sample by reactive oxygen species, and removal of heat is much faster than in air (the heat conductivity of He is high compared to air). In He, the X-ray intensity loss is also 500 times less than in air. A beryllium (Be) window separates the high vacuum from the inert atmosphere. For

TABLE 4.3 Facilities in the United States Where Access to Synchrotron Radiation Is or Will Soon Be Available

Facility	Institute	URL
Advanced Photon Source (APS)	Argonne National Laboratory	http://www.aps.anl.gov/
Cornell High Energy Synchrotron Source (CHESS)	Cornell University	http://www.tn.cornell.edu/
National Synchrotron Light Source (NSLS)	Brookhaven National Laboratory	http://www.nsls.bnl.gov/
Stanford Synchrotron Radiation Laboratory (SSRL)	Stanford University	http://www-ssrl.slac.stanford.edu/
Synchrotron Ultraviolet Radiation Facility (SURF)	National Institute of Standards and Technology	http://physics.nist.gov/MajResFac/SURF/
Synchrotron Radiation Center (SRC)	University of Wisconsin-Madison	http://www.src.wisc.edu/
Center for Advanced Microstructures and Devices (CAMD)	Louisiana State University	http://www.camd.lsu.edu/
Advanced Light Source (ALS)	Lawrence Berkeley Laboratory	http://www-als.lbl.gov/

wavelengths shorter than 1 nm, Be is very transparent — that is, it is an excellent X-ray window. A 25 μm thick Be window can withstand a 1 atm pressure differential across a small diameter (< 1 in). For large area exposures, windows up to 6 cm dia. have been developed. Be windows age with X-ray exposure and must be replaced periodically.

4.2.3 Access to the Technology

Today, the construction cost for a typical synchrotron totals over $30 million, restricting the access to LIGA. Obviously, a less expensive alternative for generating intense X-rays is preferred. Along this line, in Japan Ishikawajima–Harima Heavy Industries (IHI) is building compact synchrotron X-ray sources (e.g., an 800 MeV synchrotron of about 30 feet per side) (<http://www.ihi.co.jp/>).

By the end of 1993, eight nonprivately owned synchrotrons were in use in the United States. The first privately owned synchrotron was put into service in 1991 at IBM's Advanced Semiconductor Technology Center (ASTC) in East Fishkill, New York. Table 4.3 lists the eight U.S. synchrotron facilities.

Most of the facilities listed in Table 4.3 allow LIGA work. For example, Cronos Integrated Microsystems, Inc., a JDS Uniphase Company and a spin-off from MCNC (Research Triangle Park, NC), in collaboration with the University of Wisconsin-Madison, announced its first multiuser LIGA process sponsored by ARPA in September 1993 (<http://www.memsrus.com/CIMSmain2ie.html>). The Center for Advanced Microstructure Devices (CAMD), at Louisiana State University, has three beam lines dedicated exclusively to micromachining work, and the Advanced Light Source (ALS) at Berkeley has one beam line available for micromachining.

Like Cronos, Forschungszentrum Karlsruhe GmbH offers a multiuser LIGA service (LEMA, or LIGA-experiment for multiple applications). The commercial exploitation of LIGA is pursued by at least three German organizations: microParts GmbH STEAG (<http://www.microparts.de>); IMM (<http://www.imm.uni-mainz.de>); and Forschungszentrum Karlsruhe, or KfK (<http://www.fzk.de>). In the U.S., Louisiana State University's CAMD (<http://www.camd.lsu.edu/>), Baton Rouge, and the associated start-up Mezzo Systems, Inc. (now International Mezzo Technologies, Inc., <http://www.mezzotech.biz/>), are promoting the technology.

4.3 LIGA and LIGA-Like Process Steps

4.3.1 X-Ray Masks

4.3.1.1 Introduction

X-ray mask production is one of the most difficult aspects of X-ray lithography. To be highly transmissive to X-rays, the mask substrate by necessity must be a low-Z (atomic number) thin membrane. X-ray

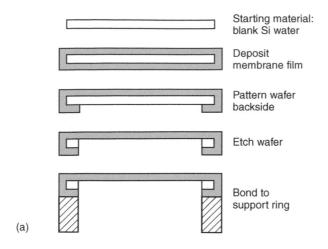

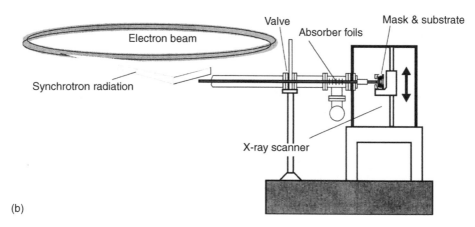

FIGURE 4.5   (a) Schematic of a typical X-ray mask and (b) mask and substrate assembly in an X-ray scanner. (The latter reprinted with permission from IMM [1995] brochure.)

masks should withstand many exposures without distortion, be alignable with respect to the sample, and be rugged. A possible X-ray mask architecture and its assembly with a substrate in an X-ray scanner are shown in Figure 4.5. The mask shown here has three major components: an absorber; a membrane, or mask blank; and a frame. The absorber contains the information to be imaged onto the resist. It is made up of a material with a high atomic number (Z), often Au, patterned onto a membrane material with a low Z. The high-Z material absorbs X-rays, whereas the low-Z material transmits X-rays. The frame lends robustness to the membrane/absorber assembly so that the whole can be handled confidently.

The requirements for X-ray masks in LIGA differ substantially from those for the IC industry. A comparison is presented in Table 4.4 [Ehrfeld et al., 1986]. The main difference lies in the absorber thickness. To achieve high contrast (>200), very thick absorbers (>10 μm vs. 1 μm) and highly transparent mask blanks (transparency > 80%) must be used because of the low resist sensitivity and the great depth of the resist. Another difference focuses on the radiation stability of membrane and absorber. For conventional optical lithography, the supporting substrate is a relatively thick, optically flat piece of glass or quartz that is highly transparent to optical wavelengths. It provides a highly stable (>10^6 μm) basis for the thin (0.1 μm) chrome absorber pattern. In contrast, the X-ray mask consists of a very thin membrane (2 to 4 μm) of low-Z material carrying a high-Z thick absorber pattern [Lawes, 1989]. A single exposure in LIGA results in an exposure dose 100 times higher than in the IC case.

TABLE 4.4 Comparison of Masks in LIGA and the IC Industry

	Semiconductor lithography	LIGA process
Transparency	$\geq 50\%$	$\geq 80\%$
Absorber thickness	$\pm 1\,\mu m$	$10\,\mu m$ or higher
Field size	$50 \times 50\,mm^2$	$100 \times 100\,mm^2$
Radiation resistance	$=1$	$=100$
Surface roughness	$<0.1\,\mu m$	$<0.5\,\mu m$
Waviness	$<\pm 1\,\mu m$	$<\pm 1\,\mu m$
Dimensional stability	$<0.05\,\mu m$	<0.1–$0.3\,\mu m$
Residual membrane stress	$\sim 10^8\,Pa$	$\sim 10^8\,Pa$

Source: Reprinted with permission from Ehrfeld, W. et al. (1986) "Mask Making for Synchrotron Radiation Lithography," *Microlectron. Eng.* 5, pp. 463–70.

TABLE 4.5 Comparison of Membrane Materials for X-Ray Masks

Material	X-ray transparency	Non-toxicity	Dimensional stability	Remark
Si	0 (50% transmission at 5.5 μm thickness)	++	0 (thermal exp coefficient 2.6°C^{-1} 10^{-6}) Young's modulus = 1.3	Single crystal Si, well developed, rad hard, stacking faults cause scattering, material is brittle
SiN$_x$	0 (50% transmission at 2.3 μm thickness)	++	(thermal exp coefficient 2.7°C^{-1} 10^{-6}) Young's modulus = 3.36	Amorphous, well developed, rad hard if free of oxygen, resistant to breakage
SiC	+ (50% transmission at 3.6 μm thickness)	++	(thermal exp coefficient 4.7°C^{-1} 10^{-6}) Young's modulus = 3.8	Poly and amorphous, rad hard, some resistance to breakage
Diamond	+ (50% transmission at 4.6 μm thickness)	++	++ (thermal exp coefficient 1.0°C^{-1} 10^{-66}) Young's modulus = 11.2	Poly, research only, highest stiffness
BN	+ (50% transmission at 3.8 μm thickness)	++	0 (thermal exp coefficient 1.0°C^{-1} 10^{-6}) Young's modulus $=$ 1.8	Not rad hard, i.e., not applicable for LIGA
Be	++	–	++	Research, especially suited for LIGA, even at 100 μm the transparency is good, 30 μm typical, difficult to electroplate, toxic material
Ti	–	++	0	Research, used for LIGA, not very transparent, films must not be more than 2 μm to 3 μm thick

We will look into these different mask aspects separately before detailing a process with the potential of obviating altogether the need for a separate X-ray mask, through the use of conformal or transfer masks.

4.3.1.2 X-Ray Membrane (Mask Blank)

The low-Z membrane material in an X-ray mask must have a transparency for rays with a critical wavelength λ_c from 0.2 to 0.6 nm of at least 80% and should not induce scattering of those rays. To avoid pattern distortion, the residual stress σ_r in the membrane should be less than 10^8 dyn/cm^2. Mechanical stress in the absorber pattern can cause in-plane distortion of the supporting thin membrane, requiring a high Young's modulus for the membrane material. Humidity or high deposited doses of X-ray might also distort the membrane directly. During one typical lithography step, the masks may be exposed to 1 MJ/cm^2 of X-rays. Since most membranes must be very thin for optimal transparency, a compromise has to be found among transparency, strength, and form stability. Important X-ray membrane materials are listed in Table 4.5. The higher radiation dose in LIGA prevents the use of BN and compound mask blanks that incorporate a polyimide layer. Those mask blanks are perfectly appropriate for classical IC lithography work but will not do for LIGA work. Mask blanks of metals such as titanium (Ti) and beryllium were specifically developed for LIGA applications because of their radiation hardness [Ehrfeld et al., 1986; Schomburg et al., 1991] In comparing titanium and beryllium membranes, beryllium can have a much greater membrane thickness d and still be adequately transparent. For example, a membrane transparency of 80%, essential for adequate exposure of a 500 μm thick PMMA resist layer, is obtained with a thin 2 μm titanium film, whereas with beryllium, a thick 300 μm membrane achieves the same result. The thicker beryllium membrane permits easier processing and handling. In addition, beryllium has a greater Young's modulus E than titanium (330 vs. 140 kN/mm^2) and, since it is the product of $E \times d$ that determines the amount of mask distortion, distortions due to absorber stress should be much smaller for beryllium blanks [Schomburg et al., 1991; Hein et al., 1992]. Beryllium thus comes forward as an excellent membrane material for LIGA because of its high transparency and excellent damage resistance. Such a mask should be good for up to 10,000 exposures and may cost $20,000 to $30,000 ($10,000 to $15,000 in quantity). Stoichiometric silicon nitride (Si_3N_4) used in X-ray mask membranes may contain numerous oxygen impurities, absorbing X-rays and thus producing heat. This heat often suffices to prevent the use of nitride as a good LIGA mask. Single-crystal silicon masks have been made (1 cm square and 0.4 μm thick, and 10 cm square and 2.5 μm thick) by electrochemical etching techniques. Nanostructures, Inc. (<http://www.nanostructures.com/services.htm>) is one of the companies that make such thin Si masks. For Si and Si_3N_4, the Young's modulus is quite low compared with CVD-grown diamond and SiC films, with a Young's modulus as much as three times higher. These higher stiffness materials are more desirable because the internal stresses of the absorbers, which can distort mask patterns, are less of an issue. Unfortunately, diamond and SiC membranes are also the most difficult to produce.

4.3.1.3 Absorber Materials

The requirements on the absorber are high attenuation (>10 dB), stability under radiation over extended periods of time, negligible distortion (stress $< 10^8$ dyn/cm^2), ease of patterning, repairability, and low defect density. Typical absorber materials are listed in Table 4.6. Gold is used most commonly, and some groups are looking at the viability of tungsten and other materials. In the IC industry, an absorber thickness of 0.5 μm might suffice, whereas LIGA deals with thicker layers of resist, requiring a thicker absorber to maintain the same resolution.

Figure 4.6 illustrates how X-rays, with a characteristic wavelength of 0.55 nm, are absorbed along their trajectory through a Kapton preabsorber filter, an X-ray mask, and resist [Bley et al., 1991]. The low-energy portion of the synchrotron radiation is absorbed mainly in the top portion of the resist layer because absorption increases with increasing wavelength. The Kapton preabsorber filters out much of the low-energy radiation to prevent overexposure of the top surface of the resist. The X-ray dose at which the resist gets damaged D_{dm} and the dose required for development of the resist D_{dv} as well as the "threshold dose" at which the resist starts dissolving in a developer D_{th} are all indicated in Figure 4.6. In the areas under the absorber pattern of the X-ray mask, the absorbed dose must stay below the threshold dose D_{th}. Otherwise, the structures partly dissolve, resulting in poor feature definition. From Figure 4.6, we can deduce that the height of the gold absorbers must exceed 6 μm to reduce the absorbed radiation dose of the resist under the gold pattern to below the threshold dose D_{th}. In Figure 4.7, the necessary thickness of the gold absorber patterns of an X-ray mask is plotted as a function of the thickness of the resist to be

TABLE 4.6   Comparison of Absorber Materials for X-Ray Masks

Material	Remark
Gold	Not the best stability (grain growth), low stress, electroplating only, defects repairable (thermal exp coefficient 14.2°C^{-1} 10^{-6}) (0.7 μm for 10 dB)
Tungsten	Refractory and stable, special care is needed for stress control, dry etchable, repairable (thermal exp coefficient 4.5°C^{-1} 10^{-6}) (0.8 μm for 10 dB)
Tantalum	Refractory and stable, special care is needed for stress control, dry etchable, repairable
Alloys	Easier stress control, greater thickness needed to obtain 10 dB

patterned; the Au must be thicker for thicker resist layers and for shorter characteristic wavelengths λ_c of the X-ray radiation. To pattern a 500 µm high structure with a λ_c of 0.225 nm, the gold absorber must be more than 11 µm high.

Exposure of yet more extreme photoresist thicknesses is possible if proper X-ray photon energies are used. At 3000 eV, the absorption length in PMMA roughly measures 100 µm, which enables the above-mentioned 500 µm exposure depth [Guckel et al., 1994]. Using 20,000 eV photons results in absorption lengths of 1 cm. PMMA structures up to 10 cm thick have been exposed this way [Siddons and Johnson, 1994a]. A high-energy mask used by Guckel for these high-energy exposures has an Au absorber 50 µm thick and a blank membrane of 400 µm thick Si. Guckel obtained an absorption contrast of 400 when exposing a 1000 µm thick PMMA sheet with this mask. An advantage of using such thick Si blank membranes is that larger resist areas can be exposed, since one does not depend on a fragile membrane/absorber combination [Guckel et al., 1994b].

4.3.1.4 Absorber Fabrication

4.3.1.4.1 Single-Layer Absorbers

To make a mask with gold absorber structures of a height above 10 µm, one must first succeed in structuring a resist of that thickness. The height of the resist should in fact be a bit higher (say, 20%) than the absorber itself so as to accommodate the electrodeposited metal fully in between the resist features. Currently, no means exists to structure a resist of that height with sufficient accuracy and perfect verticality of the walls, unless X-rays are used. Different procedures for producing X-ray masks with thicker absorber layers using a two-stage lithography process have been developed.

The KfK solution calls for first making an intermediate mask with photo or electron-beam lithography. This intermediate mask starts with a 3 µm thick resist layer, in which case the needed line-width

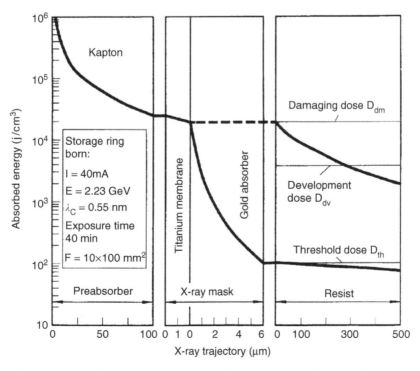

FIGURE 4.6 Absorbed energy along the X-ray trajectory including a 500 µm thick PMMA specimen, X-ray mask, and a Kapton preabsorber. (Reprinted with permission from Bley, P. et al. [1991] "Application of the LIGA Process in Fabrication of Three-Dimensional Mechanical Microstructures," presented at the 1991 International MicroProcess Conference, July 15–18, 1991, Kanazawa, Japan.)

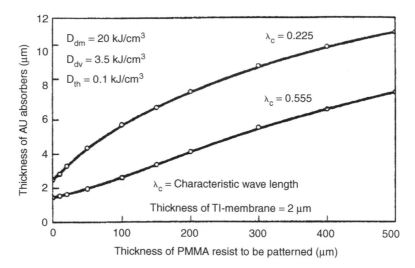

FIGURE 4.7   Necessary thickness of the gold absorbers of an X-ray mask. (Reprinted with permission from Bley, P. et al. [1991] "Application of the LIGA Process in Fabrication of Three-Dimensional Mechanical Microstructures," in *4th International Symposium on MicroProcess Conference*, Kanazawa, pp. 384–89, July 15–18, 1991.)

accuracy and photoresist wall steepness of printed features are achievable. After gold plating between the resist features and stripping of the resist, this intermediate mask is used to write a pattern with X-rays in a thicker resist, say 20 μm thick. After electrodepositing and resist stripping, the actual X-ray mask (that is, the master mask) is obtained.

Because hardly any accuracy is lost in the copying of the intermediate mask with X-rays to obtain the master mask, it is the intermediate mask quality that determines the ultimate quality of the LIGA-produced microstructures. The structuring of the resist in the intermediate mask is handled with optical techniques when the requirements of the LIGA structures are less stringent. The minimal lateral dimensions for optical lithography in a 3 μm thick resist typically measure about 2.5 μm. Under optimal conditions, a wall angle of 88° is achievable. With e-beam lithography, a minimum lateral dimension of less than 1 μm is feasible. The most accurate pattern transfer is achieved through reactive ion etching of a tri-level resist system. In this approach, a 3 to 4 μm thick polyimide resist is first coated onto the titanium or beryllium membrane, followed by a coat of 10 to 15 nm titanium deposited with magnetron sputtering. The thin layer of titanium is an excellent etch mask for the polyimide; in an optimized oxygen plasma, the titanium etches 300 times slower than the polyimide. To structure the thin titanium layer itself, a 0.1 μm thick optical resist is used. Because this top resist layer is so thin, excellent lateral tolerances result. The thin Ti layer is patterned with optical photolithography and etched in an argon plasma. After etching the thin titanium layer, exposing the polyimide locally, an oxygen plasma helps to structure the polyimide down to the titanium or beryllium membrane. Lateral dimensions of 0.3 μm can be obtained in this fashion. Patterning the top resist layer with an e-beam increases the accuracy of the three-level resist method even further. Electrodeposition of gold on the titanium or beryllium membrane and stripping of the resist finishes the process of making the intermediate LIGA mask. To make a master mask, this intermediate mask is printed by X-ray radiation onto a PMMA-resist-coated master mask. The PMMA thickness corresponds to a bit more than the desired absorber thickness. Since the resist layer thickness is in the 10 to 20 μm range, a synchrotron X-ray wavelength of 10 Å is adequate for making the master mask. A further improvement in LIGA mask making is to fabricate intermediate and master mask on the same substrate, greatly reducing the risk for deviations in dimensions caused, for example, by temperature variations during printing [Becker et al., 1986]. The ultimate achievement would be to create a one-step process to make the master mask. Along this line, Hein et al. (1992) investigated the direct patterning of 10 μm-high resist layers with a 100 kV e-beam.

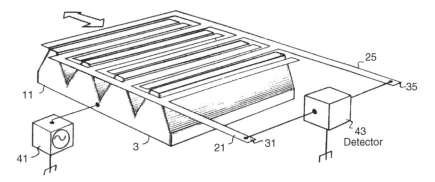

FIGURE 4.8 Mask alignment system in X-ray lithography. Conductive fingers on the mask and ridges on the Si are used for alignment. (From U.S. Patents 4,654,581 [1987] and 4,607,213 [1986].)

4.3.1.4.2 Stepped Absorbers
In principle, stepped absorber structures may result in stepped LIGA structures by means of X-ray lithography with a single mask. In this manner, variable dose depositions can be achieved at the same resist heights. The variable dose results in different molecular weights and, hence, in a different developing behavior. This technique unfortunately leads to rounded features and poor step-height control.

4.3.1.4.3 CNC Machined Absorbers
A rather unexpected approach to pattern absorber layers for X-ray masking is being pursued by Friedrich et al., from the Michigan Technological University (<http://www.me.mtu.edu/~microweb/>). Friedrich et al. are exploring X-ray mask fabrication by traditional machining methods, such as micromilling, micro-EDM, and lasers [Friedrich, 1994]. Using micromilling, this group succeeded in making mask features up to 62 μm deep and walls down to 4 μm thick and 10 μm high. The milling is carried out with a 22 μm end mill, itself fabricated using a 20 keV gallium ion beam. The advantages of this approach are rapid turn-around (less than one day per mask), low cost, and flexibility (almost any type of material can be machined) because no intermediate steps interfere. Disadvantages are less dimensional edge acuity and nonsharp interior corners, as well as much less absolute tolerance.

4.3.1.5 Alignment of X-Ray Mask to Substrate
The mask and resist-coated substrate must be properly registered to each other before they are put in an X-ray scanner. Alignment of an X-ray mask to the substrate is problematic because no visible light can pass through most X-ray membranes. To solve this problem, Schomburg et al. (1991) etched windows in their Ti X-ray membrane. Diamond membranes have a potential advantage here, as they are optically transparent and enable easy alignment for multiple irradiations without a need for etched holes.

Figure 4.8 illustrates an alternative X-ray alignment system involving capacitive pickup between conductive metal fingers on the mask and ridges on a small substrate area; Si, in this case (U.S. Patents No. 4,607,213 [1986] and 4,654,581 [1987]). When using multiple groups of ridges and fingers, two-axis lateral and rotational alignment become possible.

Another alternative may involve liquid nitrogen-cooled Si (Li) X-ray diodes as alignment detectors, eliminating the need for observation with visible light [Henck, 1984].

4.3.1.6 Conformal, Transfer, or Self-Aligned Mask for High-Aspect-Ratio Microlithography
Vladimirsky et al. [Vladimirsky et al., 1995; Malek et al., 1996] developed a procedure to eliminate the need for an X-ray mask membrane. Unlike conventional masks, the so-called X-ray transfer mask does not treat a mask as an independent unit. The technique is based on forming an absorber pattern directly on the resist surface forming a conformal, self-aligned, or transfer mask. An example process is shown in Figure 4.9. In this sequence, a transfer mask plating base is first prepared on the PMMA substrate plate

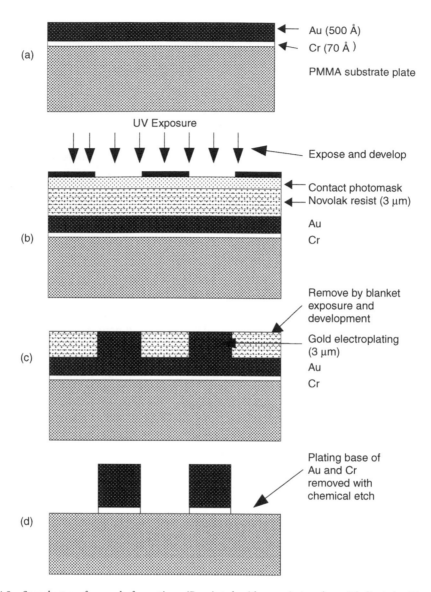

FIGURE 4.9 Sample transfer mask formation. (Reprinted with permission from Vladimirsky, Y. et al. [1995] "Transfer Mask for High Aspect Ratio Micro-Lithography," presented at Microlithography '95, Santa Clara, CA.)

by evaporating 70 Å of chromium (as adhesion layer) followed by 500 Å of gold using an electron beam evaporator. A 3 μm thick layer of standard Novolak-based AZ-type resist S1400-37 (Shipley Co.) is then applied over the plating base and exposed in contact mode through an optical mask using an ultraviolet exposure station. Three micrometers of electroplated gold on the exposed plating base further complete the transfer mask. A blanket exposure and subsequent development remove the remaining resist. The 500 Å of Au plating base is dissolved by a dip of 20 to 30 s in a solution of KI (5%) and I (1.25%) in water; the Cr adhesion layer is removed by a standard chromium etch (from KTI, Chemicals Inc., Sunnyvale, Calif.). Fabrication of the transfer mask can thus be performed using standard lithography equipment available at almost any lithography shop. Depending on the resolution required, the X-ray transfer mask can be fabricated using known photon, e-beam, or X-ray lithography techniques. The patterning of the PMMA resist with a self-aligned mask is accomplished in multiple steps of exposure and development. An example of a cylindrical resonator made this way is shown in Figure 4.10. Each exposure/development

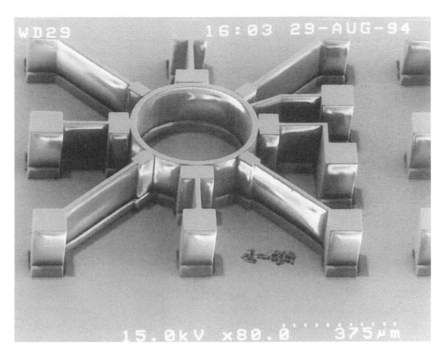

FIGURE 4.10 SEM micrograph of a cylindrical PMMA resonator made by the transfer mask method and multiple exposure/development steps. (Reprinted with permission from Vladimirsky, Y. et al., "Transfer Mask for High Aspect Ratio Micro-Lithography," presented at Microlithography '95, Santa Clara, CA, 1995, Courtesy of Dr. V. Saile, vol. 2437.)

step involves an exposure dose of about 8 to 12 J/cm². Subsequent 5 min development steps remove ~30 μm of PMMA. In seven steps, a self-supporting 1.5 mm thick PMMA resist is patterned to a depth of more than 200 μm. The resist pattern shown in Figure 4.10 is 230 μm thick and exhibits a 2 μm gap between the inner cylinder and the pickup electrodes (aspect ratio is 100:1). The resonator pattern was produced using soft (= 10 Å) X-rays and a 3 μm thick Au absorber only.

Vladimirsky et al. (1995) suggest that the forming of the transfer mask directly on the sample surface creates several additional new opportunities; besides in situ development, etching, and deposition, these include exposure of samples with curved surfaces and dynamic deformation of a sample surface during the exposure (hemispherical structures for lenses are possible this way). Elegant and cost-saving innovations like these could help mainstream LIGA.

Shih et al. further developed and fine-tuned the conformal mask method for LIGA. In one very attractive embodiment of the technology, a PMMA layer on an Al substrate is coated with a Cu-foil (17.5 μm) by cold pressing [Shih et al., 1998]. The copper foil is further laminated with a dry photoresist foil (48 μm Hitachi H-N650). After exposure and development of the dry resist, gold or tin/lead absorber patterns are electroplated on the exposed Cu foil. After stripping the dry resist in a 3 wt% NaOH solution at around 50°C and the underlying Cu foil in a $Cu(NH_3)_4Cl_2$ solution at 45°C, the conformal mask is ready for use. Using this approach, 1000 μm high structures with an aspect ratio of 6.25 were fabricated with a double-exposure development cycle only. Using a transfer mask approach, this same research group made LIGA dies for spinnerets (see Example 2) [Cheng et al., 1999]. The authors summarize the advantages of the transfer mask method as follows:

- Alleviates the difficulty in fabricating fragile mask membranes
- Avoids alignment requirements during successive exposure steps
- Reduces exposure time and absorber thickness for the same exposure source
- Enhances pattern transfer fidelity, since there is almost no proximity gap

- Avoids thermal deformation caused by exposure heat
- Increases photoresist development rate by step-wise elevated exposure dose

4.3.2   Choice of Primary Substrate

In the LIGA process, the primary substrate, or base plate, must be a conductor or an insulator coated with a conductive top layer. A conductor is required for subsequent electrodeposition. Some examples of primary substrates that have been used successfully are Al [Shih et al., 1998], austenite steel plate, Si wafers with a thin Ti or Ag/Cr top layer [Michel et al., 1993], and copper plated with gold, titanium, or nickel [Becker et al., 1986]. Other metal substrates as well as metal-plated ceramic, plastic, and glass plates have been employed [Rogner et al., 1992]. It is important that the plating base provide good adhesion for the resist. For that purpose, prior to applying the X-ray resist on copper or steel, the surface sometimes is mechanically roughened by micro grit blasting with corundum. Micro grit blasting may lead to an average roughness R_a of 0.5 μm, resulting in better physical anchoring of the microstructures to the substrate [Mohr et al., 1988a]. In the case of a polished metal base, chemical preconditioning may be used to improve adhesion of the resist microstructures. During chemical preconditioning, a titanium layer, sputter-deposited onto the polished metal base plate (e.g., a Cu plate), is oxidized for a few minutes in a solution of 0.5 M NaOH and 0.2 M H_2O_2 at 65°C. The oxide produced typically measures 30 nm thick and exhibits a micro rough surface instrumental to securing resist to the base plate. The Ti adhesion layer may further be covered with a thin nickel seed layer ($\sim$150 Å) for electroless or electroplating of nickel. When using a highly polished Si surface, adhesion promoters need to be added to the resist. A substrate of special interest is a processed silicon wafer with integrated circuits. Integrating the LIGA process with IC circuitry on the same wafer will open up additional LIGA applications (see Figure 4.49).

The back of electrodeposited microdevices is attached to the primary substrate but can be removed from the substrate if necessary. In the latter case, the substrate may be treated chemically or electro-chemically to intentionally induce poor adhesion. Ideally, excellent adhesion exists between substrate and resist, and poor adhesion exists between the electroplated structure and plating base. Achieving these two contradictory demands is one of the main challenges in LIGA.

Thick resist plates can act as plastic substrates themselves. For example, using 20,000 eV rather than the more typical 3000 eV radiation, Guckel et al. [Guckel et al., 1994a; Siddons and Johnson, 1994] exposed plates of PMMA up to 10 cm thick.

4.3.3   Resist Requirements

An X-ray resist ideally should have high sensitivity to X-rays, high resolution, resistance to dry and wet etching, thermal stability of greater than 140°C, and a matrix or resin absorption of less than 0.35 μm^{-1} at the wavelength of interest [Moreau, 1988]. These requirements are only those for IC production with X-ray lithography [Lingnau et al., 1989]. To produce high-aspect-ratio microstructures with very tight lateral tolerances demands an additional set of requirements. The unexposed resist must be absolutely insoluble during development. This means that a high contrast (γ) is required. The resist must exhibit very good adhesion to the substrate and be compatible with the electroforming process. The latter imposes a resist glass transition temperature (T_g) greater than the temperature of the electrolyte bath used to electrodeposit metals between the resist features remaining after development (say, at 60°C). To avoid stress-induced mechanical damage to the microstructures during development, the resist layers should exhibit low internal stresses [Mohr et al., 1989]. If the resist structure is the end product of the fabrication process, further specifications depend on the application itself; for example, optical transparency and refractive index for optical components or large mechanical yield strength for load-bearing applications. PMMA, for instance, exhibits good optical properties in the visible and near-infrared range and lends itself to the making of all types of optical components [Gottert et al., 1991].

Due to excellent contrast and good process stability known from e-beam lithography, PMMA is the preferred resist for deep-etch synchrotron radiation lithography. Two major concerns with PMMA as a

FIGURE 4.11   Cracking of PMMA resist. Method to test stress in thick resist layers. The onset of cracks in a pattern of holes with varying size (say 1 to 4 µm) in a resist is shifted toward smaller hole diameter the lower the stress in the film. The SEM picture displays extensive cracking incurred during development of the image in a 5-µm-thick PMMA layer on an Au covered Si wafer. The 5-µm-thick PMMA layer resulted from five separate spin-coats. Annealing pushed the onset of cracking toward smaller holes until the right cycle was reached and no more cracks were visible. (Reprinted with permission from Madou, M., and Murphy, M. (1995) "A Method for PMMA Stress Evaluation," unpublished results.)

TABLE 4.7   Properties of Resists for Deep X-Ray Lithography

	PMMA	POM	PAS	PMI	PLG
Sensitivity	−	+	+ +	0	0
Resolution	+ +	0	− −	+	+ +
Sidewall smoothness	+ +	− −	− −	+	+ +
Stress corrosion	−	+ +	+	− −	+ +
Adhesion on substrate	+	+	+	−	+

Note: PMMA = poly(methylmethacrylate), POM = polyoxymethylene, PAS = polyalkensulfone, PMI = polymethacrylimide, PLG = poly(lactide-co-glycolide). + + = excellent; + = good; 0 = reasonable; − = bad; − − = very bad.
Source: Reprinted with permission from Ehrfeld, W. et al. (1994) "LIGA at IMM," Banff, Canada.

LIGA resist are a rather low lithographic sensitivity of about $2 \, J/cm^2$ at a wavelength λ_c of 8.34 Å and a susceptibility to stress cracking. For example, even at shorter wavelengths, $\lambda_c = 5$ Å, over 90 min of irradiation are required to structure a 500 µm thick resist layer with an average ring storage current of 40 mA and a power consumption of 2 MW at the 2.3-GeV ELSA synchrotron (Bonn, Germany) (see Figure 4.6) [Wollersheim et al., 1994]. The internal stress arising from the combination of a polymer and a metallic substrate can cause cracking in the microstructures during development, a phenomenon to which PMMA is especially prone, as illustrated in the scanning electron microscope (SEM) in Figure 4.11.

To make throughput for deep-etch lithography more acceptable to industry, several avenues to more sensitive X-ray resists have been pursued. For example, copolymers of PMMA were investigated: methyl–methacrylate combined with methacrylates with longer ester side chains show sensitivity increases of up to 32% (with tertiary butylmethacrylate). Unfortunately, a deterioration in structure quality was observed [Mohr et al., 1988b]. Among the other possible approaches for making PMMA more X-ray sensitive, we can count on the incorporation of X-ray absorbing high-atomic-number atoms or the use of chemically amplified photoresists. X-ray resists explored for LIGA applications include poly(lactides), for example, poly(lactide-co-glycolide) (PLG); polymethacrylimide (PMI); polyoxymethylene (POM); and polyalkensulfone (PAS). PLG is a new positive resist developed by BASF AG that is more sensitive to X-rays by a factor of 2 to 3 compared with PMMA. Its processing is less critical, but it is not commercially available yet. From the comparison of different resists for deep X-ray lithography in Table 4.7, PLG emerges as the most promising LIGA resist. POM, a promising mechanical material, may also be suited

TABLE 4.8   A Few Common Resists for E-Beam and X-Ray Lithography. (For comparison, typical numbers for a common optical resist are given as well.)

Novolak-based resist	Tone	EBL sens (μC/cm^2)	EBL contrast	XRL sens (mJ/cm^2)	XRL contrast
PMMA	+	100	2.0	6500	2.0
PBS	+	1	2.0	170	1.3
EBR-9	+	1.2	3.0		
Ray-PF	+			125	*
COP	−	0.5	0.8	100	1.1
GMCIA		7.0	1.7		
DCOPA	−			14	1.0
Novolak based	*	200–500	2–3	750–2000	~2

* Indicates that the value is process dependent.

Source: Reprinted with permission from Campbell, S.A. (1996) *The Science and Engineering of Microelectronic Fabrication,* Oxford University Press, New York.

for medical applications given its biocompatibility. All of the resists shown in Table 4.7 exhibit significantly enhanced sensitivity compared to PMMA, and most exhibit a reduced stress corrosion [Wollersheim et al., 1994]. Negative X-ray resists have inherently higher sensitivities than positive X-ray resists, although their resolution is limited by swelling. Poly(glycidyl methacrylate-co-ethyl acrylate) (PGMA), a negative e-beam resist (not shown in Table 4.7), has also been used in X-ray lithography. In general, resist materials sensitive to e-beam exposure also display sensitivity to X-rays and function in the same fashion; materials positive in tone for e-beam radiation typically are also positive in tone for X-ray radiation. A strong correlation exists between the resist sensitivities observed with these two radiation sources, suggesting that the reaction mechanisms might be similar for both types of irradiation. IMM in Germany started developing a negative X-ray resist 20 times more sensitive than PMMA, but the exact chemistry has not yet been disclosed [IMM, 1995]. More common X-ray resists from the IC industry are reviewed in Table 4.8.

4.3.4   Methods of Resist Application

4.3.4.1   Multiple Spin Coats

Different methods to apply ultra-thick layers of PMMA have been studied. In the case of multilayer spin coating, high interfacial stresses between the layers can lead to extensive crack propagation upon developing the exposed resist. For example, Figure 4.11 is an SEM picture of a 5 μm thick PMMA layer deposited in five sequential spin coatings. Development results in the cracked riverbed mud appearance, with the most intensive cracking propagating from the smallest resist features. The test pattern used to expose the resist consisted of arrays of holes ranging in size from 1 to 4 μm. Annealing the PMMA films shifted the cracking toward holes with smaller diameter compared with the unannealed film shown in Figure 4.11 [Madou and Murphy, 1995]. CAMD (see Table 4.3) has demonstrated that multiple spin coating of PMMA can be used for up to 15 μm thick resist layers and that applying the appropriate annealing and developer (see below) eliminates cracking [Vladimirsky, 1995]. Later in this chapter, we will learn that a prerequisite for low stress and small lateral tolerances in PMMA films is a high mean molecular weight. The spin-coated resist films in Figure 4.11 do not have a high enough molecular weight to lead to adequate selectivity between radiated and nonradiated PMMA during a long development process.

4.3.4.2   Commercial PMMA Sheets

High molecular weight PMMA is commercially available as prefabricated plate (e.g., GS 233; Rohm GmbH; Darmstadt, Germany), and several groups have employed freestanding or bonded PMMA resist sheets for producing LIGA structures [Guckel et al., 1994b; Mohr et al., 1988a]. After overcoming the initial problems encountered when attempting to glue PMMA foils to a metallic base plate with adhesives, this has become the preferred method in several labs [Mohr et al., 1988a]. Guckel used commercially available thick PMMA

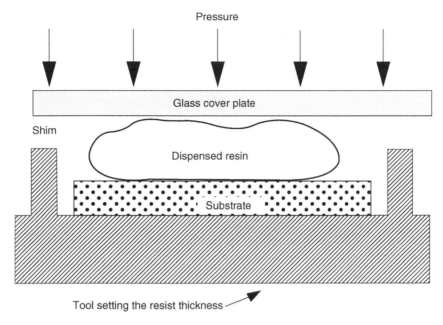

FIGURE 4.12   Principle of in situ polymerization of a thick resist layer on a metal substrate.

sheets (thickness > 3 mm), XY-sized and solvent-bonded them to a substrate and, after milling the sheet to the desired thickness, exposed the resist plate without cracking problems [Guckel et al., 1994b].

4.3.4.3   Casting of PMMA

PMMA also can be purchased in the form of a casting resin, for example, Plexit 60 (PMMA without added cross-linker) and Plexit 74 (PMMA with cross-linker added) from Rohm GmbH in Darmstadt, Germany. In a typical procedure, PMMA is polymerized in situ from a solution of 35 wt% PMMA of a mean molecular weight of anywhere from 100,000 g/mol up to 10^6 g/mol in methylmethacrylate (MMA). Polymerization at room temperature takes place with benzoyl peroxide (BPO) catalyst as the hardener (radical builder) and dimethylaniline (DMA) as the starter or initiator [Mohr et al., 1988a, 1988b]. The oxygen content in the resin, which inhibits polymerization, and gas bubbles, which induce mechanical defects, are reduced by degassing while mixing the components in a vacuum chamber at room temperature and at a pressure of 100 mbar for 2 to 3 min.

In a practical application, resin is dispensed on a base plate provided with shims to define pattern and thickness and subsequently covered with a glass plate to avoid oxygen absorption. The principle of polymerization on a metal substrate is schematically represented in Figure 4.12. Due to the hardener, polymerization starts within a few minutes after mixing of the components and ends within five minutes. The glass cover plate is coated with an adhesion preventing layer (e.g., Lusin L39; Firma Lange u. Seidel, Nurnberg). After polymerization, the antiadhesion material is removed by diamond milling, and a highly polished surface results. In situ polymerization and commercial-cast PMMA sheets top the list of thick-resist options in LIGA today. Plasma polymerization of PMMA in layers >100 μm was discussed by Guckel (1988).

4.3.4.4   Resist Adhesion

Adhesion promotion by mechanically or chemically modifying the primary substrate was introduced above under *Choice of Primary Substrate*. Smooth surfaces such as Si wafers with an average roughness R_a smaller than 20 nm pose additional adhesion challenges that are often solved by modifying the resist itself. To promote adhesion of resist to polished untreated surfaces, such as metal-coated Si wafers, coupling agents must be used to chemically attach the resist to the substrate. An example of such a coupling agent

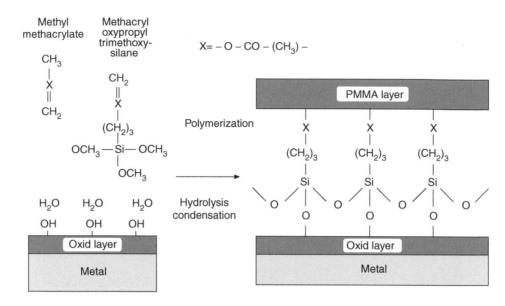

FIGURE 4.13 Schematic presentation of the adherence mechanism of methacryloxypropoyl trimethoxy silane (MEMO). (Reprinted with permission from Mohr, J. et al. [1988] "Requirements on Resist Layers in Deep-Etch Synchrotron Radiation Lithography," *J. Vac. Sci. Technol.* **B6**, pp. 2264–67.)

is methacryloxypropyl trimethoxy silane (MEMO). With 1 wt% of MEMO added to the casting resin, excellent adhesion results. The adherence is brought about by a siloxane bond between the silane and the hydrolyzed oxide layer of the metal. As illustrated in Figure 4.13, the integration of this coupling agent in the polymer matrix is achieved via the double bond of the methacryl group of MEMO [Mohr et al., 1988b]. Hydroxyethyl methacrylate (HEMA) can improve PMMA adhesion to smooth surfaces, but higher concentrations are needed to obtain the same adhesion improvement. Silanization of polished surfaces prior to PMMA casting, instead of adding adhesion promoters to the resin, did not seem to improve the PMMA adhesion. In the case of PMMA sheets, one option is solvent bonding of the layers to a substrate. In another approach, Galhotra et al. (1996) simply mechanically clamped the exposed and developed self-supporting PMMA sheet onto a 1.0 mm thick Ni sheet for subsequent Ni plating. Rogers et al. (1996) have shown that cyanoacrylate can be used to bond PMMA resist sheets to a Ni substrate and that it can be lithographically patterned using the same process sequence used to pattern PMMA. For a 300 μm thick PMMA sheet on a sputtered Ni coating on a silicon wafer, a 10 μm thick cyanoacrylate bonding layer was used. Such a thick cyanoacrylate layer caused some problems for subsequent uniform electrodeposition of metal. The dissolution rate of the cyanoacrylate is faster than the PMMA resist, resulting in metal posts with a wide profile at the base.

4.3.4.5 Stress-Induced Cracks in PMMA

The internal stress arising from the combination of a polymer on a metallic substrate can cause cracking in the microstructures during development. To reduce the number of stress-induced cracks (see Figure 4.11), both the PMMA resist and the development process must be optimized. Detailed measurements of the heat of reaction, the thermomechanical properties, the residual monomer content, and the molecular weight distribution during polymerization and soft baking have shown the necessity to produce resist layers with a high molecular weight and with only a very small residual monomer content [Mohr et al., 1988a; 1988b]. Figure 4.14 compares the molecular weight distribution determined by gel permeation chromatography of a polymerized PMMA resist (two hardener concentrations were used) with the molecular weight distribution of the casting resin. The casting resin is unimodal, whereas the polymerized resist layer typically shows a bimodal distribution with peak molecular weights centered around

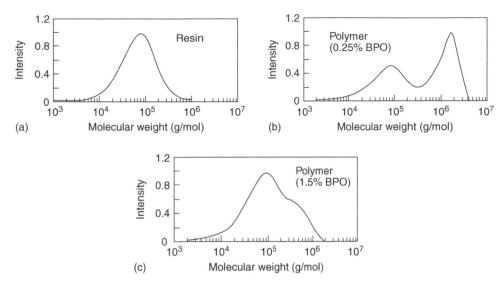

FIGURE 4.14 Molecular weight distribution of (a) the casting resin, (b) a resist layer polymerized at low hardener content, and (c) a resist layer polymerized at high hardener content determined by gel permeation chromatography. (Reprinted with permission from Mohr, J. et al. [1988] "Requirements on Resist Layers in Deep-Etch Synchrotron Radiation Lithography," *J. Vac. Sci. Technol.* **B6**, pp. 2264–67.)

90,000 and 300,000 g/mol. The first low-molecular-weight peak belongs to the PMMA oligomer dissolved in the casting resin, and the second molecular-weight peak results from the polymerization of the monomer. The molecular weight distribution is constant across the total resist thickness except for the boundary layer at the base plate, where the average molecular weight can be significantly higher (~450,000 g/mol) [Mohr et al., 1989]. The amount of the high-molecular-weight portion in the poly-merized resist depends on the concentration of the hardener. A low hardener content leads to a high-molecular-weight dominance and vice versa (see Figure 4.14) [Mohr et al., 1988b]. Since high molecular weight is required for low stress, a hardener concentration of less than 1% benzoyl peroxide (BPO) must be used. Ideally, for a low-stress resist, the residual monomer content should be less than 0.5%. The resid-ual monomer content decreases with increasing hardener content, and >1% BPO is needed to reduce the residual monomer content below 0.5%. The problem resulting from these opposite needs can be over-come by the addition of 1% of a cross-linking dimethacrylate (triethylene glycol dimethacrylate, TEDMA) to the resin. In such cross-linked PMMA, a smaller amount of BPO suffices to suppress the residual monomer content; crack-free PMMA can be obtained with 0.8% of BPO [Mohr et al., 1988a].

For solvent removal, and to further minimize the defects caused by stress, the polymerized resin is cured at 110°C for 1 hr (soft bake). The measurement of the reaction enthalpy shows that post-polymerization reactions occur at room temperature and during heating to the glass transition tempera-ture [Mohr et al., 1988a]. The rate of heating up to that temperature is 20°C/hr; after curing, the samples are cooled from 110°C to room temperature at a very low rate of 5 to 10°C/hr [Mohr et al., 1988a; Mohr et al., 1989]. The soft-bake temperature is slightly below the glass transition temperature, measured to be 115°C.

Another important factor that reduces stress in thick PMMA resist layers is the optimization of the developer. Stress-induced cracking can be minimized with solvent mixtures whose dissolution parame-ters lie near the boundary of the PMMA solubility range; that is, a nonaggressive solvent is preferred. This is discussed in more detail below, under development. Small amounts of additives such as described above for reducing stress or to promote adhesion do not influence the mechanical stability of the microstructures or the sensitivity of the resist.

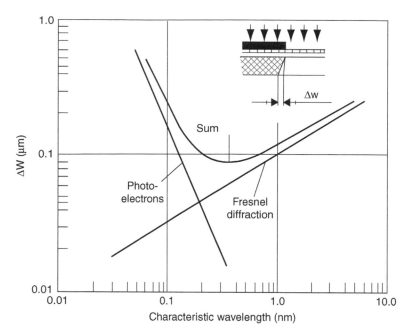

FIGURE 4.15 Fresnel diffraction and photoelectron generation as a function of characteristic wavelength λ_c and the resulting lateral dimension variation (ΔW). (Reprinted with permission from Menz, W., and Bley, P. [1993] *Mikrosystemtechnik fur Ingenieure*, VCH Publishers, Germany.)

4.3.5   Exposure

4.3.5.1   Optimal Wavelength

For a given polymer, the lateral dimension variation in a LIGA microstructure in principle could result from the combined influence of several mechanisms. These include Fresnel diffraction, the range of high energy photoelectrons generated by the X-rays, the finite divergence of synchrotron radiation, and the time evolution of the resist profiles during the development process. The theoretical manufacturing precision obtainable by deep X-ray lithography was investigated by computer simulation of both the irradiation step and the development step by Becker et al. (1984) and by Munchmeyer (1984). Results were further tested experimentally and confirmed by Mohr et al. (1988a). The theoretical results demonstrate that the effect of Fresnel diffraction (edge diffraction), which increases as the wavelength increases, and the effect of secondary electrons in PMMA, which increases as the wavelength decreases, lead to minimal structural deviations when the characteristic wavelength ranges between 0.2 and 0.3 nm (assuming an ideal development process and no X-ray divergence). To fully utilize the accuracy potential of a 0.2 to 0.3 nm wavelength, the local divergence of the synchrotron radiation at the sample site should be less than 0.1 mrad. Under these conditions, the variation in critical lateral dimensions likely to occur between the ends of a 500 μm high structure due to diffraction and secondary electrons is estimated to be 0.2 μm. The estimated Fresnel diffraction and secondary electron scattering effects are shown as a function of characteristic wavelength in Figure 4.15.

Using cross-linked PMMA, or linear PMMA with a unimodal and extremely high molecular weight distribution (peak molecular weight greater than 1,000,000 g/mol), the experimentally determined lateral tolerances on a test structure as shown in Figure 4.16a are 0.055 μm per 100 μm resist thickness, in good agreement with the 0.2 μm over 500 μm expected on a theoretical basis [Mohr et al., 1988a]. These results are obtained only when a resist/developer system with a ratio of the dissolution rates in the exposed and unexposed areas of approximately 1000 is used.

The use of resist layers, not cross-linked and displaying a relatively low bimodal molecular weight distribution, as well as the application of excessively strong solvents such as used to develop thin PMMA

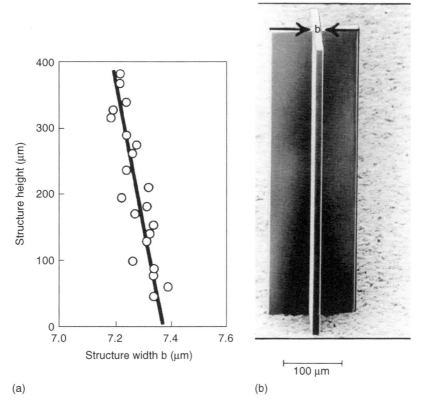

Structure height (μm)

Structure width b (μm)

(a)

100 μm

(b)

FIGURE 4.16 Structural tolerances. (a) SEM micrograph of a test structure to determine conical shape. (b) Structural dimensions as a function of structure height. The tolerances of the dimensions are within 0.2 mm over the total structure height of 400 mm [Mohr et al., 1988b]. (Courtesy of the Karlsruhe Nuclear Research Center.)

resist layers in the IC industry, leads to more pronounced conical shape in the test structure of Figure 4.16b. An illustration of the effect of molecular weight distribution on lateral geometric tolerances is that linear PMMA with a peak molecular weight below 300,000 g/mol shows structure tolerances of up to 0.15 μm/100 μm [Mohr et al., 1988]. To obtain the best tolerances requires a PMMA with a very high molecular weight, also a prerequisite for low stress in the developed resist. Finally, if the synchrotron beam is not parallel to the absorber wall but at an angle greater than 50 mrad, greater coning angles may also result [Mohr et al., 1988a].

4.3.5.2 Deposited Dose

As shown in Figure 4.17 depicting the average molecular weight of PMMA as a function of radiation dose, the X-ray irradiation of PMMA reduces the average molecular weight [Menz and Bley, 1993]. For one-component positive resists, this lowering of the average molecular weight causes the solubility of the resist in the developer to increase dramatically. The average molecular weight making dissolution possible is a sensitive function of the type of developer used and the development temperature. It can be observed from Figure 4.17 that above a certain dose (15 to 20 kJ/cm^3), the average molecular weight does not decrease any further.

The molecular weight distribution, measured after resist exposure, is unimodal with peak molecular weights ranging from 3000 g/mol to 18,000 g/mol, dependent on the dose deposited during irradiation. The peak molecular weight increases nearly linearly with increasing resist depth; that is, with decreasing absorbed dose [Mohr et al., 1989]. At the bottom of the resist layer, the absorbed dose must be higher than the development dose D_{dv} while, at the top of the resist the absorbed dose must be lower than the

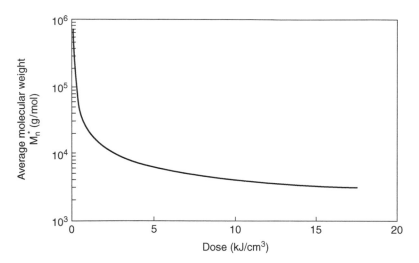

FIGURE 4.17 Average molecular weight M_n versus X-ray radiation dose. (Reprinted with permission from Menz, W., and Bley, P. [1993] *Mikrosystemtechnik fur Ingenieure*, VCH Publishers, Germany.)

damaging dose D_{dm}. In Figure 4.6, where the absorption of X-rays along the path from source to sample was illustrated, the exposure time and the preabsorber were chosen so that the bottom of a 500 μm thick PMMA layer received the necessary development dose D_{dv}, while the dose at the top of the layer stayed well below D_{dm}. Exposure of PMMA with longer wavelengths results in correspondingly longer exposure times and can lead to an overexposure of the top surface, where the lower energy radiation is mainly absorbed.

Menz and Bley (1993) describe the influence of the radiation dose on the quality of the resulting LIGA structures in a slightly different manner. Following their approach, Figure 4.18a illustrates a typical bimodal molecular weight distribution of PMMA before radiation, exhibiting an average molecular weight of 600,000. The shaded region in this figure indicates the molecular weight region where PMMA readily dissolves; that is, below the 20,000 g/mol level for the temperature and developer used. Since the fraction of PMMA with a 20,000 molecular weight is very small in nonirradiated PMMA, the developer hardly attacks the resist at all. After irradiation with a dose D_{dv} of 4 kJ/cm^3, the average molecular weight becomes low enough to dissolve almost all of the resist (Figure 4.18b). With a dose D_{dm} of 20 kJ/cm^3, all of the PMMA dissolves swiftly (Figure 4.18c). At a dose above D_{dm}, the microstructures are destroyed by the formation of bubbles. It follows that to dissolve PMMA completely and to make defect-free microstructures, the radiation dose for the specific type of PMMA used must lie between 4 and 20 kJ/cm^3. These two numbers also lock in a maximum value of 5 for the ratio of the radiation dose at the top and bottom of a PMMA structure. To make this ratio as small as possible, the soft portion of the synchrotron radiation spectrum is usually filtered out by a preabsorber (for example, a 100 μm thick polyimide foil [Kapton]) to reduce differences in dose deposition in the resist.

X-ray exposure equipment developed by KfK has a vibration-free bedding; the exposure chamber is under thermostatic control ($\pm 0.2°$C) and includes a precision scanner for the periodic movement of the sample through the irradiation plane. The polyimide window isolates the vacuum of the accelerator from the helium atmosphere (200 mbar), which serves as coolant for substrate and mask in the irradiation chamber [Becker et al., 1986]. IMM, in collaboration with Jenoptik, GmbH, developed an X-ray scanner for deep lithography, enabling irradiation of up to 1000 μm thick resists. A mask-to-resist registration within ± 0.3 μm is claimed [Editorial, 1994].

4.3.5.3 Stepped and Slanted Microstructures

For many applications, stepped or inclined resist sidewalls are very useful — consider, for instance, the fabrication of multilevel devices or prisms or, more basic yet, angled resist walls to facilitate the release of

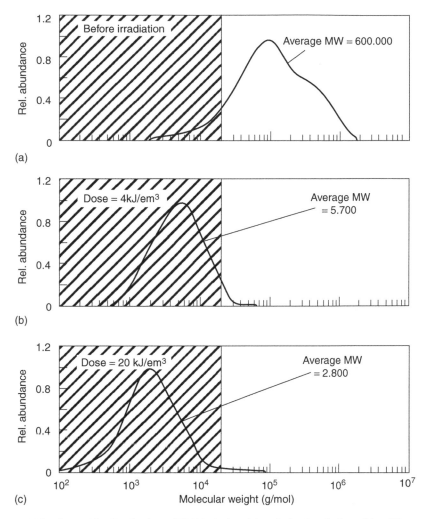

FIGURE 4.18 Molecular weight distribution of PMMA before (a) and after irradiation with 4 (b) to 20 kJ/cm³ (c). The black lines (stripes) indicate the domain in which PMMA is minimally 50% dissolved (at 38°C in the LIGA developer described in the text). (Reprinted with permission from Menz, W., and Bley, P. [1993] *Mikrosystemtechnik fur Ingenieure*, VCH Publishers, Germany.)

molded parts. Using stepped absorber layers on a single mask to make stepped multilevel microstructures was discussed above under Absorber Fabrication. We mentioned that structures made this way are not always very well resolved. To make better-resolved stepped features, one can first relief print a PMMA layer, for example, by using a Ni mold insert made from a first X-ray mask. Subsequently, the relief structure may be exposed to synchrotron radiation to further pattern the polymer layer through a precisely adjusted second X-ray mask. To carry out this process, a two-layer resist system needs to be developed consisting of a top PMMA layer that fulfills the requirement of the relief printing process and a bottom layer that fulfills the requirements of the X-ray lithography [Harmening et al., 1992]. The bottom resist layer promotes high molecular weight and adhesion, while the top PMMA layer is of lower molecular weight and contains an internal mold-release agent. This process sequence, combining plastic impression molding with X-ray lithography, is illustrated in Figure 4.19. The two-step resist then facilitates the fabrication of a mold insert by electroforming, which can be used for the molding of two-step plastic structures. Extremely large structural heights can be obtained from the additive nature of the individual microstructure levels.

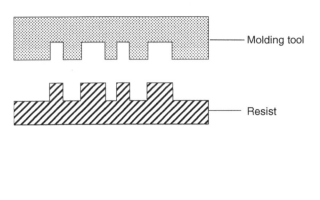

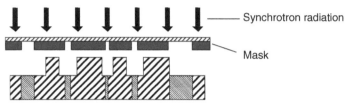

FIGURE 4.19 Stepped microstructures. (Reprinted with permission from Mohr, J. et al. [1990] "Movable Microstructures Manufactured by the LIGA Process as Basic Elements for Microsystems," in *Microsystem Technologies 90*, Reichl, H., ed., pp. 529–37, Springer, Berlin.)

There are several options for achieving miniaturized features with slanted walls. It is possible to modulate the exposure/development times of the resist, fabricate an inclined absorber, angle the radiation, or following Tabata et al. (2000), move the mask during exposure in so-called moving mask deep X-ray lithography.

To make a slanted absorber, a slab of material can be etched into a wedge by pulling it at a linear rate out of an etchant bath. Changing the angle at which synchrotron radiation is incident upon the resist, usually 90°, also enables the fabrication of microstructures with inclined sidewalls [Bley et al., 1991]. This way, slanted microstructures may be produced by a single oblique irradiation or by a swivel irradiation. One potentially very important application of microstructures incorporating inclined sidewalls is the vertical coupling of light into waveguide structures using a 45° prism [Gottert et al., 1992b]. Such optical devices must have a wall roughness of less than 50 nm, making LIGA a preferred technique for this application. The sharp decrease of the dose in the resist underneath the edge of the inclined absorber and the resulting sharp decrease of the dissolution of the resist as a function of the molecular weight in the developer result in little or no deviation of the inclination of the resist sidewall over the total height of the microstructure.

4.3.6 Development

To fully utilize the accuracy potential of synchrotron radiation lithography, it is essential that the resist/developer system have a ratio of dissolution rate in the exposed and unexposed areas of approximately 1000 (see above). The developer empirically arrived at by KfK consists of a mixture of 20 vol% tetrahydro-1,4-oxazine (an azine), 5 vol% 2-aminoethanol-1 (a primary amine), 60 vol% 2-(2-butoxyethoxy)ethanol (a glycolic ether), and 15 vol% water [Mohr et al., 1988a; Mohr et al., 1989; Ghica and Glashauser, 1982]. This developer causes an infinitely small dissolution of unexposed, high-molecular-weight, cross-linked PMMA and achieves a sufficient dissolution rate in the exposed area. It also exhibits much less stress-induced cracking than developers conventionally used for thin PMMA resists.

Systematic investigation of different organic solvents and mixtures of the above developer systems shows that solvents with a solubility parameter at the periphery of PMMA's solubility range dissolve

TABLE 4.9 Comparison of Micromolds

Parameters	LIGA	DUV	DRIE	LASER	CNC	EDM
Aspect ratio	100	22	10–25	<10	14 (drilling)	Up to 100
Wall roughness	<20 nm	~1 μm	~2 μm	1 μm–100 nm	Several microns	0.3–1 μm
Accuracy	<1 μm	2–3 μm	<1 μm	A few microns	See Madou (2002, Chapter 7, Figure 7.2 [x, y only])	Some microns
Mask needed?	Yes	Yes	Yes	No	No	No
Maximum height	A few 100 μm up to 1 cm	A few 100 μm	A few 100 μm	A few 100 μm	Unlimited	Microns to millimeters

Source: Reprinted with permission from Weber, L. et al. (1996) "Micro-Molding: A Powerful Tool for the Large Scale Production of Precise Microstructures," in *SPIE, Macromachining and Microfabrication Process Technology II*, Austin, Texas pp. 156–67.

exposed PMMA slowly but selectively and without stress-induced cracking or swelling of unexposed areas. Solvents with a solubility parameter close to those of MMA show a much higher dissolution rate but cause serious problems related to cracking and swelling. As we have already seen, the application of excessively strong solvents also leads to more pronounced conical shapes in the test structures shown in Figure 4.16. An improved developer found in the above systematic investigation is a mixture of tetrahydro-1,4-oxazine and 2-aminoethanol-1. Its sensitivity is 30% higher, but the process latitude is much narrower compared with the developer described above [Mohr et al., 1989].

KfK has built a machine for the development process that enables the continuous and homogeneous transport of developing and rinsing agents into deep structure elements and the removal of the dissolved resist from these structures. Several substrates are arranged vertically on a rotor, with each structure surface facing to the outside. During development, the developing agent flows toward the resist surface being developed, circulates, and is filtered continuously, and the temperature remains controlled at 35°C. To stop development, less-concentrated developer solutions are applied to prevent the precipitation of already dissolved resist. Three independent fluid circuits are available for immersion and spraying processes [Becker et al., 1986].

After development, the microstructures are rinsed with deionized water and dried in a vacuum. Alternatively, drying may be done by spinning and blasting with dry nitrogen. At this stage, the devices can be the final product (for example, as micro-optical components), or they can be used for subsequent metal deposition.

4.3.7 Comparison of Master Micromold Fabrication Methods

The high cost of X-ray lithography caused miniaturization engineers to search for alternative means of fabricating high-aspect-ratio metal or polymer micro masters.

Micromold inserts (or micro masters) can be fabricated by a variety of alternate techniques such as CNC machining, silicon wet bulk micromachining, precision EDM, thick deep UV resists, DRIE, excimer layer ablation, and e-beam writing (see Figure 4.2). In Table 4.9, LIGA metal molds are compared with metal masters fabricated by other means. For example, comparing metal mold inserts made by spark erosive cutting and X-ray lithography, the latter proves far superior [Hagmann and Ehrfeld, 1988]. LIGA PMMA features as small as 0.1 μm are replicated in the metal shape with almost no defects. The electroformed structures have a superior surface quality with a surface roughness R_a of less than 0.02 μm [Hagmann et al., 1987].

DRIE and thick deep UV-sensitive resists such as polyimides, AZ-4000, and SU-8 are recent contenders for micro master mold fabrication. With respect to dry etching, higher and higher aspect ratio features are being achieved; especially when using highly anisotropic etching conditions as in cryogenic DRIE, and in the Bosch process remarkable results are obtained. Wall roughness, causing form locking, remains

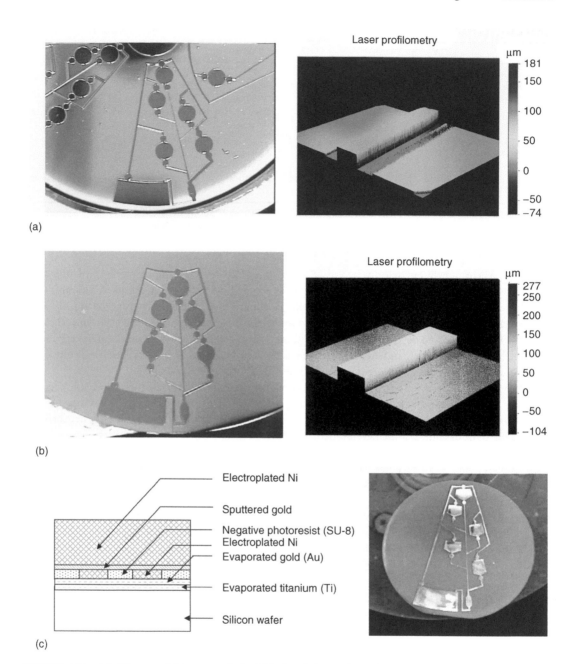

(a)

Laser profilometry

(b)

Laser profilometry

(c)

FIGURE 4.20   (a) Silicon mold insert made by UV photolithography and deep reactive ion etching (DRIE) at Burstein Technologies/UCLA/OSU. (b) Photoresist (SU-8) mold insert made by UV lithography at OSU. (c) Nickel mold insert made by UV photolithography and electroplating (OSU). The depth of the etched channel is measured by a laser profilometer.

a problem with DRIE; the dry etching process was optimized for speed, not for demolding. For small-quantity production, where the lifetime of mold inserts is not crucial, a silicon wafer etched by deep reactive-ion etching (DRIE) can be utilized directly as a mold insert for anywhere from 5 to 30 molding cycles [Madou, 2002; Wimberger-Friedl, 1999]. Figure 4.20a shows such a Si mold for building a two-point calibration fluidic device on a compact disc (lab CD). Wet etching of Si leads to much smoother surfaces than DRIE and is therefore the preferred method for making master molds from Si. For much longer-lasting molds, metallizing the Si structure and using the metal as mold is preferred. Photoresist structures on a silicon substrate have also been tested as a mold insert in plastic molding because of the simplicity and low cost of the process. Figure 4.20b shows an SU-8 photoresist mold for the same lab CD platform shown in Figure 4.20a. In low-pressure molding processes, such mold inserts do work for a limited number of runs (applying a thin metal layer over the top of the resist may further extend the lifetime of the mold), but their applicability in high-pressure processes needs to be further verified. A better approach is to use deep UV-photosensitive resists for electroplating to yield a metal tool, usually nickel or nickel–cobalt. Figure 4.20c shows a nickel mold insert made using SU-8 in our lab for the two-point calibration microfluidic platform. Both DUV and DRIE are more accessible than LIGA and will continue to improve, taking more opportunities away from LIGA. Like LIGA, both alternative techniques can be coupled with plating, but neither technique can yet achieve the extreme low surface roughness and vertical walls of LIGA.

Other competing technologies for making metal masters are laser ablation methods and ultraprecision CNC machining. The latter two methods are serial processes and rather slow, but since we are considering the production of a master only, these technologies might well be competitive for certain applications.

Laser microablation produces minimum features of about 10 μm width and aspect ratios of 1:10. Challenges include taper and surface finish control. Recast layers around the laser drilled features cause form locking and infidelity in the replication. Femtosecond pulse lasers promise thinner — or even the absence of — recast layers and excellent resolution, and they should be investigated further [Momma et al., 1997].

For large features (>50 μm) with tolerances and repeatability in the range of about 10 μm, traditional CNC-machining of materials like tool steel and stainless steel is often accurate enough for making metal mold inserts. The advantage of this technique is that the tool materials used are the same as those in conventional polymer molding, so their design, strength, and service life are well established. Complicated 3-D structures can also be machined easily. The main drawbacks are that it is difficult to make sharp corners or right angles, and the surface quality is usually poor (surface roughness is around several μm) [Madou et al., 2000]. In contrast, lithographic methods can produce molds with excellent surface quality (surface quality < 0.1 μm) and sharp corners or right angles. However, they cannot be used on conventional tool materials like steel. Diamond-based micromilling and microdrilling [Warrington, 1999] reduce the surface roughness to 1 μm or less [Roberts et al., 1997]. While diamond-based methods can achieve features smaller than 10 μm, they are applicable only to "soft" metals such as nickel, aluminum, and copper.

Table 4.9 indicates that a significant potential application of LIGA remains the fabrication of those metal molds that cannot be accomplished with other techniques because of the tight wall roughness tolerances, small size, and high aspect ratios. From the table, it is obvious that LIGA micromolds excel in both very low surface roughness and close accuracy.

Summarizing, the requirements for an optimal mold insert fabrication technique are as follows [Becker and Gätner, 2000]:

- The master has to be removed from the molded structure, so the ease of release through wall inclination control is crucial (undercuts, for example, cannot be tolerated because they cause form locking).
- The most important parameters, including master lifetime and achievable aspect ratios, depend very strongly on the surface quality of the master.
- The interface chemistry between master and polymer is critical and must be controlled.

4.3.8 Metal Deposition

4.3.8.1 Introduction

LIGA and pseudo-LIGA depend critically on metal deposition to replicate the master photoresist mold. This additive process might involve electroless metal deposition or electrodeposition.

4.3.8.2   Electroless Metal Deposition

4.3.8.2.1   Introduction

To continuously build thick metal deposits by chemical means without applying a voltage and without consuming the substrate (i.e., electroless plating), it is essential that a sustainable oxidation reaction be employed instead of the dissolution of the substrate as occurs in the case of immersion plating. In Figure 4.21, the difference between immersion plating and electroless deposition is illustrated by comparing deposit thickness vs. time. Immersion plating is self-limiting; electroless plating is not [Mallory and Hadju, 1999]. The reduction of Ni ions (Reaction [4.1]) in electroless plating is fed electrons by an oxidation reaction such as that of hypophosphite (Reaction [4.2]) resulting in an overall reaction for Ni deposition given by Reaction (4.3).

Reduction:

$$Ni^{2+} + 2e^- \rightarrow Ni \qquad \text{Reaction (4.1)}$$

Oxidation:

$$H_2PO_2^- + H_2O \rightarrow H_2PO_3^- + 2H^+ + 2e^- \qquad \text{Reaction (4.2)}$$

Overall reaction:

$$Ni^{2+} + H_2PO_2^- + H_2O \rightarrow Ni + H_2PO_3^- + 2H^+ \qquad \text{Reaction (4.3)}$$

Reaction (4.1) continues (in principle) until all the hypophosphite in solution is consumed (Reaction [4.2]). In immersion deposition, Reaction (4.1) stops as soon as the whole substrate surface is covered with Ni metal and there is no further oscillation/dissolution of uncovered substrate possible. In parallel to Reaction (4.1), more or less severe hydrogen reduction occurs:

$$2H^+ + 2e^- \rightarrow H_2 \qquad \text{Reaction (4.4)}$$

Copious hydrogen evolution can upset the quality of the depositing metal film and should be avoided. The hydrogen evolution rate is not directly related to that of the metal deposition and mainly originates from the reductant molecules. Stabilizers (i.e., catalytic poisons) are needed in electroless deposition baths because the solutions are thermodynamically unstable; deposition might start spontaneously onto

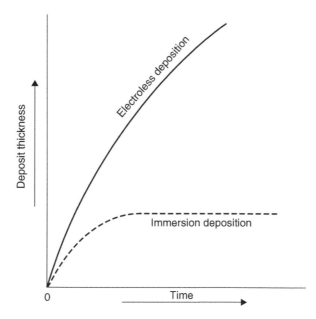

FIGURE 4.21   Thickness versus time comparison between electroless and immersion deposition. (Based with permission on Mallory, G.O., and Hadju, J.B., eds. (1990) *Electroless Plating: Fundamentals and Applications*, Orlando, FL, AESF.)

the container walls. Poisons for hydrogenation catalysts such as thiourea, Pb^{2+}, and mercaptobenzothiazole function as stabilizers in such electroless baths.

Besides stabilizers, the metal salt, and a reducing agent, electroless solutions may contain other additives such as complexing agents, buffers, and accelerators. Complexing agents exert a buffering action and prevent the pH from decreasing too fast. They also prevent the precipitation of metal salts and reduce the concentration of free metal ions. Buffers keep the deposition reaction in the desired pH range. Accelerators, also termed exaltants, increase the rate of deposition to an acceptable level without causing bath instability. These exaltants are anions, such as CN^-, thought to function by making the anodic oxidation process easier. In electroless copper, for example, compounds derived from imidazole, pyrimidine, and pyridine can increase the deposition rate to $40\,\mu m\,hr^{-1}$. The electroless deposition must occur initially and exclusively on the surface of an active substrate and subsequently continue to deposit on the initial deposit through the catalytic action of the deposit itself. Since the deposit catalyzes the reduction reaction, the term *autocatalytic* is often used to describe the plating process. Electroless plating is an inexpensive technique enabling plating of conductors and nonconductors alike (plastics such as ABS, polypropylene, Teflon, polycarbonate, etc., are plated in huge quantities). A catalyzing procedure is necessary for electroless deposition on nonactive surfaces such as plastics and ceramics. The most common method for sensitizing those surfaces is by dipping into $SnCl_2/HCl$ or immersion in $PdCl_2/HCl$ [Mallory and Hadju, 1990]. This chemical treatment produces sites that provide a chemical path for the initiation of the plating process.

Metal alloys such as nickel–phosphorus, nickel–boron, cobalt–phosphorus, cobalt–boron, nickel–tungsten, copper–tin–boron, and palladium–nickel can be produced by codeposition. In the case of Ni deposits, with or without phosphorus and boron incorporated, different electroless solutions are used to obtain optimal hardness, effective corrosion protection, or optimal magnetic properties. Recent experimental results show that composite material also can be produced by codeposition. Finely divided, solid particulate material is added and dispersed in the plating bath. Electroless Ni with alumina particles, diamond, silicon carbide, and PTFE have reached the commercial market. Table 4.10 lists electroless plating baths [Mallory and Hadju, 1990].

TABLE 4.10 Typical Electroless Plating Baths

Component	Concentration (per liter)	Application/Remark	pH	Temp (°C)
Au	1.44 g $KAu(CN)_2$, 6.15 g KCN, 8 g NaOH, 10.4 g KBH_4	Plate beam leads on silicon ICs, ohmic contacts on n-GaAs	13.3	70
Co-P	30 g $CoSO_4.7H_2O$, 20 g $NaH_2PO_2.H_2O$, 80 g $Na_3citrate.2H_2O$, 60 g NH_4Cl, 60 g NH_4OH	Magnetic properties	9.0	80
Cu	10 g $CuSO_4.5H_2O$, 50 g Rochelle salt, 10 g NaOH, 25 ml. conc. HCHO (37%)	Printed circuit boards	13.4	25
Ni-Co	3 g $NiSO_4.6H_2O$, 30 g $CoSO_4$. $7H_2O$, 30 g $Na_2malate.1/2H_2O$, 180 g $Na_3citrate.2H_2O$, 50 g $NaH_2PO_2.H_2O$		10	30
Ni-P	30 g $NiCl_2.6H_2O$, 10 g $NaH_2PO_2.H_2O$, 30 g glycine	Corrosion and wear resistance on steel	3.8	95
Pd	5 g $PdCl_2$, 20 g Na_2EDTA, 30 g Na_2CO_3, 100 mL NH_4OH (28% NH_3), 0.0006 g thiourea, 0.3 g hydrazine	Plating rate is $0.26\,\mu m/min$		80
Pt	10 g $Na_2Pt(OH)_6$, 5 g NaOH, 10 g ethylamine, 1 g hydrazine hydrate (added now and then to maintain this concentration)	Plating rate is $12.7\,\mu m/hr$		35

Source: Based with permission on Mallory, G.O., and Hadju, J.B., eds. (1990) *Electroless Plating: Fundamentals and Applications*, American Electroplaters and Surface Finishers Society (AESF), Orlando, and Romankiw, L.T. (1976) "Pattern Generation in Metal Films Using Wet Chemical Techniques," in Etching for Pattern Definition, Washington, DC.

TABLE 4.11   Electrochemical Applications in the IC Industry and Micromachining.

Photocircuit boards

- Single and multilayer epoxy boards
- Flexible circuit boards
- Electrophoretically glazed steel boards

Contacts and connectors
- Beam leads
- Contacts (pins and sockets)
- Reed switches, etc.

Cabinets and enclosures
- Corrosion protective surfaces
- Electromagnetic shielding
- Decorative purposes (anodization, electrophoretic painting)

Auxiliary equipment used in device fabrication
- Paste-screening masks
- Metal evaporation masks
- X-ray lithography masks
- Diamond saws and cutting tools

Active elements
- IC chips
- Magnetic recording heads
- Recording surfaces
- Displays
- Wear-resistant surfaces

Chip carriers and packages
- Chip in tape packages
- Surface mount boards
- Dual in-line packages
- Pin-grid array packages
- Multilayer ceramic packages
- Hybrid packages

Source: Based with permission on Romankiw, L.T., and Palumbo, T.A. (1987) "Electrodeposition in the Electronic Industry," *Proc. Symp. on Electrodeposition Technology, Theory, and Practice*, San Diego and Romankiw, L.T. et al. (1984) "Electrochemical Technology in Electronics Today and Its Future: A Review," *Oberflache-Surface* 25, 238–47.

4.3.8.2.2   *Electroless Plating in the IC Industry and in Micromachining*

The IC industry applies electroless metal deposition for a wide variety of applications, some of which are incorporated in Tables 4.11 and 4.12. In Figure 4.22, we single out the schematic for the electroless Cu deposition of a buried conductor as an illustrative example of an IC application. The fabrication of this buried conductor illustrates all the processes we have studied so far: (a) lithographic patterning of a dielectric material, (b) pattern transfer by anisotropic dry etching of a groove in the dielectric substrate, (c) evaporative deposition of an Al seed layer, (d) seed layer patterning with lift-off, (e) trench filling with electroless Cu, and (f) spin-on glass (SOG) dielectric layer deposition for planarization.

In micromachining applications, electroless deposition is used for the same purposes as in the IC industry as well as to make higher-aspect-ratio structural microelements from a wide variety of metals, metal alloys, and even from composite materials. As we will see, these applications introduce new challenges in terms of uniformity of plating rates and removal of the metal form from the master photoresist mold. Electroless plating may be chosen over electrodeposition due to the simplicity of the process because no special plating base (electrode) is needed. This represents a major simplification for combining LIGA and pseudo-LIGA structures with active CMOS electronics where a plating base could short out the active electronics. In addition, electroless Ni exhibits less stress than electrodeposited Ni, a fact of considerable importance in most mechanical structures. The major concern is the temperature of the electroless

TABLE 4.12 Situations for Which an Aqueous Electrochemical or Chemical Technique is the Only Means to Obtain the Desired Result

- Uniform plating on irregular surfaces
- Leveling of rough surfaces
- Plating of via holes and blind vias
- Formation of alloys of metastable phases
- Compositionally modulated structures
- Amorphous metal films
- Maskless, high-speed, selective deposition of metals and alloys
- Most faithful reproduction of features of polymeric masks
- Smallest dimension features with highest height-to-width aspect ratio (highest packing density of conductors)
- Metal deposits with incorporated particles (diamond, WC, TiC, Al_2O_3, Teflon, oil, etc.)
- Wear-resistant surfaces
- Corrosion-resistant surfaces
- Uniform thickness nonporous anodic films
- Anodic films with porous structure
- Very uniform, thick, pore-free, polymer films on irregular surfaces
- Very uniform thickness, glazed, and devitrified glass surfaces
- Shaping of almost any metal or semiconductor and electromachining without introducing stress
- Electropolishing of metal surfaces to mirror bright finishes

Source: Based with permission on Romankiw, L.T., and Palmbo, T.A. (1997) "Electrodeposition in the Electronic Industry," *Proc. Symp. on Electrodeposition Technology*, Theory and Practice, San Diego and Romankiw, L.T. et al. (1984) "Electrochemical Technology in Electronics Today and Its Future: A Review," *Oberflache-Surface* **25**, 238–47.

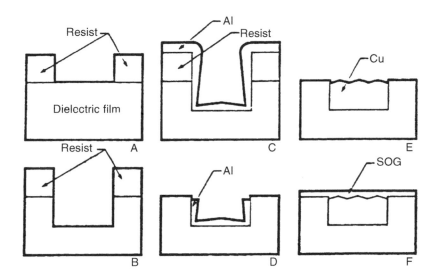

FIGURE 4.22 Schematic of a buried conductor process. (a) Photoresist pattern over a dielectric. (b) Pattern transfer by anisotropic dry etching. (c) Deposition of a thin Al seed layer. (d) Seed metal patterning with lift-off. (e) Trench filling with electroless Cu. (f) Spin-on glass (SOG) dielectric layer for planarization.

plating processes, which is often considerably higher than for electrochemical processes. Few studies of electroless plating in LIGA molds have been reported [Harsh et al., 1988].

4.3.8.2.3 *Measuring Thickness of Plated Layers*

Methods to determine the electroless deposition rate can be split into two categories: electrochemical and nonelectrochemical. Electrochemical techniques include Tafel extrapolation, DC polarization, AC impedance, and anodic stripping. Nonelectrochemical methods include weight gain, optical absorption, resistance probe, and acoustic. For more details on each of these techniques, see Ohno, 1988, and references therein [Ohno, 1988].

4.3.8.3 Electrodeposition

4.3.8.3.1 Introduction

In a typical electrodeposition through a polymer mask in IC applications, metal conductors must be deposited on a dielectric substrate. The dielectric is usually made conductive by first sputtering a thin adhesion metal layer (Ti, Cr, etc.) and then a conductive seed layer (Au, Pt, Cu, Ni, NiFe, etc.). The thickness of the thin refractory metal adhesion layer may be as small as 50 to 100 Å, while the thickness of the conducting seed layer can range from 150 to 300 Å [Romankiw et al. 1973]. The key requirement for the seed layer is that it is electrically continuous and offers low sheet resistance. After forming a pattern in a spin-coated polymer by UV exposure, e-beam, or X-ray radiation and developing away the exposed resist, contact is made to the seed layer and electrodeposition is carried out.

The IC industry tends to avoid wet chemistry techniques, but both IC and micromachining needs are forcing reconsideration of electrochemical techniques as a viable solution [Romankiw et al., 1976]. In micromachining, for example, electrochemical deposition enables the metal replication of high-aspect-ratio resist molds while maintaining the highest fidelity. Specifically in the electroforming of microdevices with LIGA (see Figure 4.1), a conductive substrate carrying the resist structures serves as the cathode. The metal layer growing on the substrate fills the gaps in the resist configuration, thus forming a complementary metal structure. The use of a solvent-containing development agent ensures a substrate surface completely free of grease and ready for plating. The fabrication of metallic relief structures is a well known art in the electroforming industry. This technology is used, for example, to make fabrication tools for compact discs where structural details in the submicron range are transferred. Because of the extreme aspect ratio (several orders of magnitude larger than in the crafting of CD masters), electroplating in LIGA poses new challenges.

The near ultimate in packing density, the extreme height-to-width aspect ratio, and the fidelity of reproduction of electrodeposition were first demonstrated in Romankiw's precursor work to LIGA in which gold was plated through polymeric masks generated by X-ray lithography. Romankiw, using X-ray exposure, developed holes in PMMA resist features of 1 μm width and depths of up to 20 μm and electroplated gold in such features up to thicknesses of 8 μm [Spiller et al., 1976]. By 1975, the same author and his team had used the electroplating approach to create micro wires 300×300 atoms in cross-section [Romankiw et al., 1993].

4.3.8.3.2 Electroplating

Electroplating takes place in an electrolytic cell. The reactions involve current flow under an imposed bias. As an example, we will consider the deposition of Ni from $NiCl_2$ in a KCl solution with a graphite anode (not readily attacked by Cl_2) and an Au cathode (inert surface for Ni deposition). With the cathode sufficiently negative and the anode sufficiently positive with respect to the solution, Ni deposits on the cathode and Cl_2 evolves at the anode. The process differs from electroless Ni deposition in that the anodic and cathodic processes occur on separate electrodes and in that the reduction is affected by the imposed bias rather than a chemical reductant. Important process parameters are pH, current density, temperature, agitation, and solution composition. The amount of hydrogen evolving and competing with Ni deposition depends on the pH, the temperature, and the current density. Since one of the most important causes of defects in metallic LIGA microstructures is the appearance of hydrogen bubbles, these three parameters need very precise control. Pollutants cause hydrogen bubbles to cling to the PMMA structures, resulting in pores in the nickel deposit, so the bath must be kept clean, for example, by circulating through a membrane filter with 0.3 μm pore openings [Harsch et al., 1988]. Besides typical impurities such as airborne dust or dissolved anode material, the main impurities are nickel hydroxide formed at increased pH values in the cathode vicinity and organic decomposition products from the wetting agents. The latter two can be avoided to some degree by monitoring and controlling the pH and by adsorption of the organic decomposition products on activated carbon.

4.3.8.3.3 Diffusion-Limited Reactions

Species in the electrolyte must be transported to and from the electrode before electrode reactions can occur. As species are being consumed or generated at the electrode surface via electrochemical reactions, the concentration of these species at the electrode surface will become smaller or larger, respectively, than in the bulk of the electrolyte. Suppose we are dealing with a species of concentration C_0 in the bulk of the electrolyte ($x = \infty$) being consumed at the electrode. The concentration gradient at the electrode is then given as:

$$\frac{dc}{dx} = \frac{C^0_{x=\infty} - C_{x=0}}{\delta} \tag{4.5}$$

where $C_x = 0$ is the concentration of the species at the electrode surface, and d represents the boundary layer thickness. From the thermodynamic relationship between the potential and concentration, we conclude that the concentration difference leads to an overpotential:

$$\eta_c = \frac{RT}{nF} \ln \frac{C_{x=0}}{C^0_{\infty}} \tag{4.6}$$

That is, the expression for the concentration polarization η_c where n is the number of electrons involved in the reaction. On the basis of Faraday's law, we can rewrite Equation (4.5) in terms of the current density as:

$$i = nFD_0 \frac{C^0_{\infty} - C_{x=0}}{d} \tag{4.7}$$

with D_0 ($\text{cm}^2\,\text{s}^{-1}$) representing the diffusion coefficient of the electro-active species and n the number of electrons transferred. At a certain potential η_c, all of the species arriving at the electrode are immediately consumed, and from that potential on, the concentration of the electro-active species at the surface falls to 0, that is, $C_{x=0} = 0$, and we reach the limiting current density i_l

$$i_l = nFD_0 \frac{C^0_{\infty}}{\delta} \tag{4.8}$$

This equation shows that the limiting current is proportional to the bulk concentration of the reacting species. On this basis, classical amperometric (current-based) sensors are used as analytical devices. We can equate $C_{x=0}/C^0_{\infty}$ in Equation (4.6) with $1 - i/i_l$ or:

$$i = i_l \left(1 - e^{\frac{nF\eta_c}{RT}} \right) \tag{4.9}$$

This expression is illustrated for a cathodically diffusion-limited reaction in Figure 4.23. The cathodic current quickly becomes diffusion limited, whereas the activation-controlled current continues rising.

The concentration overpotential and the activation overpotential have different origins. In the case of activation overpotential, the slope of i vs. η increases with η; whereas, in the transport limiting case, the slope of i vs. η decreases with increasing η and effectively becomes zero for sufficiently large values of the overpotential.

As shown in Figure 4.23, stirring of the electrolyte strongly affects the limiting current. Higher stirring rates promote convective transport of the reacting species toward the electrode and result in a smaller boundary layer δ.

4.3.8.3.4 Nonlinear Diffusion Effects on Microelectrodes
Electrochemical reactions such as electrodeposition and electrodissolution are influenced by the electrode size relative to the thickness of the diffusion layer. With microstructures of the dimensions of the diffusion layer, macroscale electrochemistry theory breaks down, and we can exploit some unexpected beneficial effects afforded through miniaturization.

The total diffusion-limited current I_l on a large substrate of area A is based on Equation (4.8) and given by $I_l = i_l \times A$. For a small spatially isolated electrode with its size reduced to the range comparable with the thickness of the diffusion layer, nonlinear diffusion caused by curvature effects must be taken into account. Figure 4.24 illustrates diffusion effects for various types of microelectrode shapes. The analysis shows that as the curvature effects become more pronounced more diffusion of species from all directions takes place, thus increasing the supply of the reactants to the electrode.

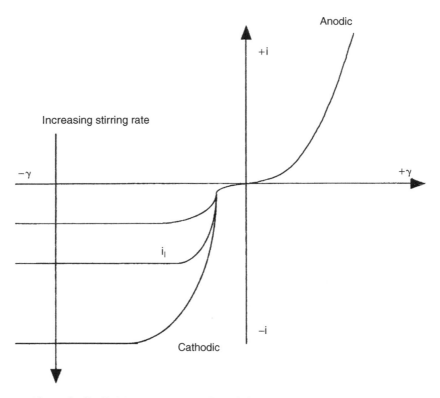

FIGURE 4.23 The cathodic limiting current is indicated for different stirring rates. The cathodic limiting current appears as a horizontal straight line limiting the current that can be achieved at any large negative value of the overpotential.

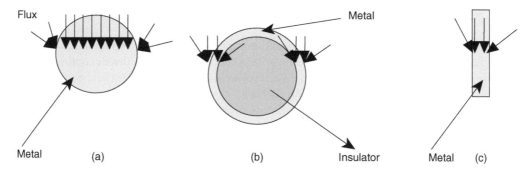

FIGURE 4.24 Convergent flux to small circular (a), ring (b), and band (c) electrodes. Left: Side view; right: plane view of the electrodes.

The diffusion layer thickness arising from linear diffusion is time dependent and given by [Madou and Morrison, 1989; Fleischmann et al., 1987]:

$$\delta = (\pi D_0 t)^{\frac{1}{2}} \tag{4.10}$$

Substituting Equation (4.10) in Equation (4.8), we obtain the "Cottrell equation":

$$I_l = nFAC_\infty^0 \left(\frac{D_0}{\pi t} \right)^{\frac{1}{2}} \tag{4.11}$$

This equation represents the current-vs.-time response on an electrode after application of a potential step sufficient to cause the surface concentration of electro-active species to reach zero. This equation, at

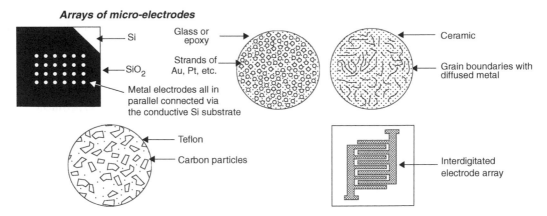

FIGURE 4.25 Arrays of microelectrodes.

short times after the potential step application, is appropriate regardless of electrode geometry and rate of solution stirring, as long as the diffusion layer thickness is much less than the hydrodynamic boundary layer thickness. Nonlinear diffusion at the edges of microelectrodes results in deviation from the simple Cottrell equation at longer times. The total current time relation with correction term becomes:

$$I_l = nFAC_\infty^0 \left(\frac{D_0}{\pi t}\right)^{\frac{1}{2}} + AnFD_0 \frac{C_\infty^0}{r} \qquad (4.12)$$

At longer times and for small electrodes, Equation (4.12) predicts that the correction term can become significant. In practice, a study on ultrasmall electrodes often starts with a linear regression of the measured current vs. $t^{-1/2}$ after application of a potential step. The intercept gives the steady state term. This term, the diffusion-limited current, $I_{l,m}$, at sufficiently long times on ultrasmall microelectrodes for some important electrode shapes, is given by [Fleischmann et al., 1987]:

$$\begin{aligned} I_{l,m} &= \pi rnFD_0C_\infty^0 \text{ (disc)} \\ I_{l,m} &= 2\pi rnFD_0C_\infty^0 \text{ (hemisphere)} \\ I_{l,m} &= 4\pi rnFD_0C_\infty^0 \text{ (sphere)} \end{aligned} \qquad (4.13)$$

In the correction term of Equation (4.13), the electrode surface area A is divided by the radius r. Hence, the principal location of charge transfer appears to be on the outer edge of the electrode. This constitutes a very favorable scaling. Phenomena scaling with the linear dimension to the first power become important in the microdomain. Surface tension exemplifies another law scaling with the first power as a very dominant force in the microdomain. In the current case, since $I_{l,m}$ is proportional to the electrode radius ($\sim r_1$) while the background current I_c (associated with the charging current of the Helmholtz capacitance) is proportional to the area ($\sim r^2$), the ratio of the Faradaic current (the charging current) to background currents should increase with decreasing electrode radius (l/r). For electrodeposition, this means that smaller features will be plated faster; for sensor applications, this translates into a higher S/N ratio or an improved sensitivity. The latter makes possible amperometric sensing with microelectrodes at unprecedented sensitivities.

With a single microelectrode, the analytical current remains small and the analytical gain is relative. However, using an array of microelectrodes (Figure 4.25) with all connected in parallel and separated by an insulating layer provides an elegant solution to this problem. In arrays of ultrasmall electrodes, the electrodes behave as an equivalent number of individual microelectrodes when the spacing between them is sufficiently large. This configuration enables analytical currents in a higher range compared to the single-electrode case. This gain is temporary because the diffusion layers around the individual microelectrodes expand across the insulating surface at a rate given by Equation (4.10). The time at which the overlap occurs is a function of both electrode size and spacing in the array, resulting in a decrease in the steady state current. In an analytical experiment, short compared to the time frame of overlap, more sensitive measurements at reasonably high currents become possible. Beside higher currents and improved S/N ratio through higher mass transport, another gain results from using an array because the signal can be averaged over many electrodes in parallel (an improvement in S/N with n the number of electrodes). Other advantages of using microelectrodes include the following:

- High mass transfer rates exist at ultrasmall electrodes, making it possible to experiment with shorter time scales (faster kinetics).
- An array of closely spaced ultrasmall electrodes can collect electrogenerated species with very high efficiency.
- Electrochemical measurements in high-resistivity media possibly including air become feasible [Madou and Morrison, 1991].

Micromachining small reaction chambers in which L, the outer boundary of the electrochemical cell, is made smaller than the diffusion layer thickness δ also increases the sensitivity of an electrochemical sensor as in Equation (4.8). δ is now replaced by a smaller L resulting in a higher diffusion limited current I_l. The latter finds its application in the design of thin-layer channel-type flow cells.

When considering a small metal disc embedded in a photoresist, nonlinear diffusion can be neglected for large substrates (r large or $I_{l,m}$ small). If r becomes small (i.e., of the dimension of and smaller than δ), the contribution of $I_{l,m}$ to the total current becomes larger than I_l; that is, an enhanced mass-transport to small electrodes occurs compared with large area electrodes. Consequently, we can expect that nonlinear behavior increases the plating rate of small features over large features.

Another factor influencing the deposition rate on a small substrate area is the level or position of the conductor surface with respect to surrounding insulating surfaces. The correction term in Equation (4.12) was derived for an inlaid microdisc electrode for which the electrochemically active substrate coincides with the dielectric insulation level (Figure 4.26a). If the conductor disc is recessed within an insulating

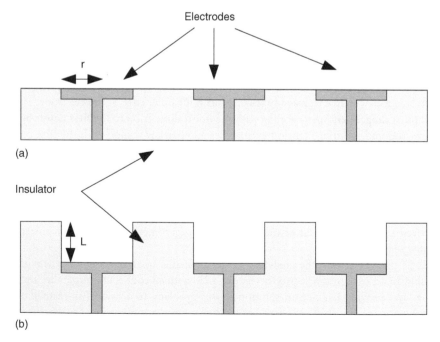

(a)

(b)

FIGURE 4.26 Nonlinear diffusion on inlaid and recessed electrodes. (a) Inlaid microelectrodes. (b) Recessed microelectrodes.

medium (Figure 4.26b), the diffusion inside the hole must be considered as well, leading to a slight modification of the correction term $I_{l,m}$:

$$I_{l,m} = AnFD_0 \frac{C_\infty^0}{r + L} \qquad (4.14)$$

in which L stands for the depth of the recession. Mass transport is only enhanced as compared with large area substrates for small values of L. Van der Putten et al. (1993a) found that the metal deposition rate at smaller recessed areas (with $r \leq L$) was higher than on the large areas, since overfilling was obtained faster for the smaller recessed features. When plating features in LIGA, we may deal with situations where a metal layer is deeply recessed ($r \ll L$) in a layer of insulating PMMA. As very high-aspect-ratio features (L/r large) can be made with LIGA, the thickness of the diffusion layers is increased artificially by the microstructured polymer layer, since it is more difficult for convective flow to reach the bottom of the deep gaps.

A microelectrode of the same size or less than the diffusion layer thickness, as shown in Figure 4.26, could be expected to plate faster than a larger electrode because of the extra increment of current due to the nonlinear diffusion contribution (Equation [4.13]). On the other hand, as derived from Equation (4.14) describing the current to a metal electrode recessed in a resist layer, the nonlinear diffusion contribution increases with decreasing radius r but decreases with increasing resist thickness L. High-aspect-ratio features consequently will plate slower than low-aspect-ratio features. Moreover, the consumption of hydrogen ions in the high-aspect-ratio features causes the pH to increase locally. As no intense agitation is possible in these crevices in the case of Ni plating an isolating layer of nickel hydroxide might form preventing further metal deposition. This all contributes to making the deposition rate, important for an economical production, much lower than the rates expected from the linear diffusion model of current density in large, low-aspect-ratio structures.

A major cause of defects in electroplating in high-aspect-ratio features is an incomplete wetting of the microchannels in the resist structure. The contact angle between PMMA and the plating electrolyte at 50°C lies between 70 and 80°. A wetting agent is thus indispensable for wetting the surface of the plastic structures for the electrolyte to penetrate into the microchannels. With a wetting agent, the contact angle between PMMA and the plating electrolyte can be reduced from 80 to 5° [Harsch et al., 1988]. In the electroforming of microdevices, a much higher concentration of wetting agent is necessary than in conventional electroplating. A dramatic illustration of this wetting effect can be seen in Figure 4.27. In the case illustrated in Figure 4.27a, where only 2.5 mL/L of a wetting agent is added to the sulfamate nickel deposition solution (see Table 4.13), nickel posts with a diameter of 50 μm often fail to form. In the same experiment, no posts with a diameter of 5 and 10 μm form at all. Increasing the wetting agent concentration to 10 mL/L results in the perfectly formed nickel posts with a diameter of 5 μm shown in Figure 4.27b.

4.3.8.3.5 Finishing Metal Parts and Metal Mold Inserts

Slight differences in metal layer thickness cannot be avoided in the electroforming process. Finish-grinding of the metal samples with diamond paste is used to smooth microroughness and slight variations in structural height. After finish grinding an electroplated LIGA or pseudo-LIGA workpiece, a primary metal shape results from removing the photoresist by ashing in an oxygen plasma or stripping in a solvent. In the case of cross-linked PMMA, the resist is exposed again to synchrotron radiation, guaranteeing sufficient solubility before being stripped.

If the metal part needs to function as a mold insert, metal is plated several millimeters beyond the front faces of the resist structures to produce a monolithic micromold (see Figure 4.28). In the latter case, to avoid damage to the mold insert when separating it from the plating base, an intermediate layer sometimes is deposited on the base plate ensuring adhesion of the resist structures while facilitating the separation of the electroformed mold insert from its plating base. In addition, it helps to prevent burrs from forming at the front face of the mold insert as a result of underplating of the resist structures.

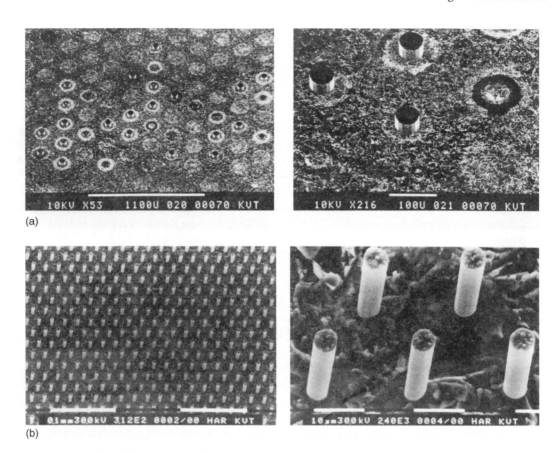

FIGURE 4.27   Electrodeposition of nickel posts of varying diameter. (a) Nickel posts with diameter of 50 μm. Only 2.5 ml/l wetting agent added to the nickel sulfamate solution. Many 50-μm posts are missing and posts with a diameter of 10 or 5 μm do not even form. (b) Nickel posts with diameter of 5 μm 10 ml/l wetting agent was added; all posts developed perfectly. (Courtesy of the Karlsruhe Nuclear Research Center.)

TABLE 4.13 Composition and Operating Conditions of Nickel Sulfamate Bath

Parameter	Value
Nickel metal (as sulfamate)	76–90 g/L
Boric acid	40 g/L
Wetting agent	2–3 mL/L
Current density	1–10 A/dm^2
Temperature	50–62°C
pH	3.5–4.0
Anodes	Sulfur depolarized

Underplating can occur because the resist does not adhere well to the substrate, allowing electrolyte solution to penetrate between the two, or because the plating solution might attack the substrate/resist interface. Finally, microcracks at the interface might contribute to underplating. Burrs are easily eliminated then by dissolving the thin auxiliary metal layer with a selective etchant, removing the mold without the need for a mechanical load. In view of the observed underplating problems, it is surprising that Galhotra et al. (1996) obtained good plating results by simply mechanically clamping the exposed and developed PMMA sheet to a nickel plating box.

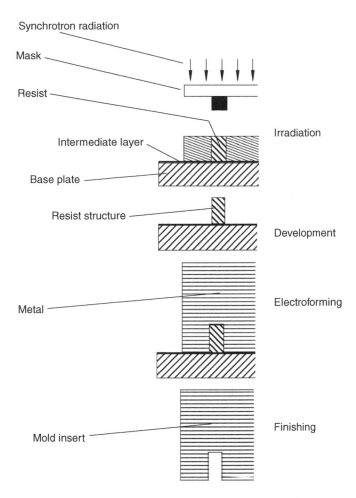

FIGURE 4.28 Fabrication of a LIGA mold insert. (Reprinted with permission from Hagmann, P. et al. [1987] "Fabrication of Microstructures with Extreme Structural Heights by Reaction Injection Molding," presented at First Meeting of the European Polymer Federation, European Symposium on Polymeric Materials, Lyon, France.)

4.3.8.3.6 *Nickel Electrodeposition*

Physical Properties of Electrodeposited Nickel. Nickel, along with copper, is one of the most widely used metals in electroplating LIGA and pseudo-LIGA structures. A nickel sulfamate bath composition optimized for Ni electrodeposition was presented in Table 4.13. In addition to nickel sulfamate and boric acid as a buffer, a small quantity of an anion-active wetting agent is usually added. Sulfur-depolarized nickel pellets are used as anode materials (sulfur-depolarized nickel has sulfur cast into it to aid in anode corrosion and keep the anode from going passive) and are held in a titanium basket [Maner et al., 1988]. Table 4.13 also lists operational parameters. Nickel sulfamate baths produce low internal stress deposits without the need for additional agents, thus avoiding a cause of more defects. The bath is operated at 50 to 62°C and at a pH value between 3.5 and 4.0. Metal deposition is carried out at current densities up to 10 A/dm². Growth rates vary from 12 (at 1 A/dm²) to 120 µm/hr (at 10 A/dm²).

The hardness of the Ni deposits can be adjusted from 200 to 350 Vickers by varying the operating conditions. Hardness decreases with increasing current density. To reach a high hardness of 350 Vickers, the electroforming must proceed at a reduced current of 1 A/dm². Also, for low compressive stress of 20 N/mm² or less, a reduced current density must be used. Internal stress in the Ni deposits is influenced not only by current density but also by the layer thickness, pH, temperature, and solution agitation. In the case of pulse plating (see below), pulse frequency has a distinct influence as well [Harsch et al., 1988].

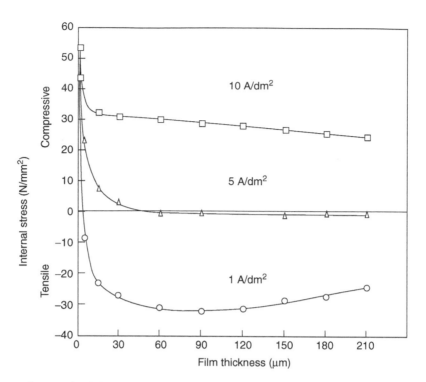

FIGURE 4.29   Influence of nickel layer thickness on internal stress. The electrolyte used is described in Table 4.13 (pH = 4; bath temperature = 52°C). (Courtesy of the Karlsruhe Nuclear Research Center.)

From Figure 4.29 we derive that for thin Ni deposits the stress is high and decreases very fast as a function of thickness. For thick Ni deposits, the stress as a function of thickness reaches a plateau. At a current density of $10\,A/dm^2$, these thick Ni films are under compressive stress; at $1\,A/dm^2$, they are under tensile stress; and at $5\,A/dm^2$, the internal stress reduces to practically zero. Stirring of the plating solution reduces stress dramatically, indicating that mass transport to the cathode is an important factor in determining the ultimate internal stress. Consequently, since high-aspect-ratio features do not experience the same agitation of a stirred solution as bigger features do, stress concentration is higher in the smallest features of the electroplated structure. Since the stress is most severe in the thinnest Ni films, Harsch et al. (1988) undertook a separate study of internal stresses in $5\,\mu m$ thick Ni films. For three plating temperatures investigated (42°, 52°, and 62°), $5\,\mu m$ thick films were found to exhibit minimal or no stress at a current density of $2\,A/dm^2$. At 62°C, the $5\,\mu m$ thick films show no stress at $2\,A/dm^2$ and remain at a low compressive stress value of about $20\,N/mm^2$ for the whole current range (1 to $10\,A/dm^2$). The internal stress at $2\,A/dm^2$ is minimal at a pH value between 3.5 and 4.5 of a sulfamate electrolyte (Table 4.13). Ni concentration (between 76 and $100\,g/L$) and wetting agent concentration do not seem to influence the internal stress. The higher the frequency of the pulse in pulse plating (see below), the smaller the internal stress.

The long-term mechanical stability of Ni LIGA structures was investigated using electromagnetic activation of Ni cantilever beams by Mohr et al. (1992). The number of stress cycles N necessary to destroy a mechanical structure depends on the stress amplitude S and is determined from fatigue curves or $S–N$ curves (applied stress on the x-axis and number of cycles necessary to cause breakage on the y-axis). Experimental results show that for Ni cantilevers produced by LIGA the long-term stability reaches the range of comparable literature data for bulk annealed and hardened nickel specimens. Usually, stress leads to crack initiation, which often starts at the surface of the structure. Since microstructures have a higher surface-to-volume ratio, the $S–N$ curves might be expected to differ from macroscopic structures, but so far this has not been observed. To the contrary, it seems that the smaller structures are more stable.

Pulse Plating of Ni. For the fabrication of micro devices with high deposition rates exceeding $120 \, \mu m/hr$, the internal Ni stress can be reduced only by raising the temperature of the bath or by using alternative electrodeposition methods. Raising of the bath temperature is not an attractive option, but using alternative electrochemical deposition techniques deserves further exploration. For example, using pulsed ($= 500 \, Hz$) galvanic deposition instead of a DC method can influence several important properties of the Ni deposit. Properties such as grain size, purity, and porosity can be manipulated this way without the addition of organic additives [Harsch et al., 1988]. In pulse plating, the current pulse is characterized by three parameters: pulse current density i_p; pulse duration t_d; and pulse pause t_p. These three independent variables determine the average current density i_a, which is the important parameter influencing the deposit quality and is given by:

$$i_a = \frac{t_d}{t_d + t_p} \, i_p \tag{4.15}$$

Pulse plating leads to smaller metal grain size and smaller porosity due to a higher deposition potential. Because each pulse pause allows some time for Ni^{2+} replenishment at the cathode (Ni^{2+} enrichment) and for diffusion away of undesirable reaction products that might otherwise get entrapped in the Ni deposit, a cleaner Ni deposit results. The higher frequency of the pulse leads to the smaller internal stress in the resulting metal deposit [Harsch et al., 1988].

Pulse plating represents only one of many emerging electrochemical plating techniques considered for microfabrication. For more background on techniques such as laser-enhanced plating, jet plating, laser-enhanced jet plating, and ultrasonically enhanced plating, refer to the review of Romankiw et al. (1987) and references therein.

4.3.8.3.7 Copper Electroplating

Copper in Printed Circuit Boards. Today's printed circuit boards may consist of as many as five or six layers of metals and dielectrics. Printed circuit board fabrication employs either an all-additive electroless copper process or a combination of electroless and electroplating processes. The continuing effort to make active devices smaller and faster, to minimize the length of connecting wire, and to make them narrower and thicker (with smaller spaces in between) necessitates more layers per board and smaller diameter holes with a larger ratio of hole length to diameter. A hole length-to-diameter ratio of 10:1 is routine, and 20:1 is feasible. Without electrochemical technology, such a degree of integration would not have been possible [Romankiw and Palumbo, 1987].

Copper in ICs: The Damascene Process. Copper wiring on IC chips was introduced by IBM in September 1997 with the "damascene process" [Editorial, 1997]. The expertise with Cu plating at IBM dates to the late 1960s at Watson, when Romankiw succeeded in electroplating narrow wires of copper onto thin film read/write heads for memory using a masking method that deposited the copper only in circuit patterns where it was needed (see also below). Copper wires conduct electricity with about 40% less resistance than traditional aluminum wires, leading to faster microprocessors. Copper wires are also less prone to the electromigration that ultimately induces wiring failure. Called damascene in reference to the metallurgists in Damascus who produced the finest polished swords in medieval times, the technique was initially used to form vias linking separate layers of wiring in chips.

In damascene patterning, the pattern of wires or vias is first formed by etching the silicon oxide layer. The metal is deposited second, and the excess is removed by polishing. Electrodeposition was the metal deposition method of choice despite the fact that some people thought the copper patterns would be filled with bubbles or that the process was too dirty. It turned out to have a faster rate of deposition, and the evenness of the copper film was better than in the case of electroless deposition. Also, copper CVD, an early favorite option, ran into other severe problems. To polish the copper at an acceptable rate while controlling corrosion, erosion, and other defects in the patterns, IBM pioneered a special chemical-mechanical process (CMP) that proved critical to the copper technology. To overcome copper's tendency to diffuse in silicon, a metal diffusion barrier preventing atoms from migrating out of a copper wire into surrounding chip material was developed. As in conventional patterning, the damascene process is repeated many times to form the alternating layers of wires and vias that form the complete wiring system of a silicon chip.

Cu in Thin Film Inductive Magnetic Heads. Earlier, in another pioneering IBM effort, Romankiw used plating-through mask technology to make thin film heads as shown in Figure 4.30 [Romankiw et al., 1970]. The development of batch-produced, thin film heads is an excellent example of how micromachining has

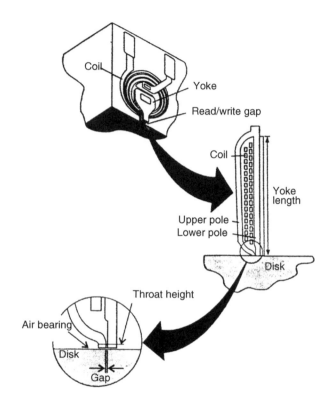

Read/write thin film head

(a)

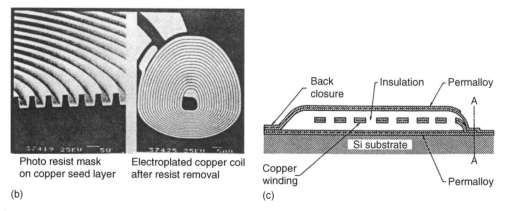

Photo resist mask on copper seed layer   Electroplated copper coil after resist removal

(b)   (c)

FIGURE 4.30 Thin film head. (a) Schematic of a multiturn thin film head with inset of pole tip structure on the air-bearing surface. (b) Resist mold and electroplated copper coil after resist removal. (c) Schematic cross-section of the head. (Reprinted with permission from Romankiw, L.T. et al. [1970] "Batch-Fabricated Thin-Film Magnetic Recording Heads," *IEEE Trans. Magn.* **MAG-6**, pp. 597–601.)

been gainfully practiced in industry — with a definite, practical, well specified application in mind and using the machining tools best suited for the problem at hand.

Read/write heads, which initially were horseshoe magnets hand wound with insulated copper wire, are of utmost importance in magnetic storage. The fabrication of traditional ferrite heads reached its limit more than a decade ago, and its further extension was hard to imagine. IBM's objective was to build the next generation read/write heads using batch fabrication and lithography techniques as much as possible. To develop a multiturn head as shown in Figure 4.30a, it was necessary to develop a technology that would handle dimensions between those of printed circuit boards and semiconductor devices. Plating of copper conductors through thick resist masks (Figure 4.30b) and of thin films of nickel–iron alloys of 80:20 nominal composition (Permalloy) (Figure 4.30c) made these thin film heads possible. The Cu coils in Figure 4.30b carry considerable current and have to be at least 2 to 3 µm thick and nearly square in cross-section. The more turns in a given length, the higher the writing field at the pole tip gap and the higher the read-back signal. The head in Figure 4.30c has 8 turns; the 3380-K IBM head has 32 turns (Figure 4.30a). The Permalloy plating in the thin film head is particularly challenging. Due to the very high sensitivity of composition to agitation of the plating solution, a plating cell had to be developed that assured uniform agitation over the entire part. This was achieved by a specially shaped paddle that moves at a predetermined frequency with a precisely defined separation from the surface being plated. With the right choice of conditions, it was eventually possible to achieve both thickness and composition uniformity better than ±5% [Romankiw et al., 1989]. Hard-baked AZ photoresist was used as a dielectric, greatly simplifying the fabrication process. Si was used as a substrate in the early days; today $Al_2O_3.TiC$ is employed.

Without these new micromachined thin film heads, the large increases in the areal densities of magnetic media in the last few years would not have been possible. According to the data storage market research firm Peripheral Research, Inc., market demand for thin film heads was projected to grow nearly 75% from 913 million units in 1998 to more than 1.5 billion units by 2001. This feat makes it probably the most successful micromachined device to date.

As the need for still more memory capacity increases, thin film inductive heads are being replaced by newer technologies such as magnetoresistive (MR) heads (introduced in 1991 by IBM) and giant magnetoresistive (GMR) heads (introduced in 1997). In MR heads, the read-back function is now performed by an MR sensor. The inductive portion of the head is used only for the write portion of the data-recording operation [Andricacos and Robertson, 1998].

4.3.8.3.8 Gold Electroplating

Whereas photocircuit boards use electroless or electroplated Cu, contacts and connectors employ electroplated Au and Au alloys or, alternatively, low-cost precious metal substitutes such as Pd, PdNi, PdAg, NiP, NiAs, and NiB with a thin gold overcoat. Separable, low-force, low-voltage contact applications require a contact finish whose resistance is stable for the projected contact life and that has sufficient wear resistance to withstand the projected number of insertions. The surface of one of the two mating parts is usually pure soft gold, while the other is a hard gold alloy. In less-critical applications, tin and tin–lead alloys are used as contact materials. Connecting chips with chip carriers or packages also often involves Au contacts. Gold does not adhere directly to silicon dioxide or silicon nitride, and Ti is typically used as an adhesion layer. Since Au and Ti form intermetallics, a Pt barrier is usually deposited between Ti and Au.

4.3.8.4 Other Plating Applications

Plating-through mask technology is successfully used in volume production of beam leads and bumps on IC wafers, fabrication of thin film magnetic heads, fabrication of PC boards, X-ray lithography mask gratings, and diffraction gratings. Electrodeposition in general has been used extensively in the electronics industry in many stages of the manufacturing process, from the device stage, chip carriers, and PC boards to corrosion protection and electromagnetic shielding of the electronic enclosures (see Tables 4.11 and 4.12). Processes include electrodeposition and electroless deposition of copper, nickel, tin, tin–lead alloys, and precious metals such as gold, gold alloys, palladium, and palladium alloys as well as NiFe, CoP, NiCoP, and other magnetic alloys [Romankiw and Palumbo, 1987].

4.3.8.5 Plating Automation

An automated electroplating facility used by KfK is shown in Figure 4.31 [Reinhard Kissler, 1994]. This setup includes provisions for on-line measurement of each electrolyte constituent in flow-through cells and concentration corrections when tolerance limits are exceeded. A computer-controlled

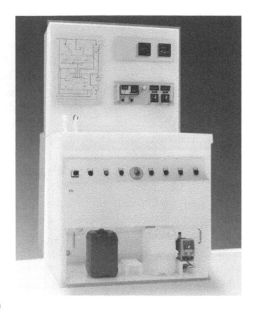

(a)

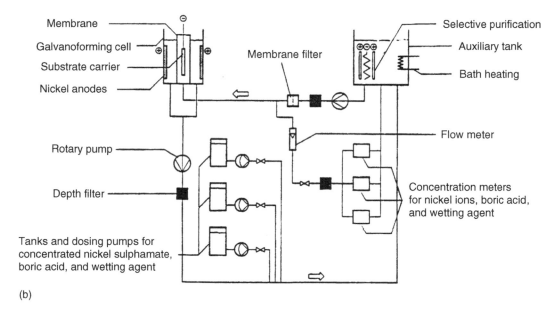

(b)

FIGURE 4.31   (a) The μGALV 750 comprises the galvanoforming cell with nickel anodes and substrate carrier and an auxiliary tank incorporating heating and purification auxiliary equipment. Three concentration meters measure nickel ions, boric acid, and the concentration of the wetting agent. The concentrations are adjusted automatically by adding via metering pumps [GmbH, 1994]. (Reprinted with permission from Kissler, R.K. (1994) "μGALV 750," Sales Brochure, GmbHDaimlerstrasse 8, Speyer 67346.) (b) Schematic drawing of a galvanoforming unit. (Reprinted with permission from Maner, A. et al. [1988] "Mass Production of Microdevices with Extreme Aspect Ratios by Electroforming," *Plating and Surface Finishing*, pp. 60–65.)

transport system moves the individual plating racks holding the microdevice substrates through the process stages whereby the substrates are degreased, rinsed, pickled, electroplated, and dried and are then returned to a magazine that can accommodate up to seven racks. The facility is designed as clean-room equipment, since contamination of the microstructures must be prevented [Maner et al., 1988]. An instrument as described here can be bought, for example, from Reinhard Kissler (the μGLAV 750) [Reinhard Kissler, 1994]. Although this type of automation is necessary for the eventual commercialization of the LIGA technique, at this exploratory stage overautomation may be counterproductive.

4.3.8.6 Plating Issues

Two major sources of difficulty associated with plating of tall structures in photoresist molds are chemical and mechanical incompatibility. Chemical incompatibility means that the photoresist mask may be attacked by the plating solution; mechanical incompatibility occurs when film stress in the plated layer causes the plated structure to lose adhesion to the substrate.

If the plating solution attacks the photoresist even slightly, considerable damage to the photoresist layer may have occurred by the time a 200 μm thick structure has been created. The limiting thickness of a plated structure due to a chemical interaction is therefore dependent on both the photoresist chemistry and the plating bath itself. In general, the plating bath must have a pH in the range of 3 to ~9.5 (acidic to mildly alkaline), a fairly wide range that can accommodate many of the commercially available photoresists. A surprising number of plating baths do not fall into that range, however. Baths that are either very strongly acidic or more than mildly alkaline tend to attack and destroy photoresist.

The plated structure must have extremely low stress to avoid cracking or peeling during the plating process. In addition, if the structure contains narrow features, the stress must be tensile because any compressive stress would result in buckling of the structure. The primary concern with regard to stresses is the incorporation of brightening agents into the plating solution. These can be selenium, arsenic, thallium, and ammonium ions, as well as others. The brightening materials exhibit different atomic configurations and sizes than most materials that we would like to plate, such as Cu, Au, Ni, FeNi, Pt, Ag, NiCo, etc. This is the source of a good deal of the stress intrinsic to the plating process. Ag and FeNi are notorious for their high stresses. The Ni-sulfamate bath, presently used extensively in LIGA, is used without brighteners, leading to Ni deposits with extremely low stress. Electroless plating baths also generally lead to low stress. These autocatalytic baths coat indiscriminately but cause very low stress. They should be compatible with the LIGA process for certain applications, such as mold insert formation where significant overplating is required.

A third difficulty in finding plating baths compatible with LIGA is that most plating baths are not intended for use in the semiconductor industry. Information (including normal deposition parameters, uniformity, stress, compatibility with various semiconductor processes, and particulate contamination) normally supplied by a vendor is not available. Since the majority of the semiconductor industry does not incorporate these kinds of thin film processes, it may be difficult to determine whether a plating bath will be suitable simply from conversations with the vendor. In many instances, trial and error decide.

4.3.8.7 Conclusions on Electroless and Electroplating through Polymer Masks

The need for higher-aspect-ratio structures ensures that the role of electrochemistry will become even more important in MEMS and the IC industry in years to come. The technology provides higher resolution, better pattern definition, and much higher-aspect-ratio patterns than chemical or even plasma ion etching can achieve.

Electrochemistry through polymer masks indeed represents one of the most powerful techniques available for formation of very high density patterns and circuits with extremely large height-to-width ratios. Most other methods, such as chemical isotropic etching, sputter etching, ion milling, and reactive ion etching, cannot be used to produce metal patterns with better than 1:1 height-to-width aspect ratios. To understand this contrasting behavior, we must look into the fundamental nature of the two different processes. In nonelectrochemical technologies, a polymeric or inorganic mask is used to cover the existing metal while either a chemical solution or an active gas phase removes the metal from the open areas.

Hence, the limitation in the height-to-width aspect ratio, the maximum circuit density, and the fidelity of reproduction of the features are caused by shadowing, which is the mechanism by which the patterning takes place. In contrast, with electrochemical techniques the exact replication of the recesses in the mask is obtained [Romankiw et al., 1984]. Thus, electrodeposition reproduces the finest features of the mask with the greatest fidelity. This is not surprising, considering that, in plating, metal ions discharge from solution present inside the mold. In so doing, the metal displaces the solution from the mold atom by atom, conforming to the smallest features that exist in the mold [Romankiw and Palumbo, 1987]. A comparison of conventional subtractive etching with electrochemical additive processes was presented as early as 1973 by Romankiw et al. (1973). For the fabrication of highly miniaturized magnetic bubble memory devices, this IBM group showed clearly that the additive plating process is superior to dry-etching-based techniques.

Figure 4.32 features a list of choices for material deposition employing electrochemical and electroless techniques [Ehrfeld et al., 1994]. The first metal LIGA structures consisted of nickel, copper, or gold electrodeposited from suitable electrolytes [Maner et al., 1988]. Nickel–cobalt and nickel–iron alloys were also tried. A nickel–cobalt electrolyte for deposition of the corresponding alloys has been developed especially for the generation of microstructures with increased hardness (400 Vickers at 30% cobalt) and elastic limit [Harsch et al., 1991]. The nickel–iron alloys permit tuning of magnetic and thermal properties of the crafted structures [Harsch et al., 1991; Thomes et al., 1992]. From Figure 4.32, it is obvious that many more materials could be combined with LIGA molds.

Beyond the deposition of a wide variety of materials in LIGA molds, additional shaping and manipulation of electroless or electrodeposited metals during or after deposition will bring a new degree of freedom to micromachinists. Of some interest in this respect is *bevel plating* from van der Putten (1993b), which introduced anisotropy in the electroless Ni plating process resulting in beveled Ni microstructures. Combining van der Putten's bevel plating with LIGA, one can envision LIGA-produced contact leads with sharp contact points. Other novel electrochemical microfabrication techniques such as laser-enhanced and jet plating and jet etching make electrochemical technology particularly useful for replicating high-aspect-ratio, micron, submicron, and nanometer mask patterns — a whole new set of micromachining tools hardly explored today in conjunction with LIGA and pseudo-LIGA.

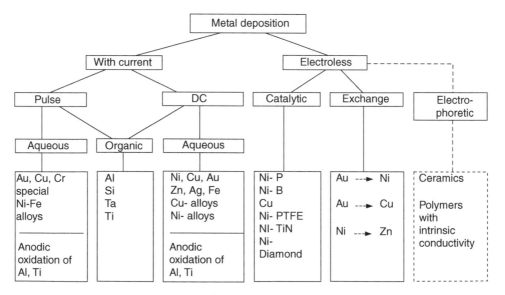

FIGURE 4.32 Processes and materials for electrochemical deposition of LIGA structures. (Reprinted with permission from Ehrfeld, W. [1994] "LIGA at IMM," notes from handouts, Banff, Canada.)

4.3.9 Molding

4.3.9.1 Introduction

LIGA and LIGA-like processes can be used directly to fabricate prototype microdevices, but for commercial-scale manufacture of microdevices, replication by molding from a LIGA or LIGA-like mold is required. If the feature sizes are larger than 50 μm and several levels of feature depths are involved, even computer numerical control (CNC) milling may be employed to manufacture the prototype device. For example, in Figure 4.33, we show a CNC machined plastic (polycarbonate) fluidic CD platform. Compared with LIGA or LIGA-like processes, CNC machining does not provide as good a surface finish or dimensional control. It does not, however, have any material limitation; various metals and plastics can be used. For feature size smaller than 50 μm (but larger than several microns), micromilling or laser ablation can be used either alone or in conjunction with CNC machining. These prototype processes, just like LIGA and LIGA-like processes, are all intensive, slow, and costly, and for these microfabrication techniques to become economically viable, microstructures must be able to be replicated successively in an "infinite loop" without going back to lithography or any other slow fabrication step to remake the primary structures. For example, in the plastic molding process, a metal structure produced perhaps by metal deposition serves as a mold insert and is used repeatedly without remaking the metal master. For mass-produced plastic devices, one of the following molding techniques is suitable: liquid resin molding, thermoplastic injection molding (IM), and compression (embossing/imprinting) molding. Casting of an elastomeric resin (e.g., polydimethylsiloxane — PDMS) against a patterned photoresist on Si for rapid prototyping purposes has also been used widely because of its simplicity.

The miniature features and the high aspect ratios of the mold inserts generated from LIGA and LIGA-like methods present new problems compared with the molding and demolding processes of macrocomponents and with the production of small but low-aspect-ratio products such as compact discs. A variety of the special processes developed to accommodate these new needs are reviewed below.

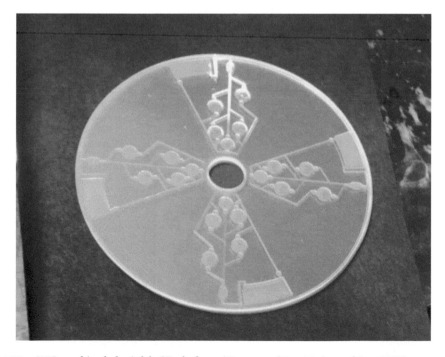

FIGURE 4.33   CNC-machined plastic lab CD platform. (Courtesy of Drs. Madou and Lee, OSU.)

4.3.9.2 Liquid Resin Molding Techniques

4.3.9.2.1 *Introduction*

Since mold inserts made by photolithographic techniques typically are limited to rather soft metals like nickel or to brittle materials such as silicon and glass, micromolding based on low-viscosity liquid resins (instead of high-viscosity polymer melts) is a very attractive approach. During liquid resin molding, the low-viscosity reactive polymer components are mixed shortly before injection into the mold cavity, and polymerization takes place during the molding process. Under liquid resin molding techniques, we distinguish between reaction injection molding, transfer molding, and casting.

4.3.9.2.2 *Reaction Injection Molding (RIM) and Transfer Molding*

Two typical setups for reaction injection molding and transfer molding are shown in Figures 4.34a and b respectively [Lee et al., 1999]. In the setup shown in Figure 4.34a, two or more highly reactive liquid resins impinge and mix inside a mixer and are injected into the mold. In reaction injection molding (RIM), the polymerization is usually mixing activated, so it can react quickly at room temperature or slightly above and convert from liquid to solid. The transfer molding setup for a thermally activated system in Figure 4.34b is simpler and less expensive. Here, the reactive species have been premixed at room temperature, and the mixture is poured into the transfer pot. A plunger squeezes the mixture through the sprue into the mold cavity. To cure the mixture, the mold is heated or radiated, depending on the resin system. With the lowest-viscosity resins, vacuum-assisted molding is used, which is even less expensive than transfer molding.

Two complications with RIM are that an internal mold release agent must be used, which tends to reduce the thermomechanical properties of molded polymer and polymer shrinkage during polymerization. External mold release agents are unsuitable, since they are difficult to introduce into the microfeatures of the mold insert. Since the resin must be fully cured at high temperatures, the cycle time is typically more than 10 min. Hanemann et al. (1998) developed a photoinitiated reaction injection molding process

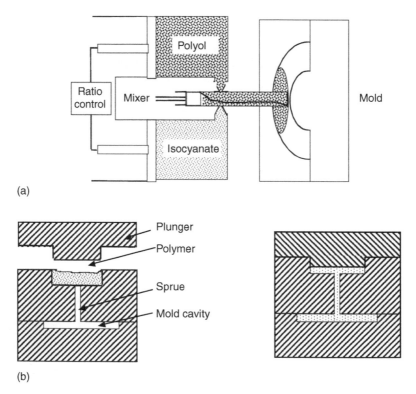

(a)

(b)

FIGURE 4.34 (a) Reaction injection molding (RIM); (b) transfer molding (TM). (Courtesy of Drs. Madou and Lee, OSU.)

by using the MMA resin. They found that the cycle time can be greatly reduced (to 2 to 3 min), but other disadvantages such as internal mold release agent and polymerization shrinkage still exist.

4.3.9.2.3 LIGA and LIGA-Like Applications of RIM

A laboratory-size reaction injection molding setup as shown in Figure 4.35 was used by the KfK group for their LIGA work [Hagmann and Ehrfeld, 1988]. The setup consists of a container for mixing the various reactants, a vacuum chamber for evacuation of the mold cavity, the molding tool, and a hydraulic clamping unit to open and close the vacuum chamber and the molding tool. After the vacuum chamber has been closed and evacuated, the tool is closed by the clamping unit. The mixed reagents are degassed in the materials container and, under a gas overpressure of up to 3 MPa, pushed through the opened inlet valve into the tool holding the evacuated insert mold. To compensate for shrinkage due to polymerization, an overpressure of up to 30 MPa must be applied during hardening of the casting resin. If the holding overpressure is too low, sunken spots appear in the plastic due to shrinkage of the polymer. For PMMA, the volume shrinkage is about 21%. If the mold material is not degassed and the mold cavity is not evacuated, bubbles develop in the molded piece resulting in defects and possible partial filling of the mold. To harden the mold material and anneal material stress, the reaction injection molding machine is operated at temperatures up to 150°C.

To fabricate RIM-based micro products, a variety of resins have been tried, including epoxy resins, silicone resins, and acrylic resins. The most promising results are obtained with resins on a methyl–methacrylate base to which an internal mold release agent is added to reduce adhesion of the molded piece to the walls of the metal mold. The mold insert in the evacuated tool is covered by means of an electrically conductive perforated plate called the gate plate (Figure 4.36a). For filling the mold, injection holes are positioned above large, free spaces in the metal structure, and the low-viscosity reactants fill the smaller sections laterally. After hardening of the molded polymer, a form-locking connection between the produced part and the gate plate is established at the injection holes, permitting the demolding of the part from the insert (Figure 4.36b). No damage to the mold insert is observed after up to 100 mold–demold cycles, even at the level of a scanning electron microscope picture. The secondary plastic structures formed are exact replicas of the original PMMA structures obtained after X-ray irradiation and development. If the final product is plastic, the molding process can be simplified considerably by omitting the

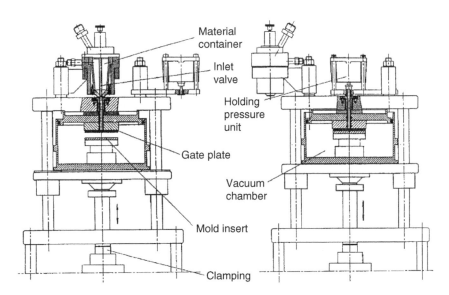

FIGURE 4.35 Schematic presentation of a vacuum molding setup. With minor changes, this setup can be used for reaction injection molding, thermoplastic injection molding, and compression molding. (Reprinted with permission from Hagmann, P., and Ehrfeld, W. [1988] "Fabrication of Microstructures of Extreme Structural Heights by Reaction Injection Molding," *J. Polymer Process. Soc.,* **4**, pp. 188–95.)

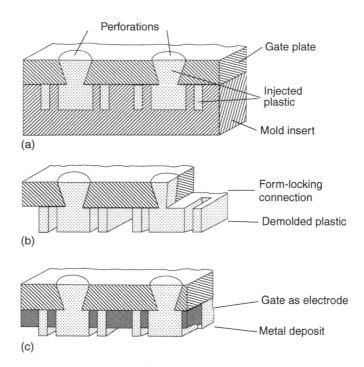

FIGURE 4.36 Molding process with a perforated gate plate. (a) Filling of the insert mold with plastic material. (b) Demolding. (c) Electrodeposition with the gate plate as plating base. (Reprinted with permission from Menz, W., and Bley, P. [1993] *Mikrosystemtechnik für Ingenieure*, VCH Publishers, Germany.)

gate plate. The mold insert is then cast over with the molding polymer in a vacuum vessel, building a stable gate block over the mold insert after hardening. The polymeric gate block is used to demold the part from the insert.

The secondary plastic templates usually are not the final product but may be filled with a metal by electrodeposition as with the primary molds. In this case, the secondary plastic templates must be provided with an electrode or plating base. The gate plate can be used directly as the electrode for the deposition of metal in the secondary plastic templates (see Figure 4.36c). It is important that the mold insert be pressed tight against the gate plate so that no uniform isolating plastic film forms over the whole interface between the gate plate and the mold insert; otherwise, electrodeposition might be impossible. Safe sealing between mold insert and gate plate is achieved by use of soft-annealed aluminum gate plates into which the insert is pressed when closing the tool. It was found that the condition to be met for perfect deposition of metal does not require that the gate plate be entirely free of plastic [Hagmann and Ehrfeld, 1988]. It suffices that a number of electrical conducting points emerge from the plastic film. Transverse growth of the metal layer produced in electrodeposition allows a fault-free continuous layer to be produced. The secondary metal microstructures are replicas of the primary metal structures, and a comparison of, for example, a master separation nozzle mold insert with a secondary separation nozzle structure did not reveal any differences in quality. After the metal forming, the top surface must be polished, and the plastic structures must be removed. Depending on the intended use of the secondary metal shape, it can remain connected to the gate plate or the gate plate can be selectively dissolved or mechanically removed.

Gate plates can be applied only if the microstructures are interconnected by relatively large openings. The injection holes in the gate plate are produced by mechanical means with openings of about 1 mm in diameter and must be aligned with large openings in the mold insert. Many desirable microstructures cannot be built this way. For example, in Figure 4.37, we show a Ni honeycomb structure with a cell diameter of $80\,\mu m$, a wall thickness of $8\,\mu m$, and a height of $70\,\mu m$ [Harmening et al., 1991]. The holes in a

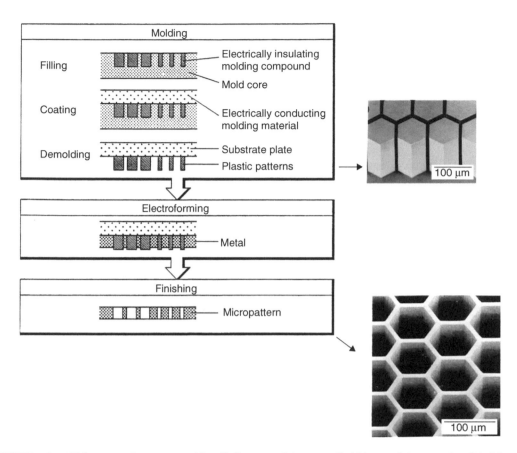

FIGURE 4.37 Ni honeycomb structure with cell diameter of 80 μm, wall thickness of 8 μm, and wall height of 70 μm. A structure like this could not be produced using a gate plate; a conductive plastic is used instead. The Ni plating occurs on the conductive plastic carrier plate, which is fused with the insulating plastic of the microstructure. (Reprinted with permission from Harmening, M. et al. [1991] "Molding Plateable Micropatterns of Electrically Insulating and Electrically Conducting Poly(Methyl Methacrylate)s by the LIGA Technique," *Makromol. Chem. Macromol. Symp.* **50**, pp. 277–84.)

gate plate are too large to inject plastic into this structure, and the interconnections in the honeycomb are too small to provide a good transverse movement of the polymerizing resin.

4.3.9.2.4 Casting

For new product development, an attractive way of rapid prototyping is casting, as demonstrated by Whitesides' PDMS process. This method, called soft lithography, has been used by many researchers because of its simplicity. The long cycle time and limitation exclusively to PDMS rubber, however, make it difficult for mass production in most industrial applications. Figure 4.38 compares two CD platforms made by two different PDMS casting processes carried out in our lab. The one based on a photoresist (SU-8) developed from a chrome-coated photomask shows much better feature shape and surface smoothness than that developed from a transparency photomask. The latter, however, is much less expensive (a few dollars vs. several hundred dollars). For feature sizes larger than 20 μm, the low-cost transparency photomask is an economical way for the initial design and testing of BIOMEMS devices.

4.3.9.2.5 Summary: Pros and Cons of RIM

The pros and cons of RIM are summarized in Table 4.14. Typical materials are optically clear reactive liquid resins such as epoxies, urethanes, silicone rubbers, cross-linkable acrylics, and hydrogels.

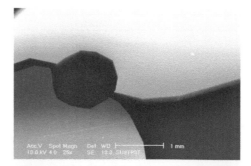
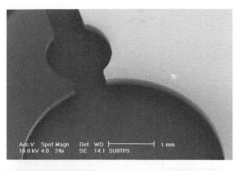

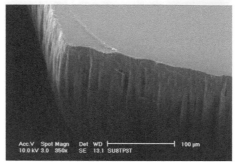
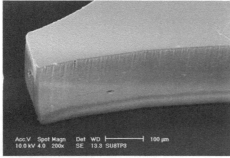

Developed from a chrome-coated
photomask

Developed from a transparency
photomask

FIGURE 4.38 SEM photos of 2 CD platforms made by PDMS casting. (Courtesy of Drs. Madou and Lee, OSU.)

TABLE 4.14 Pros and Cons of Liquid Resin Molding

Pros	Cons
Ease of mold filling (low η)	Long cycle time (minutes sometimes even hours)
Low stress on master (low η)	Polymerization shrinkage
High chemical and thermal resistance (because of the cross-linking)	Contamination (resin residue if the reaction is not 100%)
Replication extremely good for small, high-aspect-ratio, and 3-D features	Production cost medium to high

4.3.9.3 Injection Molding (IM)

4.3.9.3.1 Introduction

Injection molding (IM) is based on heating a thermoplastic material until it melts, thermostatting the mold parts, injecting the melt with a controlled injection pressure into the mold cavity, and cooling the manufactured goods. Injection molding is probably the most widely used technique in macroscopic production of polymer parts. Two types of injection molding, both good candidates for microfabrication, are detailed here: injection compression molding and thin-wall injection molding [Lee et al., 1999]. A typical injection compression molding setup is shown in Figure 4.39. The mold insert in IM, as in RIM, must have extremely smooth walls in order to demold the plastic structures. Such walls cannot be produced with a classical technique such as spark erosion. For some polymers, such as PMMA, the polymer pellets need to be dried for 4 to 6 hr at 70° before use in injection molding because PMMA absorbs water in air (up to 0.3%) leading to poorer wall quality.

4.3.9.3.2 Conventional Injection Molding

In conventional injection molding of micro parts the mold remains closed, and the walls are kept at a uniform temperature above the glass transition temperature (T_g), so that the injected molten plastic mass does not harden prematurely. To be able to fill very small mold features with aspect ratios higher than 10, the temperature of the tool holding the insertion mold should be higher than what is typically used in

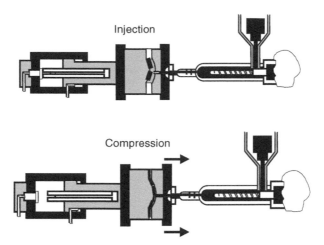

FIGURE 4.39 Typical injection/compression molding cycle sequence. (Courtesy of Drs. Madou and Lee, OSU.)

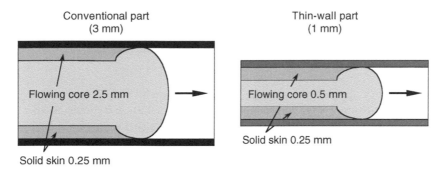

FIGURE 4.40 Schematics of solid skin buildup in conventional and thin-wall injection molding. (Courtesy of Drs. Madou and Lee, OSU.)

injection molding applications of larger parts. One should stay well below the melting temperature though (e.g., 170°C for PMMA with a melting temperature of 240°C), since the microstructure might show temperature-induced defects at a temperature closer to the melting temperature [Eicher et al., 1992].

One of the most important factors for replication fidelity in IM is the thickness of the skin, which forms on the mold surface during filling of the cavity [Lee et al., 1999] (see Figure 4.40). The injection pressure must penetrate this skin so as to replicate the shape of the master accurately. The faster the injection, the thinner the skin on the insert mold surfaces. The upper shapes of the master, forming the bottom of the parts, are usually perfectly formed. It is the bottom of the master and especially the inner corners that cause problems. Because the mold in conventional injection molding is kept closed during the entire mold-filling process, high clamping forces on the mold parts are required, and the method is used for relatively thick substrates with high replication demands and low cost. Parts made this way often have high internal stress.

Conventional thermoplastic injection molding is the technique used, for example, to make highly demanding CD formats such as CD-R involving features about 0.1 μm depth and minimum lateral dimensions of 0.6 μm; that is, an aspect ratio of 0.16. CDs are made by injection molding of polycarbonate in a cavity formed between a mirror block and a stamper (i.e., mold or master). Conventionally, nickel stampers are generated by electroforming from a glass master. The principal attributes required of a stamper substrate are toughness, thermal shock resistance, thermal conductivity, and hardness. For IM in particular, toughness, thermal shock resistance, and thermal conductivity are critical. Thermal shock resistance is the product of a material's modulus of rupture or tensile strength and its thermal conductivity, divided by the product of its coefficient of thermal expansion and its modulus of elasticity. In Table 4.15, we compare some materials for their merit as stamper substrates.

TABLE 4.15   Materials Selection for Stamper Substrates

Substrate	Knoop hardness (kg/mm^2)	Fracture toughness (MPa m$^{1/2}$)	Thermal shock resistance (W/m)	Thermal conductivity (W/m K)	Ion machinability (surface quality)
Nickel	100	~100	7138	80	Poor
Al$_2$O$_3$	2100	4	3225	30	Very good
SiC	2500	3	19,149	90	Excellent
Glassy carbon	500	2	135,517	120	Excellent
Corning 9647 (glass)	450	1.3	546	2.5	Excellent

Source: Reprinted with permission from Bifano, T.G., Fawcett, H.E., and Bierden, P.A. (1997) "Precision Manufacture of Optical Disc Master Stampers," *Precision Engineering* 20, pp. 53–62.

TABLE 4. 16   Typical Conditions for Injection Molding

Mold temperature	85°C
Polycarbonate temperature	330°C
Clamping force	60 tons
Injection time	1 s
Cooling time	2 s

Nickel compact disk stampers are normally 138 mm in diameter and 0.3 mm thick with an average roughness lower than 10 nm. For ceramic stampers, a thickness of 0.9 mm is chosen in order to avoid fracture in injection molding. Typical conditions for injection molding are summarized in Table 4.16.

Injection molding of parts with small features and low aspect ratios (e.g., CDs) has been widely applied. The main challenge in MEMS replication attempts is to extend this technique to the fabrication of components with the same small feature size but the much larger aspect ratios needed in many medical and biochemical applications. Ehrfeld and his coworkers at the Institut für Mikrotechnik in Mainz (IMM), Germany [Ehrfeld and Lehr, 1995; Dunke et al., 1998], used precision IM machines like the ones commonly used for the fabrication of CDs to mold MEMS components based on mold inserts made by LIGA. Another group at the Karlsruhe Institut für Materialforschung, Germany [Fahrenberg et al., 1995; Ruprecht et al., 1995; Goll et al., 1997; Piotter et al., 1997; Piotter et al., 1999], used CNC-machined and laser ablated metal molds in microinjection molding. Wimberger-Friedl in the Netherlands [Wimberger-Friedl, 1999] fabricated submicrometer optical grating elements by injection molding. He used 2 mm thick mold inserts that were made by e-beam lithography, RIE in fused quartz, and nickel electroplating. Kelly, at Louisiana State University [Kelly, 1999], also used LIGA-produced nickel molds and injection molding to make devices such as micro-heat-exchangers. In general, these studies show that the molds need to fill rapidly to prevent early freezing. A mold temperature above the "no-flow" temperature guarantees a complete filling of the mold. Shape deviation and damage of the fragile mold walls occur quite easily, possibly due to shrinkage of the polymer or defective filling and release. Since the mold cavity is filled at a mold temperature that exceeds the melting point T_g of the polymer, the mold needs to be cooled to obtain a sufficient green strength before part ejection. In addition, conventional venting of the cavity is not feasible due to the presence of micro features in the mold inserts. Therefore, prior evacuation of the mold cavity is needed. As a result, cycle time amounts to 5 min or longer, including the time needed for evacuation, heating, and cooling of the mold. Molding of micro features with large aspect ratios or the use of materials with higher viscosity leads to even longer cycle times.

In Figure 4.41, we show our own IM results on a CD-based fluidic platform (about 1 mm thick) [Wimberger-Friedl, 1999; Madou et al., 2000; Lee et al., 1999]. Only at injection speeds of 50 mm per second or higher does the quality of the replication become adequate. A birefringence technique was used to qualitatively examine the stress in the polymer parts as demonstrated in Figure 4.42. Samples showing

12.7 mm/sec   25.4 mm/sec

50.8 mm/sec   101.6 mm/sec

FIGURE 4.41   Microfluidic devices made at different injection rates (OQPC) [Madou et al., 2000; Lee et al., 1999].

| 12.7 mm/sec | 25.4 mm/sec | 50.8 mm/sec | 101.6 mm/sec |

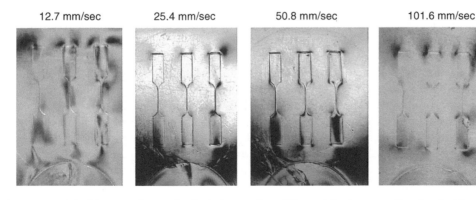

FIGURE 4.42   Birefringence of microfluidic devices made at different injection rates. (Reprinted with permission from Madou, M. et al. [2000] "A Novel Design on a CD Disc for 2-Point Calibration Measurement," *Proceedings: Solid-State Sensor and Actuator Workshop*, pp. 191–94, Hilton Head, SC, and Lee, L.J. et al. [1999] "Macro- and Micro-Molding of Polymeric Materials," *Novel Microfabrication Options for BioMEMS Conference Workshop*, The Knowledge Foundation, San Francisco.)

Reactive casting
(epoxy resin)

Hot embossing
(OQ PC)

Injection molding
(OQ PC)

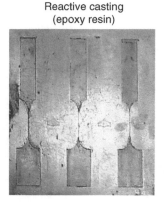
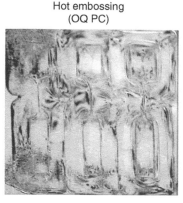
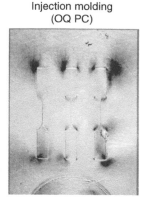

FIGURE 4.43 Comparison of birefringence patterns of microfluidic platforms made by different replication methods. (Reprinted with permission from Lee, L.J. et al. [1999] "Macro- and Micro-Molding of Polymeric Materials," *Novel Microfabrication Options for BioMEMS Conference Workshop*, The Knowledge Foundation, San Francisco.)

sharp color contrast have large molded-in stresses. At lower flow rates the residual stress is higher because during the flow the polymer has been stretched and solidified, meaning that there is a lot of flow-induced stress. Increasing the flow-rate decreases the amount of residual stress because the polymer can quickly fill in the mold and relax before solidification occurs. Nevertheless, there is some remaining stress that may cause warpage or less chemical resistance. By comparison, in casting there is essentially no flow; it is stress-free.

A comparison of sample birefringence from different replication methods is given in Figure 4.43 [Lee et al., 1999]. Conventional IM machines used for CD fabrication are expanded with special features for the molding of micro-sized parts or parts with high-aspect-ratio, micro-sized features. The IM machine periphery includes a vacuum unit for the evacuation of the mold cavity in the molding tool and a temperature control unit for the molding tool. It is desirable to use one temperature control cycle on each tool surface. The flow temperatures are adapted to the respective half of the tool or the molding and demolding process to achieve a relatively short cycle time and a homogeneous tool temperature. To keep the cycle time as short as possible, the thermal mass of the tool sections to be heated is minimized and thermally insulated from the adjacent assemblies. The guiding mechanisms of the tool halves and the ejector system should not be fitted too loosely because small transverse movements might damage the microstructures during demolding. The machines and tools to be evacuated have to comply with tolerances in the micrometer range [Dunke et al., 1998]. Demolding of the microstructures takes place by means of ejector pins, which are located at the sole, the runner system, or directly at swellings of the microstructure [Dunke et al., 1998]. Under laboratory conditions, the microstructures are removed manually. Small series production uses a handling robot that may be equipped with a feed picker. In addition, the machines used for small series and mass production are provided with a granulate drying, supply, and metering system such that fully automatic molding in personnel-free shifts is possible [Fahrenberg et al., 1995].

Commercial injection molding machines designed for micro-sized parts or parts with micro-sized features are available from Battenfeld (<www.battenfeld.de>), Boy Machines (<www.dr-boy.de/rightmachine.htm>), Engel (<www.engel.at>), Ferromatik Milacron (<www.ferromatik.com>), Murray (<www.murrayeng.com>), and Nissei (<www.nisseiamerica.com>). Figure 4.44 is a photograph of Battenfeld's Microsystem [Bley et al., 1991].

4.3.9.3.3 Injection Compression Molding

In injection compression, injection starts while the mold is closing and stops 1 mm before the mold is closed completely. This way, the part can cool with a "low" clamping force applied, and parts typically are less stressed than with conventional injection molding. This method is used for thin-wall structures with

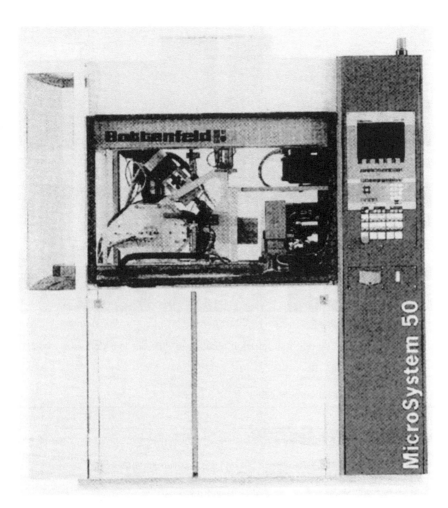

FIGURE 4.44   Battenfeld Microsystem 50 injection molding machine.

medium replication demands and low cost. DVD manufacture uses injection compression due to requirements for flatness and thickness.

4.3.9.3.4   Thin-Wall Injection Molding

Thin-wall injection molding is basically high-speed injection molding [Lee et al., 1999; Yu et al., 2000]. Speeds can be ten times faster than in conventional injection molding, reducing the thickness of the polymer skin on the mold. The polymer melt can be injected quickly into the mold to fill all the detail structure of the cavity before solidification occurs.

In our own thin-wall injection molding experiments for replication of BIO-MEMS devices such as the fluidic CD shown in Figure 4.41, we have used a Sumitomo 200-ton high-pressure and high-speed machine [Yu et al., 2000]. Another example of thin-walled structures made this way is shown in Figure 4.45. High injection speed is needed to completely fill the mold cavities for a thin-wall relay case (0.45 mm thick, flow length 50.4 mm).

4.3.9.3.5   Summary: Pros and Cons of IM

The pros and cons of injection molding are summarized in Table 4.17. Example materials in IM include polycarbonate (PC), poly(methylmethacrylate) (PMMA), polystyrene (PS), polyvinylchloride (PVC), polypropylene (PP), and polyethylene (PE).

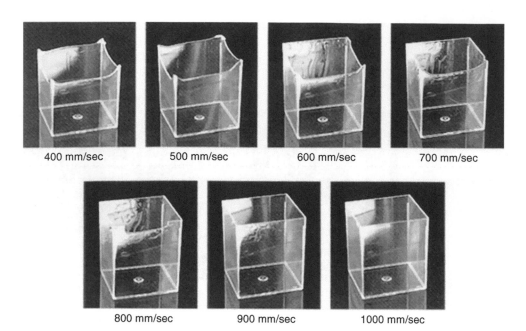

| 400 mm/sec | 500 mm/sec | 600 mm/sec | 700 mm/sec |

| 800 mm/sec | 900 mm/sec | 1000 mm/sec |

FIGURE 4.45 Relay case made by thin-wall injection molding at different injection rates (Nissei). (Courtesy of Dr. Jim Lee, OSU.)

TABLE 4.17   Pros and Cons of Injection Molding

Pros	Cons
Good for small structures with low aspect ratio, e.g., CD and DVD	Only low molecular weight polymers (may reduce mechanical and thermal strength)
Good for large, high aspect ratio, and 3-D features	More expensive equipment
Excellent dimensional control	Cyclic process only
Short cycle time (as low as 10 s)	High stress on master
High productivity	High residual stresses on molded parts
Closed mold process enables packing pressure application	

4.3.9.4   Compression Molding

4.3.9.4.1   Introduction

Compression molding (also relief imprinting or hot embossing) [Ramos et al., 1996; Becker and Dietz, 1998] provides several advantages over injection molding, such as relatively low costs for embossing tools, a simple process, and high replication accuracy for small features. The basic principle of compression molding (embossing, imprinting) is that a polymer substrate is first heated above its glass transition temperature T_g (or softening temperature). A mold (or master) is then pressed against the substrate, fully transferring the pattern onto it (embossing). After a certain time of contact between the mold and the substrate, the system is cooled to below T_g, followed by separation of the mold and the substrate (deembossing). Importantly, the hot embossing process can be achieved in either a cyclic or continuous process. A typical setup for each is shown in Figure 4.46 [Lee et al., 1999]. In a cyclic process (Figure 4.46a), a metal master is put in a hydraulic press, and a heated polymer sheet is stamped by applying the appropriate force, thus replicating the structure from the master to the polymer. This constitutes a low-cost method for making prototypes. For mass production, a continuous process is preferred (Figure 4.46b). Here, a polymer sheet stretches through a temperature chamber, and several masters mounted on a conveyor belt continuously produce parts. The process also may incorporate a lamination station to enclose certain features. This type of process is very much in keeping with our vision for continuous manufacture of MEMS devices.

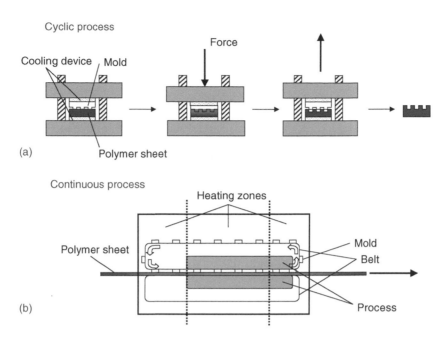

FIGURE 4.46 Schematic of cyclic and continuous embossing processes. (Reprinted with permission from Lee, L.J. et al. [1999] "Macro- and Micro-Molding of Polymeric Materials," *Novel Microfabrication Options for BioMEMS Conference Workshop*, The Knowledge Foundation, San Francisco.)

4.3.9.4.2 Hot Embossing

As seen in Figure 4.46a, hot embossing takes place in a machine frame similar to that of a press. The force frame delivers the embossing force — on the order of 5 to 20 tons. An upper boss holds the molding tool, and the lower boss holds the substrate. Processing parameters include thermal cycle, compression force, and compression speed. The temperature difference between embossing and deembossing determines the thermal cycle time, typically from 25 to 40°C. After hot embossing, in principle one could cool the whole device to room temperature before deembossing or, at the other extreme, one could deemboss just below or at the glass transition temperature. A compromise is needed; the quality of the replication may not be good if the master is removed when the polymer is still soft, while cooling all the way to room tempera-ture takes too long. A small temperature cycle leads to lower induced thermal stress. Such a smaller tem-perature cycle also reduces replication errors due to the different thermal expansion coefficients of tool and substrate. By actively heating and cooling the upper and lower bosses, cycle times of about five minutes can be obtained. A vacuum for hot embossing ensures a longer lifetime for the mold, absorbs any water released during the embossing process from the polymer material, and prevents bubbles from entrapped gases. A Jenoptik hot embossing machine is shown in Figure 4.47.

Replication of micro- and nano-size structures has been successfully achieved using hot embossing [Becker and Dietz, 1998; Kopp et al., 1997; Chou et al., 1996; Schift et al., 1999; Jaszewski et al., 1998; Casey et al., 1997; Gottschalch et al., 1999]. Adding an antiadhesive film to reduce the interaction between the mold and the replica during embossing has also been studied [Jaszewski et al., 1997; Jaszewski et al., 1999]. Instead of the conventional nickel molds, the possibility of using silicon molds has been demon-strated due to its excellent surface quality and easy mold release [Becker and Heim, 1999; Lin et al., 1996]. Also, the use of a plastic mold in the embossing process was recently illustrated [Casey et al., 1999]. In our own efforts [Lee et al., 1999; Juang et al., 2000], we used a cyclic process in a 30-ton Wabash press to fabricate a CD-based fluidic platform (see Figure 4.48). We have worked with optical quality PC (OQPC) (GE Plastics, Lexan $T_g = 135°C$) and regular PC (GE Plastics, Lexan, $T_g = 145°C$). The variables explored involve the compression force (from 2 to 25 tons) and the embossing and deembossing temperature. In

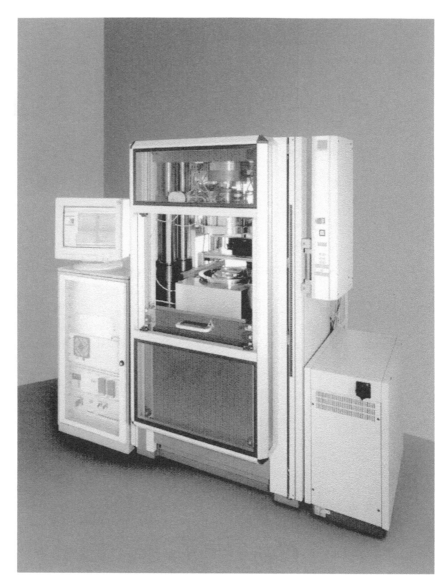

FIGURE 4.47   JENOPTIK HEX 03 hot embossing system manufactured by Jenoptik Mikrotechnik, GmbH.

Figure 4.48, we plot the depths of replicated channels in PC vs. the applied force. The smallest channels can be replicated at a relatively low force and temperature. For the deeper channels, a higher temperature and larger force must be used. This means a longer cycle time and a larger residual stress. That is why embossing is good for small structures and low aspect ratios but difficult for large structures and features. For 3-D structures (e.g., channels with different depth), the embossing pressure can be very high. Correspondingly, the molded-in stresses are also very large (see Figure 4.43).

Compression molding or relief printing is the method of choice for incorporating LIGA structures on a processed Si wafer. In a first step, the mold material is applied by polymerization onto the processed wafer with an electrically isolating layer and overlayed with a metal plating base (Figure 4.49a). At the glass transition temperature, the plastic is in a viscoelastic condition suitable for the impression molding step. Above this temperature (about 160°C for PMMA), the mold material is patterned in vacuum by impression with the molding tool. To avoid damaging the electronic circuits by contact with the molding tool, the mold material is not completely displaced during molding. A thin, electrically insulating residual layer is left

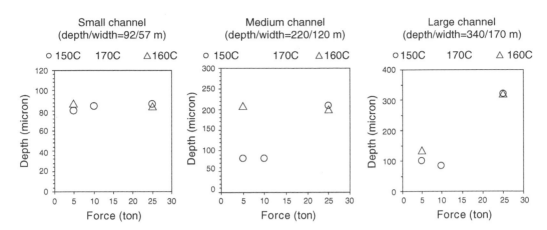

FIGURE 4.48 Depth of replicated channels (multiple-depth mold insert) versus embossing force and temperature (mold temperature: 150°C (T_g + 5°C), 160°C (T_g + 15°C), and 170°C (T_g + 25°C); compression time: 5 minutes at mold temperature and 2-hour cooling). (Reprinted with permission from Lee, L.J. et al. [1999] "Macro- and Micro-Molding of Polymeric Materials," *Novel Microfabrication Options for BioMEMS Conference Workshop*, The Knowledge Foundation, San Francisco.)

between the molding tool and the plating base on the processed wafer (Figure 4.49b). The residual layer is removed by reactive ion etching (RIE) in an oxygen plasma etch (Figure 4.49c), freeing the plating base for subsequent electrodeposition. The oxygen RIE process is as anisotropic as possible so that the side-walls of the structures do not deteriorate, and the amount of material removed from the top of the plastic microstructures is small compared with the total height. After electrodeposition on the plating base, the metal microstructures are laid bare by dissolution of the mold material (Figure 4.49d). Finally, using argon sputtering or chemical etching, the plating base between the metal microstructures is removed so that the metal parts do not short-circuit (Figure 4.49e). In the case of chemical etching, a very fast etch should be employed to avoid etching of the plating base from underneath the electrodeposited structures [Michel et al., 1993; Eicher et al., 1992].

At KfK, hot embossing is used mainly for small series production, whereas injection molding is applied to mass production. Both techniques are applied to amorphous (PMMA, PC, PSU) and semicrystalline thermoplastics (POM, PA, PVDF, PFA). Reaction injection molding is used for molding of thermoplastics (PMMA, PA), duroplastics, (PMMA), or elastomers (silicones) [Ruprecht et al., 1995].

4.3.9.4.3 *Summary: Pros and Cons of Compression Molding*
The pros and cons of compression molding are summarized in Table 4.18.

4.3.10 Demolding

4.3.10.1 Introduction

Even high-aspect-ratio metal structures can be molded and demolded with polymers quite easily, as long as a polymer with a small adhesive power and rubber–elastic properties, such as silicone rubber, is used. However, rubber-like plastics have low shape stability and would not be adequate. On the other hand, shape-preserving polymers after hardening require a mold with extremely smooth inner surfaces to prevent form-locking between the mold and the hardened polymer. Mold release has a geometric aspect (i.e., smooth and slightly inclined mold walls) and a chemical aspect (i.e., mold release agents). A mold release agent may be required for demolding. Using external mold release agents is difficult because of the small dimensions of the microstructures to be molded, as typically they are sprayed onto the mold. Consequently, internal mold release compounds need to be mixed with the polymer without significantly changing the polymer characteristics.

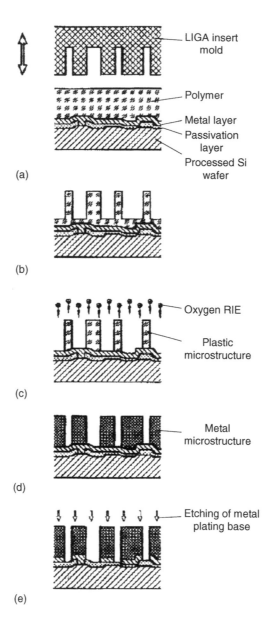

FIGURE 4.49 Fabrication of microstructures on a processed Si wafer. (a) Patterned isolation layer, conductive plating base, and in situ polymerization of PMMA. (b) Impression molding using a LIGA primary insert mold. (c) Removal of remaining plastic from the plating base by a highly directional oxygen etch. (d) Electrodeposition of metal shape. (e) Removal of plating base from between the metal structures by argon sputtering. (Reprinted with permission from Menz, W., and Bley, P. [1993] *Mikrosystemtechnik für Ingenieure,* VCH Publishers, Weinheim, Germany.)

4.3.10.2 Demolding in LIGA and LIGA-Like Applications

The demolding step in LIGA is carried out by means of a clamping unit at preset temperatures and rates (see Figure 4.35). Demolding is facilitated by slightly inclined mold walls and an internal mold release agent such as PAT 665 (Wurtz GmbH, Germany), normally employed with polyester resins [Hagmann and Ehrfeld, 1988]. The optimal yield with this agent and Plexit M60 PMMA occurs at 3 to 6 wt%. The yield drops very quickly below 3 wt% as adhesion between the molded piece and the mold insert becomes stronger. These adhesive forces are estimated by qualitatively determining the forces necessary to remove the plastic structures from the mold insert. An upper limit is 10 wt%, where the MMA does not polymerize

TABLE 4.18   Pros and Cons of Compression Molding (imprinting or hot embossing)

Pros	Cons
Low polymer flow	More difficult for structures with high aspect ratio (near T_g processing)
High molecular weight polymers (with better mechanical and thermal properties)	Less dimensional control (open mold process)
Simple process	Planar features only
Continuous or cyclic (see Figure 4.46)	High residual stresses on molded parts
Good for small structures	Difficult for large parts and multiple feature depth (too high a pressure and temperature are required)

anymore. Above 5 wt%, the Young's modulus and, hence, the mechanical stability decrease, and above 6 wt%, pores start forming in the microstructures. The internal mold release agent also has a marked influence on the optimal demolding temperature. The demolding yield decreases quickly above 60°C for a 4 wt% PAT internal mold release agent. With 6 wt%, one can obtain good yields only at 20°C [Hagmann et al., 1987; Hagmann and Ehrfeld, 1988]. The molding process initially led to a production cycle time of 120 min. For a commercial process much faster cycle times are needed, and may be attained, for instance, by optimizing the mold release agent. With that in mind, the KfK group started to work with a special salt of an organic acid, leading to a 100% yield at a release agent content of 0.2 wt% only and a temperature of 40°C (at 0.05 wt%, a 95% yield is still achieved). At 80°C, a 100% demolding yield was obtained and, significantly, a cycle time of 11.5 min was reached. During these experiments, the mold was filled at 80°C and heated to 110°C within 7.5 min. As the curing occurs at 110°C, the material needs to be cooled to 80°C for demolding. Moreover, the 0.2 wt% of the "magic release agent" did not impact the Young's modulus and the glass transition temperature of Plexit M60 [Hagmann and Ehrfeld, 1988].

A major concern of polymer-based MEMS, especially in microfluidic devices, is how to bond parts. This aspect of polymer micromachining is covered in Madou (2002, chapter 8). A good recent review paper on polymer micromolding, including a section on polymer bonding, is by Becker et al. (2000).

4.3.10.3   Sacrificial Layers

As in surface micromachining, sacrificial layers make it possible to fabricate partially attached and freed metal structures in the primary mold process [Burbaum et al., 1991]. The ability to implement these features leads to assembled micromechanisms with submicron dimension accuracies, opening many additional applications for LIGA, especially in the field of sensors and actuators. The sacrificial layer may be polyimide, silicon dioxide, polysilicon, or some other metal [Guckel et al., 1991a]. The sacrificial layer is patterned with photolithography and wet etching before polymerizing the resist layer over it. At KfK, a several micron thick titanium layer often acts as the sacrificial layer because it provides good adhesion of the polymer and it can be etched selectively against several other metals used in the process. If for exposure the X-ray mask is adjusted to the sacrificial layer, some parts of the microstructures will lie above the openings in the sacrificial layer, whereas other parts will be built up on it. These latter parts will be able to move after removal of the sacrificial layer [Mohr et al., 1990]. The fabrication of a movable LIGA structure is illustrated in Figure 4.50.

4.3.11   Alternative Materials in LIGA

Alternative X-ray resist materials and a variety of other metals for electroplating besides Ni were discussed above. In the following, we will discuss some alternative molding materials. Besides PMMA and POM, used in a commercially available form, semicrystalline polyvinylidenefluoride (PVDF), a piezoelectric material when stretched, has been used to make polymeric microstructures [Harmening et al., 1992]. The optimal molding temperature of PVDF is 180°C, and PVDF structures can be molded without using mold release agents. Fluorinated polymers such as PVDF also will enable higher-temperature

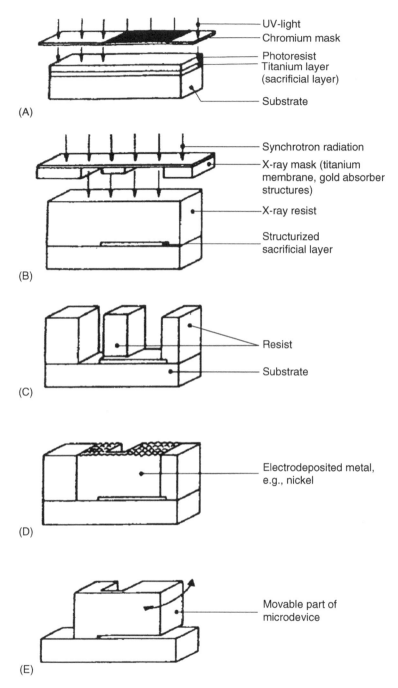

FIGURE 4.50 Movable LIGA microstructures. (Reprinted with permission from Mohr, J. et al. [1990] "Movable Microstructures Manufactured by the LIGA Process as Basic Elements for Microsystems," in *Microsystem Technologies 90*, Reichl, H., ed., pp. 529–37, Springer, Berlin.)

applications than PMMA. PC, well known for molding compact discs, is under development and should perform well [Rogner et al., 1992].

Other materials that can flow or that can be sintered, such as glasses and ceramics, can be incorporated in the LIGA process as well. In the case of ceramics, for example, plastic microstructures are used as disposable or lost molds. The molds are filled with a slurry in a slip-casting process. Before sintering at high temperatures, the plastic mold and the organic slurry components are removed completely in a burnout process at lower temperatures. First results have been obtained for zirconium oxide and aluminum oxide. An important application for ceramic microstructures is the fabrication of arrays of piezoceramic columns embedded in a plastic matrix. Because the performance of these actuators is linked to the height and width ratio of the individual ceramic columns as well as to the distance between the columns in the array, LIGA's tremendous capacity for tall, dense, and high-aspect-ratio features makes it an ideal fabrication tool [Preu et al., 1991; Lubitz, 1989].

An interesting method to make LIGA ceramic or glass structures in the future may be based on sol–gel technology. This technique involves relatively low temperatures and may enable LIGA products such as glass capillaries for gas chromatography (GC) or possibly even high-T_c superconductor actuators. Sol–gel techniques should work well with LIGA-type molds if they are filled under a vacuum to eliminate trapped gas bubbles. In the sol–gel technique, a solution is spun on a substrate that is then given an initial firing at around 200°C, driving off the solvent in the film. Subsequently, the substrate is given a high-temperature firing at 800 to 900°C to drive out the remaining solvents and crystallize the film. A major issue is the large shrinkage of sol–gel films during the initial firing. For example, the maximum thickness of high-T_c superconductor films currently achievable measures approximately 2 µm due to the high stress in the film caused during shrinkage. The latter is related to the ceramic yield or the amount of ceramic in the sol–gel compared with the amount of solvent. The sol–gel technique in general suits the production of thicker films well, as long as high-ceramic-yield sol–gels are used. As LIGA-style films tend to be thick, a reduction in the amount of solvents to allow a high ceramic yield after firing would be needed to make sol–gel compatible with LIGA. This would require some development. Also, lead–zirconium–titanate (PZT) devices could be made with sol–gel technology and LIGA, most likely by applying the metal-plated structure rather than the direct PMMA template for the creation of the device, because PZT must be processed in several elevated temperature steps. PZT sol–gel contains no particles that would prevent flow into small channels in a plated mold. The sol–gel contains high-molecular-weight polymer chains that hold the constituent metal salts, which are further processed into the final ceramic film. Thus, the sol–gel could be formed into the mold in a process much like reaction–injection molding, with the mold being filled with the chemistry under vacuum.

4.4 Examples

Example 1. Electromagnetic Micromotor

LIGA makes better electrostatic actuators than other Si micromachining techniques; the same is true for electromagnetic actuators. Most Si micromachined motors today produce negligible amounts of torque. In practical situations, what is needed are actuators in the millimeter range delivering torques of 10^{-6} to 10^{-7} Nm. Such motors can be fabricated with classical precision engineering or a combination of LIGA and precision engineering.

The performance with respect to torque and speed of a traditional miniature magnetic motor with a 1 mm diameter to 2 mm long permanent magnetic rotor was demonstrated in practice to be incomparably better than the surface micromachined motors discussed in chapter 3 [Goemans et al., 1993]. The small torque of the surface micromachined electrostatic motors is due largely to the fact that they are so flat. The magnetic device made with conventional three-dimensional metalworking techniques has an expected maximum shaft torque of 10^{-6} to 10^{-7} Nm. The torque depends on the Maxwell shearing stress on the rotor surface integrated over the area of the latter; the longer magnetic rotor easily outstrips the flat electrostatic one with the same rotor diameter [Goemans et al., 1993]. In both electric and magnetic microactuators, force production is proportional to changes in stored energy. The amount of force that may be generated per unit substrate area is proportional to the height of the actuator. Large-aspect-ratio structures are therefore preferred.

With LIGA techniques, large-aspect-ratio magnetic motors as shown in Figure 4.51 can be built [Ehrfeld et al., 1994]. Essentially, a soft magnetic rotor follows a rotating magnetic field produced by the currents in the stator coils. The motor manufacture is an example of how micromachining and precision

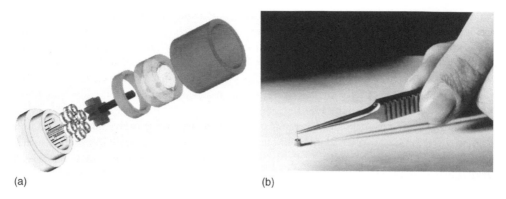

(a)   (b)

FIGURE 4.51   Large aspect ratio magnetic motor. (Courtesy of IMM, Germany.)

engineering can be complementary techniques for producing individual parts that have to be assembled afterward. Only components with the smallest features are produced by means of microfabrication techniques; other parts are produced by traditional precision mechanical methods.

The University of Wisconsin team [Guckel et al., 1991b] has also demonstrated a magnetic LIGA drive consisting of a plated nickel rotor that is free to rotate about a fixed shaft. Large plated poles (stators) come close to the rotor. The rotor is turned by spinning a magnet beneath the substrate. Further work by the same group involved enveloping coils, used to convert current to magnetic field in situ rather than externally [Guckel et al., 1992]. It is necessary to plate two different materials — a high-permeability material such as nickel and a good conductor such as gold — to accomplish this.

The motor also can be used as electrostatic motor. By applying voltages to the stator arms, the rotor (which is grounded through the post about which it rotates) is electrostatically attracted and can be made to turn by poling the voltages. Wallrabe et al. (Wallrabe et al., 1994) present design rules and tests of electrostatic LIGA micromotors. Minimum driving voltages needed were measured to be about 60 V; optimized design torques of the order of some mNm are expected. The cost of electromagnetic LIGA micromotors will often urge the investigation of LIGA-like technologies or hybrid approaches (LIGA combined with traditional machining) as more accessible and adequate machining alternatives.

Example 2. LIGA Spinneret Nozzles

Profiled capillaries (nozzles) in a spinneret plate for spinning synthetic fibers from a molten or dissolved polymer, as shown in Figure 4.52, normally are produced by micro-EDM. This process establishes a practical lower limit of 20 to 50 μm for the minimal characteristic dimension, and the method cannot satisfy the requirements to produce complexly shaped, high-aspect-ratio spinnerets at a low cost. Smaller, more precise, and high-aspect-ratio nozzles can be produced by LIGA [Maner et al., 1987].

Spinneret nozzles lend themselves to LIGA from the technical point of view. Compared with fabrication by micro-EDM, the minimum characteristic dimensions with LIGA can be reduced by an order of magnitude, and a high capillary length with excellent surface finish can be obtained easily. Moreover, the LIGA process makes all the nozzles in parallel, while micro-EDM is a serial technique. The market might focus on niche applications such as specialty, multilumen catheters. Also, medical use of atomizers for dispensing drugs is projected to increase dramatically in the coming years; this might create a demand for a wide variety of precise, inexpensive micronozzles.

Shew et al. used a conformal mask to fabricate a LIGA die with capillaries 2 mm deep and 70 μm wide for the mass production of electroplated spinnerets for the spinning of polyester fibers [Shew et al., 1999]. The conformal mask eliminates the need for the alignment steps accompanying the multiple exposure processes required in the deep X-ray lithography used by these authors. As illustrated in Figure 4.53a, a thin sputtered copper layer covers a thick PMMA layer on an Al substrate [a]. A resist layer (JSR 137N, Japan) is then patterned on the copper layer by UV lithography, followed by plating Au absorbers [b]. After stripping the resist and the copper layer [c], the conformal mask is ready for multiple X-ray exposure and

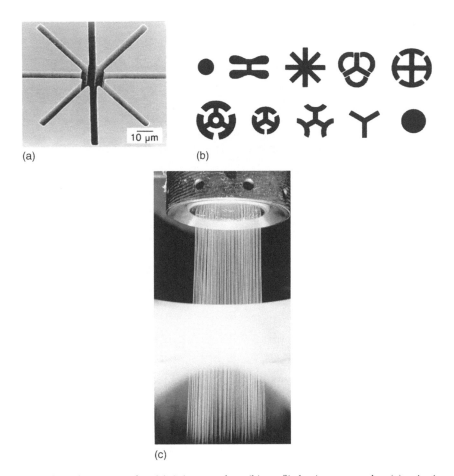

FIGURE 4.52 LIGA spinneret nozzles. (a) Spinneret plate; (b) profiled spinneret nozzles; (c) spinning synthetic fiber. (Courtesy of IMM, Germany.)

development steps [d] and die electroforming [e]. The die [f] is made by electroforming of a NiCo alloy on the Al substrate from a Ni/Co sulfamate bath at current densities between 1 and 5 A/dm². The addition of increasing amounts of cobalt sulfamate increases the hardness of the deposit to about 420 Vickers. At a concentration higher than 20 g/L Co sulfamate, the hardness saturates. With increasing hardness also comes increased residual stress, which must be carefully controlled. Polypropylene (PP) was used to duplicate the NiCo microstructures by injection molding through a stainless steel tool as shown in Figure 4.53b. After the LIGA die is removed, the result is a field of polymer microstructures standing up from the stainless steel tool. The stainless steel substrate is subsequently used for plating the spinneret structures. Polyester fibers typically melt at 260°C and are extruded at speeds of over 3000 m/min. During this fiber extrusion, the spinneret suffers high-temperature excursions of up to 280°C, and high-temperature wear resistance becomes a problem when using a NiCo. To improve the high-temperature hardness of Ni spinnerets, Shew et al. plated a Ni/SiC composite [Shew et al., 1999]. With the addition of SiC nanopowder to the sulfamate plating electrolyte, stable composite hardness of 500 Hv (Vickers hardness) at temperatures as high as 500°C was demonstrated. The authors suggest that these new types of spinnerets will enable the production of a new generation of fibers with ultrafine sizes and new functionalities at a low cost.

Example 3. LIGA Fiber-Chip Coupling

LIGA elements for coupling monomode fibers with integrated optical chips have been developed. These prealignment arrays may utilize fixed nickel guiding structures in combination with leaf springs to ensure a precise alignment of the optical fibers relative to the optical chip with an accuracy in the

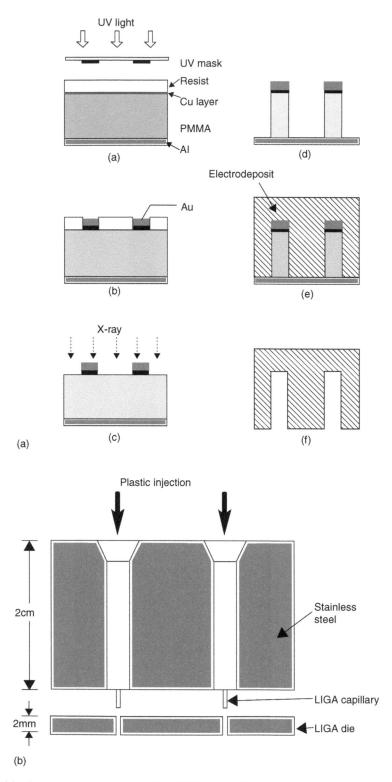

FIGURE 4.53 (a) Schematic diagram of the modified LIGA process (dimensions not to scale). (b) Duplication process of LIGA spinneret. (Based with permission on Cheng, Y. [2000] "Professional Activity Report in Micromachining, 1994–2000" Synchrotron Radiation Research Center, Hicnchu, Japan.)

Fiber-chip coupling

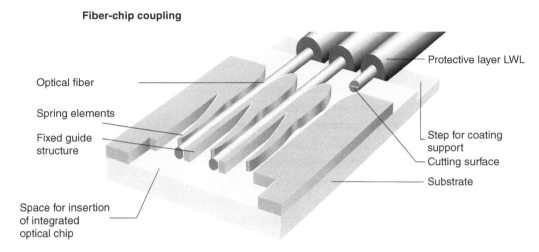

Optical fiber

Spring elements

Fixed guide
structure

Space for insertion
of integrated
optical chip

Protective layer LWL

Step for coating
support

Cutting surface

Substrate

FIGURE 4.54   Coupling for monomode fibers with nickel spring elements. (Courtesy of IMM, Germany.)

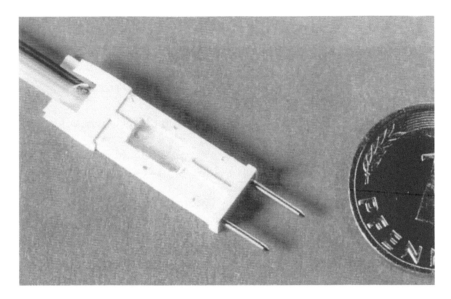

FIGURE 4.55   Pull-push LIGA connector. (Courtesy of IMM, Germany.)

submicron range as shown in Figure 4.54. The thermal expansion coefficient of the substrate material is matched to the optical chip material — for example, a glass substrate is used for coupling fibers to glass chips — and the use of spring elements simplifies the handling of the fibers [Rogner et al., 1991]. Alternatives involve the use of optical adhesives and silicon V-groove arrays (see chapter 3). Both technologies are more labor intensive and suffer from thermal expansion problems.

The use of micro-optical LIGA components is especially attractive for the coupling of multimode fibers; here, the capability of exact lithographic positioning of mechanical mounting supports is used advantageously to position the multimode fibers very precisely with respect to the micro-optical components without the need for any additional adjusting operations [Gottert et al., 1992a, 1992b].

A pull-push LIGA connector for single-mode fiber ribbons, shown in Figure 4.55, incorporates a set of precision microsprings coupling up to 12 fibers spaced at $250\,\mu m$. It is now on the market and is one of the LIGA products with mass-market appeal. To obtain good coupling efficiency, the fibers are positioned horizontally with a precision of $1\,\mu m$. LIGA is an excellent method to provide that precision. These connectors were developed at IMM [Editorial, 1994]. In line with the micromanufacturing philosophy presented in this book (i.e., to optimize the use of each micromachining technique for optimal cost/performance application), those parts of the mold insert that do not require such high accuracy are fabricated using other methods of precision machining such as EDM.

References

Andricacos, P.C., and Robertson, N. (1998) "Future Directions in Electroplated Materials for Thin-Film Recording Heads," *IBM J. Res. Dev.* **42**, pp. 671–80.

Becker, E.W., Ehrfeld, W., Hagmann, P., Maner, A., and Munchmeyer, D. (1986) "Fabrication of Microstructures with High Aspect Ratios and Great Structural Heights by Synchrotron Radiation Lithography, Galvanoforming, and Plastic Molding (LIGA Process)," *Microelectron. Eng.* **4**, pp. 35–56.

Becker, E.W., Ehrfeld, W., and Munchmeyer, D. (1984) "Untersuchungen zur Abbildungsgenauigkeit der Rontgentiefenlitographie mit Synchrotonstrahlung," KfK, Report No. 3732, Karlsruhe, Germany.

Becker, E.W., Ehrfeld, W., Munchmeyer, D., Betz, H., Heuberger, A., Pongratz, S., Glashauser, W., Michel, H.J., and Siemens, V.R. (1982) "Production of Separation Nozzle Systems for Uranium Enrichment by a Combination of X-ray Lithography and Galvanoplastics," *Naturwissenschaften* **69**, pp. 520–23.

Becker, H., and Dietz, W. (1998) "Microfluidic Device for μ-TAS Applications for Fabrication by Polymer Hot Embossing," *Proc. SPIE* **3515**, pp. 177–82.

Becker, H., and Gärtner, C. (2000) "Polymer Microfabrication Methods for Microfluidic Analytical Applications," *Electrophoresis* **21**, pp. 12–26.

Becker, H., and Heim, U. (1999) "Silicon as Tool Material for Polymer Hot Embossing," *Proceedings of 12th IEEE International Conference on MEMS*, pp. 228–31.

Bifano, T.G., Fawcett, H.E., and Bierden, P.A. (1997) "Precision Manufacture of Optical Disc Master Stampers," *Precis. Eng.* **20**, pp. 53–62.

Bley, P., Einfeld, D., Menz, W., and Schweickert, H. (1992) "A Dedicated Synchrotron Light Source for Micromechanics," in *EPAC92: Third European Particle Accelerator Conference*, March 24–28, 1992, pp. 1690–92, Berlin.

Bley, P., Gottert, J., Harmening, M., Himmelhaus, M., Menz, W., Mohr, J., Muller, C., and Wallrabe, U. (1991) "The LIGA Process for the Fabrication of Micromechanical and Microoptical Components," in *Microsystem Technologies '91*, Krahn, R., and Reichl, H., eds., pp. 302–14, VDE-Verlag, Berlin.

Bley, P., Menz, W., Bacher, W., Feit, K., Harmening, M., Hein, H., Mohr, J., Schomburg, W.K., and Stark, W. (1991) "Application of the LIGA Process in Fabrication of Three-Dimensional Mechanical Microstructures," in *4th International Symposium on MicroProcess Conference*, pp. 384–89, Kanazawa, Japan, 1991.

Burbaum, C., Mohr, J., Bley, P., and Ehrfeld, W. (1991) "Fabrication of Capacitive Acceleration Sensors by the LIGA Technique," *Sensor. Actuator. A.* **A25**, pp. 559–63.

Campbell, S.A. (1996) *The Science and Engineering of Microelectronic Fabrication*, Oxford University Press, New York.

Casey, B.G., Cumming, D.R.S., Khandaker, I.I., Curtis, A.G.S., and Wilkinson, C.D.W. (1999) "Nanoscale Embossing of Polymers Using a Thermoplastic Die," *Microelectron. Eng.* **46**, pp. 125–28.

Casey, B.G., Monaghan, W., and Wilkinson, C.D.W. (1997) "Embossing of Nanoscale Features and Environment," *Microelectron. Eng.* **35**, pp. 393–96.

Cheng, Y., Shew, B.-Y., Lin, C.-Y., Wei, D.-H., and Chyu, M.K. (1999) "Ultra-Deep LIGA Process," *J. Micromech. Microeng.* **9**, pp. 58–63.

Cheng, Y., Shew, B.-Y., Lin, C.-Y., Wei, D.-H., and Chyu, M.K. (2000) "Professional Activity Report in Micromachining, 1994–2000," Synchrotron Radiation Research Center, Hicnchu, Japan.

Chou, S.Y., Krauss, P.R., and Renstrom, P.J. (1996) "Imprint Lithography with 25-Nanometer Resolution," *Science* **272**, pp. 85–87.

Dukovic, J.O. (1993) "Feature-Scale Simulation of Resist Patterned Electrodeposition," *IBM J. Res. Dev.* **37**, pp. 125–40.

Dunke, K., Bauer, H.-D., Ehrfeld, W., Hobfeld, J., Weber, L., Horcher, G., and Muller, G. (1998) "Injection-Molded Fiber Ribbon Connectors for Parallel Optical Links Fabricated by the LIGA Technique," *J. Micromech. Microeng.* **8**, pp. 301–6.

Editorial (1994) "X-ray Scanner for Deep Lithography," Commercial Brochure, IMM, 1994.

Editorial (1994) "Fibre Ribbon Ferrule Insert Made by LIGA," Commercial Brochure, IMM, 1994.

Editorial (1997) "Back to the Future: Copper Comes of Age," *IBM Research* **35**.

Ehrfeld, W., Glashauer, W., Munchmeyer, D., and Schelb, W. (1986) "Mask Making for Synchrotron Radiation Lithography," *Microelectron. Eng.* **5**, pp. 463–70.

Ehrfeld, W., Glashauer, W., Munchmeyer, D., and Schelb, W. (1994) "LIGA at IMM," notes from handouts, Banff, Canada.

Ehrfeld, W., and Lehr, H. (1995) "Deep X-ray Lithography for the Production of Three-Dimensional Microstructures from Metals, Polymers, and Ceramics," *Radiat. Phys. Chem.* **45**, pp. 349–65.

Eicher, J., Peters, R.P., and Rogner, A. (1992) "VDI-Verlag," Report No. VDI-Bericht 960, Dusseldorf.

Fahrenberg, J., Bier, W., Mass, D., Menz, W., Ruprecht, R., and Schomburg, W.K. (1995) "A Microvalve System Fabricated by Thermoplasting Molding," *J. Micromech. Microeng.* **5**, pp. 169–71.

Fleischmann, M., Pons, S., Rolison, D.R., and Schmidt, P.P. (1987) "Ultramicroelectrodes," presented at *the Utah Conference on Ultramicroelectrodes*, Utah.

Friedrich, C. (1994) "Complementary Micromachining Processes," notes from handouts, Banff, Canada.

Galhotra, V., Marques, C., Desta, Y., Kelly, K., Despa, M., Pendse, A., and Collier, J. (1996) "Fabrication of LIGA Mold Inserts Using a Modified Procedure," in *SPIE: Micromachining and Microfabrication Process Technology II*, pp. 168–73, Austin.

Ghica, V., and Glashauser, W. (1982) "Verfahren für die Spannungsrissfreie Entwicklung von Bestrahlthen Polymethylmethacrylate-Schichten," in Deutsche Offen-legungsschrift, Germany, Patent 3039110.

Goedtkindt, P., Salome, J.M., Artru, X., Dhez, P., Maene, N., Poortmans, F., and Wartski, L. (1991) "X-ray Lithography with a Transition Radiation Source," *Microelectron. Eng.* **13**, pp. 327–30.

Goemans, P.A.F.M., Kamerbeeek, E.M.H., and Klijn, P.L.A.J. (1993) "Measurement of the Pull-Out Torque of Synchronous Micromotors with PM Rotor," in *6th International Conference on Electrical Machines and Drives*, pp. 4–8, Oxford.

Goll, C., Bacher, W., Bustgens, B., Maas, D., Ruprecht, R., and Schomburg, W.K. (1997) "An Electrostatically Actuated Polymer Microvalve Equipped with a Movable Membrane Electrode," *J. Micromech. Microeng.* **7**, September 8–10, pp. 224–26.

Gottert, J., Mohr, J., and Muller, C. (1991) "Mikrooptische Komponenten aus PMMA, Hergestellt Durch Roent-gentiefenlithographie Werkstoffe der Mikrotechnik-Bais für neue Producte," *VDI Berichte*, pp. 249–63.

Gottert, J., Mohr, J., and Muller, C. (1992a) "Coupling Elements for Multimode Fibers by the LIGA Process," in *Proceedings: Micro System Technologies '92*, pp. 297–307, Berlin.

Gottert, J., Mohr, J., and Muller, C. (1992b) "Examples and Potential Applications of LIGA Components in Microoptics," in *Integrierte Optik und Mikrooptik mit Polymeren*, Mainz, Germany.

Gottschalch, F., Hoffman, T., Torres, C.M.S., Schulz, H., and Scheer, H.-C. (1999) "Polymer Issues in Nanoimprinting Technique," *Solid State Electron.* **43**, pp. 1079–83.

Guckel, H., Christenson, T.R., Earles, T., Klein, J., Zook, J.D., Ohnstein, T., and Karnowski, M. (1994a) "Laterally Driven Electromagnetic Actuators," *Technical Digest: 1994 Solid State Sensor and Actuator Workshop*, Hilton Head Island, SC, June 13–16, pp. 49–52.

Guckel, H., Christenson, T.R., Earles, T., Klein, J., Zook, J.D., Ohnstein, T., and Karnowski, M. (1994b) "Deep Lithography," notes from handouts, Banff, Canada.

Guckel, H., Christenson, T.R., and Skrobis, K. (1992) "Metal Micromechanisms via Deep X-ray Lithography, Electroplating, and Assembly," *J. Micromech. Microeng.* **2**, pp. 225–28.

Guckel, H., Skrobis, K.J., Christenson, T.R., Klein, J., Han, S., Choi, B., and Lovell, E.G. (1991a) "Fabrication of Assembled Micromechanical Components via Deep X-ray Lithography," in *Proceedings: IEEE Micro Electro Mechanical Systems (MEMS '91)*, January 30th–February 2nd, pp. 74–79, Nara, Japan.

Guckel, H., Skrobis, K.J., Christenson, T.R., Klein, J., Han, S., Choi, B., Lovell, E.G., and Chapma, T.W. (1991b) "Fabrication and Testing of the Planar Magnetic Micromotor," *J. Micromech. Microeng.* **4**, pp. 40–45.

Guckel, H., Uglow, J., Lin, M., Denton, D., Tobin, J., Euch, K., and Juda, M. (1988) "Plasma Polymerization of Methyl Methacrylate: A Photoresist for 3D Applications," *Technical Digest: 1988 Solid State Sensor and Actuator Workshop*, Hilton Head Island, SC, pp. 9–12.

Hagmann, P., and Ehrfeld, W. (1988) "Fabrication of Microstructures of Extreme Structural Heights by Reaction Injection Molding," *J. Polym. Process. Soc.* **4**, pp. 188–95.

Hagmann, P., Ehrfeld, W., and Vollmer, H. (1987) "Fabrication of Microstructures with Extreme Structural Heights by Reaction Injection Molding," in *First Meeting of the European Polymer Federation, European Symposium on Polymeric Materials*, September, pp. 241–51, Lyon, France.

Hanemann, T., Ruprecht, R., and Haubelt, J.H. (1998) "Photomolding in Microsystem Technology," *Polym. Preprints* **39**, pp. 657–58.

Harmening, M., Bacher, W., Bley, P., El-Kholi, A., Kalb, H., Kowanz, B., Menz, W., Michel, A., and Mohr, J. (1992) "Molding of Three-Dimensional Microstructures by the LIGA Process," in *Proceedings: IEEE Micro Electro Mechanical Systems (MEMS '92)*, Travemunde, Germany, February 4–7, pp. 202–7.

Harmening, M., Bacher, W., and Menz, W. (1991) "Molding Plateable Micropatterns of Electrically Insulating and Electrically Conducting Poly(Methyl Methacrylate)s by the LIGA Technique," in *Makromol. Chem. Macromol. Symp.*, pp. 277–84.

Harsch, S., Ehrfeld, W., and Maner, A. (1988) "Untersuchungen zur Herstellung von Mikrostructuren grosser Struktur-hohe durch Galvanoformung in Nickel–sulfamatelek–trolyten," KfK, Report No. 4455, Karlsruhe, Germany.

Harsch, S., Munchmeyer, D., and Reinecke, H. (1991) "A New Process for Electroforming Movable Microdevices," in *Proceedings: 78th AESF Annual Technical Conference (SUR/FIN '91)*, Toronto, June.

Hein, H., Bley, P., Gottert, J., and Klein, U. (1992) "Elektro-nenstrahllithographie zur Herstellung von Rontgen-masken für das LIGA-Verfahren," *Feinw. Tech. Messtech.* **100**, pp. 387–89.

Henck, R. (1984) "Detecteurs au Silicium Pour Electrons et Rayons X, Principes de Fonctionnement, Fabrication et Performance," *J. Microsc. Spectrosc. Electron.* **9**, pp. 131–33.

Hill, R. (1991) "Symposium on X-ray Lithography in Japan," *National Academy of Sciences.*

IMM (1995) "The LIGA Technique," Commercial Brochure, IMM.

Jaszewski, R.W., Schift, H., Gobrecht, J., and Smith, P. (1998) "Hot Embossing in Polymers as a Direct Way to Pattern Resist," *Microelectron. Eng.* **41/42**, pp. 575–78.

Jaszewski, R.W., Schift, H., Groning, P., and Margaritondo, G. (1997) "Properties of Thin Anti-Adhesive Films Used for the Replication of Microstructures in Polymers," *Microelectron. Eng.* **35**, pp. 381–84.

Jaszewski, R.W., Schift, H., Schnyder, B., Schneuwly, A., and Groning, P. (1999) "The Deposition of Anti-Adhesive Ultra-Thin Teflon-Like Films and Their Interaction with Polymers during Hot Embossing," *Appl. Surf. Sci.* **143**, pp. 301–8.

Juang, Y.J., Lee, L.J., and Koelling, K.W. (2000) "Viscoplastic Analysis of Hot Embossing in Microfabrication," in *Proceedings: SPE ANTEC*, pp. 1032–36, Orlando.

Kelly, K.W. (1999) "Molded Microdevice Manufacture for Medical Applications via the LIGA Process," in *Novel Microfabrication Options for BioMEMS Conference (Proceedings)*, The Knowledge Foundation, San Francisco.

Kissler, R.K. (1994) "μGALV 750," sales brochure, R. Kissler, GmbH, Daimlerstrasse 8, Speyer 67346.

Kopp, M.U., Crabtree, H.J., and Manz, A. (1997) "Developments in Technology and Applications in Microsystems," *Curr. Opin. Chem. Biol.* **1**, 410–19.

Lawes, R.A. (1989) "Sub-Micron Lithography Techniques," *Appl. Surf. Sci.* **36**, pp. 485–99.

Lee, L.J., Shih, C.-H., Juang, Y.-J., Garcia, J., Madou, M.J., and Koelling, K.W. (1999) "Macro- and Micro-Molding of Polymeric Materials," in *Novel Microfabrication Options for BioMEMS Conference (Workshop)*, The Knowledge Foundation, San Francisco.

Lehr, H., and Schmidt, M. (1995) "The LIGA Technique," commercial brochure, IMM Institut fur Mikrotechnnik, GmbH, Mainz-Hechtsheim.

Leyendecker, K., Bacher, W., Stark, W., and Thommes, A. (1994) "New Microelectrodes for the Investigation of the Electroforming of LIGA Microstructures," *Electrochim. Acta* **39**, pp. 1139–43.

Lin, L.-W., Chiu, C.-J., Bacher, W., and Heckele, M. (1996) "Microfabrication Using Silicon Mold Inserts and Hot Embossing," in *Seventh International Symposium on Micro Machine and Human Science*, October 2 to 4, pp. 67–71, Nagoya, Japan.

Lingnau, J., Dammel, R., and Theis, J. (1989) "Recent Trends in X-ray Resists: Part I," *Solid State Technol.* **32**, pp. 105–12.

Lubitz, K. (1989) "Mikrostrukturierung von Piezokeramik," *VDI-Tagungsbericht* **796**, pp. 35–49.

Madou, M.J. (2002) *Fundamentals of Microfabrication*, 2nd ed., CRC Press, Boca Raton.

Madou, M.J., Lu, Y., Lai, S., Juang, Y.-J., Lee, L.J., and Daunert, S. (2000) "A Novel Design on a CD Disc for 2-Point Calibration Measurement," in *Proceedings: Solid-State Sensor and Actuator Workshop*, June 4–8, pp. 191–94, Hilton Head Island, SC.

Madou, M.J., and Morrison, S.R. (1989) *Chemical Sensing with Solid State Devices*, Academic Press, New York.

Madou, M.J., and Morrison, S.R. (1991) "High-Field Operation of Submicrometer Devices at Atmospheric Pressure," in *6th International Conference on Solid-State Sensors and Actuators (Transducers '91)*, June 24–27, pp. 145–49, San Francisco.

Madou, M.J., and Murphy, M. (1995) A Method for PMMA Stress Evaluation, unpublished results.

Malek, C.K., Vladimirsky, Y., Vladimirsky, O., Scott, J., Craft, B., and Saile, V. (1996) "X-ray Microfabrication Activities at the Center for Advanced Microstructures and Devices (CAMD)," *Rev. Sci. Instrum.* **67**, pp. 1–6, 1996.

Mallory, G.O., and Hadju, J.B. (1990) *Electroless Plating: Fundamentals and Applications*, American Electroplaters and Surface Finishers Society (AESF), Orlando.

Maner, A., Harsch, S., and Ehrfeldg, W. (1987) "Mass Production of Microstructures with Extreme Aspect Ratios by Electroforming," in *Proceedings: 74th AESF Annual Technical Conference (SUR/FIN '87)*, July, pp. 60–65, Chicago.

Maner, A., Harsch, S., and Ehrfeldg, W. (1988) "Mass Production of Microdevices with Extreme Aspect Ratios by Electroforming," *Plat. Surf. Finish.* **75**, pp. 60–65.

Masuko, N., Osaka, T., and Ito, Y., eds. (1996) *Electrochemical Technology: Innovation and New Developments*, copublished by Kodansha and Gordon and Breach Science Publishers, Japan and the Netherlands.

Menz, W., and Bley, P. (1993) *Mikrosystemtechnik für Ingenieure*, VCH Publishers, Weinheim, Germany.

Michel, A., Ruprecht, R., and Bacher, W. (1993) "Abformung von Mikrostrukturen auf Prozessierten Wafern," KfK, Report No. 5171, Karlsruhe, Germany.

Mohr, J., Burbaum, C., Bley, P., Menz, W., and Wallrabe, U. (1990) "Movable Microstructures Manufactured by the LIGA Process as Basic Elements for Microsystems," in *Microsystem Technologies 90*, pp. 529–37, Reichl, H., ed., Springer, Berlin.

Mohr, J., Ehrfeld, W., and Munchmeyer, D. (1988a) Report No. 4414.

Mohr, J., Ehrfeld, W., and Munchmeyer, D. (1988b) "Requirements on Resist Layers in Deep-Etch Synchrotron Radiation Lithography," *J. Vac. Sci. Technol.* **B6**, pp. 2264–67.

Mohr, J., Ehrfeld, W., Munchmeyer, D., and Stutz, A. (1989) "Resist Technology for Deep-Etch Synchrotron Radiation Lithography," *Makromol. Chem. Macromol. Symp.* **24**, pp. 231–51.

Mohr, J., and Strohmann, M. (1992) "Examination of Long-Term Stability of Metallic LIGA Microstructures by Electromagnetic Activation," *J. Micromech. Microeng.* **2**, pp. 193–95.

Momma, C., Nolte, S., Chichkov, N., Alvensleben, B.V., and Tunermann, F.A. (1997) "Precise Laser Ablation with Ultrashort Pulses," *Appl. Surf. Sci.* **109/110**, pp. 15–19.

Moreau, W.M. (1988) *Semiconductor Lithography*, Plenum Press, New York.

Munchmeyer, D. (1984) Ph.D. thesis, University of Karhsruhe, Germany.

Muray, J.J., and Brodie, I. (1991) "Study on Synchrotron Orbital Radiation (SOR) Technologies and Applications," Report No. 2019, SRI International, Menlo Park, CA.

Ohno, I. (1988) "Methods for Determination of Electroless Deposition Rate," in *Proceedings: Symposium on Electroless Deposition of Metals and Alloys*, pp. 129–41, Honolulu.

Piotter, V., Benzler, T., Hanemann, T., Wollmer, H., Ruprecht, R., and Haubelt, J. (1999) "Innovative Molding Technologies for the Fabrication of Components for Microsystems," *Proc. SPIE* **3680**, pp. 456–63.

Piotter, V., Hanemann, T., Ruprecht, R., Thies, A., and Haubelt, J. (1997) "New Development of Process Technologies for Microfabrication," *Proc. SPIE* **3223**, pp. 91–99.

Preu, G., Wolff, A., Cramer, D., and Bast, U. (1991) "Microstructuring of Piezoelectric Ceramic," in *Proceedings: Second European Ceramic Society Conference (2nd ECerS '91)*, pp. 2005–9, Augsburg, Germany.

van der Putten, A.M.T., and de Bakker, J.W.G. (1993a) "Geometrical Effects in the Electroless Metallization of Fine Metal Patterns," *J. Electrochem. Soc.* **140**, pp. 2221–28.

van der Putten, A.M.T., and de Bakker, J.W.G. (1993b) "Anisotropic Deposition of Electroless Nickel–Bevel Plating," *J. Electrochem. Soc.* **140**, pp. 2229–35.

Ramos, B.L., Choquette, S.J., and Nicholas, F.F. (1996) "Embossable Grating Couplers for Planar Waveguide Optical Sensors," *Anal. Chem.* **68**, pp. 1245–49.

Roberts, M.A., Rossier, J.S., Bercier, P., and Girault, H. (1997) "UV Laser Machined Polymer Substrates for the Development of Microdiagnostic Systems," *Anal. Chem.* **69**, pp. 2035–42.

Rogers, J., Marques, C., and Kelly, K. (1996) "Cyanoacrylate Bonding of Thick Resists for LIGA," in *SPIE: Microlithography and Metrology in Micromachining II*, pp. 177–82, Austin.

Rogner, A., Ehrfeld, W., Munchmeyer, D., Bley, P., Burbaum, C., and Mohr, J. (1991) "LIGA-Based Flexible Microstructures for Fiber-Chip Coupling," *J. Micromech. Microeng.* **1**, pp. 167–70.

Rogner, A., Eichner, J., Munchmeyer, D., Peters, R.-P., and Mohr, J. (1992) "The LIGA Technique: What Are the New Opportunities?" *J. Micromech. Microeng.* **2**, pp. 133–40.

Romankiw, L.T., Croll, I.M., and Hatzakis, M. (1970) "Batch-Fabricated Thin-Film Magnetic Recording Heads," *IEEE Trans. Magn.* **MAG-6**, pp. 597–601.

Romankiw, L.T., Croll, I.M., and Hatzakis, M. (1976) "Pattern Generation in Metal Films Using Wet Chemical Techniques," in *Etching for Pattern Definition*, pp. 137–39, Washington, DC.

Romankiw, L.T., Croll, I.M., and Hatzakis, M. (1984) "Electrochemical Technology in Electronics Today and Its Future: A Review," *Oberflache-Surface* **25**, pp. 238–47.

Romankiw, L.T., Croll, I.M., and Hatzakis, M. (1989) "Thin Film Inductive Heads: From One to Thirty One Turns," in *Proceedings: Symposium on Magnetic Materials, Processes, and Devices*, pp. 39–53, Hollywood, FL.

Romankiw, L.T., Croll, I.M., and Hatzakis, M. (1993) "Think Small: One Day It May Be Worth a Billion," *Interface*, Summer, pp. 17–57.

Romankiw, L.T., Krongelb, S., Castellani, E.E., Powers, J., Pfeiffer, A., and Stoeber, B. (1973) "Additive Electroplating Technique for Fabrication of Magnetic Devices," in *International Conference on Magnetics-ICM-73*, pp. 104–11.

Romankiw, L.T., and Palumbo, T.A. (1987) "Electrodeposition in the Electronic Industry," in *Proceedings: Symposium on Electrodeposition Technology, Theory and Practice*, pp. 13–41, San Diego.

Ruprecht, R., Bacher, W., Haubelt, T., and Piotter, V. (1995) "Injection Molding of LIGA and LIGA-Similar Microstructures Using Filled and Unfilled Thermoplastics," *Proc. SPIE* **2639**, pp. 146–57.

Schift, H., Jaszewski, R.W., David, C., and Gobrecht, J. (1999) "Nanostructuring of Polymers and Fabrication of Integrated Electrodes by Hot Embossing Lithography," *Microelectron. Eng.* **46**, pp. 121–24.

Schomburg, W.K., Baving, H.J., and Bley, P. (1991) "Ti- and Be-X-ray Masks with Alignment Windows for the LIGA Process," *Microelectron. Eng.* **13**, pp. 323–26.

Shew, B.-Y., Cheng, Y., Lin, C.-H., Ma, W.-P., Huang, G.-J., Kuo, C.-L., Tseng, S.-C., Lee, D.-S., and Chang, G.-L. (1999) "Manufacturing Process for LIGA Spinnerets," *Sensor. Mater.* 11, pp. 329–37.

Shih, W.-P., Cheng, Y., Lin, C.-Y., and Hwang, G.-J. (1998) "Low-Cost X-ray Conformal Mask Using X-ray Dry FILM Resist," *Microelectron. Eng.* 40, pp. 43–50.

Siddons, D.P., and Johnson, E.D. (1994) "Precision Machining Using Hard X-rays," *Synchrotron Radiat. News* 7, pp. 16–18.

Spiller, E., Feder, R., Topalian, J., Castellani, E., Romankiw, L., and Heritage, M. (1976) "X-ray Lithography for Bubble Devices," *Solid State Technol.*, April pp. 62–68.

Tabata, O., You, H., Shiraishi, H., Nakanishi, H., Nishimoto, T., Yamamoto, K., and Baba, Y. (2000) "μ-CE Chip Fabricated by Moving Mask Deep R-Ray Lithography Technology," in *Micro Total Analysis Systems 2000: Proceedings of the μTAS 2000 Symposium*, pp. 143–46, Kluwer Academic, Enschede, Netherlands.

Thomes, A., Stark, W., Goller, H., and Liebscher, H. (1992) "Erste Ergebnisse zur Galvanoformung von LIGA Mikrostrukturen aus Eisen-Nickel Legierungen," in *Symp. Mikroelektrochemie: Friedrichsroda.*

Urisu, T., and Kyuragi, H. (1987) "Synchrotron Radiation-Excited Chemical-Vapor Deposition and Etching," *J. Vac. Sci. Technol.* B5, pp. 1436–40.

Vladimirsky, O. (1995) "Spin Coating PMMA," private communication in May, Louisiana State University.

Vladimirsky, Y., Vladimirsky, O., Saile, V., Morris, K., and Klopf, J.M. (1995) "Transfer Mask for High Aspect Ratio Micro-Lithography," in *Microlithography '95 (Proceedings of the SPIE)*, 2437, pp. 391–96, Santa Clara, CA.

Waldo, W.G., and Yanof, A.W. (1991) "0.25 Micron Imaging by SOR X-ray Lithography," *Solid State Technol.* 34, pp. 29–31.

Wallrabe, U., Bley, P., Krevet, B., Menz, W., and Mohr, J. (1994) "Design Rules and Test of Electrostatic Micromotors Made by the LIGA Process," *J. Micromech. Microeng.* 4, pp. 40–45.

Warrington, R.O. (1999) "An Overview of Micromechanical Machining Processes for BioMEMS," in *Novel Microfabrication Options for BioMEMS Conference (Proceedings)*, The Knowledge Foundation, San Francisco.

Weber, L., Ehrfeld, W., Freimuth, H., Lacher, M., Lehr, H., and Pech, B. (1996) "Micro-Molding: A Powerful Tool for the Large Scale Production of Precise Microstructures," in *SPIE: Micromachining and Microfabrication Process Technology II*, Austin, pp. 156–67.

Wimberger-Friedl, R. (1999) "Injection Molding of Sub-μm Grating Optical Elements," in *Proc. SPE ANTEC*, pp. 476–80.

Wollersheim, O., Zumaque, H., Hormes, J., Hoessel, J.L., Haussling, L., and Hoffman, G. (1994) "Radiation Chemistry of Poly(lactides) as New Polymer Resists for the LIGA Process," *J. Micromech. Microeng.* 4, pp. 84–93.

Yu, L., Juang, Y.-J., Koelling, K.W., and Lee, L.J. (2000) "Thin Wall Injection Molding of Thermoplastic Microstructures," in *Proceedings: SPE ANTEC*, Orlando, pp. 468–73.

5

X-Ray–Based Fabrication

Todd Christenson
HT MicroAnalytical, Incorporated

5.1 Introduction

Originally conceived for the fabrication of smaller microelectronic features, X-ray lithography also has attributes of great utility in micromechanical fabrication. In contrast to the many micromachining processes that have been developed from microelectronic processing, however, X-ray based approaches may be performed largely without a tightly controlled clean-room environment. The mode of X-ray based microfabrication most commonly used places this type of processing in the additive category where a sacrificial mold is used to define the desired structural material. As a result, this technique lends itself to a very rich and ever-expanding material base including a variety of plastics, metals, and glasses, as well as ceramics and composites. The idea of using X-rays to define molds extends from the 1970s when its precedent involved defining high-density coils for magnetic recording read/write heads and high-density magnetic bubble memory overlays. This was where the use of X-rays for Very Large Scale Integration (VLSI) lithography was initially investigated [Romankiw et al., 1970, 1995; Spiller et al., 1976; Spears and Smith, 1972]. The distinction from VLSI X-ray lithography is that the mold or photoresist thickness for micromachining interests is generally much greater than 50 microns and may be well over 1 millimeter. X-ray processing at these thicknesses has prompted the nomenclature *deep X-ray lithography*, or *DXRL*, based microfabrication.

The primary utility of DXRL processing extends from its ability to precisely and accurately define a mold. Consequent component definition via mold filling thus is determined directly by mold acuity and stability. Exceptional definition in this regard is possible with highly collimated X-rays that may be

TABLE 5.1 Synchrotron Radiation Facilities in the United States with Active DXRL Devoted Beamlines

Storage ring	Ring parameters (energy, critical wavelength, current)	Site
Aladdin	0.8, 1 GeV 22.7 Å, 11.6 Å 260 mA, 190 mA	Synchrotron Radiation Center (SRC), Stoughton, WI
CAMD (Center for Advanced Microstructure Devices)	1.3, 1.5 GeV 7.4 Å, 4.8 Å 400, 200 mA	Louisiana State University (LSU), Baton Rouge, LA
ALS (Advanced Light Source)	1.5, 1.9 GeV 8.2 Å, 4.1Å 400 mA	Lawrence Berkeley Laboratory (LBL), Berkeley, CA
NSLS (National Synchrotron Light Source)	X-ray ring: 2.584, 2.8 GeV 2.2 Å, 1.7 Å 300, 250 mA VUV Ring: 0.808 GeV 20 Å 1000 mA	Brookhaven National Laboratory (BNL), Upton, NY
SPEAR3 (Stanford Positron Electron Accelerating Ring)	3.0 GeV 1.4 Å 500 mA	Standord Synchrotron Radiation Laboratory (SSRL), Stanford, CA
APS (Advanced Photon Source)	7.0 GeV 0.64 Å 100 mA	Argonne National Laboratory (ANL), Argonne, IL

obtained via *synchrotron radiation* from a *storage ring*. Such X-rays, in addition to possessing atomic rather than optical absorption character and thereby eliminating diffraction and standing wave effects, possess collimation on the order of 0.1 mrad. Combined with a developer that has high selectivity between unexposed and exposed photoresist, this exposure capability yields mold sidewall definition with less than 0.1 micron run-out for a thickness of several hundred microns. This radical form of lithography has resulted in a new type of foundry service to provide synchrotron radiation access. In the United States, for example, these services may be obtained from the facilities listed in Table 5.1. Many more facilities throughout the world are actively engaged in this activity.

Two distinct X-ray based microfabrication philosophies exist. That relatively thick ($\gg$ 100 microns) X-ray lithography was possible and could subsequently be applied to precision molding was first realized by Ehrfeld for application to separation nozzle fabrication for uranium isotope enrichment [Becker et al., 1982, 1986; Ehrfeld et al., 1987, 1988a, 1988b]. The process was given the German acronym "*LIGA*" representing the three basic processing steps of deep X-ray *LI*thography; mold filling by *G*alvanoformung, or electroforming; and injection molding replication, or *A*bformung. The approach defined by the LIGA process aims to become mostly independent from a synchrotron radiation X-ray exposure step by defining a master metal mold insert that is then used to replicate numerous further plastic molds via plastic injection molding. Recently improved storage ring access and the ability to expose multiple samples simultaneously has resulted in another option of using DXRL to form all sacrificial molds. The typical DXRL process sequence is outlined in Figure 5.1. In either case, the ultimate result is the ability to batch fabricate prismatically shaped components with nearly arbitrary in-plane geometry at thicknesses of several hundred microns to millimeters while maintaining submicron dimensional control. This result translates to 100 ppm accuracy for millimeter and submillimeter dimensions. Precision, or repeatability, is obtained by the batch nature of this lithographically based process.

The novel implications of this precision are manifold. For precision-engineered componentry, an increase in resolution is possible over such conventional machining techniques as stamping, or fine blanking, or electrodischarge machining (EDM), while the throughput associated with batch or parallel fabrication is facilitated. Results of using DXRL-based molding for the fabrication of EDM electrodes

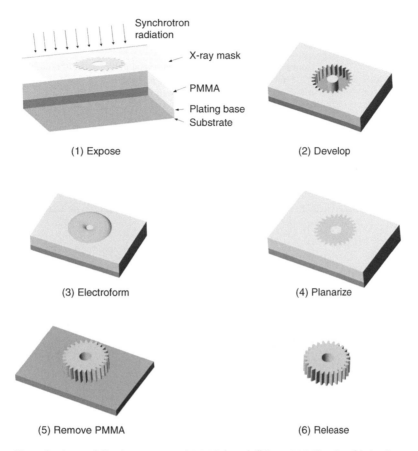

Synchrotron radiation

X-ray mask

PMMA

Plating base
Substrate

(1) Expose

(2) Develop

(3) Electroform

(4) Planarize

(5) Remove PMMA

(6) Release

FIGURE 5.1 (**See color insert following page 9-22.**) DXRL-based ("direct LIGA") microfabrication process.

have also shown that an arrayed type of tool may be used for batch plunge EDM of many precision parts in parallel. The conclusion is that DXRL-based processing is a tool appropriate for miniature precision piece-part, or component manufacture when material property requirements can be met through an additive deposition procedure or sequence of multiple molding steps.

A fundamental issue in fabricating MEMS is the ability to batch fabricate complex three-dimensional mechanisms with appropriately scaled tolerances. The three-dimensional requirement is of particular concern for constructing microactuators where force output scales directly with the volume of stored energy, as well as for micromachined sensors such as seismic sensors, for example, in the definition of inertial proof masses. Fabrication based on planar processing is immediately challenged in this regard, and a so-called high *aspect-ratio* (HAR) process is needed. Furthermore, the difficulty with process integration in realizing a MEM system often results from temperature effects. Most additive mold filling used in DXRL-based processing takes place at less than 100°C and is therefore appropriate for postprocessed components. By appropriately accommodating possible semiconductor X-ray damage, the possibility of integrating microelectronics and microsensors with postprocessed scaled precision metal mechanisms becomes particularly attractive. The flexibility of the X-ray based approach is also becoming apparent in its application to optical MEMS or Micro Opto Electro Mechanical Systems (MOEMS). Exposures at arbitrary angles with the substrate are possible, and multiple-angled exposures can be accommodated. Table 5.2 provides a summary of DXRL processing attributes.

The fundamental fabrication issues involved in DXRL processing center on X-ray mask fabrication, thick photoresist application, deep X-ray exposure, and selective development. Subsequent process issues pertaining to typical device interests include mold filling, planarization, multilayer processing, assembly and mechanism construction, and device integration. Applications arising from this fabrication technique are summarized in the categories of precision components and sensors and actuators.

TABLE 5.2 DXRL-Based Processing Attributes

Type of structural geometry accommodated	Prismatic with arbitrary 2-D shape and sidewall angle
Structural thickness	Commonly 200–800 microns up to several millimeters (10 centimeters demonstrated)
Lateral run-out	<0.1 micron per 100 microns of vertical length
Minimum critical dimension	A few microns typical — function of photoresist stability
Surface roughness	10–20 nm RMS typical, as low as several nm
Thickness control	With conventional lapping: ± a few microns typical, as low as 0.5 micron across 4 inch diameter
Materials (electroformable metals)	Ni, Cu, Au, Ag, NiFe, NiCo, NiFeCo,…
Materials (pressed powders, embossed, hot forged)	Alumina, PZT, Ferrites, NdFeB, SmCo, variety of plastics and glasses

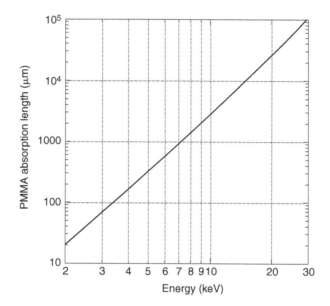

FIGURE 5.2 Absorption length of PMMA ($C_5H_8O_2$, density = 1.19 g/cc) as a function of X-ray photon energy.

5.2 DXRL Fundamentals

5.2.1 X-Ray Mask Fabrication

The fidelity of the *X-ray mask* pattern largely determines the results obtained with DXRL-assisted processing. An X-ray mask has two components: a supporting substrate or membrane with sufficient X-ray transparency and an absorber layer patterned upon it. The required X-ray transmission character for the supporting mask substrate material may be found by considering the X-ray absorption behavior of the X-ray photoresist plotted in Figure 5.2 for PMMA (poly[methyl methacrylate]), the most common X-ray photoresist. This plot depicts the X-ray photon energies required to practically expose a given thickness of PMMA to a dose sufficient to make the PMMA susceptible to dissolution by a suitable solvent (the developer). For typical DXRL layer thicknesses of 100 micron to 1 millimeter, these energies range from 3.5 to 7 keV (or wavelength from 3.5 to 1.7Å). Possible mask substrates include Be, C, and Si slices with thicknesses near 100 μm and diamond, Si, SiC, SiN, and Ti membranes with thicknesses of 1 to 2 μm

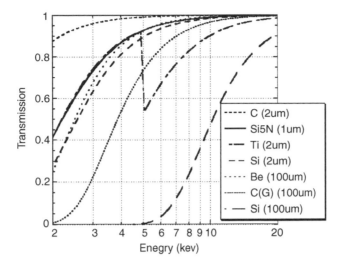

FIGURE 5.3   X-ray transmission of various X-ray mask substrates as a function of X-ray photon energy. The non-stoichiometric form of silicon nitride is a low stress LPCVD (low pressure chemical vapor deposition) form used in membrane fabrication. For the carbon mask at 100 μm thickness, graphite with a density of 1.65 g/cc was used.

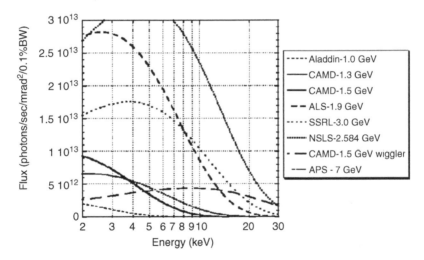

FIGURE 5.4   Synchrotron radiation spectra for U.S. storage rings listed in Table 5.1. The spectral curve for the APS source, which has a peak flux near $1(10)^{14}$ photons/sec/mrad^2/0.1%BW at 16 keV, does not appear on the graph.

[Sekimoto et al., 1982; Visser et al., 1987; Guckel et al., 1989]. The transmission behavior for these mask substrates is plotted in Figure 5.3. What is needed is a mask substrate that is stable to large X-ray flux and that also has sufficient thickness or support to allow easy handling while maintaining the transmission necessary to readily achieve the required exposure dose. For PMMA the required dose is near 3 kJ/cm^3. A corresponding source of X-ray radiation is also required. The nearly ideal light source for this task is found in the synchrotron radiation (SR) that is emitted by charged particle storage rings. The properties of light that result from accelerating charged particles at relativistic energies include highly intense radiation over a broad spectral range from UV to X-ray wavelengths and collimation of the order of 0.1 mrad. The SR beam is emitted in a narrow horizontal stripe of light through a vacuum line (*beamline*) extending from the storage ring. This requires the mask and photoresist combination to be scanned vertically through the beam, which additionally determines a local dose rate in the photoresist. The SR spectra for U.S. synchrotron radiation facilities are plotted in Figure 5.4. Frequently used SR equations are listed in Table 5.3

TABLE 5.3   Frequently Used Synchrotron Radiation Formulas

Wavelength/energy conversion	$\lambda(\text{Å}) = \dfrac{12398.5}{h\nu(\text{eV})}$
Bending magnet radius	$\rho(m) \cong \dfrac{3.336E(\text{GeV})}{B(\text{T})}$
Critical wavelength	$\lambda_C(\text{Å}) \cong \dfrac{18.64}{B(\text{T})E^2(\text{GeV})}$

$h = 4.136(10)^{-15}$ eVs, ν = frequency, E = energy of circulating charge particle, B = bending magnet field

[Margaritondo, 1988]. Low energy (<2 keV) flux creates thermal problems generated by a locally peaked dose at the surface of the photoresist; this requires a filter to limit the dose to a value that avoids the eventual photoresist damage that occurs in PMMA near a dose of 20 kJ/cm^3 [Ehrfeld, 1988b]. Filtering is achieved by inserting slices of lower atomic number materials in the SR beamline including a window that vacuum isolates the storage ring from the exposure beamline. Thus, a mask substrate material with high thermal conductance is needed to help dissipate the heat from radiation absorbed by the mask substrate and absorber. The data in Figures 5.3 and 5.4 suggest that Be and C in the form of graphite can be used as X-ray mask substrates at thicknesses of 100 μm for all SR sources. The cost and hazards associated with Be processing make it less attractive, however, and graphite has proven to be a convenient mask material, particularly when its surface is treated to decrease the roughness due to bulk porosity and thereby facilitate accurate lithography. Appropriate types of graphite substrate material include fully dense glassy carbon or porous graphite which is filled via pyrolytic carbon deposition and subsequently lapped and polished.

Requirements for the X-ray absorber are set by the minimum required exposure contrast; this may be defined as the exposure dose at the PMMA bottom surface in the mask transmission areas divided by the exposure dose delivered to the PMMA top surface under the absorber regions. Mask contrast is a function of the X-ray source, mask substrate, and mask absorber and exposed photoresist thickness. Absorption calculation results reveal two categories of X-ray masks. One X-ray mask type is particularly well suited for low energy exposures of PMMA up to several hundred micron thicknesses. It requires a mask substrate with high transmission of flux near 4 keV, such as membranes from one to a few microns thick of strain controlled silicon nitride, silicon carbide, or silicon, for example. Atomic number and density determine the figure of merit for an absorber material. Thus, materials such as platinum, tantalum, tungsten, and gold are chosen, with gold being most prevalent because it is readily electroplated. The corresponding absorber for the low energy mask is several microns of gold. Accurate dimensional control again requires an additive technique in which a photoresist several microns thick is defined — usually via a UV optical microlithographic mask transfer — to provide for through mask electroplating of gold. Vertical sidewall photoresist pattern transfer is essential to maintain accurate X-ray exposure definition because an absorber sidewall taper leads directly to an X-ray exposed and developed sidewall taper that also results from an insufficient absorber thickness. To achieve vertical photoresist patterns, a bilayer deep UV (DUV) photoresist technique using spun-on PMMA has been implemented. It takes advantage of the highly selective developing system used with DXRL PMMA exposures [Lin, 1975; Guckel et al., 1990]. Poly(methyl methacrylate), as a DUV photoresist exhibits minimal out-gassing during exposure. This is of particular concern for membrane contact lithography, in which mask-membrane separation may consequently arise. Figure 5.5a. shows the vertical sidewalls of an 8 μm thick PMMA layer. Subsequent gold electroplating and photoresist removal yields an absorber structure such as shown in Figure 5.5b. A number of recent commercial photoresist additions have also proven suitable for low energy X-ray mask absorber definition. These include Shipley SJR5740 and 220 developed for through-mask electroplating applications for magnetic read/write heads [Romankiw, 1995]. Thicker UV-exposed photoresists have also been applied to through mask plating for less critical applications [Allen, 1993; Loechel and Maciossek, 1995; Despont et al., 1997]. Figure 5.6 shows a typical membrane-type low-energy X-ray mask.

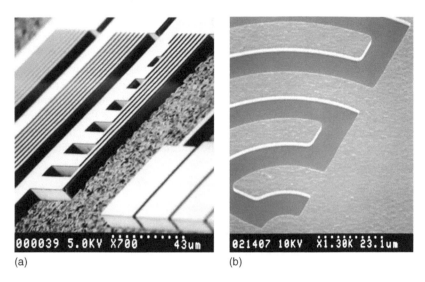

(a) (b)

FIGURE 5.5 (a) Eight-micron thick PMMA patterned with a conformal portable mask (CPM), or bilayer process. A flood deep UV source exposes the PMMA that was subsequently developed using G–G developing solution. (b) Resulting 6-μm-thick gold absorber pattern after PMMA stripping.

FIGURE 5.6 Frontside (left) and backside view of 1-μm thick silicon nitride membrane X-ray mask 5 × 7 cm in area supported by a 4″ diameter silicon wafer with 8-μm thick patterned and electroplated gold absorber layer.

Exposures of PMMA thicknesses from millimeters to over 1 centimeter entail substantially different X-ray masks. These exposures involve X-ray photon energies over 10 keV and for sufficient contrast require gold absorber thicknesses over 20 microns. Because of the difficulty in precisely patterning this thickness of photoresist to submicron tolerances with UV lithography, a two-step mask fabrication sequence is commonly employed. Consequently, a low energy X-ray mask with 2 μm gold absorber is used to pattern several 10s of microns of PMMA directly upon the high energy X-ray mask substrate. In order to maintain a maximum PMMA top dose below the damage threshold high energy SR exposures of thicker photoresist must provide for increased filtering of softer X-rays. Some or all of the increased filtering required may be readily provided by the X-ray mask substrate itself leading to a thicker X-ray mask substrate with increased mask mechanical stiffness. Thus, the high energy mask substrate may be a relatively thick layer of low Z material, such as 100 μm thick silicon, for example. Water cooling of the X-ray mask and exposure substrate also becomes a necessity with higher energy exposures due to the increased overall delivered-power during exposure. Another convenient means to enhance the contrast of a low energy X-ray mask is to provide a thick negative X-ray resist on the backside of what becomes a high energy X-ray mask. When exposed through from the frontside thin absorber pattern with SR, the backside negative resist maintains the same polarity as the original frontside absorber pattern and is then used as a plating mold for additional absorber deposition.

An X-ray mask manufacturability issue results from the constraints of using a membrane-based low-energy mask to achieve high-energy exposures. Figure 5.3 reveals that carbon as graphite at 100 μm thickness may serve both roles. Graphite is an inexpensive rugged mask substrate capable of being practically used for low- and high-energy DXRL exposures; it has good thermal conductivity, and it avoids the hazards and expense of beryllium and other low Z compounds and their associated processing. A carbon X-ray mask example is depicted in Figure 5.7.

Since X-rays emitted in synchrotron radiation are well suited for lithography of submicron critical dimensions, appropriate X-ray masks for relatively thick submicron pattern transfer have been fabricated. Figure 5.8 shows a gold absorber pattern suitable for replication of sub–0.2 µm features into PMMA as thick as 10 µm. The pattern was realized with direct electron beam writing of PMMA on a preformed silicon nitride X-ray mask membrane. Challenges in X-ray patterning of thick (≫1 µm) submicron features, however, lie largely with X-ray photoresist stress, adhesion, and resulting mechanical stability, which will be discussed in the next section.

FIGURE 5.7   X-ray mask patterned on a 3″ diameter graphite support 100-µm thick with 20-µm thick gold absorber pattern.

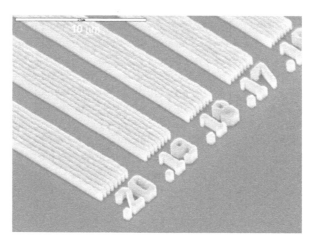

FIGURE 5.8   Submicron gold absorber pattern with approximately 1.0-µm thickness residing on SiN X-ray mask membrane.

5.2.2 Thick X-ray Photoresist

As with many MEMS technologies, processing is particularly sensitive to internal and process-induced strain as well as to adhesion issues. The control of these two areas substantially determines DXRL-based manufacturability. Process stability in this regard mostly concerns the application of a thick, stress-free photoresist with exceptional adherance to a metallized substrate for electrolytic deposition or other substrate material that will facilitate further processing. The difficulty in most photoresist application methods is avoiding large internal tensile stress, which can lead to crazing during development. A number of X-ray photoresist materials suitable for thick application have been examined [Ehrfeld et al., 1997]. Because it possesses relatively good mechanical stability and high resolution, PMMA has been the primary X-ray photoresist. The application procedure initially conceived uses direct polymerization with a casting resin [Mohr et al., 1988]. A cross-linking agent is typically added in the process to further increase the yield strength and thereby provide additional resistance to crazing. To improve adhesion to a metallized substrate, an adhesion-promoting chemistry or intermediate layer may be used [Khan Malek et al., 1998a; DeCarlo et al., 1998]. The nearly 20% volume contraction during polymerization results in a tensile-stressed film; this may be partially alleviated with annealing but cannot be completely avoided due to thermal mismatch with typical substrate materials that have lower thermal expansion coefficients. The consequences of residual photoresist stress are constraints on component geometry design rules and limits on processing procedures.

An alternative approach uses a precast low-stress PMMA sheet [Guckel et al., 1996a]. In this method, a linear high-molecular-weight or cross-linked PMMA sheet is used to obtain PMMA plate cutouts commensurate with a particular substrate geometry. Typical sheet thicknesses range from 500 μm to 1 cm. These sheet cutouts are then bonded to a metallized substrate by means of a technique with no significant temperature variation or induced chemical or mechanical perturbation. One well proven technique uses a thin prespun layer of high-molecular-weight (>1 million) PMMA combined with the monomer methyl methacrylate (MMA) as the bonding solvent. A plate of PMMA is placed on the desired substrate with a spun PMMA layer typically 1 μm thick. MMA applied between the PMMA surfaces wets the interface via capillary action. Since the diffusion of MMA through PMMA is rapid even at room temperatures, curing may take place without a bake-out at elevated temperature. Figure 5.9 shows an example of the result. This is a highly repeatable procedure for obtaining thick, low-stress PMMA films with arbitrary thickness. The bonded sheet may additionally be thinned via precision milling or fly-cutting. Use of a linear PMMA polymer allows dissolution of the PMMA subsequent to mold filling with a solvent such as methylene chloride.

The polymerization of organic vapors in a glow discharge is another method used to produce thin polymer films [Goodman, 1960; Biederman, 1987]. Methyl methacrylate (MMA), in particular, has been studied for use as a plasma deposited photoresist [Morita et al., 1980; Guckel et al., 1988]. The result is a highly conformal deposited film that has excellent adhesion to semiconductor materials and can uniformly fill deep trenches. The plasma polymerized methyl methacrylate (PPMMA) film is highly cross-linked, and its stability depends on reactor pressure, power, and temperature. Practically, deposition rates as high as

FIGURE 5.9   Bonded 1.5-mm thick PMMA on 4″-diameter substrate.

140 angstroms/minute may be achieved, and films as thick as 10 microns have been deposited. Increased power levels result in further decomposition of MMA and films that are less like PMMA.

Another X-ray resist is a negative-working epoxy-based photoresist formulated with SU-8 resin, which is also used for e-beam or deep near UV lithography and yields very high resolution [Lee et al., 1995; Khan-Malek, 1998b]. The EPON® resin SU-8 (SU-3 and SU-2.5 are also available), from Shell Chemical, possesses excellent high temperature stability with a glass transition temperature of more than 200°C when cross-linked [Shaw et al., 1997]. This is due to a high epoxide group functionality (with an average near eight) that also leads to an exceptionally high sensitivity as a photoresist. Deep X-ray lithography results indicate a sensitivity near 52 J/cm^3 [Bogdanov and Peredkov, 2000], or roughly 40 times greater than PMMA, to as low as 20 J/cm^3 [Jian et al., 2001]. Critical exposure dose has been found to be highly dependent on the residual solvent content [Singleton et al., 2001] and concentration of photo acid generator (PAG) [Ling et al., 2003]. A number of practical issues regarding the X-ray process use of SU-8 are critical for its success. Adhesion, for example, depends on substrate material, plating base material, and the use of adhesion promoters [Barber et al., 2004]. Additional issues concern SU-8 X-ray contrast, which dictates minimum absorber thickness [Shew et al., 2003a] and oxygen relaxation [Shew et al., 2003b]. Following exposure and post-exposure baking, the resulting SU-8 cross-linked epoxy also possesses a high optical refractive index (near 1.65) and thus has advantages as an optical lens material. What remains to be found is an easy means to chemically dissolve the SU-8 photoresist after cross-linking. Several techniques have had various degrees of success; these include solvent cracking, downstream etching, and molten salt baths [Dentinger et al., 2002]. In addition, sacrificial polymers may be prepared before SU-8 application to aid in removal.

Another negative X-ray photoresist that is amenable to dissolution after postexposure bake has been used. From Japan Synthetic Rubber Co. (JSR), it is the line of NFR photoresists that may be spun cast to several tens of microns. It employs standard TMAH-based developers; examples of structures are shown in Figure 5.10. Minimum X-ray dose levels of 100–300 J/cm^3 have been measured, and the resist may readily be dissolved in n-methyl pyrolidinone (NMP) at 60–80°C. New formulations of negative chemically amplified X-ray sensitive photoresists with marked photosensitivity improvements of 40 times greater than PMMA have also been presented [Sakai, 2003].

5.2.3   DXRL Exposure (The Direct LIGA Approach)

With its ability to expose and apply PMMA layers over 1 cm in thickness while most applications being pursued require 150 to 500 μm (10 cm exposures have been demonstrated [Siddons et al., 1994; Guckel, 1996b]), the question arises of how to use deep X-ray lithography most beneficially. What becomes immediately apparent is that multiple sheets of PMMA may be stacked and exposed simultaneously [Guckel, 1998a, 1999a; Siddons et al., 1997]. Accounting for the issues associated with the bonding or attachment of exposed PMMA sheets to a substrate accommodating development and mold filling, this procedure has significant throughput advantages. Exposures obtained with the 7.5 Tesla wiggler insertion device [Manohara, 1998] at the Center for Advanced Microstructure Devices (CAMD), at Louisana State University, reveal these advantages. For a 1 cm thick PMMA stack with a limited delivered top dose of 10 kJ/cm^3 and bottom dose of 3 kJ/cm^3, an exposure time slightly greater than 4 hours results for an exposure area of

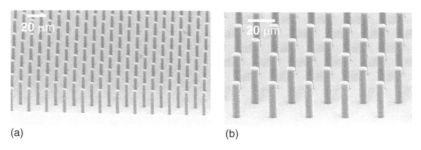

(a)   (b)

FIGURE 5.10   (a) 6 μm and (b) 8 μm diameter cylindrical NFR-015 photoresist pillars, 40 μm tall, created with 125 J/cm^3 X-ray dose.

12.5 cm × 5 cm. This may be compared to the time (depending on storage ring beam current) of roughly 20 to 30 minutes needed to perform the same area exposure at 250 μm thickness using 1.5 GeV CAMD bending magnet radiation. The enhanced high energy performance results in a throughput for a 250 μm PMMA thickness of over 200 substrates per day per beamline. The result reveals a practical approach for direct LIGA with the use of PMMA or using DXRL as the primary tool for precision mold fabrication. Similar opportunities exist for the negative chemically amplified X-ray resists, and convenient batch X-ray exposure end-stations ultimately will help to enable higher throughput.

5.2.4 Development

Although PMMA sensitivity, at roughly 500 mJ/cm², is much lower than other photoresists, the combination of high-intensity well-collimated X-ray flux and the existence of a highly selective PMMA developing system together facilitate exploitation of the high resolution of PMMA in a batch process for obtaining microminiature ultraprecision molds. A developer composed of an aqueous mixture of solvents (80% diethylene glycol monobutyl ether, 20% morpholine, 5% ethanolamine, and 15% water), commonly referred to as the G–G developer after its patent holders, has been formulated to achieve very high contrast PMMA development [Ghica and Glashauser, 1983; Stutz, 1986]. The G–G developer is typically used at a temperature of 35°C, but may be also used successfully at temperatures down to 15°C with decreased development rates yet higher selectivity. The G–G developer solution also etches copper, which may require a protective layer if the material is desired as a plating base. Figure 5.11 shows development rate curves as a function of X-ray dose [McNamara, 1999]. Development-rate temperature dependence has also been noted [Tan et al., 1998; Pantenburg et al., 2003]. Additionally, the quality of PMMA definition has been found to be dependent on developer temperature and on whether the PMMA is linear or cross-linked [Pantenburg et al., 1998].

Detailed discussions of PMMA photochemistry and pattern limits exist in several references [Ouano, 1978; Mohr, 1986; Münchmeyer and Ehrfeld, 1987; Pantenburg and Mohr, 1995; Feiertag et al., 1997a; Griffiths, 2004]. Modeling of fluorescence radiation has revealed that a dose of magnitude 500 J/cc may be deposited underneath absorber regions at areas adjacent to the mask absorber edge and at the PMMA-substrate interface due to emission of fluorescence radiation at the substrate (in this case titanium) surface [Feiertag et al., 1997b]. This behavior may readily lead to PMMA undercutting during development and ultimately to PMMA adhesion loss. The consequence is the possibility of overdevelopment and

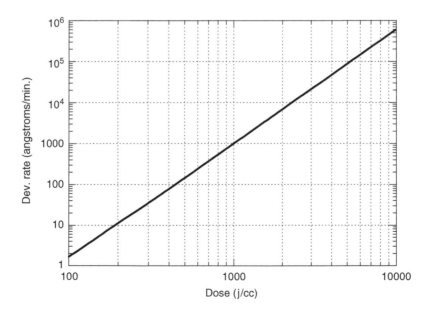

FIGURE 5.11 PMMA development rate as a function of X-ray exposure dose for ICI CQ grade PMMA with G–G developer at 35°C.

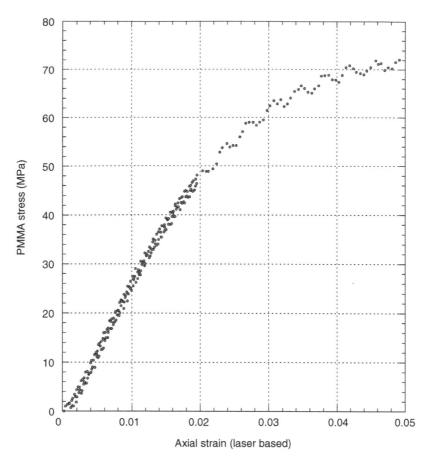

FIGURE 5.12   Tensile curve of PMMA (ICI CQ grade) measured with load frame and laser extensometer system for a DXRL patterned tensile coupon geometry with 750 μm gauge width and 150 μm thickness.

a resulting defect "foot" in the electroform. Electroplating a few microns of sacrificial metal before the structural material can resolve this situation in some circumstances. Further study of G–G developer behavior has shown that this alkaline solvent mixture induces a chemical reaction with exposed PMMA prior to dissolution and that development depends strongly on PMMA stereochemistry [Schmalz et al., 1996]. Another related issue concerns the origins and tailoring of PMMA sidewall roughness that has direct implications for optical components and tribological behavior. Results from exposures at the CAMD facility in Baton Rouge, Louisiana, have shown PMMA sidewall roughnesses as high as 25 nm and as low as 5 nm RMS with typical results near 12 nm. This result seems to concur with the typical resolution quoted for PMMA of roughly 50 to 100 Å [Van der Gaag and Sherer, 1990].

PMMA development is substantially improved via high-frequency acoustic agitation or megasonic agitation [Ehrfeld, 1991]. The development rate is increased, and redeposition of dissolved yet marginally high-molecular-weight photoresist is alleviated, thereby decreasing sidewall particulates and roughness. Analysis indicates an acoustic streaming effect likely is responsible for enhanced development rates in photoresist with structure sizes over a few micrometers [Nilson and Griffiths, 2002]. Studies of transport processes for PMMA trench development indicate that forced convective transport is effective in increasing development rate only for trench structures with an aspect ratio of less than about five [Griffiths and Nilson, 2002].

5.2.5   PMMA Mechanical Properties

The origin of design rules in the DXRL-based molding process rests on the mechanical limitations of the mold material, PMMA. A tensile curve for a high-molecular-weight linear PMMA material is shown in Figure 5.12; it was measured with a milliscale tensile pulling technique. Table 5.4 provides a summary of properties

TABLE 5.4   Properties of PMMA (nom. at 20°C)

E – bulk modulus	~3.0 GPa
G – shear modulus	~1.7 GPa
v – Poisson's ratio	0.40
α – linear thermal coefficient of expansion	~7(10)$^{-5}$/°C
Water absorption (by wt. %)	
(during immersion)	
– 24 hrs.	0.2%
– 7 days	0.5%
– 21 days	0.8%
– 48 days	1.1%
Glass transition temperature	105°C
σ_Y – tensile strength	~70 Mpa
δ – density	1.19 g/cc
κ – thermal conductivity	0.193 W/m · K
Heat capacity	1.42 J/g · K
n – refractive index at 365 nm, 1014 nm	1.514, 1.483
Abbe number	58.0
ε_r – dielectric constant	
– 60 Hz	3.5
– 1 kHz	3.0
– 1 MHz	2.6
– 30 GHz	2.57

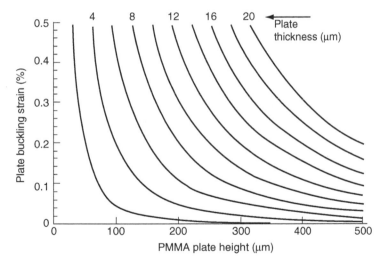

FIGURE 5.13   Buckling strain of PMMA plates typical of DXRL-defined vertical geometry with bottom side attached to a substrate and top side free as a function of plate height for various plate thicknesses (for plate length ≫ plate height).

for PMMA. Many of these parameters are also dependent on the molecular weight or degree of cross-linking, as well as on common additives for enhanced UV absorption and stabilization. The large thermal coefficient of expansion leads to significant concerns of geometry distortion and strain-induced buckling. Similar effects also can occur with water absorption. For example, in a PMMA sheet bonded to a silicon substrate, a 10°C temperature change leads to a 0.11% strain in the PMMA. This situation typically manifests itself via buckling of PMMA plates defined by three built-in edge conditions at the substrate interface and side edges and a free edge condition at the top edge. The solution for this plate-buckling problem may be used to generate design curves that indicate mechanically limiting regions of geometry and thereby establish minimum PMMA design rules [Christenson, 1995a]. One such graph is provided in Figure 5.13. The net effect is that this behavior constitutes the major restriction on aspect-ratio and

FIGURE 5.14   DXRL exposed and developed 200-µm thick PMMA mold form.

process integration. This same effect, on the other hand, may be used to quantify residual PMMA strain via patterned vertical plates used as buckling strain test structures.

5.3   Mold Filling

The material base that may be accessed largely determines the application areas that may benefit from DXRL fabrication. Deep X-ray lithography-produced PMMA mold forms are used most commonly to provide an electroforming mask as depicted, for example, in Figure 5.14. In the case of electroforming, the DXRL mold substrate must be provided with a plating base suitable for an electrolyte of interest. An example plating base is a three-layer film of Ti/Cu/Ti or Ti/Au/Ti in which the lower titanium layer provides adhesion to a substrate such as silicon or glass and the top titanium layer provides for photoresist (PMMA) adhesion while also protecting the electroplating seed layer, which in this case is copper or gold. The top titanium layer may be removed after PMMA development with dilute (100:1) hydrofluoric acid.

A large number of elemental metals and alloys may be electroplated, but this capability alone does not lead to their use as engineering or structural materials or allow electrodeposition into high-aspect-ratio mold forms. Many electroplating procedures, for example, require high temperatures or electrolytes that will dissolve or react chemically with PMMA molds, but the main hindrance in employing a greater number of electroplated materials to microelectroforming is internal strain. Electrodeposit strain can impart delamination stresses onto adjacent PMMA mold material and eventually lead to adhesion loss of the deposit itself. Geometry and process control cannot be attained without minimizing deposit stress to an acceptable level. An integrated means to obtain local measurement of internal electrodeposit stress may be accomplished with patterned beam structures as depicted in Figure 5.15 [Guckel et al., 1992]. This approach employs arrays of clamped-clamped beams and ring-and-beam structures of various beam lengths that are sensitive to lateral buckling. The structures reveal internal strain levels by visual determination of a critical buckling beam length; they are fabricated with a lower sacrificial layer etched away to free the beam structure from the substrate.

FIGURE 5.15 DXRL patterned and electroformed nickel ring and beam structure used to measure tensile internal strain via mechanical buckling. The beam cross member 2 µm wide and 18 µm thick was originally straight after electroforming and has now buckled after release via a sacrificial etch as a result of tensile internal strain. Together with the outer and inner ring radii of 100 µm and 80 µm, respectively, the critical buckling geometry of this structure represents a 0.08% internal tensile strain.

TABLE 5.5 Common Electroforming Bath Compositions and Low-Stress Plating Conditions

Nickel		Copper	
$Ni(NH_2SO_3)_2 \cdot 4H_2O$	440 g/l	$CuSO_4 \cdot 5H_2O$	68 g/l
Ni (as metal)	80 g/l	Cu (as metal)	17 g/l
Boric acid	48 g/l	Sulfuric acid	170 g/l
Wetting agent	0.4%/vol	Chloride ion	70 ppm
Temperature	50°C	Proprietary brightener, carrier and leveler additives	per manufacturer
pH	3.8–4.0	Temperature	25°C
Current density	50 mA/cm²	Current density	25 mA/cm²
Anode	Sulfur depolarized nickel in Ti basket w/anode bag	Anode	phosphorized copper (0.05% P) w/anode bag

Considerable research into how stress arises during electrodeposition has been carried out [Marti, 1966; Weil, 1970; Harsch et al., 1988]. In the course of this research, certain electrolytes have proven to be particularly well suited for low-stress electroforming. The most notable of these is the sulfamate-based electroforming bath that has made intricate electroforming of nickel possible. This bath formulation, which is specified in Table 5.5, is used for most LIGA electroforming due to the relative ease and rapidity with which thick low-stress electroforms may be obtained. Example resultant electroformed nickel structures subsequent to PMMA dissolution are shown in Figure 5.16. Another material particularly convenient for electroforming is copper, in which case the formulation listed in Table 5.5 is particularly well suited for

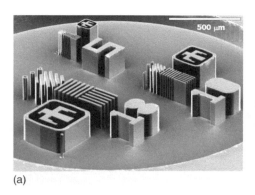

(a)   (b)

FIGURE 5.16   (a) Electroformed nickel test structures 200-μm tall. (b) Close-up of 5-μm wide nickel lines and spaces.

FIGURE 5.17   Electroplated nickel in PMMA craze resulting from PMMA stress concentration test structure. The resulting nickel wall is 20 μm in height and 0.12 μm in width.

high aspect-ratio electroforming. An increase in the sulfuric-acid-to-copper ratio has been observed to increase plating solution conductivity and throwing power, which is advantageous for high aspect-ratio geometry. Complications arising from electroforming through varied geometry and component dimensions arise due to locally varying current density, which results from current crowding and limited electrolyte conductivity in addition to mass transport variation of the electrolyte to the electrodeposit surface [Mehdizadeh et al., 1993]. These issues may be partially alleviated with techniques such as pulse-plating and electrolyte-agitation schemes including jet or fountain plating and paddle cells [Andricacos et al., 1996]. Extremely high aspect-ratio cavities may be readily electroplated with high accuracy, as demonstrated by the structure in Figure 5.17. The electroform replication of the PMMA mold is also essentially perfect to the subnanometer length scale [Hall, 2000]. Other electroformable materials that have been demonstrated with direct LIGA processing include gold, silver, indium, platinum, lead–tin, and alloys of nickel including nickel–iron, nickel–cobalt, nickel–manganese, and nickel–iron–cobalt.

The ability to electroform ferromagnetic materials facilitates implementing a large class of magnetic devices. The nickel–iron alloy 78 Permalloy (78% Ni/22% Fe), for example, may be electroformed to 500 μm

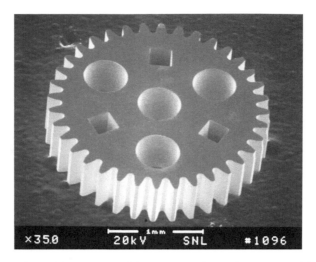

FIGURE 5.18 Gear electroformed from 78/22 nickel/iron to a thickness of 500 μm. In addition to possessing desirable soft ferromagnetic response, this material has been found to perform well as part of mechanisms where tribology is of concern, due to its high hardness and yield strength.

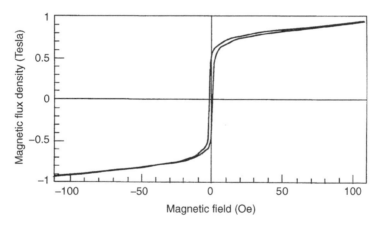

FIGURE 5.19 B–H response of an electroplated 78 permalloy sample. The initial permeability is near 4000, and the magnetic saturation flux density is 1 Tesla.

thicknesses as shown in Figure 5.18. This soft (low coercivity) ferromagnetic material has near zero magnetostriction and a B–H behavior as shown in Figure 5.19. Since microactuator force is developed from the equivalent mechanical pressure generated in a magnetic field as defined by Equation (5.1),

$$p_M = \frac{B^2}{2\mu_o} \ (\text{Pa}) \tag{5.1}$$

where B is the magnetic flux density (T) and μ_o is the permeability of free space $(4\pi(10)^{-7}\,\text{Hy/m})$, a need for soft ferromagnetic materials that can accommodate as large a B field as possible is evident. Recent work has demonstrated electroplated NiFeCo material with magnetic saturation flux density of 2.1 Tesla [Osaka et al., 1998]. Such materials have great potential if they can be incorporated into DXRL processing.

Another important magnetic material for micromechanisms is identified by considering electromagnetic scaling issues for micromagnetic devices that show the advantage of using permanent magnets for efficient miniature electromagnetic transducers. Permanent magnets that possess the best qualities for microactuator applications are rare-earth-based and include the samarium cobalt (SmCo) and neodymium iron

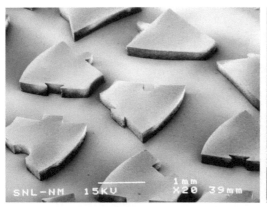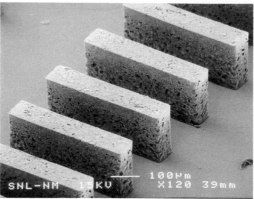

FIGURE 5.20 SEM photographs of bonded $Nd_2Fe_{14}B$ permanent magnets formed with DXRL-defined PMMA molds. Powder particle size ranges from 1 to 5 µm, and the binder is a methylene chloride resistant epoxy. These images show magnetized structures, and thus mild distortion is present due to magnetic perturbations of the electrons. Maximum energy products of 9 MGOe have been obtained with this process.

boron (NdFeB) families. These materials may be directly processed in a bonded form containing isotropic rare-earth-based powder mixed with a binder material such as epoxy. The figure of merit of interest, maximum energy product, varies for these types of magnets from 6 to 10 MGOe (mega-gauss-oersted, $1 MGOe = 100/4\pi$ kJ/m^3) for bonded varieties to over 40 MGOe in fully dense anisotropic forms. Work with bonded rare-earth permanent magnet (REPM) material has demonstrated the direct molding of bonded REPM from DXRL PMMA mold forms [Christenson et al., 1999a]. This has enabled the batch fabrication of a multitude of prismatic REPM geometry with feature sizes as small as 5 µm and dimensional tolerances of 1 µm.

Examples of bonded permanent magnet geometry are shown in Figure 5.20; they were fabricated with the process outlined in Figure 5.21. An unmagnetized mixture of REPM powder and epoxy is applied by calendering and pressing to a substrate with a DXRL-defined PMMA mold while in a low viscosity state. After curing in a press at pressures near 10 ksi, the substrate with pressed composite is planarized. The entire substrate of permanent magnets is then subjected to a magnetizing field of at least 35 kOe in the desired magnetization orientation, and the magnets are subsequently released from the substrate by dissolving the PMMA and etching an underlying metal layer. A related result using a blended form of photoresist (epoxy based SU-8) and Sm_2Co_{17} particle film uses direct UV lithography to pattern permanent magnets with 2.8 MGOe energy density [Dutoit et al., 1999].

Fabrication of REPM structures with greater maximum energy product requires higher temperature processing, for which PMMA is not well suited. An intermediate mold approach has therefore been examined. Direct slurry casting of alumina from PMMA mold forms has yielded replicated geometry with a minimum of distortion (<2 µm) by careful firing sequences [Garino et al., 1998]. Figure 5.22 depicts example glass/alumina composite structures. Other ceramic materials have been demonstrated by this technique including ferrites and PZT [Ritzhaput-Kleissl et al., 1996; Pioter et al., 1997]. The resulting alumina mold may be used to perform high-temperature processing with powder metals. The complications of anisotropic REPM powder processing have prompted development of an approach that does not involve powders [Christenson et al., 1999b]. In a technique similar to hot embossing, a bulk anisotropic slab of REPM material is placed over the alumina mold, which is decorated with a copper release layer. By vacuum hot pressing at 700°C and 2000 psi, strain rates for die upset $Nd_2Fe_{14}B$ material of 22% are realized thereby allowing filling of the alumina mold. This approach has achieved maximum energy products of 23 MGOe in submillimeter permanent magnet shapes; test geometry results are shown in Figure 5.23.

Further micromold filling techniques that have been explored involve approaches ranging from glass castings [Lee and Vasile, 1998] and embossing to flame spray [Christenson, 1998a]. A promising batch microfabrication technique from the standpoint of dramatically widening the available material base is the use

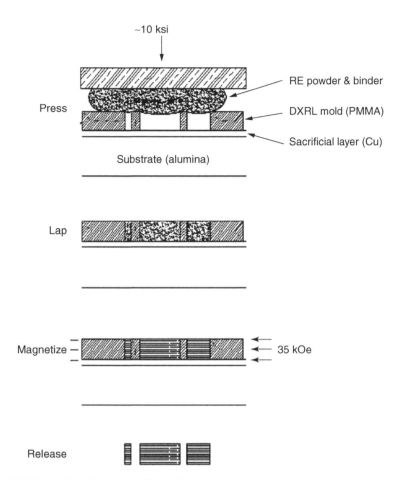

FIGURE 5.21 DXRL-based bonded REPM fabrication sequence.

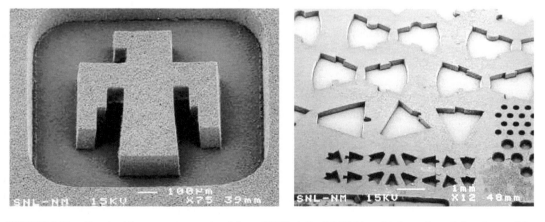

FIGURE 5.22 Alumina/glass structures created from DXRL-defined PMMA mold forms intended to be used for further high temperature mold processing.

of DXRL-patterned electrodes for electrodischarge machining (EDM) [Takahata et al., 1999]. The batch micro-EDM approach uses an electrode pattern defined with an array of components to simultaneously plunge cut a number of parts from a conductive material. Although currently not as accurate as electroforming, these latter techniques may suit applications requiring batch microfabrication of a particular material.

FIGURE 5.23   Hot forged $Nd_2Fe_{14}B$ permanent magnet test bars extracted from intermediate DXRL-formed alumina mold.

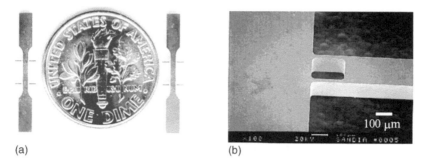

(a)   (b)

FIGURE 5.24   (a) DXRL-fabricated metal tensile specimen and (b) close-up of laser displacement tab attachment.

5.4   Material Characterization and Modification

The material properties of DXRL-fabricated components are unique in that they depend on the processing parameters used in their fabrication. Thus a means of in situ metrology is needed to generate data reflecting lot-to-lot and intrasubstrate variation. This methodology is based on the drop-in test die approach used in integrated circuit planar processing. For ultraprecise high aspect-ratio DXRL-defined components, commensurate routine dimensional metrology remains a challenge. Mechanical property measurement techniques are readily available, however, and have revealed material property deficiencies to which MEMS are particularly sensitive. Postprocess treatments show some promise for accommodating these deficiencies.

Strain diagnostic structures, mentioned previously, readily provide localized stress data as a function of process conditions, as with current density during electroplating, for example. A measure of tensile mechanical behavior has been achieved with a milliscale torsion tester comprising a tabletop servohydraulic load frame instrumented with a laser displacement measurement system [Christenson et al., 1998b]. This system is compatible with common DXRL structural dimensions and accommodates the tensile specimen geometry depicted in Figure 5.24. The specimen has a gauge section roughly $3000\,\mu m$ long with a cross-section defined by a $762\,\mu m$ width and a height set by the electrodeposition thickness that may be from $50\,\mu m$ to $1\,mm$. Two pairs of extruding optical marker bars that are read by the laser displacement system define the gauge length. In order that these markers will only negligibly interfere with the measurement and will not rotate, a pair of spars $25\,\mu m$ wide attach the marker to the gauge section. The specimen shape may be used as a drop-in test structure on any mask definition to allow tensile property measurement on every processed substrate.

The tensile testing procedure has aided in building a mechanical property database. Particularly useful information that has been extracted includes sensitivity to plating current density and temperature cycle effects. Data for nickel is listed in Figure 5.25 and Table 5.6, and results for copper are given in Figure 5.26 and Table 5.7. Nickel–iron deposits are found to have substantially higher tensile strength, as has been

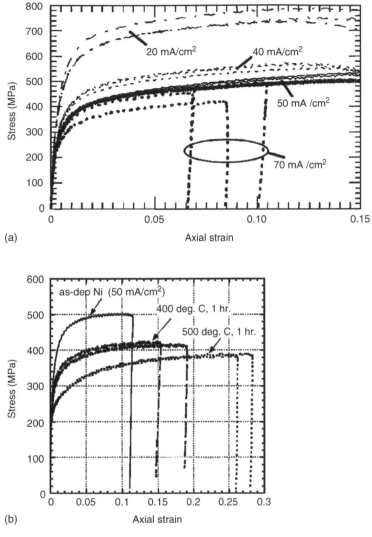

FIGURE 5.25 Tensile data for electrodeposited nickel from a sulfamate bath as a function of (a) current density and (b) anneal cycles.

TABLE 5.6 Electrodeposited Nickel Mechanical Property Current Density Dependence (from Figure 5.23)

Current density (mA/cm²) {no. of tests}	Elastic modulus (Gpa)	0.2% proof stress (Mpa)	Maximum stress (Mpa)
20 {3}	156 ± 9	441 ± 27	758 ± 28
40 {3}	155 ± 11	305 ± 12	562 ± 9
50 {3}	160 ± 20	277 ± 8	521 ± 19
70 {4}	131 ± 13	275 ± 18	460 ± 31

previously noted [Safranek, 1986]. Figure 5.27 shows the tensile data for 78/22 Ni–Fe deposited from a sulfate/citrate bath at room temperature and a pH of 4.8. The high yield strength of the Ni–Fe deposit that is also common to other electrodeposited alloys renders these materials well suited for use in spring fabrication.

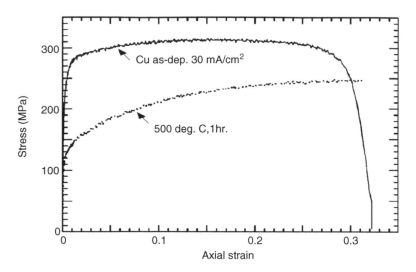

FIGURE 5.26   Tensile data for electrodeposited copper from a copper sulfate bath.

TABLE 5.7   Electrodeposited Copper Mechanical Properties (from Figure 5.24)

Anneal conditions	Elastic modulus (Gpa)	0.2% proof stress (Mpa)	Maximum stress (Mpa)
As-deposited @ 30 mA/cm^2	110 ± 10	239 ± 5	377 ± 5
500°C, 1 hr anneal	76 ± 10	122 ± 5	327 ± 5

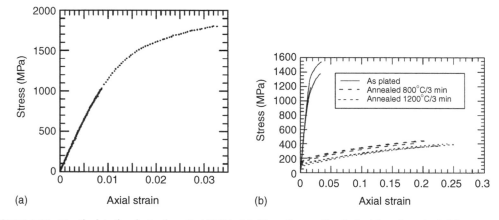

FIGURE 5.27   Tensile data for electrodeposited 78/22 nickel/iron from sulfate bath: (a) as-deposited; (b) temperature dependence.

A mechanical property complication in electroformed deposits is apparent, however, from deposit cross-section micrographs and grain orientation distribution analyses. For direct-current nickel deposits, for example, the initial deposit is found to have a fine grain structure that evolves into a coarser lenticular grain structure oriented parallel to the deposition direction. Figure 5.28 shows examples of this structure. Indentation data reveals corresponding anisotropic and varying elastic and hardness properties throughout the electrodeposited structure [Buchheit et al., 1998]. Further analysis by electron backscatter diffraction (EBSD) microtexture measurement has revealed local spatial variations in as-electrodeposited nickel crystallographic texture [Buchheit et al., 2002]. Another result of the morphology variation is an internal strain gradient through the thickness of the deposit usually with an initially compressive deposit [Harsch et al., 1988]. Means to resolve this nonuniformity include the use of pulse plating and bath chemical additions.

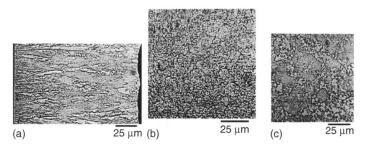

FIGURE 5.28 (a) Metallurgical cross-section photograph of electrodeposited nickel with deposition direction from left to right nucleating on an evaporated copper plating base. (b) Nucleating surface. (c) Top surface.

TABLE 5.8 Surface Mechanical Properties of Implanted Electrodeposited Nickel

	As-deposited Ni	Ti-C implanted
E (elastic modulus) (GPa)	230	400
H (hardness) (GPa)	6.3	13.6
σ_Y (yield strength) (GPa)	1.4	4.8

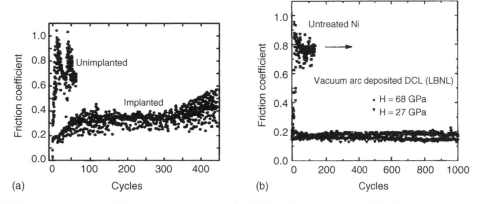

FIGURE 5.29 Friction coefficients of electroformed nickel and comparison with (a) Ti/C treatment and (b) diamond-like carbon deposition. Data was obtained in laboratory ambient with unidirectional sliding against a 1.6 mm radius 44°C steel ball with 9 gf load.

Fatigue study of electroformed nickel has also been carried out in the context of X-ray patterned specimens with $26 \times 260\,\mu m$ cross-section [Boyce, 2003]. An endurance limit of 35–40% of ultimate tensile strength was measured similar to what is found in bulk nickel.

Since scaled micromechanisms fabricated with DXRL-based processing contain components with much larger surface-area-to-volume ratios, surface behavior may substantially affect or dominate mechanism dynamics. In the limiting case, situations of metal component surfaces in contact over time can lead to a static friction situation that prohibits mechanism operation. As a result of these surface sensitivities and material base limitations, surface treatments that improve tribology and wear for existing electrodeposited material have been explored. Important requirements for these treatments are that they do not substantially alter dimensional tolerances and that they do not introduce additional adhesion or shedding problems. Dual implantation of titanium and carbon, a method that has been used previously for iron and steels, has been very successful in this regard [Myers et al., 1997]. The procedure involves implanting Ti and C to a dose of $2(10)^{17}/cm^2$ at respective energies of 180 and 45 keV. The result for electroplated nickel is similar to iron and steel in that an amorphized region formed near the nickel surface possesses significantly enhanced mechanical properties. Table 5.8 shows a comparison summary of these properties. Furthermore, the implanted surface enhances tribological properties as Figure 5.29 explains. Fixturing to address the line-of-sight nature of the implant is needed; this allows articulation of a substrate with attached DXRL components to yield an approximately even dose over all part topology while providing a batch wafer treatment.

Another approach to improving as-electroplated metal tribology has been explored with the use of diamond-like carbon coatings [Ager et al., 1998]. Experiments showed that carbon deposited via a pulsed vacuum-arc process has the advantage of providing partially conformal deposits. These films were deposited on nickel with thicknesses of 100 to 200 nm and possess hardnesses over 30 GPa. The resulting coatings provide similar enhancements in tribological performance to Ti/C implantation as the test results in Figure 5.29 show.

Further surface enhancements to X-ray lithographically molded electroformed nickel have been made by use of wafer scale plasma-enhanced chemical vapor deposition (PECVD) of diamond-like nanocomposite material (Dylyn) from Bekaert Advanced Coating Technologies [Prasad et al., 2002]. These films, which consist of a diamond-like network of C:H and a second network of Si:O, provide excellent adhesion to nickel-based components and considerable protection of nickel surfaces to wear induced deformations that have been found to occur with pure nickel and have been investigated via electron backscattered diffraction (EBSD) technique [Prasad et al., 2003].

5.5   Planarization

The demonstration of submicron in-plane tolerances for DXRL-fabricated components have resulted in similar requirements on thickness control for out-of-plane tolerances. Leveling during electroforming may be aided by bath composition, agitation, and pulsed-current techniques. Controlling several-hundred-micron-thick electrodeposits to micron tolerances has proved impractical for the variety of component geometries of interest, and in this case a concession is needed. An alternative approach is to let the electrodeposit form over the top of the photoresist mold and then planarize the plastic/metal geometry matrix to a desired thickness.

The planarization technique that has proven particularly appropriate for these materials is a diamond lapping procedure that has been referred to as *nano-grinding* [Christenson and Guckel, 1995b; Gatzen et al., 1996]. Machining operations involving fly-cutting and milling are found to generate considerable impact loading and shear stress on microelectroformed parts. Standard lapping procedures involving loose alumina abrasive slurry on a glass plate, on the other hand, are found to preferentially remove the softer PMMA material leaving a situation where the matrixed metal is prone to smearing that results in loss of edge definition. A locally rounded metal surface is also a consequence of the rolling abrasive. In the nano-grinding approach, a diamond slurry is applied to a composite soft metal plate, into which it is embedded via a conditioning plate or ring. The fixed diamond loaded surface is then used as a grinding wheel to cut the metal/PMMA composite layer on a substrate held with a precision vacuum mounted fixture. The metal plate flatness may be maintained to 0.0001″ over a 15″ diameter by the proper use of diamond cutting rings, and thus a lapped surface can be held to a thickness within one micron over a 4″ region. Results for a copper part are shown in Figure 5.30. Nearly any material may be accommodated with this technique. Such results are also a prerequisite for the precise incorporation of multiple layer processing.

5.6   Angled and Reentrant Geometry

Considerable flexibility of X-ray mask/substrate orientation relative to the SR beam is available due to the highly collimated beam and the absence of reflection outside grazing incidence angles. The structure possibilities in this regard are enhanced greatly [Feiertag et al., 1997c]. The first straightforward result is a prismatic angled exposure as shown in Figure 5.31. In this case, both mask and substrate are oriented at an angle relative to the SR beam and scanned across the X-ray mask area. X-ray mask contrast potentially becomes a problem at the edge of the absorber definition where a gradient in contrast exists across

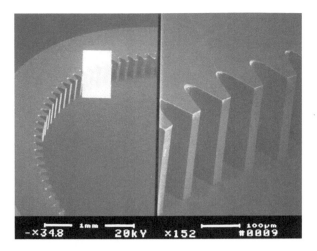

FIGURE 5.30 Copper internal gear that was planarized via the diamond nano-grinding process. Typical RMS surface roughness of less than 2 nm RMS is possible.

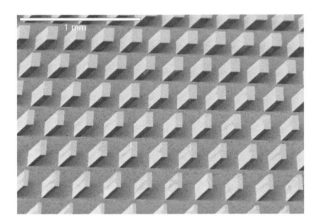

FIGURE 5.31 Copper structures fabricated from a 45 degree DXRL exposure in PMMA. Metal thickness is 200 microns normal to the substrate.

the subtended length defined by the absorber thickness. Thus for critical applications, the X-ray mask absorber may require patterning at the desired structure angle. A reduced mask contrast, however, may be useful in intentionally generating tapered geometry such as that shown in Figure 5.32 [Guckel, 1996b]. The PMMA flank is generated by a partial exposure underneath the X-ray mask absorber region. For adjacent exposed areas with a relatively much higher delivered dose, the flank profile is essentially constant. A transition to a parabolic flank dependence occurs in deeper regions of lower dose where diffusion transport limited development begins to contribute.

High-energy ultra-deep X-ray exposures have yielded extremely thick photoresist exposures and structures [Siddons et al., 1997]. Results from the NSLS depicted in Figure 5.33 reveal over 3 millimeter thick structures. Extending this effort still further, using 35 keV X-rays has enabled a 110 millimeter deep lithography result [Siddons et al., 1994].

Extending angled exposures to two directions can induce reentrant patterns. In particular, at ±45° exposure angles of the X-ray source relative to the X-ray mask and substrate combination, PMMA structures such as shown in Figure 5.34 are generated. In this type of procedure, photoresist exposure latitude becomes critical because top to bottom X-ray doses can become exaggerated.

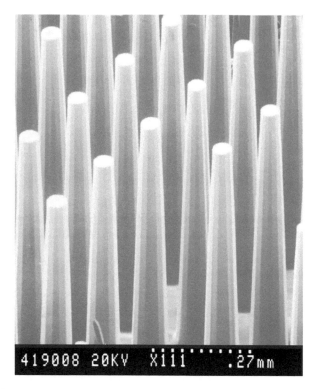

FIGURE 5.32   Tapered PMMA test structures realized with a low-contrast X-ray mask and NSLS exposure. PMMA thickness is 1.5 mm.

FIGURE 5.33   A 3-mm-thick PMMA gear patterned with high energy NSLS exposure.

5.7   Multilayer DXRL Processing

The use of multiple masks to form multilevel stepped prismatic structures makes possible such geometry as flow channels, stacked gears, and gears with shafts as well as integrated packaging structures. Several means have been explored to achieve a multilayer DXRL process. Direct plating has proven to be one viable approach; its results are depicted in Figure 5.35 [Massoud-Ansari et al., 1996]. In this technique, subsequent to planarization of the first layer, a second level of PMMA is bonded to the first composite

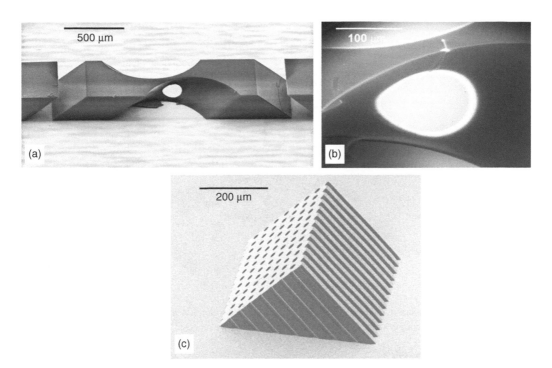

FIGURE 5.34   (a) Circular hole in PMMA created with two intersecting X-ray exposures at ±45°, (b) close-up of circular hole, (c) perforated 45° PMMA prism shape.

FIGURE 5.35   Two-level nickel DXRL-formed test structures produced via direct electroforming.

metal/PMMA layer, X-ray exposed with an aligned X-ray mask, developed, electroformed, and finally planarized. Consecutive layer geometry is typically restricted not to overhang previous geometry.

Another multilayer process uses a batch diffusion bonding and release procedure [Christenson and Schmale, 1999c]. The fabrication sequence is shown in cross-section in Figure 5.36. Two planarized substrates decorated with DXRL-patterned and electroformed layers and with PMMA removed are aligned face-to-face with gauge pins that are press-fit into complementing alignment structures. The substrates are joined in a

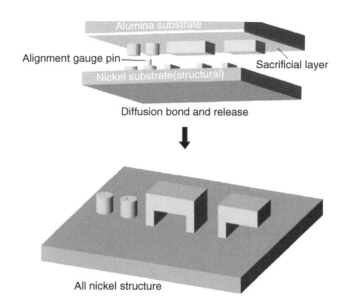

FIGURE 5.36   (See color insert following page 9-22.) Batch wafer-scale diffusion bonding procedure for multilevel fabrication of DXRL-patterned geometry. This sequence may be directly repeated to integrate further levels.

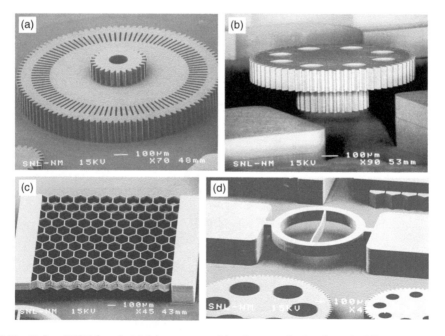

FIGURE 5.37   Various DXRL-based nickel structures resulting from two-level wafer-scale diffusion bonding. (a) Small gear on large gear, (b) large gear on small gear, (c) hexagonal structured cantilever, (d) ring and beam tensile strain measurement structure.

vacuum hot press, and the sacrificial layer on one substrate is etched entirely away leaving one layer bonded to the other. Figure 5.37 reveals results for a test mask pattern. This procedure is compatible with arbitrary overlapping geometry and may be repeated indefinitely. Careful attention must be given to the temperature cycles that have the potential to degrade mechanical properties substantially. For this reason, a milliscale mechanical torsion tester suited for batch measurement of shear bond strength of high aspect-ratio structures

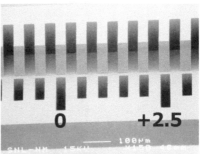

FIGURE 5.38 Vernier indicators of two-level diffusion bonding alignment that reveal an alignment accuracy of less than 1 μm. Each vernier mark indicates 0.5 μm misalignment.

FIGURE 5.39 Assembled nickel gear on nickel shaft and 10× close-up showing 1-μm journal bearing clearance at 120-μm structural thickness.

has been used to optimize diffusion bonding parameters [Christenson et al., 1998c]. Using this approach, a diffusion bond cycle of 450°C at 7 ksi for 4 hours with electrodeposited nickel was measured to yield a 130 MPa bond strength that is suitable for many applications. Alignment accuracy is measured with vernier alignment structures (Figure 5.38) that indicate less than 1 μm alignment error over a 4″ diameter substrate.

Local photodegredation has been shown to be a convenient means of creating multiple layer PMMA components without using solvents or adhesives [Henzi et al., 2003]. This procedure uses UV-induced side and main chain scission of PMMA within an absorption length of the surface with 240 nm UV light and the resulting depressed glass transition temperature in this layer to bond PMMA structures at low temperatures. Depressed PMMA glass transition temperatures to 40°C are rendered by a 240 nm irradiation dose of 10–20 J/cm^2.

5.8 Sacrificial Layers and Assembly

The multilayer fabrication schemes in the previous section aim to substantially alleviate the assembly burden. Many devices, however, still require some form of micro assembly. The reason for the reliance on assembly lies in the high aspect-ratio nature of these types of components. Maintaining intercomponent tolerances of one micron or below is not feasible directly at very high aspect-ratios, and thus assembly may be used to allow biasing of the 1 micron or less between two separately fabricated components that are then assembled to realize an acceptable precision fit. Figure 5.39 is an image of a resulting combination. This

FIGURE 5.40   Partially released nickel electrostatic comb-drive.

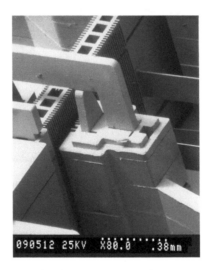

FIGURE 5.41   Five-layer stacked 78 permalloy linear actuator mechanism over 1 mm in height.

microassembly sequence realizes released components via a sacrificial layer etch that may also be used to partially release structures such as the one in Figure 5.40. The microassembly of DXRL-fabricated structures has been extended to more elaborate mechanisms via stacking (Figure 5.41) [Fischer and Guckel, 1998] and multistage element assembly (Figure 5.42) [Christenson, 1998a]. Individual microcomponent assembly techniques that have been employed include the use of magnetic probes for ferromagnetic materials, liquid surface tension to pick up certain geometry, and parallel robotic schemes [Feddema and Christenson, 1999]. Additive wafer transfer schemes such as batch processes incorporating wafer-scale component transfer with wafer-scale bonding and sacrificial-layer release sequences ultimately have the most appeal.

5.9   Application Examples

Two general application categories that have benefited significantly from X-ray based fabrication include precision components and microactuators or volume controlled devices.

FIGURE 5.42   Assembled two-layer gear train and rack mechanism fabricated from 78 permalloy. The aluminum substrate was machined conventionally to accept press-fit tool-steel gauge pins on which the keyed bushings and gears are assembled.

FIGURE 5.43   Spiral spring with 3-micron wide flexural element and 100-micron structural height fabricated from nickel.

Example 1. Precision Components

X-ray based fabrication is particularly well suited for high-volume manufacturing of components requiring micron to submicron dimensions and tolerances. The type of components in this category include precision screens, filters and nozzles, optical components and alignment structures, miniature connectors and springs, and RF components. Another adjunct category of precision components involves those requiring sustained operation at high pressures where the material base and scalable part dimensions afforded by deep X-ray lithography and molding are appropriate. Such components include miniature reactor vessels, chromatography columns, and custom packages, for example.

An additional emerging category of precision microstructure components has been enhanced by the ability of deep X-ray lithography to print extruded two-dimensional features in a multitude of directions. This category consists of unique topology for which no alternative fabrication approach exists. It is exemplified by devices with radically different means of achieving engineering functions and commensurately unique performance characteristics.

A number of precision component examples will be described to illustrate the diversity of applications. The first (Figure 5.43) reveals a spiral spring with 3 micron feature size and 0.15 N/m lateral spring

FIGURE 5.44   One hundred–micron thick released nickel gears with 7–micron involute gear tooth thickness.

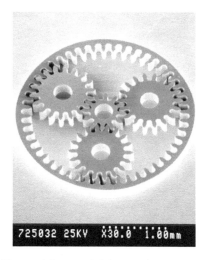

FIGURE 5.45   Planetary gear set fabricated from nickel for RMB Corporation by the Microelectronics Center of North Carolina (MCNC).

constant. Such a component would require prohibitively difficult conventional machining. Because cantilevered springs of this type possess a stiffness proportional to the cube of their thickness in the direction of flexure, a consistent spring stiffness requires submicron tolerances. The ability to pattern fine features within relatively large geometry allows accurate scaling of gears as depicted in Figure 5.44. Arbitrary gear teeth profiles are possible, and thus any pressure angle for an involute geometry, for example, may be accommodated. These components have been applied to miniature precision planetary gear box construction (Figure 5.45). Micro cycloid-gear systems have also been fabricated through multiply exposed and aligned DXRL sequences resulting in gear ratios of 18 [Hirata et al., 1999].

A large number of additional piece part applications have centered on grid structures or arrayed apertures. The screen in Figure 5.46 illustrates a typical example. This rather large nickel component contains 50 µm wide spars with less than 0.25 µm deviation through its thickness of 150 ± 1 µm. In this case, repeatability is of prime importance; the lithographic dimensional control afforded by DXRL-based processing is well suited for this. A related structure is used in a "lobster-eye" optic for X-ray astronomy [Peele, 1999]. It requires an array of square apertures that reflect X-ray flux at a grazing angle off of their orthogonal sidewalls. These reflections are directed into a central focus area if the axis of each aperture is radial

FIGURE 5.46   Precision screen part fabricated from 150-μm thick nickel.

FIGURE 5.47   Nickel X-ray "lobster-optic" test structure 200 μm thick with 50-μm apertures and 3-μm sidewalls. The outer sidewalls were patterned at 1.5-μm thickness where partial development occurred.

to the center of a sphere. A desired material for this X-ray optic structure is nickel with a figure of merit that is strongly dependent on sidewall surface roughness, the goal for which in this application is a few nanometers. Figure 5.47 depicts a test structure of the type needed for the optic element. Sidewall roughness for these structures is typically 10 nm and has been measured as low as 7 nm RMS. Thermal reflow of X-ray photoresist such as PMMA may reduce sidewall roughness to under 3 nm.

The complementary structure to a grid also may be fabricated readily with X-ray-based technology as shown in the test structure of Figure 5.48, which has relatively small gaps surrounded by larger areas of

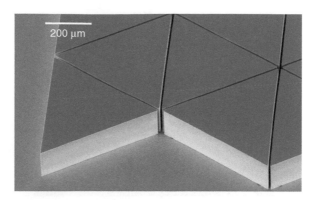

FIGURE 5.48   Close-spaced copper triangular shapes, 150 μm thick with 5-μm gap spacing.

FIGURE 5.49   PMMA cylindrical columns, 300 μm tall, 30 μm in diameter on 33 μm pitch.

material. An intermediate form of an arrayed gap construction is illustrated in Figure 5.49, which shows a pillar array with fine separation.

Another device category requiring arrayed channels includes heat exchangers. Reduced heat exchanger dimensions of 100 microns to 1 mm result in more effective heat transfer due to decreased thermal diffusion lengths. A DXRL-based fabrication method for cross-flow micro heat exchangers has been realized [Harris et al., 2000; Kelly et al., 2001]. The scheme uses a DXRL-patterned two-layer nickel mold insert (Figure 5.50a) to hot emboss half of the heat exchanger panel (Figure 5.50b) [Despa et al., 1999]. Two panels are then placed face-to-face and bonded to realize the automobile-like radiator shown in Figure 5.50c. Estimated performance for this device realized in PMMA with an air-pressure drop of 175 Pa and coolant-pressure drop of 5 kPa is a heat dissipation of 6.0 W/cm^2 or 33.3 W/cm^3, which is a substantial increase over larger traditionally scaled devices.

The primary X-ray photoresist material, PMMA, has excellent optical properties, and thus it may be patterned and used directly for optical components. Figure 5.51 reveals an integrated fiber alignment guide and prismatic focusing optic structure fabricated with PMMA [Holswade, 1999]. Two support sprues hold together two components, which are removed after pin-press fit mounting. The structure accommodates input and output fibers oriented at 90° to each other. The light from one fiber is focused into a line across a reflective encoded surface that reflects a signal back into the receiving fiber.

This type of cylindrical optic has been extended one dimension further to approximate a spherical two-surface lens. In this approach, two crossed plano-cylindrical surfaces are microfabricated in place

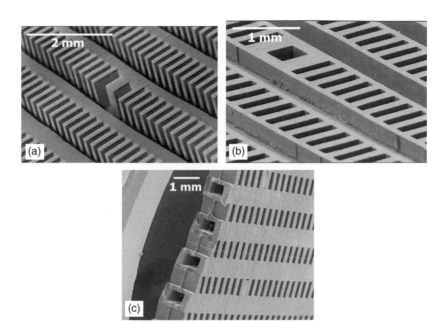

FIGURE 5.50 Cross-flow micro heat exchanger realized by bonding two PMMA halves embossed from a nickel mold [Harris, 2000]. (a) Heat exchanger nickel mold insert, (b) one side of embossed PMMA heat exchanger, (c) assembled and bonded plastic heat exchanger showing cross-section of coolant channels.

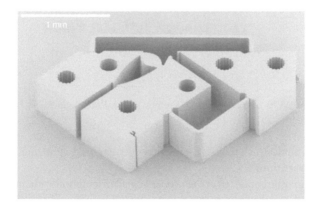

FIGURE 5.51 PMMA optical fiber alignment structure and focusing lens at 500 μm thickness. This structure, which provides for a minimum sized spot to optically monitor a moving surface, is press fit over 200-μm diameter pins and subsequently has its two retaining breakable beams detached, thereby providing a sprue function. The lens is a best-form aspheric designed for a 0.16 NA single-mode fiber at 632.8 nm with a center thickness of 200 μm.

to allow for focusing in two perpendicular directions across a lens aperture. The term *X-Optics* has been applied to this scheme, which uses deep X-ray lithography to form "crossed" micro lens systems from X-ray photopolymers [Christenson and Sweatt, 2001]. Additional materials may also be incorporated with this approach by using a molding sequence where an intermediate metal mold created from the initial photopolymer, for example, supports molding of other polymers and glasses. The process depicted in Figure 5.52 exposes X-ray photoresist at ±45 degrees relative to the substrate perpendicular through an X-ray mask with synchrotron radiation. This renders highly accurate and arbitrary lens surfaces that may, for example, be either positive or negative in shape. A single X-ray mask is used whereby the mask and substrate are not moved relative to each other between exposures. Thus the surfaces created during the two exposures are aligned to one another to lithographic accuracy. In addition, as long as

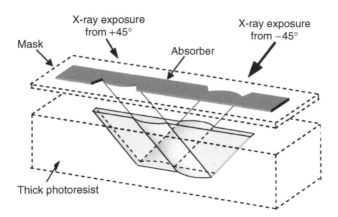

FIGURE 5.52   Exposure procedure used to form X-Optics.

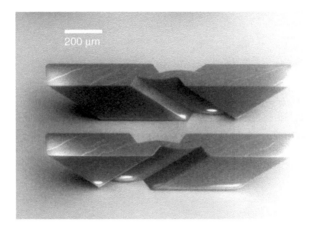

FIGURE 5.53   Pair of PMMA bi-convex (positive) X-Optic lenses. PMMA thickness is 300 μm, and effective lens aperture is 150 μm.

the X-ray mask is flat, optical axis registration across the exposed substrate is ensured regardless of mask-to-substrate gap variations.

X-ray mask contrast may be enhanced at the proper angled orientation with an intermediate mask realized by appropriate tilted absorber exposures. A convenient means to achieve this is the use of a frontside X-ray mask absorber of gold at 1 μm thickness, for example, to expose at an angle or several angles a negative X-ray resist that may be 10 to 20 μm thick on the back of an X-ray mask. This backside resist may then be used as an absorber electroplating mold to generate sharply defined edges at specific exposure angles. With the lens optic axis oriented parallel to the substrate, a multitude of lenses may be patterned in place to directly form precisely aligned multielement lenses. Within the same process, optical stops, prisms, diffractive elements, and optical fiber couplers may be constructed in place to allow simultaneous fabrication of monolithic optical subsystems.

Figure 5.53 shows an X-Optics result for two bi-convex lens elements patterned to form a 1:1 focusing test optic. Characterization of this test optic has revealed the ability to focus a 9 μm diameter core single-mode fiber output to a 10–12 μm spot with a lens aperture of 150 μm and a focal length of 300 μm.

With the same ±45 degree X-ray exposure approach, an alignment structure suitable for an optical fiber may be fabricated as shown in Figure 5.54. When viewed end-on this structure consists of a sequence of narrowing square apertures tilted at 45 degrees. As an optical fiber is inserted from the end of the structure, a slight press-fit is obtained to orient and clamp the fiber in place. For a centered optical fiber, the fiber's center axis will be held in line with the optical axis of subsequent optical components when this

FIGURE 5.54   PMMA optial fiber holder and alignment structure for 125-μm diameter optical fiber fabricated with ±45° X-Optic exposures.

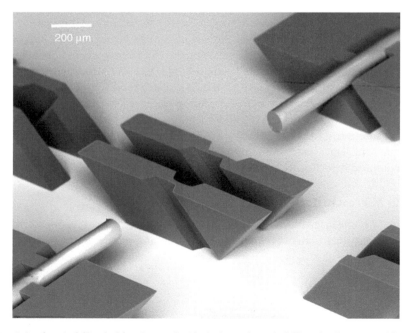

FIGURE 5.55   Pair of optical fiber holders inserted with single mode optical fibers in alignment with X-Optic lenses. All patterned material is PMMA at 300 μm thickness, and the substrate is glass.

structure is fabricated simultaneously with adjacent optical elements (Figure 5.55). This aids in micro-optical packaging and connecting.

Extending multiple tilted X-ray exposures yet one more step has been useful as a means to accurately fabricate three-dimensional photonic band gap structures or photonic crystals [Feiertag et al., 1997d; Cuisin

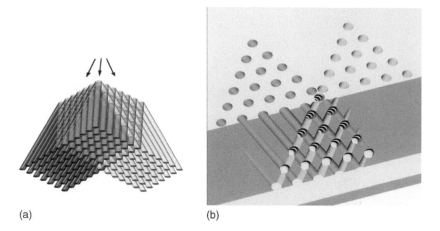

(a)   (b)

FIGURE 5.56   (a) Yablonovite X-ray exposure construction, (b) Yablonovite fcc stucture.

et al., 1999, 2000, 2002; Kiriakidis et.al., 2000]. The "three-cylinder" photonic band gap precursor or its complement, the "slanted-pore" photonic band gap structure, which has been most commonly fabricated with X-ray lithography, involves projecting a planar triangular lattice of circles at a tilt of 35.26° with respect to the plane of the circles and then rotating this projection by 120° with respect to the azimuth to form three exposures. The resulting slanted pore fcc structure has been coined "Yablonovite" by its discoverer, Prof. E. Yablonovitch [Yablonovitch et al., 1991]. The basic construction and its complement are shown in Figure 5.56. To form a material with a complete photonic band gap, a material with sufficiently high refractive index is also required. Since X-ray photoresists do not fulfill this role, other materials must fill exposed and developed photoresist molds. Nickel, for example, has a very large index of refraction in the infrared. With a positive photoresist such as PMMA, therefore, one possible fabrication sequence proceeds as follows. The slanted pore structure is exposed via DXRL in PMMA and filled via copper electrodeposition. The copper structure (Figure 5.57) is then used as a mold for the electrodeposition of nickel. After the copper is selectively etched from the composite copper–nickel structure, the result is Yablonovite formed from nickel as shown in Figure 5.58. With a large area X-ray scanner this fabrication approach is scalable to large substrate area production [Sweatt et al., 2002]. Additional structures of this kind have been suggested having larger band gaps that are also particularly well suited to the deep X-ray lithography approach outlined here [Toader et al., 2003].

Example 2. Microactuators

The ability to integrate magnetic material with batch DXRL-based microactuator fabrication has made possible high-pressure microactuators with low driving point impedance. Both linear and rotary devices have profited in accord with a figure of merit for prismatic-based actuators defined by the force generated in a working volume,

$$F_x = \rho_E h \frac{\partial A}{\partial x} \tag{5.2}$$

defined by a height h and area A tangent to the axis of motion in the x direction. The energy density multiplier ρ_E is limited for a magnetic actuator by the magnetic flux saturation density B_{sat}, as was previously described or for an electrostatic actuator by the electric breakdown field E_{bkdn} via,

$$\rho_{E_m}(\text{max}) = \frac{B_{sat}^2}{2\mu_o} \tag{5.3}$$

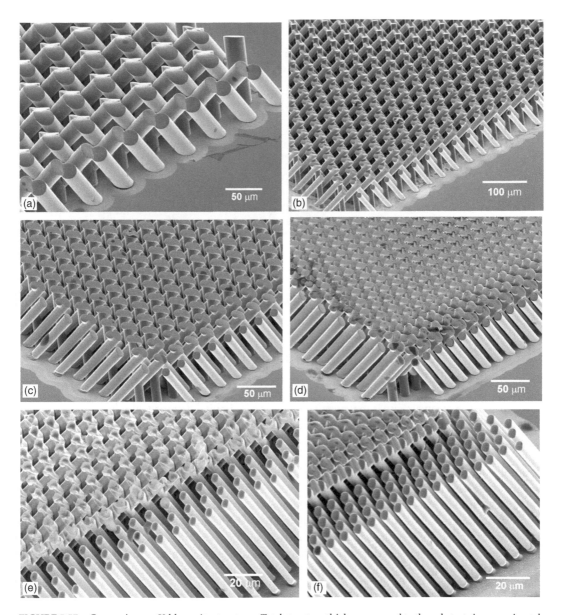

FIGURE 5.57 Copper inverse Yablonovite structures. Total structure thickness normal to the substrate is approximately 100 μm. (a) 30-μm cylinders on 60-μm pitch, (b) 15-μm cylinders on 45-μm pitch, (c) 15-μm cylinders on 30-μm pitch, (d) 15-μm cylinders on 22.5-μm pitch, (e) 6-μm cylinders on 15-μm pitch, (f) 6-μm cylinders on 12-μm pitch.

$$\rho_{E_e}(\max) = \frac{1}{2}\,\varepsilon_o E_{\text{bkdn}}^2, \tag{5.4}$$

where $\mu_o = 4\pi(10)^{-7}\,\text{Hy/m}$ and $\varepsilon_o = 8.854(10)^{-12}\,\text{F/m}$ are the vacuum permeability and permittivity respectively. In terms of driving electrical parameters, ni (ampere-turns) for magnetics and V (volts) for electrostatics, and working volume air gap d, the relationships for actuator pressure become,

$$\rho_{E_m} = \frac{1}{2}\,\mu_o\left(\frac{ni}{d}\right)^2 \tag{5.5}$$

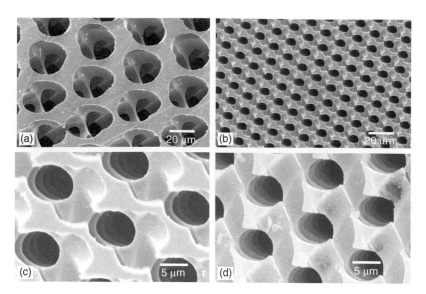

FIGURE 5.58   Nickel Yablonivite structures replicated from copper intermediate forms. (a) 30 μm pores, (b) 10 μm pores, (c) 10 μm pores close-up, (d) 6 μm pores.

$$\rho_{E_e} = \frac{1}{2} \, \varepsilon_o \left(\frac{V}{d} \right)^2. \tag{5.6}$$

The caveat in the magnetic case is that an efficient magnetic circuit is assumed with high permeability or soft ferromagnetic material. The plot of these equations in Figure 5.59 reveals the impedance-pressure tradeoff for various working gaps.

Linear magnetic microactuators have been fabricated with three basic components. A movable plunger constrained by a folded spring flexure operates into a magnetic circuit portion that supports the transfer of flux to the working gap. Both components are fabricated from a soft ferromagnetic material such as 78/22 nickel/iron or 78 Permalloy. The third component, a coil, has been most effectively realized by conventional miniature coil winding techniques using a DXRL-defined mandrel that may be press-fit into the magnetic circuit. Figure 5.60 shows a typical construction for such a magnetic *variable reluctance* (VR) linear microactuator. Typical performance of this microactuator type at a structural thickness of 150 μm is a force output in excess of 1 mN, with a stroke of 450 μm for an input current of 40 mA and corresponding input power of 10 mW. The resonant characteristics include a resonant frequency near several hundred Hz and an amplitude of ±200 μm at an input power of 200 μW with a quality factor in air near 200. With an inductance change of 1 μHy/μm, the possibility of using inductance sensing to control microactuator position has been exploited [Guckel et al., 1996c, 1998b; Stiers and Guckel, 1999]. The device depicted in Figure 5.61 uses a sense coil as part of an oscillator circuit whose resonant frequency is compared to an input frequency representing the desired actuator position. A phase detector circuit that integrates the difference in drive frequency and sense frequency yields an effective phase shift signal that constitutes a positional error. Despite a significantly reduced electrical quality factor for microscale inductance based oscillators, with the use of an integrating phase detection scheme a positional accuracy of 260 nm was realized with a settling time of roughly 100 ms.

Magnetic rotational microactuators have also been fabricated using a process similar to that of linear types, with the difference being the substitution of a rotor for the linear armature. Figure 5.62 shows a rotational magnetic micromotor integrated with silicon rotor position sensing p–n junction photodetectors; this demonstrates the ability to postfabricate DXRL-based structures readily on semiconductor substrates with microelectronics [Guckel et al., 1993]. Maximum achieved rotor speeds for 140 μm diameter rotors are 150,000 rpm for a closed-slot stator design that minimizes torque ripple in this variable reluctance design.

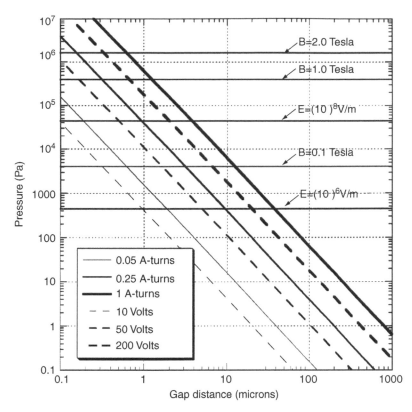

FIGURE 5.59 Plot of magnetic and electric pressure generated for various drive levels. Solid lines are magnetic with *ni* in ampere-turns, and dashed lines are electric with *V* in volts as in Equations (5.5) and (5.6). Maximum pressures for limiting B and E fields are also indicated.

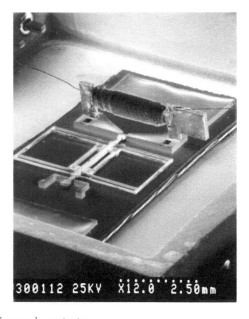

FIGURE 5.60 Packaged VR linear microactuator.

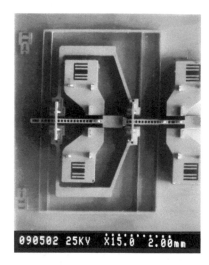

FIGURE 5.61 Linear magnetic microactuator with sensing for closed loop control. All components are 78 Permalloy. Wound coil forms remain to be inserted into the serrated magnetic pad areas.

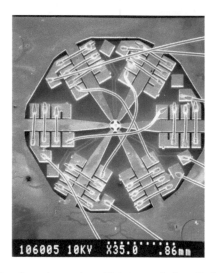

FIGURE 5.62 DXRL-based VR stepping micromotor with integrated photodiode position sensors.

These types of micromotors have been operated continuously for over three months at over $(10)^{10}$ rotational cycles and reveal no change in operating characteristics, which for a journal bearing is quite acceptable. The low wear rate is suspected to be partially due to a vertical variable reluctance rotor suspension (a consequence of a thinner rotor than stator) that maintains rotor rotation off of the substrate. This category of micromotors with integrated wire bonded coils is rather inefficient with output torques of the order of 10 nNm. A more efficient coil for rotational magnetic micromotors has been implemented with conventionally wound soft magnetic mandrels providing a large air-gap flux and resulting in submillimeter diameter 200 µm thick rotors capable of providing 0.2 µNm of torque [Christenson et al., 1996].

Much greater magnetic microactuator efficiency may be obtained by incorporating permanent magnets. Magnetization is independent of scale, which is an attribute making permanent magnet use at the microscale much preferable to an electromagnet. An electromagnetic coil possesses a magnetic field production scaling dependence that is linear and is brought about by current density constraints that are in turn due to thermal and electromigration limitations. Coils utilizing superconductors share the same scaling relationship, which arises due to a similar critical current density limit. The use of rare-earth-based permanent

FIGURE 5.63 Brushless DC motor with 5 mm diameter four-pole rotor fabricated with DXRL-formed bonded $Nd_2Fe_{14}B$ magnets. The rotor is mounted on a 320 μm diameter shaft that extends into a 1.6 mm diameter ball bearing residing under the rotor. The outer magnetic return path is 78 Permalloy with an 8 mm outer diameter. Nine coils with 120 windings per slot are arranged on a DXRL-defined PMMA coil form.

FIGURE 5.64 Overview of tunable IR filter with integrated three-phase magnetic VR stepping linear microactuator. The IR filter is seen on the left with a 2.5 × 2.5 mm area and is supported by the frame surrounding the device. Three coils are assembled into the stator forms located to the right [Ohnstein, 1996].

magnets, in particular, is attractive because of their high magnetization and high demagnetization resistance, which makes them amenable to nearly arbitrary aspect-ratio. The incorporation of DXRL-fabricated rare-earth permanent magnets has been demonstrated with a miniature brushless DC motor as depicted in Figure 5.63, which shows a low profile brushless DC motor with a 5 mm diameter four-pole radially anisotropic bonded $Nd_2Fe_{14}B$ rotor [Christenson et al., 1999d]. The slotless four-pole, nine-slot configuration rotates at 15,000 rpm with 2 volt 5 mA three-phase signals and generates a maximum torque near 10 μNm at 300 μm rotor thickness. The planar configuration of this motor is suitable for such tasks as are present in low profile miniature hard-disk drives, for example.

Example 3. Microactuator Applications

Linear actuators fabricated via DXRL approaches have found many uses in optical devices. A spectrometer application, for example, uses a three-phase magnetic VR linear stepping motor to stretch and compress a tunable IR filter for miniaturized gas analyzers and multispectral IR systems [Ohnstein et al., 1995, 1996]. Figure 5.64 depicts an overview of the device, which includes a DXRL-fabricated wire-grid IR filter

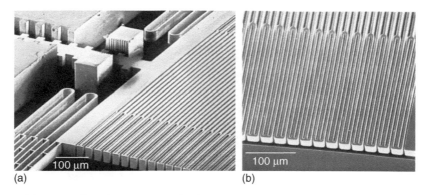

(a)   (b)

FIGURE 5.65   (a) IR wire filter shown connected to linear stepping microactuator armature; (b) close-up of Ni/Fe IR wire filter.

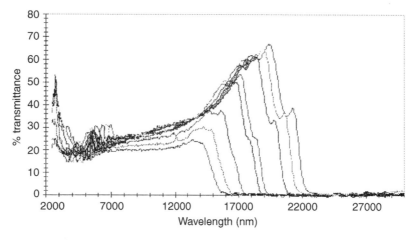

FIGURE 5.66   Transmission curves of the tunable IR filter.

that is also part of a distributed flexure attached to an armature of a linear stepping motor. All components are fabricated from electroplated 78 Permalloy. The IR filters, shown in detail in Figure 5.65, have an active area ranging from 1×1 mm to 2.5×2.5 mm commensurate with an IR focal plane array and a wire spacing that may be deformed to yield cut-off between wavelengths of 8 and 32 μm. The variable filter response is shown in Figure 5.66. Total filter displacements of 1.7 mm were achieved with a stepping actuator providing an average force of 2.4 mN/step at a drive current of 45 mA; this is more than sufficient for the filter spring constants at roughly 1 mN/mm.

A fiber optic switching device has also been developed with a magnetic VR linear microactuator. A 1×2 single-mode fiber switch consisting of a two-phase VR linear spring constrained microactuator is depicted in Figure 5.67 [Guckel et al., 1999b]. Switch performance for a 125 μm diameter fiber with an 8 μm core includes a switching speed under 1 ms at a power dissipation of below 20 mW with as low as 0.5 dB insertion loss in air. Index matching fluids may also be used with this switch, which may additionally incorporate permanent magnets for a latched state with zero power dissipation.

An application that uses the simplicity of a single-layer electrostatic drive is shown in Figure 5.68. This component consists of an electrostatic comb-drive that resonates an amplitude grating to render structured illumination in a miniature microscope [Lee et al., 2003]. The intended use of the miniature microscope is to image tissue morphology and biochemistry in vivo to provide better identification of tumor cells. The resonant amplitude grating scans a transverse amplitude of ± 100 μm with a resonant frequency of near 30 Hz and a quality factor in air of approximately 50. The 2×2 mm grating with 75 lines/mm is attached to the body of the comb-drive with an integral spring.

FIGURE 5.67   Assembled single mode 1 × 2 optical fiber switch.

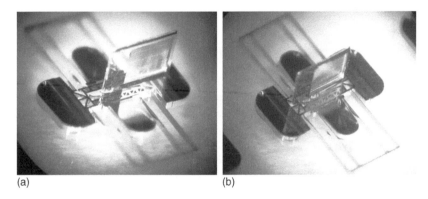

(a)   (b)

FIGURE 5.68   (a) Resonant amplitude grating device before grating engagement with holding springs. All materials are comprised of electroplated 78/22 Ni/Fe. Electrostatic drive consists of 50 electrode fingers 250 μm thick with 10 μm electrode gaps. Fabrication substrate is glass. (b) Resonant amplitude grating device with grating inserted into holding springs.

These examples illustrate that X-ray fabrication provides a unique tool set for milli- and microscale optical components particularly when form factor and precision alignment are crucial [Mohr, 1997; Mohr et al., 2003].

Example 4. Other Applications

Deep X-ray lithography fabricated gears have been used to realize a miniature in-line gear pump [Deng et al., 1998]. A magnetic coupling consists of a 78 Permalloy bar embedded in a 1.5 mm diameter PMMA gear that is driven by an external motor and permanent magnet. It drives two counter-rotating gears in a PMMA cavity. Via a maximum calculated coupling torque of 12.4 μNm and a rotation speed of 3500 rpm with 200 μm thick gears, pumping rates of 70 μl/min were achieved through 696 μm diameter inlet and outlet ports. Such a pump is self-priming and in this case could pull a maximum of 45 mm of water.

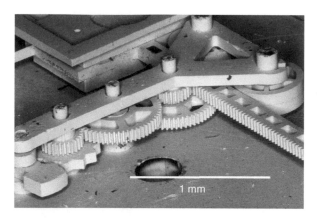

FIGURE 5.69 Assembled escapement mechanism used to provide damping for a spring-mass accelerometer structure via a rack-and-pinion coupling.

Combining the high density of electroplated materials such as gold with high aspect-ratio DXRL structures addresses two main issues for miniaturized inertial sensors. Figure 5.69 reveals an additional possibility of integrating mechanical damping to provide an acceleration-time response [Polosky et al., 1999]. In this device, a proof-mass with extended rack feature when accelerated is driven into a step-up gear train that drives an escapement and consequently provides damping. The resulting integrating accelerometer with total dimensions of 4×6 mm can sense 4 to 8 g accelerations.

Additional applications in instrumentation have been pursued. To precisely define quadrupole mass filters, DXRL-defined hyperboloid poles have been fabricated [Wiberg et al., 2003]. Additionally, to address vacuum pumping necessary for instrumentation, precision scroll pump geometry has been realized.

The precision afforded by DXRL processing combined with materials such as copper and ceramics is also enabling new high frequency RF components. Sources above 100 GHz that require cavity precision beyond that of conventional machining are under investigation [Ives et al., 2003]. Other RF component examples include high frequency passive elements such as bandpass filters and transmission lines that have been constructed within severe precision and tolerance constraints [Willke, T.L., 1997; Park et al., 2001].

5.10   Conclusions

Tools for the microfabrication of mechanical structures continue to be aided by advancing microelectronic concerns. Many applications of micromechanics ultimately depend in turn on its integration with microelectronics. A key aspect of micromachining is achieving precision at the submillimeter scale. Highly collimated X-rays enable commensurate lithographic precision at this scale while providing a unique high aspect-ratio capability. The utility of this batch fabrication approach is additionally enhanced by the additive nature of mold-based processing that lends itself to accommodating a large material base. The result is a generalized prismatic-based microfabrication tool set that can realize many useful microelectromechanical components and ultimately devices through successive additive batch microfabrication steps. Consequent microsystem integration and packaging remain challenges that this technology is well positioned to address via its low-temperature character. Access to appropriate synchrotron radiation sources continues to improve as applications demanding high microscale precision and accuracy become more prevalent. The impending technological impact of these applications with entirely new characteristics looms large.

Acknowledgments

This work was performed in part at Sandia National Laboratories, a multiprogram laboratory operated by Sandia Corp., a Lockheed Martin company, for the U.S. Department of Energy under contract DE-AC04-84AL85000.

For Further Information

An international workshop devoted to high aspect ratio microstructure technology (HARMST) was first held in 1995 with a particular emphasis on DXRL processing and LIGA technique. The meeting is held directly following the international transducers conference every other year. HARMST '95 was held in Karlsruhe, Germany; HARMST '97 in Madison, WI, USA; HARMST '99 in Kisarazu, Japan; HARMST '01 in Baden-Baden, Germany; and HARMST '03 in Monterey, CA, USA. A book of abstracts is published with the workshop, and selected papers from the workshop are published in the *Microsystem Technologies* research journal published by Springer.

Information on VLSI X-ray lithography with some DXRL subject matter is covered in the international conference on Electron, Ion, and Photon Beam Technology and Nanofabrication (EIPBN) with papers published in the *Journal of Vacuum Science and Technology B*.

Book chapters devoted to the topic of LIGA may be found in M. Madou (1997) *Fundamentals of Microfabrication*, CRC Press 1997, and P. Rai-Choudhury, ed. (1997) *Handbook of Microlithography, Micromachining, and Microfabrication*, vol.2, *Micromachining and Microfabrication*, SPIE Press, 1997, in chapters entitled "Plating Techniques" and "High Aspect Ratio Processing" (vol. 1 also contains a comprehensive review chapter on "X-ray Lithography").

Many LIGA related journal articles may be found in the IoP Journal of Micromechanics and Microengineering: Structures, Devices, and Systems.

Websites where additional information on LIGA may be found include: <http://www.fzk.de/stellent/groups/imt/documents/published_pages/en_imt_index.php>, <http://www.imm-mainz.de/>, and <http://www.ca.sandia.gov/liga/>

References

Ager III, J.W., Monteiro, O.R., Brown, I.G., Follstaedt, D.M., Knapp, J.A., Dugger, M.T., and Christenson, T.R. (1998) "Performance of Ultra Hard Carbon Wear Coatings on Microgears Fabricated by LIGA," *Materials Science of MEMS Devices — MRS Symposium Proc. Series* 546, pp. 115–20.

Allen, M.G. (1993) "Polyimide Based Processes for the Fabrication of Thick Electroplated Microstructures," *Proc. Transducers '93*, Yokohama, June 7–10, 1993, pp. 60–65.

Andricacos, P.C. et.al. (1996) "Vertical Paddle Plating Cell," U.S. Patent 5,516,412.

Barber, R.L., Ghantasala, M.K., Divan, R., Mancini, D.C., and Harvey E.C. (2004) "An Investigation of SU-8 Resist Adhesion in Deep X-ray Lithography of High-Aspect-Ratio Structures," *Proc. SPIE Device and Process Technologies for MEMS, Microelectronics, and Photonics III* 5276, pp. 85–91.

Becker, E.W, Betz, H., Ehrfeld, W., Glashauser, W., Heuberger, A., Michel, H.J., Münchmeyer, D., Pongratz, S., and Siemens, R.V. (1982) "Production of Separation Nozzle Systems for Uranium Enrichment by a Combination of X-ray Lithography and Galvoplastics," *Naturwissenschaften* 69, pp. 520–23.

Becker, E.W, Ehrfeld, W., Hagmann, P., Maner, A., and Münchmeyer, D. (1986) "Fabrication of Microstructures with High Aspect Ratios and Great Structural Heights by Synchrotron Radiation Lithography, Galvanoformung, and Plastic Molding (LIGA Process)," *Microelectron. Eng.* 4, pp. 35–56.

Biederman, H. (1987) "Polymer Films Prepared by Plasma Polymerization and their Potential Application," *Vacuum* 37, pp. 367–73.

Bogdanov, A.L., and Peredkov, S.S. (2000) "Use of SU-8 Photoresist for Very High Aspect Ratio X-ray Lithography," *Microelectron. Eng.* 53, pp. 493–96.

Boyce, B.L. (2003) "Fatigue of LIGA Nickel," *Proc. SPIE Reliability, Testing and Characterization of MEMS/MOEMS II* 4980, pp. 175–82.

Buchheit, T.E., Christenson, T.R., Schmale, D.T., and LaVan, D.A. (1998) "Understanding and Tailoring the Mechanical Properties of LIGA Fabricated Materials," *Materials Science of MEMS Devices — MRS Symposium Proc. Series* 546, pp.121–26.

Buchheit, T.E., LaVan, D.A., Michael, J.R., Christenson, T.R., and Leith, S.D. (2002) "Microstructural and Mechanical Properties Investigation of Electrodeposited and Annealed LIGA Nickel Structures," *Metall. Mat. Trans. A* 33A, pp. 539–54.

Christenson, T.R. (1995a) Micro-Electromagnetic Rotary Motors Realized by Deep X-ray Lithography and Electroplating, PhD Thesis, University of Wisconsin, Madison.

Christenson, T.R., and Guckel, H. (1995b) "Deep X-ray Lithography for Micromechanics," *Proceedings of SPIE Micromachining and Microfabrication Process Technology* 2639, pp. 134–45.

Christenson, T.R., Guckel, H., and Klein, J. (1996) "A Variable Reluctance Stepping Microdynamometer," *Microsystem Technologies* 2, no. 3, pp. 139–43.

Christenson, T.R. (1998a) "Advances in LIGA-Based Post Mold Fabrication," *Proc. of SPIE Micromachining and Microfabrication Process Technology IV* 3511, pp. 192–203.

Christenson, T.R., Buchheit, T.E., Schmale, D.T., and Bourcier, R.J. (1998b) "Mechanical and Metallographic Characterization of LIGA Fabricated Nickel and 80%Ni–20%Fe Permalloy," *Microelectromechanical Structures for Materials Research — Mater. Res. Proc.* 518, pp. 185–90.

Christenson,T.R., Buchheit, T.E., and Schmale, D.T. (1998c) "Torsion Testing of Diffusion Bonded LIGA Formed Nickel," *Materials Science of MEMS Devices — MRS Symposium Proc. Series* 546, pp. 127–32.

Christenson, T.R., Garino, T.J., and Venturini, E.L. (1999a) "Deep X-ray Lithography Based Fabrication of Rare-Earth Based Permanent Magnets and their Applications to Microactuators," *Electrochem. Soc. Proc.* 98-20, pp. 312–23.

Christenson,T.R., Garino, T.J., and Venturini, E.L. (1999b) "Microfabrication of Fully-Dense Rare-Earth Permanent Magnets via Deep X-ray Lithography and Hot Forging," *HARMST Workshop*, Kisarazu, Japan, Book of Abstracts, pp. 82–83.

Christenson, T.R., and Schmale, D.T. (1999c) "A Batch Wafer Scale LIGA Assembly and Packaging Technique via Diffusion Bonding," *Proc. of IEEE MEMS Workshop*, Orlando, FL, Jan. 17–21, 1999, pp. 476–81.

Christenson, T.R., Garino, T.J., Venturini, E.L., and Berry, D.M. (1999d) "Application of Deep X-ray Lithography Fabricted Rare-Earth Permanent Magnets to Multipole Magnetic Microactuators," *Digest of Technical Papers — 10th Int. Conf. On Solid-State Sensors and Actuators — Transducers'99*, Sendai, Japan, June 7–10, 1999, pp. 98–101.

Christenson, T.R., and Sweatt, W.C. (2001) "Micro-Optomechanical Uses of Deep X-ray Lithography (X-Optics)," *HARMST Workshop*, Baden-Baden, Germany, Book of Abstracts, June 17–19, 2001, pp. 3–4.

Cuisin, C., Chen, Y., Decanini, D., Chelnokov, A., Carcenac, F., Madouri, A., Lourtioz, J.M., and Launois, H. (1999) "Fabrication of Three-Dimensional Microstructures by High Resolution X-ray Lithography," *J. Vac. Sci. Technol. B* 17, pp. 3444–48.

Cuisin, C., Chelnokov, A., Lourtioz, J.-M., Decanini, D., and Chen, Y. (2000a) "Submicrometer Resolution Yablonovite Templates Fabricated by X-ray Lithography," *Appl. Phys. Lett.* 77, no. 6, pp. 770–72.

Cuisin, C., Chelnokov, A., Lourtioz, J.-M., Decanini, D., and Chen, Y. (2000b) "Fabrication of Three-Dimensional Photonic Structures with Submicrometer Resolution by X-ray Lithography," *J. Vac. Sci. Technol. B* 18, pp. 3505–9.

Cuisin, C., Chelnokov, A., Decanini, D., Peyrade, D., Chen, Y., and Lourtioz, J.M. (2002) "Sub-Micrometre Dielectric and Metallic Yablonovite Structures Fabricated from Resist Templates," *Opt. Quant. Electron.* 34, pp. 13–26.

DeCarlo, F., Song, J.J., and Mancini, D.C. (1998) "Enhanced Adhesion Buffer Layer for Deep X-ray Lithography Using Hard X-rays," *J. Vac. Sci. Technol. B* 16, pp. 3539–42.

Deng, K., Dewa, A.S., Ritter, D.C., Bonham, C., and Guckel, H. (1998) "Characterization of Gear Pumps Fabricated by LIGA," *Microsyst. Technol.* 4, no. 4, pp. 163–67.

Denginger, P.M., Clift, W.M., and Goods, S.H. (2002) "Removal of SU-8 Photoresist for Thick Film Applications," *Microelectron. Eng.* 61–62, pp. 993–1000.

Despa, M.S., Kelly, K.W., and Collier, J.R. (1999) "Injection Molding of Polymeric LIGA-HARMS," *Microsyst. Technol.* 6, no. 2, pp. 60–66.

Despont, M., Lorenz, H., Fahrni, N., Brugger, J., Renaud, P., and Vettiger, P. (1997) "High-Aspect Ratio, Ultrathick, Negative-Tone Near-UV Photoresist for MEMS Applications," *Proc. IEEE MEMS Workshop*, Nagoya, Japan, IEEE Press, Jan. 26–30, 1997, pp. 518–22.

Dutoit, B.M., Besse, P.-A., Blanchard, H., Guérin, L., and Popovic, R.S. (1999) "High Performance Micromachined Sm$_2$Co$_{17}$ Polymer Bonded Magnets," *Sensor. Actuator.* **77**, pp. 178–82.

Ehrfeld, W., Bley, P., Götz, F., Hagmann, P., Maner, A., Mohr, J., Moser, H.O., Münchmeyer, D., Schelb, W., Schmidt, D., and Becker, E.W. (1987) "Fabrication of Microstructures Using the LIGA Process," *Proc. of IEEE Micro Robots and Teleoperators Workshop*, Hyannis, MA, Nov. 9–11, 1987, pp. 1–11.

Ehrfeld, W., Götze, F., Münchmeyer, D., Schelb, W., and Schmidt, D. (1988a) "LIGA Process: Sensor Techniques via X-ray Lithography," Technical Digest, IEEE Solid State Sensor and Actuator Workshop, Hilton Head Isl., SC, June 6–9, 1988, pp. 1–4.

Ehrfeld, W., Bley, P., Götz, F., Mohr, J., Münchmeyer, D., Schelb, W., Baving, H.J., and Beets, D. (1988b) "Progress in Deep-Etch Synchrotron Radiation Lithography," *J. Vac. Sci. Technol. B* **6**, no. 1, pp. 178–85.

Ehrfeld, W. (1991) Univ. of Wisconsin, private communication.

Ehrfeld,W., Hessel, V., Lehr, H., Lowe, H., Schmidt, M., and Schenk, R. (1997) "Highly Sensitive Resist Material for Deep X-ray Lithography," *Proc. of SPIE — Intl. Society for Optical Engineering Conference* 3049, SPIE Press, Bellingham, WA, pp. 650–58.

Ehrfeld, W., and Schmidt, A. (1998) "Recent Developments in Deep X-ray Lithography," *J. Vac. Sci. Technol. B* **16**, pp. 3526–34.

Feddema, J.T., and Christenson, T.R. (1999) "Parallel Assembly of LIGA Components," *Tutorial on Modeling and Control of Micro and Nano-Manipulation, IEEE Intl. Conf. On Robotics and Automation.*

Feiertag, G., Ehrfeld, W., Lehr, H., Schmidt, A., and Schmidt, M. (1997a) "Calculation and Experimental Determination of the Structure Transfer Accuracy in Deep X-ray Lithography," *J. Micromech. Microeng.* **7**, Detroit, MI, May 11, 1999, pp. 323–31.

Feiertag, G., Ehrfeld, W., Lehr, H., Schmidt, A., and Schmidt, M. (1997b) "Accuracy of Structure Transfer in Deep X-ray Lithography," *Microelectron. Eng.* **35**, pp. 557–60.

Feiertag, G., Ehrfeld, W., Lehr, H., and Schmidt, M. (1997c) "Sloped Irradiation Techniques in Deep X-ray Lithography for 3D Shaping of Microstructures," *Proc. SPIE* 3048, *Emerging Lithographic Technologies*, SPIE Press, Bellingham, WA, pp. 136–45.

Feiertag, G., Ehrfeld, W., Freimuth, H., Kolle, H., Lehr, H., Schmidt, M., Sigalas, M.M., Soukoulis, C.M., Kiriakidis, G., Pedersen, T., Kuhl, J., and Koeneig, W. (1997d) "Fabrication of Photonic Crystals by Deep X-ray Lithography," *Appl. Phys. Lett.* **71**, pp. 1441–43.

Fischer, K., and Guckel, H. (1998) "Long Throw Linear Magnetic Microactuators Stackable to One Millimeter of Structural Height," *Microsyst. Technol.* **4**, No. 4, pp. 180–83.

Garino, T.J., Christenson, T.R., and Venturini, E.L. (1998) "Fabrication of MEMS Devices by Powder-Filling into DXRL-Formed Molds," *Materials Science of MEMS Devices — MRS Symposium Proc. Series* 546, pp. 195–200.

Gatzen, H.H., Maetzig, J.C., and Schwabe, M.K. (1996) "Precision Machining of Rigid Disk Head Sliders," *IEEE Trans. Magn.* **32**, pp. 1843–49.

Ghica, V., and Glashauser, W. (1983) U.S. Patent No. 4,393,129, "Method of Stress-Free Development of Irradiated Polymethylmethacrylate".

Goodman, J. (1960) "The Formation of Thin Polymer Films in the Gas Discharge," *J. Polym. Sci.* **44**, pp. 551–52.

Griffiths, S.K., and Nilson, R.H. (2002) "Transport Limitations on Development Times of LIGA PMMA Resists," *Microsyst. Technol.* **8**, pp. 335–42.

Griffiths, S.K. (2004) "Fundamental Limitations of LIGA X-ray Lithography: Sidewall Offset, Slope and Minimum Feature Size," *J. Micromech. Microeng.* **14**, pp. 999–1011.

Guckel, H., Uglow, J., Lin, M., Denton, D., Tobin, J., Euch, K., and Juda, M. (1988) "Plasma Polymerization of Methyl Methacrylate: A Photoresist for 3D Applications," *IEEE Solid-State Sensor and Actuator Workshop*, Hilton Head Island, SC, pp. 9–12.

Guckel, H., Burns, D.W., Christenson, T.R., Tilmans, H.A.C. (1989) "Polysilicon X-ray Masks," *Microelectron. Eng.* **9**, June 6–9, 1988, pp. 159–61.

Guckel, H., Christenson, T.R., Skrobis, K.J., Denton, D.D., Choi, B., Lovell, E.G., Lee, J.W., Bajikar, S.S., and Chapman, T.W. (1990) "Deep X-ray and UV Lithographies for Micromechanics," *Technical Digest — IEEE Solid State Sensor and Actuator Workshop*, pp. 118–22.

Guckel, H., Burns, D., Rutigliano, C., Lovell, E., and Choi, B. (1992) "Diagnostic Microstructures for the Measurement of Intrinsic Strain in Thin Films," *J. Micromech. Microeng.* **2**, pp. 86–95.

Guckel, H., Christenson, T.R., Skrobis, K.J., Klein, J., and Karnowsky, M. (1993) "Design and Testing of Planar Magnetic Micromotors Fabricated by Deep X-ray Lithography and Electroplating," *Digest of Technical Papers — 7th Int. Conf. on Solid-State Sensors and Actuators — Transducers '93*, Yokohama, pp. 76–9.

Guckel, H., Christenson, T.R., and Skrobis, K. (1996a) U.S. Patent No. 5,576,147, "Formation of Microstructures using a Preformed Photoresist Sheet."

Guckel, H. (1996b) "LIGA and LIGA-like Processing with High Energy Photons," *Microsyst. Technol.* **2**, pp. 153–56.

Guckel, H., Earles, T., Klein, J., Zook, J.D., and Ohnstein, T. (1996c) "Electromagnetic Linear Actuators with Inductive Position Sensing," *Sensor. Actuator. A* **53**, pp. 386–91.

Guckel, H. (1998a) "High Aspect-Ratio Micromachining Via Deep X-ray Lithography," *Proceedings of the IEEE* **86**, pp. 1586–93.

Guckel, H., Fischer, K., and Stiers, E. (1998b) "Closed-Loop Controlled Large Throw, Magnetic Linear Microactuator with 1000 μm Structural Height," *Proc. of IEEE MEMS Workshop*, Heidelberg, Germany, Jan 25–29, 1998, pp. 414–18.

Guckel, H., Fischer, K., Stiers, E., Chaudhuri, B., McNamara, S., Ramotowski, M., Johnson, E.D., and Kirk, C. (1999a) "Direct, High Throughput LIGA for Commercial Applications: A Progress Report," Book of Abstracts, *HARMST Workshop*, Kisarazu, Japan, June 13–15, 1999, pp. 2–3.

Guckel, H., Fischer, K., Chaudhuri, B., Stiers, E., McNamara, S., and Martin, T. (1999b) "Single Mode Optical Fiber Switch," *HARMST '99 Workshop*, Kisarazu, Japan, June 13–15, 1999, pp. 146–47.

Hall, A. (2000) Sandia National Laboratories, private communication.

Harris, C., Despa, M., and Kelly, K. (2000) "Design and Fabrication of Cross-Flow Micro Heat Exchanger" *J. Microelectromech. Syst.* **9**, no. 4, pp. 502–08.

Harsch, S., Ehrfeld, W., and Maner, A. (1988) "Untersuchungen zur Herstellung von Mikrostructuren grosser Strukturhöhe durch Galvanoformung in Nickel-sulfamatelektrolyten," Kernforschungszentrum Karlsruhe, Germany, Report No. 4455.

Henzi, P., Truckenmüller, R., Herrmann, D., Achenbach, S., Wallrabe, U., and Mohr, J. (2003) "Bonding of LIGA PMMA Microstructures by UV Irradiation and Low Temperature Welding," *HARMST '03 Workshop*, Monterey, CA, June 15–17, 2003, pp. 197–98.

Hirata, T., Chung, S.-J., Hein, H., Akashi, T., and Mohr, J. (1999) "Micro Cycloid-Gear System Fabricated by Multiexposure LIGA Technique," *Proc. SPIE Materials and Device Characterization in Micromachining II* 3875 pp. 164–71.

Holswade, S. (1999) Sandia National Laboratories, private communication.

Ives, L., Kory, C., Read, M., and Booske, J. (2003) "Application of HARMST to RF Source Development," *HARMST '03 Workshop*, Monterey, CA, June 15–17, 2003, pp. 199–200.

Jian L., Aigeldinger, G., Desta, Y.M., and Goettert, J. (2001) "SU-8 as Negative Photoresist in Deep X-ray Lithography," *HARMST '01 Workshop*, Baden-Baden, Germany, June 17–19, 2001, pp. 79–80.

Kelly, K.W., Harris, C., Stephens, L.S., Marques, C., and Foley, D. (2001) "Industrial Applications for LIGA-Fabricated Micro Heat Exchangers," *Proc. SPIE MEMS Components and Applications for Industry, Automobiles, Aerospace and Communication* 4559, SPIE Press, Bellingham, WA, pp. 73–84.

Khan Malek, C.G., and Das, S.S. (1998a) "Adhesion Promotion between Poly(Methylmethacrylate) and Metallic Surfaces for LiGA Evaluated by Shear Stress Measurements," *J. Vac. Sci. Technol. B* **16**, pp. 3543–46.

Khan Malek, C. (1998b) "Mask Prototyping for Ultra-Deep X-ray Lithography: Preliminary Studies for Mask Blanks and High-Aspect-Ratio Absorber Patterns," *Proc. of SPIE Conference on Materials and*

Device Characterization in Micromaching 3512, Sept. 21–22, Santa Clara, CA, SPIE Press, Bellingham, WA, pp. 277–85.

Kiriakidis, G., and Katsarakis, N. (2000) "Fabrication of 2-D and 3-D Photonic Band-Gap Crystals in the GHz and THz Regions," *Mater. Phys. Mech.* 1, pp. 20–26.

Lee, K.Y., LaBianca, N., Rishton, S.A., Zolgharnain, S., Gelorme, J.D., Shaw, J., and Chang, T.H.-P. (1995) "Micromachining Applications of a High Resolution Ultrathick Photoresist," *J. Vac. Sci. Technol. B* 13, pp. 3012–16.

Lee, C.-K.T., and Vasile, M.J. (1998) "A Direct Molding Technique to Fabricate Silica Micro-Optical Components," *Materials Science of MEMS Devices — MRS Symposium Proc. Series* 546, pp. 189–94.

Lee, J., Rogers, J.D., Descour, M.R., Hsu, E., Aaron, J.S., Sokolov, K., and Richards-Kortum, R.R. (2003) "Imaging Quality Assessment of Multi-Modal Miniature Microscope," *Opt. Express* 11, pp. 1436–51.

Lin, B.J. (1975) "Deep UV Lithography," *J. Vac. Sci. Technol.* 12, pp. 1317–20.

Ling, Z.-G., and Lian, K. (2003) "Expansion of SU-8 Application Scope by PAG Concentration Modification," *Proc. SPIE Micromachining and Microfabrication Process Technology VIII* 4979, pp. 402–9.

Loechel, B., and Maciossek, A. (1995) "Surface Micro Components Fabricated by UV Depth Lithography and Electroplating," *Proc. of SPIE Micromachining and Microfabrication Process Technology* 2639, SPIE Press, Bellingham, WA, pp. 174–84.

Manohara, H. (1998) "Development of Hard X-ray Exposure Tool for Micromachining at CAMD," *Proc. of 15th Intl. Conf. on Application of Accelerators in Research and Industry*, part 2, AIP Conf. Proc., Duggan, J.L., and Morgan, I.L. eds., no. 475, Springer–Verlag, AIP Press, Secaucus, NJ, pp. 618–21.

Margaritondo, G. (1988) *Introduction to Synchrotron Radiation*, Oxford University Press, New York.

Marti, J.L. (1966) "The Effect of Some Variables upon Internal Stress of Nickel as Deposited from Sulfamate Electrolytes," *Plating* Jan., pp. 61–71.

Massoud-Ansari, S., Mangat, P.S., Klein, J., and Guckel, H. (1996) "A Multi-Level, LIGA-Like Process for Three-Dimensional Actuators," *Proc. of IEEE MEMS Workshop*, San Deigo, CA, Feb. 11–15, 1996, pp. 285–89.

McNamara, S. (1999) U. of Wisconsin, private communication.

Mehdizadeh, S., Kukovic. J., Andricacos, P.C., Romankiw, L.T., and Chen, H.Y. (1993) "The Influence of Lithographic Patterning on Current Distribution in Electrochemical Microfabrication," *J. Electrochem. Soc.* 140, p. 3497.

Mohr, J. (1986) Analyse der Defektursachen und der Genauigkeit der Strukturübertragung bei der Röntgentiefenlithographie mit Synchrotronstrahlung, dissertation, Universität Karlsruhe.

Mohr, J., Goettert, J., Mueller, A., Ruther, P., and Wengeling, K. (1997) "Micro-Optical and Optomechanical Systems Fabricated by the LIGA Technique," *Proc. SPIE Miniaturized Systems with Micro-Optics and Micromechanics II* 3008, SPIE Press, Bellingham, WA, pp. 273–78.

Mohr, J., Ehrfeld, W., and Münchmeyer, D. (1988) "Requirements on Resist Layers in Deep-Etch Synchrotron Radiation Lithography," *J. Vac. Sci. Technol. B* 6, pp. 2264–67.

Mohr, J., Hollenbach, U., Last, A., and Wallrabe, U. (2003) "Microoptics Made by LIGA Technology," *HARMST '03 Workshop*, Monterey, CA, June 15–17, 2003, pp. 169–70.

Morita, S., Tamano, J., Hattori, S., and Ieda, M. (1980) "Plasma Polymerized Methyl-Methacrylate as an Electron-Beam Resist," *J. Appl. Phys.* 51, pp. 3938–41.

Münchmeyer, D., and Ehrfeld, W. (1987) "Accuracy Limits and Potential Applications of the LIGA Technique in Integrated Optics," *Proc. SPIE 803 — Micromachining of Elements with Optical and Other Submicrometer Dimensional and Surface Specifications*, Weck, M. ed., SPIE Press, Bellingham, WA, pp. 72–79.

Myers, S.M., Follstaedt, D.M., Knapp, J.A., and Christenson, T.R. (1997) "Hardening of Nickel Alloys by Implantation of Titanium and Carbon," *Mat. Res. Soc. Symp. Proc.* 444, pp. 99–104.

Ohnstein, T.R. et.al. (1995) "Tunable IR Filters Using Flexible Metallic Microstructures," *Proc. IEEE MEMS*, Amsterdam, Jan. 29–Feb. 2, 1995, pp. 170–74.

Ohnstein, T.R., Zook, J.D., French, H.B., Guckel, H., Earles, T., Klein, J., and Mangat, P. (1996) "Tunable IR Filters with Integral Electromagnetic Actuators," *Tech. Digest of the 1996 Solid-State Sensor and Actuator Workshop*, Hilton Head Island, SC, June 2–6, 1996, pp. 196–99.

Osaka, T., Takai, M., Hayashi, K., and Sogawa, Y. (1998) "New Soft Magnetic CoNiFe Plated Films with High B_s = 2.0-2.1 T," *IEEE Trans. Magn.* **34**, pp. 1432–34.

Ouano, A.C. (1978) "A Study on the Dissolution Rate of Irradiated Poly(Methyl Methacrylate)," *Polym. Eng. Sci.* **18**, pp. 306–13.

Pantenburg, F.J., and Mohr, J. (1995) "Influence of Secondary Effects on the Structure Quality in Deep X-ray Lithography," *Nucl. Instrum. Methods Phys. Res. B, Beam Interact. Mater. At.* June 15–17, 2003, B97, pp. 551–56.

Pantenburg, F.J., Achenbach, S., and Mohr, J. (1998) "Influence of Developer Temperature and Resist Material on the Structure Quality in Deep X-ray Lithography," *J. Vac. Sci. Technol. B* **16**, pp. 3547–51.

Pantenburg, F.J., Bankert, M.A., Domeier, L.A., Griffiths, S.K., and Wepfer, K.C. (2003) "Development Rates of LIGA PMMA X-ray Resists," *HARMST '03 Workshop*, Monterey, CA, June 15–17, 2003, pp. 9–10.

Park, K.-Y., Lee, J.-C., Kim, J.-H., Lee, B., Kim, N.-Y., Park, J.-Y., Kim, G.-H., Bu, J.-U., and Chung, K.-W. (2001) "A New Three-Dimensional 30 GHz Bandpass Filter Using the LIGA Micromachined Process," *Microwave Opt. Tech. Lett.* **30**, no. 3, pp. 199–201.

Peele, A. (1999) "Investigation of Etched Silicon Wafers for Lobster-Eye Optics" *Rev. Sci. Instrum.* **70**, pp. 1268–73.

Pioter, V., Hanemann, T., Ruprecht, R., Thies, A., and Haußelt, J. (1997) "New Developments of Process Technologies for Microfabrication," *Proceedings of SPIE Micromachining and Microfabrication Process Technology III* 3223, SPIE Press, Bellingham, WA, pp. 91–99.

Polosky, M.A., Christenson, T.R., Jojola, A.J., and Plummer, D.W. (1999) "LIGA Fabricated Environmental Sensing Device," *HARMST '99 Workshop*, Kisarazu, Japan, June 13–15, 1999, pp. 144–45.

Prasad, S.V., Christenson, T.R., and Dugger, M.T. (2002) "Tribological Coatings for LIGA MEMS," *Adv. Mater. Process.* **160**, pp. 30–34.

Prasad, S.V., Michael, J.R., and Christenson, T.R. (2003) "EBSD Studies on Wear Induced Subsurface Regions in LIGA Nickel," *Scripta Mater.* **40**, pp. 255–60.

Ritzhaupt-Kleissl, H.-J., Bauer, W., Günther, E., Laubersheimer, J., and Haußelt, J. (1996) "Development of Ceramic Microstructures," *Microsyst. Technol.* **2**, no. 3, pp. 130–34.

Romankiw, L.T., Croll, I.M., and Hatzakis, M. (1970) "Batch-Fabricated Thin-Film Magnetic Recording Heads," *IEEE Trans. Magn.* **MAG-6**, pp. 597–601.

Romankiw, L.T. (1995) "Evolution of the Plating Through Lithographic Mask Technology," *ECS Proceedings* 95-18, Romankiw, L.T., Herman, D., Jr. eds., Pennington, New Jersey, pp. 253–73.

Safranek, W.H. (1986) *The Properties of Electrodeposited Metals and Alloys*, 2nd ed., AESF Society, Orlando.

Sakai, N., Tada, K., Utsumi, Y., and Hattori, T. "Advanced Ultra-Thick Negative Photoresist with High Sensitivity for X-ray and UV Deep Lithography," *HARMST '03 Workshop*, Monterey, CA, June 15–17, 2003, pp. 15–16.

Schmalz, O., Hess, M., and Kosfeld, R. (1996) "Structural Changes in Poly(Methyl Methacrylate) during Deep-Etch X-ray Synchrotron Radiation Lithography: Part 3. Mode of Action of the Developer," *Angew. Makromol. Chem.* **239**, pp. 93–106.

Sekimoto, M., Yoshihara, H., and Ohkubo, T. (1982) "Silicon Nitride Single-Layer X-ray Mask," *J. Vac. Sci. Tech.* **21**, p. 1017–21.

Shaw, J.M., Gelorme, J.D., LaBianca, N.C., Conley, W.E., and Holmes, S.J. (1997) "Negative Photoresists for Optical Lithography," *IBM J. Res. Dev.* **41**, no. 1/2, pp. 81–94.

Shew, B.-Y., Hung, J.-T., Huang, T.-Y., Liu, K.-P., and Chou, C.-P. (2003a) "High Resolution X-ray Micromachining Using SU-8 Resist," *J. Micromech. Microeng.* **13** pp. 708–13.

Shew, B.-Y., Huang, T.-Y., Liu, K.-P., and Chou, C.-P. (2003b) "Oxygen Quenching Effect in Ultra-Deep X-ray Lithography with SU-8 Resist," *J. Micromech. Microeng.* **14**, pp. 410–14.

Siddons, D.P., Johnson, E.D., and Guckel, H. (1994) "Precision Machining with Hard X-rays," *Synchrotron Radiat. News* **7**, no. 2, pp. 16–18.

Siddons, D.P., Johnson, E.D., Guckel, H., and Klein, J.L. (1997) "Method and Apparatus for Micromachining Using X-rays," U.S. Patent No. 5,679,502.

Singleton, L., Bogdanov, A.L., Peredkov, S., Wilhelmi, O., Schneider, A., Cremers, C., Megtert, S., and Schmidt, A. (2001) "Deep X-ray Lithography with the SU-8 Resist," *Proc. SPIE Emerging Lithographic Technologies V* 4343, pp. 182–92 June 13–15, 1999, p. 83.

Spears, D.L., and Smith, H.I. (1972) "High-Resolution Pattern Replication Using Soft X-rays," *Electron. Lett.* **8**, no. 4, pp. 102–4.

Spiller, E., Feder, R., Topalian, J., Castellani, E., Romankiw, L., and Heritage, M. (1976) "X-ray Lithography for Buble Devices," *Solid State Technol.* April, pp. 62–68.

Stiers, E., and Guckel, H. (1999) "A Closed-Loop Control Circuit for Magnetic Microactuators," *HARMST '99 Workshop*, Kisarazu, Japan, June 13–15, 1999, p. 83.

Stutz, A. (1986) Untersuchungen Zum Entwicklungsverhalten eines Röntgenresist aus Vernetztem Polymethylmethacrylat, Diplomarbeit, Institut fur Kernverfahrenstechnik des Kernforschungszentrums Karslruhe und der Universität Karlsruhe.

Sweatt, W.C., Christenson, T., and Lin, S.Y. (2002) "Low Cost, Large Area Fabrication of Yablonovite in the Infrared," *Photonic and Electromagnetic Crystal Structures (PECS IV) — Abstract Book*, Los Angeles, IPAM, Univ. of California — Los Angeles, p. 154.

Sweatt, W.C., Schiller, F., and Christenson, T.R. (2003) "Fabricating Microoptical Systems with Deep X-ray Lithography," SPIE Conference, Micromachining Technology for Microoptics, 4984, San Jose, CA, Jan. 28–29, 2003, p. 4.

Takahata, K., Shibaike, N., and Guckel, H. (1999) "A Novel Micro Electro-Discharge Machining Method Using Electrodes Fabricated by the LIGA Process," *Technical Digest — IEEE MEMS '99*, Orlando, FL, Jan. 17–21, 1999, pp. 238–43.

Tan, M.X., Bankert, M.A., and Griffiths, S.K. (1998) "PMMA Development Studies Using Various Synchrotron Sources and Exposure Conditions," *SPIE Conference on Materials and Device Characterization in Micromachining* 3512, September, pp. 262–70.

Toader, O., Berciu, M., and John, S. (2003) "Photonic Band Gaps Based on Tetragonal Lattices of Slanted Pores," *Phys. Rev. Lett.* **90**, pp. 233901-1–233901-4.

Van der Gaag, B.P., and Sherer, A. (1990) "Microfabrication below 10 nm," *Appl. Phys. Lett.* **56**, p.481.

Visser, C.C.G., Uglow, J.E., Burns, D.W., Wells, G., Redaelli, R., Cerrina, F., and Guckel, H. (1987) "A New Silicon Nitride Mask Technology for Synchrotron Radiation X-ray Lithography: First Results," *Microelectron. Eng.* **6**, pp. 299–304.

Weil, R. (1970) "The Origins of Stress in Electrodeposits," *Plating*, AES Research Project 22, pp. 1231–37.

Wiberg, D.V., Myung, N.V., Eyre, B., Shcheglov, K., Orient, O.J., Moore, E., and Munz, P. (2003) "LIGA-Fabricated Two-Dimensional Quadrupole Array and Scroll Pump for Miniature Gas Chromatograph/Mass Spectrometer," *Proc. SPIE First Jet Propulsion Laboratory In Situ Instruments Workshop* 4878, SPIE Press, Bellingham, WA, pp. 8–13.

Willke, T.L., and Gearhart, S.S. (1997) "LIGA Micromachined Planar Transmission Lines and Filters," *IEEE Trans. Microwave Theory Tech.* **45**, pp. 1681–88.

Yablonovitch, E., Gmitter, T.J., and Leung, K.M. (1991) "Photonic Band Structure: The Face-Centered-Cubic Case Employing Nonspherical Atoms," *Phys. Rev. Lett.* **67**, pp. 2295–98.

6

EFAB™ Technology and Applications

Ezekiel J. J. Kruglick, Adam
L. Cohen and Christopher
A. Bang
Microfabrica Inc.

6.1 Introduction

EFAB™ technology [Cohen, 1999] is a unique micromanufacturing process based on multilayer electrodeposition of materials that is capable of producing complex, truly three-dimensional microscale components, devices, and systems, from prototypes through volume production. Figures 6.1 through 6.5 show some typical devices produced using EFAB technology. The technology was invented at the University of Southern California and licensed to Microfabrica Inc. (formerly MEMGen Corporation), a venture-funded company founded in 1999 and located in the 'Los Angeles, California area'.

In brief, the EFAB process generates multilayer devices by repeating a basic process flow several times in sequence and comprises three primary steps: (1) patterned layer deposition, (2) blanket layer deposition, and (3) planarization. The process is illustrated in Figure 6.6. In the first step (top left), a layer is patterned as deposited. A second, blanket deposition (center left) fills the spaces remaining after pattern deposition. The topography is then fully planarized (bottom left), resulting in a flat surface suitable for the next patterned deposition. A completed layer produced with EFAB technology thus consists of both a structural and a sacrificial material, as illustrated in the lower left portion of Figure 6.6. The sacrificial material in which EFAB-built devices are temporarily embedded serves the same purpose as sacrificial material in surface micromachining, i.e., mechanical support of structural material. Moreover, since both materials are conductive and platable, a layer can be deposited over another regardless of the particular cross-sectional geometry of either one. This fact eliminates virtually all restrictions on microstructure geometry,[1] allowing the structural material on a layer to overhang — or even be disconnected from — that of the previous

---

[1]The only restriction relates to the removal of sacrificial material: a hollow structure with no access to the interior cannot be fabricated, and certain geometries may include gaps, holes, and slots that are difficult to clear of sacrificial material, especially if these features are long and narrow.

FIGURE 6.1   36-layer springs made with EFAB technology.

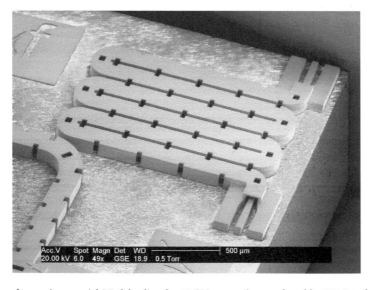

FIGURE 6.2   1-cm long microcoaxial RF delay line for 30 GHz operation produced by EFAB technology.

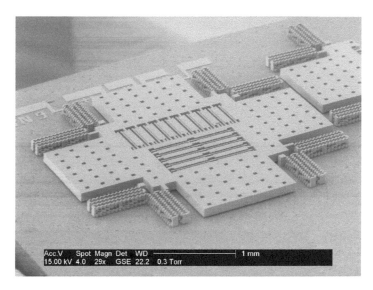

FIGURE 6.3 Two-axis accelerometer produced with EFAB technology.

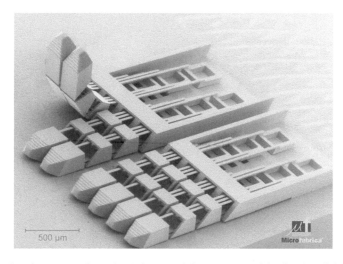

FIGURE 6.4 Micro-hands 1 mm wide with triple-jointed fingers actuated by "tendons," fabricated using EFAB technology.

layer. Such geometrical freedom also makes possible monolithically fabricated "assemblies" of discrete, interconnected parts (e.g., a gear train) without the need for actual assembly.

The layer cycle (selective deposition, blanket deposition, and planarization) is repeated to produce additional layers until the full desired height of the device is achieved (upper right). Finally, after the build is completed, the substrate is placed in a release etchant which removes the sacrificial material, leaving behind the free-standing device (lower right). The number of layers is virtually arbitrary, as is the thickness of each individual layer and, as already noted, the geometric pattern on each layer; these facts allow for a high degree of design flexibility and device variation within a single process. In a typical process flow, the sacrificial material (e.g., electroplated copper) is selectively deposited as a first step to produce a patterned deposit, while the structural material (e.g., electroplated nickel) is blanket deposited as a second step. However, this process may be inverted such that the structural material is selectively deposited and the sacrificial material is blanket deposited.

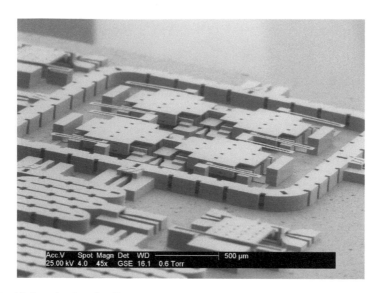

FIGURE 6.5 Co-fabricated suite of millimeter wave-components (delay lines, 30 GHz hybrid couplers, switches) produced with EFAB technology.

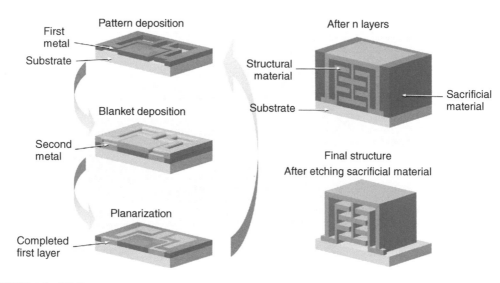

FIGURE 6.6 EFAB process steps.

EFAB technology is now qualified and routinely performed on 100-mm dielectric wafers using 150 mm-capable equipment. The process produces excellent interlayer adhesion and layer flatness. EFAB technology yields features down to 10 μm — with a path to 2–5 μm, and materials are well-characterized. Devices with over 50 layers have been built with a total height exceeding 1000 μm. Finally, a set of design rules is available for the process.

6.2   Why EFAB?

As a microfabrication technology, EFAB offers several potential benefits that make it worth considering when evaluating one's options for producing a new microscale product.

6.2.1   Flexibility

The first of these benefits is geometrical flexibility; i.e., the ability to create devices with a wide variety of truly three-dimensional geometries. It needs to be noted that true 3-D is different than what has been sometimes called "2.5-D," the latter being the "extruded" shapes commonly produced by LIGA (Lithographie (lithography), Galvanoformung (electroforming), and Abformung (molding)) or deep reactive ion etching DRIE. True 3-D becomes possible when the number of layers from which a device can be fabricated becomes significant (e.g., several dozen) and a sacrificial material is provided to support the structural material during fabrication. Extremely complex geometries — which may include overhanging, reentrant, and internal features — are possible with EFAB technology. Three-D has several advantages for microdevices. Performance can often be improved and optimized (e.g., more powerful actuators, larger sense capacitances, fully isolated microwave transmission lines, radio frequency (RF) devices with higher Q, larger proof masses). New and higher levels of functionality can be achieved (e.g., more sophisticated spring designs, microwave filters, completely new actuator geometries). Packages can be co-fabricated along with devices, and the need for assembly (e.g., of inkjet printheads) can be minimized through the use of more complex shapes, both resulting in cost savings. Many different versions of a particular design can be cofabricated together on the same substrate, allowing quicker and more complete optimization of a design. Electrostatic and electromagnetic isolation can be more easily achieved. Finally, simply having access to a 3-D process and designing in 3-D (vs. in 2-D) allows the device designer to focus on achieving product performance objectives, with much less attention to process constraints.

6.2.2   Versatility

EFAB technology allows many different types of devices to be built using the same process, much like the complementary metal oxide semiconductor (CMOS) process or, looking at the world of macroscale manufacturing, where computer numerical control (CNC) machining, injection molding, sand casting are generic, standard manufacturing processes, in the sense that a very large variety of products can be made using them. Versatility is, in fact, a prerequisite to the existence of a standard process, and the minimal versatility among most MEMS (microelectromechanical systems) processes seems to be a major reason why they have not evolved as standard ways of making things at the microscale. Not having to develop a custom process for a new design clearly helps to reduce time-to-market and development cost and promotes the development of mature, well-characterized manufacturing infrastructure. Moreover, the ability to co-fabricate different devices on a single substrate (as shown in Figure 6.5) enables cost-effective (e.g., without the need for microscale assembly) fabrication of subsystems and entire systems. Such systems make possible higher value-added contributions to commercial products, in other words, MEMS that are truly systems and not just components. System-level fabrication also leverages the cost and development-time benefits of design reuse, which enables tested libraries of component designs to be "plugged in" to create the desired result.

6.2.3   Accessibility

The fact that EFAB design is carried out with familiar 3-D CAD (computer assisted design) tools, with the manufacturing process flow automatically derived from the design, makes EFAB technology particularly accessible — even to nonexperts in MEMS design and fabrication. Indeed, the design and fabrication experience is largely what-you-see-is-what-you-get (WYSIWYG). Thus, EFAB technology makes microfabrication relatively easy and intuitive. Engineers with extensive domain-specific experience (in RF, medical devices, and so forth) but little MEMS experience can single-handedly design successful devices, a capability that has been demonstrated recently.

6.2.4   Materials

Whereas silicon, especially single-crystal silicon, has several desirable properties as an engineering material for MEMS, the availability of pure metals and alloys in EFAB technology can be advantageous in a large number of applications. Metals (e.g., Cu, Ag) that have far greater electrical conductivity and optical reflectivity (e.g., Au at infrared wavelengths) than Si can be plated. Some metals (e.g., Pt, Rh) offer high-temperature performance while others (e.g., Pt, Au) offer improved biocompatibility. Other metals (e.g., Ni-Fe) offer magnetic properties, catalytic properties (e.g., Pt), and so on. In general, metals also offer ductility and impact resistance not provided by Si.

EFAB also allows a wide variety of substrates to be used, including metals, ceramics, glass, polymers, and semiconductor wafers such as CMOS, providing an extra dimension of materials flexibility. Part of this flexibility in substrates derives from the fact that EFAB technology is based on low-temperature electrodeposition from aqueous baths, making it feasible to fabricate devices directly onto heat-sensitive substrates.

6.2.5   Size and Aspect Ratio

EFAB technology allows devices to be fabricated over a wider range of sizes than most conventional MEMS processes, from tens of microns to millimeters. Moreover, not all microdevices are best fabricated at the scale of typical Si MEMS (e.g., with heights in the range of $10\,\mu m$) but are preferably made larger to increase robustness, maximize performance (e.g., small pressure drops in microfluidic channels or large optical aperture in moving mirror devices), etc. EFAB technology helps fill a gap in available device size between Si MEMS and more macroscale manufacturing processes. Functional devices that are large in the plane of the substrate ($X–Y$) typically cannot be produced unless there is an ability to make them proportionately large in the out-of-plane (Z) direction; otherwise, the low aspect ratio leads to mechanically "floppy" structures. While LIGA, ultraviolet LIGA (UV-LIGA), and DRIE offer high aspect ratios, the 2.5-D nature of these processes limits their usefulness. Si surface micromachining, although capable of more complex geometry, usually is limited to just a few (e.g., five, in the case of Sandia's Summit V process), thin (e.g., $1–2\,\mu m$) layers. EFAB technology, however, allows for extremely complex geometry coupled with high aspect ratio, and thus it enables devices that can range in size from those typically made from Si MEMS to much larger devices.

6.2.6   Speed

Time-to-market can be as important for MEMS products as it is for macroscale products. Indeed, since MEMS are often incorporated into larger products as components or subsystems, their time-to-market must be significantly less than the higher-level product or they will be "too late" to be useful.

Thus, it is desirable to accelerate the design, prototyping, and design optimization cycles as much as possible, and EFAB technology can be beneficial in this regard. EFAB design is normally faster than design for conventional MEMS processes, since the design is far less constrained by manufacturing limitations, and design rules are few and easy to keep in mind; the design is done in 3-D instead of 2-D, eliminating the need to try to visualize the geometry that will ultimately be produced; 3-D designs may be directly simulated using finite element analysis packages, some of which are integrated directly into 3-D design software packages; many design variations can be built and tested together; and preexisting designs can more easily be reused.

EFAB prototyping is also typically faster than that of conventional MEMS. Unless the device requires development of a new material or a modification of the standard process, no time is spent developing and qualifying the fabrication process; a production-qualified process already exists through which devices can be built and tested with high confidence that they will behave in production the same way they behaved as prototypes. Secondly, prototyping is faster, since there are fewer steps to be performed per layer and deposition rates for electroplating are normally higher than for chemical vapor deposition (CVD) or physical vapor deposition (PVD). Typical lead times for prototypes with 20 layers are three to four weeks from design tape-out, subject to available capacity. Since MEMS are so small, rework and trimming to correct a design problem is very difficult, and redesign and refabrication are usually needed; to allow for this obstacle, MEMS arguably need to be prototyped even more quickly than most macroscale devices.

Finally, EFAB facilitates design optimization by speeding up the design and prototyping cycles and, as already noted, by allowing an unusually wide variety of design variations to be fabricated in a single prototyping run, which helps to make design optimization more of a parallel vs. a strictly serial process.

6.2.7 Cost

Last, but hardly least, the cost of devices fabricated with EFAB can be less than with conventional processes, for several fundamental reasons. First, the cost of developing a manufacturing process (which ultimately becomes part of the cost per device) is eliminated, so long as the standard EFAB process and materials set can be used unchanged. At this writing, many practical devices can already be fabricated without additional development, and as EFAB process capabilities grow and more materials are added to the palette, the need to do any process or materials development will be very infrequent.

Second, the equipment and materials required for EFAB manufacturing are less costly than for most other MEMS processes. For example, electrodeposition equipment costs less than CVD or PVD equipment, and Ni plating baths and anodes of the required purity cost less per kilogram than single crystal silicon or the materials used in depositing polysilicon.

Third, the 3-D nature of EFAB technology enables reduced assembly and — as will be discussed in the following section — packaging costs, the latter through the use of wafer-scale self-packaging.

6.3 EFAB™ Technology

Figure 6.7 illustrates the overall high-level process flow for EFAB technology, from design through completed devices. The process normally begins with 3-D CAD design and (optional) simulation, resulting in a file describing the geometry of the device or assembly in the STL file format. The STL file is imported into a Microfabrica-developed software package called Layerize™, which performs a mathematical cross-sectioning operation on the geometry. Layerize outputs one or more GDSII files with the geometry of each cross-section of the device, as well as a "FAB" file containing fabrication-related parameters. Layerize can also import 2-D design and layout files in GDSII format if one prefers to perform either function in the 2-D domain. GDSII files are then used to produce one or more photomasks using conventional pattern generation equipment. These photomasks may then be used to fabricate Instant Masks™ if these are to be used for selective deposition, or else the photomasks are directly provided to the EFAB layer process if photoresist will be used in selective deposition. The EFAB layer process (consisting of a set of steps repeated for every layer) is then carried out, with photomasks or Instant Masks, substrates, consumable materials, and the FAB file as inputs. Figure 6.8 depicts the steps in the layer process for Instant Masking-based EFAB technology and photoresist-based EFAB technology. After all layers have been fabricated, postprocessing is then performed on the resulting "build" to remove the sacrificial material, singulate devices built together on the wafer, and if required, perform packaging.

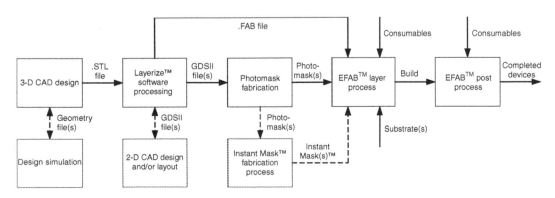

FIGURE 6.7 EFAB technology process flow, from design through finished devices.

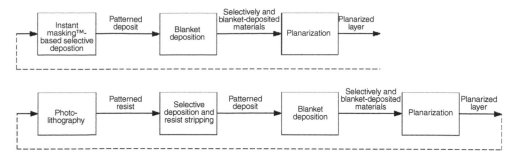

FIGURE 6.8   EFAB layer process for Instant Mask processing or photoresist processing.

6.3.1   Detailed Process

6.3.1.1   3-D CAD Design and Simulation

Design for EFAB can generally be accomplished using any of a wide variety of 3-D mechanical CAD software packages. Among the most commonly used packages are Solidworks® and Pro/ENGINEER®; however, virtually all 3-D packages are able to output geometry in the STL file format that is the *de facto* industry standard for driving rapid prototyping systems such as those based on stereolithography, 3-D printing, and fused deposition modeling. The designer works with a 3-D solid model, continuously observing the microdevice geometry on the monitor from any desired angle and immediately seeing the results of manipulations (e.g., adding a boss, forming a slot, changing a dimension). Because the ultimate microdevice will be built from multiple layers, the designer normally keeps in mind the effect (if any) that layering will have on the final design and may set dimensions along the Z (building) axis to be equal to the layer thicknesses specified later. During or after the design process, the 3-D geometry can also be exported (unless the package is integrated into the CAD software) in a number of 3-D formats to 3-D analysis packages — among them ANSYS® and COSMOSWorks — for simulation of physical behavior. Packages are available which can simulate a variety of physical domains including mechanical, thermal, electromagnetic, electrostatic, and fluid dynamic, including, in some cases, the interaction among these domains. Normally the design is iterated according to simulation results, and eventually a final design is obtained. At this point, one or more STL files is exported, representing a single device or system, or possibly several devices intended to be co-fabricated.

6.3.1.2   Layerize

The STL file is read into the Windows®-based Layerize software package, and the designer then specifies several parameters, among them the device orientation relative to the build (Z) axis; the thickness for each layer (the thickness of each layer can be set independently of the others, within the acceptable range); the number of copies of the device to be built; and the dicing lane width. Layerize then orients the CAD geometry and calculates the cross-sectional geometry for each layer at the specified thickness. The software displays the cross-sectioned geometry, allowing the user to check for design rule violations and other errors.[2] Based on measured input, Layerize can also modify the geometry to compensate for accuracy errors in the X–Y plane. It then automatically lays out the wafer (including ID label) across the available photomask area, replicating the calculated geometry as desired, taking into account the specified dicing lane width, and displaying the resulting photomask layout. Finally, Layerize exports files of two different types. The first is one or more GDSII files that represents the geometry of each photomask required to fabricate the build. For these files, Layerize handles the complex topology that may be inherent in a device design in a format that is compatible with the GDSII language. The second type is the FAB file, which contains information used in the EFAB layer process, including the thickness of each layer and the order of photomasks to be used.

---

[2]A "light" version of Layerize known as LayerView™, which similarly allows for visualization of the cross-sectioned geometry, is available at no charge from Microfabrica at www.microfabrica.com/layerview/.

6.3.1.3 Photomask Fabrication

GDSII files are sent to photomask service bureaus (e.g., Toppan Photomasks, Photo sciences), where they are used to drive laser or e-beam-based pattern generation equipment that produces standard photomasks, typically a chrome pattern on a quartz substrate.

6.3.1.4 Instant Masks

If photoresist processing will be used, no further steps are taken at this time. However, if Instant Mask processing (see below) is planned, then Instant Mask tools are fabricated using a photolithography-based process using photomasks. One method of fabricating Instant Masks is based on micromolding. For example, a photomask can be used to expose a layer of thick photoresist deposited on a substrate, which, once developed, yields a two-level mold corresponding to the photomask geometry. A liquid elastomer can then be applied to the mold, and a metal anode is then placed in contact with the elastomer and pressed to squeeze out excess material. The elastomer can thereafter be hardened, after which it can be de-molded. Using this method, Instant Masks can be fabricated from a photomask within two days for all the required masks and then installed in an EFAB cluster tool.

6.3.1.5 Selective Deposition Using Photoresist Processing

Selective deposition can be done using conventional photoresist processing combined with through-mask electroplating. Liquid or dry film photoresists may be used. The steps involve substrate cleaning, application of resist, edge bead removal (for liquid resists), soft baking (for liquid resists), exposure through a photomask, developing, and hard baking (for some liquid resists). Some resists may also require plasma treatment to remove residues of photoresist. Metal is then plated onto the substrate through the apertures in the resist, and the resist is then stripped, leaving a patterned deposit of the metal. Considerations using photoresist for EFAB processing include sidewall quality (e.g., angle, smoothness), feature size, and maximum resist thickness.

6.3.1.6 Selective Deposition via Instant Mask Processing

Instant Mask processing is similar to photoresist processing, in that it uses photolithography to create a mask which defines the desired pattern of deposition and allows material to be deposited simultaneously over an entire layer. Yet unlike photoresist processing, Instant Mask processing (i) allows the photolithographic processing to be performed completely *separate* from the device-building process; (ii) allows all of the necessary masks to be produced as a batch *prior to* part generation, rather than *during* part generation; (iii) and potentially allows the mask to be reused.[3] By eliminating the need to repeat photoresist processing for every layer, Instant Mask processing enables an automated, self-contained EFAB cluster tool, the discussion of which follows.

The Instant Masking process is somewhat analogous to the printing of ink on paper. With the latter process, a printing plate which has a surface pattern that is the mirror image of the desired distribution of ink on paper is fabricated. The plate is inked and pressed against the paper, then removed, resulting in a patterned deposit of ink. With Instant Masking, an Instant Mask tool which has a surface pattern that is the mirror image of the desired distribution of metal on a substrate is fabricated.[4] The Instant Mask tool is immersed in an electrodeposition bath and pressed against a substrate (Figure 6.9(a)). Once the substrate and tool are mated, an electric current, of a magnitude calculated to provide the desired current density, is passed between the substrate and the anode of the tool, for a time calculated to selectively deposit the desired thickness of material (somewhat greater than the intended layer thickness) (Figure 6.9(b)). The tool is then removed, leaving behind a micropatterned deposit of metal (Figure 6.9(c), Figure 6.10).

The Instant Mask tool is a two-component device, consisting of a layer of an insulating elastomeric mask material that is patterned with apertures and attached to an anode (Figure 6.9(b)). The elastomeric

---

[3]As opposed to patterned photoresist, which is chemically removed and disposed of after electrodeposition.

[4]The term "substrate" signifies either the actual substrate upon which a microstructure is fabricated, or equivalently, the previous layer of a multilayer structure, which itself behaves as a substrate for the next layer.

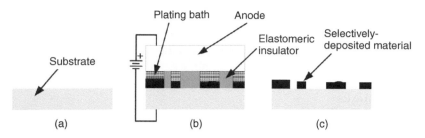

FIGURE 6.9   Instant Mask processing for selective electrodeposition.

FIGURE 6.10   Patterned copper deposit produced by an Instant Mask.

mask conforms to the topography of the substrate and prevents deposition of material other than within the region of the apertures. The anode serves its usual role in an electrodeposition cell, but it also provides mechanical support for the elastomer layer, maintaining isolated "islands" of mask material in precise alignment.

Key performance issues with Instant Mask processing are (i) how selective the deposition process really is (i.e., is there any extraneous deposit, or "flash," in the masked areas); (ii) how small a feature can be patterned with sharply defined edges and without flash; (iii) how thick a deposit can be obtained (too thin a layer is not acceptable, since layers in the range of 2–10 μm or thicker would normally be desired for micromachining); (iv) whether or not the process is reliable; and (v) whether robust Instant Mask tools can be produced. In brief, performance can be summarized as follows. Flash can be largely avoided by proper design and fabrication of the Instant Mask, adequate parallelism between mask and substrate, and application of the correct mating pressure. Features down to 30 μm in width can be generated routinely, with 20 μm features demonstrated and smaller features possible. Deposits up to 15 μm[5] in thickness with feature sizes in the range of 30 μm can be produced. Potential shorting between mask anode and substrate (cathode) associated with deposit nonuniformities must be managed. Finally, Instant Mask tools able to survive many days of immersion in plating baths have been routinely fabricated.

6.3.1.7 Blanket deposition

After cleaning, the patterned substrate is then placed in a plating system for blanket deposition. The deposition is normally performed until the deposit is somewhat thicker than the desired layer thickness.

---

[5]During Instant Mask deposition, there can be only a tiny volume (e.g., picoliters) of plating bath within the electrochemical cell formed by the anode, cathode, and patterned elastomer. This volume is only capable of supplying enough ions to deposit a few nanometers of material. The problem is solved by making the Instant Mask anode from a metal that is electrochemically dissolved during deposition so as to provide "feedstock."

After blanket deposition, all apertures that were formed during the selective deposition are filled in with the second material, completing the material deposition for the layer.

6.3.1.8 Planarization

Planarization has several key objectives. The first is to remove the blanket-deposited material overlying the selectively deposited material. The second is to establish a precise vertical (Z) dimension for the layer (either total thickness, distance between the top of the layer and another reference point, such as the substrate). Planarization thus can overcome (by depositing more thickly than needed and planarizing back) the local or cross-wafer nonuniformities in thickness that are unavoidable with a plating process. The last objective is to generate a flat, smooth surface. Such surfaces are, in general, desirable for microdevices, and in addition, if Instant Mask processing is used, these surfaces are needed to achieve good (flash-free) processing of the next layer. Planarization therefore also overcomes the surface roughness that sometimes results from plating processes.

Planarization may be performed using a variety of lapping, polishing or chemical–mechanical polishing plates, pads, and abrasive slurries (e.g., alumina or diamond). Parameters include pressure applied to the substrate, and rotational speed of the plate or pad. A means of end-point detection must also be provided, so that planarization can be stopped when the layer reaches the proper thickness. Depending on the materials and parameters used, surface roughness can be obtained on the order of 100 nm rms or better with very good planarity (e.g., submicron over a 100 mm wafer). Polishing, when required by some applications (e.g., optical mirrors or valve seats), can yield even smoother surfaces. Certain combinations of materials may exhibit differential planarization rates, causing one material to become recessed below the surface of the other; however, by proper selection of planarization materials and parameters, these effects are largely controllable.

6.3.1.9 Singulation

After all layers have been fabricated, the EFAB wafer can be singulated using processes similar to those used to process semiconductor wafers, resulting in individual device die. Singulation may be performed either before or after release of the sacrificial material.

6.3.1.10 Release

The structural material, in general, is ultimately released from the sacrificial material that surrounds and embeds it. This process may be carried out by wet chemical etching, though other methods are also possible. With copper as the sacrificial material, etching rates of several hundred microns per hour can be obtained with, for example, selectivity to nickel exceeding over 500:1. Material can be removed from fairly high aspect ratios; for example, copper has been successfully etched from a volume with an exposed cross-section of $25 \times 150\,\mu m$ to a depth of 3 mm.

If the sacrificial material is to be removed, it is necessary to create openings in any otherwise fully enclosed region to allow access for the release etchant. A successful release etch is dependent on the amount of sacrificial material to be removed, the etch path length, and the size of the opening through which the release fluid must travel. The etch path length defines a released region from the point of access to the main volume of etchant, to the point farthest from this to be released. As a general rule, devices are most easily released if designed as "openly" as possible. If achieving two-material structures is desired, sacrificial material can be embedded within structural material by fully enclosing it in the latter, thus trapping it.

6.3.2 Materials

6.3.2.1 Substrates

A flat, reasonably smooth conductive material, to which the deposited materials are capable of adhering, is needed as the substrate for the EFAB process. Substrates for the EFAB process may be composed of a wide variety of materials, including metals, ceramics (e.g., alumina), glass, semiconductors (e.g., silicon and GaAs), and piezoelectric materials. In the case of nonconductive substrates, seed layers may be applied to

TABLE 6.1   Properties of EFAB Nickel Structural Material

Property	Value
Chemical composition	>99.5% Ni
Modulus of elasticity	190 GPa
Poisson's ratio	0.31
Hardness	XX
Yield Strength	>800 MPa
Tensile Strength	>800 MPa
Fatigue life	infinite @ <100 MPa
Thermal expansion coefficient	13.4e-6/K
Electrical resistivity (in plane)	6.84×10^{-6} Ω-cm
Electrical resistivity (85 by 85 m interlayer junction)	<50
Layer-to-layer adhesion	Near bulk strength

render them conductive; such layers are ultimately removed, except where needed to anchor the devices, to avoid shorting of electrical devices. At this writing, the substrate most commonly used for EFAB devices is metallized alumina. By providing a layer of sacrificial material between a substrate and the bottommost layer of the device, the device can be built free–standing, i.e., completely free of a substrate. EFAB devices can also be built inverted (last layer deposited first) and then bonded (e.g., by adhesive) to another substrate, either before or after release.

6.3.2.2   Structural Materials

EFAB technology can be used to form microstructures from many materials that can be deposited by electrolytic or electroless deposition from aqueous solutions at temperatures typically ranging from 15 to 95°C. Such materials include many pure metals, e.g., nickel, copper, gold, silver, platinum, lead, zinc, iron, and cobalt; a variety of metal alloys, e.g., nickel–iron, nickel–cobalt, copper–zinc (i.e., brass); certain semiconductors, e.g., lead sulfide, bismuth telluride; composites formed from a particulate co-deposited with a metal e.g., Ni-Teflon®, Ni-diamond; and potentially, conductive polymers [Schlesinger and Paunovic, 2000]. One constraint is that there must be a depositable sacrificial material available that can be selectively removed (e.g., by etching) after all layers are formed. Other considerations are reasonable stress and deposition rate, good interlayer adhesion, and ability to be planarized.

The structural material that has received the greatest attention so far is nickel. Nickel has good strength and stiffness and is reasonably temperature and corrosion resistant. Properties of the Ni structural material used currently for EFAB processing have been well characterized and are shown in Table 6.1.

In addition to Ni, Ni–Co is a commercially-available material for EFAB devices. Structures made from silver and copper have been demonstrated, but these materials are not yet available for use.

An EFAB-like process under development fabricates devices from Cu and a polymer dielectric. It should be noted that most devices can be successfully built without this capability, since both the dielectric substrate and air can usually be used for electrical isolation.[6] Incorporating dielectric using a backfill approach after removal of sacrificial material is another option. Finally, as with LIGA, EFAB structures may be used as tools to shape (e.g., by injection molding or embossing) polymers and other materials.

6.3.2.3   Sacrificial Materials

The sacrificial material that has received the greatest attention so far is copper, used as a sacrificial material for nickel and several other materials. Cu is easily plated and can be etched at a reasonable rate with respect to Ni with excellent etch selectivity. Other considerations in choosing copper are reasonable deposit stress and deposition rate and the ability to be planarized. In general, etch rate, etch selectivity, stress, plating rate, and ability to be planarized are major criteria in selecting a sacrificial material.

---

[6]It should be recalled that most surface micromachining processes (e.g., MUMPS, Summit V) also do not offer a permanent (vs. sacrificial) structural dielectric, other than the substrate itself.

TABLE 6.2 Performance of Commercial EFAB Technology and Roadmap

Characteristic	2005	2006	2007
Number of layers	No known limit (50 demonstrated)		
Overall device height	No known limit (1000 µm demonstrated)		
Single layer thickness (µm)	2–50 (see Figure 6.10)	2–50	1–100
Layer thickness/Z dimensional tolerance (µm)	+/−0.6	0.5	0.3
Layer-to-layer registration (µm)	+/1.0–1.5	+/−0.5	+/−0.3
X/Y accuracy (µm)	+0/−1.0	+0/−0.4	+0/−0.3
Minimum feature size (µm)[a]	10	5	2
Substrate size (mm)	100	150	200

[a]In the X–Y layer plane only (parallel to the substrate). Some limitations apply to stand-alone structures and thick layers.

6.3.3 EFAB™ Technology Performance

Table 6.2 summarizes the performance of commercially available EFAB technology and provides a roadmap through 2006. Design rules detailing the current state of the technology are available from Microfabrica.

6.3.4 EFAB™ Technology vs. Alternative Approaches

Table 6.3 compares the major microfabrication processes with EFAB technology. Whereas EFAB technology offers certain potential advantages over these other processes in some areas, there are, of course, some applications for which it is not the best choice among alternative methods.

Like any manufacturing technology, EFAB technology has certain limitations and shortcomings.

6.3.4.1 Size

Compared with most microfabrication processes, EFAB technology can produce very tall structures. However, it is not economical for the production of devices taller than several millimeters[7] unless there is no alternative fabrication technology that can provide the required geometrical complexity or material properties.

6.3.4.2 Geometry

Compared with most microfabrication processes, EFAB technology is extremely flexible in geometry. However, because of the sacrificial material, geometries are not entirely unrestricted. The need to remove this material imposes some constraints on the design, though there are far fewer constraints than if sacrificial material were not used to support the structural material.

6.3.4.3 Materials

When compared with surface micromachining, an electrodeposition-based process is more restricted in the choice of materials, since it is possible to deposit, using physical or chemical vapor deposition, an extremely wide variety of materials. However, EFAB technology allows electrodeposition and nonelectrodeposition processes to be combined in a hybrid fashion, enabling the integration, where needed, of additional materials (e.g., oxides, piezoelectrics, shape memory alloys).

6.3.4.4 Minimum features

Compared with some alternative microfabrication processes, EFAB technology's minimum X–Y features at this writing (Q2 2005) are — at $10 \times 10\,\mu m$ — several times larger. However, the EFAB roadmap will reduce

---

[7]Unless, perhaps, if the X–Y extents are very small, allowing many devices to be made simultaneously.

TABLE 6.3   EFAB Compared with Other Micromachining Processes

Characteristic	EFAB	Bulk μmach.	Surface μmach.	LIGA
True 3-D geometry (unlimited number of layers) without assembly	Y	N	N	N
Devices hundreds of μm in height without a synchrotron	Y	N	N	N
Layer thickness of any layer	50	n/a	0.1–3[a]	2–2000
Device design at the 3-D CAD level, with automatic 2-D layout and process planning	Y	N	N	n/a
Co-fabrication of packages	Y	Y (difficult)	N	N
Metal structural materials	Y	N	Y	Y
Dielectric structural materials	N (Y for EFAB-like process in development)	N	Y	N (only by molding)
Other structural materials (e.g., piezoelectrics, shape memory alloys)	Y[b]	N	Y	N
Substrates	Metal, ceramic, glass, semiconductors, etc.	Si	Si typically	Metal, ceramic, glass, etc[c]
Vertical wall smoothness	1.2–3.0 μm (in Q2 '05; 0.3–2.0 in future)	<20 nm[d]	n/a	30–50 nm[e]
Minimum features <10 μm	N (2 μm on EFAB roadmap for '07)	Y	Y	N (on a commercial basis)

[a] Sometimes not easily specified or varied on a layer-by-layer basis.
[b] By combining electrodeposition with CVD, PVD, or other deposition processes.
[c] Semiconductor substrates are problematic due to exposure to high-energy X-rays.
[d] [Vargo et al., 2000]
[e] http://greenmfg.me.berkeley.edu/LIGA/right.htm

minimum feature size to 2 μm by 2007. Meanwhile, it should be noted that since EFAB processing can deposit multiple thick layers, device footprints can be larger than usual without loss of stiffness, thus obviating, in many cases, the need for smaller features. Small gaps between structures can be achieved, however, when necessary by exploiting the multilayer nature of EFAB. For example, structures can be fabricated on different layers in a prealigned fashion, and then deformed in a massively parallel manner to render them coplanar.

6.3.5   EFAB™ Equipment

The EFAB process can be performed in a conventional cleanroom environment using a set of discrete process tools (for plating, planarization, inspection, etc.). Alternatively, if Instant Masks are used, the process may be performed in a fully automated, self-contained cluster tool that can execute all the steps under computer control. Several EFAB cluster tools have been built by Microfabrica, the latest of which is shown in Figure 6.12. The tool pictured is designed for prototyping and accommodates a 21 mm wafer and several 100 mm Instant Masks. It features a class 100 enclosure and is installed on a factory floor, serving essentially as a "cleanroom in a box."

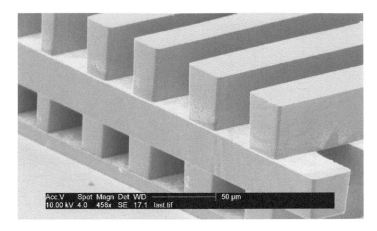

FIGURE 6.11 EFAB structure with thick layers (from bottom to top: 12, 30, 40, and 50 μm thick).

FIGURE 6.12 Self-contained, automated EFAB cluster tool.

6.3.6 Future EFAB™ Technology Capabilities

The continued development of EFAB technology includes several new capabilities in addition to process improvements and new materials already discussed, several of which are in development by Microfabrica. New capabilities include the following:

6.3.6.1 Multiple Materials

It is desirable for some applications to make structures from more than a single structural material. In the simplest case, different structural materials can be used on different layers; this procedure is fairly straightforward to implement. In the more sophisticated case, more than one structural material can be used on the same layer; this process normally requires that there be one additional selective deposition step for such a layer, for each additional material desired. For example, if one desires three structural

FIGURE 6.13   A multilayer capacitor co-fabricated with package walls.

materials (along with one sacrificial material) on a given layer, three selective depositions and one blanket deposition can be used.

6.3.6.2   Self-packaging

As a 3-D process, EFAB technology can be used to cofabricate a package — or at least part of a package — along with and surrounding a device. Figure 6.13 shows a simple example of this capability, in which the walls of a package have been fabricated along with a capacitor inside them. Such a package typically requires another element, such as a lid, to complete it. EFAB "self-packaging" can be performed at a wafer scale, with each die containing its own packaged device, offering a path to significant reduction in total product cost.

6.3.6.3   IC Integration

EFAB devices can be fabricated directly onto semiconductor wafers such as CMOS and GaAs by virtue of the low-temperature processing involved. Benefits of on-chip integration vs. the use of wire bonding or flip-chip interconnection to a separate die include reduced package size; lower capacitance, inductance, and noise; improved scalability to large arrays of devices; and, potentially, higher reliability (fewer interconnects). Other methods of integrated semiconductor die with EFAB devices are also possible.

6.4   EFAB™ Applications

By now it should be clear that EFAB has almost arbitrary aspect ratios. Tens of layers that are 10 or more microns high can be used to generate structures with heights ranging from hundreds of microns to several millimeters, if desired. These structures, taller even than many LIGA structures, can have micron-level geometric precision in different patterns on each layer. In general, this procedure allows for mixtures of scales that operate outside the intuition of a designer who is accustomed to either bulk or surface micromachining. Inertial devices can combine extremely large proof masses with extremely densely packed capacitor plates; actuators can generate forces much higher than usually associated with MEMS; RF-systems can be built with fully three-dimensional coaxial interconnects and passives; and all of these devices, and more, can be co-fabricated next to one another. Designers have the flexibility to consider the "ideal" shape for a device, rather than the usual shape.

Let us consider the actuation density available from electrostatic comb actuators, commonly used to provide displacement-invariant force in MEMS. Figure 6.14 shows comb drives implemented in various processes and Figure 6.15 shows the resulting force as a function of voltage. The implementations shown are: surface micromachining, which has little depth but typically very small gaps; bulk DRIE silicon etching,

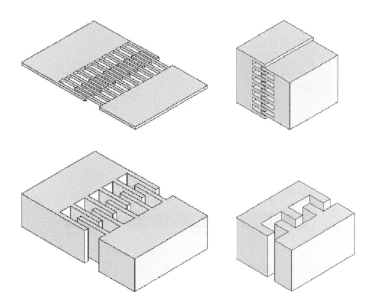

FIGURE 6.14 (See color insert following page 9-22.) A variety of comb drive actuator implementations, clockwise from top left: surface micromachining, DRIE, LIGA, and EFAB. A comparison of the available force (or capacitive sensitivity) is shown on the chart, demonstrating the high power densities available with EFAB.

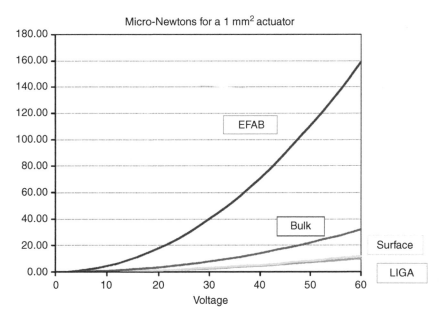

FIGURE 6.15 (See color insert following page 9-22.) Calculated forces as a function of voltage for comb drive actuators having a 1×1 mm footprint for different fabrication processes.

shown with moderate gaps and moderate depth; LIGA, which has very deep features but relatively large gaps; and EFAB technology, which mixes extremely high aspect ratios with small gaps. The skeptical reader may wonder if EFAB technology's advantages are only manifested at high voltages (the right side of the chart), but all of these actuators have force proportional to voltage squared; therefore, the ratio of forces is the same at all voltages and the shapes on the chart do not vary with voltage range (only the

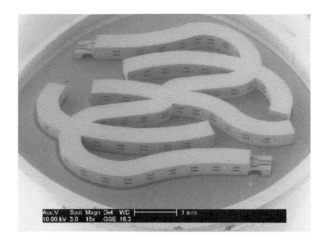

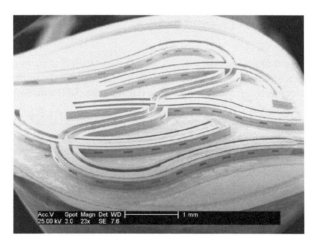

FIGURE 6.16 A three-pole Ka-band filter implemented in microcoaxial construction shown (top) with and (bottom) without the upper layers, the latter to illustrate the center conductor. The whole filter fits in an 8-mm diameter package.

magnitude of the force does). The higher force can be used in a wide range of applications, from actuating electrical relays to powering micropumps or surgical tools.

These capacitive actuators also illustrate an example of sensor flexibility available with EFAB technology. The capacitors that provide on the order of twenty times greater force provide similar increases in sensitivity. It has been realized several times during an EFAB device design review that the capacitive output change will swamp previously used circuitry, caused by too much signal from the sensor!

For inertial sensors there are similar sets of advantages. The proof mass can be extremely large, often larger than even bulk etched devices (caused by greater height and a material nearly four times denser than silicon), and this can be combined with the high capacitances illustrated in the previous example. Both of these elements can also coexist with springs of almost arbitrary shape and dimension, leaving it up to the designer to decide where to place things and how to shape the various components. The EFAB process offers the additional opportunity to build the supports, sense, drive, or other elements within the proof mass if so desired, which can be useful for controlling mass distribution. Additionally it is usually feasible to build lateral and vertical accelerometers at the same time, which is often a challenge with other fabrication technologies. Figure 6.3 shows a multiple-axis EFAB accelerometer.

Pressure sensors and the like can also benefit from the mix of scales available. A simple membrane pressure-sensing element can be coupled to finely detailed sensors to vastly increase sensitivity or, alternately, reduce compliance to widen sense range or provide more robust structures. The more complex constructions available also enable additional features such as self-test or local monitoring of induced stress to compensate the device.

Radio frequency MEMS can benefit from the geometrical freedom of EFAB by building complex, fully three-dimensional coaxial structures such as transmission lines, couplers, filters, phase shifters, and the like. Demonstrated and published designs have shown high performance in small sizes [R. Chen, 2004, and Reid, 2004]. Figure 6.16 shows an example of a three-pole Ka-band filter implemented in EFAB technology. It is shown complete, and also without the top layers so that the internal conductor can be seen. The whole device fits within an 8-mm diameter circle.

These examples have illustrated just a few of the ways in which changing to a fully three-dimensional, arbitrary aspect ratio process gives the designer tools to increase performance by large factors. There are also examples of devices which simply could not be produced by previous processes and can be produced by the EFAB process. One example is complex meso-scale mechanics such as the helical spring seen in Figure 6.1. Devices such as this one (and some much more complex spring assemblies) can be constructed at scales up to millimeters with features accurate to microns. Thus, devices that bridge the scale gap between macro- and microsystems can be built. Another example is construction of magnetic microdevices which include high-conductivity coils and high-magnetic permeability elements. Examples of such devices are motors, solenoid actuators, and microscale linear variable differential transformers (LVDTs) for displacement sensing.

Acknowledgments

The authors wish to express their appreciation to the Defense Advanced Research Projects Agency (DARPA) Defense Sciences Office and Microsystems Technology Office, for seed funding of EFAB technology development under the Mesoscale Machines and MEMS Programs.

EFAB™, Instant Masking™, Instant Mask™, and LayerView™ are trademarks of Microfabrica, Inc Solid Works and COSMOSWorks are registered trademarks of SolidWorks Corporation. Pro/ENGINEER is a registered trademark of Parametric Technology Corporation. ANSYS is a registered trademark of ANSYS, Inc. Windows® is a registered trademark of Microsoft Corporation.

Defining Terms

Build: The result of the EFAB layer process: one or more layers deposited on a wafer or other substrate

EFAB™ technology: A multilevel 3-D microfabrication process based on electrochemical deposition and planarization to define the geometry of the fabricated device on each layer

Instant Masking™: A micropatterning method involving a conformable patterned mask, an electrically active bath, and a source of current. When used for electrodeposition, the mask is mated with a substrate in the presence of an electrodeposition bath, and current is applied to deposit material onto the substrate in regions not contacted by the mask.

Layerize™: Special-purpose software, developed for use with EFAB technology, that imports a 3-D geometry file produced by standard 3-D CAD systems and exports a 2-D geometry file in a format acceptable to standard photomask pattern generators.

Substrate: Either the actual substrate upon which a microstructure is fabricated, or equivalently, the previous layer of a multilayer structure, which itself behaves as a substrate for the next layer.

References

Chen, R.T., Brown, E.R., and Bang, C.A. (2004) "A Compact Low-Loss Ka-Band Filter Using 3-Dimensional Micromachined Integrated Coax," in *Proc. IEEE International Conference on Micro Electro Mechanical Systems*, 25–29 January, pp. 801–4.

Cohen, A.L., Zhang, G., Tseng, F.G., Frodis, U., Mansfeld, F., and Will, P.M. (1999) "EFAB: Rapid, Low-Cost Desktop Micromachining of High Aspect Ratio True 3-D MEMS," *Technical Digest, 12th IEEE*

International Conference on Microelectromechanical Systems, 17–21 January, Orlando, Florida, pp. 244–51.

Reid, J., and Webster, R. (2004) "A 60 GHz Branch Line Coupler Fabricated Using Integrated Rectangular Coaxial Lines," in *IEEE MTT-S International Microwave Symposium*, 6–11 June, pp. 441–44.

Schlesinger, M., and Paunovic, M., eds. (2000) *Modern Electroplating*, John Wiley & Sons, New York.

Vargo, S.E, Wu, C., and Turner, T. (2000), "Reactive Ion Etching of Smooth, Vertical Walls in Silicon," in NASA Tech Briefs, 24, no. 10.

For Further Information

Please visit Microfabrica Inc.'s Web site at www.microfabrica.com.

7

Single-Crystal Silicon Carbide MEMS: Fabrication, Characterization, and Reliability

Robert S. Okojie
NASA Glenn Research Center

7.1 Introduction

Pressure monitoring during deep-well drilling and in automobiles and jet engines requires pressure sensors and electronics that can operate reliably at temperatures between 200°C and 600°C [Alexander's Gas and Oil Connections, 2003, Matus et al., 1991]. Conventional silicon semiconductor pressure transducers increasingly suffer instability and failure as the operation is extended toward higher-temperature regimes. The failures include degradation of metal/semiconductor contacts and weakening of wire-bonds caused by temperature-driven intermetallic diffusion [Khan et al., 1988]. Robust architecture based on Silicon-On-Insulator (SOI) technology has extended device operation to near 400°C [Tyson and Grzybowski, 1994]. However, at 500°C the onset of thermoplastic deformation of silicon becomes the ultimate factor limiting silicon-based microelectromechanical systems (MEMS) [Huff et al., 1991]. Several commercially available silicon and piezoceramic pressure transducers employ complex and costly packaging schemes (e.g., plumbing for air or water cooling) to help maintain continuous and reliable operation at high temperature. Added to the complexity and cost is the increase in size and weight. Therefore, to meet the increasing need for sensors to operate reliably at higher temperature and at relatively lower cost, new and innovative devices made from materials more robust than silicon are needed.

Technological advancement in the growth of wide-bandgap semiconductor crystals such as silicon carbide (SiC) has made it possible to extend the operation of solid-state devices and MEMS beyond 500°C. SiC has long been viewed as a potentially useful semiconductor material for high-temperature applications [Neudeck et al., 2002]. Its excellent electrical characteristics — wide-bandgap, high-breakdown electric field, and low intrinsic carrier concentration — make it a superior candidate for high-temperature electronic applications [Pearson et al., 1957].

Table 7.1 shows the comparison of relevant electrical and mechanical properties between 6H-SiC and silicon. The fact that SiC exhibits excellent thermal and mechanical properties at high temperature, combined with its fairly large piezoresistive coefficients, makes it well-suited for use in the fabrication of high temperature electromechanical sensors. There already are efforts to develop pressure sensors based on SiC [Okojie et al., 1996, Berg et al., 1998, Mehregany et al., 1998].

SiC appears in various crystal structures called polytypes. The polytypes most frequently available are the hexagonal 4H-SiC and 6H-SiC and the cubic-SiC (also referred to as 3C-SiC or β-SiC). The prominent physical differences between the 4H- and the 6H-SiC are the stacking sequences of the Si-C atomic bi-layers, the number of atoms per unit cell, and the lattice constants [Park, 1998]. The hexagonal crystals are grown in large boules, mostly by the sublimation (Acheson) process, and then sliced into wafers; currently, the largest commercially available single crystal size is 76 mm in diameter. Homoepitaxial growth by chemical vapor deposition (CVD) can be performed on the hexagonal single crystal substrate to obtain epilayers of various thicknesses and doping levels (typically nitrogen doping for n-type and aluminum or boron for p-type), as desired for various device applications.

Electronically, 4H- and 6H-SiC have bandgaps of 3.2 eV and 2.9 eV, respectively. The bulk-grown 3C-SiC polytype of device quality is not readily commercially available because of the nonexistent effective growth technology. Aggressive research is ongoing in this area; the payoff is the harnessing of its relatively high electron mobility to produce high frequency device switching. Heteroepitaxial CVD of 3C-SiC on silicon substrate is generally applied to grow epilayers a few microns thick. However, the ~20% lattice mismatch that exists at the 3C-SiC/Si heterojunction induces dislocation defects in the 3C-SiC heteroepilayer, which greatly degrades the performance of electronic devices fabricated in it. Such defects in 3C-SiC and its relatively low bandgap (2.3 eV) have largely limited its attractiveness for broader, high-temperature mechanical sensor applications.

The discussions in this chapter will focus on the technological progress made in recent years in terms of implementing MEMS-based SiC piezoresistive pressure sensors, as representative of accelerometers and flow sensors fabricated in SiC that apply the same sensing principles. These SiC sensing devices are critically needed for closer proximity sensing in the applicable harsh environment. This chapter's five sections will provide readers with in-depth understanding of the present and future challenges that confront this emerging technology, and of the efforts being made to surmount these challenges.

TABLE 7.1   Comparison of Properties of α-(6H)-SiC with Silicon

Properties	Silicon	6H-SiC
Bandgap (eV)	1.12	2.9
Melting Point (°C)	1420	>1800
Breakdown Voltage ($\times 10^6\,\mathrm{Vcm}^{-1}$)	0.3	2.5
Young's Modulus of Elasticity (GPa)	165	448
Thermal Conductivity [W(cm-C)$^{-1}$]	1.5	5.0
Electron Saturation Velocity ($\times 10^7\,\mathrm{cms}^{-1}$)	1	2
Maximum Operating Temperature (°C)	300	1240

The basic fabrication principles of single-crystal silicon carbide are presented in Section 7.2, in which SiC device fabrication by photoelectrochemical etching (PECE) and conventional electrochemical etching (ECE) are discussed. Important etching parameters leading to the fabrication of resistor and diaphragm structures for sensing applications are also discussed in this section. Relevant SiC fabrication processes using more recently developed deep reactive ion etching (DRIE) are discussed in another chapter, hence they will only be discussed here in context. In strain sensor technology, the output of the device is primarily the function of the gauge factor (*GF*) and resistance. Section 7.3 will discuss the characterization of SiC *GF* and resistance, and their temperature dependencies. Contact metallization that is stable at the desired operating temperature is recognized as critical for successful implementation of high-temperature sensors. A successful approach to realizing thermally stable ohmic contacts to SiC electronics and sensors is presented in Section 7.4, along with the deleterious effects of oxygen. Section 7.5 discusses sensor testing and performance characteristics, and finally, Section 7.6 will present the most recent reliability evaluation of SiC pressure sensors that were fabricated using the DRIE process [Beheim and Salupo, 2000], improved metallization, and packaging.

7.2 Photoelectrochemical Fabrication of 6H-SiC

The difference between PECE and ECE is basically the photo-generation of electron–hole pairs (EHPs) that occurs during the application of the former. The ECE process is a conventional anodization process in which high voltage or current is applied to etch away material. In both cases, the etched substrate acts as the anode electrode. The PECE process is adopted where shallow and anisotropic etching is critical. For example, in the PECE of resistors, it is important to minimize isotropic etching in order to maintain uniform resistor geometry with minimum undercutting across the wafer. Nonuniform geometry of the resistors typically leads to increased circuit imbalance if configured in the Wheatstone bridge form, and reproducibility becomes difficult.

Unlike silicon technology, processes associated with the fabrication of deep structures in single-crystal SiC are limited generally to ECE and DRIE. This situation is the result of the near–inert surface chemistry of SiC that makes conventional wet chemical etching tenuous at room temperature. The fabrication of piezoresistor-based sensors in single-crystal SiC requires at least one intentionally doped epilayer, usually n-type, in which the sensing element is fabricated; this epilayer is grown on a p-type substrate in which the cavity is fabricated to create a diaphragm. However, for several reasons, it is often necessary to have two epilayers grown on an n-type substrate, as shown in Figure 7.1. The first epilayer is p-type, followed by a second epilayer of n-type conductivity. The resistor sensing elements are fabricated in the n-type layer, whereas the p-type epilayer primarily functions as an electrochemical etch-stop during PECE of the n-type resistors.

The ECE process is applied with high voltage or current, which enables high etch rates in fabricating cavities that are as deep as 200 μm to create a diaphragm in the n-type substrate. The fabrication of the piezoresistor for use as a strain sensor requires a thorough investigation of the etching characteristics of the n$^+$-type SiC epilayer. In SiC piezoresistive strain sensor technology, the preference for the highly

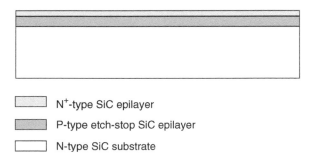

  N$^+$-type SiC epilayer

  P-type etch-stop SiC epilayer

  N-type SiC substrate

FIGURE 7.1   Typical SiC substrate with two epilayers for use in photoelectrochemical etching (PECE).

doped n$^+$-type epilayer over the lower-doped n-type epilayers for use as the sensing element is due to the associated low resistivity and lesser sensitivity to temperature across the range of operating temperature. However, this preference leads to a trade-off in lower strain sensitivity due to reduction in the *GF*.

In order to fabricate a SiC diaphragm transducer, a controllable etching process for thinning and forming of deep cavities in the backside is required. For diaphragm fabrication, the desirable etch rates should be in the range of 1 μm/min. It has been shown [Okojie, PhD Thesis, 1996] that by using ECE in dilute Hydrofluoric acid (HF), cavities with depth greater than 100 μm in both p- and n-type 6H-SiC can be fabricated. In many cases using dilute HF during the ECE process, a porous layer is formed; this layer has a texture that depends on the concentration of HF in water and the voltage or current density applied. At low HF concentration, the anodization produces a soft, highly porous SiC layer, whereas anodization at higher HF concentrations produces a hard layer of low porosity. Because these porous layers have large crystal surface areas, enhanced thermal oxidation occurs within them relative to the bulk. Subsequent dipping in buffered hydrofluoric (BHF) acid removes the now-oxidized porous material to reveal the etched cavity.

In n-type SiC, ECE is possible only when the positive anode voltage (with the SiC as the anode electrode) is sufficiently high to cause the semiconductor space charge layer at the SiC/electrolyte interface to break down. Under this condition, electrons are injected from the SiC/HF interface into the bulk of the semiconductor, resulting in dissolution reactions. The voltage required for the breakdown depends on the doping level of the semiconductor and on the concentration of the electrolyte.

Two ECE experiments were conducted with n-type SiC: (a) anodization in a 0.625% HF in a two-electrode cell (SiC-anode with platinum counterelectrode) and subsequent thermal oxidation to remove the porous layer, and (b) anodization in a 0.625% HF in a three-electrode configuration (SiC-anode, platinum-counterelectrode, and reference standard calomel electrode) and oxidation to remove the porous layer. Both experiments demonstrated rapid etching of n-type SiC. In the first experiment, the average etch rate was 0.3 μm/min, and in the second, an etch rate of 0.8 μm/min was obtained. At an average current density ($J_{av} \sim 100$ mA/cm^2) used in both experiments, higher etch-rates should have been obtained if etching was the only process that occurred. The actual etch rates obtained from the experiments suggested that other processes were occurring (e.g. water decomposition), which would use up part of the anodization current. Gaseous products released during the anodization process acted as masks to decrease the reaction rate, thereby reducing the etch rate.

Materials must possess certain properties to be used as a mask in electrochemical etching. In a dilute HF, the mask should be inert in the concentrations used, and no electrolytic reaction should occur during anodization that might interfere with the etching process. Two masking materials were investigated. Platinum, which is very resistant to HF, generated gas bubbles at the high anodic potentials used, resulting in nonuniform etching. Silicon nitride did not form bubbles because it is an insulator. However, silicon nitride is not completely resistant to HF.

Because the process of deep etching to form a diaphragm takes a long time (200 minutes to etch a 200-μm deep cavity at an average of 1 μm/min), a certain minimum thickness of nitride is required to survive the dilute HF electrolyte during the anodization process; thus, the complete stripping of the nitride mask and consequent undesired etching of the entire SiC surface can be prevented. The investigations briefly described led to the following important experimental conclusions [Okojie, PhD Thesis, 1996], that at a fixed voltage of 12 V:

1. Etching in 2.5% HF occurs at higher rates than in 0.625% HF.
2. Etching in 2.5% HF produced deep etch pits and/or hillocks on the surface of SiC, whereas no etch pits or hillocks were observed on surfaces etched in 0.625% HF. The pits are attributed to the enhanced electrochemical etching around the micropipes in the SiC substrate.
3. The surface of the etched region was smoother and the depth more uniform when etched in 0.625% HF.

In an attempt to further optimize the etching process for forming n-type SiC diaphragms, an additional series of experiments was conducted with different etching parameters and concentrations of HF.

Hillocks between 10 μm and 50 μm formed randomly on the diaphragm surface. The presence of uncontrollable discontinuities of this size results in nonuniform stress distributions that adversely affect the operation of a sensor. Although the cause of the hillock formation was not conclusively determined, there was evidence that the formations were associated with bubbles generated during the high-rate etching.

The existence of micropipes in the SiC wafers, which could lead to high current–density concentrations in localized areas, could result in selectively high etch rates around the defect sites, increasing the possibility of etch pit formation. Although the etching potential of highly doped n-type SiC using the ECE process is much higher than that of p-type SiC, it is possible to stop the etching process at the np-junction if the ohmic contact used for anodization potential control is made only to the n^+-SiC epilayer (refer to Figure 7.1). The positive anodic potential on the n^+-SiC layer will cause the junction to be reverse-biased, thereby preventing the flow of current through the underlying p-SiC epilayer. In order for the etch-stop to be effective, the breakdown voltage of the np-junction must be higher than the etching potential of the n^+-SiC epilayer. Also, the p-type layer should have a doping level significantly lower than that of the n^+-SiC epilayer to minimize the possibility of tunneling current [Shor et al., 1993].

The process of using PECE of SiC for the purpose of fabricating well-defined resistor structures is described as follows. As shown in Figure 7.1, the starting wafer is an n-type 6H-SiC substrate upon which a 5-μm thick, lightly doped (3×10^{18} cm^{-3}) p-type epilayer is grown by CVD, followed by a 2 μm thick n^+-type 6H-SiC epilayer. An ohmic contact metallization, preferably nickel, is deposited and patterned into a circular shape on the top n^+-SiC epilayer to enable control of the anodization potential during the PECE process. Platinum is sputtered onto the top of the wafer, covering the ohmic contact metal and the entire n^+-SiC epilayer. The platinum in direct contact with the epilayer is then patterned into the shape of serpentine resistor elements. This platinum acts as an etch mask, so that the serpentine resistor patterns can be transferred onto the n^+-SiC during the PECE process. Contact electrode wire is wire-bonded on the section of the platinum mask that is in direct contact with the nickel ohmic contact. A thin layer of black wax is then applied over areas covered by the ohmic contact electrode. The wafer is then immersed in dilute HF electrolyte, with the side to be etched facing up. This face is then exposed to a UV light source, with the anodization potential set at 1.7 V_{SCE} (SCE means Standard Calomel Electrode, which is the reference electrode against which the anode voltage is measured). Under this condition, the exposed sections of the n^+-SiC epilayer are photoelectrochemically etched, and the sections under the serpentine-shaped platinum etch mask are unetched. After the anodization process, the wax is stripped in acetone and the platinum mask and nickel ohmic contacts are stripped in *aqua regia* and nitric-hydrochloric acid in a 50:50 mixture, respectively. After stripping, the resistor patterns transferred to the n^+ epilayer are revealed.

The current vs. time curves of a typical PECE are shown in Figure 7.2. For the epilayer thickness used, the photocurrent rises to a first maximum in the first five minutes, and then drops rapidly because of the blocking action of the bubbles during the release of gaseous products. A second maximum appears in the curve after twelve minutes, when the bubbles deflate, after which time the current density gradually

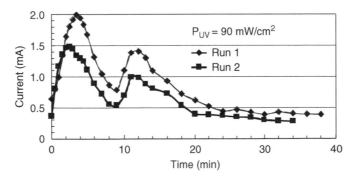

FIGURE 7.2 Current density vs. time during photoelectrochemical etching of n-type 6H-SiC in dilute HF electrolyte. Two anodization I–t characteristics indicate the reproducibility of the process as long as ohmic contact is present.

decreased. The initial rise of the current is due to the rapid generation of electron-hole pairs (EHPs) during the initial stages of the etching process. These carriers are available near the surface of the SiC to participate in the chemical reaction process.

A combination of various mechanisms leads to the gradual decay of the current, as shown in Figure 7.2. As the etching progresses, the reacting species will need to diffuse through the porous SiC layer being formed; thus, they can get to the bulk SiC surface just as the products of the reaction need to diffuse out. Therefore, as the porous SiC thickness increases, the mass transport process is slowed down, resulting in a reduced rate of reaction. The porous SiC will also shade the bulk SiC from the UV intensity, resulting in a drop in the EHPs needed to sustain the reaction.

Once the etching reaches the underlying p-type epilayer, the abrupt change in epilayer conductivity from n^+-SiC to that of the underlying p-type SiC causes the etching to stop. Since most of the electric field is confined within the n^+-SiC and the applied $1.7\ V_{SCE}$ anode potential causes the np-junction to be reverse-biased, only an insignificant leakage current will flow through the p-epilayer and be transported to participate in the etching process at the electrolyte interface.

The current–time curves are repeatable between two etching runs, as can also be seen in Figure 7.2. It should be noted that the etch rates of the n^+-SiC epilayer is dependent on the UV intensity and the doping level of the epilayer. By increasing the UV intensity, which causes more EHPs to be generated, the n^+-SiC epilayer can be etched more rapidly and more selectively between n- and p-type, because fewer carriers are present at the p-epilayer to participate in the etching.

SiC etch rates greater than $20\ nm/min$ have been achieved. The wafer is thermally oxidized and then dipped in 49% HF to remove the porous layer that formed during anodization. After the process of selective etching, the np-junction is sometimes not well-isolated electrically, which leads to unsteady outputs during device operation. To prevent this situation from occurring, the etching time of the n^+ piezoresistor epilayer is increased, so that the PECE process partially occurs in the p-epilayer. Poor electrical isolation can also be avoided if the junction isolation is verified immediately after PECE before the mask is stripped.

Usually, a second thermal oxidation is carried out to ensure that the np-junction of all the devices is well-isolated. The metal contact via is opened in the oxide by conventional photolithographic process and buffered-HF wet etching to expose sections of the resistors. Subsequently, ohmic contact metallization is deposited and patterned over the via to form the electrical contacts. This process is followed by the deposition and patterning of the diffusion barrier to form the final device shown in the cross section of Figure 7.3(a). The top view drawing of the sensor is shown in Figure 7.3(b).

Another process that can be used to pattern n^+-SiC epilayers is the ECE. As stated earlier, when the small positive anodic potential is applied through the ohmic contact on the n^+-SiC, current cannot flow between the n^+-SiC epilayer and the underlying p-SiC substrate. However, etching selectivity may not be as high as in the PECE, since leakage current will increase for the voltage needed to perform a high etching rate with dark current.

After the platinum mask is removed, the surface of the underlying resistor is sometimes pitted. The pitting is the result of the pinholes in the platinum, which allow electrolyte to seep through the mask and etch the otherwise protected n^+-SiC surface. Therefore, it is important to ensure that the etch mask used for the pattern transfer is free of pinholes.

Undercutting of the piezoresistors during etching may also occur as a result of the small dark current between the edge of the etching n^+-SiC and the electrolyte under bias of the anodic potential. Although this potential is made small enough to avoid reverse-bias current flow between the n-type epilayer and the underlying p-type epilayer, lateral current conduction between the expanding side of the n-type SiC epilayer and the electrolyte allows undercutting etching to proceed during the PECE process.

The above problems typically lead to undesirable nonuniformity in the resistor patterns. Using n-type epilayers with lower doping reduces dark etching at anodization potentials of about $1.7\ V_{SCE}$ and can minimize these effects. This result is due to the fact that, at that anodic potential, fewer carriers from the dark current are injected into the electrolyte relative to the EHPs generated by the UV light source. However, this approach means the gauge factor and the temperature effects of the piezoresistor will change since gauge factor is dependent on doping level.

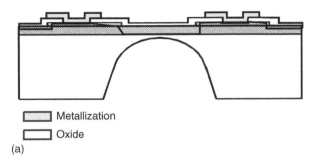

(a)

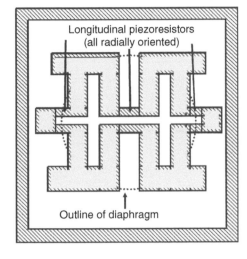

(b)

FIGURE 7.3   (a) Cross section view of 6H-SiC after PECE of the top n^+-epilayer and ECE of the backside cavity. Notice the curvature in the cavity, which is characteristic of the ECE process. (b) Top view of patterned piezoresistors in n-type 6H-SiC.

By substituting another appropriate mask, such as polyimide or polyimide on silicon nitride, for platinum, anodization reaction will occur only at the appropriate areas. The polyimide can prevent pinhole formation, and the silicon nitride can minimize the effect of undercutting. Polyimide is highly conformal, and therefore it will plug the pinholes. Silicon nitride is not conductive and is, therefore, electrically and chemically inactive during PECE.

One effective method used to neutralize pinholes in platinum is by double-layer deposition. After the deposition of the first platinum layer, the film is sputter-etched, and subsequently a second platinum layer is deposited. This process significantly reduces the pinhole formation and pitting associated with the use of a platinum etch mask. In many cases, the p-type SiC layer is not a fully effective etch-stop. This effect was observed in p-type SiC with low doping levels. Apparently, in lightly doped material, the electric field in the space–charge region is not high enough to prevent all the photo-generated carriers from reaching the surface to cause etching. In addition, the UV light incident on the np-junction causes higher leakage currents across the junction than higher-doped, p-type SiC. Although the anodic voltage is applied only through the ohmic contact on the top n-type SiC epilayer, the light-induced current through the junction leads to etching of the p-type SiC. To avoid etching of the p-type SiC epilayer, the reference voltage (V_{SCE}) must be reduced to a level that curtails the drifting of photo-carriers assisted by electric field when the p-epilayer is eventually exposed to the electrolyte. This fabrication procedure can be adopted to produce resistors in n-type epilayers with any doping level.

The characteristics of a diaphragm-based pressure sensor device are determined by the piezoresistors and by the dimensions of the diaphragm. Two key dimensions that characterize any circular diaphragm are thickness and radius. Since the radius is generally a fixed value determined by the pressure range and package specification, thickness is left as the main controlling variable. Therefore, the process of etching of the diaphragm to achieve a desired thickness and shape is of primary importance in diaphragm-based pressure sensor technology.

Until recently, the application of SiC diaphragms as pressure sensing devices was almost nonexistent. Fortunately, SiC microfabrication technology has advanced significantly over the last decade, largely because of the ability to perform high-energy plasma etching of various types of structures. Some structures have been selectively etched in SiC using reactive ion etching (RIE), but the etch rates reported were too low for practical use in the wafer thinning and shaping [Palmour et al., 1987] needed for pressure sensor fabrication. Rapid progress has been made in DRIE with etch rates up to 1 μm/min already reported [Khan and Adesida, 1999].

7.3   Characterization of 6H-SiC Gauge Factor

The piezoresistive effect is associated with changes in resistance as a result of an applied mechanical stimulus. The resistance can be measured by incorporating the gauge within a Wheatstone bridge circuit configuration, and the resultant output voltage is related to the applied force. The piezoresistive effect is rather small in metals, but in semiconductors it is much more pronounced. The explanation can be found in Equation (7.1) for the gauge factor,

$$GF = \frac{dR}{R\varepsilon} = 1 + 2v + \frac{d\rho}{\rho\varepsilon} \qquad (7.1)$$

where R = electrical resistance (Ω); ρ = electrical resistivity (Ω-cm) of the gauge material; v = Poisson ratio (a geometrical effect); and ε = applied strain.

The change in the resistance with strain is dependent on two terms: one is associated with the geometrical piezoresistive effect, and the second originates from the strain dependency of the resistivity, as shown in Equation (7.1). In metals, this latter term is zero, whereas it is significantly nonzero in semiconductors. In semiconductors, the contribution of the geometrical term to the GF is of the same order as in metals. Therefore, the gauge factor in semiconductors is related to the large strain sensitivity of the resistivity. This fact can be explained on the basis of the energy-band structure of semiconductors. In n-type semiconductors such as 6H-SiC that have multiple valleys in the conduction band, the piezoresistive effect is associated with a change in the relative energy positions of the multivalley minima under applied stress. This effect causes the electrons to transfer among the valleys, causing a net change in the mobility, which in turn has a dominant effect on the strain (stress) dependency of the resistivity [Herring and Vogt, 1956].

In order to measure the GF of 6H-SiC, resistors were etched in 6H-SiC epilayers (p- and n-type). The configuration of the resistors was such that two of the elements were transversely oriented, whereas the other two were oriented longitudinally. The substrate was then diced into rectangular chips, each containing a pair of Wheatstone bridge circuits, one of which is shown in Figure 7.4(a). The equivalent circuit of the Wheatstone bridge is shown in Figure 7.4(b). It depicts the transverse and longitudinal piezoresistors arranged alternately. With current flowing parallel to their length, the transverse resistors R_1 and R_4 would experience a perpendicular strain (strain perpendicular to current flow), and the longitudinal resistors R_2 and R_3 would experience a parallel strain (strain parallel to current flow).

The purpose of using two Wheatstone bridges (only one shown here) on a single chip is to simultaneously compare the measured responses from both the tensile and the compressive states of the bridges. The rectangular chip was then attached to a machined metal diaphragm made of Incoloy™ with high-temperature glass, as shown in the cross section schematic of Figure 7.5. One set of Wheatstone bridges was placed at the edge of the metal diaphragm, and the other one was on the push rod. Pressure was applied to the diaphragm through a back port (not shown), which caused the push rod of the metal diaphragm to deflect the SiC beam. One set of piezoresistors, closer to the center of the metal diaphragm, was placed in tension, whereas the other, closer to the periphery of the diaphragm, was in compression. The ΔR of each

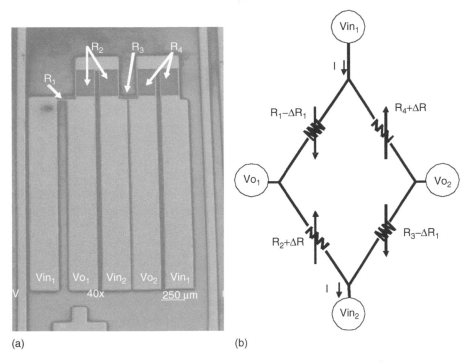

FIGURE 7.4 (a) Top view SEM picture of 6H-SiC beam with resistors and (b) equivalent Wheatstone bridge configuration. R_1 and R_4 are the transverse piezoresistors while R_2 and R_3 are the longitudinal piezoresistors.

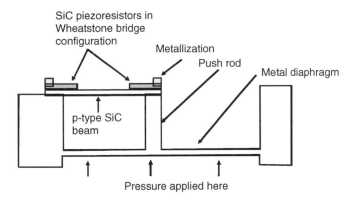

FIGURE 7.5 SiC cantilever beam transducer integrated to a metal diaphragm and push rod and used for gauge factor characterization.

arm of the bridge was measured to obtain the longitudinal and transverse gauge factors. All of the beams were uniaxially deformed perpendicular to within 4° of the 6H-SiC crystal's basal plane, which is the [0001] direction. Because the beam is integrated with the metal diaphragm, it is necessary to calculate the strain and stresses on the surface of the beam and to analyze their distribution across the beam. The problem is solved as a superposition of two systems, namely: (1) one with an edge-fixed diaphragm, and (2) one with the beam, with one edge of the beam fixed (clamped) and the second edge guided. Deflection, w_p, at the center resulting from uniform loading (pressure) on the front of the metal diaphragm is expressed as:

$$w_p = \frac{Pa^4}{64D_m} \tag{7.2}$$

where D_m is the flexural rigidity (N–m) of the metal membrane material and is expressed as:

$$D = \frac{E_m t_m^3}{12(1 - v_m^2)} \tag{7.3}$$

where E_m represents the Young's modulus (Pa) of the diaphragm, t_m (m) is the metal diaphragm thickness, and v_m is the metal Poisson constant. Deflection of the metal diaphragm, w_f, at the center resulting from a concentrated load at the center

$$w_f = \frac{Fa^2}{16\pi D_m} \tag{7.4}$$

where F (N) is the concentrated load or the contact force on the boss section of the metal diaphragm. The net deflection resulting from combined loading, when the concentrated load acts in the opposite direction of the applied pressure, is determined by:

$$w_{net.} = w_p - w_f = \frac{a^2}{16D_m}\left(\frac{Pa^2}{4} - \frac{F}{\pi}\right) \tag{7.5}$$

Deflection of a beam fixed (clamped) on one end, "guided" on the other (guided means that the slope at the guided point is always zero, but slight deflection is allowed and may change). Applying the Castigliano (1966) method allows the beam deflection to be expressed as:

$$w_b = \frac{Fl^3}{12E_b I_b} \tag{7.6}$$

where w_b = beam deflection (m); l = length of the beam (m); E_b = modulus of elasticity of the beam (Pa); and I_b = moment of inertia of the beam, expressed as:

$$I_b = \frac{bh^3}{12} \tag{7.7}$$

where b = beam width (m) and h = beam thickness (m). During loading, the deflection of the beam will be equal to that of the diaphragm; therefore, Equations (7.5) and (7.6) can be set equal to solve for the contact force, F (which is the contact force between the diaphragm boss and the beam). Because the radius of the diaphragm and the length of the beam are, for all intents and purposes, equal, the equation for F simplifies to:

$$F = \frac{3}{4}\left(\frac{Pa^2 \pi E_b bh^3}{4aE_m t_m^3 \pi + 3E_b bh^3}\right) \tag{7.8}$$

However, the maximum strains (ε_{max}) occur at either end of the beam and have opposite signs.

$$\varepsilon_{max} = \frac{Fat}{4E_b I_b} \tag{7.9}$$

The maximum stress, σ_{max}, at the edge of the beam can be calculated from Hook's Law:

$$\sigma_{max} = \varepsilon_{max} E_b \tag{7.10}$$

Based on the dimensions of the bossed metal diaphragm and the beam (diaphragm thickness, t_m = 0.2 mm; Young's modulus, E_d = 207 GPa, diaphragm radius, a = 4 mm, beam thickness, h = 305 μm; beam length, l = 4 mm, beam width, b = 1.52 mm; beam Young's modulus, E_b = 448 GPa, the strain in the SiC beam calculated from the dimensions of the metal diaphragm, was approximately 1 *nano*strain/Pa (~7 μstrains/psi).

Based on the above analysis, it becomes possible to calculate the *GF* of the 6H-SiC material, because there is a direct relationship between the applied strain and the gauge factor, as indicated in Equation (7.1).

The *GF* and Temperature Coefficient of Resistance (TCR) of n-type 6H-SiC were analyzed in the basal (0001)-plane. The characteristics of 6H-SiC piezoresistors were established individually in a four-arm Wheatstone bridge. The load on the beam was applied through pressure exerted on the metallic diaphragm previously described and transmitted by the boss acting as a push rod to the beam. Patterned strain gauges were fabricated in homoepitaxially grown, 2 μm, n-type epilayers on p-type 6H-SiC substrates with two n-type doping levels, namely: 2×10^{17} cm^{-3} and 3×10^{18} cm^{-3}. In hexagonal crystals, because the piezoresistance tensor is isotropic in the basal (0001)-plane, the gauge can be rotated about the *c*-axis without affecting the piezoresistivity [Davis et al., 1988]. Longitudinal and transverse gauges were measured, which yielded results corresponding to the piezoresistive coefficients π_{11} and π_{12}, respectively. Figure 7.6 shows normalized change in the resistance vs. applied strain in longitudinal n-type 6H-SiC gauges for the two n-type doping levels. The measured *GF*s were approximately -25 for the lower doping level and -20 for the higher doping level. Measurements conducted on the transverse piezoresistors ($N_d = 3 \times 10^{18}$ cm^{-3}) yielded a gauge factor of approximately 11. The longitudinal and the transverse piezoresistance coefficients, π_{11} and π_{12}, can be calculated from the relationship [Rapatskaya et al., 1968]:

$$\frac{dR}{R} = \frac{dV}{V} = \pi_{11}\sigma \qquad (7.11)$$

where *dV* represents the bridge output of the Wheatstone bridge circuit, and *V* is the input voltage. By using the stress obtained from Equation (7.18), the longitudinal gauge factor was found to be about -5.12×10^{-12} cm²/dyne for the SiC gauge with the lower doping level, and -4.3×10^{-12} cm²/dyne for the SiC gauge with the higher doping level. One initial conclusion is that an increase in doping level results in a decrease of the *GF*. Since the piezoresistive effect is an energy band transport phenomenon, factors such as impurities, donor ionization energy, mobility, defect density, and overall quality of the crystal may influence it substantially.

7.3.1 Temperature Effect on Gauge Factor

The piezoresistive properties of a 6H-SiC beam transducer described above were measured between 25°C and 250°C. Measurements were carried out on a beam consisting of longitudinal and transverse n-type 6H-SiC (doping level $N_d = 2 \times 10^{19}$ cm^{-3}) beams fabricated as described previously. Figure 7.7(a) shows the relative

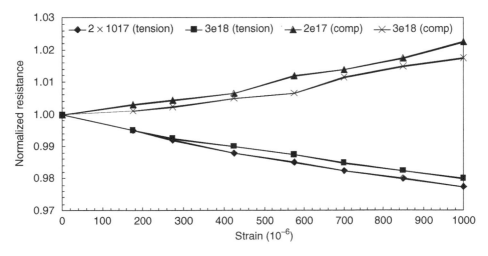

FIGURE 7.6 Normalized changes in resistivity vs. applied strain in longitudinal n-type 6H-SiC gauges with two doping levels.

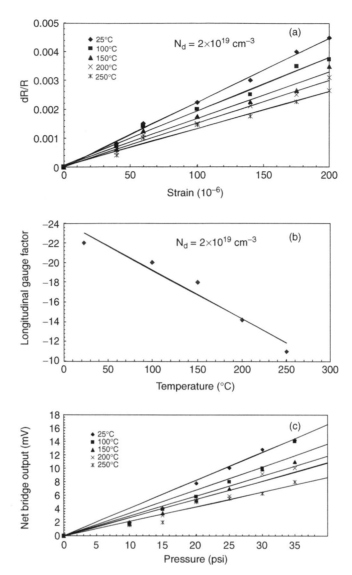

FIGURE 7.7   (a) Relative change in resistance of the logitudinal piezoresistors as a function of strain at different temperatures (n$^+$ 6H-SiC, $N_d = 2 \times 10^{19}\,\mathrm{cm}^{-3}$). (b) Longitudinal guage factor as a function of temperature. At 250°C the gauge factor is approximately 60% of its room temperature value (n$^+$ 6H-SiC, 2 µm thick epilayer, $N_d = 2 \times 10^{19}\,\mathrm{cm}^{-3}$). (c) Net bridge output as a function of pressure at five different temperatures ($N_d = 2 \times 10^{19}\,\mathrm{cm}^{-3}$). Bridge input voltage is 5 V.

change in resistance of the longitudinal piezoresistors as a function of strain at different temperatures. At all temperatures, a linear relationship is observed between $\Delta R/R$ and strain, but the strain sensitivity decreases with increasing temperature. Figure 7.7(b) shows the longitudinal gauge factor as a function of temperature. At 250°C the gauge factor is approximately 60% of its room temperature resistance for this doping level.

The bridge output as a function of pressure and temperature is shown in Figure 7.7(c). The decrease in output due to increasing temperature is explained in terms of intravalley carrier transport; the external applied heat energy leads carriers to acquire more energy, which enables more of them to be transported and occupy other energy minima. Therefore, when strain is applied under heat, only a few electrons can be transported to the energies not yet occupied. As a result of the fewer electrons available for intravalley exchange, the piezoresistance decreases.

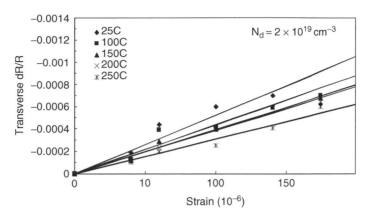

FIGURE 7.8 Relative change in resistance of the transverse piezoresistors as a function of strain at different temperatures ($N_d = 2 \times 10^{19}\,\text{cm}^{-3}$).

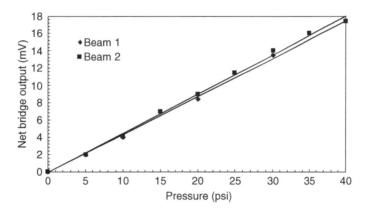

FIGURE 7.9 Net bridge output as function of pressure of two different beam sensors. In both cases the dependence is linear ($N_d = 2 \times 10^{19}\,\text{cm}^{-3}$).

The bridge *GF* decreases linearly with temperature, as seen in Figure 7.7(b). The relative change in resistance versus strain of the transverse piezoresistors is shown in Figure 7.8, from which the transverse *GF* can also be calculated. In order to check the reproducibility of the measurements, another beam transducer structure was assembled, and the bridge output as a function of pressure was measured. For comparison, Figure 7.9 shows the results obtained on both beams. In both cases, the dependence between the bridge output and the applied pressure is linear; however, one of the beams exhibited a slightly lower sensitivity. This lowered sensitivity was probably a result of either a geometrical factor (i.e., the metal diaphragm in both cases did not have exactly the same dimensions) or the mounting procedure, or both. Random temperature variations made it difficult to measure the *GF* of the individual resistor elements, especially the transverse resistors, which exhibited very small changes in resistance with pressure.

7.3.2 Temperature Effect on Resistance

Another important consideration in the selection of a resistor for high or low temperature applications is how the resistor's electrical resistance changes with temperature. Resistance variation with temperature is usually expressed as a TCR, which is defined as:

$$\beta = \frac{1}{R_o} \frac{R_f - R_o}{T_f - T_o} \qquad (7.12)$$

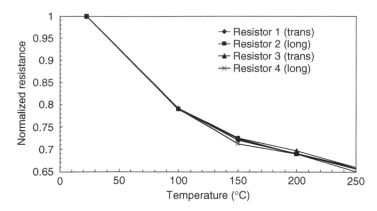

FIGURE 7.10   Change in normalized resistance of four individual gauges in a transducer as function of temperature. All measured resistances decrease as the temperature increases (n^+ 6H-SiC, $N_d = 2 \times 10^{19}/cm^3$).

where R_o = resistance at room or reference temperature (Ω); R_f = resistance at operating temperature; T_o = room or reference temperature (usually 25°C); and T_f = operating temperature. The TCR may be positive or negative and is usually expressed either in ppm/°C or in %/°C. Practically, the TCR can be influenced by resistor structure, as well as by processing conditions such as uniformity of the resistivity across the wafer.

To evaluate the TCR behavior of SiC, the resistances of four individual gauges ($N_d = 2 \times 10^{19} cm^{-3}$) in a transducer were measured and plotted as a function of temperature (Figure 7.10). All measured resistances in this sample decreased with temperatures up to 250°C, caused by increasing ionization of the donors in the heavily doped SiC. In contrast, the initial resistance measurements carried out with the lower-doped n-type 6H-SiC ($1.8 \times 10^{17} cm^{-3}$) decreased with temperature in the range between −60° to 25°C. Above 25°C, the resistance increased. Using Equation (7.12), the average TCR value for the range 25 to 625°C was found to be 0.56%/°C. In the $3 \times 10^{18} cm^{-3}$ doped samples, the resistance was observed to decrease up to 100°C and then begin to increase. The average TCR value for this sample in the range 100 to 625°C was found to be 0.28%/°C.

The decrease in resistance with increase of temperature below a certain temperature limit, which typically lies between 0 and 25°C, is associated with the increasing ionization of dopant impurities. In this temperature range, the semiconductor resistance is primarily controlled by carrier ionization. Once most dopant impurities have become ionized, carrier phonon-related lattice scattering increases with the temperature to increase the resistance. This observed behavior is consistent with well-known semiconductor carrier transport physics. In highly doped n-type SiC, the impurity ionization is completed at higher temperature because of the large number of impurities and the wide bandgap.

7.4   High Temperature Metallization

SiC-based technology appears to be the most mature wide-bandgap semiconductor material with the proven capability to function at temperatures above 500°C [Jurgens, 1982 and Palmour et al., 1991]. However, the contact metallization of SiC typically undergoes severe degradation beyond this temperature because of enhanced thermochemical reactions and microstructural changes. The causative factors of contact failures include interdiffusion among layers, oxidation, and compositional and microstructural changes. These mechanisms are potential device killers by way of contact failure. Liu et al. (1996) and Papanicolaou et al. (1998) have demonstrated stable ohmic contacts at 650°C for up to 3000 hours and 850°C for a short duration in vacuum. Vacuum aging is, however, not representative of the environmental condition in which SiC pressure sensor devices are expected to operate.

In order to fabricate any high-temperature electronic device, it is essential to have ohmic contacts and diffusion barriers that are capable of withstanding the device's operational temperatures. It is necessary

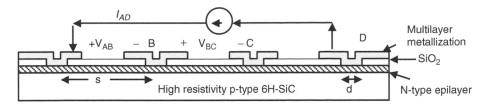

FIGURE 7.11 Cross-sectional view of multilayer metallization contact on n-type 6H-SiC epilayer for contact resistivity measurements.

to identify metals and alloys that form acceptable ohmic contact to 6H-SiC. For the ohmic contacts on n-type SiC, the metals include titanium and its alloys of nickel–titanium and titanium–tungsten. It has been shown previously that titanium contacts deposited on 3C-SiC withstand 20 hours at 650°C [Zeller et al., 1987]. Nickel is also known to be a good ohmic contact to n-type SiC, but it exhibits severe adhesion problems. Nickel–titanium alloys may combine the favorable electrical properties of nickel with the high reactivity and adhesion of titanium. Titanium–tungsten exhibits good diffusion barrier properties and may also be ohmic. Several groups have also demonstrated the effectiveness of TiN as a diffusion barrier. This section will start with the examination of a recently investigated alternative scheme that uses a TaSi$_2$ diffusion barrier, which shows promising performance characteristics to support stable device operation at 600°C in air.

7.4.1 General Experimental and Characterization Procedure

Several (0001)-oriented, Si-face 6H-SiC substrates (tilt angle 3.5°) with n-type, 1 μm-thick epilayers of different doping levels were used. The wafers were initially cleaned in equal volume of H$_2$O$_2$ and H$_2$SO$_4$ solution and dipped in 49% HF for five seconds, after which they were rinsed with deionized water and blown dry with nitrogen. They were then thermally oxidized in dry oxygen ambient at 1150°C for four hours to yield an oxide thickness of about 60 nm. This oxide was then stripped in 49% HF and the samples were rinsed and dried again. A second thermal oxidation, for five hours at 1150°C, was performed, yielding a cleaner thermal oxide layer. Photoresist was applied and patterned into circular patterns. Circular contact holes were then etched through the oxide with buffered HF (BHF) to expose circular sections of the epilayer surface. The contact holes consisted of twelve rows, with each row made up of four circular contact holes of the same diameter, d, with 225 μm equidistant separations among their centers as shown in Figure 7.11. The diameter of the contacts ranged from 6 to 28 μm. After stripping the photoresist, the samples were RCA-cleaned again, but with no HF dip, to ensure that a very clean epilayer surface with a monolayer of oxide was formed. The samples were immediately transferred into the sputtering chamber for metal depositions. A 300°C dehydration process, in vacuum for 20 minutes, followed, to remove any water trapped within micropipes in the wafers. The Current–Voltage (I–V) characteristic measurement of epilayer contacts was conducted by probing two adjacent contacts from the same row. The contact resistivity was measured using the modified four-point method of Kuphal (1981), expressed as

$$r_{cs} = \frac{A}{I_{AD}}\left[V_{AB} - V_{BC}\frac{\ln\left(\left(3\frac{s}{d}\right)-\left(\frac{1}{2}\right)\right)}{2\ln 2}\right] [\Omega\text{-cm}^2] \tag{7.13}$$

where V_{AB} is the voltage measured across contacts A and B in Figure 7.11, whereas V_{BC} measures the voltage between contacts B and C (epilayer resistance); A = contact area (μm); s = distance between adjacent contacts (μm); and d = diameter of the contact (μm). Current, I_{AD}, was passed through the circular contact pad, A, as shown in Figure 7.11, and out through pad D. Unless otherwise specified, the applied current was set at 1 mA. Measurements were made with four probes coming into contact with the Shockley pads that extend over the oxide. Therefore, the spreading and contact resistance, R_s and R_c, respectively, can be calculated by

$$R_s + R_c = \frac{V_{AB} - V_{BC}}{I_{AD}} \tag{7.14}$$

The resistance of the probes in these measurements was negligible. Since the main parameter of interest in these measurements was the overall change in resistance under the contact, R_c and R_s were lumped together in determining the contact resistivity. As a result of the lumping together of the spreading and contact resistances, the result was considered to be on the high end of the average specific contact resistivity.

7.4.2 Characterization of Ti/TiN/Pt Metallization

Several (0001)-oriented, highly resistive, Si-face, p-type 6H-SiC substrates, each with n-type epilayers (1 μm thick) of different doping levels ranging between $3.3 \times 10^{17}\,cm^{-3}$ and $1.9 \times 10^{19}\,m^{-3}$, were purchased from Cree Research, Inc. The wafers were initially cleaned by modified RCA method and dipped in 49% HF for five seconds, followed by rinsing and blow-drying. An *ex situ* dehydration process, at 200°C in nitrogen ambient for 20 minutes to desorb water trapped within the micropipes, followed this cleaning process. Depositions of Ti (50 nm)/TiN (50 nm)/Pt (100 nm) were made on the samples by sputtering without breaking vacuum. Titanium nitride was obtained by reactive sputtering of titanium in 20% nitrogen/argon ambient. The top platinum layer was etched in light *aqua regia* to form rectangular and circular probe pads that overlapped the field oxide. The exposed TiN/Ti on the field oxide was selectively etched in 1:1 EDTA:H_2O_2 to electrically isolate. The pads offered total coverage of the contact regions and facilitated broad area probe contact during testing. In the as-deposited state, the titanium contact on the n-type epilayer was ohmic for the sample with the highest doping level ($1.9 \times 10^{19}\,cm^{-3}$). The contact resistance using Equation (7.13) was found to be approximately $1 \times 10^{-5}\,cm^{\mp}$. In order to obtain ohmic contact to n-type 6H-SiC with lower doping levels (3.3×10^{17}–$10^{18}\,cm^{-3}$), high-temperature annealing was required.

The experimental results of the Ti/TiN/Pt ohmic contact are summarized in Table 7.2. The I–V characteristics of the as-deposited metallization on all samples were rectifying, except for the highest doped sample ($1.9 \times 10^{19}\,cm^{-3}$). After 30 to 60 seconds of rapid thermal anneal at 1000°C in argon ambient, ohmic contact was achieved on all samples except for the lightest doped, which remained rectifying after three and a half minutes of annealing. The average barrier height before annealing was obtained from the forward I–V characteristic curve using the thermionic emission model:

$$J = J_s\left[e^{\left(\frac{qV}{nkT}\right)} - 1 \right] \tag{7.15a}$$

where J is the forward current density (A/cm^2); V is the applied voltage; q is the electronic charge; k is the Boltzman constant; T is the temperature (K); and n is the ideality factor that models the deviation from

TABLE 7.2 Summary Results of Electrical Characteristics of Ti/TiN/Pt Metallization on N-type 6H-SiC Epilayers

Sample No.	Conc. (cm^{-3})	As-Deposited	Annealed	Total Time (min.)	SBH$_{as-dep}$ (eV)	r_{cs} ($10^{-4}\,\Omega cm^2$)
A	3.3×10^{17}	Rectifying	Rectifying	3.5	0.84	
B	1.4×10^{18}	"	Ohmic	0.50	0.82	3.42
C	1.5×10^{18}	"	Ohmic	1.00	0.74	2.50
D	1.7×10^{18}	"	Ohmic	0.50	0.82	2.10
E	2.7×10^{18}	"	Ohmic	0.50	0.80	1.50
F	1.9×10^{19}	Ohmic	Ohmic	0.50	n/a	0.15

the theoretical ideal I–V characteristic, depending on the integrity of the metal/epilayer interface. The saturation current density, J_s, is expressed as:

$$J_s = A^*T^2 e^{\left(-\frac{q\phi_B}{kT}\right)} \tag{7.15b}$$

where A^* is the effective Richardson constant ($Acm^{-2}K^{-2}$), and ϕ_B (V) is the Schottky Barrier Height (SBH) between the metal in intimate contact with the 6H-SiC epilayer. The ideality factors before and after annealing was ranged from 1 to 1.05 and J_s in the range of 9.44×10^{-8} to 4.4×10^{-3} (Acm^{-2}). In this work, the effective Richardson constant was estimated by

$$A^* = 120(m_e^*/m_o) \; [Acm^{-2}K^{-2}] \tag{7.15c}$$

where m_e^*/m_o is the ratio of the effective electron mass to the electronic rest mass. For a value of $0.45m_o$, A^* was calculated to be 54 $Acm^{-2}K^{-2}$. The obtained average SBH values ranged from 0.54 to 0.84 eV. The ohmic contact obtained after annealing is believed to be a result of the barrier-lowering effect caused by the change at the metal/SiC interface during annealing.

The Auger Electron Spectroscopy (AES) depth profile of Figure 7.12(a) indicated distinct boundaries of the as-deposited metals on the SiC epilayer. However, in Figure 7.12(b), intermixing and zone reactions after annealing are evident, and a new layer consisting mainly of Pt, Ti, Si, and C atoms is observed in

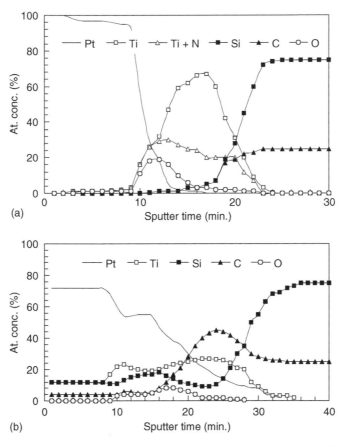

(a)

(b)

FIGURE 7.12   (a) Auger Electron Spectroscopy depth profile of as-deposited Ti/TiN/Pt metallization on n-type 6H-SiC. (b) Auger Electron Spectroscopy of Ti/TiN/Pt after rapid thermal anneal at 1000°C for thirty seconds in argon atmosphere. Synchronous 2:1 ratio tracking between titanium and carbon suggested the formation of TiC_{1-x} on the epilayer surface.

direct contact with the epilayer. The synchronous tracking of Ti and C atoms at a constant ratio of almost 1:2 (discounting the primary Ti–N signal that was difficult to distinguish) strongly suggested the formation of a TiC species. Several groups had previously confirmed the formation of TiC_{1-x} and Ti_5Si_3 species for wafers annealed in the range of 500 to 1200°C. The decrease in the SBH could be associated with the low work function of TiC (3.35 eV) compared to that of titanium (4.1 eV) (1976).

The contact resistivity plotted against the impurity concentration, which exhibited the characteristic exponential dependence of contact resistivity on impurity concentration, is shown in Figure 7.13. In order to estimate the specific contact resistivity, r_{cs}, the spreading resistance, R_s, and the contact resistance, R_c, were decoupled. TiC was assigned R_s on the assumption that it was the new layer in contact with the 6H-SiC epilayer. Therefore, R_s was evaluated with respect to its thickness by using the method of Cox and H. Strack (1967):

$$R_s = \rho_{TiC} \frac{t}{\pi\left(\frac{d}{2}\right)^2} \qquad (7.16)$$

where d is the contact diameter (μm), ρ_{TiC} is the resistivity (μΩ-cm) of the assumed TiC layer, and t is the thickness (~100nm). Substituting for R_s in (7.16) in Equation (7.14), we have:

$$\rho_{TiC} \frac{t}{\pi\left(\frac{d}{2}\right)^2} + R_c = \frac{1}{I_{AD}} [V_{AB} - V_{BC}] \qquad (7.17)$$

For a TiC resistivity 200 μΩ-cm [Toth, 1971], the contact resistance, R_c, and the specific contact resistivity, ρ_c, was then evaluated. The values obtained were such that $R_s \ll R_c$. A comparison of the Figure 7.12 Auger depth profile of the annealed samples to the as-deposited samples revealed the reactions at TiN/Pt and 6H-SiC/Ti interfaces. The oxygen content (at. 17%) between the former interfaces was an artifact of the deposition system. The degree of its effect on the electrical characteristics was not known, but the 6H-SiC/Ti interface was relatively free of oxygen contamination. The surface of the top platinum layer exhibited a faint brown coloration after annealing, indicating the appearance of titanium species, which was confirmed by Auger surface spectral analysis. The average contact resistivity, ranging from 1.5×10^{-5} to 3.42×10^{-4} Ω-cm², and Schottky barrier height, between 0.8–0.84 eV, were in reasonable agreement with previously published results [Waldrop et al., 1993]. Interlayer delamination, sometimes attendant in a pre-processed titanium layer, was not observed. Auger Electron Spectroscopy (AES) revealed the out-diffusion of titanium-silicon species. The continued presence of the TiN layer after annealing at 1000°C suggested its partial survivability, but it did not offer a full barrier against platinum diffusion.

In order for this multilayer, high-temperature metallization to be applicable, the oxygen contamination in the metal must be kept to less than 3 at. % oxygen. High levels of oxygen contamination at the

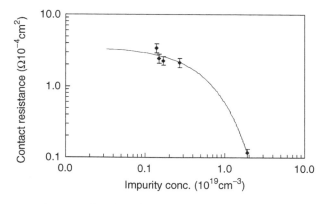

FIGURE 7.13   Bulk contact resistance as function of doping level.

interfaces of Pt/TiN or TiN/Ti could be the result of full or partial decomposition of the titanium nitride layer at high temperature and replacement by a layer of titanium oxide, caused by the high affinity of titanium toward oxygen.

Formation of titanium oxide results in two deleterious effects: (a) it greatly reduces the effectiveness of the diffusion barrier, and (b) it forms a dielectric layer, which leads to rectification and failure of the ohmic contact. Another destructive effect is penetration of oxygen through the outer platinum layer. Inspection with scanning electron microscope (SEM) indicated that the deposited platinum layer contained a high density of pinholes. At high temperature, oxygen would diffuse through the pinholes, thereby degrading the titanium nitride diffusion barrier. Oxygen also reacts with the titanium beneath the barrier, forming a solid titanium oxide layer with undesirable rectifying properties that extends to the 6H-SiC surface. For the metallization scheme to work effectively in air, the issue of oxygen contamination must be resolved.

7.4.3 Ti/TaSi$_2$/Pt Scheme

Oxygen contamination generally posed a big problem for the Ti/TiN/Pt scheme, but the problem is minimized if the device with such metallization is hermetically sealed. However, the diffusion barrier integrity of the TiN layer against platinum diffusion to the semiconductor interface is also undesirable. A robust, high-temperature metallization scheme must have, at a minimum, the following attributes: (a) ohmic contact with reasonably low contact resistance relative to the bulk epilayer; (b) long-term contact stability in the harsh environment; (c) compatibility with SiC large-scale integrated electronics fabrication technology; (d) good wire-bond strength; and (e) compatibility with high-temperature interconnect and packaging technology.

In order to meet these criteria, it was necessary to identify metallization schemes that will both form an ohmic contact on n-type SiC and offer an excellent diffusion barrier against both oxygen penetration and migration of any top layer metallization, such as platinum, toward the contact SiC interface. In addition, such a scheme should have a top surface that is essentially wire-bondable. In developing this new scheme, thermodynamic and thermochemical issues were taken into consideration with the recognition that at 600°C, the activation energies of several metals are enough to promote reactions or intermixing among metals. Metal layers with low mutual diffusivities were identified in order to keep intermixing at a minimum. In the case where they mix, however, the alloys formed must be thermodynamically, mechanically, and electrically stable. They must maintain excellent diffusion barrier characteristics. By combining the ability of titanium to form ohmic contact on n-type SiC, the diffusion barrier characteristics of TaSi$_2$, and the relative stability of the interface of the two layers, a new scheme was developed with a result that proved far superior to the Ti/TiN scheme.

A sequential deposition of Ti (100 nm)/TaSi$_2$ (200 nm)/Pt (300 nm) multilayer contact was performed in a three-gun, UHV/load-lock sputtering system. Details of the deposition parameters are shown in Table 7.3. A 2-μm aluminum layer was deposited by e-beam evaporation and used as an etch mask during reactive ion etching (RIE) to pattern the multilayer metallization into Shockley probe pads over the contacts. Following the RIE, the etch mask was selectively removed with an aluminum etchant to expose the underlying platinum layer. The specific contact resistance, ρ_{cs}, was calculated using the modified four-point probe measurement method described earlier.

TABLE 7.3 Process Parameters for the Deposition of Ti/TaSi$_2$/Pt

| Layer | Thickness (nm) | Deposition Conditions | | | | |
		Pressure (mTorr)	Power (W)	Gas Flow (sccm)	Time (min.)	Deposition Method
Ti	100	6	200 R.F.	50 Ar	16.5	Sputtering
TaSi$_2$	200	6	100 R.F.	50 Ar	33.3	Sputtering
Pt	300	9	75 D.C.	50 Ar	6.3	Sputtering

The specific contact resistances of the sample sets as a function of time at 500°C and 600°C in air atmosphere are shown in Figures 7.14(a) and 7.14(b), respectively. The samples treated at 500°C, shown in Figure 7.14(a), exhibited higher contact resistance values initially that remained high for the first 40 hours. The contact resistance dropped thereafter and remained practically constant for the entire period between 70 and 600 hours. The results of the samples tested at 600°C in air for 150 hours are shown in Figure 7.14(b). The specific contact resistance also increased after an initial 30-minute anneal at 600°C in forming gas, but subsequent heat treatments in air saw a nearly exponential decrease in the contact resistance values, which appeared to taper off after 100 hours. There is an obvious difference in the contact resistance values between the two sample sets in the first 40 hours, as depicted in Figure 7.14. The observed differences could perhaps be attributed to one or a combination of three things: (a) oxygen contamination of samples treated at 500°C in air; (b) the probable existence of surface states; and (c) incomplete reaction product formation at the SiC interface after the initial 30-minute anneal at 600°C in forming gas, and the reaction was accelerated by the subsequent heat treating at 600°C. Since the heat

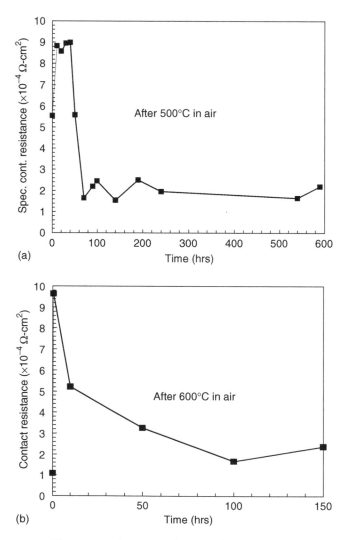

FIGURE 7.14 (a) Average specific contact resistance as a function of time after 500°C in air. The high contact resistance in the first few hours may be attributed to conditions stated in the text. (b) Average specific contact resistance as a function of time after 600°C in air. The high contact resistance in the first few hours may be attributed to conditions stated in the text.

treatments in air between the sample sets are different, variations in activation energies may cause differences in product formation, thereby leading to variations in electrical characteristics. After the contrast in results for times less than 100 hours between the sample sets, the average specific contact resistance for both sets leveled off to values around 2–3 $\times$ 10^{-4} Ω-cm^2.

To begin understanding the active mechanisms, we examined the relationship between the electrical and thermochemical characteristics within the context of diffusion barrier formation and interfacial reactions. The Auger profiles after annealing at 600°C for 30 minutes in forming gas, and after 50 hours at 500°C in air, are shown in Figures 7.15(a) and 7.15(b), respectively. Generally, the two figures are similar in terms of phenomenological changes taking place within the layers. Figure 7.15(a) shows the preferential, unidirectional migration of silicon into platinum and toward the surface. This action is of significant importance, as it forms the basis for the diffusion barrier characteristics of the metallization. This migration creates a silicon-depleted zone inside the metallization, as depicted in Figure 7.15(b). The AES profile shows a buildup of silicon, at a nearly consistent platinum-to-silicon ratio of 2:1, within the platinum layer. Since we anticipate stable titanium silicide and titanium carbide as the reaction products between titanium and SiC, no new source of silicon exists that will migrate to the surface. Although no strong indications of titanium carbide and titanium silicide signals were observed at the epilayer interface, the extension of the contact boundary, a few nanometers into the epilayer, strongly suggests an underlying physical and chemical reaction.

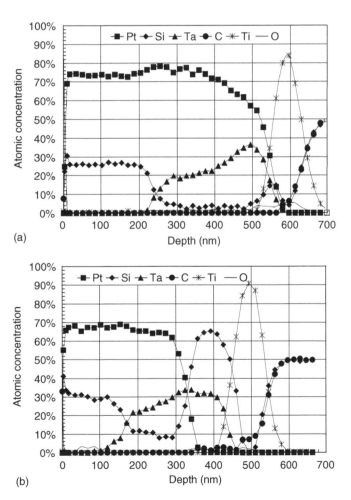

FIGURE 7.15 (a) Auger depth profile of 6H-SiC/Ti/TaSi$_2$/Pt after heat treatments at: 600°C anneal in H$_2$ (5%)/ N$_2$ 30 minutes. (b) Auger depth profile of 6H-SiC/Ti/TaSi$_2$/Pt after heat treatments at 500°C treatment in air for fifty hours.

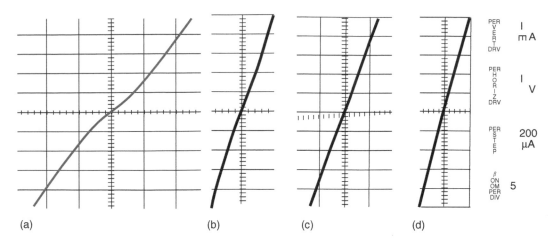

(a)   (b)   (c)   (d)

FIGURE 7.16   I–V characteristics of samples measured at different stages: (a) as-deposited, (b) 600°C annealed in H_2 (5%)/N_2 forming gas for 30 minutes. (c) sample treated at 500°C in air for 630 hours, and (d) sample treated at 600°C in air for 150 hours. The current and voltage scales are the same for all I–V plots.

The representative I–V characteristics after various conditions for both sets of samples are shown in Figure 7.16(a–d). The I–V curve shown in Figure 7.16(a) is that of the as-deposited condition, which was the same for both sample sets. The observed weak rectification could be related to low-level oxidation issues previously discussed in this chapter. The entire sample set exhibited linear I–V characteristics after annealing at 600°C in H_2 (5%)/N_2 and forming gas for 30 minutes, as shown in Figure 7.16(b). The slight curvature observed was attributed to the ongoing reaction at the SiC interface. Figure 7.16(c) shows that the representative I–V characteristic of the sample set after 630 hours at 500°C in air remained practically unchanged relative to Figure 7.16(b). The observed slight change in the I–V slope was an artifact of the oxidation process occurring on the Shockley pads; scratching the surface with the probe tips was required to get good electrical contact between the pad and the probe tip. For the sample set treated at 600°C in air, the slope of the I–V characteristic after 150 hours, shown in Figure 7.16(d), compares very well with that of Figure 7.16(b) and Figure 7.16(c), given that they were all from the same wafer (net epilayer doping level, $N_d = 2 \times 10^{19}\,\text{cm}^{-3}$). This result is an indication of the improved thermal stability of the new SiC interface.

An understanding of these chemical reactions can be drawn from previous works by Bellina (1987) and Chamberlain (1980). The temperatures in which we performed the heat treatments were in a range consistent with the work of Chamberlain, who identified the formation of titanium carbide and silicide at similar temperatures. Applying Chamberlain's parabolic reaction rate relation gives an approximation of the minimum thickness of reaction products:

$$x^2 = x_o + 2K(t - t_o)\ [\text{cm}^2] \tag{7.18a}$$

$$K = K_o \exp\left(\frac{-262.7\,\text{KJmol}^{-1}}{RT} \right)\ [\text{cm}^2\text{s}^{-1}] \tag{7.18b}$$

where k_o is the reaction rate constant (cm^2s^{-1}); K is the temperature dependence of the rate constant; x_o and t_o are initial distance and time constants, which in our case were set to zero; and R is the universal gas constant ($8.314\,\text{J-mol}^{-1}\text{K}^{-1}$). Without correction for pressure, a layer of no less than 2 nm of new products would have formed at the epilayer interface after 30 minutes of anneal at 600°C. Thermodynamic and reaction-limited kinetics are proposed to be acting in sequence, such that the decomposition of $TaSi_2$ in the presence of platinum proceeds with a final reaction:

$$4Pt + TaSi_2 \rightarrow Ta + 2Pt_2Si \tag{7.19}$$

If this reaction proceeds to the right, the limiting reactant would be silicon, since it is fully consumed by platinum. This proposition is made more likely, given that the heat of formation of Pt_2Si ($-\Delta H_f \sim 10\,\text{kcal mol}^{-1}$) is less than that of $TaSi_2$ ($-\Delta H_f \sim 28\,\text{kcal mol}^{-1}$) [Andrews and Phillips, 1975].

It is important to examine the oxygen concentrations in Figures 7.15(a) and 7.15(b). Both parts of the figure show very little migration of oxygen into the metallization. This finding implies a much more stable contact structure than the Ti/TiN/Pt approach [Okojie et al., 1999], which was affected by significant oxidation of the metals in the contact. Various groups have extensively studied the oxidation kinetics of the silicides. Murarka (1988) concluded that the oxidation mechanisms for most silicides essentially have the same heats of formation as those of normal oxides. Lie (1984) and Razouk (1982) confirmed that the oxidation kinetics of tantalum disilicide has a parabolic rate. This result implies that it would take an appreciable length of time for oxygen to diffuse through the entire contact.

7.5 Sensor Characteristics

Generally, the design of devices that sense physical phenomena and provide electrical readouts calls for the interpolation of two or more kinds of mathematical relationships. In the design of high-temperature pressure sensors, the Equations that describe the physical phenomena (i.e., pressure and diaphragm deflection) are interpolated with the electrical Equations that express the resulting output voltage.

The Equations that model deflecting diaphragms are classified into two main categories. One category models maximum diaphragm deflections that are less than the diaphragm thickness (linear case), whereas the other supports diaphragms with maximum deflections greater than the diaphragm thickness (nonlinear case). If the maximum deflection of the membrane is less than its thickness, as occurs in applications of short-range pressure measurement, there is generally a reasonable degree of linearity of diaphragm deflection in response to applied pressure. For larger pressure measurements where deflection is equal to or greater than the thickness of the diaphragm, the deflection of the diaphragm in response to applied pressure is no longer linear [Timoshenko and Woinowsky-Krieger, 1959]. For the device to be used continuously over a long period of time, the membrane must be capable of repeatedly deflecting under applied pressure with precision and little hysteresis. To achieve this aim, the membrane must retain its elastic property after it is subjected to maximum applied pressures. To that effect, there is a need to choose materials with an appreciable linear region on the Stress-Strain curve. For the diaphragm to retain its elastic integrity, the stress induced by pressure must not exceed the yield or fracture point. In essence, the maximum operating stress should be at a point below the yield and fracture stress limit. If the operating stress reaches the elastic or fracture limit, there is a very strong likelihood that the diaphragm would lose its elasticity, become permanently deformed, and possibly fracture.

A resistor on a circular diaphragm arranged tangentially, with current flowing parallel to the resistor, will experience a longitudinal stress induced by the tangential strain component. It will also experience a transverse (radial) strain component (strain perpendicular to resistor length), which usually inserts a negative piezoresistance coefficient. On the other hand, as depicted in Figure 7.3(b), the radially oriented resistor, with current flowing parallel along its length, will be dominated by the stress induced by the longitudinal strain component. The transverse effect is introduced via the tangential stress, with its corresponding negative piezoresistance coefficient. These findings have been extensively verified and used in silicon. The output of the sensor is strongly affected by the orientation of the resistors. Therefore, the resistor geometry and orientation should be such that only one strain component exists and the other is suppressed.

When the maximum deflection, w, of a clamped circular plate is less than its thickness, the equation that describes it is expressed as [Timoshenko and Woinowsky-Krieger, 1959]:

$$w = 0.89\frac{Pa^4}{64D} \tag{7.20}$$

where P is the applied pressure (Pa), a is the radius of the diaphragm (μm), and D is the flexural rigidity (N–m) of the membrane material. D is expressed as:

$$D = \frac{Et^3}{12(1 - v^2)} \qquad (7.21)$$

where E is the Young's modulus (Pa) and t = membrane thickness (m). The total stress on the membrane associated with such a small deflection at the clamped edge is expressed as:

$$(\sigma_r)_r = \frac{3}{4}\frac{Pa^2}{t^2} \qquad (7.22)$$

The choice of large or small deflection equation, as stated before, is basically dictated by device application. A circular diaphragm can be easily mounted, and in the case of materials with high elastic moduli, high pressures can be applied on diaphragms with reasonable diameter-to-thickness ratios. According to the theory of plates and shells by Timoshenko and Woinowsky-Krieger (1959), the radial and tangential stresses (σ_r and σ_t, respectively) at any point on the front side of a circular diaphragm with fixed edges can be related to the applied pressure, P, on the front of the diaphragm; its thickness, t; radius, a; and the distance, r, from the center of the point of interest as:

$$\sigma_r = -\frac{3}{4}\frac{Pa^2}{t^2}\left[(v + 3)\frac{r^2}{a^2} - (v + 1)\right] \qquad (7.23)$$

and

$$\sigma_t = -\frac{3}{4}\frac{Pa^2}{t^2}\left[(1 + 3v)\frac{r^2}{a^2} - (v + 1)\right] \qquad (7.24)$$

The same set of Equations can be used to define stresses on the back of the diaphragm, when pressure is applied there, but with negative sign. A simple analysis of the radial and tangential stress distribution on the front side of the diaphragm shows that both stresses change (as a result of applied pressure from the front side), at a certain distance from the center of the diaphragm, from compressive to tensile. When pressure is applied to the front side of the diaphragm, any piezoresistor on the back or on the front side of the diaphragm will be subjected to parallel and perpendicular stresses, depending on its location. Therefore, the functional relationship between the fractional change in electrical resistance ($\Delta R/R$) of the piezoresistor and the perpendicular and parallel stress components is given by:

$$\frac{\Delta R}{R} = \pi_t\sigma_\parallel + \pi_r\sigma_\perp \qquad (7.25)$$

where π_t and π_r are parallel and perpendicular piezoresistive coefficients, respectively, and $\sigma_\parallel$ and $\sigma_\perp$ are parallel and perpendicular stress components, respectively. For tangentially and radially oriented piezoresistors on a circular diaphragm whose radius is large as compared with the resistor dimensions, this fractional change in resistance can be expressed as follows:

for tangential resistors:

$$\left(\frac{\Delta R}{R}\right)_t = \pi_t\sigma_t + \pi_r\sigma_r \qquad (7.26)$$

and for radial resistors:

$$\left(\frac{\Delta R}{R}\right)_r = \pi_t\sigma_r + \pi_r\sigma_t \qquad (7.27)$$

In the case of semiconductors with hexagonal crystal structure (such as 6H-SiC), the problem is much more complicated, in terms of resolving the piezoresistive constants in different directions. Earlier attempts to characterize the piezoresistance of 6H-SiC as a function of crystallographic orientation yielded significant differences in the obtained values [Rapatskaya et al., 1968, Azimov et al., 1974, Guk et al., 1974a, and Guk et al., 1974b]. Possible reasons for the discrepancies were the differences and imperfections of the Lely (1955) platelets, which were the only crystals available at that time, and also the inconsistencies and quality of the metal ohmic contacts. Recent advances in SiC technology have led to more reproducible and better quality SiC wafers and ohmic contacts, making it possible to obtain more reliable results.

On the backside of the wafer, a circular cavity mask was aligned with each set of piezoresistors that form a Wheatstone bridge network. Electrochemical etching (ECE), as previously described, was used to fabricate the circular diaphragm cavities. The irregularity of the sidewalls and base of the cavity indicates that a consistent control of the electrochemical etching process is required. This batch of sensors had a diaphragm thickness of about 50 μm and a chip area of 1.48 mm^2. The above process was followed by a 20-hour wet oxidation at 1150°C to ensure complete p–n junction isolation and passivation of the active elements. Contact holes were etched through the oxide using BHF to expose sections of the resistor elements. This procedure was followed by *in vacuo* sputter deposition of the Ti/TaSi$_2$/Pt high-temperature metallization and patterning of the metallization by the method described earlier. The wafer was then diced into chips and individually mounted on specially designed pressure sensor headers. Gold wires were bonded from the sensor to the header pins to facilitate external electrical connection.

The performance characteristics of the three sensors shown in Table 7.4 are representative of multiple batch fabrication and diaphragm thickness. The room temperature net output voltage, as a function of

TABLE 7.4 Performance Characteristics of 6H-SiC Pressure Sensor (Epilayer Doping Level $N_d = 2 \times 10^{19}\,\text{cm}^{-3}$)

Temp (°C)	Characteristics	Sensor 5	Sensor 6	Sensor 10
23	Full-scale output (mV)	41.70	39.51	66.42
	Linearity (%)	0.604	0.420	0.900
100	Full-scale output (mV)	37.50	34.47	57.23
	Linearity (%)	0.400	0.800	0.005
	TCGF (%/°C)	−0.13	−0.17	−0.20
	TCR (%/°C)	−0.24	−0.23	−0.23
200	Full-scale output (mV)	32.20	27.92	47.50
	Linearity (%)	0.93	0.86	0.12
	TCGF (%/°C)	−0.13	−0.17	−0.16
	TCR (%/°C)	−0.17	−0.17	−0.17
300	Full-scale output (mV)	28.20	24.56	36.54
	Linearity (%)	0.39	0.86	0.93
	TCGF (%/°C)	−0.12	−0.14	−0.16
	TCR (%/°C)	−0.12	−0.12	−0.12
400	Full-scale output (mV)	21.77	19.61	28.33
	Linearity (%)	0.95	0.30	0.41
	TCGF (%/°C)	−0.13	−0.13	−0.15
	TCR (%/°C)	0.04	0.04	0.05
500	Full-scale output (mV)	18.11	15.73	25.04
	Linearity (%)	0.57	0.20	0.12
	TCGF (%/°C)	−0.12	−0.13	−0.13
	TCR (%/°C)	0.07	0.07	0.06
600	Full-scale output (mV)	18.69	10.66	23.00
	Linearity (%)	0.65	0.94	0.16
	TCGF (%/°C)	−0.10	−0.13	−0.11
	TCR (%/°C)	0.07	0.07	0.08
23	Full-scale output (mV)	41.50	39.46	66.34

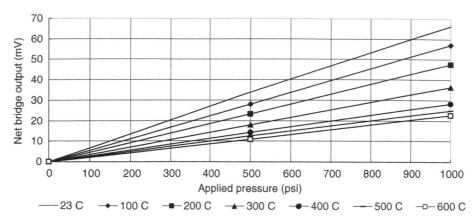

FIGURE 7.17   Net bridge output voltage of 6H-SiC pressure sensor as function of pressure at various temperature regime.

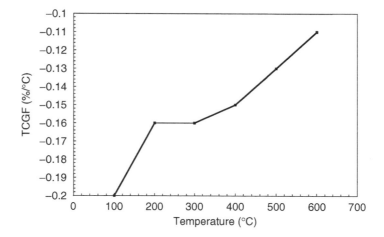

FIGURE 7.18   Temperature coefficient of gauge factor of 6H-SiC (calculated over 100°C increments) as function of temperature (epilayer doping level, $N_d = 2 \times 10^{19}\,\mathrm{cm}^{-3}$).

applied pressure at various temperatures, is shown in Figure 7.17 for sensor #10. With a bridge input of 5V, the full-scale output (FSO) was 66.42 mV at room temperature for an applied pressure of 1000 psi, indicating a sensitivity of 0.013 mV/V/psi. Very low hysterisis of 0.7% FSO and nonlinearity of −0.9% FSO were obtained. The 600°C output of 25.04 mV indicated a 62% output drop from the room temperature value. The characterization of the *GF*, described in Section 7.2, showed a linear drop in *GF* with increased temperature. The output was observed to decrease as temperature increased, but it became gradually insensitive to temperature as the temperature approached 600°C. Keyes (1960) had previously predicted this behavior in silicon. The temperature coefficient of gauge factor (TCGF), a measure of the output sensitivity to temperature, is defined here as:

$$\gamma = \frac{1}{V_{(T_o)}} \frac{V_{(T)} - V_{(T_o)}}{T - T_o} 100 \; [\%/°C] \tag{7.28}$$

where $V_{(T_o)}$ and $V_{(T)}$ are the full scale outputs at room temperature and final temperature. The TCGF (calculated over 100°C increments), shown in Figure 7.18, indicated an initial pronounced sensitivity that approached smaller (less-negative) values as the temperature increased. The TCGF response is expected to be lower in magnitude for doping levels greater than $2 \times 10^{19}\,\mathrm{cm}^{-3}$. The effect of temperature on resistance

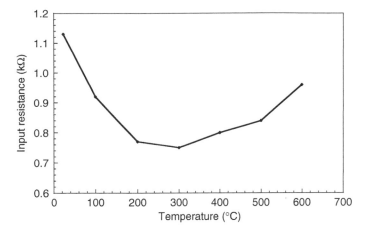

FIGURE 7.19 Bridge resistance of 6H-SiC piezoresistive pressure sensor as function of temperature.

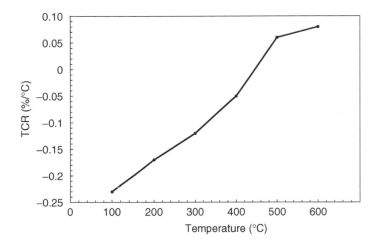

FIGURE 7.20 Temperature coefficient of resistance of 6H-SiC (calculated over 100°C increments) as function of temperature (epilayer doping level, $N_d = 2 \times 10^{19} \, cm^{-3}$).

is shown in Figure 7.19. It indicates a gradual decrease from a room temperature bridge resistance value of $1.13 \, k\Omega$ to about $750 \, \Omega$ at 300°C caused by carrier ionization. The upward swing of the resistance is associated with the growing dominance of the lattice scattering mechanism [Streetman, 1990]. From this result, the TCR from Equation (7.20) was calculated over 100°C increments and is shown in Figure 7.20. The negative TCR characteristic, relative to the room temperature resistance, was consistent with an n-type 6H-SiC epilayer of this doping level ($2 \times 10^{19} \, cm^{-3}$). For more heavily doped crystals, the negative TCR will extend to higher temperatures, thereby allowing for a less-complex compensation scheme.

7.6 Reliability Evaluation

Recently, Masheeb et al. (2002) reported the demonstration of a leadless (no wire-bond), SiC-based pressure transducer at 500°C. The elimination of the gold bonds, and the protection of the metallization from the harsh environment, offer the potential for long-term survival of SiC pressure transducers at high temperature. These developments have increased the possibility of direct insertion of uncooled SiC pressure sensors into high-temperature environments. However, for SiC pressure sensor technology to transition from the laboratory to commercial production, several reliability challenges, including package-induced stress and contact degradation, must be resolved. To be commercially viable, these sensors must undergo appropriate reliability testing standards that are specific for harsh-environment microsystems. However, the existing Joint Electron Device Engineering Council standards are only applicable to conventional-environment semiconductor microsystems [JEDEC Standards]. Because of the absence of a testing standard for high temperature devices, we recently developed and employed an Accelerated Stress Test (AST) protocol to evaluate the reliability performance of packaged SiC pressure transducers up to 300°C in air for over 140 hours under cyclic pressure and temperature [Savrun et al., 2004].

7.6.1 Reliability by Package Design

The SiC sensor (2.1 × 2.1 mm²) is mounted on an aluminum nitride (AlN) header (0.25 in. diameter) by the direct chip attach (DCA) method, as shown in Figure 7.21, so that only the sensor's circular diaphragm is free to deflect in and out of the reference cavity [Okojie, patent pending]. The sealing glass is applied only into and above the narrow gap (<10 μm) between the SiC sensor and the inner walls of the AlN, thereby providing a reference pressure in the cavity. This DCA approach eliminates the need for wire-bonding, thereby eliminating failure mechanisms associated with high-temperature diffusion of gold bond wires [Khan et al., 1994]. Because the sealing glass is applied at $T_\delta = 750$°C, the SiC/glass interface is assumed to be in a relaxed, stress-free state. Upon cooling to room temperature, the mismatches in the coefficients of thermal expansion (CTE) of the contracting components induce a net stress on the SiC sensor. This package-induced stress is a primary cause of transducer instabilities during cyclic temperature excursions and must be significantly minimized. Therefore, having components with equal CTEs will, ideally, eliminate the package-induced stress. AlN is selected because its CTE ($\alpha_{AlN} = 4.1$ ppm/°C) is close to that of SiC ($\alpha_{SiC} = 3.7$ ppm/°C). The CTE of the sealing glass (~4.1 ppm/°C) is also close to that of SiC. Thus, the resulting induced net strain on the SiC sensor as a result of the packaging process can be estimated as:

$$\varepsilon_\delta = (\alpha_{glass} - \alpha_{SiC})(T_\delta - T) \tag{7.29}$$

This calculation approximates a lateral residual strain of about 300 μstrains on the SiC sensor solid surface at room temperature. Because the distance separating the SiC sensor and the kovar tube is greater than 12 mm, the large thermomechanical stress between the kovar and AlN does not propagate to the SiC sensor.

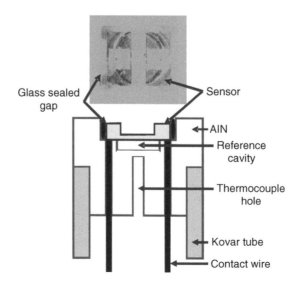

FIGURE 7.21 Top and cross-sectional views of MEMS-DCA package featuring direct wire contact to sensor and thermocouple access hole for temperature compensation, and calibration.

7.6.2 Transducer Parametric Analysis

The resistors are arranged in a Wheatstone bridge configuration, which, ideally, will have a pressure off-set voltage, V_{oz}, of zero if all resistors are exactly the same. Deviation from perfect resistor symmetry results in a non-zero V_{oz} across the bridge that is expressed as:

$$V_{oz} = \frac{V_{in}}{2}\left(\frac{R_2 - R_1}{R_1 + R_2} + \frac{R_4 - R_3}{R_3 + R_4}\right) \tag{7.30}$$

where R_1, R_2, R_3, and R_4 are the bridge resistor elements (Ω) and V_{in} is the input voltage (V). Another influence on V_{oz} is the thermomechanical package-induced stress discussed above. The relative change in each resistor element due to externally applied and residual strain leads to the modification of equation (7.1) to be expressed as:

$$\Delta R = R(\varepsilon + \varepsilon_\delta)GF \tag{7.31}$$

The TCR of a resistor approximates the resistor value via the expression:

$$R(T) = R_o(1 + \beta\Delta T) \tag{7.32}$$

where R_o is the resistance at the reference temperature (room temperature), and ΔT (°C) is the temperature change from the reference temperature. The temperature effect of the gauge factor is similarly modeled by:

$$G(T) = GF(1 + \gamma\Delta T) \tag{7.33}$$

It is important to note that under an unstrained condition (i.e., no applied pressure and sensor not packaged), the bridge output is governed only by Equations (7.30) and (7.32). With the sensor attached to the package at elevated temperature and cooled down to room temperature, it experiences a package-induced residual strain, as expressed in Equation (7.29), thus creating a change in the resistance and V_{oz}. When pressure and temperature are applied simultaneously, each resistor element experiences a relative change, which is expressed by combining Equations (7.29) and (7.31–7.33) as:

$$\Delta R(T) = R_oG_o[1 + (\beta + \gamma)\Delta T + \gamma\beta(\Delta T)^2](\varepsilon + \varepsilon_\delta) \tag{7.34}$$

Therefore, with the sensor fully packaged and no applied pressure ($\varepsilon = 0$), the change in resistance with temperature results in change in V_{oz}. As the operating temperature increases, ε_δ will decrease correspondingly (relaxation), and the resistance change becomes largely governed by β and less by γ parameters. Equation (7.34) is particularly important for understanding the overall behavior of the transducer, especially with regard to temperature compensation and transducer calibration. It also reveals the effect of thermomechanically induced stress on the system and its deleterious effect on the long-term output stability.

7.6.3 AST Protocol

The AST protocol used is shown in Figure 7.22. We evaluated the stability and reliability of 6H-SiC pressure transducers at temperatures starting from 25°C up to the temperature of instability onset, $T_{unstable}$. This is the temperature at which the V_{oz} fluctuates such that no reliable pressure measurement can be performed.

The initial pressure tests at room temperature were performed in Step 1, where 20 cycles of 0 psi → P_{max} → 0 psi were applied at steps of 10%P_{max} and held for ten seconds per step. The pressure, P_{max}, was predetermined to be 100 psi from previously validated finite element modeling and burst pressure analysis. The failing transducers after this step are removed for failure analysis (FA) to determine the failure mechanisms. The passing transducers after Step 1 were thus validated to handle the maximum pressure.

In Step 2, the transducers were heated from room temperature to 100°C and cycled 20 times between 0 and 100 psi. This 20-cycle pressurization was repeated at increments of 100°C following one-hour, zero-pressure

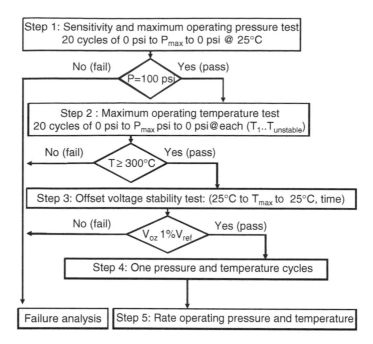

FIGURE 7.22   The AST protocol for evaluating the long-term reliability of SiC pressure sensors at high temperature.

stabilization at each temperature increment. This intermittent procedure continued at every temperature until instability was observed at $T_{unstable}$. Thus, Step 2 ascertained the maximum temperature T_{max} at which the V_{oz} remained stable. Devices that exhibited instability in Step 2 at temperatures below 300°C were screened out.

The transducers that exhibited $T_{max} \geqslant 300$°C proceed to Step 3, where they were subjected to their corresponding T_{max} as a function of time. Prior to the start of Step 3, the room temperature reference offset voltage, V_{ozref} (25°C) and the V_{ozref} (T_{max}) were recorded. From then on, the offset voltage values at the two temperatures (25°C and T_{max}) were recorded after each cycle of heating and cooling over time. The dwell time at T_{max} was arbitrarily chosen. The deviations of zero pressure offset voltages from the V_{ozref} values at these two temperature extremes determined the degree of stability of the transducer.

Finally, in Step 4, another round of pressurization and temperature treatment (one cycle of pressure and temperature ramp-up and ramp-down) was performed on the passing transducers to rate operating sensitivity, pressure, and temperature.

7.6.4   Stability of Transducer Parameters

Only six of the twelve 6H-SiC transducers evaluated survived to Step 4 of the AST. The results of the failure analyses on the failed transducers will be the subject of a future publication. For the passing transducers, three were evaluated at T_{max} of 300°C while the other three were evaluated at 400°C. The $V_{oz}(T)$ of a representative transducer from the 400°C batch is shown in Figure 7.23(a), as recorded at the end of Step 3 of the AST protocol. It shows the V_{oz} ($P = 0$, T), for an applied bridge input voltage $V_{in} = 5$ V. At 0 psi, the output resistance of the bridge changed from 476.63 to 479.39 Ω between 25 and 400°C, respectively. The plot of the full-scale (100 psi) response as a function of temperature is also included, to show the dynamic operating range of the transducer and the effect of temperature. From the plots, the difference between the V_{oz} ($P = 0$ psi, T) and V_o ($P = 100$ psi, T) becomes slightly smaller with increasing temperature. This finding is shown from a different perspective in Figures 7.23(b) and 7.23(c) in terms of the dynamic range (net bridge output) and sensitivity as a function of pressure at various temperatures; these figures show the characteristic drop in net pressure, net output voltage, strain and sensitivity with increasing

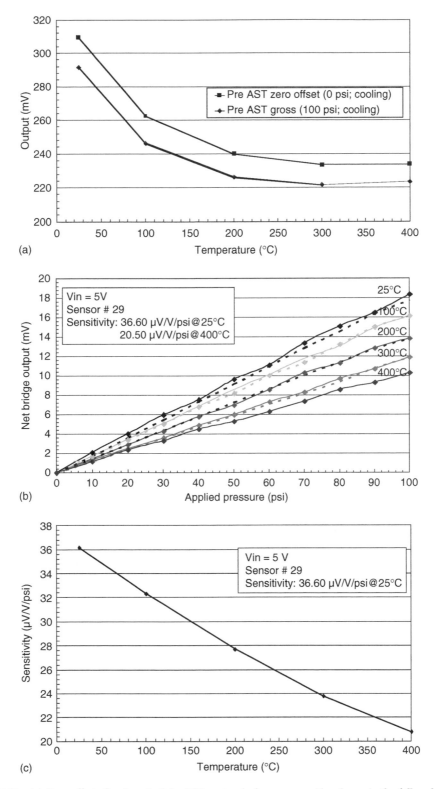

FIGURE 7.23 (a) Zero offset after Step 3 of the AST protocol of one sensor. Also shown in the full-scale output at 100 psi. (b) Net voltage after Step 3. Solid and dashed plots represent heating and cooling excursions, respectively. (c) Thermal stability of sensitivity after Step 3 of the AST.

temperature, respectively. This behavior is characteristic of piezoresistive sensors seen in literature. The sensitivity at 400°C, as plotted in Figure 7.23(c), shows a drop of about 40% from the room temperature sensitivity. Although not shown, the drift in the full-scale sensitivity as a function of time at the two temperature extremes remains within $\pm 1\,\mu V/V/psi$ of the reference values, which is equivalent to errors of 2.9% and 4.8% at 25°C and 400°C, respectively.

7.6.5 Long-Term Stability

The long-term stability of the transducer V_{oz} during Step 3 was evaluated by cyclic heating and cooling of the two batches separately from 25°C to their respective T_{max} (Step 3) over time. For the $T_{max} = 300°C$ sub-batch, Figure 7.24(a) shows that after 145 hours, the maximum drift of V_{oz} (25°C, t) from V_{ozref} (25°C) is 0.5 mV. Similarly, the maximum drift in V_{oz} (300°C, t) was approximately 0.6 mV from V_{ozref} (300°C). With regard to the 400°C sub-batch shown in Figure 7.24(b), the maximum drifts of V_{oz} (25°C, t) and V_{oz} (400°C, t) from their corresponding references were 1.9 mV and 2.0 mV, respectively. In both cases, the reference values were set after a 24-hour transducer burn-in period. In both sub-batches, the drifts in the V_{oz} are observed to fluctuate around the reference values, as opposed to irreversible degradation of the transducer.

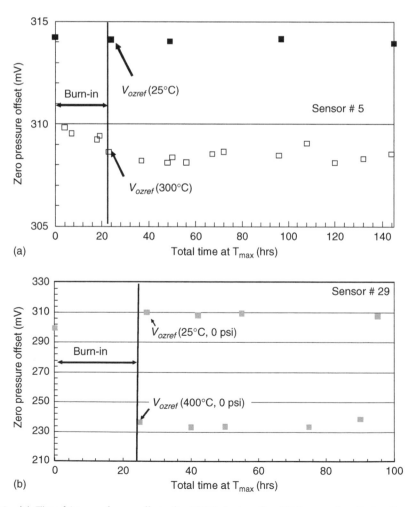

FIGURE 7.24   (a) Time history of zero offset of a 300°C device after 145 hours of cyclic heating at 300°C and cooling. (b) Zero pressure offset from a 400°C device after 90 hours of cyclic heat heating to 400°C and cooling to room temperature.

Summary

It is noteworthy that the development of SiC sensor technology is fundamentally motivated by the need to perform instrumentation in environments of extreme temperature, extreme vibration, and harsh chemical media. Accurate measurement in such environments, therefore, requires a new generation of sensors that can survive in such environments. In that respect, this chapter has presented critical technology issues that have been investigated in recent years as part of the effort toward realizing a new generation of microelectromechanical systems (MEMS) in single-crystal silicon carbide. Four important areas were addressed, namely: electrochemical etching methodologies for resistors and diaphragm structures; piezoresistance characterization in terms of strain and pressure; high-temperature metallization to support sensor and electronic operation at high temperatures; and, finally, an initial evaluation of the long-term reliability of SiC pressure sensors.

Photoconductive selectivity, as a method of fabrication of structures on either n-type or p-type SiC, was demonstrated. This principle was then used to fabricate piezoresistors in n-type epilayers by applying photoelectrochemical etching. The p-type epilayer beneath the n-type epilayer served as an etch-stop. The newly developed process for resistor fabrication can be applied, with minor adjustments, to fabricate resistors with any n-type doping level. This work also demonstrated preferential oxidation of porous SiC to facilitate good pattern definition and fast removal of residues. The thinning and the cavity etching to form the diaphragm was carried out by applying dark etching on the back side of the n-SiC wafer. The resulting cavities are relatively free of etch-pits and hillocks. The average etch-rates were found to be 0.6–0.8 μm/min. Thermally stable Ti/TaSi$_2$/Pt ohmic contact was demonstrated on n-type SiC, which is the conductivity of choice for piezoresistive SiC sensors. As a result of these efforts, a first-generation, batch-microfabricated 6H-SiC diaphragm-based piezoresistive pressure sensor was produced. This progress has since led to significant improvements in fabrication and packaging technologies to produce SiC pressure transducers that could soon be ready for commercialization.

Acknowledgments

The first five sections of this chapter were the outcome of research work performed at Kulite Semiconductor Products, Leonia, NJ, and funded by NASA Glenn Research Center. My thanks go to Dr. Anthony D. Kurtz, the chairman of Kulite, for thrusting upon me the challenge of implementing the objectives of this project, the bulk of which culminated in my doctoral thesis. Thanks to Dr. William N. Carr, my doctoral adviser, for giving me the opportunity to be his graduate student. At Israel Institute of Technology, Technion, many thanks go to Drs. Ben Z. Weiss and Ilana Grimberg for their technical support. Also, at Kulite, the full support of Alex N. Ned is greatly appreciated. I am grateful to the entire staff of department 200 with special reference to Gary Provost (formerly at Kulite) for metallization/dielectric depositions and other equipment support; to Mahesh Patel for support in testing the sensors; and to Scott Goodman for material support in testing the sensors. Here at NASA Glenn Research Center, where the more recent advanced work on Ti/TaSi$_2$/Pt metallization in Section 7.4 and the reliability evaluation in Section 7.6 were performed, I thank the various efforts of the technician cadre and Drs. Phillip G. Neudeck and Lawrence G. Matus for their critical reviews of this chapter. The reliability evaluation of Section 7.6 was performed in collaboration with Dr. Ender Savrun and Vu Nguyen of Sienna Technologies, Inc. This portion of the work was funded under the Technology Transfer Project at NASA Glenn and by the Glennan Microsystems Initiative.

References

Alexander's Gas and Oil Connections (2003) "DOE Adds Three New Projects to Deep Reservoir Drilling Program," *News and Trends: North America* **8**(13).

Andrews, J.M., and Phillips, J.C. (1975) "Chemical Bonding and Structure of Metal-Semiconductor Interfaces," *Phys. Rev. Lett.* **35**, pp. 56–59.

Azimov, S.A., Mirzabaev, M.M., Reifman, M.B., Uribaev, O.U., Khairullaev, Sh., and Shashkov, Yu.M. (1974) "Investigation of the Influence of Uniaxial Elastic Deformation on the Galvanomagnetic Properties of Hexagonal SiC," *Sov. Phys. Semicond.* **8**(11), pp. 1427–1428.

Beheim, G., and Salupo, C.S. (2000) "Deep RIE Process for Silicon Carbide Power Electronics and MEMS," *Proc. MRS 2000 Spring Meeting*, San Francisco, May 24–28, MRS Proc. 622, paper T8.9.

Bellina, J.J., Jr., and Zeller, M.V. (1987) "Novel Refractory Semiconductors," *Mater. Res. Soc. Symp. Proc.* 97, D. Emin, T.L. Aselage, and C. Wood, eds., Pittsburgh, PA., p. 265.

Berg, J. von, Ziermann, R., Reichert, W., Obermeier, E., Eickhoff, M., Krötz, G., Thoma, U., Cavalloni, C., and Nendza, J.P. (1998) "Measurement of the Cylinder Pressure in Combustion Engines with a Piezoresistive β-SiC-on-SOI Pressure Sensor," *Tech. Proc., 4th Int. High Temp. Electron. Conf.*, pp. 245–249.

Carrabba, M.M., Li, J., Hachey, J.P., Rauh, R.D., and Wang, Y. (1989) *Electrochem. Soc. Extended Abstr.* 89-1, p. 727.

Castigliano, A. (1966) *"The Theory of Equilibrium of Elastic Systems and its Applications,"* Dover, New York.

Chamberlain, M.B. (1980) *"Thin Solid Films,"* **72**, pp. 305–311.

Cox, R.H., and Strack, H. (1967) "Ohmic Contacts for GaAs Devices," *Solid State Electron.* **10**, 1213.

Davis, R.F., Sitar, Z., Williams, B.E., Kong, H.S., Kim, H.J., Palmour, J.W., Edmond, J.A., Ryu, J., Glass, J.T., and Carter, C.H., Jr. (1988) "Critical Evaluation of the Status of the Areas for Future Research Regarding the Wide Band Gap Semiconductors Diamond, Gallium Nitride and Silicon Carbide," *Mater. Sci. Eng.* **B1**, pp. 77–104.

Herring, C., and Vogt, E. (1956) *Phys. Rev.*, **101**, p. 944.

Huff, M.A., Nikolich, D., and Schmidt, M.A. (1991) "A Threshold Pressure Switch Utilizing Plastic Deformation of Silicon," *Transducers '91, Int. Conf. Solid State and Actuators Digest Tech. Papers*, p. 177.

Guk, G.N., Usol'tseva, N.Ya., Shadrin, V.S., and Mundus-Tabakaev, A.F. (1974) "Piezoresistance of α-SiC under Hydrostatic Pressures," *Sov. Phys. Semicond.* **8**(3), pp. 406–407.

Guk, G.N., Lyubimskii, V.M., Gofman, E.P., Zinov'ev, V.B., and Chalyi, E.A. (1974) "Temperature Dependence of the Piezoresistance π_{11} of n-Type SiC(6H)," *Sov. Phys. Semicond.* **9**, 104.

JEDEC, See www.jedec.org for applicable standards.

Jurgens, R.F. (1982) *IEEE Trans. Ind. Electron.* 1E-29, pp. 107–111.

Keyes, R.W. (1960) *"Solid State Physics,"* vol. 11, Academic Press, New York.

Khan, M., Fatemi, H., Romero, J., and Delenia, E. (1988) "Effect of High Thermal Stability Mold Material on the Gold-Aluminum Bond Reliability in Epoxy Encapsulated VLSI Devices," *Proc. 26th Int. Reliability Physics Symp.*, April, 12–14, pp. 40–49.

Khan, F.A., and Adesida, I. (1999) "High Rate Etching of SiC Using Inductively Coupled Plasma Reactive Ion Etching in SF6-based Gas Mixtures," *Appl. Phys. Lett.* **75**(15), pp. 2268–2270.

Kuphal, E. (1981) "Low Resistance Ohmic Contacts to n-Type and p-InP," *Solid State Electron.* **24**, 69–78.

Lely, J.A. (1955) *Bericht Deutsche Keram. Gesel.* 32: 229.

Lie, L.N., Tiller, W.A., and Saraswat, K.C. (1984) "Thermal Oxidation of Silicides," *J. Appl. Phys.* **56**(7), pp. 2127–2132.

Liu, S., Reinhardt, K., Severt, C., Scofield, J., Ramalingam, M., and Tunstall, C., Sr. (1996) "Long-Term Thermal Stability of Ni/Cr/W Ohmic Contacts on N-Type SiC," *Proc. 3rd Int. High Temp. Electron. Conf.*, pp. VII (9–13).

Masheeb, F., Stefanescu, S., Ned, A.A., Kurtz, A.D., and Beheim, G. (2002) "Leadless Sensor Packaging for High Temperature Applications," *5th IEEE Int. Conf. Micro Electro Mechanical Systems*, pp. 392–395.

Matus, L.G., Powell, J.A., and Salupo, C.S. (1991) "High-Voltage 6H-SiC p-n Junction Diodes," *Appl. Phys. Lett.* **59**, pp. 1770–1772.

Mehregany, M., Zorman, C.A., Rajan, N., and Wu, C.H. (1998) "Silicon Carbide MEMS for Harsh Environments," *Proc. IEEE* **86**(8), pp. 1594–1609.

Murarka, S.P. (1980) "Refractory Silicides for Integrated Circuits," *J. Vac. Sci. Technol*, **17**(4), pp. 775–792.

Neudeck, P.G., Okojie, R.S., and Chen, L.Yu. (2002) "High-Temperature Electronics—A Role for Wide Bandgap Semiconductors?" *Proc. IEEE* **90**(6), pp. 1065–1076.

Okojie, R.S., Ned, A.A., Kurtz, A.D., and Carr, W.N. (1996) "α(6H)-SiC Pressure Sensors for High Temperature Applications," *Proc. 9th Int. Workshop Micro Electro Mechanical Systems*, pp. 146–149.

Okojie, R.S. (1996) "Characterization and Fabrication of α(6H)-SiC as a Piezoresistive Pressure Sensor for High Temperature Applications," Ph.D. Thesis, New Jersey Institute of Technology, Department. Electrical and Computer Engineering, Newark, NJ.

Okojie, R.S., Ned, A.A., and Kurtz, A.D. (1997) "Operation of Alpha 6H-SiC Pressure Sensor at 500°C." *Int. Conf. Solid State Sensors and Actuators*, 1997, vol. **2**, pp. 1407–1409.

Okojie, R.S., Ned, A.A., Kurtz, A.D., and Carr, W.N. (1999) "Electrical Characterization of Annealed Ti/TiN/Pt Contacts on N-Type 6H-SiC Epilayer," *IEEE Trans. Electron. Devices* **46**(2), pp. 269–274.

Okojie, R.S. (2005) "MEMS Direct Chip Attach Packaging Methodologies and Apparatuses for Harsh Environments," US Patent #6,845.664.

Palmour, J.W., Davis, R.F., Astell-Burt, P., and Blackborow, P. (1987) "Science and Technology of Microfabrication," *Mat. Res. Soc. Symp.*, R.E. Howard, E.L. Hu, S. Namba, and S.W. Pang, eds., Pittsburgh, PA, p. 185.

Palmour, J.W., Kong, H.S., Waltz, D.G., Edmond, J.A., and Carter, C.H., Jr. (1991) "6H-Silicon Carbide Transistors for High Temperature Operation," *Proc. 1st Int. High Temp. Electron. Conf.*, pp. 229–236.

Papanicolaou, N.A., Edwards, A.E., Rao, M.V., Wickenden, A.E., Koleske, D.D., Henry, R.L., and Anderson, W.T. (1998) "A High Temperature Vacuum Annealing Method for Forming Ohmic Contacts on GaN and SiC," *Proc. 4th Int. High Temp. Electron. Conf.*, pp. 122–127.

Park, Y.S., ed. (1988) "*SiC Materials and Devices*." Academic Press, New York.

Pearson, G.L., Read, W.T., and Feldman, W.L. (1957) *Acta Met.* **5**, p. 181.

Rapatskaya, I.V., Rudashevskii, G.E., Kasaganova, M.G., Iglitsin, M.I., Reifman, M.B., and Fedotova, E.F. (1968) "Piezoresistance Coefficients of n-Type α-SiC," *Sov. Phys. Solid State* **9**(12), pp. 2833–2835.

Razouk, R.R., Thomas, M.E., and Pressaco, S.L. (1982) "Oxidation of Tantalum Disilicide/Polycrystalline Silicon Structures in Dry O_2," *J. Appl. Phys.* **53**(7), pp. 5342–5344.

Savrun, E., Nguyen, V., and Okojie, R.S. (2004) "A Silicon Carbide Pressure Sensor for High Temperature (600°C) Applications," Presented at the IMAPS Int. High Temperature Electronics Conf., Santa Fe, NM, May 17–20.

Shor, J.S., Zhang, X.G., and Osgood, R.M. (1992) *J. Electrochem. Soc.* **139**, p. 1213.

Shor, J.S., Okojie, R.S., and Kurtz, A.D. (1993) *Institute of Physics Conference Series* No. 137, Chapter 6, pp. 523–526.

Smithells, C.J. ed. (1976), "*Metals Reference Book*," 5th ed., Butterworth & Co., London.

Streetman, B.E. (1990) "*Solid State Electronic Devices*," 3rd ed., Prentice Hall, Englewood Cliffs, NJ.

Timoshenko, S., and Woinowsky-Krieger, S. (1959) "*Theory of Plates and Shells*," 2nd ed., McGraw Hill, New York.

Ting, C.Y., and Wittmer, M. (1982) "The Use of Titanium-Based Contact Barrier Layers in Silicon Technology," *Thin Solid Films*, **96**, pp. 327–345.

Toth, L.E. (1971) "*Transition Metals Carbides and Nitrides*." Academic Press, New York.

Tyson, S.M., and Grzybowski, R.R. (1994) "High Temperature Characteristics of Silicon-on-Insulator and Bulk Silicon Devices at 500°C," *Trans. 2nd Int. High Temperature Electron. Conf.*, Omni Charlotte Hotel, Charlotte, NC., p. 9.

Waldrop, J.R., and Grant, R.W. (1993) "Schottky Barrier Height and Interface Chemistry of Annealed Metal Contacts to Alpha 6H-SiC: Crystal Face Dependence," *Appl. Phys. Lett.* **62**(21), pp. 2685–2687.

Zeller, M.V., Bellina, J., Saha, N., Filar, J., Hargraeves, J., and Will. H. (1987) *Mat. Res. Soc. Symp. Proc.* **97**, pp. 283–288.

8

Deep Reactive Ion Etching for Bulk Micromachining of Silicon Carbide

Glenn M. Beheim and
Laura J. Evans
NASA Glenn Research Center

8.1   Introduction

It is often desired to insert microsensors and other MEMS into harsh, e.g., hot or corrosive, environments. Silicon carbide (SiC) offers considerable promise for such applications, because it can be used to fabricate both high-temperature electronics and extremely durable microstructures. One of the attractive characteristics of SiC is the compatibility of its process technologies with those of silicon, which allows for the cofabrication of SiC and silicon MEMS. However, a very important difference in the processing of these semiconductors arises from the chemical inertness of SiC, a characteristic which makes it attractive for use in corrosive environments, but which also causes it to be very difficult to micromachine.

Realization of the full potential of SiC MEMS will require the development of a set of micromachining tools for SiC which is comparable to the tool set available for silicon. Micromachining methods are generally classified as bulk, in which the wafer is etched, or surface, in which deposited surface layers are patterned. Surface micromachining methods for deposited SiC layers have been developed to a high level [Song, 2001]. Silicon carbide can be readily etched to the required depths of just several μm using reactive ion etching (RIE) processes [Yih, 1997]. Further work remains to be done, however, in developing RIE processes with greater selectivity for SiC. Present RIE processes lack the selectivity needed to etch entirely through a SiC layer, while minimally modifying an underlying silicon or silicon dioxide layer. This limitation has motivated the development of a micromolding method in which SiC is deposited into molds formed by RIE of silicon or silicon dioxide [Yasseen, 1999].

The emphasis here is on bulk micromachining of SiC for the fabrication of SiC microstructures with vertical dimensions from approximately ten μm to several hundred μm. Three methods for bulk micromachining of SiC have been developed: wet etching, micromolding, and deep reactive ion etching (DRIE). Each of these has an important role to play in SiC MEMS fabrication.

Conventional wet etching of SiC is difficult to accomplish because silicon and carbon have a high bond energy. Temperatures greater than 600°C are required to obtain significant etch rates [Faust, 1960]. Electrochemical etching, despite the requirement for specialized equipment, is a more practical means to micromachine SiC [Shor, 1994]. It can provide a high rate (>1 μm/min); it has a high selectivity to other materials, so the etch can be effectively masked; and it allows for the use of p-n junction etch stops for precise depth control. Etch directionality, however, is poor, resulting in significant undercutting of the mask. Highly directional (or anisotropic) etch processes, which provide vertical etch sidewalls, are needed to accurately transfer the mask pattern deeply into the wafer.

Bulk micromolding methods have been developed which circumvent the limitations of available deep-etch processes for SiC [Lohner, 1999]. Here, silicon molds are fabricated using highly developed silicon DRIE processes. The molds are filled with polycrystalline SiC using chemical vapor deposition (CVD) and then removed by wet etching. Thick, finely featured SiC microcomponents have been fabricated using this approach. The molded polycrystalline SiC, however, does not have the excellent mechanical and electronic properties of single-crystal SiC. In particular, deep etching of single-crystal SiC wafers is required if high-quality SiC electronics are to be integrated on-chip with thick SiC microstructures.

Conventional, parallel-plate RIE is not well-suited for deep etching of SiC (to depths greater than about 10 μm) because it provides low etch rates, high rates of mask erosion, and relatively poor directionality (although superior to that which can be achieved using electrochemical etching). The use of advanced high-density plasma (HDP) reactors can alleviate all these difficulties. High-density plasma has enabled the DRIE process for silicon, and its usefulness for deep etching of SiC is becoming well-established.

8.2 Fundamentals of High-Density Plasma Etching

High-density plasma systems typically have a plasma density (ions per unit volume) that is two orders of magnitude higher than that of conventional RIE. This higher plasma density produces a much greater flux of ions and reactive species to the substrate surface, which yields much higher etch rates. Also, HDP etching is performed at lower pressures, which helps minimize bowing of the etch sidewalls caused by scattering of ions. Lower pressures also facilitate the transport of etchants and etch products into and out of narrow trenches. In addition, the low operating pressure helps to provide smoother surfaces, because sputtered mask materials are less likely to be scattered back onto the etched surfaces, which can cause micromasking and the formation of "grass" or other etch residues. Another important advantage of most HDP systems is the capability to independently control the plasma density and the energy with which ions bombard the substrate. This capability allows for the adjustment of the chemical and mechanical components of the etch process to give a satisfactory trade-off between the etch rate and the erosion rate of the mask. The ratio of the substrate to mask etch rates, which is called the mask selectivity, is an important parameter for deep etching because it determines the maximum depth that can be etched with a given mask thickness.

There are a number of different HDP systems, including magnetically enhanced RIE (MERIE), electron cyclotron resonance (ECR), helicon, and inductively coupled plasma (ICP) systems. MERIE systems use a magnetic field to confine electrons to the plasma. This situation enhances the plasma density but usually negatively affects the uniformity. In an ECR system, plasma electrons are confined using strong magnets and excited at the electron cyclotron frequency using microwaves. A field of 875 G is required to achieve resonance at the commonly used microwave frequency of 2.45 GHz. In a helicon reactor, a helicon wave is excited using an RF-driven antenna (typically 1–50 MHz) in conjunction with a relatively weak magnetic field (typically 20 to 200 G). This system creates a high-density, uniform plasma, but the chamber has a large aspect ratio. Detailed information about different HDP systems has been provided

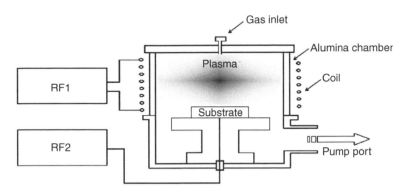

FIGURE 8.1　Schematic of an inductively coupled plasma (ICP) etching system. The RF generators are labeled RF1 and RF2.

by Lieberman [1994]. Inductively coupled plasma has come to dominate the HDP market, largely as a result of lower complexity and cost. An ICP reactor will be discussed in further detail.

A schematic of an ICP plasma etching system is shown in Figure 8.1. Typically, a halogenated gas, such as SF_6, is supplied to the reactor through a mass flow controller. A high throughput pumping system maintains the system at a low pressure and enables a high gas flow to rapidly replenish depleted etchants. A plasma is generated by supplying RF power to a coil wrapped around the ceramic walls of the chamber. Power supplied to the coil produces a time-varying axial magnetic field inside the chamber. This situation induces an azimuthal electric field that accelerates the electrons to high energies. The circumferential electric field helps to confine the electrons to the plasma, which results in the much higher plasma densities characteristic of ICP. Because confinement of the electrons increases the probability that an electron will undergo an ionizing collision with a gas molecule before it leaves the plasma, the ionization rate is increased. The plasma can, therefore, be sustained at lower pressures (as low as 1 mTorr), which increases the mean path length the electrons travel between collisions. This increased path length enables the acceleration of the electrons to higher energies, which increases the likelihood that a collision with a gas molecule will break the molecule apart. Relatively inert, fluorinated gases like SF_6 are thereby dissociated to produce extremely reactive radicals like atomic fluorine. A greater quantity of highly reactive species is produced in ICP, in comparison with conventional RIE.

At the same time the substrate is subjected to a high flux of reactive radicals, it can be bombarded with energetic ions. A second generator supplies RF power to an electrode, onto which the wafer is clamped. Upon application of the AC voltage, electrons and ions are alternately attracted. The less massive electrons are considerably more mobile, so the electrode acquires a negative charge that gives it a time-averaged negative potential with respect to the plasma. This negative bias potential acts to repel electrons, preventing further net accumulation of charge. The potential gradient between the plasma and the substrate electrode causes ions drifting out of the plasma to be accelerated across the dark space (or sheath) between the plasma and substrate. Typically, the ions are accelerated to energies from several tens of eV to several hundred eV. The damage which these ions cause to the substrate, e.g., by creating highly reactive dangling bonds, can dramatically increase the etch rate. The ions strike the substrate at normal incidence, provided they do not undergo scattering collisions with gas molecules in the sheath. Inductively coupled plasma enables highly anisotropic etching, since the low operating pressures of ICP ensure that ions strike only the horizontal wafer surfaces and are not scattered to the etch sidewalls.

8.3　Fundamentals of SiC Etching

Various halogen etch chemistries (such as chlorine-, fluorine-, and bromine-based) have been studied for plasma etching of SiC [Leerungnawarat, 2001]. Fluorine-based chemistries are most easily implemented since nontoxic feed gases can be used. The plasma dissociates the relatively inert fluorinated gas (e.g., SF_6)

to produce highly reactive radicals such as atomic fluorine. For DRIE of silicon, a fluorine etch chemistry is almost always used, in part because of convenience, but primarily because fluorine provides the high etch rates needed for economical deep etching. Fluorine plasmas provide the highest etch rates for SiC, as well. For these reasons, basic principles of fluorine plasma etching of SiC will be discussed.

The reactivity of SiC in a fluorine plasma contrasts strongly with that of silicon. Exposure of silicon to atomic fluorine produces rapid etching with an isotropic profile, as shown in Figure 8.2. Silicon reacts spontaneously with atomic fluorine, producing volatile etch products, e.g., SiF_4 and SiF_2, which rapidly desorb from the surface. Etching proceeds laterally as well as vertically, which results in an isotropic profile. Because of the high bond energy of SiC, both physical and chemical processes are required for dry etching of SiC in plasma. The etch rate of SiC, when exposed to atomic fluorine, is extremely low unless energy is supplied to the surface. In reactive ion etching, this energy is provided by ion bombardment, which results in an anisotropic etch profile, as shown in Figure 8.3. Energetic ions damage the SiC lattice (i.e., breaking Si−C bonds), which promotes reactions with atomic fluorine to produce volatile SiF_x and CF_x species. If the ions are well collimated, only the horizontal surfaces are etched, producing the desired anisotropy.

In fluorine-based DRIE of silicon, the required anisotropic profile is produced by modifying the process to cause the formation of a passivating layer on the etch sidewall. Most commonly, a time-multiplexed etch/passivate process is employed, in which the reactor is programmed to switch back and forth between the etching of silicon and the deposition of an etch-resistant fluorocarbon polymer [Bhardwaj, 1995]. Low-energy ion bombardment is used during the etch step to remove the polymer from the horizontal surfaces, but it is left intact on the etch sidewalls. Proper balancing of the etch and passivation steps produces a vertical sidewall, which, at high magnification in a scanning electron microscope, has a scalloped appearance.

In DRIE of SiC, the low reactivity of the substrate makes the etch profile inherently anisotropic. However, the low reactivity of SiC creates a number of other problems, including low etch rate, low selectivity to the mask, and a tendency for the formation of residues on the etched surfaces. The net result is that DRIE of SiC is decidedly more difficult than DRIE of silicon. It is the high reactivity of silicon with fluorine that enables the silicon DRIE process to attain the outstanding attributes of high-rate, one-step etch masking, and tunable sidewall slope. Because the silicon DRIE process uses low ion energies, it provides a high selectivity with respect to photoresist, the most economical etch mask. Etch processes for SiC, however, generally etch photoresist at a comparable rate, so other masks must be employed. The time-multiplexed process for DRIE of silicon provides excellent control of the etch profile, since the etch and passivation

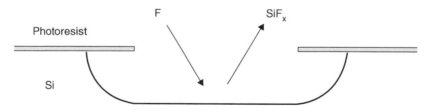

FIGURE 8.2   Etching of silicon in a fluorine plasma using a photoresist etch mask.

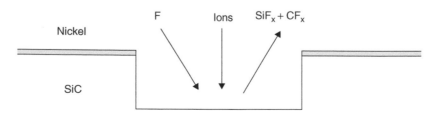

FIGURE 8.3   Etching of silicon carbide in a fluorine plasma using a nickel etch mask.

processes can be independently controlled. With inherently anisotropic etch processes like DRIE of SiC, it may prove more difficult to obtain the same high aspect ratios (etch depth divided by minimum feature size) as can be realized in silicon.

Dry etching of SiC is usually performed using a primary source of fluorine radicals, with secondary gases added to control or enhance the process. Some primary gases that have been investigated include CF_4, SF_6, NF_3, CHF_3, and C_2F_6. Additives include O_2, Ar, H_2, and N_2. The primary gases NF_3 and SF_6 exhibit the highest etch rates, as a result of their rapid dissociation in plasma. Although NF_3 tends to etch SiC at a faster rate [Leerungnawarat, 2001], SF_6 is more desirable for use as a feed gas because it is less expensive and safer. C_2F_6 has lower perfluorinated compound (PFC) emissions than both NF_3 and SF_6. Combinations of primary gases have also been investigated to enhance etch rates or reduce residues. The addition of oxygen to the plasma acts to enhance the generation of reactive fluorine atoms and to facilitate removal of carbon in the form of CO_x [Chabert, 2001]. Oxygen also helps to maintain a clean system by reducing sulfur-based deposition (from SF_6) on the walls of the reactor [Chabert, 2001]. If too much oxygen is added, the primary gas becomes diluted, and the etch rate is lowered; also, the etch rate may be further reduced by the formation of SiO_n on the surface [Khan, 2001]. Hydrogen additives are used as metal scavengers, e.g., forming volatile aluminum hydride compounds (AlH_z) [Yih, 1997]. The addition of hydrogen also results in decreased etch rates (caused by gas-phase reactions with fluorine atoms), greater risk of contamination, and reduced aspect ratios (increased polymer formation can lead to outward sloping sidewalls that can limit trench depth). Physical sputtering of the substrate surface can be increased by adding nonreactive gases such as argon, helium, or nitrogen. The addition of such gases can reduce the formation of residues in the etched areas by sputtering away etch-resistant materials that would otherwise interfere with the etch process.

The high ion energies needed to etch SiC at acceptable rates make it impractical to use photoresist as an etch mask. A number of different materials (metals and dielectrics) have been used as etch masks. Masks that have been studied include Al, Cu, Ni, indium tin oxide (ITO), Cr, Ti, SiO_2, SnO_2, and Si_3N_4. High selectivities (greater than 40) for SiC with respect to the mask material are desired in order to perform deep etching. Nickel has generally been found to have good selectivity, along with exhibiting less micromasking effects than other metals. Some groups have reported extremely high selectivity for aluminum in oxygen-containing plasma as a result of the conversion of the Al surface to Al_2O_3, which has a lower sputter yield [Cho, 2001]. Aluminum, however, also shows great tendencies towards micromasking. Copper has been reported to have higher selectivity compared to nickel and aluminum during etching in SF_6, because a sputter-resistant metal-fluoride layer forms to a greater thickness on a copper mask [Kim, 2004].

The formation of residues on the etched surface is a significant problem in SiC etching. The highly energetic ions cause significant erosion of the etch mask, and sputtered mask materials that redeposit on the SiC surface can act as micromasks. Sputter erosion of the electrode surface or the walls of the chamber can also cause micromasking. Because the etch process is highly anisotropic, micromasks are not undercut, and pillar-like features ("grass") and other residues can result. High pressures increase the likelihood of residue formation, since sputtered mask materials are more likely to be scattered back onto the SiC surface. Contamination from the reactor can be prevented by covering surfaces with graphite, Kapton®, Si, SiO_2, or Teflon® [Yih, 1997].

Trenching is another problem that is frequently encountered when etching SiC. The flux is often enhanced at the base of the sidewall, which increases the etch rate and produces a trench. The trenching effect can be lessened by reducing the ion energies or increasing the pressure. However, both these changes generally reduce the SiC etch rate.

8.4 Applications of SiC DRIE

8.4.1 Review

Throughput is a key economic factor, and much work has focused on the development of high-rate etch processes for SiC. Various types of HDP reactors have been used to demonstrate SiC etch rates greater than 1 μm/min, which compares fairly well with the 2 μm/min rates that are typical of silicon DRIE. In SiC, however, the attainment of such a high etch rate necessitates highly energetic ion bombardment, which dramatically increases the sputter erosion of the etch mask. This increase can severely limit the etch depths that can be obtained using feasible mask thicknesses. In addition, high-rate etching often does not provide the smooth etched surfaces needed, as a result of the increased tendency for micromasking. A summary of high etch rates obtained by different groups using conventional RIE and HDP was provided by Leerungnawarat [2001].

Several groups have demonstrated deep reactive ion etching of SiC. In one case, a parallel-plate RIE was used to etch a via through an 80-μm thick substrate at a rate of 0.23 μm/min [Sheridan, 1999]. The feed gas was NF_3, and the chamber pressure was relatively high, 225 mTorr, which provided a high concentration of reactive species. The high operating pressure also resulted in relatively low ion energies, which minimized sputtering of the mask. Since the selectivity to the electroplated nickel-alloy mask was only 25, thick masks were needed for deep etching of SiC. Anisotropy was quite good for the roughly 100-μm diameter via, but scattering of ions will limit the aspect ratios that can be achieved at such a high pressure.

Etching of 4H-SiC was reported using a helicon HDP reactor [Chabert, 2001]. The feed gas was SF_6 and 25% O_2. It was stated that the addition of O_2 did not affect the etch rate but prevented the deposition of sulfur compounds on the chamber walls. A 50-μm thick nickel sheet was used as a shadow mask to etch a via through a 330-μm thick 4H-SiC substrate. The wafer was etched through in approximately six hours, giving an etch rate of 0.9 μm/min. The selectivity with respect to the nickel mask was 40:1. Etch rates as high as 1.35 μm/min were obtained, but rough surfaces were produced at the higher rates.

A 140-μm deep via was etched through a 6H-SiC wafer using an ICP reactor [Khan, 2001]. A combination of SF_6 with 20% O_2 was used at a chamber pressure of 8 mTorr and total flow rate of 10 sccm. The ICP coil power was 900 W and the bias voltage was varied between −300 V and −390 V. A high etch rate of 820 nm/min was maintained by increasing the bias voltage during the etch, to compensate for the reduction in transportation kinetics as the hole depth increased. Selectivity to the electroplated nickel mask was approximately 40.

A combination of SF_6 with 25% O_2 was used to etch vias up to 100 μm deep in 4H-SiC using an ICP reactor [Cho, 2001]. An etch rate of approximately 0.6 μm/min was obtained at a pressure of 5 mTorr and a total flow rate of 15 sccm. Selectivity to the 3-μm thick plated Al mask was more than 50. The addition of O_2 to the feed gas mixture was found to increase the selectivity to Al through oxidation of the Al surface to produce the more sputter-resistant Al_2O_3. SEM micrographs show rough sidewalls that are almost vertical, along with a clean, although rounded, bottom surface. Clean surfaces could be attributed to the narrow width (less than 50 μm) of the etched features.

A magnetically enhanced ICP reactor was used for deep etching of sintered SiC substrates in mixtures of SF_6 and O_2 [Tanaka, 2001]. Selectivity to an electroplated nickel mask was approximately 27. Addition of 5% O_2 to SF_6 was found to produce trenches with no residue, whereas larger percentages of O_2 and no O_2 created grass-like residues. Etching in 5% O_2, at 1.8 mTorr, 150 W source power, and 150 W stage power yielded a depth of 216 μm, with a rate of 0.51 μm/min. Microloading effects were shown (decreased etch depth with decreased mask opening width), along with different residues generated for different mask widths. For widths of less than 100 μm and greater than millimeters, the bottom surfaces of the trenches were residue-free, while residues were observed for an intermediate trench width of 170 μm. Tapering of the upper parts of trenches, caused by regression of the Ni mask, also was observed.

The dependence of etch rate on mask width or percentage of exposed wafer for deep etching using SF_6 was experimentally determined by Leerungnawarat [2001]. Etch rates fell with decreasing via diameter and increasing exposed wafer area. Diameters up to 150 μm and wafer exposures from 20 to 100% were studied. Vias as deep as 100 μm were etched.

A magnetically enhanced ICP reactor was used in conjunction with a 3-μm thick copper mask to demonstrate 20 μm DRIE in 6H-SiC substrates [Kim, 2004]. The etch parameters were 1500 W source power, −150 V bias, 10 mTorr pressure, and 50 sccm of SF_6 feed gas flow. Etch rate was approximately 1150 nm/min. Selectivity to the copper mask was reported to be infinite.

Using an ICP reactor with SF_6, very smooth etched surfaces were obtained for an etch depth of 45 μm, and a via was etched through a 100-μm thick 6H-SiC wafer [Beheim, 2000]. At a rate of 0.3 μm/min, the selectivity with respect to a nickel mask was 80. Since the presentation of these results, this laboratory has made substantial progress in SiC DRIE, as will be described in the next section.

8.4.2 Applications of SiC DRIE: Experimental Results

This section will present experimental results for DRIE of SiC using an inductively coupled plasma reactor (STS Multiplex ICP). The reactor has an automated loadlock and a 1000 L/sec turbopump with a dry backing pump. Helium is supplied to the backside of the electrostatically clamped wafer to provide effective heat transfer to a water-cooled chuck. The SiC substrates, which range in size from 10 mm square to 50 mm in diameter, are attached to 100 mm silicon carrier wafers using a thin layer of photoresist. The use of a sacrificial carrier wafer (which etches at approximately 2 μm/min) helps to minimize roughness caused by the sputtering of etch-resistant materials onto the SiC surface.

Lift-off of electron-beam evaporated films is a convenient means of fabricating both nickel and ITO etch masks. Lift-off of ITO masks as thick as 3.5 μm was readily accomplished, while film stresses limited the evaporated nickel masks to thicknesses no greater than 2500 Å. For extremely deep etching, nickel masks with thicknesses up to 15 μm were fabricated by selective electroplating.

A series of experiments was performed to determine the effects of various process parameters on the SiC etch rate and selectivity to the mask. During these experiments, the coil power was 800 W and the feed gas was 100% SF_6. The gas flow was maintained as high as was practical, given the constraints of the pumping system and the mass flow controller. The 10-mm square, n-type 4H-SiC substrates were masked using evaporated nickel.

Figure 8.4 shows the etch rate of 4H-SiC as a function of pressure for a fixed platen power of 75 W. The highest etch rate of 0.30 μm/min was obtained at a pressure of 3 mTorr. The etch rates for 4H-SiC are not significantly different from those which were reported for the 6H polytype [Beheim, 2000]. The dopant type also has been found to have little effect on the etch rate.

Figure 8.5 shows the SiC etch rate and the selectivity to the nickel mask (ratio of the SiC and Ni etch rates) as functions of the power supplied to the substrate electrode or platen. The platen power determines the kinetic energy of the ions that bombard the substrate. Selectivity was found to decrease with increasing ion energies, because sputtering of the mask was increased. For a platen power of 75 W, the etch rate was 0.30 μm/min and the selectivity to the nickel mask was 125.

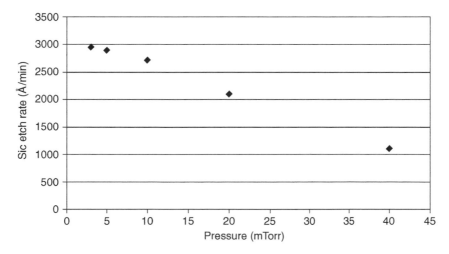

FIGURE 8.4 Etch rate of n-type 4H-SiC as a function of pressure for 800 W coil power, 75 W platen power, and 100% SF_6.

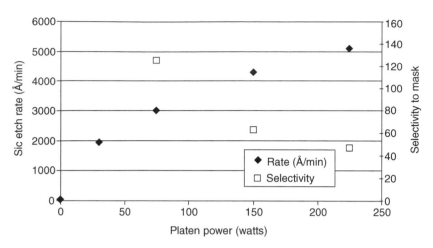

FIGURE 8.5   Etch rate of n-type 4H-SiC and selectivity to a nickel mask as functions of platen power for 800 W coil power, 5 mTorr pressure, and 100% SF_6.

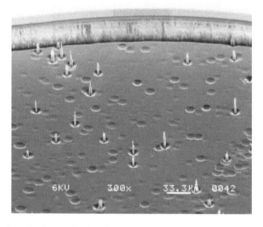

FIGURE 8.6   6H-SiC (n-type) etched to a depth of 45 μm using the baseline 100% SF_6 process. The SiC substrate was cleaned in hot sulfuric acid prior to etching. The electroplated nickel etch mask has not been stripped.

An investigation was performed to determine the effects of the various parameters on the surface morphology of deeply etched features. It was found that 5 mTorr pressure and 75 W platen power, in conjunction with 800 W coil power and 55 sccm SF_6, provides a deep etch that is satisfactory for many applications. Relative to these baseline process parameters, higher platen powers give lower selectivity to the mask and higher pressures promote residue formation, whereas lower pressures cause increased trenching at the base of the sidewall.

The surface morphology subsequent to etching was found to be strongly influenced by the cleanliness of the surface at the start of the etch process. A number of cleaning procedures (solvents, hot sulfuric acid etch, oxidation followed by etch in hydrofluoric acid) were tried, but none was found to be satisfactory. Figure 8.6 shows typical results obtained by cleaning a 6H-SiC substrate using standard methods prior to loading it into the reactor. After etching to a depth of 45 μm, the etched surface is covered with dimples and spike-shaped residues.

An in situ plasma cleaning procedure was found to be effective at providing smooth surfaces for etch depths of approximately 75 μm. Figure 8.7 shows the result when the SiC surface was sputter-cleaned in argon immediately prior to a 45-μm deep etch using the baseline process. A disadvantage of the argon

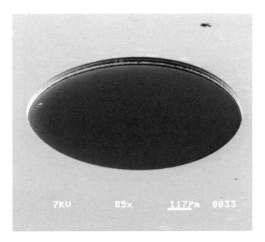

FIGURE 8.7 A 1-mm diameter well etched to a depth of 45 μm in n-type 6H-SiC using the baseline 100% SF_6 etch process after an in situ sputter-cleaning in argon. The ITO etch mask has been stripped.

FIGURE 8.8 A 1-mm diameter well etched to a depth of 160 μm in p-type 6H-SiC using the baseline 100% SF_6 etch process after an in situ oxygen plasma-clean using energetic ion bombardment. The ITO etch mask has been stripped.

sputter clean is the significant mask erosion that it causes. The ten-minute sputter-etch in argon was found to remove 1 μm of ITO, or 2500 Å of nickel, but it removed only 800 Å of SiC. An oxygen plasma cleaning process, with moderately energetic ion bombardment (75 W platen power), was found to be as effective as the argon sputter-cleaning, with much less mask erosion. In the case of ITO masks, however, the oxygen plasma pretreatment was found to leave etch-resistant residues on the SiC surface, unless the openings in the mask were quite wide (>100 μm).

When the SiC surface is plasma-cleaned prior to deep etching, the etched surface is initially smooth, but with increasing etch depth, first dimples, and then rough residues, gradually appear. For etch depths less than approximately 75 μm, the surfaces are generally smooth, but for greater depths, the etched surface becomes increasingly rough. Figure 8.8 shows a 160 μm deep etch of a p-type 6H-SiC substrate that was oxygen–plasma cleaned prior to the start of the baseline etch process. Auger analysis of the rough residues showed small amounts of aluminum. The residues are apparently the result of micromasking caused by aluminum sputtered from the interior components of the vacuum chamber, in particular, the aluminum oxide uniformity shield that surrounds the wafer and covers the outer area of the 200-mm diameter electrode.

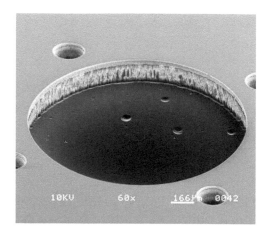

FIGURE 8.9   A 1-mm diameter well etched to a depth of 245 μm in p-type 6H-SiC using a SF$_6$ + 85% Ar etch process after an in situ sputter-cleaning using argon. The electroplated nickel mask has not been stripped.

The rough residues are found only if dimples are also present. Because a dimple is caused by a local enhancement of etch rate, it is plausible that a dimple may react more readily with aluminum than with the untextured surface; the trapped aluminum would then serve as a micromask. If no dimples are present, aluminum may remain mobile on the surface until it is sputtered off, causing no roughening of the surface. Experimentally, residual hydrocarbons and water were found to greatly increase the density of dimples in the absence of in situ plasma cleaning prior to etching. As determined by X-ray photoelectron spectroscopy (XPS), the etched SiC surface was found to be covered with a thin ($<$10 Å) fluorocarbon layer, which mediates the etch reaction. Hydrogen is known to reduce the thickness and passivating properties of such a fluorocarbon layer on an etched silicon surface, so it is feasible that hydrogen-bearing contaminants may cause the local enhancement of etch rate that results in the observed dimples. The texture-inducing contaminants are apparently found in the background vacuum of the chamber, since the dimples gradually appear with continued etching, despite an initial in situ cleaning of the surface.

Smooth surfaces, even for extremely deep etching, were obtained by diluting the SF$_6$ etchant gas with argon to cause continuous sputter-cleaning of the surface throughout the etch process. In a series of experiments, the argon concentration was gradually increased until smooth etched surfaces were obtained at 85% Ar and 15% SF$_6$. Other parameters were unchanged from the baseline process (5 mTorr pressure, 800 W coil and 75 W platen powers). Dilution with 85% Ar reduces the SiC etch rate from 0.30 to 0.22 μm/min. The selectivity to nickel is reduced from 125 to 55. The selectivity to ITO is only 20, which is insufficient for deep etching. For DRIE using the Ar-diluted SF$_6$ process, thick nickel masks were fabricated by selective electroplating. Figure 8.9 shows the smooth surface that resulted after etching p-type 6H-SiC to a depth of 245 μm using SF$_6$ and 85% argon. The mask in this case was 6-μm thick electroplated nickel.

A key attribute of the silicon DRIE process is the ability to accurately transfer finely detailed features deeply into the wafer, as quantified by the aspect ratio. In DRIE of silicon, aspect ratios of 30 or greater can be achieved by using a time-multiplexed etch/passivate process which is highly anisotropic. In this process, an etch step is alternated with a passivation step that coats the substrate with a conformal polymer film. The etch step employs low-energy ion bombardment that quickly removes the polymer from the horizontal surfaces. However, the polymer on the etch sidewalls remains, which inhibits lateral etching.

Deep RIE processes suitable for the fabrication of high-aspect-ratio SiC microstructures are still in the very early stages of development. Although the etch rate of SiC in the absence of ion bombardment is very small, it is still sufficient to roughen the sidewalls during lengthy etches, as can be seen in Figure 8.9. In an effort to minimize lateral etching, a time-multiplexed etch/passivate process has been employed. Preliminary attempts to use this approach to etch moderately high aspect ratio structures in SiC have

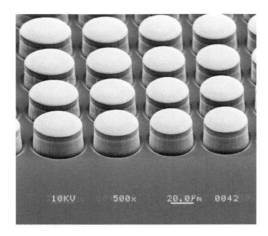

FIGURE 8.10   Moderate aspect ratio features etched in 6H-SiC to a depth of 26 μm using a multiplexed etch/ passivate process after an in situ sputter-cleaning using argon. The electroplated nickel mask has not been stripped.

FIGURE 8.11   Trench etched to a depth of 50 μm in 6H-SiC using a multiplexed etch/passivate process. The nickel mask is beginning to recede from the edge, which is causing etching to occur behind the polymer coating on the sidewall.

yielded promising results, as shown by Figures 8.10 and 8.11. For this nonoptimized etch process, the etch step was 30 sec in duration, with a gas flow of 70 sccm SF_6; the passivation step was 7 sec, with a gas flow of 80 sccm C_4F_8. The coil power was 800 W continuously, and 75 W was applied to the platen only during the etch step. The pressure control valve was maintained at 15% closed. The etch rate for this process was found to be 0.24 μm/min. The multiplexed etch/passivate process was preceded by a ten-minute sputter etch in argon. This sputter etch provided for clean etched surfaces, but it also rounded the edges of the electroplated nickel mask, as can be seen from Figures 8.10 and 8.11. In order to etch deeper than 50 μm, the duration of the argon sputter etch must be reduced so that the nickel mask retains a more nearly square profile. This profile would allow for deeper etching before the nickel recedes from the edge of the sidewall, as is evidenced in Figure 8.11. This process produced sidewalls that sloped slightly outwards. The width of the 50-μm deep trench shown in Figure 8.11 decreases from 10 μm at the top to 8 μm at the bottom. Trenching at the base of the sidewalls is evident in both Figures 8.10 and 8.11. Work is ongoing to refine this process, in order to increase the verticality of the etch sidewall, to reduce the sidewall roughness, to reduce the trenching, and to increase the etch depth.

FIGURE 8.12   (See color insert following page 9-22.) Packaged SiC pressure sensor chip. Through the semitrans-parent SiC can be seen the edge of the well, which has been etched in the backside of the wafer to form the circular diaphragm. The metal-covered n-type SiC which connects the strain gauges is highly visible, but the n-type SiC strain gauges are more faintly visible. There is a U-shaped strain gauge over the edge of the diaphragm at top and bottom, and there are two vertically oriented linear gauges in the center of the diaphragm.

8.4.3   Applications of SiC DRIE: Fabrication of a Bulk-Micromachined SiC Pressure Sensor

This section will describe a practical application for SiC DRIE, specifically, a pieozoresistive SiC pressure sensor, that can be used at temperatures as high as 500°C. Silicon piezoresistive pressure transducers were some of the first MEMS sensors and now represent a large fraction of the MEMS devices manufactured. Silicon pressure sensors with precisely controlled dimensions can be inexpensively fabricated using highly anisotropic wet etch processes. At moderate temperatures, silicon is a nearly ideal elastic material. However, at elevated temperatures, silicon deforms plastically; the highest possible operating temperature for a silicon pressure sensor has not yet been determined, but it is probably in the range of 600–700°C. Silicon carbide is a superior material for high-temperature pressure sensors because SiC maintains its excellent mechanical properties at very high temperatures (>1000°C). A production-worthy DRIE process for SiC is a key to the successful commercialization of high-temperature SiC pressure sensors, since DRIE is the only bulk-micromachining method for SiC which provides the required dimensional control.

A packaged SiC pressure sensor chip is shown in Figure 8.12 [Ned, 2001]. The glass used to attach the chip limits the operating temperature to 500°C. A SiC diaphragm is micromachined by using DRIE to etch a 1-mm diameter well to a depth of about 60 μm in the backside of a 120 μm thick 6H-SiC wafer. The trench at the base of the sidewall is then removed by using an isotropic electrochemical etch to remove a few μm of SiC. The trench must be removed because it creates a stress concentration that weak-ens the diaphragm. On the front side of the diaphragm, four SiC strain gauges are fabricated in an n-type epitaxial layer. An underlying p-type epitaxial layer provides electrical isolation between the strain gauges. The strain gauges are defined using a photo-electrochemical etch process which automatically stops when the p-type layer is reached. The strain gauges are configured in a Wheatstone bridge, with two gauges positioned over the edge of the diaphragm and the other two gauges positioned in the center of the diaphragm. The gauges on the edge of the diaphragm have the greatest sensitivity to pressure, while the sensitivity of the gauges in the center of the diaphragm is smaller and opposite in sign. The output of the Wheatstone bridge responds to the difference in the resistances of the edge-mounted and center-mounted gauges. The temperature-induced resistance changes, therefore, are compensated, since they are

FIGURE 8.13   (See color insert following page 9-22.) SiC pressure sensor chip in leadless package. The backside of the SiC chip is exposed to the measurement environment. The metallization on the front side of the chip, visible through the semitransparent SiC, is protected from oxidation in a hermetically sealed cavity.

largely the same for all four gauges. A three-layer metallization system, consisting of titanium, tantalum silicide, and platinum, is used to electrically contact the SiC piezoresistors.

The highly anisotropic DRIE process for SiC produces diaphragms that are uniform in thickness and have precisely positioned edges. These diaphragms maximize the sensitivity and provide a high uniformity of response from one sensor to the next, since the sensitivity is strongly influenced by the positions of the strain gauges in relation to the edge of the diaphragm.

Recently, an improved packaging method has been developed [Masheeb, 2002]. In this leadless sensor, shown in Figure 8.13, the pressure sensor chip is inverted so that the metal contacts are no longer exposed to the harsh measurement environment. In addition, the fragile leads are eliminated by using conductive glass to connect the pins of the package to the pads on the chip. Current work is directed at increasing the operating temperature beyond 500°C and providing a means for temperature compensation.

Summary

Deep RIE of SiC is a key to enabling the development of a wide range of SiC MEMS for use in harsh environments. High-density plasma makes DRIE of SiC possible, because it provides the required high chemical reactivity, low operating pressure, and independent control of ion energy. Deep RIE of SiC is quite different from DRIE of silicon because SiC is chemically inert. The need for energetic ion bombardment to produce an appreciable etch rate causes RIE of SiC to be inherently anisotropic. Energetic ion bombardment necessitates the use of a nonreactive etch mask such as nickel or ITO. The attainable etch depth is typically limited by sputter erosion of the etch mask. Often, higher ion energies can be used to increase the SiC etch rate, but this method can produce an unacceptable reduction in the selectivity with respect to the mask. The etch depths for which smooth, residue-free surfaces are obtained can be limited by background contaminants in the reaction chamber. Argon can be introduced into the chamber to provide continuous sputter-cleaning of the surface during the etch process, but the addition of argon has detrimental effects on etch rate and mask selectivity. At present, SiC DRIE processes have been developed which are well-suited for the fabrication of low-aspect-ratio microstructures such as pressure sensors. However, high-aspect-ratio DRIE processes are still in the early stages of development.

Defining Terms

Anisotropy: Degree of directionality of an etch process. In reactive ion etching, an ideal anisotropic process provides negligible etching in the lateral direction.

Aspect ratio: Etch depth divided by lateral feature size. Highly anisotropic etch processes are needed to fabricate microstructures with high aspect ratios.

Deep reactive ion etching (DRIE): Highly anisotropic plasma etching to depths greater than about 10 μm.

High-density plasma (HDP): A reactor which employs some means of electron confinement to produce a plasma density (ions per unit volume) that is typically two orders of magnitude greater than that of a conventional parallel plate RIE.

Inductively coupled plasma (ICP): A type of high-density plasma reactor in which power is coupled to the plasma using a coil. The time-varying magnetic field induced by the coil helps to confine electrons to the plasma, which increases the plasma density.

Mask selectivity: The substrate etch rate divided by the etch rate of the mask.

Micromasking: The formation of residues caused by the masking effect of materials sputtered from the etch mask onto the etched surface. In some reactors, sputtering of the substrate electrode can also cause micromasking.

References

Beheim, G., and Salupo, C.S. (2000) "Deep RIE Process for Silicon Carbide Power Electronics and MEMS," in *Proc. MRS 2000 Spring Meeting*, MRS Proc. 622, paper T8.9, 24–28 May, San Francisco.

Bhardwaj, J.K., Ashraf, H. (1995) "Advanced Silicon Etching Using High Density Plasmas," *Micromachining and Microfabrication Process Technology*, K.W. Markus, ed., Proc. SPIE 2639, pp. 224–33.

Chabert, P. (2001) "Deep Etching of Silicon Carbide for Micromachining Applications: Etch Rates and Etch Mechanisms," *J. Vac. Sci. Technol. B*, **19**, pp. 1339–45.

Cho, H., Lee, K.P., Leerungnawarat, P., Chu, S.N.G., Ren, F., Pearton, S.J., Zetterling, C.M. (2001) "High Density Plasma Via Hole Etching in SiC," *J. Vac. Sci. Technol. A,* **19**, pp. 1878–81.

Faust, J.W. (1960) *The Etching of SiC*, O'Connor, J.R., and Smiltens, J., eds., Pergamon Press, London/Oxford.

Khan, F.A., Roof, B., Zhou, L., Adesida, I. (2001) "Etching of Silicon Carbide for Device Fabrication and through Via-Hole Formation," *J. Electron. Mater.*, **30**, pp. 212–19.

Kim, D.W., Lee, H.Y., Park, B.J., Kim, H.S., Sung, Y.J., Chae, S.H., Ko, Y.W., Yeom, G.Y. (2004) "High Rate Etching of 6H-SiC in SF_6-Based Magnetically-Enhanced Inductively Coupled Plasmas," *Thin Solid Films*, **447–448**, pp. 100–04.

Leerungnawarat, P., Hays, D.C., Cho, H., Pearton, S.J., Strong, R.M., Zetterling, C.M., Ostling, M. (1999) "Via-Hole Etching for SiC," *J. Vac. Sci. Technol. B*, **17**, pp. 2050–54.

Leerungnawarat, P., Lee, K.P., Pearton, S.J., Ren, F., Chu, S.N.G. (2001) "Comparison of F_2 Plasma Chemistries for Deep Etching of SiC," *J. Electron. Mater.*, **30**, pp. 202–06.

Lieberman, M., Lichtenberg, A. (1994) *Principles of Plasma Discharges and Materials Processing*, John Wiley, New York.

Lohner, K.A., Chen, K.-S., Ayon, A.A., and Spearing, S.M. (1999) "Microfabricated Silicon Carbide Microengine Structures," in *Proceedings of the MRS 1998 Fall Meeting-Symposium AA, Materials Science of MEMS*, MRS Proc. 546, pp. 85–90, 1–2 December, Boston.

Masheeb, F., Stefanescu, S., Ned, A.A., Kurtz, A.D., and Beheim, G. (2002) "Leadless Sensor Packaging for High Temperature Applications," in *15th IEEE International Conference on Micro Electro Mechanical Systems (MEMS)*, pp. 392–95, 20–24 January, Las Vegas.

Ned, A.A., Kurtz, A.D., Masheeb, F., and Beheim, G. (2001) "Leadless SiC Pressure Sensors for High Temperature Applications," in *Proc. ISA 2001, Instrument Society of America Annual Conference*, 10–13 September, Houston, TX.

Sheridan, D.C., Casady, J.B., Ellis, C.E., Siergiej, R.R., Cressler, J.D., Strong, R.M., Urban, W.M., Valek, W.F., Seiler, C.F., and Buhay, H. (1999) "Demonstration of Deep (80 (m) RIE Etching of SiC for MEMS and MMIC Applications," in *Proc. International Conference on Silicon Carbide and Related Materials 1999*, pp. 1053–56, 10–15 October, Research Triangle Park, NC.

Shor, J.S., and Kurtz, A.D. (1994) "Photoelectrochemical Etching of 6H-SiC," *J. Electrochem. Soc.* **141**, pp. 778–81.

Song, X., Rajgolpal, S., Melzak, J.M., Zorman, C.A., Mehregany, M. (2001) "Development of a Multilayer SiC Surface Micromachining Process with Capabilities and Design Rules Comparable with

Conventional Polysilicon Surface Micromachining," in *Technical Digest of the International Conference on Silicon Carbide and Related Materials 2001*, 28 October–2 November, Tsukuba, Japan.

Tanaka, S., Rajanna, K., Abe, T., Esashi, M. (2001) "Deep Reactive Ion Etching of Silicon Carbide," *J. Vac. Sci. Technol. B*, **19**, pp. 2173–76.

Yasseen, A.A., Zorman, C.A., Mehregany, M. (1999) "Surface Micromachining of Polycrystalline SiC Films Using Microfabricated Molds of SiO_2 and Polysilicon," *J. MEMS*, **8**, pp. 237–42.

Yih, P.H., Saxena, V., and Steckl, A.J. (1997) "A Review of SiC Reactive Ion Etching in Fluorinated Plasmas," *Phys. Stat. Sol.* (b) **202**, pp. 605–42.

9

Polymer Microsystems: Materials and Fabrication

Gary M. Atkinson
Virginia Commonwealth University

Zoubeida Ounaies
Texas A&M University

Mr. McGuire: "I just want to say one word to you … just one word."

Benjamin: "Yes, sir."

Mr. McGuire: "Are you listening?"

Benjamin: "Yes, I am."

Mr. McGuire: "Plastics."

Benjamin: "Exactly how do you mean?"

Mr. McGuire: "There's a great future in plastics. Think about it. Will you think about it?"

Benjamin: "Yes, I will."

(From "The Graduate," Embassy Pictures Corporation, 1967)

9.1   Introduction

In the last decade, there has been an explosion of microsystems and micromachining technology that, on the one hand, has clearly been driven by a "technology push," where multidisciplinary research in material science and chemical, mechanical, electrical, and biomedical engineering has uncovered advanced materials that enhance the functionality of microsystems. On the other hand, there has also been a strong market "pull" by new, emerging applications in commercial, military, medical, and industrial products. Market pull requires new technology to successfully achieve the required overall performance at an economically viable market price. Success implies a profitable product and a successful business venture that draws this microtechnology explosion forward. Clearly, this fact has been demonstrated for a number of well-known, successful microsystems technologies, such as airbag chips, pressure sensors, and inject printing heads, to name a few.

Both the technology push and the market pull have had a dramatic effect on micromachining technology. What began as a predominantly silicon-based technology, born out of the integrated circuit industry with "standard" integrated circuit materials (silicon, silicon dioxide, nitride, and aluminum), has expanded to include compound semiconductors (both III–V and IV–IV), a wide range of dielectrics, magnetic materials, biosensitive materials, and indeed, nearly every available metal or alloy. Additionally, fabrication techniques have also multiplied as the techniques for controlling stress during depositions, achieving extremely high aspect ratios, releasing micromechanical structures, forming sealed cavities, and packaging have proliferated. The combination of material advancements and fabrication improvements has driven this technology explosion, expanding the types of transducer structures that can be manufactured, as well as decreasing dimensions and tolerances, and improving their performance.

More recently, two promising new areas have expanded the materials and fabrication techniques even further. These areas are advanced organic polymers and nanotechnology. The combination of these areas promises to expand the range of applications that are possible, or, more precisely, accessible to, microsystems technology. Polymers have been used in integrated circuits for decades, primarily as the photosensitive layers for microlithography, and subsequently the etch mask, in a pattern-transfer step. This use extended itself to early micromachining technologies, as well. Polymers, particularly photoresists, have been used as the sacrificial layers in a number of micromachining technologies [Atkinson, 2003; Fan, 2004; Courcimault, et al., 2004] where the polymer layer provides the separation between structural components and is removed in the back end of the process. The primary disadvantage of using photoresist as a sacrificial layer is that it cannot tolerate the excessive temperatures of many deposition processes, and thus it limits the processing temperatures of the back end of the process once it is applied.

The second area of explosive growth in materials and fabrication techniques has been nanotechnology, seemingly acclaimed as the panacea of all things technical. Nanotechnology encompasses nanostructured materials, nanopowders, nanoparticles, nanolayers, and nanocoatings. Already, nanotechnology has heralded a new generation of fabrication techniques to precisely control and form nanoscale features and structures in a variety of materials, from metals to carbon, and from ceramics to polymers. Obviously, the scope of nanotechnology is beyond that of the present chapter, but the crosspollination of polymer technology with nanotechnology, particularly with the advent of high performance "smart" polymer-nanostructure composites, is important to include in this discussion.

Polymer materials offer distinct advantages over silicon and other traditional engineering materials used in MEMS. Polymers have increased fracture strength; low Young's modulus; high elongation at break; and relatively low material costs. Furthermore, polymers can be inert and biocompatible, making them extremely attractive for use in biological and chemical applications. Polymer fabrication techniques can be simple and quite varied. Polymers can be deposited on various types of substrates, and they are available in a variety of molecular structures. Thus, films can be produced with various physical and chemical properties. They can be optically transparent to opaque, and thermally and chemically resistant. There are more and more applications where polymers remain as functional layers in the completed device structure. These layers may be adhesive layers for bulk micromachined parts; membranes and cantilevers for sensor structures; deformable membranes for microfluidic valve or pump structures; channel walls in microfluidic systems; membrane-sealing layers; temperature- or humidity-sensitive layers; chemically or biologically sensitive layers; actuator mechanisms; and even the substrates themselves. There are many more variations and types of polymer structures than there are elements in the periodic table. For instance, polymers can be made to be insulating, conductive, electroactive, inert, and with varying degrees of stiffness and strength. Of course, there are challenges and disadvantages in using polymers, as well. Polymers are typically sensitive to moisture and swelling, and they generally have a low temperature tolerance relative to traditional microfabrication materials such as silicon and its oxides. Successful applications of polymer-based microtechnology will be in areas where manufacturing issues can be solved at costs that meet the market demand.

The purpose of this chapter is to review the state-of-the-art in polymer microsystems technology. It is important first to review the myriad of polymer materials that are available and have been explored for microfabrication. The focus of this chapter is on devices in which polymers perform as key structural elements of the completed device. We will next review the properties of these various polymers, specifically in reference to the types of usage or application to which they lend themselves. Particularly with microsystems, it is important to review the types of actuation and sensing mechanisms that are available from the current toolbox of polymer films. Also for microfabrication applications, one must review the techniques by which these polymeric materials can be deposited and patterned. These techniques include spin casting, vacuum deposition, vapor deposition, electrodeposition, beam-induced CVD, molding, injection molding, and embossing. We will cover essential assembly and integration techniques, fairly specific to polymers, for bonding, adhering, and combining polymer films and systems with themselves and with other nonpolymer systems. Next, we will discuss current polymer devices, specifically in which polymer films are used as the key structural or functional layers of the completed device. Finally, we will speculate on the future trends and directions in polymer microsystems.

9.2 Polymer Materials in MEMS

Microelectromechanical systems (MEMS) is the integration of mechanical elements, sensors, actuators, and electronics on a common substrate through the use of microfabrication technology. In the last decade, owing to the many attributes polymers offer over metallic and inorganic materials, there has been great interest in applying polymers in MEMS and further, in fabricating an all-polymer MEMS. In polymer-based MEMS, we envision that the polymer would actively participate in the device function, rather than play a passive role such as a coating, substrate, or electrical connection. Active polymer components take advantage of polymer properties to increase the functionality of the device. For example, electroactive polymers can be used as actuators in pumps and valves; optoelectronic polymers can function as lenses and filters; pyroelectric polymers can be used as temperature and humidity sensors. The functionality is even greater if multiple polymer materials with different properties are used in a single, multifunctional device. Owing to the great variety of polymer structures, they are able to do this better than silicon. Discussion follows on various types of polymers and polymer nanocomposites that are currently used or are being considered for use in MEMS. Where appropriate, sensing and actuation functionalities are identified and discussed.

9.2.1 PVDF and Copolymers

Pioneering work in the area of piezoelectric polymers has led to the development of strong piezoelectric activity in polyvinylidene fluoride (PVDF) and its copolymers with trifluoroethylene (TrFE) and tetraflouoroethylene (TFE) [Kawai, 1969]. These semicrystalline fluoropolymers represent the state-of-the-art in piezoelectric polymers and are currently the only commercially available piezoelectric polymers. Polyvinylidene fluoride (PVDF) and its copolymers are used in applications such as pyroelectric sensors, pressure sensors, audio-frequency transducers, sonar hydrophones, and ultrasonic transducers [Gallantree, 1983].

Polyvinylidene fluoride (PVDF) is a fluorocarbon whose main use was initially as an inert lining or pipework material in chemical plants, where its combination of processability, mechanical strength, and excellent chemical resistance are advantageous [Gallantree, 1983]. PVDF can be shaped into simple or complex configurations via extrusion, solution casting, and molding to form films, sheets, and tubes. This great flexibility in design, coupled with its pliability and relative low cost, offer an attractive basis for MEMS as a substrate as well as a functional material. Interest in the electrical properties of PVDF began in 1969, when Kawai showed that thin films that had been poled exhibited a very large piezoelectric coefficient, 6–7 pCN^{-1}, a value about ten times larger than had been observed in any other polymer. As a result of Kawai's discovery, PVDF became better known for its piezoelectric properties and has been extensively investigated as a sensor and actuator since that time. Today, PVDF accounts for virtually all of the commercially significant piezoelectric polymer applications.

PVDF is semi-crystalline, with typically 50 to 60% crystallinity, depending on thermal and processing history. Generally, PVDF is synthesized by the free-radical polymerization of 1,1-difluoroethylene and has the repeating unit shown in Figure 9.1. The amorphous phase in PVDF has a glass transition that is well below room temperature ($-50°C$), hence the material is quite flexible and readily strained at room temperature. There are two conditions that must be fulfilled in order to render PVDF piezoelectric, namely: the crystal structure should possess a nonzero net dipole moment, i.e., it must be polar; and the dipoles should have the same preferred orientation.

PVDF has four polymorphic forms: α, β, γ, and δ phases (Figure 9.2). The α-phase is the most common and stable form of PVDF. Normally, the α-phase is obtained by melt extrusion. It is nonpolar as it produces an inversion of the dipole moments in alternate chains, which leads to a cancellation of the overall dipole moment. The β-phase is the most important polymorph of PVDF, because it is polar and exhibits the highest piezoelectric properties. Usually, the β-phase is obtained by mechanical stretching of the α-phase film. Generally, stretching is either uniaxial or biaxial, yielding different properties. The γ-phase is also a polar unit cell. Generally, the γ-phase is obtained by solution crystallization and readily transforms into the β-phase upon mechanical deformation. The δ-phase is the polar form of the α-phase. It is obtained by applying a high electric field to α-phase film. As a result, the unit-cell dimensions and the configuration are the same as in the α-phase. As mentioned, the β-phase is the one most commonly used for sensing and actuation because of the higher piezoelectricity. After mechanical stretching to produce the polar β-phase, the poling process begins with either heating of the film and applying a high DC electric field, or by using corona discharge (Figure 9.3).

During the poling process, the dipoles are aligned relative to the direction of the poling field. The polarization achieved depends on the applied field and on poling temperature and poling time. After poling PVDF, the room-temperature polarization stability is excellent, however, polarization and piezoelectricity degrade with increasing temperature and are erased at PVDF's Curie temperature (temperature above which PVDF ceases to be piezoelectric and ferroelectric). The β-phase has a dipole moment perpendicular to the chain axis of 2.1 D corresponding to a dipole concentration of 7×10^{-30} cm. The presence of this nonzero dipole is at the origin of the piezoelectric and ferroelectric behavior of PVDF.

A classical definition of piezoelectricity, a Greek term for "pressure electricity," is the generation of electrical polarization in a material in response to a mechanical stress. This phenomenon is known as the direct effect. Piezoelectric materials also display the converse effect, mechanical deformation upon application of electrical charge or signal.

A subset of piezoelectricity is pyroelectricity, whereby the polarization is a function of temperature. Some pyroelectric materials are also ferroelectric. Ferroelectricity is a property of certain dielectrics,

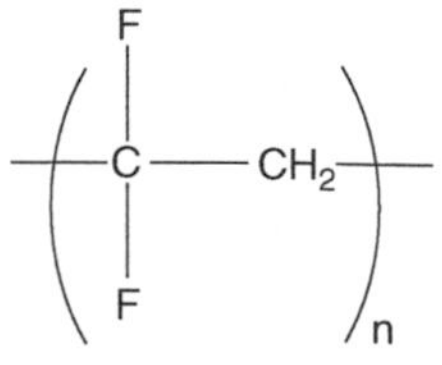

FIGURE 9.1   Structure of PVDF.

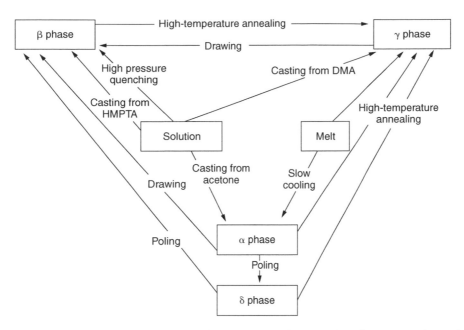

FIGURE 9.2 Conversion between various phases of PVDF. (Adapted from K. Tashiro, "Structure and Piezoelectricity of Polyvinylidene Fluoride, Ferroelectrics," (1981); G.M. Sessler, "Piezoelectricity in Polyvinylidenefluoride", *J. Acoust. Soc. Am.*, **70**(6) 1981.)

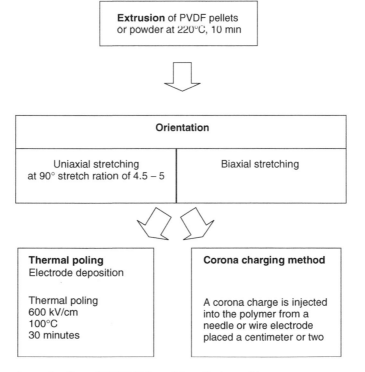

FIGURE 9.3 Processing and poling of PVDF. (Adapted from Kepler et al.)

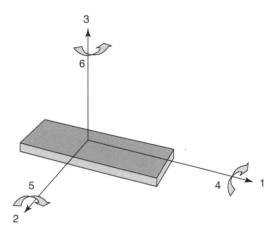

FIGURE 9.4   Tensor directions for defining the constitutive piezoelectric relations. (Adapted from Harrison et al., 2002.)

which exhibit a spontaneous electric polarization (separation of the center of positive and negative electric charge, making one side of the crystal positive and the opposite side negative) that can be reversed in direction by the application of an appropriate electric field. PVDF is piezoelectric, pyroelectric, and ferroelectric. The polarization, P, is a measure of the degree of piezoelectricity in a given material. In a piezoelectric material, a change in polarization, ΔP, results from an applied stress, X, or strain, S, under the conditions of constant temperature and zero electric field. A linear relationship exists between ΔP and the piezoelectric constants. Because of material anisotropy, P is a vector with three orthogonal components in the 1, 2, and 3 directions (Figure 9.4). The stretch direction is denoted as "1". The "2" axis is orthogonal to the stretch direction in the plane of the film. The polarization axis (perpendicular to the surface of the film) is denoted as "3". The shear planes are indicated by the subscripts "4," "5," "6" and are perpendicular to the directions "1," "2," and "3," respectively. The piezoelectric constants can be defined as

$$\Delta P_i = d_{ij} X_j \tag{9.1}$$

$$\Delta P_i = g_{ij} S_j \tag{9.2}$$

where the d_{ij} coefficient is a measure of the actuation potential of the piezoelectric material, and g_{ij} is a measure of the sensor potential. The electromechanical coupling coefficient, k_{ij}, represents the conversion of electrical energy into mechanical energy, and vice versa. The electromechanical coupling can be considered as a measure of transduction efficiency and is always less than unity, as shown below

$$k^2 = \frac{electrical\ energy\ converted\ to\ mechanical\ energy}{input\ electrical\ energy} \tag{9.3a}$$

$$k^2 = \frac{mechanical\ energy\ converted\ to\ electrical\ energy}{input\ mechanical\ energy} \tag{9.3b}$$

When the piezoelectric film is operating in an electromechanical or motor mode, the film elongates and contracts as the polarity of the alternating field changes. When operating in a mechanoelectrical or generator mode, external forces are applied that produce compressive and tensile strain. These deformations cause a change in the surface charge density, resulting in a voltage between the electrodes, whose polarity changes as the direction of the force is reversed. One limitation in applying PVDF to MEMS as the sensing and actuation structure in the device is that the use of PVDF is limited to 80°C, owing mostly to the prior stretching necessary to render the structure polar.

The electromechanical properties of PVDF have been widely investigated. For more details, the reader is referred to the wealth of literature that exists on the subjects of morphological, piezoelectric, pyroelectric, and ferroelectric properties [Furukawa, 1989; Sessler, 1981; Fukada, 2000; Davis et al., 1978; Lovinger, 1982].

TABLE 9.1 Comparison of Piezoelectric Properties of Some Semicrystalline Polymeric Materials

Polymer	Structure	T_g (°C)	T_m (°C)	Max. Use Temp. (°C)	d_{31} (pC/N)
PVDF		−35	175	80	20–28
PTrFE		32	150	90–100	12

Copolymers of polyvinylidene fluoride with trifluoroethylene (TrFE) and tetrafluoroethylene (TFE) have also been shown to exhibit strong piezoelectric, pyroelectric, and ferroelectric effects. An attractive morphological feature of the comonomers is that they force the polymer into an all-trans conformation that has a polar crystalline phase, which eliminates the need for mechanical stretching to yield a polar phase. P(VDF-TrFE) crystallizes to a much greater extent than PVDF (up to 90% crystalline), yielding a higher remanent polarization. TrFE also extends the use temperature by about 20 degrees, to nearly 100°C. Conversely, copolymers with TFE have been shown to exhibit a lower degree of crystallinity and a suppressed melting temperature when compared to the PVDF homopolymer. Although the piezoelectric constants for the copolymers are not as large as the homopolymer, the advantages of P(VDF-TrFE) associated with processability, enhanced crystallinity, and higher use temperature make it favorable for applications. Typical values of the d_{31} piezoelectric constants for PVDF and TrFE copolymers are given in Table 9.1 [Harrison et al., 2002].

Recently, researchers have reported that highly ordered, lamellar crystals of P(VDF-TrFE) can be made by annealing the material at temperatures between the Curie temperature and the melting point. These researchers refer to this material as a "single crystalline film." A relatively large single crystal P(VDF-TrFE) 75/25 mol% copolymer has been grown, exhibiting a room temperature $d_{33} = -38\,\text{pm/V}$ and a coupling factor $k_{33} = 0.33$ [Omote et al., 1997]. Zhang et al. (1998) studied the influence of introducing defects into the crystalline structure of P(VDF-TrFE) copolymer using high electron irradiation on electroactive actuation. Extensive structural investigations indicate that the electron irradiation disrupts the coherence of polarization domains (all-trans chains) and forms localized polar regions. After irradiation, the material exhibits behavior analogous to that of relaxor ferroelectric systems in inorganic materials. The resulting material is no longer piezoelectric but rather exhibits a large electric field-induced strain (5% strain) caused by electrostriction. The basis for such large electrostriction is the large change in the lattice strain as the polymer traverses the ferroelectric to para-electric phase transition and the expansion and contraction of the polar regions. Piezoelectricity can still be measured in these and other electrostrictives when a DC bias field is applied. Irradiation is typically accomplished in a nitrogen atmosphere at elevated temperatures with irradiation dosages up to 120 Mrad. The resulting polymer exhibits strains up to 7%, coupled with high stresses up to 45 MPa, making it extremely attractive as an electromechanical actuator [Zhang, 2001; Xia et al., 2003]. One disadvantage of these actuators is that extremely high actuation electric fields are required (~150 MV/m), which could correspond to high voltages (one to a few kilovolts). Madden et al. (2004) pointed out that using thin layers can reduce the high voltage requirement, which is especially relevant in MEMS; typically, polymers are 100 nm thick which requires only 15 V for actuation. Voltage reduction can also be achieved by raising the dielectric constant. One way to achieve that is by adding nanoparticles and nanopowders to the polymer [Madden et al., 2004]. The upcoming section on carbon nanotube-polyimide composites illustrates this point. Other challenges are associated with the irradiation process (it is costly and time-consuming), and the relative stiffness of the electrodes as compared to the polymer. The latter can be overcome by using more flexible electrode materials, rather than the conventional metals.

In general, PVDF-TrFE is a better candidate for MEMS if sensing and actuation functionalities are desired in the polymer material. Advantages of PVDF-TrFE over PVDF and PVDF-TFE include higher temperature use, increased mechanical and chemical resistance (especially in the irradiated case), high actuation, relatively low voltage, and ease of fabrication, i.e., no need for prior stretching or poling (again, in the case of the irradiated PVDF-TrFE).

9.2.2 Polyimides

Polyimides are a class of polymers that contains an imide group, as shown in Figure 9.5. Polyimides can have a linear or cyclic structure, generally yielding a linear aliphatic polyimide structure or an aromatic polyimide structure, respectively (Figure 9.6). Polyimides can be formulated as both thermosets and thermoplastics. They are known for their thermal and chemical stability, high mechanical strength, high modulus, radiation and solvent resistance, and excellent electrical and dielectric properties. In particular, aromatic polyimides are rigid and fairly insoluble. As a result of this attractive combination of properties, polyimides were introduced in electronics packaging to replace ceramics. Polyimide coatings are used in a variety of interconnect and packaging applications. Fluorinated polyimides are especially important in electronics applications such as integrated communications systems. They exhibit a combination of such desirable properties as low dielectric constant, high glass transition temperatures (T_g), and in some cases, optical transparency. For example, the low dielectric constant reduces the signal delays that result when an electrical signal travels from one chip to another through the packaging.

When first developed, polyimide synthesis and processing were tedious and complex. The synthesis of an aromatic polyimide was first reported in 1908; however, it was not until the late 1960s that a commercial polyimide film was introduced, paving the way for mass production and the possibility of advanced applications. This synthesis route is through a polyamic acid precursor, which then converts to the final polyimide structure (Figure 9.7). Today, because of their exceptional thermal, mechanical, and dielectric properties, polyimides are widely used as matrix materials in aircraft and as dielectric materials in microelectronic

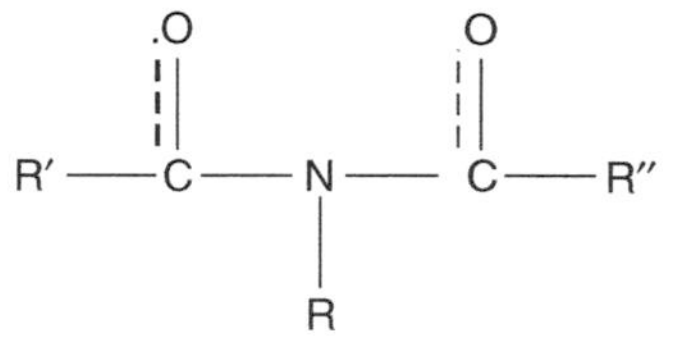

FIGURE 9.5   Structure of an imide group.

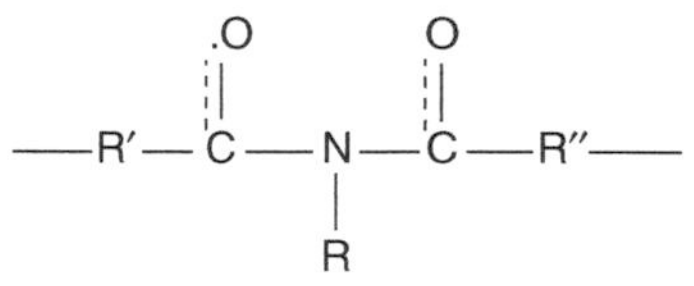

Linear polyimide

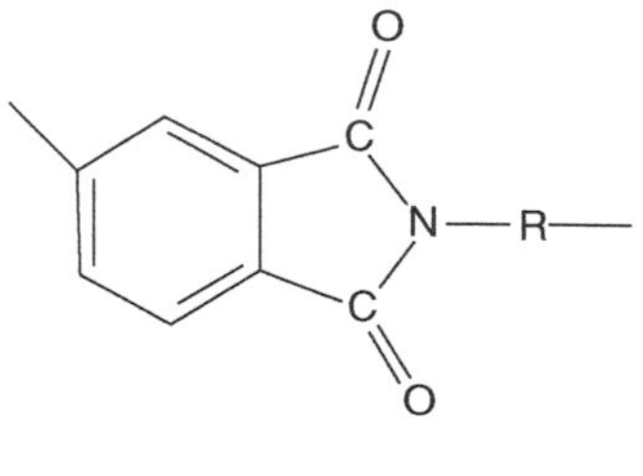

Aromatic polyimide

FIGURE 9.6   Possible polyimide structures.

devices. Their use in microelectronics dates back to the early 1970s. These properties sparked further investigation of new compositions aimed at advanced applications. Particularly interesting is the potential use of piezoelectric polyimides in MEMS devices because fluoropolymers such as PVDF, discussed above, do not possess the chemical resistance or thermal stability necessary to withstand conventional MEMS processing.

The development of high-temperature piezoelectric polymers promises to have a great impact on advanced applications such as microelectronics. Polyimides are excellent candidates because of their high temperature stability and their compatibility with MEMS processing. Recently, a series of amorphous piezoelectric polyimides containing polar functional groups has been developed at NASA Langley [Ounaies et al., 1997; Simpson et al., 1997; Park et al., 2004], based on molecular design and computational chemistry, for potential use in high-temperature applications. Specifically, nitrile-substituted polyimides are developed and investigated for the purpose of tailoring their piezoelectric response such that it is stable at temperatures in excess of 150°C. The piezoelectricity in these amorphous polymers resides entirely in the amorphous region and differs significantly from that in semicrystalline polymers and inorganic crystals [Davis, 1993]. In order to induce a piezoelectric response in amorphous systems, the presence of dipoles is necessary. A piezoelectric response is induced by applying a strong electric field (E_p) at an elevated temperature ($T_p \geqslant T_g$), which produces orientation of the molecular dipoles. The dipoles are polarized with the applied field and will partially retain this polarization when the temperature is lowered below T_g in the presence of E_p. This process is known as orientation polarization. The resulting remanent polarization, P_r, is key in developing materials with a useful level of piezoelectricity, as it is directly proportional to the material's piezoelectric response. In order to maximize piezoelectricity, the remanent polarization, P_r, must be maximized. P_r is related to the poling field by

$$P_r = \varepsilon_o \Delta \varepsilon E_p \tag{9.4}$$

where ε_o is the permittivity of space and $\Delta\varepsilon$ is the dielectric relaxation strength, which is the change in dielectric constant upon traversing the glass transition. As pointed out by Furukawa (1989), $\Delta\varepsilon$ is the parameter of greatest interest in designing amorphous polymers with large piezoelectric activity. The dielectric

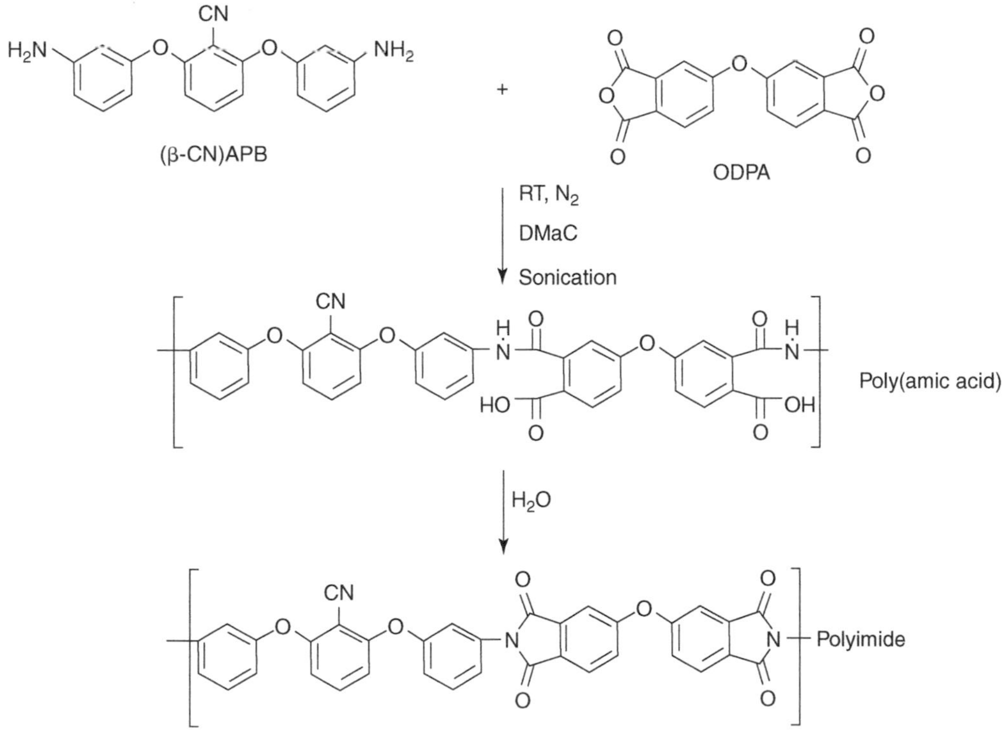

FIGURE 9.7 Polyimide synthesis route via polyamic acid (Park et al.).

relaxation strength, $\Delta\varepsilon$, is affected by dipolar mobility and concentration. Polyimide films are synthesized using two different diamines, (2,6-bis(3-aminophenoxy) benzonitrile ((β-CN)-APB) and 1,3-aminophenoxy benzene (APB), to determine the effect of the pendant nitrile group on the piezoelectric response. Each of the diamines is reacted with 4,4'-oxdiphthalic anhydride (ODPA), resulting in the two polyimides shown in Figure 9.8. The (β-CN) APB/ODPA polyimide possesses three dipole functionalities. These dipoles, along with the associated dipole moments, are presented in Table 9.2.

Typically, the functional groups in amorphous polymers are pendant to the main chain (e.g., the nitrile group). The dipoles, however, may also reside within the main chain of the polymer, such as the dianhydride units and the diphenylether group in the (β-CN) APB/ODPA polyimide. The nitrile dipole is pendant to a phenyl ring ($\mu = -4.2D$), whereas the two anhydride dipoles ($\mu = -2.34D$) are within the chain, resulting in a total dipole moment per repeat unit of 8.8 D. The importance of dipole concentration on ultimate polarization is evident from a comparison of polyacrilonitrile (PAN) and the polyimide (β-CN) APB/ODPA. PAN has a single nitrile dipole per repeat unit ($\mu = -3.5D$), resulting in a dipole concentration of $1.34\times10^{28}\,\mathrm{m}^{-3}$. This result translates into an ultimate polarization of $152\,\mathrm{mC/m}^2$ [Harrison et al., 2004]. The (β-CN) APB/ODPA polyimide, on the other hand, has a total dipole moment per monomer of 8.8D. The dipole concentration of (β-CN) APB/ODPA, however, is only $0.136 \times 10^{28}\,\mathrm{m}^{-3}$, resulting in an ultimate polarization of $40\,\mathrm{mC/m}^2$, which is less than a fourth of that of PAN. As a result, similar polyimide systems with increased nitrile concentrations were synthesized and characterized, in an attempt to produce a polyimide with a P_r as high as possible. Studies on these polymers show that increasing dipole concentration significantly increases polarization. Structure–property investigations

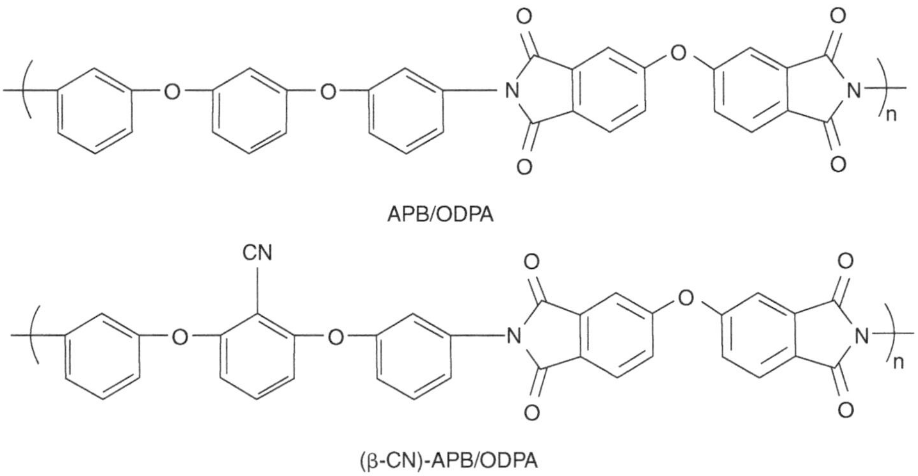

APB/ODPA

(β-CN)-APB/ODPA

FIGURE 9.8 Chemical structures of unsubstituted and nitrile-substituted polyimides (Simpson et al., 1997).

TABLE 9.2 Values of the Dipole Moments within the Nitrile-Substituted Polyimide

Dipoles	Dipole Identity	Dipole Moment (Debye)
	Pendant nitrile group	4.18
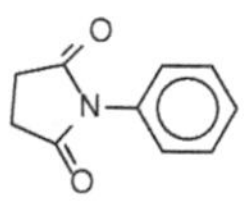	Main chain dianhydride group	2.34
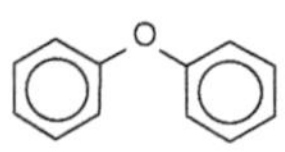	Main chain diphenylether group	1.30

designed to assess effects of these dipoles on T_g, thermal stability, and overall polarization behavior are currently being done [Klein et al., 2002].

A comparison of the properties of these piezoelectric polyimides vs. PVDF is given in Table 9.3 and in Figure 9.9. The polyimide has a lower g_{31} and d_{31} at room temperature than PVDF. However, as the temperature increases to about 75°C, the values of g_{31} and d_{31} for (β-CN) APB/ODPA increase, but those of PVDF can no longer be detected. The higher-temperature performance of these modified polyimide films allows them to maintain their piezoelectric properties at temperatures up to 150°C, twice the capability of PVDF [Simpson et al., 1997; Ounaies et al., 1997].

The piezoelectric voltage and strain coefficients at higher temperatures, coupled with the fact that polyimides resist harsh environments, make these materials ideal for high-temperature MEMS devices. However, even with their relatively high glass transition temperature ($T_g \sim 220°C$), fabrication processes must be kept at relatively low temperatures by traditional microfabrication processing standards. Low-temperature curing and in situ poling of polyimides during MEMS fabrication are two issues that are key in processing polyimide MEMS, in order to take advantage of the sensing and actuation capability of the polymer.

9.2.3 Other Electroactive Polymers (PDMS, PVC, and PMMA)

PDMS, a fairly transparent polymer with high elasticity and low stiffness, is composed of an alternating backbone of silicon and oxygen atoms with two organic functionalities attached to each silicon atom. The absence of a chromophore within alkyl-substituted polysiloxanes results in a polymer with high stability towards ultraviolet radiation. Also, the silicon–oxygen backbone gives PDMS high resistance to oxidation and chemical attack. The high torsional mobility about the silicon–oxygen bond results in the dimethyl-substituted polysiloxane having one of the lowest glass transition temperatures of any polymer ($-123°C$). Owing to the combination of the above properties, coupled with its adaptable molecular composition and excellent processability, PDMS systems attract attention for a variety of actuation applications.

TABLE 9.3   Piezoelectric Response of PVDF vs. (β-CN) APB/ODPA Polyimide

	$T = 25°C$	$T = 75°C$	$T = 150°C$
	g_{31} (mV $-$ m/N)	g_{31} (mV $-$ m/N)	g_{31} (mV $-$ m/N)
PVDF	235	376	–
(β-CN)APB/ODPA	7.6	22.0	152.7

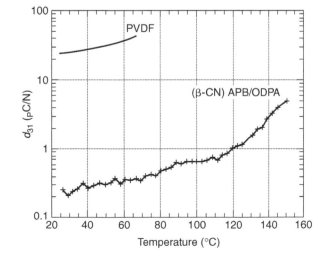

FIGURE 9.9   Piezoelectric coefficient comparison.

Electric field-induced actuation of PDMS-based systems has been reported by a number of researchers [Kornbluh et al., 1995]. The electroactive response is dominated by Maxwell stress, i.e., the interaction of the electrostatic charges on the electrodes. When a voltage is applied, charges on opposite electrodes will attract each other, squeezing the film in thickness, whereas the like charges on each electrode will tend to repel each other and expand the film in area. If the polymer film is soft, as in the case of PDMS, large motions can be produced from such a response. In general, extremely high levels of electric field are required for actuation (on the order of 50–100 MV/m).

In the early 1970s, a few studies focused on the piezoelectric behavior of PVC. The carbon–chlorine dipole in polyvinylidene chloride (PVC) can be oriented to produce a low level of piezoelectricity. The piezoelectric and pyroelectric activities generated in PVC were found to be stable and reproducible. Broadhurst et al. (1973) used PVC as a basis for understanding and studying piezoelectricity in amorphous polymers. The piezoelectric coefficient d_{31} of PVC has been reported in the range of 0.5 to 1.3 pC/N. Simultaneous stretching and corona poling of the film achieved an improved response. The enhanced piezoelectric coefficient d_{31} ranged from 1.5 to 5.0 pC/N, which is still fairly low for most actuator applications.

PMMA is the polymer of choice for photoresist materials. However, it also shows excellent charge-storage capability, which is promising for use in MEMS devices as an electret. The dipoles in PMMA are large but unstable. For this reason, poling using the method described for the nitrile-substituted polyimide does not result in a permanent polarization. Rather, the polarization decreases abruptly upon discontinuing the poling field [Harrison et al., 2002].

9.2.4 Conductive Polymers

Conductive polymers are conjugated polymers exhibiting high electrical conductivity. Polymers exhibiting electronic conduction were discovered in the late 1970s and the early 1980s [Shirakawa et al., 1977; MacDiarmid et al., 1980]. These researchers and others investigated a class of polymers that become conductive when oxidized or reduced by donor or acceptor electrons. The oxidation–reduction reaction is accompanied by addition and removal of charges from the polymer, which results in ionic mobility aimed at balancing these charges. This ionic mobility, coupled with solvent effects, produces swelling and contraction of the polymer. Several researchers have shown that the ensuing volume changes can be controlled using external stimulus such as an electric field [Marque et al., 1990; Chiarelli et al., 1994; Oguro et al., 1992].

The main response characteristics in conductive polymers arise from two sources: 1) conformational changes caused by interchain interactions pursuant to solvent swelling and absorption [Baughman, 1996; Pei et al., 1992]; and 2) dimensional changes in carrier population and electron mobility pursuant to redox reaction [Freundand et al., 1995]. The presence of both responses engenders chemoresistive, electrochemical, and electromechanical behavior in the polymers. This outcome exposes the possibility of using conductive polymers as transducers, much the same way that piezoelectric and electrostrictive polymers can display sensor and actuator functionalities. However, in contrast to the electroactive polymers described in previous sections, conductive polymers are classified as "wet" actuators, since their response is tied to ionic and solvent transport, and consequently, actuation requires the presence of an electrolyte. Oxidation and reduction of the polymer produces an exchange of ions between the electrolyte and the polymer. Properties of the exchange process, such as response speed, dimensional change, and conductivity variation, depend strongly on the doping mechanism, the diffusion properties of the polymer, and the type and size of ion that is present in the electrolyte.

One of the primary disadvantages of conductive polymers is the slow response speed. The speed depends on the phase change induced during the oxidation–reduction reaction, and the ionic and solvent diffusion across the polymer layer. Thus, scaling down the size may improve the response speed. For that reason, conductive polymers hold much promise in polymer microsystems applications. The first conductive polymer microactuators were developed successfully by Smela et al. in 1993. Since then, a number of researchers have investigated the use of conductive polymers in MEMS devices [Jager et al., 2000].

The synthesis of conductive polymers is based on polymerizing a number of commercially available monomers. The most common ones are aniline and pyrrole, yielding polyaniline and polypyrrole, respectively. A common technique for transducer fabrication is to electropolymerize a conductive polymer film onto a substrate. The properties of the polymers are very dependent on the solvent and salts used in the deposition. High tensile strengths (100 MPa), large stresses (up to 34 MPa), and stiffness of the order of 1 GPa can be obtained [Madden et al., 2004]. In addition to the superb mechanical properties, an attractive advantage is that low drive voltages are needed, usually on the order of 1–5 V. One drawback, though, is that these actuators typically exhibit low electromechanical coupling, thus requiring high power output (i.e., high current).

Strain rate, voltage, current, and mechanical properties are key parameters that govern the performance of conductive polymers as actuators. A number of researchers are investigating these materials computationally and experimentally to better control and tailor their performance [Madden et al., 2001; Madden et al., 2004; Smela et al., 1995; Smela et al., 2003]. Owing to their many advantages and the increased interest from the research community, there is no doubt that conductive polymers will become more common in MEMS applications.

9.2.5 Polymer Nanocomposites

Carbon Nanotubes (CNTs) are long, slender fullerenes. The walls of the tubes are hexagonal carbon, often capped at each end. CNT appears as a graphene sheet that has been rolled into a tube. CNTs are either single-walled nanotubes (SWNT) or multiwalled nanotubes (MWNT). A SWNT is a tube with only one wall, whereas MWNTs have many concentric tubes, held together by weak van der Walls forces. The properties of CNTs depend on chirality, diameter, and length of the tubes. The graphene sheet can be rolled up with varying degrees of twist along its length, resulting in CNTs having a variety of chiral structures. The atomic structure of CNTs is described in terms of the tube chirality, or helicity, which is defined by the chiral vector C_h and the chiral angle ϕ. Figure 9.10 shows a SWNT, a MWNT, as well as the Chiral vector and Chiral angle. Table 9.4 displays the various classifications of CNTs.

Iijima first observed MWNTs through the electric arc discharge method [sp. Iijima, 1991]. Subsequently, Iijima et al. (1993) and Bethune et al. (1993) reported the synthesis of SWNTs by the laser ablation technique. During the synthesis of the CNTs, impurities in the form of catalyst particles, amorphous carbon, and nontubular fullerenes are also produced, such that the nanotubes have to be purified to eliminate these impurities. CNTs exhibit actuation through a combination of electronic and ionic conduction since, generally, actuation is done in an electrolyte. MWNT are less effective actuators when compared to SWNTs, because their solvent-accessible surface area is typically lower than that of SWNTs [Madden et al., 2004]. Actuation is achieved by immersing nanotubes in an electrolyte and applying a potential between the nanotubes and a counter electrode. Ions are attracted to the nanotubes, leading to the accumulation of ionic charge at their surfaces, which is balanced by electronic charge within the tubes. The process is very similar to conductive polymers, where the rearrangement of the electronic structure leads to dimensional changes. Strains reach between 0.1% and approximately 1% [Baughman et al., 1999]. SWNTs have a tensile modulus that ranges from 270 GPa to 1 TPa and a tensile strength ranging from 11 to 200 GPa, which is 10 to 100 times higher than the strongest steel at a fraction of the weight [Li et al., 2000; Lu, 1997]. SWNTs are also stable up to 2800°C, and have a thermal conductivity twice as high as that of diamond. The strain behavior, combined with the excellent mechanical and thermal properties of CNTs, open up great opportunity for applications in microsystems.

Another promising technology for microsystems is combining CNTs with electroactive polymers to enhance and tailor the resulting structural and electroactive properties [Ounaies et al., 2003; Baughman et al., 1999]. In the last few years, micro and nanostructured polymer composites have generated great interest in the materials community, partly because of the promise for large gains in mechanical and physical properties with respect to traditional structural materials. Thus far, most of the research on CNT-organic composites has typically concentrated on processing CNT-reinforced polymers with substantial improvements in mechanical behavior. The basis is that CNT-polymer composites could offer modulus- and strength-to-weight ratios beyond any material currently available. Recently, Dalton et al. (2004) reported the fabrication of exceptionally strong CNT composite fibers that are tougher than either

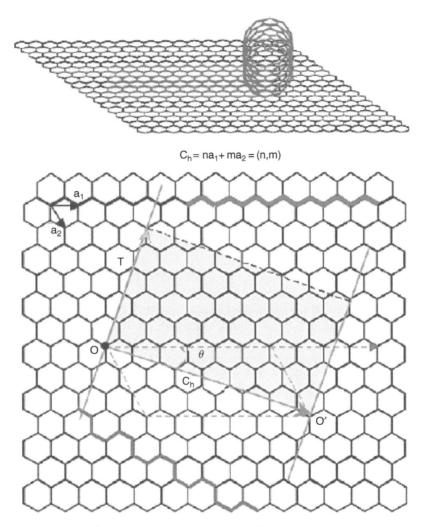

FIGURE 9.10   (a) Schematic of the roll-up of a graphene sheet to form a SWNT, (b) Definition of the chiral vector C_h translation vector T, and chiral angle θ. (Ouyang et al., 2002)

TABLE 9.4   Classification of Carbon Nanotubes

Type	ϕ	C_h
Armchair	30°	(n, n)
Zigzag	0°	(n, 0)
Chiral	0° < ϕ < 30°	(n, m)

spider silk or any fiber used for mechanical reinforcement. Velasco-Santos et al. (2003) observed an improvement in mechanical response by more than 200%, with only 1 wt% CNT added to methyl-ethyl methacrylate copolymer, substantially above other reports where large quantities of CNTs were used. Most of the researchers pursuing this route agree that while results are promising, they have yet to achieve the magnitude of property enhancement believed possible. There remain many challenges to optimize the properties of such materials, including uniform dispersion of the CNTs within the organic matrix, nanotube and material wetting and adhesion, CNT alignment, and modeling of CNT-polymer interaction.

More recently, some researchers have investigated the electrical properties of CNT-organic polymer composites using percolation theory. These studies tended to examine the electrical properties of CNT-polymer

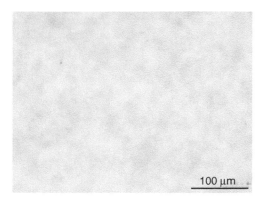

FIGURE 9.11   Optical micrograph of CNT-polyimide film.

composites with a goal of achieving a conductive composite. Notably, Pötschke et al. (2003) measured the electrical conductivity and dielectric properties of multiwall CNT-polycarbonate composites to assess the dispersion characteristics of the CNT and evaluate percolation. Valentini et al. (2004) investigated the role and interaction of CNTs as conductive fillers in an epoxy matrix. The investigators' focus was on effect of the CNTs on the curing and degree of reaction of the epoxy resin. Dang et al. (2003) also used percolation theory to study the dielectric behavior of multiwalled CNT in a barium titanate/PVDF matrix. A number of other investigations dealt with the potential of CNT-polymer composites for chemical sensing. Again, the mechanism involved percolation where the conducting paths were disrupted by the chemicals, and the response was detected as a change in resistance. Despite some of the challenges such as availability and cost of CNTs, the inherent potential suggests that further exploration is warranted. CNT-polymer composites may provide a revolutionary route to future generation multifunctional materials, owing to their exceptional thermal, mechanical, electrical and dielectric properties.

One of the challenges in developing and processing composites is the homogeneous dispersion of the inclusions in the matrix. Dispersion is an even bigger challenge in the case of CNTs, because intrinsic van der Waals attraction among the tubes, in combination with their high surface area and high aspect ratio, leads to significant agglomeration and clustering, thus preventing efficient transfer of their superior properties to the matrix. Although a number of studies have focused on dispersion of CNTs, complete dispersion of CNTs in a polymer matrix has rarely been achieved.

Using a method developed in our laboratory [Park et al., 2002; Ounaies et al., 2003], SWNT-polyimide composite films were prepared with a very high degree of SWNT dispersion, as shown in Figure 9.11. Specifically, SWNT-reinforced polyimide nanocomposites were synthesized by in situ polymerization of monomers of interest in the presence of sonication. This process enabled uniform dispersion of SWNT

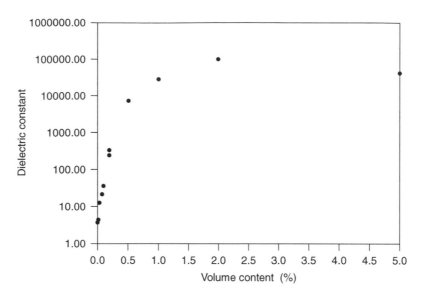

FIGURE 9.12   Temperature behavior of the dielectric constant of nitrile-substituted polyimide.

TABLE 9.5   Comparison of P_r

CNT content (vol %)	E_p MV/m	P_r mC/m^2
0	50	7
0.020	50	15
0.100	50	32

bundles in the polymer matrix. The predispersed SWNT solution remained stable throughout the reaction under sonication, producing a reasonably transparent, electrically conductive composite at very low SWNT loading. Figure 9.12 shows the dielectric constant as a function of SWNT content. The dielectric constant, ε' of the pristine polyimide was about 4. A sharp increase of the dielectric constant value was observed between 0.02 and 0.1 vol%, where ε' changed from 4 to 31. This increase continues to 1 vol% SWNT. This behavior is indicative of a percolation transition. Percolation theory predicts that there is a critical concentration, or percolation threshold, at which a conductive path is formed in the composite, causing the material to convert from a capacitor to a conductor. For CNT volume content below percolation, the conductivity is low enough that the composites can be poled (a process whereby the dipolar functionalities can be permanently oriented by application of an electric field). Orientation of the dipoles in the polyimides is induced by applying an electric field, E_p at an elevated temperature T_p $(T_p \geqslant T_g)$ where the molecular chains are sufficiently mobile to allow dipole alignment with the electric field. Partial retention of this orientation is achieved by lowering the temperature below T_g in the presence of E_p, resulting in a piezoelectric-like effect, as discussed in Section 9.2.2. Poling times need to be at least of the order of the relaxation time of the polymer at the poling temperature. The piezoelectric response can be quantified by the remanent polarization P_r. The effect of CNT addition on dipolar orientation in the polyimide films is investigated by dynamic dielectric spectroscopy. These measurements yield the dielectric relaxation strength $\Delta\varepsilon$; the larger the value of $\Delta\varepsilon$, the higher the piezoelectric performance (see Equation 9.4). Figure 9.12 shows the dielectric constant as a function of temperature at 1 KHz for the pristine polyimide, as well as for volume contents 0.02%, 0.035% and 0.10% (below percolation). The dielectric constant does not vary much with temperature below T_g. As the temperature approaches T_g however, the dielectric constant increases, since dipoles become more mobile. Increasing the amounts of CNT not only raises the room temperature dielectric constant but also increases the value of $\Delta\varepsilon$. As the content of CNT is increased to 0.1 vol%, $\Delta\varepsilon$ improves from 8 to 20. This result could indicate that the

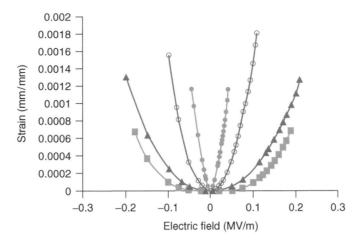

FIGURE 9.13 Strain response of CNT-polyimide composites.

TABLE 9.6 Polymer Deposition Techniques

Thin Film Deposition	Bulk Deposition
Spin casting	Casting
Vapor Deposition	Embossing
Microstamping	Micromolding
Sputtering	Offset Lithographic Printing
Beam-Induced CVD	Injection Molding
Evaporation	
Electrodeposition	
Plasma Deposition	
Self-Assembly	

presence of the CNT enhances orientation polarization by stabilizing the dipoles and preventing them from relaxing back when the field is turned off, resulting in a higher piezoelectric performance. Poling of the CNT-polyimide composites confirms this observation.

As shown in Table 9.5, three different materials were poled at 50 MV/m: the polyimide, the 0.02 vol% CNT+polyimide, and the 0.1 vol% CNT+polyimide. P_r was measured using the thermally stimulated current (TSC) method. It is observed that P_r increases with increasing content of CNT, even though the poling conditions were kept the same for all three cases. The CNTs lead to better dipolar alignment with the poling field, and possibly a higher percentage of dipoles are realigning in the direction of the field. The value of P_r for the 0.1 vol% CNT-polyimide composite is more than four times that of the pristine polyimide. The higher piezoelectric response should translate into a better sensor and actuator. Some preliminary results on actuation capability of the CNT-polyimide composites are shown in Figure 9.13 for 0.1 MV/m. The strain increases with increasing CNT volume content up to 1 vol%. The combination of mechanical, electrical, dielectric, and electroactive properties exhibited by the polyimide-carbon nanotube composites is interesting, and it promises a great deal of flexibility and tailoring between processing and properties.

9.3 Polymer Microfabrication Techniques

There are a wide variety of techniques for depositing polymer materials suitable for microfabrication. Here, we will discuss the predominant techniques that highlight the unique capabilities of polymer fabrication or show promise for manufacturing and production.

A listing of polymer deposition techniques is given in Table 9.6. These methods can be divided into two major categories, thin film techniques, such as spin casting, vapor deposition or evaporation, and bulk

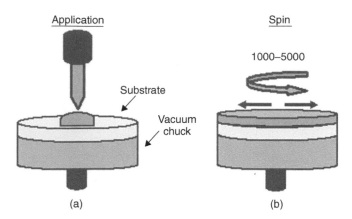

FIGURE 9.14   Spin casting of polymers: (a) polymer application and (b) dispersing through rotation.

deposition techniques, such as casting, embossing or molding procedures. Other divisions can be considered as well, such as whether or not a technique provides a finished, patterned structure, such as molding or casting, or whether or not it provides a uniform film that must be subsequently patterned. We will first discuss the thin film techniques, followed by bulk techniques and finally a discussion of lamination and bonding techniques for integration and assembly of components.

9.3.1   Thin Film Deposition

Spin casting has been used in microfabrication, as the predominant method of forming photolithographic thin films in microelectronics, for nearly a half-century. The basic technique is to apply the polymer, which is cast in a solvent solution onto the surface of the substrate, as in Figure 9.14(a). The substrate is then rotated at a rate, for optimal uniformity, that is typically between 1000 and 5000 rpm (Figure 9.14(b)). The final thickness of the resultant film is determined by:

$$t = \frac{kS^2}{\sqrt{RPM}} \tag{9.5}$$

where t is the resulting thickness, k is a proportionality constant, S is the percent solids in the solution and RPM is the spin speed. Typically, the percent solids is adjusted to give the desired thickness in the acceptable spinning range. Once the polymer is spun to the desired thickness, the casting solvent is typically baked off on a hotplate or conventional convection oven.

Although polymers generally lend themselves well to spin casting, there can be issues with adhesion to the substrate, reactions with the polymer to moisture or oxygen in the air during the spin casting process, and nonuniformity caused by surface topography. Adhesion can often be improved by driving off surface moisture through either heating the substrate above room temperature or subjecting the substrate to a vacuum or an oxygen plasma, such as in a barrel etcher or asher. Additionally, adhesion can be improved, particularly on silicon substrates, by silanating the surface with a solution such as hexamethyldisilazane (HMDS), which can be spun on and allowed to dry, or vapor-deposited.

Reaction with moisture or oxygen for some polymers, such as aromatic polyimides, can make spin casting less user-friendly. Better results can be obtained by spinning and curing the polymer films under a nitrogen ambient [Atkinson, 2003]. Once the films have been cured (even only partially), they are rendered structurally stable and can tolerate further processing in air.

Variations in surface topography, caused by the radially symmetric centrifugal force that spreads the polymer, are inherent to the spin casting process. An excellent comparison of polymer deposition methods, in this case photoresist, has been performed by Pham et al. (2004), where spin casting was compared to spray deposition and electrodeposition techniques. Pham et al. found that spray deposition yields more reproducible and controllable film thickness in most cases, though the deposition equipment and the

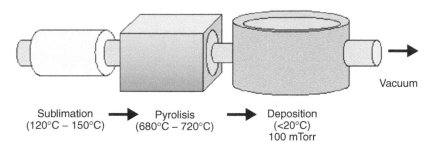

Vacuum

Sublimation → Pyrolisis → Deposition
(120°C – 150°C)   (680°C – 720°C)   (<20°C)
100 mTorr

FIGURE 9.15   (See color insert following page 9-22.) Vapor deposition of parylene.

polymer solutions themselves are more specialized, requiring optimization for the spray deposition process. Electrodeposition is found to be suitable for conductive surfaces such as metals, and it has the primary advantage of being able to coat vertical surfaces.

One of the most common fabrication steps in microfabrication is photolithography. Traditional materials such as silicon, common dielectrics and metals lend themselves quite readily to process sequences with multiple depositions of materials and multiple levels of lithography performed in between. This fact is straightforward simply because traditional microelectronic materials are not harmed by either the solvents used in applying common photoresists (e.g. ethylene glycol, monoethyl ether propylene glycol methyl ether acetate) or by the solvents/processes used to remove them (such as acetone, commercial resist strippers, or an oxygen plasma). Recently, direct photo etching of polyvinylidene fluoride was reported, in which X-rays were used for direct patterning of this versatile electroactive polymer [Manohara et al., 1999].

With the advent of polymer microsystems, however, it can be somewhat more challenging to develop a process sequence in which polymer layers that form a structural component of the completed device are not harmed by photolithography steps that follow. This is one advantage of polymers such as polyimide, which is available in a wide variety of forms and characteristics, but once cured, it is generally resistant to the solvents and processing required of photolithography. Parylene is also a useful polymer in which photolithography steps can be performed on top of it. Care must be taken to examine the tendency of commercial resist strippers to soften and delaminate existing structural polymers. Additionally, exposed underlying polymer geometries will be attacked during an oxygen plasma etch. Using acetone for removal can circumvent these issues, however it is important to keep the photoresist baking temperatures (both prebake and postbake) at 90°C or less, so that the photoresist itself can be completely removed by the acetone.

Polymers such as parylene (parylene-C) can be vapor-deposited at room temperature, and commercial systems are available for this purpose [Carlen, 2002]. A typical configuration is illustrated in Figure 9.15. Endpoint detection for parylene deposition, for more accurate film thickness control, has been investigated by Sutomo, et al. (2003).

In this process, the solid parylene dimer is sublimated in the first chamber at moderate temperatures (120–150°C). In the pyrolisis furnace, the dimer is subjected to high temperatures (680–720°C), splitting it into two monomers. The monomers then polymerize (conformally, because of the low pressure) on the cool substrate surface in the deposition chamber, held typically at 100 mTorr.

Carlen and Mastrangelo (2002, 2003) have extensively studied the deposition and characterization of parylene for microfluidic devices. In that work and others, the conformal deposition of parylene is used in the fabrication of a variety of microfluidic devices. In micromachining, parylene films are primarily used as diaphragms and sealing layers, however they have also been applied in integrated circuits as an encapsulation layer or as intermetal dielectrics.

One direct method of forming a polymer film on a substrate is by thermal evaporation. Here, a bulk polymer is vaporized by thermally heating the material in a tungsten crucible. A considerable amount of work in this area has been performed by Carlen and Mastrangelo (2002, 2003) for developing paraffin-based volumetric microactuators. Paraffin lends itself to low-temperature evaporation owing to its high vapor pressure. A custom system for paraffin deposition is shown in Figure 9.16.

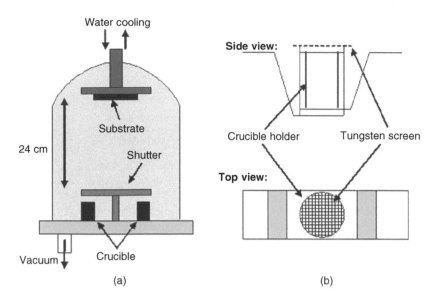

FIGURE 9.16   (a) Custom thermal evaporation system for depositing paraffin, (b) modified crucible holder eliminating polymer splashing (after Mastrangelo, 2003).

One of the key aspects of this design is that the modified crucible holder eliminates the tendency of the boiling paraffin to deposit large spheres on the thin film. This situation is prevented by the tungsten screen placed above the quartz crucible in the holder. Evaporations of Logitech paraffin were carried out at less than 5 μTorr and at a 150°C material deposition temperature, giving a deposition rate of approximately 100 nm/min and a surface roughness of 42 nm on a 4-μm-thick paraffin film.

9.3.2   Polymer Patterning Techniques

Additionally, polymers, particularly photoresists, have been found to be useful as the sacrificial layers in a number of micromachining technologies, where the advantages of photoresist as a sacrificial layer are that (1) it is readily applied at low temperatures at a wide range of thicknesses, from <1 μm to tens of microns thick; (2) it is readily patterned with standard photolithography; (3) it can be heated to moderate temperatures (typically 110°C), to round the corners; and (4) it can readily be removed in acetone, provided it has not seen temperatures in excess of 110°C. The primary disadvantage of using photoresist as a sacrificial layer is that it cannot tolerate the excessive temperatures of many deposition processes, and thus it limits the processing temperatures of the back end of the process once it is applied.

Photopatterning is also used widely in patterning SU-8 and photosensitive polyimides (HD Microsystems). These materials can act as sacrificial layers, or structural components, of a completed device. A novel approach to SU-8 fabrication is presented by Kim (2002), in which a totally dry release process is accomplished by modifying surface adhesion forces. A self-assembled monolayer (SAM) film is applied to an oxide surface (patterned or unpatterned), and metal is deposited on top of it. An SU-8 film is applied and cured, and then the SU-8 film can be peeled off. Wherever the SAM film is present on the surface, the metal film is released from the substrate and adheres to the SU-8.

One established technique for creating high-aspect-ratio microstructures in polymers is referred to as LIGA, as summarized by Kim (2000). For an extensive description of the LIGA technique, the reader is referred to Madou (1997). This technique can be used for rapid prototyping and mass production of micromolded polymer components, as shown by Kim et al. (2000). In this case, PDMS, a silicone elastomer, is patterned by casting it into electroplated micromolds fabricated through the LIGA technique. Although the original LIGA process relies on a synchrotron X-ray source to achieve high-resolution exposures in thick layers of PMMA, several others have pursued "poor man's LIGA," using the epoxy-based negative

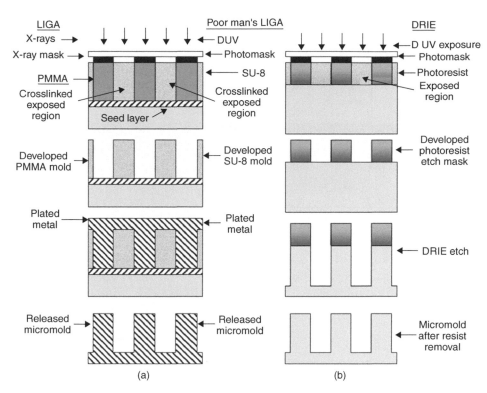

FIGURE 9.17 (See color insert following page 9-22.) Micromold fabrication techniques: (a) LIGA/"Poor Man's LIGA" and (b) DRIE micromold fabrication.

photoresist SU-8. The LIGA process provides higher aspect ratios (100:1) and thickness up to 1 mm. SU-8 based processes are typically of aspect ratios 20:1–30:1 and <100 μm in thickness. It is also possible to construct a suitable micromold in silicon by deep reactive ion etching techniques, based on an inductively coupled plasma etch with either a cryogenically cooled substrate or alternating fluorocarbon source gasses. DRIE techniques can achieve thicknesses or depths of hundreds of microns, with aspect ratios better than 30:1. Each of these approaches can provide a useful fabrication technique for the micromolding of polymer structures.

A comparison of each of these mold fabrication techniques is shown in Figure 9.17(a) and Figure 9.17(b). In Figure 9.17(a), the steps for both LIGA and Poor Man's LIGA are detailed, showing the translation from exposure to resist mold to plated metal (which is then released). The DRIE method, which also relies on standard lithography (Figure 9.17(b)), then translates the exposed resist pattern to a silicon substrate using DRIE. Note that in this case the tone of the exposure mask is the same as in the LIGA approaches, since the photoresist tone is typically positive and the mold formation is a removal process rather than an additive one.

Fabrication of PDMS microstructures from fabricated molds has been carried out, for example, by Kim (2000) and Sundararajan (2004). PDMS is cast directly or spun-cast onto the prefabricated micromolds, degassed, cured, and then peeled off. The degassing step is usually performed in a dessicator under vacuum to remove air bubbles inside the deep trenches. The curing is for one hour at 65°C. A sample of replicated micromolds in electroplated nickel and the resulting micromolded PDMS parts is shown in Figure 9.18(a) and Figure 9.18(b), respectively.

Micromolded PDMS components fabricated in this way are typically of one-piece construction, as opposed to discrete, unattached PDMS components. Because PDMS is not photodefinable, it takes additional steps to separate PDMS-fabricated parts such as by Ryu (2004). In this approach, illustrated in Figures 9.19(a–e), a photoresist mold is filled with a prepolymer PDMS solution, and a flat, smooth rubber blade is used to remove the excess solution. The PDMS remains in the recessed regions of the photoresist. The PDMS is then

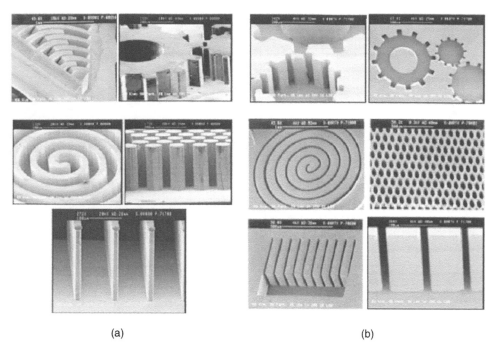

(a)   (b)

FIGURE 9.18   (a) Nickel electroplated micromolds fabricated using LIGA and (b) PDMS (silicone elastomer) replicated parts (after Kim et al. (2000)).

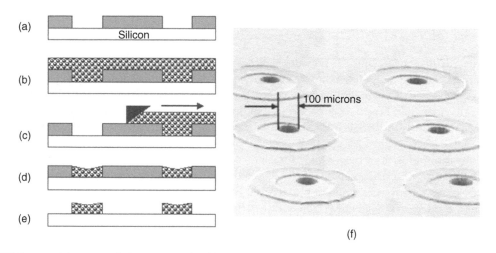

(f)

FIGURE 9.19   (a) Patterned photoresist mold (b) PDMS spin casting, (c) blade-removal of excess PDMS solution, (d) cure and polish, (e) photoresist removal in acetone and (f) SEM of completed PDMS O-rings around 100 µm through-holes in a silicon wafer (after Ryu et al. (2004)).

cured at 90°C and the photoresist is removed in acetone. Practical considerations involve accomplishing the uniform blade-removal of the PDMS solution and cleaning up residual PDMS on the surface of the resist with a light mechanical polish after curing. This process was used to fabricate the PDMS O-ring structures of Figure 9.19(f), where the elastomer O-rings are aligned with 100-µm-diameter, thru-wafer holes in a silicon substrate. This type of process has useful application in interconnect strategies for microfluidic applications, where the elasticity of the PDMS can be used to seal a microfluidic interconnect.

Various "sandwiching" techniques have been used to fabricate micromolded parts in PDMS [Anderson et al., 2000; Jo et al., 2000; Sundararajan et al., 2004]. Jo et al. used a "sandwiching" technique to fabricate

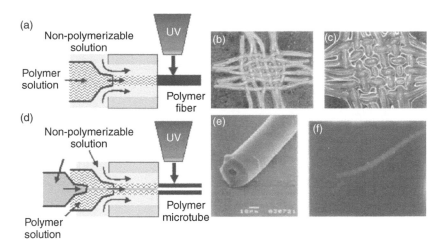

FIGURE 9.20 (See color insert following page 9-22.) (a) Photopolymerizing fiber extraction apparatus, (b) woven pH-responsive fibers, (c) pH-responsive fibers in a swollen state, (d) photopolymerizing microtube apparatus (e) SEM of a fabricated microtube, and (f) fluorescent image of glucose sensing microtube (after Jeong et al. (2004)).

thin PDMS-molded components from an SU-8 mold, which can be assembled into multilayer stacks to form complex, microfluidic components. A thicker molded PDMS layer was used as the bottom layer, thick enough to "core" with a syringe for attaching tubes for fluidic I/O. An interesting microfluidic drug-delivery system has also been fabricated using PDMS micromolding by Su (2004). SU-8 molds are used to form a reservoir and delivery channel structure over a novel microfluidic osmotic actuator that provides continual drug delivery powered by water. One key element of constructing these novel actuators was the enhanced adhesion that was found for PDMS by exposing it to an oxygen plasma. Even so, adhesion-promoting polymer layers were required to get proper adhesion of all the layers.

Several methods have been developed to pattern polymers using exposure to light, electron, or ion beams. In addition to standard photoresist polymers, photosensitive epoxy resins and polyimides are commercially available that can be patterned by exposure to DUV light and then developed in custom developing solutions. These resins allow a wide range of thicknesses (from 1 to ~100 μm) to be patterned with aspect ratios approaching 30:1. As discussed previously, these photo-patterned layers can be used as molds for both polymer applications such as PDMS, and for plating of metals. These layers can also be used directly as structural components of microsystems, such as membranes and channel walls.

Some other, rather novel, beam techniques have been implemented in which exposure can be used to pattern a polymer directly. Photopolymerization of novel polymer fibers and microtubes, as shown by Jeong et al. (2004) shows the versatility of this approach for fabricating novel three-dimensional structures in Figures 9.20(a–f). In this approach, a multiphase, laminar flow is set up using both polymerizable and nonpolymerizable solutions, as shown. As the solution is ejected, it is polymerized by a UV light source. In Figure 9.20(a), a dual-phase laminar flow results in a microstring being formed, and in Figure 9.20(b) and Figure 9.20(c), the ability to weave these fibers as well as form them from pH-responsive fibers (which induce swelling) are demonstrated. In Figures 9.20(d–f), a three-phase laminar flow with a nonpolymerizable center stream results in microtube formation. An SEM of a fabricated tube shows the hollow structure in Figure 9.20(e), and a fluorescent solution in the center of the microtube demonstrates the ability to sense glucose in Figure 9.20(f).

In a "two-photon polymerization" of a UV-light-curing optical adhesive (Norland NOA 63), Galajda and Ormos (2001) and Maruo et al. (2000) fabricated three-dimensional structures by moving the sample through an argon laser beam on a three-dimensional piezoelectric stage. The resolution of the technique is dependent on the spatial resolution of the optics focusing the argon beam, and resolution of 0.5 mm was achieved. Galajda and Ormos used this technique to create micromotors that were not only

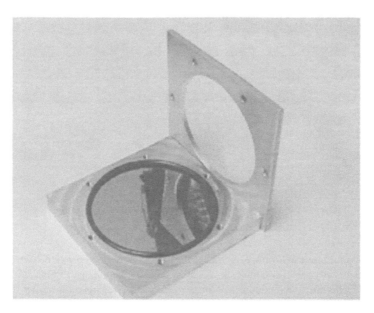

FIGURE 9.21   Aluminum mold plate and stamper for open mold casting (after Mastrangelo (2003)).

fabricated optically, but optically driven as well. Maruo et al. (1998, 2000) showed exotic, nanoscale spiral structures, tubes and funnels using both single-photon and two-photon polymerization technology.

9.3.3   Bulk Fabrication Techniques

Compared to traditional metal and semiconductor machining materials, polymers have the unique ability to be readily available in liquid form at room temperature. This fact has allowed a number of useful batch processing techniques to be developed, and extensive work in this area has been accomplished by a number of researchers, including Carlen and Mastrangelo (2002, 2003) and Sundararajan (2004). Carlen and Mastrangelo have developed and characterized a variety of techniques for the bulk fabrication of plastic microsystems, including casting, bonding, and lamination.

In order to replicate plastic parts in bulk fashion using microcasting techniques, it is first necessary to fabricate the stamp, or mold, from which the parts will be fabricated. The LIGA and DRIE approaches discussed earlier [Kim, 2000] are two methods suitable for this purpose. Mastrangelo (2003) accomplished precisely this fabrication by using a silicon wafer patterned using DRIE, in order to form a stamper for the open mold casting method depicted in Figure 9.21. The silicon wafer is placed in a recessed aluminum frame and coated with parylene-C to aid the later release of the molded plastic part. A Teflon O-ring and an aluminum top plate are then clamped down on top of the silicon wafer, and epoxy resin is poured over the stamper. The resin is cured at 80°C for three hours and is then peeled from the stamper. The aluminum mold frames used to contain the silicon stamper wafer are shown in Figure 9.22. With the open-mold casting process, it is not guaranteed that the back surface of the molded part is flat. If necessary, a closed upper plate can be used to form the part, simply necessitating an access hole for injection of the epoxy resin.

The most straightforward approach to sealing microfluidic channels and reservoirs from cast polymer components is to laminate or bond them to an unpatterned coverfilm or substrate. For a flexible coverfilm, such as Mylar, the sealing process can be performed in a desktop laminator. For sealing to a rigid coverplate, a thin adhesive layer and bonding is preferred. In this manner, flexible, patterned polymer components can readily be sealed for microfluidic applications.

Laminating polymer components can be accomplished using a roll-type laminating process, such as in a commercially available desktop unit from Kepro, as shown by Mastrangelo (2003). The molded plastic substrate is first treated with acetone and then laminated to a transparent, 50-μm-thick Mylar cover film

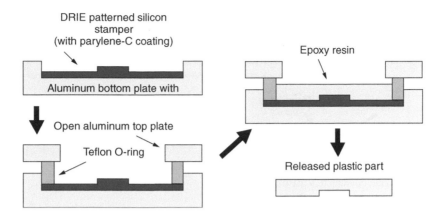

FIGURE 9.22 Open casting process (after Mastrangelo (2003)).

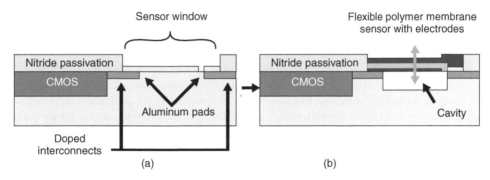

FIGURE 9.23 (**See color insert following page 9-22.**) (a) CMOS chip with sensor window in passivation and (b) CMOS electronics with polymer sensor fabricated in the sensor window.

by pressing the two components together through two rollers at 80°C. Excellent bonding can be obtained with low permeability rates and no apparent seeping at the bond interface. Rigid epoxy coverplates can be formed in the same manner as the molded epoxy components that contain channels and reservoirs also. The two epoxy substrates can then be bonded in a similar fashion to laminating, with the additional requirement of a thin adhesive layer. Mastrangelo accomplished this process using an adhesive solution of 1:4 Blanchard wax in acetone. The thinned wax solution is spun onto one of the substrates, and then the two substrates are aligned under a microscope with a z-motion stage that applies bonding pressure for two minutes. The bonding is completed by passing the substrates through the laminator system, between two hot rollers at 80°C.

9.3.4 Assembly and Integration Techniques

One of the powerful possibilities for polymer microsystems is the ease with which polymer sensors and actuators can be integrated with CMOS electronics. This ease of integration is primarily due to the fact that polymer fabrication techniques are predominantly low-temperature. The ability of completely fabricated CMOS circuits to withstand temperatures up to 400°C allows CMOS circuits to withstand nearly any polymer-based postprocessing that one might conceive. The processing power of CMOS chip technology, combined with the multifunctionality of polymers, promises to be a powerful combination for the fabrication of smart, integrated microsystems at economically viable cost. Generally, polymer-based fabrication technologies are already suitable for CMOS integration because of their low temperature deposition and fabrication techniques (<400°C). Polymers can be integrated for sensors [Atkinson et al.,

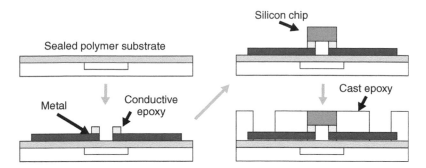

FIGURE 9.24   Flip-chip bonding to a plastic substrate.

2003], and fluidic components [Carlen and Mastrangelo, 2002, 2003], and can also provide hinges for flexible, integrated microsystems [Smela, 1999; Miyajama, 2001; and Park, 1999]. As the reverse situation, i.e. polymer microsystems subjected to typical integrated circuit processing temperatures, is not at all tolerable, there remain two approaches to integrating polymer devices and systems directly with electronics. These two approaches are post-CMOS processing and die bonding of completed integrated circuits.

In a post-CMOS processing scenario, polymer sensors and actuators are fabricated on a silicon wafer after the CMOS electronics fabrication is complete. An example of how integration is accomplished by leaving sensor regions open in the final passivation of a simple CMOS process is shown in Figure 9.23(a). In this example, doped interconnects lead to aluminum pads that have been exposed in the silicon nitride passivation. This process forms a sensor window, in which a number of sensor configurations might be fabricated using low-temperature polymer fabrication techniques, such as spin casting, evaporation, or electrodeposition.

One such structure, an integrated pressure sensor configuration, is illustrated in Figure 9.23(b), where an electroactive membrane is sandwiched in between two metallizations with a sealed cavity underneath. The sensor electrode layers can be formed by suitable low-temperature metallization techniques such as evaporation and electroplating. Processes to be integrated onto CMOS electronics have been investigated by Atkinson et al. (2003), using electroactive polyimides as the active sensor component. This work includes both suspended structures for vibration and pressure sensing as well as acoustic wave devices, for a variety of applications.

In many systems, where it would be desirable to integrate circuitry or detectors with a polymer based substrate, the area of the integrated circuit may be quite small relative to the plastic components. This situation might typically be true for a microfluidic system, for example. In this case, it is more cost-effective to fabricate the silicon chips separately, using a standard integrated circuit process, dice the components into individual chips, and then bond them onto the plastic substrates using flip-chip bonding techniques. This technique is well known for efficiently integrating silicon circuits onto printed circuit boards in the IC industry. Mastrangelo (2003) has adapted this technique for polymeric substrates, as illustrated in Figure 9.24. Beginning with a sealed plastic substrate, a metal electrode pattern is fabricated on the surface. This metal pattern will provide contact pads to the inverted silicon chip, but also can extend outward from underneath the chip to provide accessible contact pads. Using alignment marks on the backside of the silicon chip and the front side of the plastic substrate, a pick-and-place system is used to align and place the inverted silicon chip onto the conductive epoxy bonding sites. Finally, the entire substrate can be cast again to create an epoxy-sealing layer. In order to gain access to the metal electrodes, one can apply a suitable mold for the casting of the upper epoxy layer, with openings for contact pads.

The previous approaches have relied on fabricating control or detection circuitry in a traditional manner, such as in a standard CMOS integrated circuit process. One can consider an alternative approach, in the vein of "bringing the mountain to Mohamed," where a new type of electronics is developed that is more easily integrated with plastic components. One such approach has been demonstrated by Maltezos, using a novel transistor structure with mercury source and drain electrodes. These devices use not only

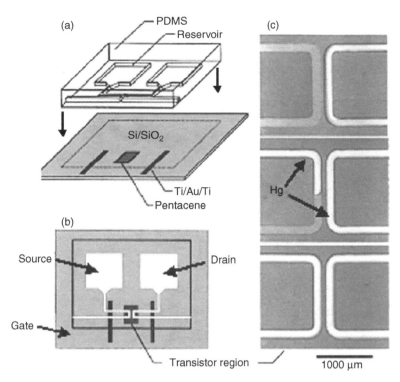

FIGURE 9.25 (a) Microfluidic source/drain substrate being integrated onto oxidized, heavily doped silicon substrate with source/drain contacts and pentacene channel material, (b) top view of integrated structure, and (c) optical micrographs of transistor region showing the ability to modulate the effective channel width using the mercury flow (after Maltezos et al. (2003)).

microfluidic technology as an integral part of the electronics, but also the microfluidic fabrication of the source and drain can be used to fabricate additional microfluidic components and systems. The basic fabrication approach is illustrated in Figure 9.25. The PDMS component of Figure 9.25(a) is formed initially by a casting the PDMS in a master mold and peeling it off. The master mold is a negative-tone relief image of the desired pattern in a 15-µm-thick layer of the photosensitive epoxy resin SU-8. The PDMS solution is cast in the SU-8 mold, cured, and then removed. A separate, highly doped silicon substrate, also shown in Figure 9.25(a), is thermally oxidized and then patterned with two shadow mask steps. A shadow mask step is used to pattern an electron beam evaporation of Ti/Au/Ti contacts that will connect to the transistor source and drain, and a separate shadow mask is used to pattern a thermal evaporation of pentacene, a p-type polymer semiconductor that will form the transistor region of the device. Droplets of mercury are placed in the reservoirs and will adhere, even when the PDMS substrate is inverted for attachment to the silicon substrate. A brief exposure to an oxygen plasma allows a good bond between the PDMS and silicon dioxide. The highly doped silicon substrate will act as the gate electrode, and the thermal oxide forms the gate dielectric. Figure 9.25(b) shows the overlapping design of the device after it is assembled. In this current work, the mercury in the reservoirs is pumped by applying a slight mechanical pressure to the PDMS reservoirs, and it then flows down the narrow channels, contacting the Ti/Au/Ti electrodes and coming in close proximity in the transistor region of the device. A unique feature of these devices is shown in Figure 9.25(c), where optical micrographs of the mercury at different levels of pressure show that the position of the mercury can be controlled, effectively allowing the gate width of the device to be tuned. Low contact resistances between the mercury and the pentacene device layers are obtained (<0.02 MΩ-cm), and threshold voltages are approximately 18 Volts. Transistor characteristics for a device where the mercury extends completely across the transistor region are shown in Figure 9.26, where the linear behavior in the inset indicates ohmic contacts between the mercury and the pentacene.

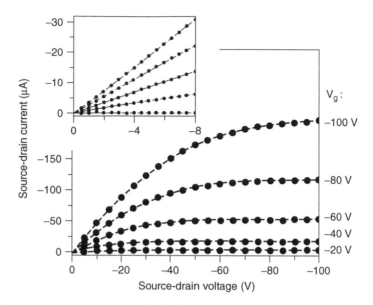

FIGURE 9.26 Transistor characteristic for a microfluidic transitor as shown in Figure 9.10, The mercury extends completely through the transistor region in this device and the gate length is 110 μm and the gate width 2.5 cm. The inset shows the device characteristics in the linear region (after Maltezos et al. (2003)).

9.4 Device Examples

The mass production of small, low-profile, smart-polymer microsystems enables polymer-based chemical, biological, or environmental monitoring systems that can be readily incorporated into a variety of products. Low-cost polymer mass production is extremely attractive for biomedical applications, such as quick turnaround laboratory analysis or remote medicine, often termed lab-on-a-chip or micro-total-analysis systems (μTAS). For a review, see Rossier (2002). Similarly, these uses may be extended to environmental monitoring for consumer, military, and industrial applications. Demand for this technology is driven by an ever-increasing need to monitor potential health hazards, toxic threats in warfare, and homeland security. Applications range from industrial and office environments, shipping and transportation systems, military environments (including wearable systems), and home consumer products. Clearly, the technology driver for success in these applications is a low-cost, flexible, and robust technology that polymer systems have considerable potential to meet. In many applications, polymers are readily employed as sacrificial layers in a microsystem process [Courcimault et al., 2004]. The advantages of polymers as sacrificial layers are: (1) a polymer is readily applied at low temperatures at a wide range of thicknesses, from <1 μm to tens of microns thick; (2) a polymer is readily patterned with standard photolithography; (3) a polymer can be heated to moderate temperatures (typically 110°C) to round the corners; and (4) a polymer can readily be removed in acetone (provided it has not seen temperatures in excess of 110°C). Polymer layers have also found application as structural layers that provide flexibility to microsystems, forming hinges and flexible interconnects for integrated systems and packaging [Smela, 1999; Park, 1999; Miyajima, 1999; Ucok, 2004]. There are applications and device examples too numerous to be covered here, illustrating the extreme versatility of this technology for microfabrication. The discussion here is limited to some notable examples in which the structural layers in the completed device are formed from polymers.

9.4.1 Tactile Sensor and Shear Sensors in Parylene Membranes

Fan (2004) demonstrated an example of a parylene membrane process applied to tactile sensors and shear sensors, as shown in Figure 9.27. In Figure 9.27(a), the photoresist layer is spin cast and patterned.

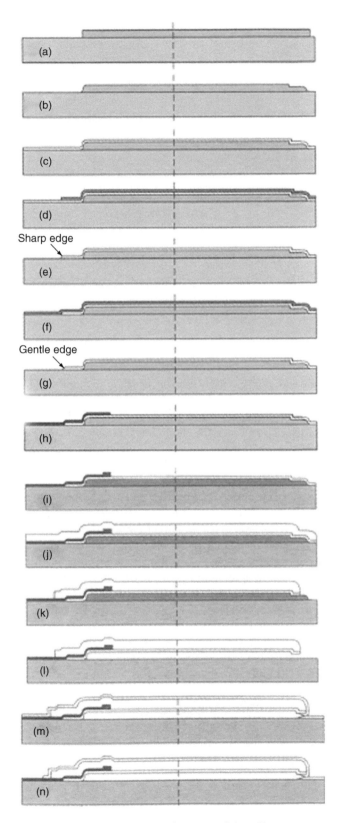

FIGURE 9.27 Process steps for a parylene membrane (after Fan et al. (2004)).

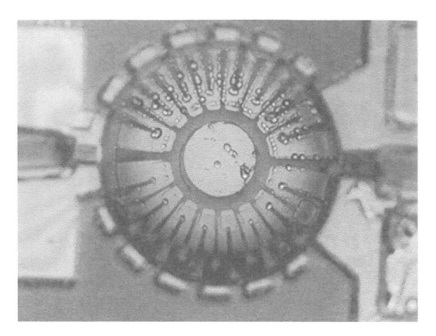

FIGURE 9.28   Tactile sensor with serpentine-gold strain-gauge resistors on a parylene membrane (after Fan (2004)).

In Figure 9.27(b), the photoresist is selectively thinned, which is accomplished by a lower dose exposure with a subsequent mask. A 1-μm-thick parylene layer is deposited (Figure 9.27(c)) and then coated with aluminum, which is patterned photolithographically (Figure 9.27(d)). The exposed parylene layer is then removed with an O_2 plasma etch and then the aluminum is removed, as shown in Figure 9.27(e). This step unfortunately leaves a sharp edge in the parylene, but this edge is removed quite cleverly by spinning an additional photoresist layer (Figure 9.27(f)) and then plasma etching is performed once again to remove the photoresist. Since the photoresist is thinner on top of the parylene and the etch selectivity is 1:1, the gentle photoresist slope is transferred to the parylene edge, as shown in Figure 9.27(g). A 200-nm gold film is deposited and patterned (Figure 9.27(h)). It is important to note that heating during this Au deposition can cause the sacrificial layer to release solvents, which when trapped under the parylene result in an increase in the height of the membrane in a dome shape. This added height can be left in place for the final structure of the device, or prevented by additional baking of the sacrificial layer or cooling during the gold deposition. In order to optimize the performance of the serpentine metal strain gauge resistors on the parylene membrane, metal liftoff is used to locally increase the thickness of the metal film (Figure 9.27(i)). The entire device is coated with a thick (8-μm) parylene film (Figure 9.27(j)) and then patterned with an Al etch mask and O_2 plasma etching as in Figure 9.27(k), in which the etch hole regions of the sacrificial layer, as well as outlying contact pads, are now exposed. The sacrificial layer is removed in acetone, which takes considerable time (several hours), and the details of this release process are not described (Figure 9.27(l)). The wafers are then dried under an infrared heat lamp. Optionally, the conformal nature of the parylene vapor deposition can be used to seal the cavity (Figure 9.27(m)), which requires that an additional metal masked oxygen plasma etch process be performed to expose the contact pads, as in Figure 9.27(n). A completely fabricated structure is shown in Figure 9.28, and these devices were successfully demonstrated as tactile sensors and similar geometries as shear stress sensors in microfluidic flow channels.

9.4.2 Microfluidic Pumps and Valves Using Paraffin Actuators

One of the promising areas of application for polymer microsystems is microfluidic systems. Here it is desirable to be able to fabricate fluidic channels, pumps, valves, and mixers. Microfluidic systems fabricated

TABLE 9.7   Actuation Pressure, P_a, for Several Low Voltage
Microactuators (after Carlen and Mastrangelo (2002))

Actuation Type	P_a (J/m^3)
Piezoelectric (PZT)	10^5
Electrostatic (comb-drive)	10^3
Electromagnetic Thermal Expansion (Ni/Si)	10^5
Thermo-Pneumatic	10^5
Solid-Liquid Phase Change (acetimide)	10^6
Shape Memory Alloy (Ni-Ti)	10^7
Solid-Liquid Phase Change (paraffin)	10^7

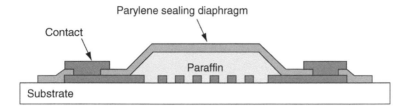

FIGURE 9.29   Paraffin actuator (after Carlen and Mastrangelo (2002)).

from polymers are expected to be applied to a variety of systems, including lab-on-a-chip systems for chemical and biological analysis, in vivo drug delivery and monitoring, and disposable sensing and screening devices. One key design issue with these polymer microsystems is achieving the actuation levels required to effectively pump and control fluid flow.

The actuation energy, E_a, for a particular electromechanical structure is defined by Zhang and Jang (1995) as the actuator force, F_a, multiplied by the maximum displacement, ε_a:

$$E_a = F_a \varepsilon_a \tag{9.6}$$

One can argue that for any particular actuation technology, the smaller the volume required to achieve a given actuation energy, the more powerful the technology. As suggested by Mastrangelo, one can estimate this actuation power by dividing the actuation energy by the volume of the actuator:

$$P_a = \frac{E_a}{V_a} = \frac{F_a \varepsilon_a}{V_a} \tag{9.7}$$

Although this might be more accurately defined as an "actuation pressure," since P_a has units of pressure, it does have the effect of normalizing the differences between actuator technologies. A comparison between the actuation pressure for well-known actuation technologies is shown in Table 9.7.

It is apparent from Table 9.7 that the solid–liquid phase change actuators have the ability to provide significantly more pressure on a microscale, and particularly, paraffin as an actuation material is quite promising. It is important to note that although shape memory alloys are also capable of providing large forces, they have limitations in terms of both the overall strain they are able to achieve, roughly 6%–8%, and the frequency of operation. In that regard, shape memory polymers have the largest strain levels, as well as the lowest frequency response [Gall, 2004; Mastrangelo, 2003]. Another interesting approach is reported by Smela et al. (1999), in which the volume change of the conjugate polymer polypyrrole, upon doping or applied voltage, is used to create an actuation pressure of 7×10^4 J/cm^3, with a polypyrrole hinge lifting a silicon plate.

A simple microactuator using paraffin was described by Carlen and Mastrangelo (2002) and is illustrated in Figure 9.29. In this piston-type actuator, the 5–20-µm-thick paraffin layer is heated through its melting point by passing current through a resistive metal heater, typically fabricated from either

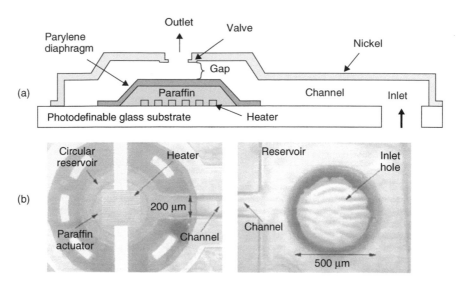

FIGURE 9.30   (**See color insert following page 9-22.**) (a) Cross-section of paraffin active blocking valve actuator and (b) optical micrograph of a nickel-plated microvalve from the backside (after Carlen and Mastrangelo (2002)).

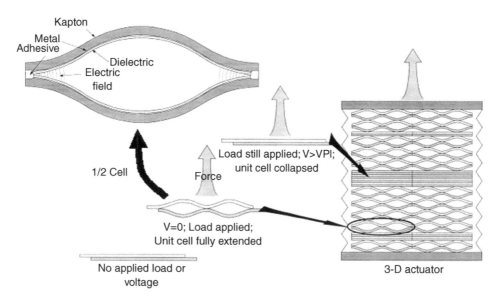

FIGURE 9.31   Biomimetic MEMS actuator (after Horning and Johnson (2002)).

aluminum or gold. The expansion of the paraffin vertically displaces the flexible parylene diaphragm. Circular microactuators were fabricated with a 200-μm radius and a 9-mm-thick paraffin film. The resulting vertical displacement was 2.6 mm for a 5-volt, 100-mW heater input. This actuator was fabricated into an active blocking microvalve by forming a channel and valve seat above the piston using photoresist and electroplated nickel, as shown in Figure 9.30(a) and Figure 9.30(b). Flow rates of $10^{-3}-10^{-1}$ sccm were obtained, with an 800-Torr pressure differential (from inlet to outlet) for paraffin actuators from 200 to 800 μm in diameter, and power consumption levels in the range of 50–150 mW. Leak rates as low as 5×10^{-4} sccm were obtained for the smallest devices. Additional structures such as a mass flow control valve, an inline microvalve, microfluidic pumps, and a polymerase chain reactor device were also fabricated, showing that paraffin microactuators can be quite readily fabricated into a variety of structures.

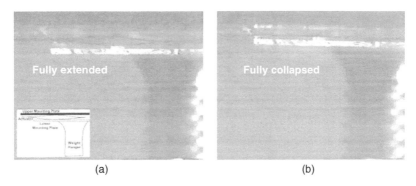

FIGURE 9.32   Biomimetic MEMS actuator cell (a) open and (b) electrostatically collapsed (after Horning and Johnson (2002)).

9.4.3   Biomimetic Muscle in Polyimide Using Electrostatic Actuation

Exciting work is being performed by Horning and Johnson (2002) in the development of biomimetic MEMS actuators for robotic applications. Based on the electrostatic actuators of Minami et al. (1993), Horning and Johnson have adapted this microactuator design to a process suitable for biomimetic muscles in robotic applications. Biological muscles characteristically have desirable properties such as a large force-to-weight ratio and high strain. They are generally robust and reliable and also have built-in fault tolerance. These MEMS actuators are designed to have these characteristics, as well. Additionally, it is desirable for a synthetic MEMS muscle to have low power consumption when the actuator is holding a position but not actually moving, owing to the relatively low power generation and storage technologies currently available.

Horning and Johnson (2002) have developed an electrostatic MEMS actuator in polyimide, as shown in Figure 9.31. From the half-cell inset, one can see that thin metal and dielectric (Kapton, a commercially available polyimide) layers form electrodes that are laminated together at their endpoints (though the electrodes are not in contact because of a thin dielectric layer not shown). When external forces are applied to the cell, it expands like a spring. When a voltage is applied between the upper and lower electrodes of the cell, the flexible electrodes are pulled together. Key to the low voltage operation of the cell are the teardrop-shaped ends of the cell, where the electrode spacing is the smallest, and very low voltages will create a high field and collapse the electrodes at the ends. This procedure has the effect of bringing the adjacent region closer together so that they will collapse as well, forming a "zipping" action that collapses the entire cell.

To form biomimetic structures, additional cells can be fabricated in parallel, to increase the force as well as the number of additional layers of cells being stacked on top of each other, which increases the total displacement. Quite elegantly, 3-D arrays of cells can be fabricated with large displacements and large forces. It is also important to note that like biological muscle tissue, these polymer-MEMS actuators only work in tension. Also like biological muscles, antagonistic pairs of muscles must be used for bidirectional motion. Also, antagonistic actuators can be used to tune the stiffness of a joint, as when an object is being held, and in this case, the only power consumption is the leakage current through the dielectrics, which has been measured in the picoamp to 10 nanoamp range. Finally, two other aspects of this actuator provide further beneficial biomimetic action. One is that the cellular nature of this polyimide MEMS actuator implies that the failure of an individual cell will not adversely affect the overall operation of a large-scale 3-D actuator and will provide a substantial amount of fault tolerance. Also, the springlike structure of this actuator also mimics the springlike nature of muscles and tendons, further aiding locomotion in robotic applications.

These actuators have been fabricated in a process adapted from printed circuit manufacturing. A sheet of Kapton is metalized with 10 nm of aluminum, which is then coated with a thin dielectric (to ultimately prevent the electrodes from shorting). Next, the electrode areas, bonding areas, and actuator boundaries are defined by photolithography, oxygen plasma, and etching the aluminum. Two sheets are placed face-to-face and then "spot-bonded" together by cutting a laser slit in each sheet and applying adhesive. Figure 9.32(a)

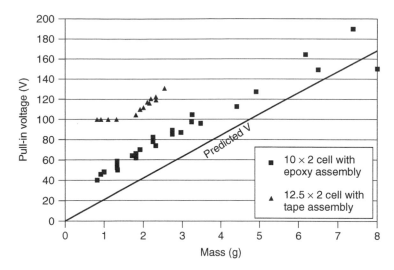

FIGURE 9.33 Experimental vs. theoretical pull-in voltage vs. load for the biomimetic MEMS actuator cell for two assembly techniques (after Horning and Johnson (2002)).

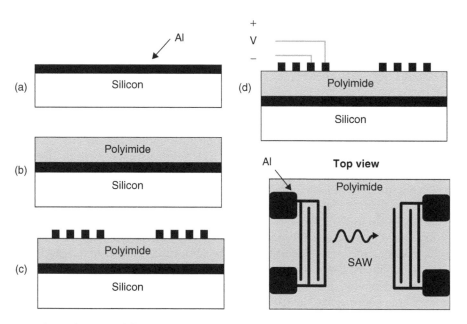

FIGURE 9.34 (**See color insert following page 9-22.**) SAW fabrication on electroactive polyimide (after Atkinson (2003)).

shows a cross-sectional photograph of a photograph of a fabricated actuator that has been pulled open, and Figure 9.32(b) shows the actuator after a voltage has been applied and the actuator has collapsed. Figure 9.33 shows a measurement of the pull-in voltage as a function of the external load, where an improved fabrication technique improved considerably the match with the theoretically predicted results.

9.4.4 Surface Acoustic Wave Devices in Polymers

An interesting use of electroactive polymers in sensor applications is to apply them to surface acoustic wave (SAW) devices. For an excellent review, the reader is referred to Varadan and Varadan (2000). One particularly useful polymer for this application is an electroactive polyimide where a cyano group dipole has been added to the β-site of an APB/ODPA polyimide [Ounaies, 2002]. This polymer exhibits electroactivity at temperatures higher than conventionally available polymers such as PVDF and can be spin cast and patterned with conventional microfabrication techniques [Atkinson et al., 2003]. Details on the structure and sensor and actuator behavior of this polyimide were presented in section 9.2.2. The ease with which polymers can be formed on an existing substrate, combined with the simple structure of SAW devices, makes it quite straightforward to fabricate SAW devices on substrates with prefabricated CMOS integrated circuits. With the variety of sensors that are possible with SAW devices, including chemical, biological, physical, and inertial configurations, it is possible to consider integrating a variety of sensors all in one straightforward postprocessing fabrication sequence. A simple process for fabricating polymer SAW devices on a CMOS substrate is illustrated in Figure 9.34 [Atkinson et al., 2003]. In this structure, the interdigitated transducers (IDTs) are used to pole the electroactive polyimide in situ. SAW devices have been fabricated with 5-mm line/space IDTs ($\lambda = 4$-mm) in a 2.25-mm-thick β-CN electroactive polyimide film. These devices show a 61 MHz transmission peak, with relatively high attenuation. The poling was accomplished with 100 V (20 MV/cm) applied to the IDT electrodes, at the glass transition temperature (215°C).

9.5 Future Directions and Challenges

The application potential for electroactive polymers and polymer nanocomposites in MEMS is immense. Typical applications that are being explored include MEMS devices in medical instrumentation, sensors, robotics, optics, computers, and ultrasonic and electroacoustic transducers. One important emerging application area for polymer MEMS is in the biomedical field, where polymers are being explored as potential artificial muscle actuators, as invasive medical robots for diagnostics and microsurgery, as actuator implants to stimulate tissue and bone growth, and as sensors to monitor vascular grafts and to prevent blockages. Polymer-based biomedical sensors, lab-on-a-chip systems, and μTAS systems will continue to proliferate. In some cases, such as in vivo, this proliferation may be owing to the unique properties of the polymer films themselves. In other cases, low-cost production methods for the fabrication of polymer microsystems allow these products to enter markets where costs must be necessarily low. This low cost requirement is especially true in applications such as disposable laboratory and medical devices or transportation sensors. In other cases, where a flexible substrate is required, such as in smart apparel and flexible catheters, the unique potential of polymers can be realized. It is evident that polymer microsystems will continue to expand and create viable markets. The versatility and flexibility of polymers can be compared to that of optical lithography in microfabrication. Optical lithography, with the myriad of optical techniques that can be applied, has always been able to dominate the microfabrication lithography market. Although historically, polymers were not the "original" material for microsystems, the enormous range of material properties and fabrication techniques may allow polymer-based systems to eventually surpass the more traditional materials in the not-too-distant future.

Still, polymer and polymer nanocomposite fabrication techniques applicable to the micro domain remain comparatively nascent, relative to traditional microfabrication methods. Techniques for high-resolution patterning, and overcoming the tendencies of polymers to swell or delaminate, are still important for many polymer systems. Improving the performance of these materials, in terms of their electroactivity, strength, and temperature tolerance, will continue to expand their applicability.

References

Anderson, J.R., Chiu, D.T., Jackman, R.J., Cherniavskaya, O., McDonald, J.C., Wu, H. Whitesides, S.H., and Whitesides, G.M. (2000) "Fabrication of Topologically Complex Three Dimensional Microfluidic Systems in PDMS by Rapid Prototyping," *Anal. Chem.*, **72**, pp. 3158–64.

Ataka, M., Omodaka, A., Takeshima, N., and Fujita, H. (1993) "Fabrication and Operation of Polyimide Bimorph Actuators for a Ciliary Motion System," *J. Microelectromech. Syst.*, **2**, pp. 146–50.

Atkinson, G.M., Pearson, R.E., Ounaies, Z., Harrison, J.S., Park, C., Dogan, S., and Midkiff, J.A. (2003) "Novel Piezoelectric Polyimide MEMS," in *Proc. 12th Annual Conference on Solid-State Sensors, Actuators and Microsystems.*, June 2003 Boston, MA, pp. 782–785.

Baughman, R.H. (1996) "Conducting polymer artificial muscles," *Synth. Met.*, **78**, pp. 339–53.

Baughman, R.H., Chanxing, C., Zakhidov, A.A., Iqbal, Z., Barisci, J.N., Spinks, G.M., Wallace, G.G., Mazzoldi, A., De Rossi, D., Rinzler, A.G., Jaschinski, O., Roth, S., and Kertesz, M. (1999) "Carbon Nanotube Actuators," *Science*, **284**, pp. 1340–44.

Bethune, D.S., et al. (1993) "Cobalt Catalyzed Growth of Carbon Nanotubes with Single Atomic Layer Walls," *Nature*, **363**, pp. 605–7.

Brei, D., and Moskalik, A.J. (1997) "Deflection Performance of a Distributed Polymeric Piezoelectric Micromotor," *J. Microelectromech. Syst.*, **6**, pp. 62–9.

Broadhurst, M.G., Malmberg, C.G., Mopsik, F.I., and Harris, W.P. (1973) "Piezo-and Pyroelectricity in Polymer Electrets," *Electrets: Charge Storage and Transport in Dielectrics*, M.M. Perlman, ed., The Electrochemical Society, Inc., Princeton, NJ, pp. 492–504.

Broadhurst, M.G., Harris, W.P., Mopsik, F.I., and Malmberg, C.G. (1973) "Piezoelectricity, Pyroelectricity and Electrostriction in Polymers," in *Polymer Preprints, Amer. Chem. Soc. Div. Poly. Chem.*, **14**, p. 820.

Carlen, E.T., and Mastrangelo, C. (2002) "Electrothermally Activated Paraffin Microactuators," *J. Microelectromech. Syst.*, **11**, pp. 165–73.

Carlen, E.T., and Mastrangelo, C. (2002) "Surface Micromachined Paraffin-Actuated Microvalve," *J. Microelectromech. Syst.*, **11**, pp. 408–20.

Chiarelli, P., De Rossi, D., Della Santa, A., Mazzoldi, A. (1994) *Polym. Gels Networks*, vol. **2**, p. 289.

Courcimault, C.G., Allen, M.G., Jayachandran, J.P., Kohl, P.A., and Bidstrup-Allen, S.A. (2004) "A Sacrificial-Polymer-Based Trench Refill Process for Post-DRIE Surface Micromachining," in *Technical Digest of the Solid-State Sensor, Actuator and Microsystems Workshop*, Hilton Head Island, SC, pp. 200–03.

Czaplewski, D.A., Kameoka, J., Mathers, R., Coates, W., and Craighead, H.G. (2003) "Nanofluidic Channels with Elliptical Cross Sections Formed Using a Nonlithographic Process," *Appl. Phys. Lett.*, **83**, pp. 4836–38.

Dalton, A.B., Collins, S., Razal, J., Munoz, E., Ebron, V.H., Kim, B.G., Coleman, J.N., Ferraris, J.P., and Baughman, R.H. (2004) "Continuous Carbon Nanotube Composite Fibers: Properties, Potential Applications, and Problems," *J. Mat. Chem.*, **14**, pp. 1–3.

Dang, Z.M., Fan, L.Z., Shen, Y., and Nan, C.W. (2003) "Dielectric Behavior Of Novel Three-Phase MWNTs/BaTiO$_3$/PVDF Composites," *Mater. Sci. Eng.*, **B103**, pp. 140–44.

Davis, G.T., McKinney, J.E., Broadhurst, M.G., and Roth, S.C. (1978) "Electric-Field-Induced Phase Changes in Poly(vinylidene fluoride)," *J. Appl. Phys.*, **49**, p. 4998.

Davis, G.T. (1993) "Piezoelectric and Pyroelectric Polymers," in *Polymers for Electronic and Photonic Applications*, C.P. Wong, ed., Academic Press, Inc., Boston, p. 435.

Dokmeci, M., and Najafi, K. (2001) "A High-Sensitivity Polyimide Capacitive Relative Humidity Sensor for Monitoring Anodically Bonded Hermetic Micropackages," *J. Microelectromech. Syst.*, **10**, pp. 197–203.

Fan, Z., Engel, J.M., Chen, J., and Liu, C. (2004) "Parylene Surface-Micromachined Membranes for Sensor Applications," *J. Microelectromech. Syst.*, **13**, pp. 484–90.

Frecker, M.I., and Aguilera, W.M. (2004) "Analytical Modeling of a Segmented Unimorph Actuator Using Electrostrictive P(VDF-TrFE) Copolymer," *Smart Mater. Struc.*, **13**, pp. 82–91.

Freundand, M.S., and Lewis, N.S. (1995) "A Chemically Diverse Conducting Polymer-Based 'Electronic Nose,'" in *Proc. National Academy of Science USA*, pp. 2652–56.

Fukada, E. (1995) "Piezoelectricity and Pyroelectricity of Biopolymers," in *Ferroelectric Polymers*, H.S. Nalwa, ed., *Chemistry, Physics and Application*, Marcel Dekker, Inc., New York.

Fukada, E. (2000) "History and Recent Progress in Piezoelectric Polymers," in *IEEE Transactions on Ultrasonics, Ferroelectrics and Frequency Control*, **47**, pp. 1277–90.

Furukawa, T. (1989) "Piezoelectricity in Polymers," in *IEEE Trans. Electr. Insul.*, **24**, pp. 375–93.

Galajda, P., and Ormos, P. (2001) "Complex Micromachines Produced and Driven by Light," *Appl. Phys. Lett.*, **78**, pp. 249–51.

Galipeau, D.W., Story, P.R., Vetelino, K.A., and Mileham, R.D. (1997) "Surface Acoustic Wave Microsensors and Applications," *Smart Mater. Struct.*, **6**, pp. 658–67.

Gall, K., Kreiner, P., Turner, D., and Hulse, M. (2004) "Shape Memory Polymers for Microelectromechanical Systems," *J. Microelectromech. Syst.*, **13**, pp. 472–83.

Gallantree, H.R. (1983) "Review of Transducer Applications of Polyvinylidene Fluoride," in *Piezoelectricity*, American Institute of Physics, New York.

Harrison, J., and Ounaies, Z. (2002) "Piezoelectric Polymers," in *Encyclopedia of Smart Materials*. John Wiley and Sons, New York.

Horning, R., and Johnson, B. (2002) "Polymer-based Microactuators for Biomimetics," *Neurotechnology for Biomimetic Robots*, Ayers, J., Davis, J., Rudolph, A., eds., MIT Press, Cambridge, MA.

Iijima, S., (1991), "Helical Microtubules of Graphitic Carbon," *Nature*, **354**, pp. 56–8.

Jager, E.W.H., Lundstrom, I., and Inganas, O. (2000) *Science*, **288**, pp. 2335–38.

Iijima, S., and Ichlhashi, T. (1993) "Single Shell Carbon Nanotubes of 1-nm Diameter," *Nature*, **363**, pp. 603–05.

Jeong, W., Mensing, G., Lee, S., and Beebe, D.J. (2004) "A Continuous Method for Manufacturing Polymer Strings and Tubes," in *Technical Digest of the Solid-State Sensor, Actuator and Microsystems Workshop*, Hilton Head Island, SC, pp. 388–89.

Jo, B-H. Van Leberghe, L.M., Motsegood, K.M., and Beebe, D.J. (2000) "Three-Dimensional Micro-Channel Fabrication in Polydimethylsiloxane (PDMS) Elastomer," *J. Microelectromech. Syst.*, **9**, pp. 76–81.

Kawai, H. (1969) "The Piezoelectricity of Poly(vinylidene fluoride)," *Jpn. J. Appl. Phys.*, **8**, p. 975.

Kepler, R.G., and Anderson, R.A. (1978) "Ferroelectricity in Polyvinylidene Fluoride," *J. Appl. Phys.*, **49**, pp. 1232–35.

Kim, G.M., Kim. B., Liebau, M., Huskins, J., Reinhoudt, D.N., and Brugger, J. (2002) "Surface Modification with Self-Assembled Monolayers for Nanoscale Replication of Photoplastic MEMS," *J. Microelectromech. Syst.*, **11**, pp. 175–81.

Kim, K., Park, S., Manohara, H., and Lee, J. (2000) "Polymethylsiloxane (PDMS) for High Aspect Ratio Three-Dimensional MEMS," in *Proc. of ISIM 2000*, Beijing, China.

Klein, D.J., Ounaies, Z., and Bryant R.G. (2002) "Synthesis and Characterization of Cyano-Containing Piezoelectric Polyimide Homopolymer and Copolymers," in *Proc. First World Congress on Biomimetics*, 9–11 December, Albuquerque, New Mexico.

Kornbluh, R., Pelrine, R., and Joseph, J. (1995) in *Proc. Third IASTED International Conference on Robotics and Manufacturing*, pp. 1–6, Cancun, Mexico.

Li, F. et al. (2000) "Tensile Strength of Single Walled Carbon Nanotubes Directly Measured from their Macroscopic Ropes," *Appl. Phys. Lett.*, **77**, pp. 3161–63.

Li, M-H., Wu, J.J., and Gianchandani, Y.B. (2001) "Surface Micromachined Polyimide Scanning Thermocouple Probes," *J. Microelectromech. Syst.*, **10**, pp. 3–9.

Lochun, D., Kilitziraki, M., Harrison, D., and Samuel, I. (2001) "Manufacturing Flexible Light-Emitting Polymer Displays with Conductive Lithographic Film Technology," *Smart Mater. Struct.*, **10**, pp. 650–56.

Lovinger, A.J. (1982) Developments in Crystalline Polymers, D.C. Basset, ed., Applied Science Publishers, London.

Lu, J.P. (1997) "Elastic Properties of Single and Multilayered Nanotube," *J. Phys. Chem. Solids*, **58**, pp. 1649–52.

Madden, J.D.W., Vandesteeg, N.A., Anquetil, P.A., Madden, P.G.A., Takshi, A., Pytel, R.Z., Lafontaine, S.R., Wieringa, P.A., and Hunter, I.W. (2004) "Artificial Muscle Technology: Physical Principles and Naval Prospects," in *IEEE J. Oceanic Eng.*, **29**, pp. 706–28.

MacDiarmid, A.G., and Heeger, A.J. (1980) "Organic Metals and Semiconductors: The Chemistry of Polyacetylene (CH)x, and Its Derivatives," *Synth. Met.*, **1**, pp. 101–18.

Madou, M. (1997) *Fundamentals of Microfabrication*, CRC Press, Boca Raton, FL.

Manahara, H.M., Morikawa, E., Choi, J., and Sprunger, P.T. (1999) "Pattern Transfer by Direct Photo Etching of Poly(vinylidene fluoride) Using X-rays," *J. Microelectromech. Syst.*, **8**, pp. 417–22.

Maruo, S., and Ikuta, K. (2000) "Three-dimensional Microfabrication by Use of Single-Photon Adsorbed Polymerization," *Appl. Phys. Lett.*, **76**, pp. 2656–58.

Maruo, S., and Kawata, S. (1998) "Two-Photon-Absorbed Near-Infrared Photopolymerization for Three-Dimensional MicroFabrication," *J. Microelectromech. Syst.*, **7**, pp. 411–15.

Mastrangelo, C.H. (2003) "Microfabrication Techniques for Plastic Microelectromechanical Systems (MEMS)," Final Technical Report, AFRL-IF-RS-TR-2003-161, AFRL/DARPA.

Matlezos, G., Nortrup, R., Jeon, S., Zaumseil, J., and Rogers, J.A. (2003) "Tunable Organic Transistors that Use Microfluidic Source and Drain Electrodes," *Appl. Phys. Lett.*, **83**, pp. 2067–69.

Marque, P., and Roncali, J. (1990) "Structural Effect on the Redox Thermodynamics of Poly(thiophenes)," *J. Phys. Chem.*, **94**, pp. 8614–17.

Minami, K., Kawamura, S., and Esashi, M. (1993) "Fabrication of Distributed Electrostatic Micro Actuator (DEMA)," *J. Microelectromech. Syst.*, **2**, pp. 121–27.

Miyajima, H., Asaoka, N., Arima, M., Minamoto, Y., Murakami, K., Tokuda, K., and Matsumoto, K. (2001) "A Durable, Shock-Resistant Electromagnetic Optical Scanner with Polyimide-Based Hinges," *J. Microelectromech. Syst.*, **13**, 418–24.

Oguro, K., Kawami, Y., and Takenaka, H. (1992) "Bending of an Ion-Conducting Polymer Film-Electrode Composite by an Electrical Stimulus at Low Voltage," *J. Micromachine Soc.*, **5**, pp. 27–30.

Omote, K., Ohigashi, H., and Koga, K. (1997) "Temperature Dependence of Elastic, Dielectric and Piezoelectric Properties of 'Single Crystalline' Films of Vinylidene Fluoride Trifluoroethylene Copolymer," *J. Appl. Phys.*, **81**, p. 2760.

Ounaies, Z., Young, J.A., Simpson, J.O., and Farmer, B.L. (1997) "Dielectric Properties of Piezoelectric Polyimides," in *Materials Research Society Proceedings: Materials for Smart Systems II*, 459, p. 59.

Park, C., Ounaies, Z., Wise, K., and Harrison, J. (2004) "In Situ Poling And Imidization Of Amorphous Piezoelectric Polyimides," *Polymer*, **45**, pp. 5417–25.

Park, K-T., and Esashi, M. (1999) "A Multilink Active Catheter with Polyimide-Based Integrated CMOS Interface Circuits," *J. Microelectromech. Syst.*, **8**, pp. 349–57.

Pei, Q., and Inganas, O. (1992) "Electrochemical Application of the Bending Beam Method. 1. Mass Transport and Volume Changes in Polypyrrole During Redox," *J. Phys. Chem.*, **96**, pp. 10, 507–14.

Pham, N.P., Boellaard, E., Burghartz, J.N., and Sarro, P.M. (2004) "Photoresist Coating Methods for Integration of Novel 3-D RF Microstructures," *J. Microelectromech. Syst.*, **13**, pp. 491–99.

Potschke, P., Dudkin, S.M., and Alig, I. (2003) "Dielectric Spectroscopy on Melt Processed Polycarbonate-Multiwalled Carbon Nanotube Composites," *Polymer,* **44**, pp. 5023–30.

Rossier, J.S., Schwarz, A., Reymond, F. Ferrigon, R., Bianchi, F., and Girault, H.H. (1999) "Microchannel Networks for Electrophoretic Separations," *Electrophoresis*, **20**, pp. 727–31.

Ryu, K.S., Wang, X., Shaikh, K., and Liu, C. (2004) "A Method for Precision Patterning of Silicone Elastomer and its Applications," *J. Microelectromech. Syst.*, **13**, pp. 568–75.

Sessler, G.M. (1981) "Piezoelectricity in Polyvinylidenefluoride," *J. Acoust. Soc. Am.*, **70**, pp. 1596–1608.

Shirakawa, H., Louis, E.J., MacDiarmid, A.G., Chiang, C.-K., and Heeger, A.J. (1977) "Synthesis of Electrically Conducting Organic Polymers: Halogen Derivatives of Polyacetylene, (CH)x," *J. Chem. Soc., Chem. Commun.*, pp. 578–97.

Simpson, J. O., Ounaies, Z., and Fay, C. (1997) "Polarization and Piezoelectric Properties of a Nitrile Substituted Polyimide," in *Materials Research Society Proceedings: Materials for Smart Systems II*, 459, p. 53.

Smela, E., Inganäs, O., Pei, Q.B., and Lundstrom, I. (1993) *Adv. Mater.*, **5**, pp. 630–32.

Smela, E., Inganäs, O., and Lundström, I. (1995) "Controlled Folding of Micrometer-Size Structures," *Science*, **268**, pp. 1735–38.

Smela, E., Kallenbach, M., and Holdenreid, J. (1999) "Electrochemically Driven Polypyrrole Bilayers for Moving and Positioning Bulk Micromachined Silicon Plates," *J. Microelectromech. Syst.*, **8**, pp. 373–83.

Smela, E. (2003) "Conjugated Polymer Actuators for Biomedical Applications," *Adv. Mater.*, **15**, pp. 481–94.

Studer, V., Pépin, A., and Chen, Y. (2002) "Nanoembossing of Thermoplastic Polymers for Microfluidic Applications," *Appl. Phys. Lett.*, **80**, pp. 3614–16.

Su, Y.-C., and Lin, L. (2004) "A Water-Powered Drug Delivery System," *J. Microelectromech. Syst.*, **13**, pp. 75–81.

Sundararajan, N., Pio, M.S., Lee, L.P., and Berlin, A. (2004) "Three-Dimensional Hydrodynamic Focusing in Polydimethylsiloxane (PDMS) Microchannels," *J. Microelectromech. Syst.*, **13**, pp. 559–67.

Sutomo, W., Wang, X., Bullen, D. Braden, S.K., and Liu, C. (2003) "Development of an End-Point Detector for Parylene Deposition Process," *J. Microelectromech. Syst.*, **12**, pp. 64–69.

Tan, W., Fan, Z.H. Qiu, C.X. Ricco, A.J., and Gibbons, I. (2002) "Miniaturized Capillary Isoelectric Focusing in Plastic Microfluidic Devices," *Electrophoresis*, **23**, pp. 3638–45.

Ucok, A.B., Giachino, J.M., and Najafi, K.N. (2004) "The WIMS Cube: A Microsystem Package With Actuated Flexible Connections and Re-Workable Assembly," in *Technical Digest of the Solid-State Sensor, Actuator and Microsystems Workshop*, pp. 117–20, Hilton Head Island, SC.

Valentini, L., Puglia, D., Frulloni, E., Armentano, I., Kenny, J.M., and Santucci, S. (2004) "Dielectric Behavior of Epoxy Matrix/Single-Walled Carbon Nanotube Composites," *Compos. Sci. Technol.*, **64**, pp. 23–33.

Varadan, V., and Varadan, V. (2000) "Microsensors, Microelectromechanical Systems (MEMS) and Electronics for Smart Structures and Systems," *Smart Mater. Struct.*, **9**, pp. 953–72.

Velasco-Santos, C., Martýnez-Hernández, A.L., Fisher, F., Ruoff, R., and Casta, V.M. (2003) "Dynamical–Mechanical and Thermal Analysis of Carbon Nanotube-Methyl-Ethyl Methacrylate Nanocomposites," *J. Phys. D: Appl. Phys.*, **36**, pp. 1423–28.

Weston, D.F., Smekal, T., Rhine, D.B., and Blackwell, J. (2001) "Fabrication of Microfluidic Devices in Silicon Using Plasma Etching," *J. Vac. Sci. Technol. B*, **19**, pp. 2846–51.

Xia, F., Li, H., Huang, C., Huang, M., Xu, H., Bauer, F., Cheng, Z.-Y., and Zhang, Q. (2003) "Poly(vinylidene fluoride-trifluoroethylene) Based High Performance Electroactive Polymers," in *Proc. SPIE Smart Structures and Materials, Electroactive Polymer Actuators and Devices*, 5051, pp. 133–42.

Zhang, Q.M., Bharti, V., and Zhao, X. (1998) "Giant Electrostriction and Relaxor Ferroelectric Behavior in Electron-Irradiated Poly(vinylidene fluoride trifluoroethylene) Copolymer," *Science*, **280**, pp. 2101–04.

Zhang, Q.M. (2001) "P(VDF-TrFE)-Based Electrostrictive Co/Ter-Polymers and Its Device Performance," *Electroactive Polymer Actuators and Devices*, Bar-Cohen, Y., ed., SPIE Press, Bellingham, WA.

Zhang, Z., and Jang, B. (1995) "Actuation Power Capacities of Polymers, Gels and Composites for Use in Smart Material Systems," in *Proceedings of the SPIE*, 2447, pp. 26–94.

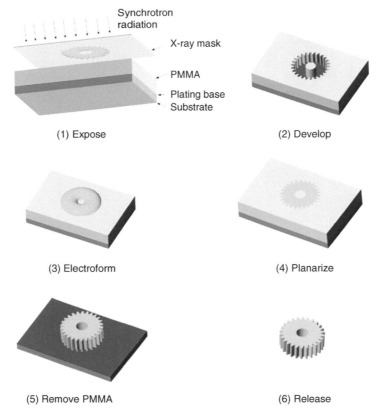

Synchrotron
radiation

X-ray mask

PMMA

Plating base
Substrate

(1) Expose

(2) Develop

(3) Electroform

(4) Planarize

(5) Remove PMMA

(6) Release

COLOR FIGURE 5.1   DXRL-based ("direct LIGA") microfabrication process.

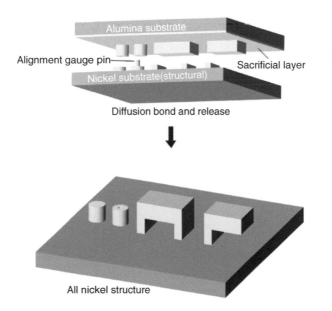

Alumina substrate

Alignment gauge pin

Sacrificial layer

Nickel substrate(structural)

Diffusion bond and release

All nickel structure

COLOR FIGURE 5.36   Batch wafer-scale diffusion bonding procedure for multilevel fabrication of DXRL-patterned geometry. This sequence may be directly repeated to integrate further levels.

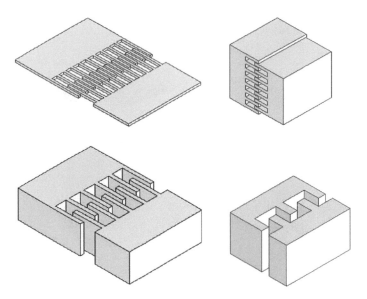

COLOR FIGURE 6.14 A variety of comb drive actuator implementations, clockwise from top left: surface micro-machining, DRIE, LIGA, and EFAB. A comparison of the available force (or capacitive sensitivity) is shown on the chart, demonstrating the high power densities available with EFAB.

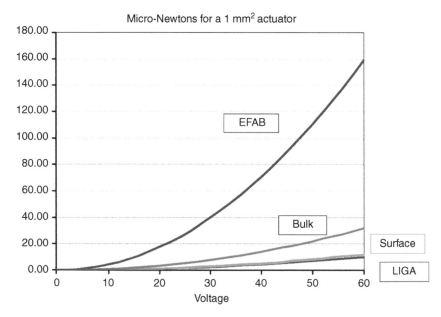

COLOR FIGURE 6.15 Calculated forces as a function of voltage for comb drive actuators having a 1×1 mm footprint for different fabrication processes.

COLOR FIGURE 8.12 Packaged SiC pressure sensor chip. Through the semitransparent SiC can be seen the edge of the well, which has been etched in the backside of the wafer to form the circular diaphragm. The metal-covered n-type SiC which connects the strain gauges is highly visible, but the n-type SiC strain gauges are more faintly visible. There is a U-shaped strain gauge over the edge of the diaphragm at top and bottom, and there are two vertically oriented linear gauges in the center of the diaphragm.

COLOR FIGURE 8.13 SiC pressure sensor chip in leadless package. The backside of the SiC chip is exposed to the measurement environment. The metallization on the front side of the chip, visible through the semitransparent SiC, is protected from oxidation in a hermetically sealed cavity.

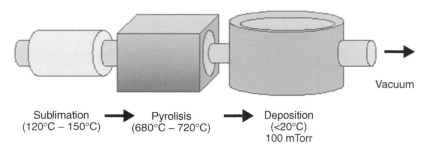

COLOR FIGURE 9.15 Vapor deposition of parylene.

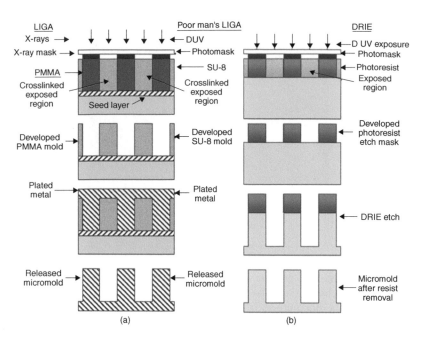

COLOR FIGURE 9.17   Micromold fabrication techniques: (a) LIGA/"Poor Man's LIGA" and (b) DRIE micromold fabrication.

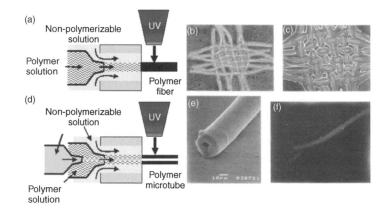

COLOR FIGURE 9.20   (a) Photopolymerizing fiber extraction apparatus, (b) woven pH-responsive fibers, (c) pH-responsive fibers in a swollen state, (d) photopolymerizing microtube apparatus (e) SEM of a fabricated microtube, and (f) fluorescent image of glucose sensing microtube (after Jeong et al. (2004)).

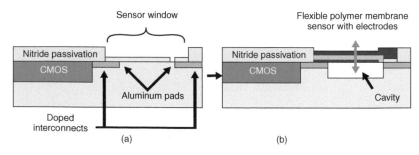

COLOR FIGURE 9.23   (a) CMOS chip with sensor window in passivation and (b) CMOS electronics with polymer sensor fabricated in the sensor window.

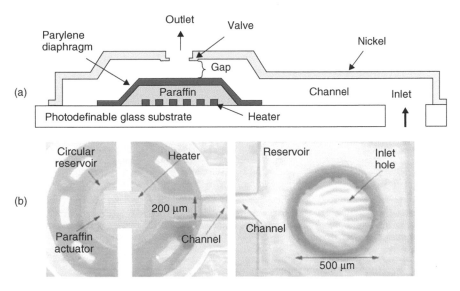

COLOR FIGURE 9.30   (a) Cross-section of paraffin active blocking valve actuator and (b) optical micrograph of a nickel-plated microvalve from the backside (after Carlen and Mastrangelo (2002)).

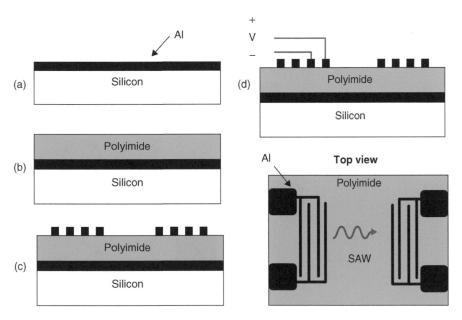

COLOR FIGURE 9.34   SAW fabrication on electroactive polyimide (after Atkinson (2003)).

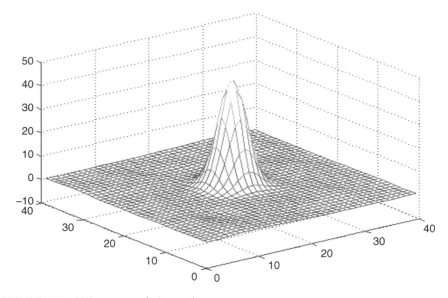

COLOR FIGURE 10.2   PIV cross-correlation peak.

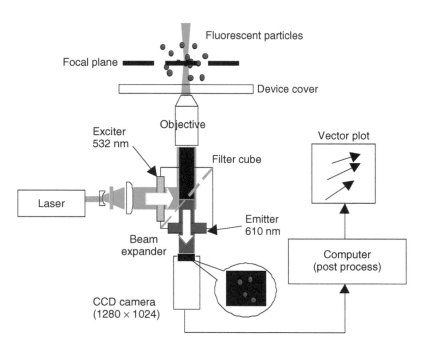

COLOR FIGURE 10.3   Schematic of a μPIV system. A pulsed Nd:YAG laser is used to illuminate fluorescent flow-tracing particles, and a cooled CCD camera is used to record the particle images.

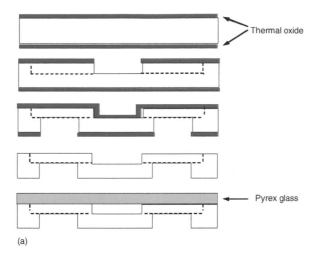

Thermal oxide

Pyrex glass

(a)

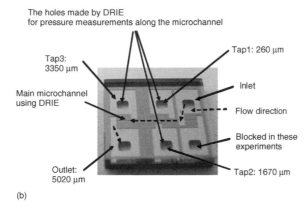

The holes made by DRIE
for pressure measurements along the microchannel

Tap3:
3350 μm

Tap1: 260 μm

Main microchannel
using DRIE

Inlet

Flow direction

Blocked in these
experiments

Outlet:
5020 μm

Tap2: 1670 μm

(b)

COLOR FIGURE 10.10   The schematics of (a) fabrication procedure and (b) photo of MC-II.

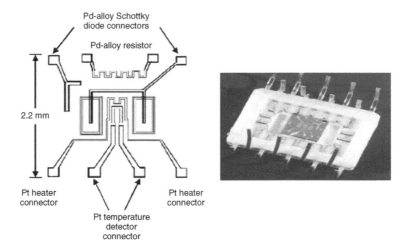

Pd-alloy Schottky
diode connectors

Pd-alloy resistor

2.2 mm

Pt heater
connector

Pt heater
connector

Pt temperature
detector
connector

COLOR FIGURE 11.1   Schematic diagram and picture of the silicon-based hydrogen sensor. The Pd alloy Schottky diode (rectangular regions) resides symmetrically on either side of a Pt heater and temperature detector. The Pd alloy Schottky diode is used for low-concentration measurements, whereas the resistor is included for high-concentration measurements.

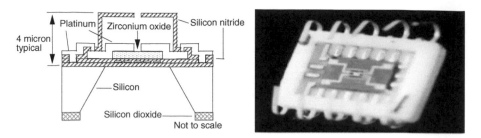

COLOR FIGURE 11.7   The structure of a microfabricated amperometric oxygen sensor and a picture of a sensor packaged without its protective orifice.

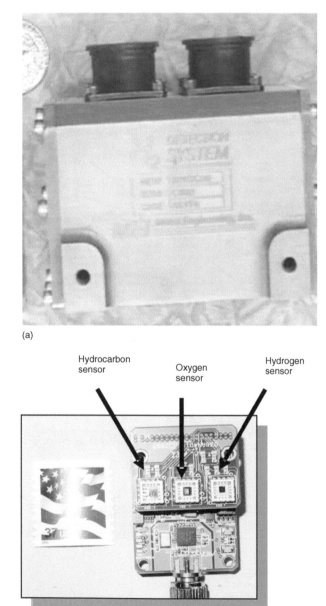

COLOR FIGURE 11.10   (a) Shuttle system hardware (H$_2$ sensor with electronics); (b) A prototype version of a "Lick and Stick" leak sensor system with hydrogen, hydrocarbon, and oxygen detection capabilities combined with power and supporting electronics (including signal conditioning and telemetry).

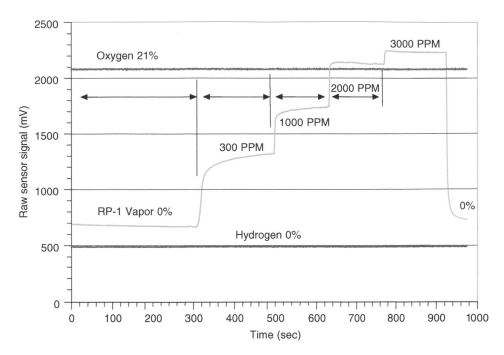

COLOR FIGURE 11.11   Response of the three sensors of this system to a constant oxygen environment and varying hydrocarbon (RP-1) concentrations. The sensor signal shown is the output from the signal conditioning electronics, which processes the measured sensor current at a constant voltage.

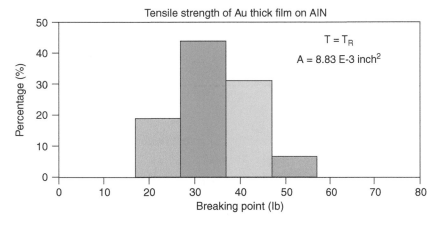

COLOR FIGURE 12.3   Tensile strength of adhesion of selected Au thick-film on AlN substrate measured by stud pull tests.

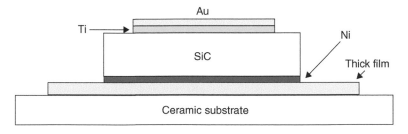

COLOR FIGURE 12.8   Schematic diagram of as-fabricated SiC device and die-attach structure.

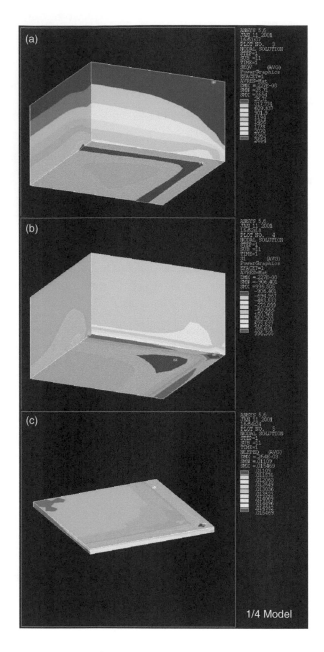

COLOR FIGURE 12.12 Thermal stress and strain distribution in SiC die and Au thick-film layer. (a) Von Mises stress distribution in SiC die. (b) Maximum principal stress distribution in SiC die. (c) Equivalent plastic strain in Au thick-film layer.

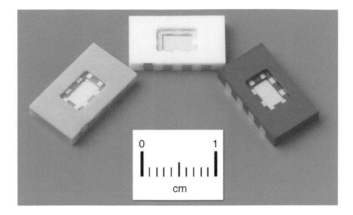

COLOR FIGURE 12.15 AlN (left), 96% Al_2O_3 (top), and 92% Al_2O_3 (right) chip level packages with Au thick-film metallization. These 8-pin (I/O) packages were developed at NASA Glenn Research Center for low-power SiC electronics and sensors for operation up to 500°C.

COLOR FIGURE 12.16 AlN substrate and Au thick-film metallization based printed circuit board designed for the 8-pin (I/O) AlN packages shown in Figure 12.15.

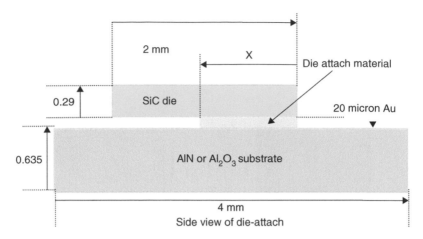

COLOR FIGURE 12.18 Side view of an example of die-attach structure with lateral stress attenuation. Where X (0 mm < X < 2 mm) is a parameter determined by device design and packaging requirements, such as mechanical strength and resonant frequencies.

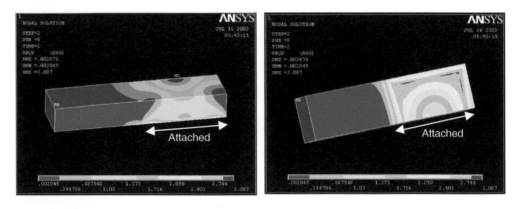

COLOR FIGURE 12.19 FEA simulation of thermal stress of the partially attached SiC die on AlN substrate using 20 μm Au. The attached area is 1 mm × 1 mm. All stress units are MPa. (a) Von-Mises stress contour plot of top half of partially attached SiC die. Attaching area is 1 mm × 1 mm. (b) Von-Mises stress contour plot of bottom half of partially attached SiC die. Attaching area is 1 mm × 1 mm.

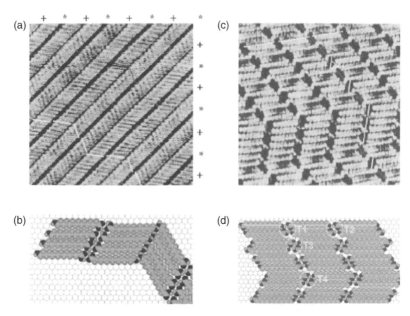

COLOR FIGURE 14.1 (a) 15-hydroxypentadecanoic acid at the interface of a 1-nonanol solution and the basal plane of graphite, imaged by STM. Scan size: $15 \times 15\,nm^2$. A single molecular length is indicated by a black or a blue bar. The asterisks (∗) mark troughs that are composed of carboxylic acid dimers, and the pluses (+) mark troughs that are composed of alcohol functional groups. The yellow bars depict the global pattern, which can be described as a super-herringbone structure. (b) Ball model of 15-hydroxypentadecanoic acid on a graphite surface: carbon atoms are green, oxygen atoms are red, hydrogen atoms are grey. The molecules are positioned in registry with the underlying graphite lattice. (c) STM images of 16-hydroxyhexadecanoic acid diluted in a 1-hexanol solution on graphite. Molecular lengths are indicated by the blue bars. The blue bars are collinear, indicating that the fatty acids occupy the same graphite lattice rows. The red and the yellow bars depict the two different orientations observed for the dark spots. Image size: $10 \times 10\,nm^2$. (d) Ball model of 16-hydroxyhexadecanoic acid on a graphite surface. (Reprinted with permission from Wintgens, D. et al. [2003] "Packing of $HO(CH_2)_{14}COOH$ and $HO(CH_2)_{15}COOH$ on Graphite at the Liquid–Solid Interface Observed by Scanning Tunneling Microscopy: Methylene Unit Direction of Self-Assembly Structures," *J. Phys. Chem. B.* **107**, pp. 173–79.)

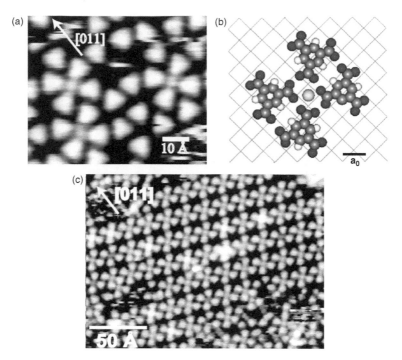

COLOR FIGURE 14.2 (a) STM image of trimesic acid molecules adsorbed on a Cu(100) surface at room temperature. The triangular shape reflects a flat-lying adsorption geometry. Cloverleaf-shaped arrangements of four tma molecules with a central protrusion represent $[Cu(tma)_4]^{n-}$ coordination compounds. (b) Model for the $[Cu(tma)_4]^{n-}$ configuration with a central Cu atom coordinated by four carboxylate ligands. Gray: carbon atoms; red: oxygen atoms; white: hydrogen atoms; yellow: Cu adatoms. The Cu substrate is represented by the orange lattice where $a_0 = 3.61$ Å. In agreement with the tunneling data the tma triangles do not point straight to the central site, which indicates a unidentate coordination of the Cu adatom. (c) Regular array of $[Cu(tma)_4]^{n-}$ complexes. (Reprinted with permission from Lin, N. et al. [2002] "Real-Time Single-Molecule Imaging of the Formation and Dynamics of Coordination Compounds," *Angew. Chem. Int. Ed.* **41**, pp. 4779–83).

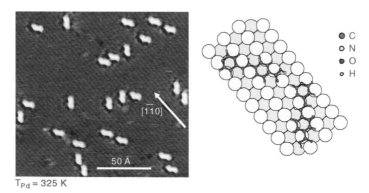

COLOR FIGURE 14.3 Left: PVBA molecules randomly distributed on the Pd(110) surface at low coverage ($\theta = 0.018$ ML). The molecules were adsorbed and imaged at a substrate temperature of 325 K and deposited with a rate of 3×10^{-5} ML/s. The atomic rows of the Pd substrate and the dog-bone internal molecular structure are resolved (image size: 19×18 nm^2: $U_t = 1.04$ V, $I_t = 1$ nA). Right: Ball model for unrelaxed PVBA molecules on Pd(110). Large light and dark grey circles: Pd atomic rows of the surface; small dark grey circles: carbon atoms; yellow: nitrogen atoms; red: oxygen atoms; and white: hydrogen atoms. The molecular axis is oriented by $\pm 35.3°$ with respect to the $[1{-}10]$ Pd rows providing optimal coordination of surface Pd and molecular subunits. (Reprinted with permission from Weckesser, J. et al. [1999] "Binding and Ordering of Large Organic Molecules on an Anisotropic Metal Surface: PVBA on Pd(110)," *Surf. Sci.* **431**, pp. 168–73.)

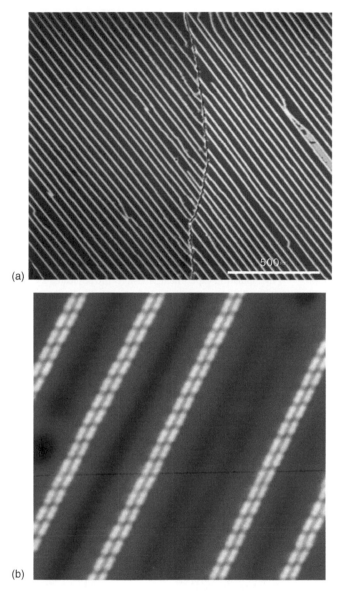

(a)

(b)

COLOR FIGURE 14.4 One-dimensional supramolecular PVBA super-grating by self-assembly mediated by H-bond formation on an Ag(111) surface at 300 K (measured at 77 K). (a) An STM topograph of a single domain extending over two terraces demonstrates ordering at the micron length scale. (b) A close-up image of the self-assembled twin chains reveals that they consist of coupled rows of PVBA molecules. (Reprinted with permission from Barth, J.V. et al. [2000] "Building Supramolecular Nanostructures at Surfaces by Hydrogen Bonding," *Angew. Chem. Int. Ed.* **39**, pp. 1230–3.)

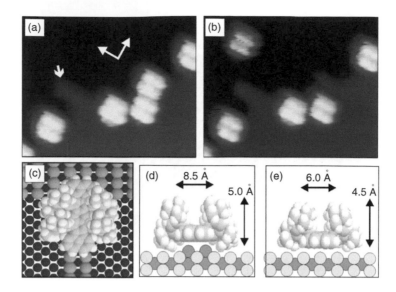

COLOR FIGURE 14.6 (a, b) Manipulation sequence of a Lander molecule labeled 1 from a step edge on Cu(110). A nanostructure from a previous manipulation experiment is visible and indicated by an arrow. (13.6 × 11.2 nm², V = −1.768 V, I = −37 nA, T = 95 K). (b) After the manipulation of the molecule along the [1–10] direction, the nanostructure underneath becomes visible (tunneling parameters for manipulation: V = −0.055 V, I = −1.05 nA). (c) Model of the Lander on the nanostructure, showing that the board is parallel to the nanostructure. For clarity, the length of the nanostructure is extended beyond the molecule. Cross-sectional side view of the Lander (d) on a nanostructure and (e) on a flat terrace. The distances between and heights of the spacer groups (from calculated STM images) are stated. (Reprinted with permission from Schunack, M. et al. [2002] "Adsorption Behavior of Lander Molecules on Cu(110) Studied by Scanning Tunnelling Microscopy," *J. Chem. Phys.* 117, pp. 6259–65.)

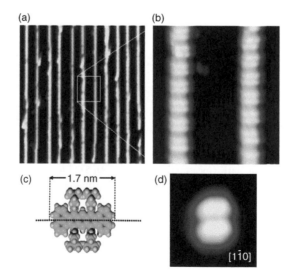

COLOR FIGURE 14.7 (a) 60 × 60 nm² STM image showing the molecular chains formed after the deposition of SL molecules on a nanopatterned Cu–O surface. The molecules adsorb exclusively on bare Cu stripes. (b) 14 × 14 nm² high-resolution STM image. The individual SL molecules can be individually resolved, and their conformation can be extracted. (c) Filled-space model of the single Lander (SL) molecule. (d) Typical 3 × 3 nm² STM image of the SL molecule on a clean Cu(110) surface revealing the SL as four lobes corresponding to the spacer legs. (Reprinted with permission from Otero, R. et al. [2004] "One Dimensional Assembly and Selective Orientation of Lander Molecules on a O–Cu Template," *Angew. Chem. Int. Ed.* 43, pp. 2092–95.)

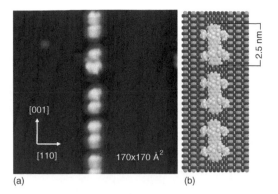

(a)　　　　　　　　　　　　　　　　(b)

COLOR FIGURE 14.8　(a) 17 × 17 nm² STM image showing the molecular chains formed after the deposition of VL on the nanopatterened O–Cu surface. The molecules adsorb exclusively on bare Cu stripes, but their orientation is no longer perpendicular to the direction of the chain. The molecules are found either perfectly aligned with the direction of the chain or forming an angle of 20° with respect to that direction. (b) Ball model of the VL adsorbed in a Cu trench. (Reprinted with permission from Otero, R. et al. [2004] "One Dimensional Assembly and Selective Orientation of Lander Molecules on a O–Cu Template," *Angew. Chem. Int. Ed.* 43, pp. 2092–95.)

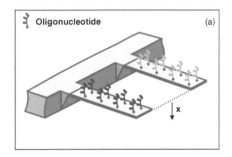

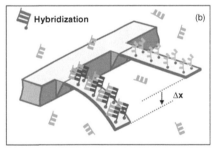

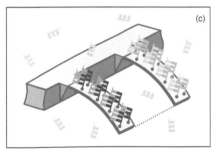

COLOR FIGURE 14.10　Simple scheme illustrating the hybridization experiment. Each cantilever is functionalized on one side with a different oligonucleotide base sequence (marked red or blue). (a) The differential signal is set to zero. (b) After injection of the first complementary oligonucleotide (green), hybridization occurs on the cantilever that provides the matching sequence (red), increasing the differential signal. (c) Injection of the second complementary oligonucleotide (yellow) causes the cantilever functionalized with the second oligonucleotide (blue) to bend. (Reprinted with permission from Fritz, J. et al. [2000] "Translating Biomolecular Recognition into Nanomechanics," *Science* 288, pp. 316–18.)

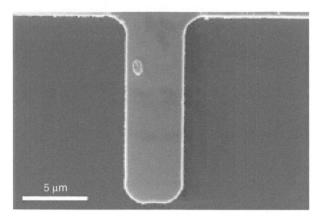

COLOR FIGURE 14.11 SEM micrograph of a single *E. coli* bacterium on an antibody-coated silicon nitride cantilever oscillator. (Reprinted with permission from Craighead, H.G. [2000] "Nanoelectromechanical Systems," *Science* 290, pp. 1532–35.)

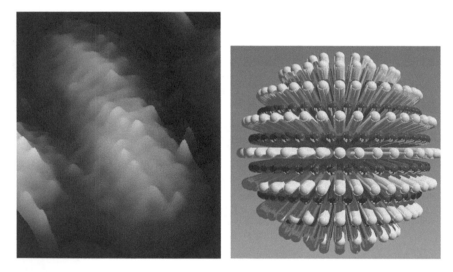

COLOR FIGURE 14.12 Decanethiol/MPA (2:1 molar ratio) self-assembled onto gold nanoparticles. The molecules separate into alternating domains that wrap around the gold core. Left: These domains appear as "ripples" in the STM images. Right: Schematic model of the MPMN. (Reprinted with permission from Jackson, A.M. et al. [2004] "Spontaneous Assembly of Subnanometre–Ordered Domains in the Ligand Shell of Monolayer-Protected Nanoparticles," *Nature Materials* 3, pp. 330–36.)

10

Optical Diagnostics to Investigate the Entrance Length in Microchannels

Sang-Youp Lee, Jaesung Jang
and Steven T. Wereley
Purdue University

10.1   Introduction

Understanding the flow in the entrance region of a microchannel is very important because the transport properties of the flow, such as the pressure gradient and the heat transfer coefficient, depend strongly on the flow region. Planar microfabrication techniques generally constrain the reservoir upstream of the microchannel to be the same height as the channel — a marked departure from the geometries used at macroscopic-length scales to study entrance length. Advanced optical diagnostics for microscale flows are essential in acquiring a spatially resolved view of the flow in the small entrance region of microchannel flows. This chapter will describe the entrance length problem in microchannels, outline several microscale fluid diagnostic techniques, and use those techniques to analyze the entrance length in a microchannel.

Entrance length is most frequently defined as the axial distance required for the centerline velocity to reach 99% of the fully developed centerline velocity, usually assuming a uniform flow at the channel entrance [Shah and London, 1978]. However, a uniform entrance velocity profile is seldom achieved in microchannels because of flow development in the inlet region [Beaver et al., 1970], the velocity overshoot caused by the abrupt velocity gradients owing to the sharp turn of the flow at the entrance [Sparrow and Anderson, 1977], and the no-slip condition at the wall [Shah and London, 1978]. For high Reynolds numbers, because the dimensionless entrance length, $L_e/(Re_D D_h)$, is approximately proportional to the Reynolds number, the entrance length correlation is calculated from a boundary-layer type governing equation, and the entrance length is independent of inlet velocity profile [Shah and London, 1978; Atkinson et al., 1969]. However, for the low Reynolds numbers typical of microchannel flows, the dependence of the entrance length on the inlet velocity profile needs to be considered, because the axial diffusion of vorticity determines the velocity development in the upstream reservoir and the entrance of the microchannel [Vrentas et al., 1966].

For a circular duct, the dimensionless entrance length L_e/D_h is given as ~6% of the Reynolds number [Fox and McDonald, 1998] when $Re_D > 400$ [Shah and London, 1978]. Entrance length correlations accounting for the nonlinear relationship between L_e/D_h and Reynolds number were proposed by Atkinson et al. (1969) and Chen (1973) for a tube and parallel plates to be

$$\frac{L_e}{D_h} = C_1 + C_2 Re_D \tag{10.1}$$

$$\frac{L_e}{D_h} = \frac{C_1}{(C_2 Re_D + 1)} + C_3 Re_D \tag{10.2}$$

where L_e is the entrance length, D_h is the hydraulic diameter, and C_1, C_2, and C_3 are appropriate constants. For Atkinson's correlation, C_1 and C_2 are 0.59 and 0.056 for a circular channel, respectively, and for parallel plates, they are 0.625 and 0.044, respectively. For Chen's correlation, C_1, C_2, and C_3 are given as 0.6, 0.035, and 0.056, respectively for circular channels, and for parallel plates, they are 0.63, 0.035, and 0.044, respectively.

The Atkinson type correlation, Equation (10.1), shows the form of a linear combination of the creeping flow and boundary-layer type solutions and shows that there exists a finite entrance length even when the Reynolds number approaches zero. Chen (1973) proposed a more accurate relation, Equation (10.2), based on the solution of Friedmann et al. (1968). Both correlations show that when the Reynolds number approaches zero, there exists a constant portion of entrance length. Chen's type of correlation indicates that the entrance length has a rational relation with Reynolds number when the Reynolds number is small. When the Reynolds number becomes large, both types of correlations are dominated by a proportional relationship between the dimensionless entrance length and the Reynolds number. Consequently, when the entrance length is scaled $L_e/(Re_D D_h)$, it is constant for large Reynolds numbers.

In a rectangular channel, four boundary layers form, one on each wall beginning at the channel inlet, similar to the boundary layer growth on a flat plate, and the layers merge at some point downstream [Shah and London, 1978]. Therefore, the flow field in rectangular channels is more complicated than in circular channels, because the flow field depends on two cross-sectional coordinates and is thus a function of aspect ratio. Owing to the nonlinear terms in the momentum equations, an exact solution does not exist, but approximate and numerical solutions are available. Han (1983) provided the approximate analytical solution for the rectangular channel and showed the entrance length as a function of aspect ratio. The entrance length decreases as the channel becomes more like two parallel plates and less like a square (aspect ratio approaching either infinity or zero). For example, $L_e/(Re_D D_h)$ is 0.0660 and 0.0427 for the aspect ratios 2.0 and 4.0, respectively. Some experimental results are also available. Sparrow et al. (1967) showed that the entrance lengths are about 0.08 for the aspect ratio 2.0 and 5.0. Fleming and Sparrow (1969) presented an approximate analytical solution for aspect ratios of 2.0 and 5.0, and the dimensionless entrance lengths are found as 0.07 and 0.052, respectively. Also, the accepted correlations for a circular channel, as shown in Equation (10.1) and Equation (10.2), can be used with the hydraulic diameter approach. Though adopting the hydraulic diameter concept is crude, this approximation gives acceptably accurate results for rectangular channels with the aspect ratios $0.25 < \alpha < 4$ [Fox and McDonald, 1998]. In spite of these extensive earlier works, the entrance length problem in microchannels is still largely unsolved because of the planar upstream reservoir geometry.

The importance of the entrance length can be emphasized further when the channel application is extended to small length scales, because the entrance length can be a significant portion of the total length of the channel. Because microfabrication techniques are typically planar processes based on film etching or developing, two important geometrical limitations exist. First, there is often a uniform depth throughout microfabricated devices. Thus, in microchannel fabrication, these planar processes result in the height of the reservoir upstream of the microchannel entrance being the same as that of the microchannel itself. Therefore, the flow conditions upstream of the channel entrance must be considered when studying entrance length in microchannels, rather than simply assuming a uniform velocity profile. A second important implication of planar microfabrication is that the cross-sections of the microchannels are usually rectangular or trapezoidal. This limitation makes entrance length estimation in microscale channels even more difficult, because the accuracy of using entrance length correlations of a circular channel for rectangular channel entrance length estimation is not known in the microscale. When both of these limitations are considered together, the results from past research on rectangular channels must be applied carefully in microchannels, since the aspect ratios H/W and W/H, equivalent in macroscale channels, result in totally different inlet geometries when using planar microfabrication techniques. Lee et al. (2002) and Wereley et al. (2002) first investigated the entrance length problem in typical microchannel geometries and showed that the entrance length estimation is not as simple as expected with the conventional entrance length correlations.

In the current study, the entrance lengths of two different aspect ratio microchannels are investigated for low Reynolds numbers using μPIV experiments. To ensure that the microchannel has sufficient length to fully develop, the microchannels have large L/D ratios. One microchannel (MC-I) is fabricated from transparent acrylic using conventional machining, a precision sawing technique, and has the dimensions 120.0 mm (L) $\times$ 0.252 mm (W) $\times$ 0.694 mm (H). The other microchannel (MC-II) is fabricated with silicon using DRIE and measures 5020 μm (L) $\times$ 104.6 μm (W) $\times$ 38.6 μm (H). The two channels yield the reciprocal aspect ratios of (H/W) $\sim$2.75 (MC-I) and $\sim$0.37 (MC-II), respectively. These aspect ratios would be considered identical in the flow from an infinite reservoir. However, because of the planar upstream reservoir geometry, these aspect ratios must be treated separately. Therefore, the aspect ratio and inlet geometry must be considered at the same time. For both microchannels, eight different Reynolds numbers are tested. They are 1, 10, 20, 30, 50, 70, and 100 for MC-I, and 5, 10, 17.5, 24, 35.7, 48.4, 62, and 76.2 for MC-II. The entrance lengths are found by comparing the widthwise velocity profile as a function of the axial distance with the fully developed velocity profile, as well as the local centerline velocity with the fully developed centerline velocity.

10.2 Optical Diagnostics Metrology in Microscale Fluid Mechanics

Optical diagnostics have played a significant role in experimental fluid mechanics, beginning with the development of Schlieren and shadowgraphy imaging of compressible flows in the early 20th century. The use of optical diagnostics was greatly accelerated in the second half of the 20th century, with the development of lasers and high-speed, high-resolution electronic cameras. Beginning in the 1990s and continuing through the present day, many of these optical diagnostic techniques were extended to microscale fluid mechanics. These techniques can be divided into pointwise and full-plane techniques. The pointwise techniques measure one spatial point at a time (although often a dense temporal stream can be acquired at each spatial point), whereas the full-plane techniques measure many spatial points simultaneously. In the present entrance-length experiments, a full-plane technique (μPIV) was used to acquire a complete picture of the entrance region with a single measurement.

10.2.1 Pointwise Methods

Laser Doppler velocimetry (LDV) has been a standard optical measurement technique in fluid mechanics since the 1970s. In the case of a dual-beam LDV system, two coherent laser beams are aligned so that they intersect at some point. The volume of the intersection of the two laser beams defines the measurement volume. In the measurement volume, the two coherent laser beams interfere with each other, producing a pattern of light and dark fringes. When a seed particle passes through these fringes, a pulsing reflection is created that is collected by a photomultiplier, processed, and turned into a velocity measurement. Traditionally, the measurement volumes of standard LDV systems have characteristic dimensions on the order of a few millimeters. Compton and Eaton (1996) used short focal length optics to obtain a measurement volume of $35 \times 66 \mu m$. Using very short focal length lenses, Tieu et al. (1995) built a dual-beam solid-state LDA system with a measurement volume of approximately $5 \times 10 \mu m$. Their micro-LDV system was used to measure the flow through a 175-μm-thick channel, producing time-averaged measurements that compare well to the expected parabolic velocity profile, except within 18 μm of the wall. Advancements in microfabrication technology are expected to facilitate the development of new generations of self-contained solid-state LDV systems with micron-scale probe volumes. These systems will likely serve an important role in the diagnosis and monitoring of microfluidic systems [Gharib et al., 2002]. However, the size of the probe volume significantly limits the number of fringes that it can contain, which subsequently limits the accuracy of the velocity measurements.

Optical Doppler tomography (ODT) has been developed to measure micron-scale flows embedded in a highly scattering medium. In the medical community, the ability to measure in vivo blood flow under the skin allows clinicians to determine the location and depth of burns [Chen et al., 1997]. ODT combines single-beam Doppler velocimetry with heterodyne mixing from a low-coherence Michelson interferometer. The lateral spatial resolution of the probe volume is determined by the diffraction spot size. The Michelson interferometer is used to limit the effective longitudinal length of the measurement volume to that of the coherence length of the laser. The ODT system developed by Chen et al. (1997) has a lateral and longitudinal spatial resolution of 5 μm and 15 μm, respectively. The system was applied to measure flow through a 580-μm-diameter conduit.

10.2.2 Full-Field Methods

Full-field experimental velocity measurement techniques generate velocities that are minimally two-component velocity measurements distributed within a two-dimensional plane. These types of velocity measurements are essential in microfluidics for several reasons. First, global measurements, such as the pressure drop along a length of channel, can reveal the dependence of flow physics upon length scale by showing that the pressure drop for flow through a small channel is smaller or larger than a flow through a large channel. However, global measurements are not very useful for pointing to the precise cause of why the physics might change, such as losing the no-slip boundary for high Knudsen number gas flows. A detailed view of the flow, such as that provided by full-field measurement techniques, is indispensable for establishing the reasons why flow behavior changes at small scales. Full-field velocity measurement techniques are also useful for optimizing complicated processes like mixing, pumping, or filtering — typical microfluidic processes. Several of the common macroscopic full-field measurement techniques have been extended to microscopic-length scales. These techniques are scalar image velocimetry, molecular tagging velocimetry, and particle image velocimetry. These techniques will be introduced briefly in this section and then discussed in detail in the following sections.

Scalar image velocimetry (SIV) refers to the determination of velocity-vector fields by recording images of a passive scalar quantity and inverting the transport equation for a passive scalar. Dahm et al. (1992) originally developed SIV for measuring turbulent jets at macroscopic-length scales. Successful velocity measurements depend on having sufficient spatial variations in the passive-scalar field and relatively high Schmidt numbers. Since SIV uses molecular tracers to follow the flow, it has several advantages at the microscale over measurement techniques such as PIV or LDV, which use discrete flow-tracing particles. For instance, the molecular tracers will not become trapped in even the smallest passages within a MEMS device. In addition, the discrete flow tracing particles used in PIV can acquire a charge and move in response to not only hydrodynamic forces but also electrical forces in a process called electrophoresis. However, molecular tracers typically have much higher diffusion coefficients than discrete particles, which can significantly lower the spatial resolution and velocity resolution of the measurements.

Paul et al. (1998) analyzed fluid motion using a novel dye that, while not normally fluorescent, can be made fluorescent by exposure to the appropriate wavelength of light. Dyes of this class are typically called *caged dyes*, because the fluorescent nature of the dye is defeated by a photoreactive bond that can be easily broken. This caged dye was used in the microscopic SIV procedure to estimate velocity fields for pressure- and electrokinetically driven flows in 75-μm-diameter capillary tubes. A 20×500-μm sheet of light from a $\lambda = 355$-nm, frequency-tripled Nd:YAG laser was used to uncage a 20-μm-thick, cross-sectional plane of dye in the capillary tube. In this technique, only the uncaged dye is excited when the test section is illuminated with a shuttered beam from a continuous-wave Nd:YVO$_4$ laser. The excited fluorescent dye is imaged using a $10\times$, NA $= 0.3$ objective lens onto a CCD camera at two known time exposures. The velocity field is then inferred from the motion of the passive scalar. We approximate the spatial resolution of this experiment to be on the order of $100 \times 20 \times 20$ μm, based on the displacement of the fluorescent dye between exposures and the thickness of the light sheet used to uncage the fluorescent dye.

Molecular tagging velocimetry (MTV) is another technique that has shown promise in microfluidics research. In this technique, flow-tracing molecules fluoresce or phosphoresce after being excited by a light source. The excitement is typically in the form of a pattern such as a line or grid written into the flow. The glowing grid lines are imaged twice, with a short time delay between the two images. Local velocity vectors are estimated by correlating the grid lines between the two images [Koochesfahani et al., 1997]. MTV has the same advantages and disadvantages, at least with respect to the flow-tracing molecules, as SIV. In contrast to SIV, MTV infers velocity in much the same way as particle image velocimetry — a pattern is written into the flow, and the evolution of that pattern allows for inferring the velocity field. MTV was demonstrated at microscopic-length scales by Maynes and Webb (2002) in their investigation of liquid flow through capillary tubes, as well as by Lempert et al. (2001) in an investigation of supersonic micronozzles. Maynes and Webb (2002) investigated aqueous glycerin solutions flowing at Reynolds numbers ranging from 600 to 5000 through a 705-μm-diameter, fused-silica tube of circular cross-section. They state that the spatial resolution of their technique is 10 μm across the diameter of the tube and 40 μm along the axis of the tube. The main conclusion of this work is that the velocity measured in their submillimeter tube agreed quite well with laminar flow theory, and that the flow showed a transition to turbulence beginning at a Reynolds number of 2100. Lempert et al. (2001) flowed a mixture of gaseous nitrogen and acetone through a 1-mm, straight-walled "nozzle" at pressure ratios ranging from highly underexpanded to perfectly matched. Because the nozzle was not transparent, the measurement area was limited to positions outside the nozzle. A single line was written in the gas, normal to the axis of the nozzle, by a frequency-quadrupled Nd:YAG (256 nm) laser. The time evolution of the line was observed by an intensified CCD camera. The investigators reported measurements at greater than Mach 1, with an accuracy of ± 8 m/s and a spatial resolution of 10 μm perpendicular to the nozzle axis.

The machine vision community developed a class of velocimetry algorithms, called *optical-flow* algorithms, to determine the motion of rigid objects. The technique can be extended to fluid flows by assuming that the effect of molecular diffusion is negligible, and requiring that the velocity field is sufficiently smooth. Since the velocity field is computed from temporal and spatial derivatives of the image field, the accuracy and reliability of the velocity measurements are strongly influenced by noise in the image field. This technique imposes a smoothness criterion on the velocity field, which effectively low-pass filters the data, and which can lower the spatial resolution of the velocity measurements [Wildes et al., 1997]. Amabile et al. (1996) applied the optical-flow algorithms to infer velocity fields from 500–1000-μm diameter microtubes by indirectly imaging 1–20-μm diameter X-ray-scattering emulsion droplets in a liquid flow. High-speed X-ray microimaging techniques were presented by Leu et al. (1997). A synchrotron is used to generate high-intensity X-rays that scatter off the emulsion droplets onto a phosphorous screen. A CCD camera imaging the phosphorous screen detects variations in the scattered X-ray field. The primary advantage of the X-ray imaging technique is that one can obtain structural information about the flow field without having optical access. Hitt et al. (1996) applied the optical flow algorithm to in vivo blood flow in microvascular networks, with diameters ~ 100 μm. The algorithm spectrally decomposes subimages into discrete spatial frequencies by correlating the different spatial frequencies to obtain flow field information. The advantage of this technique is that it does not require discrete particle images to obtain

FIGURE 10.1   The same interrogation region at two different times. Notice the displacement of the particle image pattern.

reliable velocity information. Hitt et al. (1995) obtained *in vivo* images of blood cells flowing through a microvascular network using a 20× water-immersion lens with a spatial resolution on the order of 20 μm in all directions.

Particle image velocimetry (PIV) has been used since the mid-1980s to obtain high spatial resolution, two-dimensional velocity fields in macroscopic flows. The experimental procedure is, at its heart, conceptually simple to understand. A flow is made visible by seeding it with particles. The particles are photographed at two different times. The images are sectioned into many smaller regions called *interrogation regions*, as shown in Figure 10.1. The motion of the group of particles within each interrogation region measuring *p* by *q* pixels is determined using a statistical technique called a *cross-correlation*. If the array of gray values which forms the first image is called $f(i, j)$ and the second image $g(i, j)$, the cross-correlation is given by

$$\Phi(m, n) = \sum_{j=1}^{q} \sum_{i=1}^{p} f(i, j) \cdot g(i + m, j + n) \qquad (10.3)$$

The cross-correlation for a high-quality set of PIV measurements should look like Figure 10.2. The location of the peak indicates how far the particles have moved between the two images. Curve fitting with an appropriate model is used to obtain displacement results accurate to 0.1 pixels.

A PIV bibliography by Adrian (1996) lists more than 1200 references describing various PIV methods and the problems that they have been used to investigate. For a good reference describing many of the technical issues pertinent at macroscopic-length scales, see the text [Raffel et al., 1998]. This section will provide a brief explanation of how PIV works in principle and then concentrate on how PIV is different at small-length scales.

Santiago et al. (1998) demonstrated the first μPIV system — a PIV system with a spatial resolution small enough to be able to make measurements in microscopic systems. Their system was capable of measuring slow flows — velocities on the order of hundreds of microns per second — with a spatial resolution of 6.9 × 6.9 × 1.5 μm. The system used an epifluorescent microscope and an intensified CCD camera to record 300-nm-diameter polystyrene flow-tracing particles. The particles are illuminated using a continuous Hg-arc lamp. The continuous Hg-arc lamp is chosen for situations that require low levels of illumination (e.g., flows containing living biological specimens) and where the velocity is small enough that the particle motion can be frozen by the CCD camera's electronic shutter.

Koutsiaris et al. (1999) demonstrated a system suitable for slow flows that used 10-μm glass spheres for tracer particles and a low spatial resolution, high-speed video system to record the particle images, yielding a spatial resolution of 10.2 μm. The investigators measured the flow of water inside 236-μm round glass capillaries and found agreement between the measurements and the analytical solution within the measurement uncertainty.

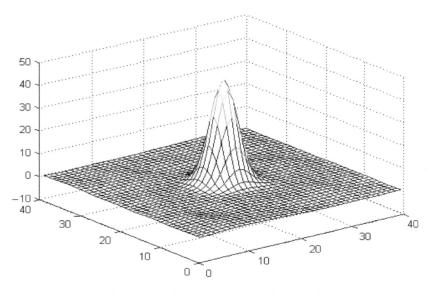

FIGURE 10.2 (See color insert following page 9-22.) PIV cross-correlation peak.

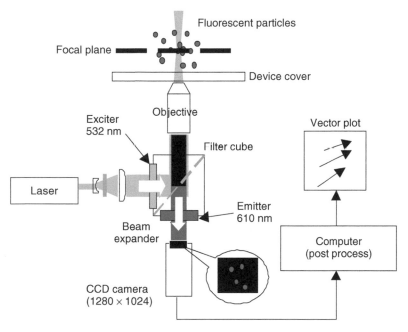

FIGURE 10.3 (See color insert following page 9-22.) Schematic of a μPIV system. A pulsed Nd:YAG laser is used to illuminate fluorescent flow-tracing particles, and a cooled CCD camera is used to record the particle images.

Later applications of the μPIV technique moved steadily toward faster flows. The Hg-arc lamp was replaced with a New Wave, two-headed Nd:YAG laser that allowed cross-correlation analysis of singly exposed image pairs acquired with sub microsecond time steps between images. At macroscopic-length scales, this short time step would allow analysis of supersonic flows. However, because of the high magnification, the maximum velocity measurable with this time step is on the order of meters per second.

Meinhart et al. (1999a) applied μPIV to measure the flow field in a 30-μm-high $\times$ 300-μm-wide rectangular channel, with a flow rate of 50 μl/hr, equivalent to a centerline velocity of 10 mm/s, or three orders of magnitude greater than the initial effort a year before. The experimental apparatus, shown in Figure 10.3, images the flow with a 60×, NA = 1.4, oil-immersion lens. The 200-nm-diameter, polystyrene flow-tracing particles were small enough so that they faithfully followed the flow and were 150 times smaller than the smallest channel dimension. A subsequent investigation by Meinhart and Zhang (2000) of the flow inside a microfabricated inkjet printer head yielded very high speed μPIV measurements. Using a slightly lower magnification (40×) and consequently lower spatial resolution, measurements of velocities as high as 8 m/s were made.

10.3   Overview of μPIV

10.3.1   Fundamental Physics Considerations of μPIV

Three fundamental problems differentiate μPIV from conventional macroscopic PIV: the particles are small compared to the wavelength of the illuminating light; the particles are small enough that the effects of Brownian motion must be addressed; and the illumination source is typically not a light sheet, but rather an illuminated volume of the flow.

10.3.1.1   Particles that are Small Compared to λ

Flow-tracing particles must be large enough to scatter sufficient light so that their images can be recorded. In the Rayleigh scattering regime, where the particle diameter d is much smaller than the wavelength of light, $d \ll \lambda$, the amount of light scattered by a particle varies as d^{-6} [Born and Wolf, 1997]. Because the diameter of the flow-tracing particles must be small enough that the particles do not disturb the flow being measured, they can frequently be on the order of 50–100 nm. Their diameters are then 1/10 to 1/5 the wavelength of green light, $\lambda = 532$ nm, and are therefore approaching the Rayleigh scattering criteria. This situation places significant constraints on the image recording optics, making it extremely difficult to record particle images.

   One solution to the imaging problem is to use epifluorescence imaging to record light emitted from fluo-rescently labeled particles, using an optical filter to remove the background light. This technique was used successfully in liquid flows to record images of 200–300-nm-diameter fluorescent particles [Adrian, 1996; Meinhart et al., 1999a]. Whereas fluorescently labeled particles are well suited for μPIV studies in liquid flows, they are not readily applicable to high-speed gas flows, for several reasons. First, commercially available, flu-orescently labeled sub micron particles are available only in aqueous solutions. Furthermore, the emission decay time of many fluorescent molecules is on the order of several nanoseconds, which may cause streaking of the particle images for high-speed flows. Presently, seeding gas flows remains a significant problem in μPIV.

10.3.1.2   Effects of Brownian Motion

When the seed particle size becomes small, the collective effect of collisions between the particles and a moderate number of fluid molecules is unbalanced, preventing to some degree, the particle from follow-ing the flow [Probstein, 1994]. This phenomenon, commonly called *Brownian motion*, has two potential implications for μPIV: one is to cause an error in the measurement of the flow *velocity*; the other is to cause an uncertainty in the *location* of the flow tracing particles — although this problem is eliminated by using pulse lasers for illumination. In order to assess the effects of Brownian motion, it is first necessary to establish how particles suspended in flows behave.

10.3.1.2.1   Flow/Particle Dynamics
In stark contrast to many macroscale fluid mechanics experiments, the hydrodynamic size of a particle (a measure of its ability to follow the flow based on the ratio of inertial to drag forces) is usually not a con-cern in microfluidic applications because of the large surface-to-volume ratios at small-length scales. A simple model for the response time of a particle subjected to a step change in local fluid velocity can be used to gauge particle behavior. Based on a simple first-order inertial response to a constant flow accel-eration (assuming Stokes flow for the particle drag), the response time τ_{p} of a particle is:

$$\tau_{\mathrm{p}} = \frac{d_{\mathrm{p}}^2 \rho_{\mathrm{p}}}{18\eta} \tag{10.4}$$

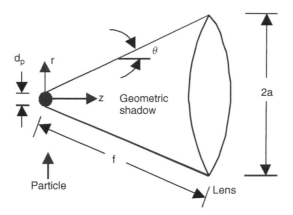

FIGURE 10.4 Geometry of a particle with a diameter d_p, being imaged through a circular aperture of radius a, by a lens of focal length f (after [Meinhart et al., 2000b]). The values r and z are the in-plane radius and the out-of-plane coordinate, respectively, with the origin located at the point source.

where d_p and ρ_p are the diameter and density of the particle, respectively, and η is the dynamic viscosity of the fluid. Considering typical μPIV experimental parameters of 300-nm-diameter polystyrene latex spheres immersed in water, the particle response time would be 10^{-9} sec. This response time is much smaller than the time scales of any realistic liquid or low-speed gas flow field.

In the case of high-speed gas flows, the particle response time may be an important consideration when designing a system for microflow measurements. For example, a 400-nm particle seeded into an air micronozzle that expands from the sonic at the throat to Mach 2 over a 1-mm distance may experience a particle-to-gas relative flow velocity of more than 5% (assuming a constant acceleration and a stagnation temperature of 300 K). Particle response to flow through a normal shock would be significantly worse. Another consideration in gas microchannels is the breakdown of the no-slip and continuum assumptions as the particle's Knudsen number Kn_p, defined as the ratio of the mean free path of the gas to the particle diameter, approaches (and exceeds) one. For the case of the slip flow regime ($10^{-3} < Kn_p < 0.1$), it is possible to use corrections to the Stokes drag relation to quantify particle dynamics [Beskok et al., 1996]. For example, a correction offered by Melling (1986) suggests the following relation for the particle response time:

$$\tau_p = (1 + 2.76 Kn_p) \frac{d_p^2 \rho_p}{18 \eta} \tag{10.5}$$

10.3.1.2.2 Velocity Errors

Santiago et al. (1998) briefly considered the effect of Brownian motion on the accuracy of μPIV measurements. Devasenathipathy et al. (2003) provided the more in-depth consideration of the phenomenon of Brownian motion necessary to completely explain its effects in μPIV. For time intervals Δt much larger than the particle inertial response time, the dynamics of Brownian displacement are independent of inertial parameters such as particle and fluid density, and the mean square distance of diffusion is proportional to $D\Delta t$, where D is the diffusion coefficient of the particle. For a spherical particle in an unbounded medium subject to Stokes drag law, the diffusion coefficient D was first given by Einstein (1905) as:

$$D = \frac{\kappa T}{3 \pi \eta d_p} \tag{10.6}$$

where d_p is the particle diameter, κ is Boltzmann's constant, T is the absolute temperature of the fluid, and η is the dynamic viscosity of the fluid.

The random Brownian displacements cause particle trajectories to fluctuate about the deterministic pathlines of the fluid flow field. Assuming the flow field is steady over the time of measurement and the local velocity gradient is small, the imaged Brownian particle motion can be considered a fluctuation about a streamline that passes through the particle's initial location. An ideal, non-Brownian (i.e., deterministic) particle following a particular streamline for a time period Δt has x- and y-displacements of

$$\Delta x = u\Delta t \tag{10.7a}$$

$$\Delta y = v\Delta t \tag{10.7b}$$

where u and v are the x and y components of the time-averaged, local fluid velocity, respectively. The relative errors, ε_x and ε_y, incurred as a result of imaging the Brownian particle displacements in a two-dimensional measurement of the x and y components of particle velocity, are given as

$$\varepsilon_x = \frac{\sigma_x}{\Delta x} = \frac{1}{u}\sqrt{\frac{2D}{\Delta t}} \tag{10.8a}$$

$$\varepsilon_y = \frac{\sigma_y}{\Delta y} = \frac{1}{v}\sqrt{\frac{2D}{\Delta t}} \tag{10.8b}$$

This Brownian error establishes a lower limit on the measurement time interval Δt because, for shorter times, the measurements are dominated by uncorrelated Brownian motion. These quantities describe the relative magnitudes of the Brownian motion and will be referred to here as Brownian intensities. The errors estimated by Equation (10.8a) and Equation (10.8b) show that the relative Brownian intensity error decreases as the time of measurement increases. Larger time intervals produce flow displacements proportional to Δt, whereas the RMS of the Brownian particle displacements grows as $\Delta t^{1/2}$. In practice, Brownian motion is an important consideration when tracing 50–500-nm particles in flow field experiments with flow velocities of less than about 1 mm/s. For a velocity on the order of 0.5 mm/s and a 500-nm seed particle, the lower limit for the time spacing is approximately 100 μs for a 20% error because of Brownian motion. This error can be reduced by averaging over several particles in a single interrogation spot and by ensemble averaging over several realizations. The diffusive uncertainty decreases as $1/\sqrt{N}$, where N is the total number of particles in the average [Bendat and Piersol, 1986].

Equation (10.8) demonstrates that the effect of the Brownian motion is relatively less important for faster flows. However, for a given measurement, when u increases, Δt will generally decrease. Equation (10.8a) and Equation (10.8b) also demonstrate that when all conditions but Δt are fixed, increasing Δt will decrease the relative error introduced by the Brownian motion. Unfortunately, a longer Δt will decrease the accuracy of the results, because the PIV measurements are based on a first-order accurate approximation to the velocity. Using a second-order accurate technique allows for a longer Δt to be used without increasing this error [Wereley and Meinhart, 2001a].

10.3.1.3   Volume Illumination of the Flow

Another significant difference between μPIV and macroscopic PIV is that, owing to a lack of optical access and significant diffraction in light sheet forming optics, light sheets are typically not a practical source of illumination for microflows. Consequently, the flow must be volume illuminated, leaving two choices for how to visualize the seed particles — with an optical system whose depth of focus exceeds the depth of the flow being measured, or with an optical system whose depth of focus is small compared to that of the flow. Both of these techniques have been used in various implementations of μPIV. Cummings (2001) used a large depth of focus imaging system to explore electrokinetic and pressure-driven flows. The advantage of the large depth of focus optical system is that all particles in the field of view of the optical system are well focused. The disadvantage of this scheme is that all depth information is lost, and the resulting velocity fields are completely depth-averaged. Cummings (2001) addressed this problem with advanced processing techniques that will not be covered here.

The second choice of imaging systems is one whose depth of focus is smaller than that of the flow domain. The optical system will then focus those particles that are within the depth of focus of the imaging system, whereas the remaining particles will be unfocused and contribute to the background noise level. Because the optical system is being used to define thickness of the measurement domain, it is important to characterize exactly how thick the depth of focus, or more appropriately, the *depth of correlation* δz_m, is. Meinhart et al. (2000b) considered this question in detail by starting from the basic principles of how small particles are imaged.

10.3.1.3.1 Depth of Field

The depth of field of a standard microscope objective lens is given by Inoué & Spring (1997) as:

$$\delta z = \frac{n\lambda_0}{\mathrm{NA}^2} + \frac{ne}{\mathrm{NA} \cdot M} \tag{10.9}$$

where n is the refractive index of the fluid between the microfluidic device and the objective lens; λ_0 is the wavelength of light in a vacuum being imaged by the optical system; NA is the numerical aperture of the objective lens; M is the total magnification of the system; and e is the smallest distance that can be resolved by a detector located in the image plane of the microscope (for the case of a CCD sensor, e is the spacing between pixels). Equation (10.9) is the summation of the depths of field resulting from diffraction (first term on the right-hand side) and geometric effects (second term on the right-hand side).

The cutoff for the depth of field caused by diffraction, i.e., first term on the right-hand side of Equation (10.9), is chosen by convention to be one-quarter of the out-of-plane distance between the first two minima in the three-dimensional point spread function, $u = \pm\pi$ in diffraction variables. Substituting $\mathrm{NA} = n \sin \theta = na/f$, and $\lambda_0 = n\lambda$ yields the first term on the right-hand side of Equation (10.9).

If a CCD sensor is used to record particle images, the geometric term in Equation (10.9) can be derived by projecting the CCD array into the flow field, and then, considering the out-of-plane distance, the CCD sensor can be moved before the geometric shadow of the point source occupies more than a single pixel. This derivation is valid for small light collection angles, where $\tan \theta \sim \sin \theta = \mathrm{NA}/n$.

10.3.1.3.2 In-Plane Spatial Resolution Limits

The overall goal of μPIV is to obtain reliable two-dimensional velocity fields in microfluidic devices with high accuracy and high spatial resolution. In this section, we will discuss the theoretical requirements for achieving both of these goals and address the relative tradeoffs between velocity accuracy and spatial resolution.

The most common mode of PIV is to record two successive images of flow-tracing particles that are introduced into the working fluid, and which accurately follow the local motion of the fluid. The two particle images are separated by a known time delay, Δt. Typically, the two particle image fields are divided into uniformly spaced interrogation regions, which are cross-correlated to determine the most probable local displacement of the particles.

High spatial resolution is achieved by recording the images of flow-tracing particles with sufficiently small diameters, d_p, so that they faithfully follow the flow in microfluidic devices. The particle should be imaged with high resolution optics and with sufficiently high magnification so that the particles are resolved with at least 3–4 pixels per particle diameter. Following Adrian (1991), the diffraction-limited spot size of a point source of light, d_s, imaged through a circular aperture is given by

$$d_s = 2.44 \, (M + 1) f^\# \lambda \tag{10.10}$$

where M is the magnification, $f^\#$ is the *f-number* of the lens, and λ is the wavelength of light. For infinity-corrected microscope objective lenses, $f^\# \approx 1/2[(n/\mathrm{NA})^2 - 1]^{1/2}$. The numerical aperture, NA, is defined as $\mathrm{NA} = n \sin \theta$, where n is the index of refraction of the recording medium, and θ is the half-angle subtend by the aperture of the recording lens. The actual recorded image can be estimated as the convolution of point-spread function with the geometric image. Approximating both these images as Gaussian functions, the effective image diameter, d_e, can be written as (Adrian and Yao, 1985)

$$d_e = [d_s^2 + M^2 d_p^2]^{1/2} \tag{10.11}$$

TABLE 10.1 Effective Particle Image Diameters When Projected Back into the Flow, d_e/M (μm) (After [Wereley, S.T., and Meinhart, C.D. 2003])

	Microscope Objective Lens Characteristics				
Particle Size, d_p	$M = 60$ NA = 1.4	$M = 40$ NA = 0.75	$M = 40$ NA = 0.6	$M = 20$ NA = 0.5	$M = 10$ NA = 0.25
0.01 μm	0.29	0.62	0.93	1.24	2.91
0.10 μm	0.30	0.63	0.94	1.25	2.91
0.20 μm	0.35	0.65	0.95	1.26	2.92
0.30 μm	0.42	0.69	0.98	1.28	2.93
0.50 μm	0.58	0.79	1.06	1.34	2.95
0.70 μm	0.76	0.93	1.17	1.43	2.99
1.00 μm	1.04	1.18	1.37	1.59	3.08
3.00 μm	3.01	3.06	3.14	3.25	4.18

The effective particle image diameter places a bound on the spatial resolution that can be obtained by μPIV. Assuming that the particle images are sufficiently resolved by the CCD array, the location of the correlation peak can be sufficiently resolved to within 1/10th the particle image diameter (Prasad et al., 1993). Therefore, the uncertainty of the correlation peak location for a $d_p = 0.2\,\mu$m diameter particle recorded with a NA = 1.4 lens is $\delta x \sim d_e/10M = 35$ nm. Table 10.1 gives effective particle diameters recorded through a circular aperture and then projected back into the flow, d_e/M.

10.3.1.3.3 Out-of-Plane Spatial Resolution
It is common practice in PIV to use a sheet of light to illuminate the flow-tracing particles. In principle, the light sheet illuminates only particles contained within the depth of focus of the recording lens. This process provides reasonably high quality, in-focus particle images to be recorded with low levels of background noise being emitted from the out-of-focus particles. The out-of-plane spatial resolution of the velocity measurements is defined clearly by the thickness of the illuminating light sheet.

Because of the small length scales associated with μPIV, it is difficult, if not impossible, to form a light sheet that is only a few microns thick, and even more difficult to align a light sheet with the object plane of a microscope objective lens. Consequently, it is common practice in μPIV to illuminate the test section with a volume of light, and rely on the depth of field of the lens to define the out-of-plane thickness of the measurement plane.

The effective particle image diameter in volume illumination can be derived as follows. Following the analysis of Olsen & Adrian (2000a), using $f^\# \approx 1/2[(n/\text{NA})^2 - 1]^{1/2}$, the effective image diameter of a particle displaced a distance z from the objective plane can be approximated by combining Equation (10.10) and Equation (10.11), and adding a third term to account for the geometric spreading of a slightly out-of-focus particle

$$d_e = \left[M^2 d_p^2 + 1.49(M + 1)^2\lambda^2\left[(n/\text{NA})^2 - 1\right] + \frac{M^2 D_a^2 z^2}{(s_o + z)^2} \right]^{1/2} \tag{10.12}$$

where s_o is the object distance, and D_a is the diameter of the recording lens aperture.

The out-of-plane spatial resolution can be determined in terms of the *depth of correlation*. The depth of correlation is defined as the axial distance, z_{corr}, from the object plane in which a particle becomes sufficiently out-of-focus, so that it no longer contributes significantly to the signal peak in the particle-image correlation function. Following the analysis of Olsen & Adrian (2000a), the expression of the depth of correlation is derived as follows:

$$z_{\text{corr}} = \left[\left(\frac{1 - \sqrt{\varepsilon}}{\sqrt{\varepsilon}} \right)\left[\frac{d_p^2[(n/\text{NA})^2 - 1]}{4} + \frac{1.49(M + 1)^2\lambda^2[(n/\text{NA})^2 - 1]^2}{4M^2} \right] \right]^{1/2} \tag{10.13}$$

The depth of correlation, z_{corr}, is strongly dependent on numerical aperture, NA, and particle size, d_p, and is weakly dependent upon magnification, M. The variable ε represents the relative contribution of a

TABLE 10.2 Thickness of the Measurement Plane for Typical Experimental Parameters, $2z_{corr}$ (μm) (After [Wereley, S.T., and Meinhart, C.D. 2003])

Particle Size, d_p	Microscope Objective Lens Characteristics				
	$M = 60$ NA = 1.4	$M = 40$ NA = 0.75	$M = 40$ NA = 0.6	$M = 20$ NA = 0.5	$M = 10$ NA = 0.25
0.01 μm	0.36	1.6	3.7	6.5	34
0.10 μm	0.38	1.6	3.8	6.5	34
0.20 μm	0.43	1.7	3.8	6.5	34
0.30 μm	0.52	1.8	3.9	6.6	34
0.50 μm	0.72	2.1	4.2	7.0	34
0.70 μm	0.94	2.5	4.7	7.4	35
1.00 μm	1.3	3.1	5.5	8.3	36
3.00 μm	3.7	8.1	13	17	49

particle displaced a distance z from the object plane, compared to a particle located at the object plane. Table 10.2 gives the thickness of the measurement plane, $2z_{corr}$, for various microscope objective lenses and particle sizes. The highest out-of-plane resolution for these parameters is $2z_{corr} = 0.36\,\mu$m for a NA = 1.4, $M = 60$ lens, and particle sizes $d_p < 0.1\,\mu$m. For these calculations, it is important to note that the effective numerical aperture of an oil-immersion lens is reduced when imaging particles suspended in fluids such as water, where the refractive index is less than that of the immersion oil.

10.3.1.3.4   Particle Visibility
The ability to obtain highly reliable velocity data depends significantly upon the quality of the recorded particle images. In macroscopic PIV experiments, it is customary to use a sheet of light to illuminate only those particles that are within the depth of field of the recording lens. This process minimizes background noise resulting from light emitted by out-of-focus particles. However, in μPIV, the small length scales and poor optical access necessitate the use of volume illumination.

Experiments using the μPIV technique must be designed so that in-focus particle images can be observed despite the background light produced by out-of-focus particles and the test section surfaces. The background light from test section surfaces can be removed by using fluorescent techniques to filter out elastically scattered light [Santiago et al., 1998].

Background light from unfocused particles is not so easily removed, but the amount can be lowered to acceptable levels by judiciously choosing proper experimental parameters. Olsen & Adrian (2000a) presented a theory to estimate peak particle visibility, defined as the ratio of the intensity of an in-focus particle image to the average intensity of the background light produced by the out of focus particles.

Assuming light is emitted uniformly from the particle, the light of a single particle reaching the image plane can be written as

$$J(z) = \frac{J_p D_a^2}{16(s_o + z)^2} \tag{10.14}$$

where J_p is total light flux emitted by a single particle. Approximating the intensity of an in-focus particle image as Gaussian,

$$I(r) = I_o \exp\left(\frac{-4\beta^2 r^2}{d_e^2}\right) \tag{10.15}$$

where the unspecified parameter, β, is chosen to determine the cutoff level that defines the edge of the particle image. Approximating the Airy distribution by a Gaussian distribution, with the area of the two axisymmetric functions being equal, the first zero in the Airy distribution corresponds to

$I/I_o = \exp(-\beta^2 = -3.67)$ [Adrian and Yao, 1985]. Since the total light flux reaching the image plane is $J = \int I(r)dA$, Equation (10.14) and Equation (10.15) can be combined, yielding (Olsen & Adrian, 2000a)

$$I(r, z) = \frac{J_p D_a^2 \beta^2}{4\pi d_e^2(s_o + z)^2} \exp\left(\frac{-4\beta^2 r^2}{d_e^2}\right) \tag{10.16}$$

Idealizing particles that are located a distance $|z| > \delta/2$ from the object plane as being out-of-focus and contributing uniformly to background intensity, and particles located within a distance $|z| < \delta/2$ as being completely in-focus, the total flux of background light, J_B, can be approximated by

$$J_B = A_v C\left\{\int_{-a}^{-\delta/2} J(z)dz + \int_{\delta/2}^{L-a} J(z)dz\right\} \tag{10.17}$$

where C is the number of particles per unit volume of fluid, L is the depth of the device, and A_v is the average cross sectional area contained within the field of view.

Combining Equation (10.14) and Equation (10.17), correcting for the effect of magnification, and assuming $s_o \gg \delta/2$, the intensity of the background glow can be expressed as (Olsen and Adrian, 2000a)

$$I_B = \frac{CJ_p LD_a^2}{16M^2(s_o - a)(s_o - a + L)} \tag{10.18}$$

Following Olsen & Adrian (2000a), the visibility of an in-focus particle, V, can be obtained by combining Equation (10.12) and Equation (10.16), dividing by Equation (10.18), and setting $r = 0$ and $z = 0$,

$$V = \frac{I(0, 0)}{I_B} = \frac{4M^2\beta^2(s_o - a)(s_o - a + L)}{\pi C L s_o^2\left(M^2 d_p^2 + 1.49(M + 1)^2\lambda^2[(n/NA)^2 - 1]\right)} \tag{10.19}$$

For a given set of recording optics, particle visibility can be increased by decreasing particle concentration, C, or by decreasing test section thickness, L. For a fixed particle concentration, the visibility can be increased by decreasing the particle diameter, or by increasing the numerical aperture of the recording lens. Visibility depends only weakly on magnification and on object distance, s_o.

An expression for the volume fraction, V_{fr}, of particles in solution that produce a specific particle visibility can be obtained by rearranging Equation (10.19) and multiplying by the volume occupied by a spherical particle

$$V_{fr} = \frac{2d_p^3 M^2\beta^2(s_o - a)(s_o - a + L)}{3VLs_o^2\left(M^2 d_p^2 + 1.49(M + 1)^2\lambda^2[(n/NA)^2 - 1]\right)} \times 100\% \tag{10.20}$$

Reasonably high quality particle-image fields require visibilities of, say, $V \sim 1.5$. For the purpose of illustration, assume that we are interested in measuring the flow at the centerline, $a = L/2$, of a microfluidic device with a characteristic depth, $L = 100\,\mu m$. It is also important to seed the flow so that the percentage of particle volume fraction of seed particles is kept below a suitable level. Thus the particle loading on the fluid is not too large. Table 10.3 shows the maximum percent volume fraction of particles that can be in the fluid, while maintaining an in-focus particle visibility $V = 1.5$, for various experimental parameters. Here, the object distance, s_o, is estimated by adding the working distance of the lens to the designed coverslip thickness.

Meinhart et al. (2000b) demonstrated these competing signal-to-noise ratio and spatial resolution issues with a series of experiments using known particle concentrations and flow depths. The measured signal-to-noise ratio was estimated from particle-image fields taken of four different particle concentrations and four different device depths. A particle solution was prepared by diluting $d_p = 200$-nm-diameter polystyrene particles in de-ionized water. Test sections were formed using two feeler gauges of known thickness

TABLE 10.3 Maximum Percent Volume Fraction of Particles, V_{fr}, While Maintaining an In-focus Visibility, $V = 1.5$, for Imaging the Center of an $L = 100\,\mu m$ Deep Device (After [Wereley, S.T., and Meinhart, C.D. 2003])

Particle Size, d_p	Microscope Objective Lens Characteristics				
	$M = 60$ $NA = 1.4$ $s_o = 0.38\,mm$	$M = 40$ $NA = 0.75$ $s_o = 0.89\,mm$	$M = 40$ $NA = 0.6$ $s_o = 3\,mm$	$M = 20$ $NA = 0.5$ $s_o = 7\,mm$	$M = 10$ $NA = 0.25$ $s_o = 10.5\,mm$
$0.01\,\mu m$	2.0E−5	4.3E−6	1.9E−6	1.1E−6	1.9E−7
$0.10\,\mu m$	1.7E−2	4.2E−3	1.9E−3	1.1E−3	1.9E−4
$0.20\,\mu m$	1.1E−1	3.1E−2	1.4E−2	8.2E−3	1.5E−3
$0.30\,\mu m$	2.5E−1	9.3E−2	4.6E−2	2.7E−2	5.1E−3
$0.50\,\mu m$	6.0E−1	3.2E−1	1.8E−1	1.1E−1	2.3E−2
$0.70\,\mu m$	9.6E−1	6.4E−1	4.1E−1	2.8E−1	6.2E−2
$1.00\,\mu m$	1.5E+0	1.2E+0	8.7E−1	6.4E−1	1.7E−1
$3.00\,\mu m$	4.8E+0	4.7E+0	4.5E+0	4.2E+0	2.5E+0

TABLE 10.4 The Effect of Background Noise on Image Quality (i.e., Signal-to-Noise Ratio) [Meinhart et al., 1999b]

Depth (μm)	Particle Concentration (by Volume)			
	0.01%	0.02%	0.04%	0.08%
25	2.2	2.1	2.0	1.9
50	1.9	1.7	1.4	1.2
125	1.5	1.4	1.2	1.1
170	1.3	1.2	1.1	1.0

sandwiched between a glass microscope slide and a coverslip. The particle images were recorded near one of the feeler gauges to minimize errors associated with variations in the test-section thickness, which could result from surface tension or deflection of the coverslip. The images were recorded with an oil-immersion $M = 60\times$, $NA = 1.4$ objective lens. The remainder of the μPIV system was as previously described.

Meinhart et al. (2000b) defined the signal-to-noise ratio of the image to be the ratio of the peak image intensity of an average in-focus particle divided by the average background intensity. The maximum possible spatial resolution is defined as the smallest square interrogation region that would, on average, contain three particle images and is directly related to the number concentration of the particle images in each PIV image.

The measured signal-to-noise ratio is shown in Table 10.4. As expected, the results indicate that, for a given particle concentration, a higher signal-to-noise ratio is obtained by imaging a flow in a thinner device. This higher ratio occurs because decreasing the thickness of the test section decreases the number of out-of-focus particles, whereas the number of in-focus particles remains constant. In general, thinner test sections allow higher particle concentrations to be used, which can be analyzed using smaller interrogation regions. Consequently, the seed-particle concentration must be chosen judiciously so that the desired spatial resolution can be obtained, while maintaining adequate image quality (i.e., signal-to-noise ratio).

10.3.2 Special Processing Methods for μPIV Recordings

When evaluating digital PIV recordings with conventional correlation-based algorithms or image-pattern tracking algorithms, a sufficient number of particle images is required in the interrogation window or the tracked image pattern to ensure reliable and accurate measurement results. However, in many cases, especially in μPIV measurements, the particle image density in the PIV recordings is not high enough. These PIV

FIGURE 10.5 One of the LID-PIV recordings. Image size: 256 × 256 pixels [Wereley et al., 2002b]. Copyright 2002, AIAA. Reprinted with permission.

recordings are called *low image density* (LID) recordings and are usually evaluated with particle-tracking algorithms. When using particle-tracking algorithms, the velocity vector is determined with only one particle, hence the reliability and accuracy of the technique are limited. In addition, interpolation procedures are usually necessary to obtain velocity vectors on the desired regular grid points from the random distributed particle-tracking results, and therefore, additional uncertainties are added to the final results. Fortunately, special processing methods can be used to evaluate the μPIV recordings, so that the errors resulting from the low-image density can be avoided [Wereley et al., 2002b]. In this section, two methods are introduced to improve measurement accuracy of μPIV: improving the evaluation algorithm and using a digital image processing technique.

10.3.2.1 Ensemble Correlation Method

For correlation-based PIV evaluation algorithms, the correlation function at a certain interrogation spot is usually represented as:

$$\Phi_k(m, n) = \sum_{j=1}^{q}\sum_{i=1}^{p} f_k(i, j) \cdot g_k(i + m, j + n) \tag{10.21}$$

where $f_k(i, j)$ and $g_k(i, j)$ are the gray value distributions of the first and second exposures, respectively, in the kth PIV recording pair at a certain interrogation spot of a size of $p \times q$ pixels. The correlation function for a singly exposed PIV image pair has a peak at the position of the particle image displacement in the interrogation spot (or window), which should be the highest among all the peaks of Φ_k. The subpeaks, which result from noise or mismatch of particle images, are usually obviously lower than the main peak (i.e., the peak of the particle image displacement). However, when the interrogation window does not contain enough particle images or the noise level is too high, the main peak will become weak and may be lower than some of the subpeaks, and consequently, an erroneous velocity vector is generated. In the laminar and steady flows often measured by μPIV systems, the velocity field is independent of the measurement time. That means the main peak of $\Phi_k(m, n)$ is always at the same position for PIV recording pairs taken at different times, whereas the subpeaks appear with random intensities and positions in different recording pairs. Therefore, when averaging Φ_k over a large number of PIV recording pairs (N), the

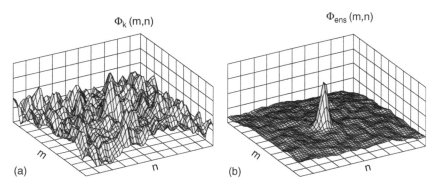

$\Phi_k\,(m,n)$ $\qquad\qquad\qquad$ $\Phi_{ens}\,(m,n)$

(a) $\qquad\qquad\qquad\qquad\qquad$ (b)

FIGURE 10.6 Effect of ensemble correlation: (a) results with conventional correlation for one of the PIV recording pairs; and (b) results with ensemble correlation for 101 PIV recording pairs [Wereley et al., 2002b]. Copyright 2002, AIAA. Reprinted with permission.

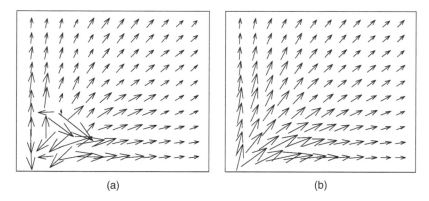

(a) $\qquad\qquad\qquad\qquad\qquad$ (b)

FIGURE 10.7 Comparison of the evaluation function of a single PIV recording pair (a) with the average of 101 evaluation functions (b) [Wereley et al., 2002b]. Copyright 2002, AIAA. Reprinted with permission.

main peak will remain at the same position in each correlation function, but the noise peaks, which occur randomly, will average to zero. The averaged (or ensemble) correlation function is given as:

$$\Phi_{ens}(m, n) = \frac{1}{N} \sum_{k=1}^{N} \Phi_k(m, n) \tag{10.22}$$

The ensemble correlation requires a steady flow. The concept of averaging correlation functions can also be applied to other evaluation algorithms such as correlation tracking and the MQD method. This method was first proposed and demonstrated by Meinhart et al. (1999b).

The ensemble correlation function technique is demonstrated for 101 LID-PIV recording pairs (Φ_{ens}) in Figure 10.6 in comparison to the correlation function for one of the single recording pairs (Φ_k). These PIV recording pairs are chosen from the flow measurement in a microfluidic biochip for impedance spectroscopy of biological species [Gomez et al., 2001]. With the conventional evaluation function in Figure 10.6(a), the main peak cannot easily be identified among the subpeaks, so the evaluation result is neither reliable nor accurate. However, the ensemble correlation function in Figure 10.6(b) shows a very clear peak at the particle image displacement, and the subpeaks can hardly be recognized.

The effect of the ensemble correlation technique on the resulting velocity field is demonstrated in Figure 10.7 with the PIV measurement of flow in the microfluidic biochip. All the obvious evaluation errors resulting from the low image density and strong background noise (Figure 10.7(a)) are avoided by

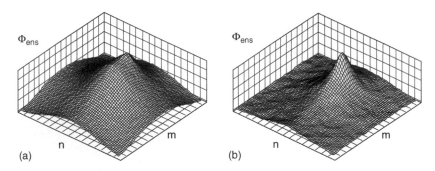

FIGURE 10.8   Ensemble correlation function for 100 image sample pairs (a) without and (b) with background removal [Wereley et al., 2002b]. Copyright 2002, AIAA. Reprinted with permission.

using the ensemble correlation method based on 101 PIV recording pairs (Figure 10.7(b)). One important note here is that because the bad vectors in Figure 10.7(a) all occur at the lower left corner of the flow domain, the removal of these bad vectors and subsequent replacement by interpolated vectors will only coincidentally generate results that bear any resemblance to the true velocity field in the device. In addition, if the problem leading to low signal levels in the lower left-hand corner of the images is systematic (i.e., larger background noise), even a large collection of images will not generate better results, because they will all have bad vectors at the same location.

10.3.2.2   Removing Background Noise

When using the ensemble correlation technique, a large number of μPIV recording pairs is usually obtained, enabling the removal of the background noise from the μPIV recording pairs. One of the possibilities for obtaining an image of the background from plenty of PIV recordings is averaging these recordings [Gui et al., 1997]. Because the particles are randomly distributed and quickly move through the camera view area, their images will disappear in the averaged recording. However, the image of the background (including the boundary, contaminants on the glass cover, and particles adhered to the wall) maintains the same brightness distribution in the averaged recording, because it does not move or change. Another method is building at each pixel location a minimum of the ensemble of PIV recordings, because the minimal gray value at each pixel may reflect the background brightness in the successively recorded images [Cowen and Monismith, 1997]. The background noise may be successfully removed by subtracting the background image from the PIV recordings.

A data set from a flow in a microchannel is used to demonstrate this point. The size of the interrogation regions is 64 × 64 pixels, and the total sample number is 100 pairs. The mean particle image displacement is about 12.5 pixels from left to right. In one particular interrogation region in the images, the particle images in a region at the left side of the interrogation region look darker than those outside this region. This situation may result from an asperity on the glass cover of the microchannel.

The ensemble correlation function for the 100 image sample pairs without background removal is given in Figure 10.8(a), which shows a dominant peak near zero displacement because the fleck does not move. When the background image is built with the minimum gray value method and subtracted from the image sample pairs, the influence of the asperity is reduced, so that the peak of the particle image displacement appears clearly in the evaluation function in Figure 10.8(b).

10.4   Entrance Length Measurement in Microchannel Flow

The μPIV technique is quite well suited for use in the entrance length phenomenon. Its highly spatially resolved measurements allow a full investigation of the velocity field as it progresses from the upstream reservoir toward the fully developed region in the microchannel. Furthermore, the high dynamic range

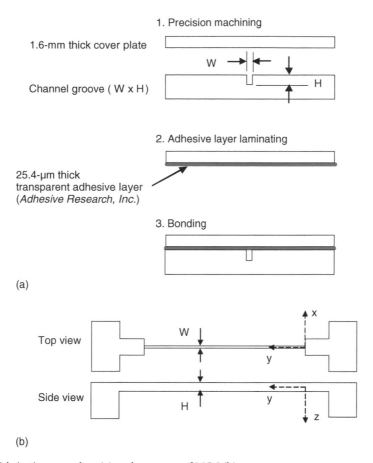

FIGURE 10.9 Fabrication procedure (a) and geometry of MC-I (b).

of the technique allows for measurement of both slow flows and fast flows (i.e., low and high Reynolds numbers) without changing the experimental setup.

10.4.1 Microchannel Fabrication

MC-I was fabricated from transparent acrylic. Figure 10.9(a) shows the fabrication procedure. The microchannel groove and the cover plate were made using conventional machining — a precision sawing technique. The 1.6-mm-thick cover plate was laminated with a 10.4-µm-thick, pressure sensitive adhesive layer (Adhesive Research, Inc.) that had an optical transmission rate over 95%. This plate was bonded on top of the microchannel groove. Further, this bonded piece was compressed by an external clamp to ensure that the channel was leakproof. Figure 10.9(b) shows the completed microchannel structure. Two deep reservoirs or plenums, leading to the shallow plenums, were built at both ends of the microchannel to ensure even distribution of the flow. The microchannel dimension was optically measured and found as $252 \pm 6\,\mu$m and $694 \pm 7\,\mu$m, respectively.

The fabrication procedure of MC-II is shown in Figure 10.10. First, 525-µm-thick, P-type (100) 4″ silicon wafers were thermally oxidized by 0.6 µm on both sides. The microchannels and pressure measurement reservoirs on the front side were defined by photolithography. Then the DRIE process (Cornell University) was used to form the microchannels and pressure measurement reservoirs. The wafers were oxidized again, and lithography using AZ 4620 was done on the backside. AZ 4620 is a thick photoresist,

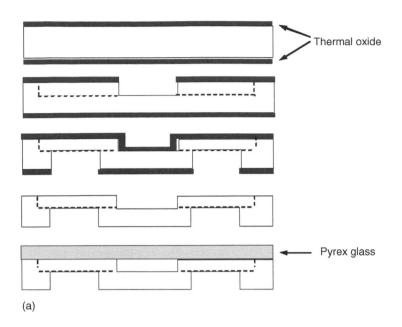

(a)

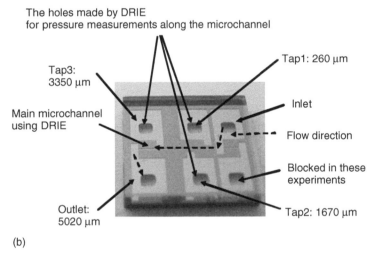

(b)

FIGURE 10.10   (**See color insert following page 9-22.**) The schematics of (a) fabrication procedure and (b) photo of MC-II.

used as a barrier layer for the backside DRIE process. Another DRIE was applied to etch through the wafers to the bottom surfaces of the reservoirs, connecting the reservoirs to the backside of the chip where tubes were epoxy-bonded to supply flow, remove flow, and measure pressure. After the remaining photoresist and oxide were removed, anodic bonding was done with 1-mm-thick, Pyrex 7740 glass wafers at 350°C and 1000 V. Finally, the bonded wafers were diced (Figure 10.10(a)).

The microchannel chip (8.1×8.1 mm^2) consists of the main microchannel, inlet and outlet reservoirs, and five reservoirs for measuring pressure along the main microchannel. It has short, narrow channels connecting the main microchannel with the three pressure reservoirs (Figure 10.10(b)). The five pressure reservoirs are located at $0\,\mu$m (inlet), $260\,\mu$m, $1670\,\mu$m, $3350\,\mu$m, and $5020\,\mu$m (outlet) from the inlet. In the current study, the $5020\,\mu$m channel was tested and the cross-section was found to be $104.6\,\mu$m (W) $\times$ $38.6\,\mu$m (H).

10.4.2 Microparticle Image Velocimetry System

Figure 10.3 shows a schematic of the μPIV system [Santiago et al., 1998; Wereley et al., 1998] used to make the measurements. A two cavity, frequency-doubled Nd:YAG laser (New Wave Inc.) was used for the illumination. The wavelength and the pulse width were 532 nm and about 3~5 ns, respectively. The laser beam was delivered into the inverted epifluorescent microscope (Nikon, TE200) through the beam expander assembly, which was located between the laser aperture and the microscope back aperture. This beam expander assembly is carefully designed for the laser beam to have characteristics similar to the original mercury lamp. A negative and a positive lens in a Galilean telescope arrangement were used to expand the beam, and a 5° diffuser was located between two lenses to disorder the collimation of the laser beam, so that laser focal points did not damage internal optics. The laser beam was guided to the flow field of the microchannel device by passing through an epifluorescence filter cube and an objective lens. The filter cube, located below the objective lens, was an assembly of the exciter, the emitter, and a dichroic mirror. Because the dichroic mirror only transmits light in the range >585 nm, the beam was redirected to the objective lens. The beam coming out the objective lens illuminated a large volume of fluid within the microchannel in which the seeding particles were suspended.

Fluorescent particles (Duke Scientific Co.) are used for flow seeding. For MC-I and MC-II, 1-μm- and 0.69-μm-diameter polystyrene were used, respectively. The fluorescent particles absorb the illuminating laser beam ($\lambda \sim 532$ nm) and emit a longer wavelength ($\lambda \sim 610$ nm). The signal from the measurement region includes the emitted light from both in-focus and out-of-focus particles, and the reflection from the background. The reflection from the background is eliminated by the emitter filter and the dichroic mirror, and both the focused and unfocused particle images are imaged on the interline transfer CCD camera. After a specified time delay, Δt, the same process as described previously is performed to acquire the second image frame for the cross-correlation based interrogation. Two different CCD cameras were used for MC-I and MC-II. For MC-I, Flow Master 3s, LaVision, Inc., was used. The camera resolution is 1280×1024 pixel^2. The field of view is 428.8×343.0 μm^2 using a 20$\times$ objective lens (Nikon, Plan Fluor, 0.45 NA). For MC-II, Imager Intense, LaVision, Inc., was used. The camera resolution is 1376×1040 pixel^2, producing a field of view of 221.9×167.7 μm^2 with a 40$\times$ objective lens (Nikon, Plan Fluor, 0.60 NA).

The particle concentration must be considered carefully in μPIV because the background noise keeps increasing as the particle concentration increases, whereas the particle image intensity remains about same value [Meinhart et al., 1999b]. Since MC-I has a large depth (694 μm), the ratio of the out-of-focus volume to focused volume is also large. The particle concentration is optimized to be 0.038% (by volume) by balancing the valid detection rate and the background noises from out-of-plane particles. For MC-II, the fluorescent particle concentration was 0.057% (by volume).

10.4.3 Experimental Procedure

The flow is pumped into the inlet of the microchannels using a syringe pump (Harvard Apparatus, 22). Because the syringe pump works with a worm gear and stepper motor, its actuation is not continuous at low speeds. Also, the friction between the piston and the syringe wall is large when flow rate is small. The velocity fluctuations for each Reynolds number showed less than 3% and 4% in transverse and axial components, respectively, but the axial velocity fluctuation for $Re_D = 1$ is slightly lower than 9%. Consequently, the pump stability is better than 3~4%, except for the slowest speed measured.

For MC-I, experiments were performed at eight Reynolds numbers, $Re_D = 1, 10, 20, 30, 40, 50, 70,$ and 100. Seventy image pairs, each showing the entire entrance region of the channel, were acquired. In the case of MC-II, eight Reynolds numbers were also chosen. These numbers were 5, 10, 18, 24, 36, 48, 62, and 76. At each Reynolds number, 125 image pairs were acquired. Optical access was available from the narrow side (i.e., x–y plane) and wide side for MC-I and MC-II, respectively, at the middle planes of the microchannel depths. The evaluation accuracy was improved by using advanced interrogation algorithms, i.e., ensemble average correlation method [Meinhart et al., 1999b] and CDIC [Wereley and Gui, 2003].

Also, a specially developed image processing technique, the μPIV image filter [Gui et al., 2002], was used. Interrogation windows 32 × 32 pixel were used.

10.4.4 Results and Discussion

10.4.4.1 Velocity Profiles

Figure 10.11 shows velocity profile development for Reynolds number (a) 1, (b) 20, (c) 40, and (d) 70 in MC-I. Each velocity profile represents the average of seven neighboring axial measurement locations. Normalization was performed with the fully developed centerline velocity. For all Reynolds numbers, the velocity fields developed quickly, and the last four profiles (filled symbols) did not show significant differences.

When the Reynolds number is small, the flow develops very quickly, because the vorticity diffusion is greater than the momentum. This fact can be seen with the profiles at $y = 17.4\,\mu m$. As the Reynolds number increases, the inlet velocity profiles become more blunt and, for $Re_D = 70$, the inlet profiles are rather close to the inviscid core flat profiles. Nevertheless, the overall development lengths in this Reynolds number range are shorter than those predicted by existing correlations, except for $Re_D = 1$. For example, the entrance lengths for Reynolds numbers from 10 to 100 are expected to vary from 0.372 to 2.0 mm with Chen's correlation, which means the entrance lengths are longer than one field of view for $Re_D > 20$. When the Reynolds number is small, $Re_D = 1$, the constant portion of the entrance length correlation dominates the total entrance length.

Figure 10.12 shows the normalized velocity profiles in MC-II from the entrance to the fully developed region at Reynolds number (a) 18, (b) 36, and (c) 62. The normalization was performed in the same manner as for Figure 10.11. The solid lines indicate the fully developed theoretical velocity profiles. The maximum velocities at the inlet are to the left of the center in both cases. This skewed inlet flow is due to the asymmetric

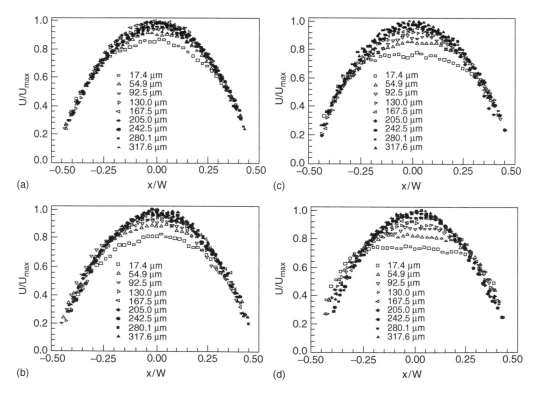

FIGURE 10.11 Normalized velocity profiles for Reynolds number (a) 1, (b) 20, (c) 40, and (d) 70 in MC-I. Last four profiles are shown with filled symbols.

flow path from the reservoir to the channel entrance. These asymmetric results at the inlet were also shown by Zhao (2003) using ANSYS simulations. The skewness decreases as the Reynolds number increases. In MC-II, a slight velocity overshoot is seen in the Reynolds number 18 case and becomes more significant as the Reynolds number increases. Fully developed profiles are observed within one field of view, as in MC-I.

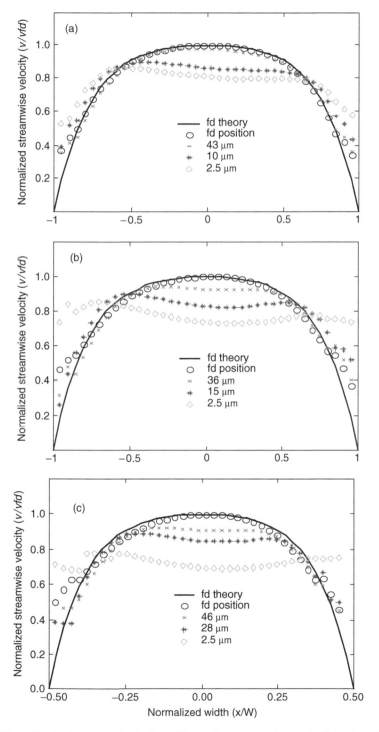

FIGURE 10.12   Normalized velocity profiles for Reynolds number (a) 18, (b) 36, and (c) 62. The solid lines indicate the theoretical fully developed velocity profiles.

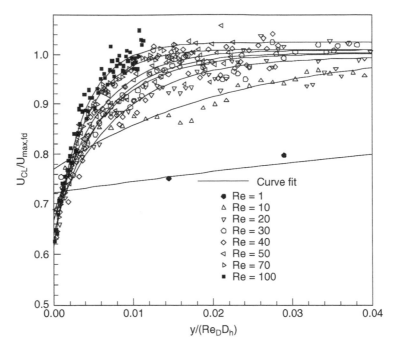

FIGURE 10.13   Normalized centerline velocities for various Re_D's in MC-I (Aspect ratio, H/W, 2.75).

10.4.4.2   Centerline Velocity

The entrance length was investigated quantitatively, using a curve fit of the normalized centerline velocity as a function of the dimensionless axial distance from the microchannel entrance. An exponential decaying function is used for curve fitting the centerline velocity development, as shown,

$$\frac{U_{CL}}{U_{CL,fd}} = N_0 - N_1 \cdot \exp\left(-N_2 \frac{y}{Re_D D_h}\right) \tag{10.23}$$

where $N_{0, 1, and 2}$ are positive constants and the dimensionless entrance lengths were found with the condition $U_{CL}/U_{CL,fd} = 0.99 N_0$ [Lee et al., 2002]. Figure 10.13 and Figure 10.14 show the centerline developments in MC-I and MC-II, respectively. The data sets for the various Reynolds numbers do not collapse onto a universal curve using the data reduction variable $y/(Re_D D_h)$ for either microchannel. Thus, the flow fields in the low Reynolds number range do not have a strong dependence on Reynolds number. In the MC-I case, the Reynolds number 1 and 10 cases show the strongest deviations, and in MC-II, in Fig. 8, Reynolds number 5 and 10 cases show the strongest deviations.

As the Reynolds number increases, the initial slopes of data sets increase. This increased slope results in the dimensionless entrance length, $L_e/(Re_D D_h)$, decreasing as the Reynolds number increases. For MC-I, for example, $L_e/(Re_D D_h) \sim 0.0705$ for $Re_D = 10$ and 0.0285 for $Re_D = 20$. Using 99% of the fully developed centerline velocity, the entrance lengths were found to be $y/D_h = 0.6194, 0.7054, 0.5695, 0.7191, 0.8604,$ 1.0080, 1.0314, and 1.358. The $Re_D = 1$ case shows large fluctuations after the dimensionless entrance length about 0.6. For the case of $Re_D = 100$, the entrance length is found about 500 μm, which is slightly out of one field of view.

Figure 10.14 shows the curve fit results for MC-II. The normalized entrance lengths, y/D_h, are found to be 1.12, 1.24, 1.26, 1.28, 1.49, 1.87, 2.1, and 2.43 for each Reynolds number. Though the Reynolds numbers do not exactly match the MC-I case, the y/D_h values for MC-II are two or three times as large as the dimensionless entrance length in the MC-I case.

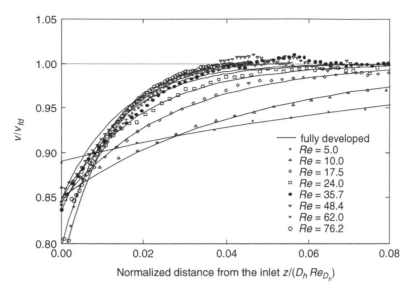

FIGURE 10.14 Normalized centerline velocity for various Re_D's in MC-II (Aspect ratio, H/W, 0.37). (a) Curve fits in the form of Atkinson's correlation. (b) Curve fits in the form of Chen's correlation.

10.4.4.3 Entrance Length Correlation

New entrance length correlations are attempted for the current experiments in the form of Atkinson's (see Equation (10.1) and Figure 10.15(a)) and Chen's (see Equation (10.2) and Figure 10.15(b)) correlations. Also, the existing entrance length correlations, Atkinson et al. (1969) and Chen (1973), are compared with the current experimental data.

Overall, Figure 10.15 shows that the existing correlations do not show good agreement with the experimental results. For Figure 10.15(a), the fitting functions have slopes of ~0.008 and ~0.018 for MC-I and MC-II, respectively, which indicates the weak Reynolds number dependence, and the constant portions are ~0.55 and ~0.96, respectively. Both slopes are much smaller than those of Atkinson's correlation (i.e., 0.056 for circular channel or 0.044 for parallel plates). The constant portion of MC-I is close to that of Atkinson's correlation (i.e., 0.59 for circular channel or 0.625 for parallel plates). For MC-II, the constant portion is found to be 0.96, which is greater than any of the established correlations. The reason that the constant entrance length in MC-II is greater than that of MC-I seems because of the mixed effect of both the aspect ratio and the planar reservoir geometry. Since the aspect ratio (H/W) of MC-II is 0.37, the widthwise vorticity propagation should take longer than in the MC-I case, whereas the vorticity should be fully developed in the depthwise direction because of the planar reservoir geometry. For Figure 10.15(b), the Reynolds number coefficients in the denominator of the first term, i.e., 0.008 and 0.018, are from the low Reynolds number linear fit of Figure 10.15(a). The estimates of the constant portion entrance lengths are close to the values from Figure 10.15(a). As the Reynolds number increases, the second terms on the right hand side, i.e., slopes, dominate the entrance length. The slopes are found to be 0.010 and 0.026 for MC-I and MC-II, respectively, compared to Chen's correlations, which give the slopes of 0.056 for circular channel and 0.044 for parallel plates.

10.4.5 Conclusion

The entrance length of microchannels with planar upstream reservoir geometries has been studied using μPIV experiments for a broad Reynolds number range, $1 < Re_D < 100$. Two microchannels with reciprocal aspect ratios of 2.75 and 0.37 were used for the experiments. The velocity field was measured using μPIV,

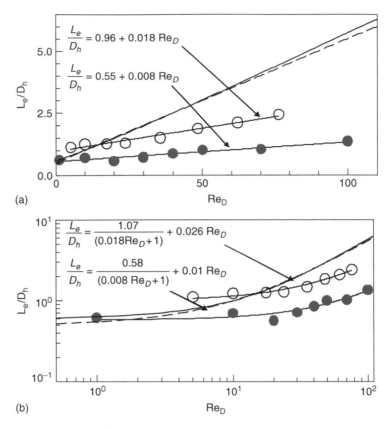

FIGURE 10.15 The comparisons of the entrance length correlations. The symbol ● and ○ indicate the experimental results of MC-I and MC-II, respectively. The symbol --- and —— indicate Atkinson's and Chen's correlations, respectively.

and the velocity profiles were extracted. The dimensionless entrance length for each Reynolds number was found using a decaying exponential function curve fit based on the entrance length definition, i.e., the axial distance required for the centerline velocity to reach 99% of the fully developed velocity. From both the velocity profiles and the centerline velocities, the entrance lengths were found to have significant differences compared to those predicted by the existing entrance length correlations.

Entrance length correlations have been found in the form of existing correlations such as Atkinson's and Chen's. For both microchannels, the correlations showed weaker dependence on Reynolds number, i.e., slope, than the existing correlations in the linear portion of the entrance length. The slopes were only 18% and 46% of Chen's correlation for MC-I and MC-II, respectively. The constant portion of entrance length agrees approximately in MC-I with the existing correlations, whereas MC-II shows a greater value than the estimates from Atkinson's and Chen's correlations. Thus, for MC-II, the experimental results are greater than the estimates when the Reynolds number is less than 15. The difference in correlations of MC-I and MC-II is no doubt due to the aspect ratio. Even if two microchannels have reciprocal aspect ratios, they must be considered completely different aspect ratios because of the planar inlet geometries.

The results indicate that the *effects* of the planar reservoir geometry and the aspect ratio on the entrance lengths have mixed *effects* and must be considered together. The planar geometry forces the predevelopment in the plenum region and thus both microchannels show the reduced entrance lengths. Also, MC-I and MC-II have the reciprocal aspect ratios so that the existing correlation can not account for the difference of these two channels because of the uniform inlet velocity profiles assumption. However, the entrance length difference between these two channels show that the vorticity diffusion in MC-II takes longer because the width of channel is greater than the height, which results in longer entrance length. The influence of aspect ratio could be resolved by another set of experiments with an aspect ratio of 1, i.e.,

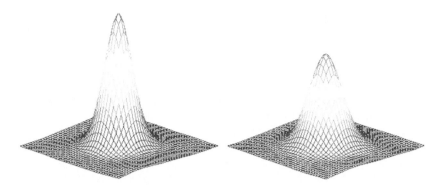

FIGURE 10.16   A pair of correlation functions demonstrates possible variations in peak width caused by Brownian motion. The autocorrelation (no Brownian motion) is on the left, and the cross-correlation (contains Brownian motion) is shown on the right. Both figures are shown in the same scale.

square channel. Further, the depthwise velocity profile at the inlet needs to be measured to investigate more thoroughly the effect of the planar plenum geometry.

10.5   Extensions of the μPIV Technique

The μPIV technique is very versatile and can be extended in several meaningful ways to make different, but related, measurements of use in characterizing microflows. One extension, called microparticle image thermometry, involves using the μPIV to measure fluid temperature in the same high spatial resolution sense as the velocity is measured. Another extension involves using a similar measurement technique but having wavelengths longer than visual, called near-infrared, to measure flows completely encased in silicon — a real benefit in the MEMS field. The third extension is to avoid the low signal-to-noise ratio caused by the volume illumination by using evanescent waves to illuminate only a very thin region near the flow boundary.

10.5.1   Microparticle Image Thermometry

Microparticle image thermometry (μPIT) is a minimally invasive, high-resolution temperature measurement technique based on the particle diffusion caused by Brownian motion. The square of the expected distance traveled by a particle with diffusivity D (recall Equation (10.6)) in some time window Δt is given by

$$\langle s^2 \rangle = 2D\Delta t \qquad (10.24)$$

Combining Equation (10.6) and Equation (10.24), it can be observed that an increase in fluid temperature, with all other factors held constant, will result in a greater expected particle displacement, $\sqrt{<s^2>}$.

Olsen and Adrian (2000b) derived analytical equations describing the shape and height of the cross-correlation function in the presence of Brownian motion for both light-sheet illumination and volume illumination (as is used in μPIV). In both cases, the signal peak in the cross-correlation has a Gaussian shape, with the peak located at the mean particle displacement. One of the key differences between light-sheet PIV and μPIV (volume illumination PIV) lies in the images formed by the seed particles. In light-sheet PIV, if the depth of focus of the camera is set to be greater than the thickness of the laser sheet, then all of the particle images will (theoretically) have the same diameter and intensity.

From their analysis of cross-correlation PIV, Olsen and Adrian (2000b) found that one effect of Brownian motion on cross-correlation PIV is to increase the correlation peak width, Δs_0 — taken as the $1/e$ diameter of the Gaussian peak (see Figure 10.16). For the case of light-sheet PIV, they found that:

$$\Delta s_{o,a} = \sqrt{2}\,\frac{d_e}{\beta} \qquad (10.25)$$

when the Brownian motion is negligible, to:

$$\Delta s_{o,c} = \sqrt{2}\,\frac{(d_e^2 + 8M^2\beta^2 D\Delta t)^{1/2}}{\beta} \tag{10.26}$$

when the Brownian motion is significant (note that in any experiment, even one with a significant Brownian motion, $\Delta s_{o,a}$ can be determined by computing the autocorrelation of one of the PIV image pairs). Here, d_e is the particle image diameter which is given in Equation (10.11) for light sheet illumination and Equation (10.16) for volume illumination. M is the magnification, D is the diffusivity, and Δt is the time delay between two frames of images. The constant β is a parameter arising from the approximation of the Airy point-response function as a Gaussian function (see Equation (10.11)). Adrian and Yao (1985) found a best fit to occur for $\beta^2 = 3.67$. It can be seen that Equation (10.26) reduces to Equation (10.25) in cases where the Brownian motion is a negligible contributor to the measurement (i.e., when $D\Delta t \rightarrow 0$).

Hohreiter et al. (2002) derived the temperature term explicitly from diffusivity by manipulating Equation (10.25) and Equation (10.26). To avoid the complicated calculations in Equation (10.11), squaring both Equation (10.25) and Equation (10.26), taking their difference, and multiplying by the quantity $\pi/4$ will convert the individual peak width (peak diameter for a three-dimensional peak) expressions to the difference of two correlation peak areas — namely, the difference in area between the auto- and cross-correlation peaks. Performing this operation and substituting Equation (10.6) in for D yields

$$\Delta A = \frac{\pi}{4}(\Delta s_{o,c}^2 - \Delta s_{o,a}^2) = C_0 \frac{T}{\mu}\Delta t \tag{10.27}$$

where C_0 is the parameter $2M^2\kappa/3d_p$. κ and d_p are Boltzmann's constant and the physical particle diameter.

Hohreiter et al. (2002) successfully demonstrated that local temperature can be measured experimentally based on the particle diffusion theory, as in Equation (10.27). An Olympus BX50 system microscope with BX-FLA fluorescence light attachment (housing the dichroic mirror/prism and optical filter cube) was used to image the particle-laden solution. All experiments were carried out with a 50× objective (NA = 0.8). A Cohu (model 4915-3000) 8-bit CCD video camera was used — the CCD array consisted of 768 (horizontal) × 494 (vertical) pixels and a total image area of 6.4 × 4.8 mm. The particles — 700-nm-diameter polystyrene latex microspheres (Duke Scientific, Palo Alto, California) — had a peak excitation wavelength at 542 nm and peak emission at 612 nm. A variable-intensity halogen lamp was used for illumination. Optical filters were used to isolate the wavelength bands most applicable to the particles — 520 to 550 nm for the incident illumination, and greater than 580 nm for the particle fluorescence.

In the experiments, the power supplied to a patch heater (Figure 10.17) was adjusted incrementally, and several successive images of the random particle motion were captured at each of several temperature steps. At each temperature, the system was allowed five minutes to come to a steady temperature, as recorded by a thermocouple. Measured temperatures varied from 20 to 50°C. Figure 10.18 shows the experimental results. The measured temperatures with correlation peak width change show a good agreement with the data from the thermocouple measurement. The average error over all the test cases is about ±3°C.

10.5.2 Infrared μPIV

Another recent extension of μPIV that has potential to benefit microfluidics research in general, and silicon MEMS research in particular, is that of infrared IR-PIV [Han and Breuer, 2001]. The main difference between the established technique of μPIV and IR-PIV is the wavelength of the illumination, which is increased from visible wavelengths to infrared wavelengths to take advantage of silicon's relative transparency at IR wavelengths. Although this difference may seem trivial, it requires several important changes to the technique while enabling important new types of measurements to be made.

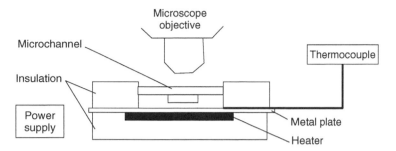

FIGURE 10.17 Microchannel, heater, and thermocouple arrangement used for experimentation. (After Hohreiter et al., 2002.)

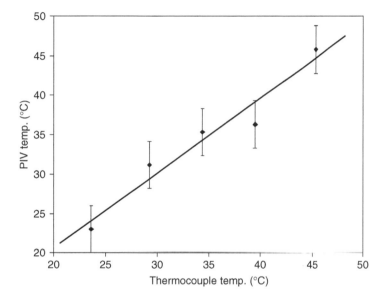

FIGURE 10.18 Temperature inferred from PIV measurements plotted versus thermocouple-measured temperature. Error bars indicate the range of average experimental uncertainty, ±3°C. (After Hohreiter et al., 2002.)

10.5.2.1 Differences between μPIV and IR-PIV

The fluorescent particles that allow the use of epifluorescent microscopes for μPIV are not available with both absorption and emission bands at IR wavelengths [Han and Breuer, 2001]. Consequently, elastic scattering must be used, in which the illuminating light is scattered directly by the seed particles with no change in wavelength. Using this mode of imaging, it is not possible to separate the images of the particles from those of the background using colored barrier filters, as in the μPIV case. The intensity of elastic scattering intensity I of a small particle of diameter d varies according to

$$I \propto \frac{d^6}{\lambda^4} \tag{10.28}$$

where λ is the wavelength of the illuminating light [Born and Wolf, 1997]. Thus, a great price is exacted for imaging small particles with long wavelengths. The main implication of Equation (10.28) is that there is a trade-off between using longer wavelengths, where silicon is more transparent, and using shorter wavelengths, where the elastic scattering is more efficient. Typically, infrared cameras are also more efficient at longer wavelengths. Han and Breuer (2001) found a good compromise among these competing factors by using 1-μm polystyrene particles and an illumination wavelength of $\lambda = 1200$ nm.

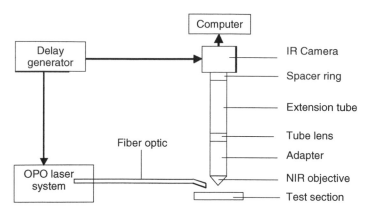

FIGURE 10.19 Schematic of the experimental apparatus for IR-PIV. (After Han and Breuer, 2001.)

An experimental apparatus suitable for making IR-PIV measurements is described by Han and Breuer (2001) and is shown in Figure 10.19. As with μPIV, a dual-headed Nd:YAG laser is used to illuminate the particles. However, in this case, the 532-nm laser light is used to drive an opto-parametric oscillator (OPO) — a nonlinear crystal system that transforms the 532-nm light into any wavelength between 300 and 2000 nm. The laser light retains its short pulse duration when passing through the OPO. The output of the OPO is delivered via fiber optics to the microfluidic system being investigated. Han and Breuer (2001) use an off-axis beam delivery, as shown in Figure 10.19, with an angle of 65° between the normal surface of the device and the axis of the beam. Alternatively, dark field illumination could be used. The light scattered by the particles is collected by a Mititoyo near infrared (NIR) microscope objective (50×, NA = 0.42) mounted on a 200-mm microscope tube and delivered to an Indigo Systems indium gallium arsenide (InGaAs) NIR camera.

The camera has a 320 × 256 pixel array with 30 μm pixels — a relatively small number of relatively large pixels compared to the high resolution cameras typically used for μPIV applications. The NIR is a video rate camera which cannot be triggered, meaning that the PIV technique needs to be modified slightly. Instead of using the computer as the master for the PIV system, the camera is the master, running at its fixed frequency of 60 Hz. The laser pulses are synchronized to the video sync pulses generated by the camera and can be programmed to occur at any point within a video frame. For high-speed measurements, a process called frame straddling is used, in which the first laser pulse is timed to occur at the very end of one frame, and the second laser pulse is timed to occur at the very beginning of the next video frame. Using the frame straddling technique, the time between images can be reduced to as little as 0.12 ms — suitable for measuring velocities on the order of centimeters per second. Higher-speed flows can be measured by recording the images from both laser pulses on a single video frame. The signal-to-noise ratio of the double-exposed images is decreased somewhat when compared to the single-exposed images, but flows on the order of hundreds of meters per second can be measured with the double exposed technique.

Liu et al. (submitted to *Exp. Fluids*, 2004) successfully made IR-PIV measurement for laminar flow of water in a circular micro capillary tube of hydraulic diameter 255 μm. Figure 10.20 shows the velocity field (a) and the velocity profile (b) in the tube, respectively, and Figure 10.20(b) shows that the experimental measurements agree very well with velocity profiles predicted from laminar theory.

10.5.3 Evanescent Wave PIV

As explained in the Section 10.3.1.3, one of the limitations of μPIV is that a volume of light is used to illuminate the flow, only a small fraction of which coincides with the depth of field of the imaging optics. This situation leads to the presence of out-of-focus particle images in the recorded images. These out-of-focus particle images reduce the signal-to-noise ratio of the measurements and limit the concentration of seed particles that can be used without completely obscuring the focused particle images. One solution to this

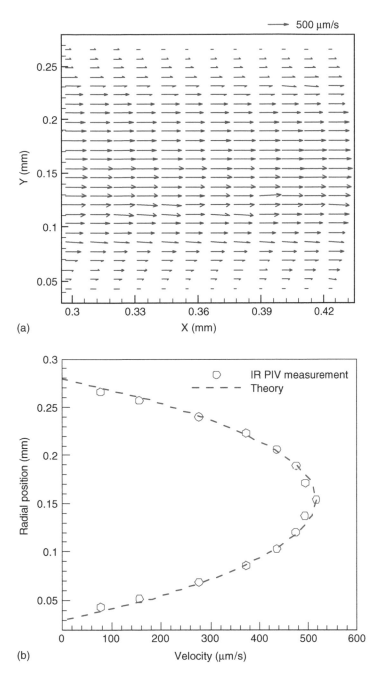

FIGURE 10.20 IR-PIV results for very low-speed flow: (a) velocity vectors, and (b) comparison of the measurements with the theoretical profile.

problem is to use evanescent wave illumination to restrict the illumination to a region within a few hundred nanometers of the wall.

While the light undergoes total internal reflection (TIR), some fraction of the incident light penetrates into the less-dense medium and propagates parallel to the surface. This parallel light wave is called *evanescent wave* [Jin et al., 2003]. In this illumination, the field amplitude decays exponentially over a distance comparable to the wavelength in the sample medium, allowing the information depth of optical signals

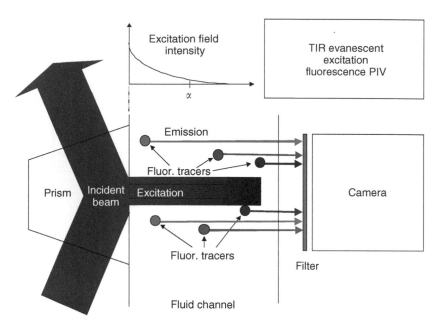

FIGURE 10.21   Schematic diagram of evanescent wave PIV system.

to be limited to scales of tens or hundreds of nanometers, a far higher resolution than can be achieved with conventional imaging optics. The penetration depth, z_p, is given as:

$$z_p = \frac{\lambda_o}{2\pi} \ (n_2^2 \sin^2 \theta - n_1^2)^{-1/2} \tag{10.29}$$

where n_1 and n_2 are the refractive indices of two mediums of $n_1 < n_2$, λ_o is the wavelength of the light in vacuum, and θ is the incident angle greater than the critical angle, $\theta_c = \sin^{-1}(n_2/n_1)$, to generate total internal reflection.

Zettner and Yoda (2003) made a PIV measurement in a rotating Couette flow using evanescent wave. The fluorescent particles of few hundreds in diameter were used, the spatial resolution in depth direction was about 380 nm, and the field of view was 200 μm × 150 μm. The results showed a reasonable agreement with the exact solution for single-phase Newtonian fluid flow.

Jin et al. (2003) used total internal reflection fluorescent microscopy (TIRFM) to measure the slip velocity on hydrophilic and hydrophobic surfaces. The penetration depth ranges from 98 to 232 nm. Fluorescent particles of 200–300 nm in diameter were used for PTV evaluation. The minimal slip difference was found between a hydrophilic surface and a hydrophobic surface. Though the measurement using evanescent wave illumination is a promising tool to measure near-surface flow, it is still difficult to measure the slip velocity for several reasons, such as the interaction between seed particles and channel surface, relatively large particle size, and surface roughness.

References

Adrian, R.J., and Yao, C.S. (1985) "Pulsed Laser Technique Application to Liquid and Gaseous Flows and the Scattering Power of Seed Materials," *Appl. Opt.*, **24**, pp. 44–52.

Adrian, R.J. (1991) "Particle-Imaging Techniques for Experimental Fluid Mechanics," *Annu. Rev. Fluid Mech.*, **23**, pp. 261–304.

Adrian, R.J. (1996) *Bibliography of Particle Image Velocimetry Using Imaging Methods: 1917–1995*, University of Illinois at Urbana-Champaign, Urbana, IL.

Amabile, M., Dunsmuir, J., Lazillotto, A.M., and Len, T. (1996) "Applications of X-ray Micro-Imaging, Visualization and Motion Analysis Techniques to Fluidic Microsystems," *Technical Digest of the IEEE Solid State Sensor and Actuator Workshop*, pp. 123–126, 3–6 June, Hilton Head Island, SC.

Atkinson, B., Brocklebank, M.P., Card, C.C.H., and Smith, J.M. (1969) "Low Reynolds Number Developing Flows," *AIChE J.*, **15**, pp. 548–53.

Beavers G.S., Sparrow E.M., and Magnuson R.A. (1970) "Experiments on Hydrodynamically Developing Flow in Rectangular Ducts of Arbitrary Aspect Ratio," *Int. J. Heat Mass Trans.*, **13**, 4, pp. 689–702.

Bendat, J.S., and Piersol, J.G. (1986) *Random Data: Analysis and Measurement Procedures*, John Wiley and Sons, New York.

Beskok, A., Karniadakis, G.E., and Trimmer, W. (1996) "Rarefaction and Compressibility," *J. Fluid. Eng.*, **118**, pp. 448–56.

Born, M., and Wolf, E. (1997) *Principles of Optics*, Pergamon Press, New York.

Chen, R-Y. (1973) "Flow in the Entrance Region at Low Reynolds Numbers," *J. Fluid. Eng.*, **95**, pp. 153–58.

Chen, Z., Milner, T.E., Dave, D., and Nelson J.S. (1997) "Optical Doppler Tomographic Imaging of Fluid Flow Velocity in Highly Scattering Media," *Opt. Lett.*, **22**, pp. 64–66.

Compton, D.A., and Eaton, J.K. (1996) "A High-Resolution Laser Doppler Anemometer for Three-Dimensional Turbulent Boundary Layers," *Exp. Fluid.*, **22**, pp. 111–17.

Cowen, E.A., and Monismith, S.G. (1997) "A Hybrid Digital Particle Tracking Velocimetry Technique," *Exp. Fluid.*, **22**, pp. 199–211.

Cowen, E.A., Chang, K.A., and Liao, Q. (2001) "A Single-Camera Coupled PTV-LIF Technique," *Exp. Fluid.*, **31**, pp. 63–73.

Cummings, E.B. (2001) "An Image Processing and Optimal Nonlinear Filtering Technique for PIV of Microflows," *Exp. Fluid.*, **29** [Suppl.], pp. 42–50.

Dahm, W.J.A., Su, L.K., and Southerland, K.B. (1992) "A Scalar Imaging Velocimetry Technique for Fully Resolved Four-Dimensional Vector Velocity Field Measurements in Turbulent Flows," *Phys. Fluids A (Fluid Dynamics)*, **4**, pp. 2191–206.

Devasenathipathy, S., Santiago, J.G., Wereley, S.T., Meinhart, C.D., and Takehara, K. (2003) "Particle Imaging Techniques for Microfabricated Fluidic Systems," *Exp. Fluid.*, **34**, pp. 504–14.

Einstein, A. (1905) "On the Movement of Small Particles Suspended in a Stationary Liquid Demanded by the Molecular-Kinetic Theory of Heat," *Theory of Brownian Movement*, Dover, New York, pp. 1–18.

Fleming, D.P., and Sparrow, E.M. (1969) "Flow in the Hydrodynamic Entrance Region of Ducts of Arbitrary Cross Section," *ASME J. Heat Trans.*, **91**, pp. 345–54.

Fox, R.W., and McDonald, A.T. (1998) *Introduction to Fluid Mechanics*, 5th ed., John Wiley and Sons, New York.

Friedmann, M., Gillis, J., and Liron, N. (1968) "Laminar Flow in a Pipe at Low And Moderate Reynolds Numbers," *App. Sci. Res.*, **19**, pp. 426–38.

Gharib, M. et al. (2002) "Optical Microsensors for Fluid Flow Diagnostics", *AIAA, Aerospace Sciences Meeting and Exhibit, 40th*, AIAA Paper 2002–0252. 14th January, Reno, NV.

Gomez, R., Bashir, R., Greng A, Bhunia, M., Ladishch, J., and Wereley, S. (2001) "Microfluidic Biochip for Impedance Spectroscopy of Biological Species," *Biomed. Microdevices*, **3**, pp. 201–309.

Guezennec, Y.G., Brodkey, R.S., Trigui, N., and Kent, J.C. (1994) " Algorithms for Fully Automated 3-Dimensional Particle Tracking Velocimetry," *Exp. Fluid.*, **17**, pp. 209–19.

Gui, L., Merzkirch, W., and Shu, J.Z. (1997) "Evaluation of Low Image Density PIV Recordings with the MQD Method and Application to the Flow in a Liquid Bridge," *J. Flow Visualization Image Process.*, **4**, pp. 333–43.

Gui, L., Wereley, S.T., and Lee, S.Y. (2002) "Digital Filters for Reducing Background Noise in Micro PIV Measurements," *Proceedings of the 11th International Symposium on the Application of Laser Techniques to Fluid Mechanics*, paper 12.4, 8–11 July, Lisbon, Portugal.

Han, L.S. (1983) "Hydrodynamic Entrance Lengths for Incompressible Laminar Flow in Rectangular Ducts," *J. App. Mech.*, **27**, pp. 273–77.

Han, G., and Breuer, K.S. (2001) "Infrared PIV for Measurement of Fluid and Solid Motion Inside Opaque Silicon Microdevices," *Proceedings of 4th International Symposium on Particle Image Velocimetry*, paper number 1146, September, Göttingen, Germany.

Hitt, D.L., Lowe M.L., and Newcomer, R. (1995) "Application of Optical Flow Techniques to Flow Velocimetry," *Phys. Fluids*, 7, pp. 6–8.

Hitt, D.L., Lowe, M.L., Tindra, J.R., and Watters, J.M. (1996) "A New Method for Blood Velocimetry in the Microcirculation," *Microcirculation*, 3, pp. 259–63.

Hohreiter, V., Wereley, S.T., Olsen, M., and Chung, J. (2002) "Cross-Correlation Analysis for Temperature Measurement," *Meas. Sci. Technol.*, 13, pp. 1072–78.

Inoué, S., and Spring, K.R. (1997) *Video Microscopy*, 2nd ed., Plenum Press, New York.

Jin, S., Huang, P., Park, J., Yoo, J.Y., and Breuer, K.S. (2003) "Near-surface Velocimetry using Evanescent Wave Illumination," *ASME International Mechanical Engineering Congress & Exposition*, paper# IMECE2003-44015, 16–21 November, Washington, DC.

Keane, R.D., Adrian, R.J., and Zhang, Y. (1995) "Super-Resolution Particle Imaging Velocimetry," *Meas. Sci. Technol.*, 6, pp. 754–68.

Koochesfahani, M.M., Cohn, R.K., Gendrich, C.P., and Nocera, D.G. (1997) "Molecular Tagging Diagnostics for the Study of Kinematics and Mixing in Liquid Phase Flows," *Developments in Laser Techniques in Fluid Mechanics*, R.J. Adrian et al., eds., pp. 125–34, Springer-Verlag, New York.

Koutsiaris, A.G., Mathioulakis, D.S., and Tsangaris, S. (1999) "Microscope PIV for Velocity-Field Measurement of Particle Suspensions Flowing Inside Glass Capillaries," *Meas. Sci. Technol.*, 10, pp. 1037–46.

Lee, S.Y., Wereley, S.T., Gui, L.C., Qu, W.L., and Mudawar, I. (2002) "Microchannel Flow Measurement using Micro Particle Image Velocimetry," *Proceedings of ASME/IMECE*, Paper #2002-33682, November, New Orleans, LA.

Lempert, W.R., Jiang, N., Sethwram, S., and Samimy, M. (2001) "Molecular Tagging Velocimetry Measurements in Supersonic Micro Nozzles," *Proceedings of 39th AIAA Aerospace Sciences Meeting and Exhibit*, pp. 2001–2044, January Reno, NV.

Leu, T.S., Lanzillotto, A.M., Amabile, M., and Wildes, R. (1997) "Analysis of Fluidic and Mechanical Motions in MEMS by Using High Speed X-ray Micro-Imaging Techniques," *Proceedings of Transducers '97, 9th International Conference on Solid-State Sensors and Actuators*, pp. 149–150, 16–19 June, Chicago.

Liu, D., Garimella, S.V., and Wereley, S.T. (2004) "Infrared Micro-Particle Image Velocimetry of Fluid Flow in Silicon-Based Microdevices," submitted to Experiments in Fluids, 2004.

Maynes, D., and Webb, A.R. (2002) "Velocity Profile Characterization in Sub-Millimeter Diameter Tubes Using Molecular Tagging Velocimetry," *Exp. Fluid.*, 32, pp. 3–15.

Meinhart, C.D., Wereley, S.T., and Santiago, J.G. (1999a) "PIV Measurements of a Microchannel Flow," *Exp. Fluid.*, 27, pp. 414–19.

Meinhart, C.D., Wereley, S.T., and Santiago, J.G. (1999b) "A PIV Algorithm for Estimating Time-Averaged Velocity Fields," *J. Fluid. Eng.*, 122, pp. 285–89.

Meinhart, C.D., and Zhang, H. (2000) "The Flow Structure Inside a Microfabricated Inkjet Printer Head," *J. Microelectromech. Syst.*, 9, pp. 67–75.

Meinhart, C.D., Wereley, S.T., and Santiago, J.G. (2000a) "Micron-Resolution Velocimetry Techniques," *Laser Techniques Applied to Fluid Mechanics*, R.J. Adrian et al., eds., pp. 57–70, Springer Verlag, New York.

Meinhart, C.D., Wereley, S.T., and Gray, M.H.B. (2000b) "Volume Illumination for Two-Dimensional Particle Image Velocimetry," *Meas. Sci. Technol.*, 11, pp. 809–14.

Melling, A. (1986) "Seeding Gas Flows for Laser Anemometry," AGARD Conference in Advanced Instrumentation for Aero Engine Components, Paper no. 339.

Olsen, M.G., and Adrian, R.J. (2000a) "Out-of-Focus Effects on Particle Image Visibility and Correlation in Particle Image Velocimetry," *Exp. Fluid.*, [Suppl.], pp. 166–74.

Olsen, M.G., and Adrian, R.J. (2000b) "Brownian Motion and Correlation in Particle Image Velocimetry," *Opt. Laser Technol.*, 32, pp. 621–27.

Ohmi, K., and Li, H.Y. (2000) "Particle-Tracking Velocimetry with New Algorithm," *Meas. Sci. Technol.*, **11**, pp. 603–16.

Paul, P.H., Garguilo, M.G., and Rakestraw, D.J. (1998) "Imaging of Pressure- and Electrokinetically Driven Flows Through Open Capillaries," *Anal. Chem.*, **70**, pp. 2459–67.

Prasad, A.K., Adrian, R.J., Landreth, C.C., and Offutt, P.W. (1992) "Effect of Resolution on the Speed and Accuracy of Particle Image Velocimetry Interrogation," *Exp. Fluid.*, **13**, pp. 105–16.

Probstein, R.F. (1994) *Physicochemical Hydrodynamics: An Introduction*, John Wiley and Sons, New York.

Raffel, M., Willert, C., and Kompenhans, J. (1998) *Particle Image Velocimetry: A Practical Guide*, Springer, New York.

Santiago, J.G., Wereley, S.T., Meinhart, C.D., Beebe, D.J., and Adrian, R.J. (1998) "A Particle Image Velocimetry System for Microfluidics," *Exp. Fluid.*, **25**, pp. 316–19.

Shah, R.K., and London, A.L. (1978) *Laminar Flow Forced Convection in Ducts — A Source Book for Compact Heat Exchanger Analytical Data, Supplement 1 to Advances in Heat Transfer Series*, Academic Press, New York.

Sparrow, E.M., Hixon, C.W., and Shavit, G. (1967) "Experiments on Laminar Flow Development in Rectangular Ducts," *ASME J. Basic Eng.*, **89**, pp. 116–24.

Sparrow, E.M., and Anderson, C.E. (1977) "Effect of Upstream Flow Processes on Hydrodynamic Development in a Duct," *ASME J. Fluids Eng.*, **99**, pp. 556–60.

Stitou, A., and Riethmuller, M.L. (2001) "Extension of PIV to Super Resolution Using PTV," *Meas. Sci. Technol.*, **12**, pp. 1398–1403.

Takehara, K., Adrian, R.J., Etoh, G.T., and Christensen, K.T. (2000) "A Kalman Tracer for Super Resolution PIV," *Exp. Fluid.*, **29**, pp. s34–s41.

Tieu, A.K., Mackenzie, M.R., and Li, E.B. (1995) "Measurements in Microscopic Flow with a Solid-State LDA," *Exp. Fluid.*, **19**, pp. 293–94.

Wereley, S.T., Meinhart, C.D., Santiago, J.G., and Adrian, R.J. (1998) "Velocimetry for MEMS Applications," *Micro-Electro-Mechanical Systems, DSC.* **66**, ASME, pp. 453–59.

Wereley S.T., and Meinhart, C.D. (2001a) "Adaptive Second-Order Accurate Particle Image Velocimetry," *Exp. Fluid.*, **31**, pp. 258–68.

Wereley S.T., Gui, L., and Meinhart, C.D. (2001b) "Flow Measurement Techniques for the Microfrontier," *AIAA, 39th Aerospace Sciences Meeting and Exhibit*, AIAA Paper 2001–0243, 8–11 January, Reno, NV.

Wereley, S.T., and Gui, L.C. (2001) "PIV Measurement in a Four-Roll-Mill Flow with a Central Difference Image Correction (CDIC) Method," *Proceedings of 4th International Symposium on Particle Image Velocimetry*, paper number 1027, September, Göttingen, Germany.

Wereley, S.T., Lee, S.Y., and Gui, L.C. (2002a) "Entrance Length and Turbulence Transition in Microchannels," *American Physics Society, Div. Fluid Dynamics Annual Meeting*, November, Dallas, TX.

Wereley, S.T., Gui, L., and Meinhart, C.D. (2002b) "Advanced Algorithms for Microscale Velocimetry," *AIAA J.*, **40**, pp. 1047–55.

Wereley, S.T., and Gui, L., (2003) "A Correlation-Based Central Difference Image Correction(CDIC) Method and Application in a Four-Roll Mill Flow PIV Measurement," *Exp. Fluid.*, **34**, pp. 42–51.

Wereley, S.T., and Meinhart, C.D. (2003) "Micron-Resolution Particle Image Velocimetry" in *Micro- and Nano-Scale Diagnostic Techniques*, Kenny Breuer, ed., Springer-Verlag, New York, *in Press.*

Wiesendanger, R. (1994) *Scanning Probe Microscopy and Spectroscopy: Methods and Applications*, Cambridge University Press, Cambridge, England.

Vrentas, J.S., Duda, J.L., and Bargeron, K.G. (1966) "Effect of Axial Diffusion of Vorticity on Flow Development in Circular Conduits: Part I. Numerical Solutions," *AIChE J.*, **15**, pp. 837–44.

Zettner, C.M., and Yoda, M. (2003) "Particle Velocity Field Measurements in a Near-Wall Flow Using Evanescent Wave Illumination," *Exp. Fluid.*, **34**, pp. 115–21.

Zhao, Y. (2003) Design and Characterization of Micro and Molecular Flow Sensors, M.S.M.E Thesis, Department of Mechanical Engineering, Purdue University, West Lafayette, IN.

11

Microfabricated Chemical Sensors for Aerospace Applications

Gary W. Hunter and
Jennifer C. Xu
NASA Glenn Research Center

Chung-Chiun Liu
Case Western Reserve University

Darby B. Makel
Makel Engineering, Incorporated

11.1 Introduction

The advent of microelectromechanical systems (MEMS) technology is important in the development and use of chemical sensor technology for a range of applications, especially those that include operation in harsh environments or that affect safety. As will be discussed in this chapter, chemical microsensors can provide unique information that can significantly improve safety and reliability while decreasing costs of a system or process. Such information can also be used to improve a system's performance and reduce its effect on the environment. Chemical sensor data also can complement that derived from physical measurements such as temperature, pressure, heat flux, etc., further improving overall knowledge of a system and expanding its capabilities.

However, the application of even traditional, macrosized chemical sensor technology can be problematic. Chemical sensors often need to be specifically designed, or tailored, to operate in a given environment. It is often the case that a chemical sensor that meets the needs of one application will not function adequately in another application. The more demanding the environment and the more specialized the requirement, the greater the need to modify existing chemical sensor technologies to meet these requirements or, as necessary, to develop new chemical sensor technologies. Four common parameters are typically cited as relevant in determining whether a chemical sensor can meet the needs of an application: sensitivity, selectivity, response time, and stability. Sensitivity refers to the ability of the sensor to detect the desired chemical species in the range of interest. Selectivity refers to the ability of the sensor to detect the species of interest in the presence of interfering gases, which also can produce a sensor response. Response time refers to the time it takes for the sensor to provide a meaningful signal (often defined as 90% of the steady-state signal) when the chemical environment is changed. Finally, stability refers to the degree to which the sensor baseline and response to a given environment change over time. Simply stated, for micro- and macrochemical sensing systems, one needs a sensor that will accurately determine the species of interest in a given environment, with a response large and rapid enough to be of use in the application, and whose response does not significantly drift over its designed operational lifetime. Depending on the application, it is often difficult to find a suitable sensor material which provides the required sensitivity, selectivity, response time, and stability.

The microfabrication of chemical sensors involves much more than just making a macrosized sensor smaller. The processing used to produce a sensor material as a macroscopic bulk pellet can change considerably when it is desired to fabricate the material as part of a miniaturized system. For example, a chemical sensor in the form of a macroscopic bulk pellet can be fabricated from the powder of a starting material, pressed into a pellet containing lead wires, and then sintered at high temperatures to form the resulting sensor. However, using the same starting material, pressing the material onto a substrate to form a smaller, or even microscopic, sensor is often not a viable option. Rather, a thin or thick film of the sensor material must be deposited onto a substrate which itself, at a minimum, must support the sensor and allow for connections to the outside world. The underlying substrate with sensor film may not survive the sintering that typically is done with the pellet sensor material.

Further, given the surface sensitive nature of many chemical sensors, the effects of miniaturization can be dramatic and can include significant changes in sensor sensitivity and response time. These effects are in part owing to the fact that the sensor film is often produced by techniques such as sputtering, which may result in different material properties than those of bulk materials. The resulting surface-to-volume ratio of a thin film is larger than that of a bulk material; surface effects which may affect only a small percentage of a sensor in the bulk form may occur within a significant larger percentage volume of a thin film sensor. This situation can strongly affect the sensor's response. For example, oxidation may occur on the surface of a sensor exposed to high temperatures. In a bulk material, this oxidation may only be a small percentage of the sensor's volume, whereas in a thin film material, the same oxidation thickness may account for a sizable percentage of the sensor's volume. If the sensor's detection mechanism relies on bulk conduction, this oxidation could significantly affect the sensor response by changing the nature of the volume of the sensor. In addition, stresses in sensor thin films that degrade sensor response or catastrophically damage the sensor structure may be less of a factor in bulk materials [Hughes, 1987; Hunter, 1998]. Therefore, new technical challenges often must be overcome as sensor technology is miniaturized.

This chapter will discuss the development and application of microfabricated chemical sensor technology for aerospace (aeronautics and space) applications. These applications are particularly challenging, because often they include operation in harsh environments or have requirements which have not been previously emphasized by commercial suppliers. Section 11.2 will discuss the chemical sensing needs of three important aerospace applications: leak detection, emission monitoring, and fire detection. For each application, the use and advantages of MEMS-based approaches will be discussed. Each application has vastly different problems associated with the measurement of chemical species. Nonetheless, the development of a common base technology can address the measurement needs of a number of applications. Section 11.3 will discuss three base technologies being used to develop chemical sensors for these applications: microfabrication techniques as applied to chemical sensors, nanomaterials, and high temperature silicon carbide (SiC) based electronics. Section 11.4 will discuss the use of these base technologies to produce chemical sensors for a range of species: hydrogen (H_2), nitrogen oxides (NO_x), oxygen (O_2), hydrocarbons (C_xH_y), carbon monoxide (CO), and carbon dioxide (CO_2). The following two sections discuss the integration, application, and future direction of chemical sensor technology in aerospace applications and provide a summary.

11.2 Aerospace Applications

Aerospace applications vary considerably in their requirements for chemical sensor technology. In general, there are three envisioned uses of chemical sensor technology in aerospace vehicle applications: 1) system development and ground testing and where the sensor provides information on the state of a system that does not fly; 2) vehicle health monitoring (VHM) which involves the long-term monitoring of a system in operation to determine the health of the vehicle system (e.g., is the engine losing performance, increasing emissions, or developing a leak in the fuel system); and 3) active control of the vehicle in a feedback mode where information from a sensor is used to change a system parameter in real time (e.g., fuel flow to the engine changed because of sensor measurements of emissions). A sensor could be used in all three applications, by qualifying the system on the ground, monitoring its health in flight, and using this information in real time to change the operation parameters. These applications have a strong impact on vehicle safety, performance, and reliability. The design of each individual sensor depends on the requirements of each application. As will be discussed, very different sensor technology is necessary depending on temperature range, ambient gas, interfering gases, etc. For example, the sensor technology used for ground testing for leaks in an air ambient is very different from that used for leak testing in an inert (no oxygen) ambient. This present section illustrates the needs of three specific aerospace applications and how MEMS-based technology can address these needs by examining a ground-based VHM application (leak detection), an in-flight VHM application (fire detection), and an application (emission monitoring) that can be included in all three aerospace application areas: ground testing, VHM, and active control.

11.2.1 Leak Detection

Launch vehicle safety requires the rapid and accurate detection of fuel leaks. One major application is the detection of low concentrations of H_2 associated with the launch and operation of the space shuttle. Hydrogen leaks can lead to hazardous situations and, unless their locations can be rapidly identified, can lead to explosive situations, delays in vehicle launches, and significant costs. In the summer of 1990, hydrogen leaks on the space shuttle while it was on the launch pad temporarily grounded the fleet until the leak source could be identified and repaired. The standard leak detection system was a mass spectrometer connected to an array of sampling tubes placed throughout the region of interest. Although able to detect hydrogen in a variety of ambient environments, the mass spectrometer leak detection approach suffered delay times, because gas had to be transported from the fuel pipe through the sampling tubes to the remotely located mass spectrometer. Pinpointing the exact location of the leak was also problematic, because often the fuel pipes were covered with insulation. A leak emitted from the pipe could travel in the insulation and exit the insulation at a location different from the leak source. Thus, the mass spectrometer leak detection approach was unable to adequately isolate the location of the hydrogen leaks leading to a delay in shuttle launches. Problems with hydrogen leak detection on the shuttle continue. As recently as 1999, the launch of STS-93 was delayed for two days because of an ambiguous signal in the mass spectrometer leak detection system. The shuttle's hydrogen leak detection system must be improved in order to allow improved safety and reduced operational costs.

However, at the time of the 1990 leaks, no commercial hydrogen sensors existed that operated satisfactorily in this and other space-related applications, primarily owing to the conditions in which the sensor must operate. The hydrogen sensor must be able to detect hydrogen from low concentrations (ppm range) through the lower explosive limit (LEL), which is 4% in air. The sensor must be able to survive, and preferably function, even during exposure to 100% hydrogen without damage or change in calibration. Further, the sensor may be exposed to gases emerging from cryogenic sources, which could cool the sensor causing undesired changes in sensor output. Thus, sensor temperature measurement and control is necessary. Operation in inert environments is also necessary, because the sensor may have to operate in areas purged with helium. To allow leak detection even in space, the ability to operate in a vacuum is preferred. Commercially available sensors, which often require oxygen (air) to operate or depend upon moisture, did not meet the needs of this application (especially operation in inert environments), and thus the development of new types of hydrogen sensors was initiated [Hunter, 1992a].

NASA has been active in efforts to improve fuel leak detection capabilities for propulsion systems. The long-term objective has been an automated detection system using MEMS-based, point-contact hydrogen sensors. The application of MEMS-based sensor technology allows the fabrication of sensors with minimal size, weight, and power consumption to beneficially decrease vehicle weight and power requirements. A number of these sensors placed in the region of interest could allow multiplexing of the signals and "visualization" of the magnitude and location of the hydrogen leak. Potential uses include placing the sensors on fuel lines, inside insulation, and throughout chambers inside the vehicle. The ability to launch a vehicle on schedule depends on a leak detection system that can monitor key regions in the vehicle and automatically locate a leak in a limited time. This process decreases cost and manpower associated with the detection and location of leaks and improves overall system reliability.

Hydrogen is not the only fuel that may be used in launch vehicle applications. Other potential fuels considered for launch vehicles include propane, methane, ethanol, and hydrazine. Leak detection sensors for these fuels will need to meet many of the same requirements as the hydrogen sensor: sensitive detection of the fuel in possibly inert or cryogenic environments; sensor survival in high concentrations of the fuel; and minimal size, weight, and power consumption. Methane, which is a naturally occurring by-product of animal wastes, already exists in small amounts in the environment. This fact increases the difficulty of its measurement in some applications. Further, the detection mechanism which may readily allow the detection of hydrogen may not work for the detection of methane or ethanol [Chen, 1996]. Hydrazine is toxic, which adds further importance to leak detection systems; the gas must be detected not only for reasons of explosion prevention, but for personnel safety reasons as well. Just as, in 1990, there was no hydrogen sensor technology that met the needs of launch vehicle safety applications, there are still no suitable fuel leak detection systems for launch vehicles that use alternate fuels to hydrogen.

11.2.2 Fire Safety Monitoring

The detection of fires on board commercial aircraft is extremely important to avoid catastrophic situations. Although dependable fire detection equipment presently exists within the cabin, detection of fire within the cargo hold has been less reliable [Grosshandler, 1997]. In aircraft where cargo hold fire detection is required, the fire detection equipment presently used in many commercial aircraft relies on the detection of smoke. These smoke detectors are either optically based or they depend on particle ionization. Although highly developed, these sensors are subject to false alarms, with estimated false alarm rates as high as 200:1. Since some cargo areas are inaccessible to the flight crew during flight, visual confirmation is not possible. The presence of false alarms decreases the pilots' confidence in these systems and may potentially cause accidents if pilots react to reported fires that may not exist. These false alarms may be caused by a number of sources, including: changes in humidity; condensation on the fire detector surface; and contamination from animals, plants, or other contents of the cargo bay.

A second, independent method of fire detection to complement the conventional smoke detection techniques, such as the measurement of chemical species indicative of a fire, will help reduce false alarms and improve aircraft safety. Although many chemical species are indicative of a fire, two species of particular interest are CO and CO_2 [Grosshandler, 1995]. Some requirements for these chemical sensors differ from those of leak detection and emission monitoring. The sensor must withstand temperatures ranging from -30 to 50°C, pressures from 18.6 kPa to 104 kPa, and relative humidities from 0 to 95%. Different types of fires produce different chemical signatures [Grosshandler, 1995]; the sensor must be able to detect the presence of a real fire and must not be affected by the presence of gases produced by contents of the cargo hold. The response must be quick, reliable, and able to provide relevant information to the pilot.

The use of MEMS-based chemical sensing technology for fire detection would be similar to that for the leak detection application above; a number of sensors placed throughout a region would give an indication of the chemical signature of a fire and its location. Similar to the leak detection application, a smaller system is preferred. The larger the sensor system, the fewer sensors that can be placed in a cargo bay, and the more likely they are of being dislodged because they stick out into the chamber. The ideal is to have a multifunctional sensor array which gives all the information necessary to determine the presence of the fire, while being easily integrated with existing fire detection systems in the cargo bay.

11.2.3 Engine Emission Monitoring

One of the NASA goals is to significantly reduce emissions of future aircraft. Control of emissions from aircraft engines is an important component in the development of the next generation of these engines. A reduction of engine emissions can be achieved, in principle, if the emissions of the engine are monitored in real time, and that information is fed back and used to modify the combustion process of the engine. This active combustion control also depends on the development of actuators that control the combustion process, as well as a control system to interpret the sensor data and appropriately adjust the engine parameters. The ability to monitor the type and amount of emissions being generated by an engine is also important in determining the operational status of the engine. For example, by monitoring the emission output of the engine and correlating changes in the emissions produced with previously documented changes in engine performance, long-term degradation of engine components might be monitored. If such an emissions monitoring system was developed, an initial use would be measurement of emissions during the development stage of the engine.

There are very few sensors available commercially which are able to measure the components of the engine's emissions in situ. The harsh conditions, such as high temperatures, inherent near reaction chambers of the engine render most sensors inoperable. A notable exception to this limitation in sensor technology is the commercially available oxygen sensor presently used in automobile engines [Logothetis, 1991]. This sensor, which is based on the properties of zirconium dioxide (ZrO_2), has been instrumental in decreasing automotive engine emissions. However, comparable sensors for other components of the gas stream do not exist; highly sensitive monitoring of emissions of nitrogen oxides, hydrogen, and hydrocarbons is not presently possible in situ with point-contact sensors placed near the engine. Even the traditional ZrO_2-based sensor has sensitivity limits as well as size, weight, and power consumption requirements that discourage its use in some applications.

Therefore, the development of a new class of sensors for emissions monitoring of aircraft engines is necessary. The difficulties inherent with larger sensors placed in the engine stream include: 1) added weight to the engine; 2) less ability to pinpoint the chemical signature of the stream at a given location if the sensors vary in position because of their size; 3) more turbulence experienced by the array in the engine flow stream, and more interference with the flow of gases in the engine; 4) a larger projectile if the sensor array is dislodged from the measuring site and emitted into the engine. Thus, an emission sensor array should be as small as possible, which suggests the use of MEMS-based technology.

The environment in which a sensor must operate in emission monitoring applications varies drastically from the environment of launch vehicle propellant leak applications. The sensor must operate at high temperatures and detect low concentrations of the gases to be measured. Although the measurement of NO_x is important in these applications, the measurement of other gases present in the emission stream, such as C_xH_y, CO, CO_2, and O_2, is also of interest. Table 11.1 provides a partial listing, identified by the Glennan Microsystems Initiative (see For Further Information), of gases of interest in an emissions

TABLE 11.1 A Partial Listing of Gas of Interest in an Emissions Stream as Identified by the Glennan Microsystems Initiative

Chemical Species	Concentration Range
CO	50–2000 ppm
CO_2	2–14%
NO_x	10–2000 ppm
O_2	0–25%
C_xH_y	5–5000 ppm

stream. The measurement ranges depend on the chemical species and the engine, but generally the detection of NO_x and C_xH_y may be necessary at sensitivities down to 5–10 ppm, CO down to 50 ppm, O_2 from 0% to near 25%, and CO_2 from 2 to 14%. Such a wide range of detection requirements demands several types of sensor technologies, each of which should have stable, or at least predictable, performance for extended periods of time. Given the limited number of appropriate sensors for this application, development of new sensor technologies is also necessary for this application.

11.3 Sensor Fabrication Technologies

In order to meet the chemical sensor needs of aerospace applications described in Section 11.2, the development of a new generation of sensor technologies is necessary. Active development of chemical sensor technology to meet these needs is taking place at NASA Glenn Research Center (NASA GRC) and Case Western Reserve University (CWRU), based on recent progress in three technology areas: 1) micromachining and microfabrication technology to fabricate miniaturized sensors; 2) nanocrystalline materials, to develop sensors with improved stability and higher sensitivity; and 3) emergence of silicon carbide (SiC) semiconductor technology to provide electronic devices and sensors operable at high temperatures. This section will give a brief overview of these technologies, and Section 11.4 discusses chemical sensor technology development and application.

11.3.1 Microfabrication and Micromachining Technology

Microfabrication and micromachining technology is derived from advances in the semiconductor industry [Madou, 1997]. A significant number of silicon-based microfabrication processes were developed for the integrated circuit (IC) industry. Of the various processing techniques, lithographic reduction, thin film metallization, photoresist patterning, and chemical etching have found extensive use in chemical sensor applications. These processes allow the fabrication of microscopic sensor structures. The ability to mass-fabricate many of these sensors on a single wafer (also known as batch processing) using presently available semiconductor processing techniques significantly decreases the fabrication costs per sensor. A large number of sensors is fabricated simultaneously with the same processing, resulting in high reproducibility in response between the sensors. The smaller sensor size also enables better sensor performance, especially in low concentration measurements where the small surface area of the sensor allows a smaller number of molecules to have a larger relative effect on the sensor output.

However, these processes produce mainly two-dimensional, planar structures, which have limited application. By combining these processes with micromachining technology, three-dimensional structures can be formed which have a wider range of application to chemical sensing technology. Micromachining technology is generally defined as the means to produce three-dimensional micromechanical structures using both bulk and surface micromachining techniques. The techniques used in micromachining fabrication include chemical anisotropic and dry etching, the sacrificial layer method, and LIGA (lithographie, galvano forming, Abformung). Chemical anisotropic etching is an etching procedure that depends on the crystalline orientation of the substrate. For silicon etching, potassium hydroxide (KOH) and tetramethyl ammonium hydroxide (TMAH) solutions are most commonly used as etching agents. Dry etching processes include ion milling, plasma etching, reactive ion etching, and reactive ion beam etching. These dry etching processes are not dictated or limited by the crystalline structure. LIGA techniques have been used to produce high aspect ratio multistructures [Madou, 1997].

The sacrificial layer method employs a deposited underlayer that can be chemically removed. This technique can be used to make a chamber electrode structure to protect the integrity of the sensor element. For many applications, temperature control is necessary. Incorporation of a microfabricated heating element and a temperature detector allows feedback control of the operating temperature. In these microstructures, a small thermal mass is desirable in order to minimize heat loss and heat energy consumption. Achieving a small thermal mass is accomplished by micromachining techniques which allow the selective removal of the underlying silicon substrate, producing a suspended diaphragm structure. This diaphragm structure underneath the sensing element, combined with a heater and a temperature detector integrated as part of the sensor, is useful as a sensor platform that has small thermal mass and quickly reaches thermal equilibration with less heat loss and energy consumption. The sacrificial layer method can also be used to make a chamber structure that surrounds the sensor and protects the integrity of the sensing element.

MEMS processes allow the mass production of sensors of minimal size, weight and power consumption that are tailored for a given application. This sensor processing can be done using silicon (Si), either as a substrate on which a structure is built, or as a semiconductor that is part of an electrical circuit. Alternate materials to Si may be used as substrates as long as they are compatible with IC processing. However, micromachining of alternate materials to Si is less developed; different processes are often necessary to produce the same structures in, for example, ceramics.

11.3.2 Nanomaterials

Nanocrystalline materials are materials whose grain size is less than 100 nanometers. Nanocrystalline materials have several inherent advantages over conventionally fabricated materials, including increased stability and sensitivity at high temperature. The improved stability is because the mechanisms involved in the material grain growth are different for nanocrystalline materials than for materials with micro- or macrosized grains. Thus, these materials are more stable than conventional materials. Drift in the properties of gas-sensitive materials such as tin oxide (SnO_2), which have long-term heating caused by grain boundary annealing, have been previously noted [Jin, 1998; Hunter, 1992a]. This drift causes changes in the sensor output with time and reduces sensor sensitivity. The use of nanocrystalline materials, such as nanocrystalline SnO_2, is being investigated to stabilize the sensor materials for long-term operation. Further, since the detection mechanism of materials such as SnO_2 depends on the number of grain boundaries [Hunter, 1992a], nanocrystalline materials have a higher density of grains, and thus, a higher sensitivity. However, while nanomaterials may provide advantages as sensor materials, their use in microsensor platforms depends on the capability to process them in a manner compatible with the processing used for the other components of the sensor structure.

11.3.3 SiC-Based High-Temperature Electronics

Silicon carbide-based semiconductor electronic devices are presently being developed for use in conditions under which conventional semiconductors cannot adequately perform. Because of its wide bandgap and low intrinsic carrier concentration, SiC operates as a semiconductor at significantly higher temperatures than are possible with silicon semiconductor technology. SiC is now available commercially, and processing difficulties associated with fabrication of SiC devices are being addressed [Neudeck, 1995; Neudeck, 2002].

Silicon carbide occurs in many different crystal structures, or polytypes, with each crystal structure having its own unique properties. In many device applications, SiC's exceptionally high breakdown field (more than five times that of Si), wide bandgap energy (more than twice that of Si), and high thermal conductivity (more than three times that of Si) could lead to substantial performance gains. Combined with other material properties, such as its superior mechanical toughness, SiC is an excellent material for use in a wide range of harsh environments.

Silicon carbide's capability to function in extreme conditions is expected to enable significant improvements to a wide range of applications and systems. These improvements include: 1) greatly improved high-voltage switching for energy savings in public electric power distribution; 2) more powerful microwave electronics for radar and communications; and 3) electronics, sensors, and controls for cleaner-burning more fuel-efficient jet aircraft and automobile engines. A potentially major application of SiC as a semiconductor is in a gas-sensing structure. One set of approaches used to produce SiC-based gas sensors is discussed in Section 11.4.4.

11.4   Chemical Sensor Development

As discussed in the previous sections, the needs of aerospace applications are driving the development of a new generation of chemical microsensor technology. While the sensor design and sensing approach depends strongly on the application, the application needs nevertheless have some common factors:

1. Optimally, the detection of gas should take place in situ. Thus, it is strongly preferred that the sensor operate in the environment of the application.
2. The sensor must have minimal size, weight and power consumption to allow placement of multiple sensors in a number of locations without, for example, significantly increasing the weight or power consumption of the vehicle.
3. Ideally, the sensors should be readily multiplexed to allow a number of sensors to be placed in a given region and the results of the measurement fed back to monitoring hardware or software. This procedure will allow the accurate monitoring of a region and by coordinating the signals from an array of sensors determine, for example, the location of a leak or fire.

This section will discuss the work at NASA GRC and CWRU using microfabrication technology, nanomaterials, and SiC semiconductor technology to develop H_2, C_xH_y, NO_x, CO, O_2, and CO_2 sensors. Three different sensor platforms are used: a Schottky diode, a resistor, and an electrochemical cell. A brief description is given of each sensor configuration and its present stage of development. The Si-based hydrogen sensor is at a relatively mature stage of development, whereas the state of development of the other sensors ranges from the proof-of-concept level to the prototype stage. Integration of the sensor technology with hardware and software, as well as system demonstrations, are also taking place in specific applications with Makel Engineering, Inc. (MEI).

11.4.1   Si-Based Hydrogen Sensor Technology

The needs of hydrogen detection for aerospace leak detection applications just discussed require that the sensor work in a wide concentration range, from the ppm range to 100% hydrogen in inert environments. The number of sensor types that work in these environments is limited, and to find one sensor to completely meet the needs of this application is difficult. For example, metal films which change resistance upon exposure to hydrogen have a response proportional to partial pressure of H_2:$(P_{H2})^{1/2}$. This dependence is due to the sensor detection mechanism, which consists of migration of hydrogen into the bulk of the metal, changing the bulk conductance of the metal [Hunter, 1992b]. The result is low sensitivity at low hydrogen concentrations, but a continued response over a wide range of hydrogen concentrations. In contrast, Schottky diodes, composed of a metal in contact with a semiconductor (MS), or a metal in contact with a very thin oxide on a semiconductor (MOS), have a very different detection mechanism. For hydrogen detection, the metal, e.g., palladium (Pd), is hydrogen-sensitive. For Pd-SiO_2-Si Schottky diodes, hydrogen dissociates on the Pd surface and diffuses to the Pd-SiO_2 interface, affecting the electronic properties of the MOS system, which results in an exponential response of the diode current to hydrogen [Lundstrom, 1989].

This exponential response has higher sensitivity at low concentrations and decreasing sensitivity at higher concentrations as the sensor saturates. Thus, by combining both a resistive sensor and a Schottky diode, sensitive detection of hydrogen throughout the range of interest can be accomplished. Further, temperature control is necessary for both sensor types for an accurate reading. MEMS-based technology enables fabrication in a limited area of both sensors (resistor and Schottky diode) with temperature control and leads to minimal size, weight, and power consumption.

NASA GRC and CWRU have developed palladium (Pd) alloy Schottky diodes and resistors as hydrogen sensors on Si substrates. Palladium alloys are more resilient than pure Pd to damage caused by exposure to high concentrations of hydrogen [Hughes, 1987]. Although the sensor signal is affected by the presence of oxygen, the sensor does not require oxygen to function. The NASA GRC/CWRU design uses a palladium chrome (PdCr) alloy because of its ability to withstand exposure to 100% hydrogen [Hunter, 1998]. The sensor structure is shown in Figure 11.1 and includes a PdCr Schottky diode, a PdCr resistor, a temperature detector, and a heater all incorporated in the same chip. The sensor dimensions are approximately 2.2 mm on a side. The combination of the Schottky diode and resistor sensing elements results in a sensor with a broad detection range; the Schottky diode provides sensitive detection of low concentrations of hydrogen up to near 1%, whereas the resistor provides sensitivity from 1% up to 100% hydrogen. However, if the sensor is only exposed to low concentrations of hydrogen, then a different alloy, palladium silver (PdAg), can be used in the Schottky diode structure for higher sensitivity.

The MEMS-based hydrogen sensors using both PdCr and PdAg alloys have been selected for use or have been demonstrated in several applications. These sensors have been used on the assembly line of the Ford Motor Company (see Commercial Uses section below) and on the space shuttle. They have also been selected for use on the safety system of the experimental NASA X-33 vehicle and the water purification system for the International Space Station. Supporting hardware and software have also been included with the sensor to provide signal conditioning and control of the sensor temperature, and to record sensor data [Hunter, 1998; Hunter, 2000]. Each application has different requirements; the whole system from sensors to supporting hardware and software was tailored for these applications. It should be noted that the successful operation of these MEMS-based sensors depends on this supporting technology and associated packaging.

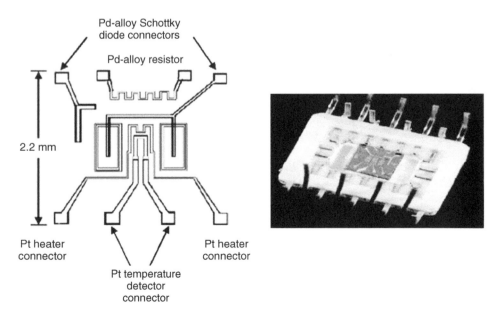

FIGURE 11.1 (See color insert following page 9-22.) Schematic diagram and picture of the silicon-based hydrogen sensor. The Pd alloy Schottky diode (rectangular regions) resides symmetrically on either side of a Pt heater and temperature detector. The Pd alloy Schottky diode is used for low-concentration measurements, whereas the resistor is included for high-concentration measurements.

This complete hydrogen detection system (two sensors on a chip with supporting electronics) flew on the STS 95 mission of the space shuttle (launched in October, 1998) and again on STS 96 (launched in May, 1999) [Hunter, 2000; Hunter, 1999a]. The hydrogen detection system was installed in the aft main engine compartment of the shuttle and was used to monitor the hydrogen concentration in that region. Presently, a mass spectrometer monitors the hydrogen concentration in the aft compartment before

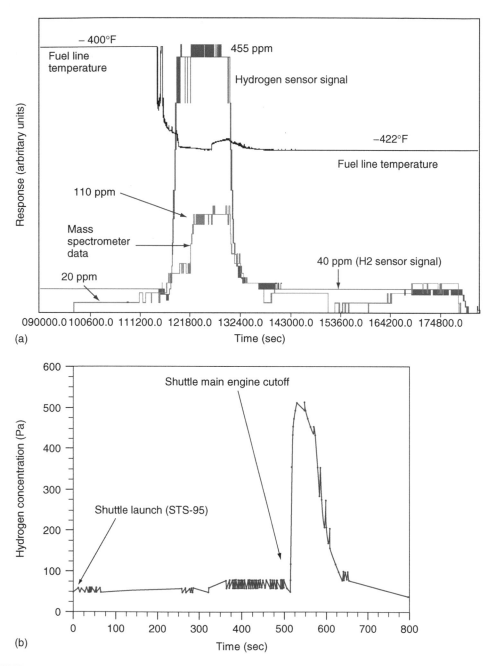

FIGURE 11.2   The response of a MEMS-based PdCr hydrogen sensor in the aft compartment of the space shuttle: (a) compared to the response of a mass spectrometer during loading of liquid hydrogen while on the launchpad. The time scale, in seconds, refers to a clock associated with the shuttle monitoring system; (b) Recorded data in flight during the launch of the shuttle. The time scale, in seconds, refers to time after launch.

launch, whereas after launch "grab" bottles are used. The insides of these "grab bottles" are initially at vacuum. During flight, the grab bottles are pyrotechnically opened for a brief period of time and the gas in the aft compartment is captured in the bottle. Several of these bottles are opened at different times during the takeoff and after the shuttle returns to earth; their contents are analyzed to determine the time profile of the gases in the aft chamber. However, this information is only available after the flight and cannot be used to monitor the shuttle condition in real time.

Data from the STS 95 mission is shown in Figure 11.2. The response of the hydrogen sensors was compared to that of the mass spectrometer and the information obtained by the grab bottles. During ground monitoring, the hydrogen sensor (Schottky diode) response was compared with that of the mass spectrometer. The sensor and the mass spectrometer intake were located at different points in the aft compartment. The results are shown in Figure 11.2a, where the hydrogen sensor signal, the mass spectrometer hydrogen signal, and the hydrogen fuel line temperature are shown as a function of time. The decrease in the fuel line temperature corresponds to the onset of fuel flow to the shuttle's hydrogen tanks. As this fueling begins, a small increase in hydrogen concentration is observed with both the hydrogen sensor and mass spectrometer. After the initial signal, the hydrogen concentration is seen to drop by both the sensor and the mass spectrometer. Overall, the hydrogen sensor response is seen to generally parallel that of the mass spectrometer, but with a larger signal and quicker response time (perhaps caused by the relative location of the each measuring device with respect to the hydrogen source). These results are for just one sensor monitoring the chamber; a number of these MEMS-based sensors placed around the chamber could be used to "visualize" the location of the leak.

The hydrogen sensor response during the launch phase of flight is shown in Figure 11.2(b). No significant sensor response is seen until the cutoff of the shuttle's main engines. Near this time, a spike in the hydrogen concentration is observed, which decreases with time back to baseline levels. These results are qualitatively consistent with the leakage of very small concentrations of unburnt fuel from the engines into the aft compartment after engine cutoff. These observations also qualitatively agree with those derived by analyzing the contents of the grab bottles. Moreover, the advantage of this microsensor approach is that the hydrogen monitoring of the compartment is continuous and, in principle, could be used for real-time health monitoring of the vehicle in flight.

11.4.2   Nanocrystalline Tin Oxide Thin-Films for NO$_x$ and CO Detection

The detection of NO$_x$ and CO can be accomplished using nanocrystalline SnO$_2$ materials deposited on a microfabricated or micromachined substrate [Hunter, 2000]. The substrate can be either Si or ceramic. Figure 11.3 shows the basic structure of the sensor design with a Si substrate. It consists of two components, Si and glass, which are fabricated separately and then bonded together. The microfabrication process allows the sensor to be small in size, with low heat loss and minimal energy consumption. Energy consumption is further reduced by depositing the sensor components (temperature detector, heater, and gas-sensing element) over a diaphragm region. This procedure minimizes the thermal mass of the sensing area, thereby decreasing both the power consumption for heating and the time to reach thermal equilibrium. The substrate used for the Si-based design is a 0.15-mm-thick glass. On one side of the glass are the heater and temperature detector, and the sputtered platinum interdigitated fingers reside on the other side. The width of the fingers and the gap between them is 30 microns each. The overall sensor dimensions are approximately 300 microns on a side, with a height of 250 microns.

The SnO$_2$ is placed across the interdigitated fingers using a sol-gel technique [Jin, 1998]. The gel containing the nanocrystalline SnO$_2$ particles is spun onto the substrate and patterned across the interdigitated finger region using conventional photolithography. This film is subsequently fired, yielding a nanocrystalline SnO$_2$ film whose conductivity is measured across the fingers. An example of a fired nanocrystalline SnO$_2$ film is shown in Figure 11.4. The size of each grain of SnO$_2$ is on the order of 10 nanometers with the porosity between the grains of roughly equal dimensions. Changes in the conductivity of SnO$_2$ across the interdigitated electrodes are measured and correlated to CO and NO$_x$ concentration. The detection of NO$_2$ has been demonstrated down to the 5 ppm level at 360°C, with the highest level of sensitivity in the

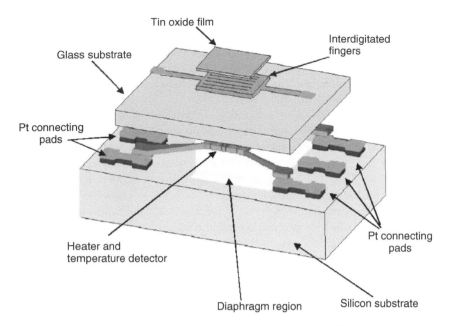

FIGURE 11.3 Structure of a SnO_2-based sensor on a Si substrate. The sensor package has Si and glass components and includes a Pt temperature detector and heater.

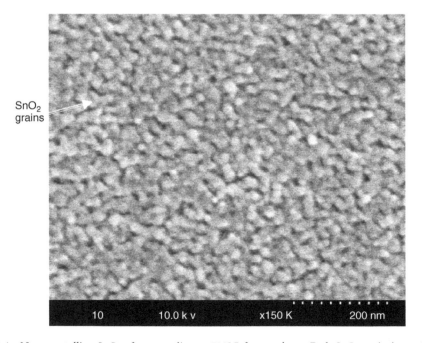

FIGURE 11.4 Nanocrystalline SnO_2 after annealing at 700°C for one hour. Each SnO_2 grain is on the order of 10 nanometers.

lower ppm with a very stable response [Hunter, 1999a]. Similar results with the detection of CO have been obtained [Jin, 1998].

Two major efforts in this development work are to optimize the SnO_2 response to a given gas and to stabilize the SnO_2 for long-term, high-temperature operation. Doping the SnO_2 can improve both the

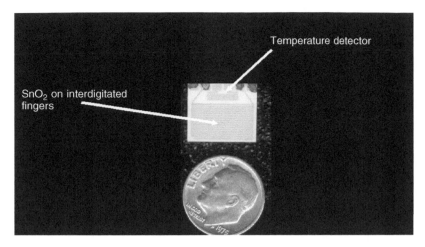

FIGURE 11.5 Tin oxide–based sensors fabricated on ceramic substrates for high-temperature operation. Shown are the interdigitated fingers on which SnO_2 is deposited and the temperature detector. The heater is on the backside of the ceramic.

selectivity of the sensor to a specific gas [Jin, 1998] as well as sensor stability. For example, the inclusion of Pt has been shown to improve sensitivity to CO, whereas the inclusion of SiO_2 has been shown to significantly decrease the grain growth of the SnO_2 [Hunter, 2000]. Another goal of the work is to fabricate the nanocrystalline SnO_2 films on alternate substrates for different applications. For example, for fire detection applications, a Si substrate is generally preferred. For emission-sensing applications, deposition on a ceramic substrate, as shown for the sensors in Figure 11.5, is often advantageous. While the configuration of interdigitated fingers, temperature detector, and heater are easily transferable to ceramics, micromachining and substrate bonding for high-temperature applications are more challenging.

A tin oxide-based NO_x sensor has been tested in a ground-based power generation turbine engine. The sensor was fabricated on an alumina substrate, as shown in Figure 11.5, and placed in a region where the exhaust flow stream from the engine is extracted and passed over the sensor. Supporting electronics were placed in a cooler region near the sensor. The temperature of the gas flowing over the sensor is near 350°C, and the sensor temperature is maintained using the built-in temperature detector and heater. The response of the sensor was compared to an industry standard continuous emission monitoring (CEM) system, which was simultaneously measuring the NO_x production. As seen in Figure 11.6, the sensor is initially saturated by the high level of NO_x. By diluting the flow into the sensor region with air by a factor of 18 (i.e., 18 parts air:1 part exhaust), the sensor response is brought back into detection range. The engine load is then increased, which correspondingly increases the NO_x production. This process is reflected by both the response of the microfabricated sensor (Figure 11.6) and the readings of the CEM. The two readings show good agreement: the sensor measured 540 ppm NO_x (after correction for the dilution) compared to the CEM measurement of 593 ppm. These results support the viability of replacing instrument-rack-sized CEM systems with MEMS-based chemical sensor technology.

11.4.3 Electrochemical Cell Oxygen Detection

A microfabricated O_2 sensor has applications both in fuel leak detection and, as demonstrated in the automotive emissions control example in Section 11.2.3, emissions control. Commercially available O_2 sensors are typically electrochemical cells using ZrO_2 as a solid electrolyte and platinum (Pt) as the anode and cathode. The anode is exposed to a reference gas, usually air, while the cathode is exposed to the gas to be detected. Zirconium dioxide becomes an ionic conductor of $O^=$ at temperatures of 600°C and above. This property of ZrO_2 to ionically conduct O_2 means that the electrochemical potential of the cell can be used to measure the ambient oxygen concentration at high temperatures [Logothetis, 1991].

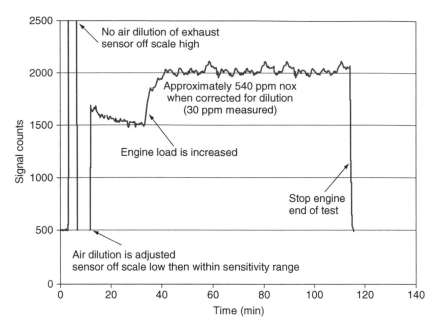

FIGURE 11.6   The response of a SnO$_2$-based NO$_x$ sensor in a ground-based turbine engine emission test. The emission stream is diluted with air by 18:1 near the beginning of the test to avoid saturation of the sensor. The sensor output is displayed as signal counts processed by the supporting electronics.

However, the operation of these commercially available sensors in this potentiometric mode limits the range of oxygen detection. This sensor is very sensitive near zero percent O$_2$ but has limited sensitivity at higher O$_2$ concentrations. Further, the current manufacturing procedure of this sensor, using sintered ZrO$_2$, is relatively costly and results in a complete sensor package with a power consumption on the order of several watts.

The objective of the research described here is to develop a ZrO$_2$ solid electrolyte O$_2$ sensor using microfabrication and micromachining techniques. As noted in previous sections, the presence of O$_2$ often affects the response of H$_2$, C$_x$H$_y$, and NO$_x$ sensors. An accurate measurement of the O$_2$ concentration would help to better quantify the response of other sensors in environments where the O$_2$ concentration is varying. Thus, the combination of an O$_2$ sensor with other microfabricated gas sensors is envisioned to optimize the ability to monitor emissions.

A schematic cross-section of the microfabricated O$_2$ sensor design is shown in Figure 11.7. As discussed in Section 11.4.2, microfabricating the sensor components onto a micromachined diaphragm region allows the sensor to be small in size, to have decreased energy consumption, and to have reduced time to reach thermal equilibrium. When operated in the amperometric (current-measuring) mode, the current of this cell is a linear function of the surrounding O$_2$ concentration. This linear response to oxygen concentration significantly increases the O$_2$ sensor detection range over O$_2$ sensors operated in the potentiometric mode. A chamber structure with a well-defined orifice can be micromachined to cover the sensing area. This orifice provides a pathway to control oxygen diffusion, which is important in amperometric measurements. A filter may also be placed over the sensor to minimize flow and provide protection. As discussed in the introduction to this chapter, fabrication of a MEMS-based version of a macrodevice is not just a matter of decreasing the dimensions of the macrodevice. For example, the fabrication of a microfabricated version of the oxygen sensor is a significant technical challenge, with obstacles not encountered in the fabrication of the macrosized sensor [Ward, 2002]. These challenges include adhesion between the various layers to each other and the substrate, pinhole electrical leaks between the layers, and diffusion effects caused by heating the sensor to high temperatures. Testing of a complete O$_2$ sensor has been accomplished, and further improvements on the design are planned.

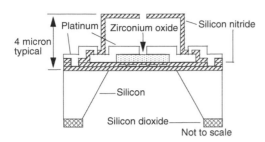

FIGURE 11.7 (See color insert following page 9-22.) The structure of a microfabricated amperometric oxygen sensor and a picture of a sensor packaged without its protective orifice.

11.4.4 SiC-Based Hydrogen and Hydrocarbon Detection

Hydrogen and hydrocarbon sensors are being developed using SiC semiconductors in order to meet the needs of leak detection and emissions monitoring applications. Like the Si-based system discussed in Section 11.4.1, the SiC-based Schottky diode does not need oxygen to operate. In contrast to the Si-based Schottky diode sensors which, are limited by the properties of the Si semiconductor to lower temperature operation (significantly below 350°C), SiC can operate at high enough temperatures (at least 600°C) as a semiconductor to allow the detection of hydrocarbons or NO_x. This capability opens the possibility of using SiC-based gas sensors in engine emission sensing applications. The development at GRC and CWRU of high-temperature, SiC-based gas sensors for use in harsh environments has focused on the development of a stable, gas-sensitive Schottky diode. One main advantage of a Schottky diode sensing structure is its high sensitivity, which is useful in emission sensing applications where the concentrations to be measured are low (see Table 11.1).

The detection mechanism of SiC-based Schottky diodes is similar to that for Si-based diodes, discussed in Section 11.4.1. For hydrogen, this detection mechanism involves the dissociation of the hydrogen on the surface of a catalytic metal. The atomic hydrogen migrates to the interface of the metal and the insulator, or the metal and the semiconductor, forming a dipole layer. Given a fixed forward bias voltage, this dipole layer affects the barrier height of the diode, resulting in an exponential change of current flow with the change in barrier height. The magnitude of this effect can be correlated with the amount of hydrogen and other gas species (especially oxygen) present in the surrounding ambient. The detection of gases such as hydrocarbons is made possible if the sensor is operated at a temperature high enough to dissociate hydrocarbons and produce atomic hydrogen. The resulting atomic hydrogen affects the sensor output in the same way as molecular hydrogen [Baranzahi, 1995; Hunter, 1995].

The major advantage of SiC as a semiconductor for Schottky diode gas sensors is shown in Figure 11.8, which shows the zero bias capacitive response of a Pd/SiC Schottky diode to one hydrocarbon, 360 ppm propylene, in nitrogen at a range of temperatures [Chen, 1996]. The sensor temperature is increased from 100 to 400°C in steps of 100°C and the sensor response is observed. At a given temperature, the sensor is exposed to air for 20 minutes, N_2 for 20 minutes, 360 ppm of propylene in N_2 for 20 minutes, N_2 for 10 minutes, and then 10 minutes of air. There are two points to note in the sensor behavior, as shown in Figure 11.8. First, the baseline capacitance in air does not change between 100 and 200°C, but decreases to a slightly lower value as the temperature is raised to 300°C and above suggesting the possibility that temperature dependent chemical reactions which affect the diode electronic properties may have an onset at temperatures above 200°C. Second, Figure 11.8 clearly shows that the magnitude of sensor response to 360 ppm propylene depends strongly on the operating temperature. A sensor operating temperature of 100°C is too low for propylene to dissociate on the Pd surface, so the device does not respond at all. The three other curves for 200°C, 300°C, and 400°C show that elevating the temperature increases the sensor's response to propylene. The presence of propylene can be detected at any of these higher temperatures, with 200°C being the minimum operating temperature determined in this study. Since the standard long-term

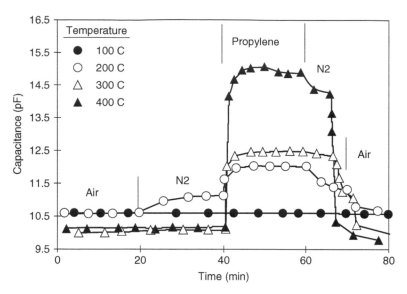

FIGURE 11.8 The temperature dependence of the zero bias capacitance of a Pd/SiC Schottky diode to various gas mixtures. The sensor response to propylene is seen to require a temperature above a minimum temperature ($\sim 200°C$) and to be strongly temperature-dependent.

operating temperature of Si is usually below 200°C, these results demonstrate the significant advantages of using SiC rather than Si in gas-sensing applications.

Several SiC-based Schottky diode structures have been investigated. Two example MS structures investigated are Pd on SiC (Pd/SiC) and PdCr on SiC (PdCr/SiC). Very sensitive detection, on the order of 100 ppm of hydrogen and hydrocarbons in both inert and oxygen-containing environments, has been demonstrated. Although the Pd/SiC sensor response is affected by prolonged high temperature heating at 400°C, the PdCr/SiC structure has been shown to be much more stable, with comparable sensitivity [Hunter, 1997]. The difference in behavior appears to be linked to reduced migration of Si into the PdCr metal from the SiC substrate and less formation of SiO_2 on the surface of the metal. More recent work shows the advantages of using atomically flat SiC semiconductor surfaces for the fabrication of chemical sensors rather than the standard SiC surfaces, which tend to have a high density of defects [Hunter, 2004]. Metal contacts to SiC for durable high-temperature operation, either as chemical sensors or as electrical device connections, are areas of continued investigation.

The third structure involves the incorporation of chemically reactive oxides into the SiC-based Schottky diode structure [Hunter, 1997; Hunter, 1999b]. A wide variety of materials, e.g., metal oxides such as SnO_2, are sensitive to C_xH_y and NO_x at high temperatures. These materials could be incorporated as a sensitive component into MOS structures. Unlike silicon, SiC-based devices can be operated at high enough temperatures for these materials to be reactive to gases such as C_xH_y and NO_x. This high-temperature operation results in a different type of gas-sensitive structure, a metal reactive-oxide semiconductor structure (MROS). The potential advantages of this type of SiC-based structure include: 1) increased sensor sensitivity because the diode responds to gas reactions caused by both the catalytic metal as well as the reactive oxide; 2) improved sensor stability, because the gas reactive oxide can act as a barrier layer between the metal and SiC, potentially stabilizing the sensor's structure against degradation at high temperature; and 3) the ability to vary sensor selectivity by varying the reactive oxide element. This MROS approach has been demonstrated using a SnO_2 film resulting in a Pd/SnO_2/SiC structure. This structure has been shown to have improved stability and significantly different responses than the Pd/SiC structure. Further, the geometry and processing conditions by which the SnO_2 is deposited affects the sensor response. Multiple types of oxide layers are envisioned to produce different selectivities for the

SiC-based Schottky diodes, which would improve the ability to selectively determine the concentration of a specific gas (see Section 11.5.4).

Development of SiC-based sensor packaging to allow sensor operation at temperatures greater than 300°C is also an active research area. For example, temperatures above 300–400°C will likely be necessary to optimally measure gases such as propylene and methane. Unlike the SnO_2 sensors in Section 11.4.2, which can be directly deposited onto a MEMS-based structure, the SiC-based sensors are semiconductor chips that are processed separately and then packaged. The type of packaging depends on the application. For leak detection applications in a room-temperature ambient, one difficulty with operating sensing elements that function at higher temperatures than ambient is that a considerable amount of power (on the order of several watts) may be necessary to properly heat the sensing element. Such power demands to achieve optimal sensing temperatures may limit sensor use in some applications. For these applications, one approach towards minimizing the heating element power demand is to mount the sensor in a MEMS-based package that thermally isolates the sensors and decreases the thermal mass. An early prototype SiC-based packaged sensor design is discussed by Hunter (1999b). For emissions applications, the objective is to package the sensor, with connections to the outside world, for operation in high-temperature engine environments. Thus, issues such as stable electrical contacts to SiC and the integrity of electrical insulation at high temperatures are of importance. Micromachining processes for SiC are being developed, for example, to enable formation of diaphragm structures that decrease the sensor's thermal mass and power consumption in the same manner as is done with Si structures. These issues are discussed in more detail elsewhere in this book (see chapter on "Deep Reactive Ion Etching for Bulk Micromachining of SiC").

11.4.5 NASICON-Based CO_2 Detection

The detection of CO_2, like the detection of oxygen, can be based on the use of solid electrolytes. However, the solid electrolyte for CO_2 sensors is NASICON (sodium super ionic conductor) rather than the zirconia used ion O_2 detection. NASICON is an ionic conductor composed of $Na_3Zr_2Si_2PO_{12}$. When combined with auxiliary electrolytes Na_2CO_3-$BaCO_3$, NASICON can detect CO_2 with high selectivity. A sol–gel process can be used to synthesize the NASICON electrolyte [Yang, 2000]. The ability to detect CO_2 over a range of concentrations has been demonstrated [Hunter, 1999a]. A microfabricated, miniaturized sensor structure can be similar to that of Figure 11.7, which can be easy to incorporate into a package with other sensors such as the CO sensor. The largest challenge to application of this material system to a miniaturized CO_2 sensor has involved incorporation of NASICON and Na_2CO_3-$BaCO_3$ onto a MEMS-based structure and eliminating drift in the sensor response caused by depletion of the chemical balance of the electrolyte during operation [Xu, 2004]. Significant progress has been made in these areas.

11.5 Future Directions, Sensor Arrays, and Commercialization

NASA's long-term goals include significantly improving safety and decreasing the cost of space travel, significantly decreasing the amount of emissions produced by aeronautic engines, and improving the safety of commercial airline travel. In order to reach these goals, the development of "smart" vehicle (a vehicle that can monitor itself and adapt to changing conditions) technology is envisioned. This monitoring, maintenance, and management must take place throughout the operational lifetime of the vehicle to ensure high safety, integrity, and reliability with decreased maintenance and costs. For self-monitoring to be effective, both physical and chemical parameters must be measurable throughout the vehicle, including the propulsion system. Thus, implementation of sensor technology in a variety of locations including harsh environments, vacuum, or in turbulent conditions will be necessary. Large, single sensors measuring a single parameter using a single detection mechanism will no longer be a viable technology approach. Rather, MEMS-based microsensor arrays which can measure several parameters of interest simultaneously and are integrated on a single device will be necessary to meet the objective of a "smart" vehicle.

One of the factors that has caused reluctance in applying sensor technology throughout a vehicle is the reliability of the sensor technologies themselves. Widespread use of sensors has been hampered, for example, by the fact that sensors can be damaged, need calibration, or can produce unreliable results as they drift from calibration. Thus, integrated hardware and software that can monitor and condition the sensor, detect problems, perform self-calibration as necessary, and provide this information to the user is also required; the accompanying electronics are what make a MEMS-based sensor system "smart." Further, these complete "smart" systems must meet the system requirements of the vehicles on which they are installed. As described here, for space-based systems, this situation may include using military grade components or being able to use the power available on the vehicle.

The sections that follow discuss ongoing work towards the development of MEMS-based chemical microsensor arrays. This discussion includes high-temperature membranes selective to individual gases, as well as examples of arrays in development with supporting electronics. The particular examples given are a leak detection array and a high-temperature electronic nose. The more advanced of these activities is the leak detection array work: some of the same techniques being developed for that system can also be applied to other array systems. Finally, the nonaerospace, commercial applications of this technology are also discussed in this section.

11.5.1   High-Selectivity Gas Sensors Based on Ceramic Membranes

Chemical sensors often lack selectivity and display similar responses to several different gaseous species [Hunter, 2000]. The selectivity of these sensors can be improved through the use of a gas-selective membrane over the sensor, to exclude other interfering gaseous species from reaching the sensing element. A selective membrane with a narrow distribution of pore sizes can filter gas molecules according to molecular size, thereby offering size selectivity to the species of interest. Because the gas molecules of interest are on the order of several angstroms, the pore size would also have to be only several angstroms in diameter.

Zeolites appear particularly well-suited for this filtering application. Zeolites are crystalline ceramic materials that contain open channels (or pores) as a natural part of their structure. Zeolites are composed primarily of aluminum, silicon, and oxygen. These elements combine to form SiO_4 and AlO_4 tetrahedra, which arrange in three-dimensional networks, yielding uniform channels through the ceramic crystal. Whereas these channels are identical in size or diameter within each crystallite of a zeolite, the diameters of these channels can vary significantly (from 3 to 12 angstroms), depending mainly on the ratio of aluminum to silicon. The effective diameters of the crystalline channels can be modified by the size of ions within the channels. Different ions may be incorporated into the crystalline channels by ion exchange. Further, as a ceramic material, zeolites are capable of withstanding high temperatures and harsh environments, which may be encountered in gas-measuring applications.

Composite membranes containing zeolites have been prepared using various processing techniques, including sol-gel and electrochemical vapor deposition methods [Hunter, 2000], to achieve desired selectivity. Figure 11.9 shows the grain structure of one of the prototype films. This filtering technology has the potential to be combined with the MEMS-based chemical sensor technology to further enhance the sensor selectivity to gas molecules of interest.

11.5.2   Leak-Detection Array

The development of an integrated smart leak-detection system, using miniaturized chemical sensor microsystems, is underway. The objective is to produce a microsensor array, which includes hydrogen, oxygen, and hydrocarbon sensors microfabricated by MEMS-based technology. Thus, a variety of potential launch vehicle fuels (hydrogen or hydrocarbons) and oxygen can be measured simultaneously. The size of the supporting electronics for this system is determined, in part, by the aerospace application. For space-based applications, electronics must often meet military specifications, must be radiation-hardened, or must function using vehicle voltage supplies that are not optimized for small-scale electronics (e.g., 28 V rather than 5 V). The smallest, most powerful electronics may not meet these specifications, or a bulky transformer may have to be added to the system to accommodate power restrictions. Results of these restrictions for the flight hardware that flew on the shuttle and included only a hydrogen sensor

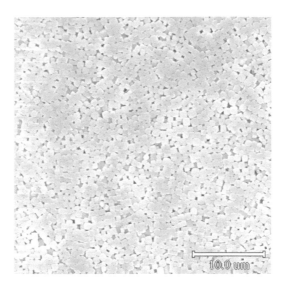

FIGURE 11.9   A scanning electron micrograph of a prototype of a selectively permeable zeolite thin-film membrane.

are shown in Figure 11.10(a). Whereas the sensor size is only 2 mm on a side and wafer thin, the associated packaging and hardware necessary to qualify a system for manned space flight is many orders of magnitude larger.

Significant progress has been made in reducing the size and power consumption of the overall system and in allowing the sensor system to be stand-alone, with signal conditioning electronics, power, data storage, and telemetry. The final system will be self-contained with the surface area comparable to a postage stamp. Thus, this postage-stamp-sized "lick and stick" type gas sensor technology can enable a matrix of leak detection sensors placed throughout a region with minimal size and weight, as well as with no power consumption from the vehicle. The sensors can detect a fuel leak from next generation vehicles, and the sensors can then combine that measurement with a determination of the oxygen concentration to ascertain if an explosive condition exists. Sensor outputs are fed to a data processing station, enabling real-time visual images of leaks, and enhancing vehicle safety.

A prototype model of the "lick and stick" sensor system has been fabricated and is shown in Figure 11.10(b) [Hunter, 2004]. The complete system has signal conditioning electronics, power, data storage, and telemetry, with hydrogen, hydrocarbon, and oxygen sensors. Figure 11.11 shows the operation of the electronics plus the three sensor systems simultaneously. In particular, the data highlights the response of the SiC-based gas sensor at various hydrocarbon fuel (RP-1) concentrations. The oxygen concentration is held constant and the hydrogen sensor signal shows no response, suggesting a lack of cross-sensitivity between the hydrogen and the hydrocarbon sensors. The hydrocarbon sensor, a $Pt/Cr_3C_2/SiC$ Schottky diode, is operated at 400°C and is able to detect fuel concentrations from 300 to 3000 ppm. The magnitude of the response to 300 ppm RP-1 fuel suggests the ability to detect concentrations well below 300 ppm. The range of the response is broad enough to detect low concentrations of gases (to at least 0.3%), but detection of up to 100% RP-1 fuel would require a complementary sensor as is done in hydrogen detection. It is envisioned that this multiple microsensor approach, combined with MEMS-based fabrication of the component parts, can allow the placement of sensors in a number of locations to better determine the safety of the region of interest.

11.5.3   High-Temperature Electronic Noise

Integration of a number of the individual high-temperature gas sensors discussed into a single platform will enable the formation of a sensor array. This array, composed of a variety of sensors, will allow for the

(a)

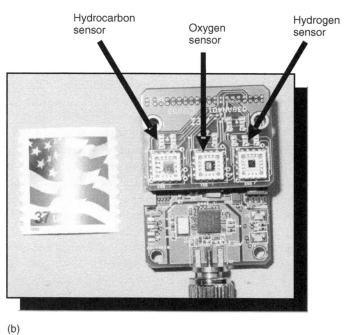

(b)

FIGURE 11.10 (See color insert following page 9-22.) (a) Shuttle system hardware (H_2 sensor with electronics); (b) A prototype version of a "Lick and Stick" leak sensor system with hydrogen, hydrocarbon, and oxygen detection capabilities combined with power and supporting electronics (including signal conditioning and telemetry).

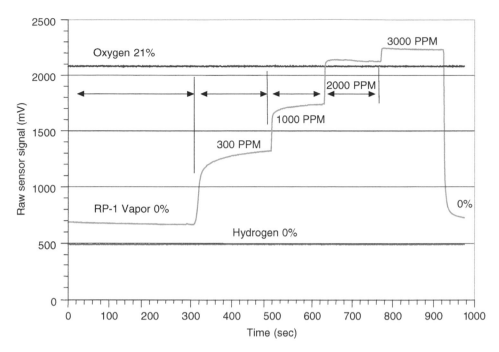

FIGURE 11.11 (**See color insert following page 9-22.**) Response of the three sensors of this system to a constant oxygen environment and varying hydrocarbon (RP-1) concentrations. The sensor signal shown is the output from the signal conditioning electronics, which processes the measured sensor current at a constant voltage.

detection of a number of gases on a single chip operable in high-temperature environments, such as in an engine. The formation of an array of the high-temperature-capable sensors discussed in this paper could detect H_2, C_xH_y, NO_x, CO, O_2, and CO_2. Development of such a microfabricated gas sensor array operable at high temperatures and high flow rates would be a dramatic step toward realizing the goal of monitoring and control of emissions produced by an aeronautic engine. Such a gas sensor array would, in effect, be a high-temperature electronic nose and be able to detect a variety of gases of interest. Several of these arrays could be placed around the exit of the engine exhaust to monitor the emissions produced by the engine. The signals produced by this nose could be analyzed to determine the constituents of the emission stream, and this information then could be used to control those emissions. Microfabrication of these sensors is necessary; a conventional, bulky system would add weight to the aircraft and impede the flow of gases leaving the engine exhaust. The component parts and packaging of this sensor array for a high-temperature environment would be fundamentally different than that used in the leak detection example discussed in the previous section.

The concept of an electronic nose has been in existence for a number of years [Gardner, 1994]. Commercial electronic noses for near-room-temperature applications presently exist, and there are a number of efforts to develop other electronic noses. However, these electronic noses often depend significantly on the use of polymers and other lower-temperature materials to detect the gases of interest. These polymers are generally unstable above 400°C and thus would not be appropriate for use in harsh engine environments. Thus, a separate development is necessary for a high-temperature electronic nose.

The development of such a high-temperature electronic nose has begun using the high-temperature sensors discussed in this chapter. In addition to the higher temperature limit of the sensors which constitute this nose, another major difference in approach compared to conventional noses is that there are three very different sensor types which constitute the high-temperature electronic nose: resistors, electrochemical cells, and Schottky diodes. As discussed earlier in this chapter, each sensor type relies on different mechanisms to detect gases so that each sensor type gives fundamentally different responses to the gas environment. For example, in general, electrochemcial cells have logarithmic responses to different gas concentrations, whereas Schottky diodes have exponential responses; resistors may have a sensitivity to many gases, but electrochemical cells can be operated to detect only gases which have certain binding energies; the time response of a Schottky diode is largely determined by migration of a chemical species into a film, whereas the time response of resistors like SnO_2 is related to surface effects. Thus, each sensor type provides very different types of information on the environment. This characteristic is in contrast to a conventional array of sensors that generally consist of elements of the same type, e.g., SnO_2 resistors doped differently for different selectivities. Each sensor in this conventional system provides information available through the SnO_2 resistors (reactions occurring on the surface of the sensor film) but do not provide information determined by electrochemical cells or Schottky diodes.

A first generation high-temperature electronic nose has been demonstrated on a modified automotive engine [Hunter, 2001]. Figure 11.12 shows the response of a tin oxide-based sensor (doped for NO_x sensitivity), an oxygen sensor, and a SiC-based hydrocarbon sensor. The figure shows the individual sensor responses during the initial start of the engine, a warm-up period, a steady state operation period, and at the engine turn-off. The sensors were operated at 400°C and the engine operating temperature was 337°C. Each sensor had a different characteristic response. The oxygen sensor showed a decrease in O_2 concentration, whereas the NO_x and C_xH_y concentrations increased at start-up. The hydrocarbon concentrations decreased as the engine warmed up to steady-state, whereas the NO_x concentration increaseed before stabilizing. The O_2, NO_x, and C_xH_y concentrations all returned to their start-up values after the engine was turned off. These results are qualitatively consistent with what would be expected for this type of engine. They also show the value of using sensors with very different response mechanisms in an electronic nose array; the information provided by each sensor was unique and monitored a different aspect of the engine's chemical behavior. It is envisioned that the high-temperature electronic nose will sense responses from this array of very different, high-temperature, MEMS-based sensors (resistors, diodes, and electrochemical cells) and integrate this information with neural net processing to allow a more accurate determination of the chemical constituents of harsh, high-temperature environments.

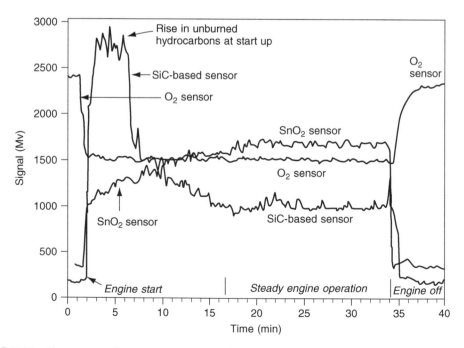

FIGURE 11.12   The response of a sensor array composed of a tin oxide based sensor (doped for NO_x sensitivity), an oxygen sensor, and a SiC-based hydrocarbon sensor in an engine environment.

11.6   Commercial Applications

The gas sensors being developed at NASA GRC and CWRU are meant for aerospace applications, but they can also be used in a variety of commercial applications as well. For example, an early design of the Pd-alloy hydrogen sensors, using PdAg, was adapted from space applications for use in an automotive application. GenCorp Aerojet Corporation, in conjunction with NASA Marshall Space Flight Center, developed hardware and software to monitor and control the NASA GRC and CWRU sensors to enable the sensors to monitor and display the condition of the tank of a natural gas-powered vehicle. Several of these systems have been purchased for use on the Ford Motor Company assembly line to test for leaks in the valves and fittings of natural gas vehicles. This complete system received a 1995 R&D 100 Award as one of the 100 most significant inventions of that year.

Likewise, the high-temperature C_xH_y, NO_x, and O_2 sensors are being developed for aeronautic applications but can also be applied to commercial applications. For example, the conditions in an aeronautic engine have similarities to those in an automotive engine. Thus, sensors that work in aeronautic engine applications may be operable in automotive engine applications. The hydrogen sensors which can detect leaks in propulsion systems can be used to detect hydrogen evolution in fuel cells. Other possible applications include combustion process monitoring, catalytic reactor monitoring, alarms for high-temperature pressure vessels and piping, chemical plant processing, polymer production, and volatile organics detection.

11.7   Summary

The needs of aerospace applications require the development of chemical gas sensors with capabilities beyond those of conventional commercial gas sensors. These requirements include operation in harsh environments, high sensitivity, and minimal size, weight and power consumption. MEMS-based sensor technology is being developed to address these requirements using microfabrication and micromachining technology, as well as SiC semiconductor technology. The combination of these technologies allows for the fabrication of a wide variety of sensor designs with responses and properties that can be tailored to the given application. Integration of these sensors with supporting packaging technology, hardware, and software is necessary for their application. Sensors designed for aerospace applications also have significant commercial applications. Although each application is different and the sensor needs to be tailored for that environment, the base technology being developed for aerospace applications can have significant impact on a broad range of fields.

Acknowledgments

The authors would like to acknowledge the contributions of Dr. Philip G. Neudeck, Mr. Gustave Fralick, Dr. Lawrence G. Matus, Dr. Daniel L. P. Ng, Mr. Paul Raitano, and Dr. Jih-Fen Lei, of NASA GRC; Dr. Liang-Yu Chen and Dr. Valarie Thomas of the Ohio Aerospace Institute; Mr. Q. H. Wu of the CWRU Electronics Design Center; Mr. William Rauch and Dr. M. Liu of Georgia Institute of Technology; Mr. José Perotti and Mr. Greg Hall of the NASA Kennedy Space Center; Dr. Benjamin Ward of Makel Engineering Inc.; and the technical assistance of Mr. Drago Androjna, Mr. Peter Lampard, Mr. Michael Artale, and Mr. Carl Salupo of Akima/NASA GRC.

Defining Terms

Ambient: The gas composition which is dominant in the surrounding environment (e.g. air, pure nitrogen, pure helium etc.).

Detection Mechanism: The chemical and/or physical reaction by which a sensor responds to a given chemical species.

Interfering gases: Gases which can cause a competing response in a sensor and can thus mask or interfere in the sensor's response to a given chemical species. For example, many sensors that respond to hydrogen can also respond to carbon monoxide; thus carbon monoxide is an interfering gas in the measurement of hydrogen.

Lithographic reduction: The reduction in size of patterns, such as those used for microfabricated structures, onto a photolithographic mask. These patterns are eventually transferred from a mask to a thin film in the formation of a sensor structure.

LIGA: Lithographie, Galvano forming, Abformung (i.e., Lithography, Plating, Molding), a German acronym signifying a combination of deep-etch X-ray lithography, electroplating, and molding.

Lower explosive limit (LEL): The lowest concentration at which a flammable gas becomes explosive. This limit depends on the flammable gas and the corresponding amount of oxidant. For hydrogen in air, this limit is a hydrogen concentration of 4%.

Micromachining technology: The fabrication of three dimensional miniature structures using processing techniques such as etching and LIGA.

Photoresist patterning: The use of a polymer film, photoresist, to create miniature features after exposure to UV radiation and removal by a developer.

Response time: The time it takes for a sensor to respond to the environment. Since some sensors never completely stabilize and reach a stable maximum value, often a value equal to 90% the steady state value is cited.

Sacrificial layer: A thin layer of material which is added to multilayer structure for the purpose of later being removed (or sacrificed) to open up that region of that structure.

Selectivity: The ability of a sensor to respond to only one chemical species.

Sensitivity: The amount of change in a sensor's output from baseline to a given chemical species.

Sol-Gel: The formation of compound through the use organic compounds and water which with drying form a gel which can be applied to a substrate. Higher temperature processing removes the organics and leaves the chemically sensitive material.

Stability: The reproducibility and repeatability of a sensor signal and baseline over time.

Thin Film: Typically, a film whose thickness is less than 1 micron.

Vehicle Health Monitoring: The capability to efficiently perform checkout, testing and monitoring of vehicles, subsystems and components before, during and after operation. This includes the ability to perform timely status determination, diagnostics and prognostics. The use of both sensors to determine the vehicle health and software to interpret the data is necessary.

References

Chen, L.-Y., Hunter, G.W., Neudeck, P.G., Knight, D., Liu, C.C., and Wu, Q.H. (1996) "SiC-Based Gas Sensors," in *Proc. Third International Symposium on Ceramic Sensors, 190th Meeting of the Electrochemical Society, Inc.*, pp. 92–105, 6–11 October, San Antonio, TX.

Gardner, J.W., and Bartlett, P.N. (1994) "A Brief History of Electronic Noses," *Sensor. Actuator. B*, **18**, pp. 211–20.

Grosshandler, W.L. (1995) "A Review of Measurements and Candidate Signatures for Early Fire Detection," NISTIR 555, Nat. Inst. of Stand., and Tech., Gaithersburg, MD.

Grosshandler, W.L., ed. (1997), "Nuisance Alarms in Aircraft Cargo Areas and Critical Telecommunications Systems," in *Proc. Third NIST Fire Detector Workshop*, NISTIR 6146, December, Gaithersburg, MD.

Hughes, R.C., Schubert, W.K., Zipperian, T.E., Rodriguez, J.L., and Plut, T.A. (1987) "Thin Film Palladium and Silver Alloys and Layers for Metal-Insulator-Semiconductor Sensors," *J. Appl. Phys.*, **62**, pp. 1074–1038.

Hunter, G.W. (1992a) "A Survey and Analysis of Commercially Available Hydrogen Sensors," NASA Technical Memorandum 105878.

Hunter, G.W. (1992b) "A Survey and Analysis of Experimental Hydrogen Sensors," NASA Technical Memorandum 106300.

Hunter, G.W., Neudeck, P.G., Chen, L.Y., and Knight, D. (1995) "Silicon Carbide-Based Hydrogen and Hydrocarbon Gas Detection," *AIAA-95-2247*, presented at the 31st AIAA Joint Propulsion Conference, 10–12 July, San Diego, CA.

Hunter, G.W., Neudeck, P.G., Chen, L.-Y., Knight, D., Liu, C.C., and Wu, Q.H. (1997) "SiC-Based Schottky Diode Gas Sensors," in *Proc. International Conference on SiC and Related Materials*, 1–7 September, Stockholm, Sweden.

Hunter, G.W., et al. (1998) "A Hazardous Gas Detection System for Aerospace and Commercial Applications," *AIAA-98-3614*, presented at *31st Joint Propulsion Conference*, 13–15 July, Cleveland, OH.

Hunter, G.W., Neudeck, P.G., Chen, L.-Y., Liu, C.C., Wu, Q.H., Sawayda, S., Jin, Z., Hammond, J.D., Makel, D.B., Liu, M., Rauch, W.A., and Hall, G. (1999) "Chemical Sensors for Aeronautic and Space Applications III," *NASA TM 1999-209450*, presented at *Sensors Expo 99*, September, Cleveland, OH.

Hunter, G.W., Neudeck, P.G., Chen, L.-Y., Makel, D.B., Liu, C.C., and Wu, Q.H. (1999) "SiC-Based Gas Sensor Development," in *Proc. International Conference on Silicon Carbide and Related Materials*, pp. 1439–42, 10–15 October, Research Triangle Park, MD.

Hunter, G.W., Neudeck, P.G., Fralick, G., Liu, C.C., Wu, Q.H., Sawayda, S., Jin, Z., Makel, D.B., Liu, M., Rauch, W.A., and Hall, G. (2000) "Chemical Microsensors For Aerospace Applications," in *Microfabricated Systems and MEMS V, Proceedings of the International Symposium*, 198th Meeting of the Electrochemical Society, Hesketh, P.J. et al., eds., pp. 122–41, 22–27 October, Phoenix, AZ.

Hunter, G.W., Neudeck, P.G., Fralick, G., Makel, D.B., Liu, C.C., Wu, Q.H., Thomas, V., and Hall, G. (2001) "Microfabricated Chemical Sensors For Space Health Monitoring Applications," in *AIAA 2001–4689, AIAA Space 2001 Conference*, August, Albuquerque, NM.

Hunter, G.W., Neudeck, P.G., Xu, J., Lukco, D., Trunek, A., Artale, M., Lampard, P., Androjna, D., Makel, D., Ward, B., and Liu, C.C. (2004) "Development of SiC-based Gas Sensors for Aerospace Applications," presented at *2004 Spring Meeting of the Materials Research Society*, San Francisco.

Jin, Z.H., Zhou, H.J., Jin, Z.L., Savinell, R.F., and Liu, C.C. (1998) "Application of Nano-crystalline Porous Tin Oxide Thin Film for CO Sensing," *Sensor. Actuator. B*, **52**, pp. 188–94.

Logothetis, E.M. (1991) "Automotive Oxygen Sensors," *Chemical Sensor Technology*, 3, ed., Kodansha Ltd., pp. 89–104.

Lundstrom, I. (1989) "Physics with Catalytic Metal Gate Chemical Sensors," *CRC Crit. Rev. Solid State Mater. Sci.*, **15**, pp. 201–78.

Madou, M. (1997) *Fundamentals of Microfabrication*, CRC Press, Boca Raton, FL.

Neudeck, P.G. (1995) "Progress in Silicon Carbide Semiconductor Electronics Technology," *J. Electron. Mater.*, **24**, pp. 283–88.

Neudeck, P.G., Okojie, R.S., and Chen, L.Y. (2002) "High-Temperature Electronics — A Role for Wide Bandgap Semiconductors?" in *Proc. of the IEEE*, 90, pp. 1065–76.

Ward, B. (2003) Thin-Film Microsensors for High-Temperature Gas Detection Arrays, Ph.D. Thesis, Case Western Reserve University.

Xu, J., Hunter, G.W., Liu, C.C., Hammond, J.W., Ward, B., Lukco, D., Artale, M., Lampard, P., and Androjna, D. (2004) "Miniaturized Thin-Film Carbon Dioxide Sensors," Abstract, AM-NET-2-2004, presented at the *American Ceramic Society Meeting*, April, Indianapolis, IN.

Yamazoe, N., Baranzahi, A., Spetz, A.L., Glavmo, M., Nytomt, J., and I. Lundstrom (1995) "Fast Responding High Temperature Sensors For Combustion Control," in *Proc, 8th International Conference on Solid-State Sensors and Actuators and Eurosensors IX*, pp. 741–744, Stockholm, Sweden.

Yang, Y., and Liu C. C. (2000), "Development of a NASICON-based Amperometric Carbon Dioxide Sensor," *Sensor. Actuator. B*, **62**, pp. 30–34.

For Further Information

For more information on the development of high-temperature sensor materials, see cism.ohio-state.edu/mse/cism/home_page.html for activities of the Center for Industrial Sensors and Measurements, the Ohio State University.

The Glennan Microsystem Initiative is a consortium of NASA, industry, and universities to address microsystems for harsh environments.

For further information of the semiconductor device technology, see Sze, S.M. (1981) *Physics of Semiconductor Devices*, 2nd ed., New York: John Wiley & Sons.

12

Packaging of Harsh Environment MEMS Devices

Liang-Yu Chen and
Jih-Fen Lei
NASA Glenn Research Center

12.1 Introduction

Microelectromechanical system (MEMS) devices, as they are defined, are both electrical and mechanical devices. Via microlevel mechanical operation, MEMS devices, as sensors, transform mechanical, chemical, optical, magnetic, and other nonelectrical parameters to electrical or electronic signals, and as actuators, MEMS devices transform electrical or electronic signals to nonelectrical or electronic operations. Therefore, MEMS devices very often interact with the environment electrically, magnetically, optically, chemically, and mechanically. In order to support these nonconventional device operations (i.e., the device's mechanical operation and the nonelectrical interactions between a MEMS device and its environment), new packaging capabilities beyond those provided by conventional integrated circuit (IC) packaging technology are required [Madou, 1997]. A chemically inert, optically dark, and electromagnetically "quiet" environment for packaging conventional ICs, provided by hermetic sealing and electromagnetic screening, is no longer suitable for packaging most MEMS devices. Because MEMS devices have very specific requirements for their immediate packaging environment, it is expected that the design of MEMS packaging will be very device-dependent. This is in contradiction to the conventional IC packaging practice, in which a universal package design can accommodate many different ICs. Compared to conventional IC packaging, the most distinct issue of MEMS packaging is to meet the requirements imposed by device mechanical operability, nonelectrical signal exchange, and thermomechanical reliability of MEMS devices.

NASA is interested in using harsh-environment operable MEMS [Hunter, 2004] and electronic devices [Neudeck, 2000; McCluskey and Pecht, 1999; Kirschman, 1999; Willander and Hartnagel, 1997] to characterize in situ combustion processes of aerospace engines and atmospheres of inner solar planets such as Venus. The operation environment of a high-temperature pressure sensor, one of the sensors most wanted by NASA and the aerospace industry for diagnosis and control of a new generation of aerospace engines, specifies the general requirements for packaging these harsh-environment MEMS. This pressure sensor must operate in an engine combustion chamber, and must be exposed to the high temperature environment, in order to measure in situ combustion chamber pressure in real time. The specifications of the high-temperature pressure sensor and the standard of in situ operation environment have all been determined by the Propulsion Instrumentation Working Group (PIWG) [PIWG, 2001]. The sensor operates at temperatures up to 500°C in a gas ambient composed of chemically reactive species such as oxygen in air, hydrocarbon or hydrogen in fuel, and catalytically poisoning species such as NO_x and SO_x in combustion products. High-temperature pressure sensors are also needed for atmosphere profiling missions for Venus, where the temperature is up to 500°C and the gas ambient is chemically corrosive. These sensor operation environments are summarized as high-temperature, high and dynamic pressure, and chemically corrosive. This is indeed a harsh environment compared to the standard operation conditions for most advanced (commercial) sensors and electronics, so the packaging materials and basic components, including substrate, metallization material(s), electrical interconnections (such as wire-bond), die-attach, and sealing, must be able to withstand an environment that is 500°C, corrosive, especially oxidizing and reducing, and with high dynamic pressure.

One of the suggested high-temperature pressure sensors is a semiconductor piezoresistive device. The sensing mechanism of this device depends on the mechanical deformation of semiconductor resistors residing on a thin diaphragm (fabricated by micromachining). This diaphragm is configured by the differential pressure on the diaphragm. Therefore, this device is very sensitive to external forces applied on the device (diaphragm). The major source of undesired external force is the thermomechanical stress from the die-attach structure caused by mismatches of coefficients of thermal expansion (CTE) of the die material (such as SiC), the substrate material, and the die-attaching material. The thermal stress of the die-attach must be suppressed in order to achieve precise and reliable device operation, because the thermally induced stress may generate unwanted device response to the changes in thermal environment, and in the extreme case, thermal stress can cause permanent mechanical damages to the die-attach.

Harsh-environment MEMS packaging offers new challenges to the device packaging field. In the next section, general material requirements for the major components of harsh-environment MEMS packaging are first analyzed by adopting the basic structure of conventional IC (chip-level) packaging, in which the die is attached to a substrate using a die-attaching material layer, and the die is electrically interconnected to the package by wire-bond. Following that, a SiC wafer-bonding method for wafer level packaging will be discussed. An electrical interconnection system that is based on ceramic substrates and thick-film metallization (both electrically and mechanically) will be addressed, followed by introducing conductive die-attach schemes compatible to the metallization schemes for chip-level packaging. The thermomechanical stress of a Au thick-film based die-attach assembly is discussed analytically, and is numerically analyzed using nonlinear finite element analysis (FEA). A low-stress die-attach structure for high-temperature MEMS packaging is introduced in Section 12.7.2.

12.2   Material Requirements

A summary of the basic requirements of the major materials needed for packaging harsh-environment (such as high-temperature) MEMS may help to establish the general guidelines for material selections.

12.2.1   Substrates

The basic function of the packaging substrate is to provide a framework (platform) for attaching the device die; metallization for electrical interconnection (such as wire-bond); mounting the leads connecting the chip to the external environment; building nonelectrical signal paths; and mechanical, and possibly electromagnetic, shielding. Plastic or polymer-type materials are not suitable for 500°C operation because of melting and depolymerization at temperatures above 350°C [McCluskey et al., 1997; Pecht et al., 1999]. Most metal and alloy materials suffer severely from corrosion, especially from oxidation at temperatures approaching 500°C in air. Ceramics, the remaining material system suitable for substrates, with excellent and stable chemical and electrical properties meet the basic requirements for substrates for operation in harsh environments, especially at high temperatures and corrosive gas ambient. After selecting the substrate material, a metallization scheme (both materials and processing), and associated sealing materials matching the substrate material, must be identified or developed simultaneously. In order to reduce the thermal stress of the die-attach, the CTE of the substrate material must match that of the device material (such as SiC). The properties of substrate surface and the interfaces formed at high temperature with other packaging materials, such as the die-attach material, also become very important. At high temperatures, the surfaces of some ceramics (nitrides and carbides) gradually react with the gas ambient, such as oxygen and water vapor, and therefore would lead to changes in properties such as surface resistivity and surface adhesiveness. Concerns such as these regarding the surface properties of "well-known" ceramic materials in high-temperature corrosive gas ambient may lead to valuable results from research into packaging materials for high-temperature MEMS.

12.2.2   Metallization and Electrical Interconnection Systems

Most metals and alloys, including some noble metals, oxidize at temperatures approaching 500°C in air. So the metals and alloys commonly used in conventional IC packaging such as Cu, Al, and Au/Ni-coated Kovar are excluded from packaging applications at 500°C and above, unless a perfectly hermetically sealed inert or vacuum condition is achievable. Intermetallic phases form at the interface of different metals such as Al and Cu at temperatures above 200°C. Intermetallic phases very often reduce mechanical strength of an interconnection system, so achieving material consistency of an interconnection system becomes extremely important to avoid thermomechanical failure at high temperatures.

Because of their chemical stability and good electrical conductivity, precious metals are considered for applications in substrate metallization and electrical interconnection. Some precious metals (e.g., palladium) react with atomic hydrogen (H) to form a hydrogen-rich alloy at elevated temperatures [Lewis, 1967]. Though this situation is desirable for gas sensing devices [Hunter et al., 1999], it is not desirable for electrical interconnection applications, because the phase transition may cause significant changes in the physical and electrical properties. Gold (Au) is widely used for both substrate metallization and wire-bond in packaging and hybridizing conventional high-frequency, high-reliability ICs. Besides the high conductivities and superior chemical stability at high temperatures, Au also has a low Young's modulus and a narrow elastic region. If Au is used as a die-attaching material, these properties of Au are helpful to reduce thermal stress generated at the Au-die and the Au-substrate interfaces caused by CTE mismatches. These features of Au are especially desirable for applications in packaging 500°C operable MEMS because a wide range of operation temperatures is of concern. It has been reported that Au thin film or wire with small grain size suffered from electro-migration of Au atoms at grain boundaries at high temperature under extreme current density ($\sim 10^6$ A/cm^2) operation [Goetz and Dawson, 1996]. Surface modifications and coatings have been suggested for the Au conductor approach to withstand high-temperature and high-current-density operation [Goetz and Dawson, 1996]. The electrical migration effect is proportional to J^{-2} (J is the current density), therefore, the obtainable lifetime of a Au conductor operated at high temperature but low current density may still be substantial, even without surface modification, as demonstrated in the next section. Developing a low-cost, highly conductive, and thermally and chemically stable conductor with excellent thermomechanical properties for the electrical interconnection is a challenging goal for packaging harsh-environment (high-temperature) devices [Grzybowski and Gericke, 1999; Harman, 1999].

12.2.3   Die-Attach

The basic function of the die-attach is to provide mechanical, electrical, and thermal support to the die. Most die-attaching (adhesive) materials used for packaging conventional ICs for operation at temperatures below 250°C are not suitable for 500°C operation. The basic failure mechanisms are depolymerization of epoxy-type materials, oxidizing and melting of eutectic-type materials, and softening and melting of glass-type materials. Generally, we expect the die-attach materials to be electrically and thermally conductive and physically and chemically stable at high temperatures, and to be cured or processed at a temperature that is not too high to damage the die. For applications in packaging MEMS devices, the thermomechanical properties of the die-attach material, such as CTE, Young's modulus, fatigue or creep properties and their temperature dependences, are very much of concern, as the die-attaching material is the material in intimate contact, both thermally and mechanically, with the die and the substrate. The material properties, especially the CTE, are expected to match those of the die and the substrate materials. If the CTE of the die is not completely consistent with that of the substrate, the die-attaching material is then ideally expected to thermomechanically compromise the CTE mismatch between the die and the substrate materials to reduce the thermal stress. Another major concern regarding the die-attaching materials for packaging high-temperature MEMS devices is long-term chemical and mechanical stabilities of the interfaces formed with the die (or the metallization of the die) and substrate (or the metallization of the substrate) materials at high temperatures. If the die-attach is expected to be electrically conductive, then the electronic properties of its interface with the die (or the metallization materials on the die) would also become critically important. Combining all these thermomechanical, chemical, and electrical requirements for the die-attaching material, it is apparent that a material system for die-attach with a suitable die-attaching process is critical to achieving success in packaging high-temperature MEMS devices.

The thermomechanical stress in the die-attach caused by the differences in physical properties (particularly CTE mismatch) of materials of the die-attach structure may cause degradation and failure of packaged devices. The extreme case of die-attach failure caused by thermal stress is cracking of the die or the attaching material. Besides this catastrophic failure caused by fatigue and creep of the die or the attaching material(s), thermal stress in die-attach can also cause the high-temperature MEMS device to have an undesired thermal response, irreproducibility of operation, and output signal drift. Therefore, thermomechanical failure caused by CTE mismatches of the materials involved in the die-attach is expected to be a common and important thermomechanical issue that must be addressed in packaging MEMS devices, especially for high-temperature MEMS devices.

12.2.4   Hermetic Sealing

The basic purpose of hermetic sealing is to create and maintain a stable and sometimes inert ambient for the packaged device. This simple function at low (room) temperature, however, is difficult to achieve at temperatures approaching 500°C. First, at such high temperatures, most soft or flexible sealing materials such as plastic or polymer-based materials can no longer operate. Second, sealing very often applies between different materials. The CTE mismatches of these materials make hermetic sealing difficult over a wide temperature range, especially under thermal cycling conditions. Third, thermal processes such as diffusion and degassing at material surfaces are activated or significantly promoted at high temperatures [Palmer, 1999], thus it becomes difficult to maintain an ambient in a small enclosure by sealing. Therefore, it is expected that creative sealing concepts are necessary to meet the sealing requirements for packaging many high-temperature MEMS devices.

12.3   Wafer-Level Packaging

Complicated wafer-level processes have been extensively used in both MEMS device fabrication and in MEMS wafer level packaging. These wafer-level processes include wafer oxidizing; wet chemical and dry etching; thin-film deposition (metal, silicon, oxides, nitrides, etc.); wafer bonding, sealing, and dicing. The materials used in wafer-level processes include oxides, ceramics, metals or alloys, and plastics. Most wafer-level packaging technologies developed so far are based on the Si wafer material system, which is not optimal for high-temperature operation. However, wafer-level metallization, passivation [Spry et al., 2004], and wafer bonding [Yushin et al., 2003] for single-crystal SiC material systems, which is electrically and mechanically suitable for high-temperature operation, have been reported recently. Wafer bonding is a key step of wafer-level packaging. In this section, a wafer-bonding method reported for single-crystal 6H-SiC wafers is summarized. Single crystal SiC wafers have been used extensively for high-temperature MEMS devices fabrication.

It has been reported that Si-face 6H-SiC wafers with on axis (0001) orientation and $3 \cdot 10^{18}/cm^3$ n-type doping have been fusion-bonded at temperatures above 800°C [Yushin, 2003]. The wafer surface roughness was 1.5 nm (RMS). The samples were cleaned in JTB-111 solution first, then in 10% hydrofluoric acid and de-ionized water. Surface oxides were desorbed in ultra-high vacuum (UHV) at 1100°C for 30 minutes prior to fusion. The fusion bonding was performed in UHV at temperatures between 800–1100°C under a uniaxial stress of 20 MPa for 15 hours. Examination by Transmission Electron Microscope (TEM) showed a 3-nm carbon interlayer at the bonding interface. The interface carbon might have resulted from high-temperature annealing [Bryant, 1999]. The bonded SiC retained its crystalline quality. Electrical measurements showed that azimuthal orientation misalignment between the wafers had significant effects on interface states. It is important to note that many single-crystal structures of SiC, such as the hexagonal structures (6H and 4H), present polar characteristics, i.e., Si terminates one side of a wafer and C terminates the other side of the wafer. The effects of polarity on SiC/SiC direct fusion bonding have not been reported. Except for the difference of the fabrication process for the surface silicon dioxide layer, the bonding process and mechanisms of SiC-SiO_2/SiO_2-SiC may be similar to those of the Si-SiO_2/SiO_2-Si system. However, it is of interest to note the effects of residual interfacial carbon, which usually accumulate at SiO_2/SiC interfaces, on interface states and the bonding quality.

Whereas 6H-SiC wafer bonding has been accomplished, electrical and mechanical, especially interfacial properties of bonded 6H-SiC wafer(s) at elevated temperatures, still need to be characterized for extreme-temperature applications.

12.4 High-Temperature Electrical Interconnection System

As discussed in the last section, ceramic materials are naturally selected as packaging substrate materials because of their superior high-temperature chemical and electrical stabilities. Aluminum oxide (Al_2O_3) is a low-cost substrate widely used for conventional IC packaging. Another advantage of using Al_2O_3 is that many thick-film and thin-film materials have been developed for metallization of Al_2O_3. In comparison to Al_2O_3, aluminum nitride (AlN) has a higher thermal conductivity and a lower CTE, which is very close to that of SiC. These AlN features make it useful for packaging high- temperature and high-power devices [Martin and Bloom, 1999]. If the basic framework of conventional IC packaging is adopted, the next step after selecting the substrate is to identify metallization material(s). In this section, we review a Au thick-film based, 500°C operable electrical interconnection system for chip-level, low-power, harsh-environment MEMS packaging [Chen et al., 2000a; 2000c].

Thick-film metallization materials are usually composed of fine metal (such as gold) powder, an inorganic binder (such as metal oxides), and an organic vehicle. The screen-printing technique is usually used for thick-film coating for thickness control and patterning. During the initial drying process (at 100 to about 150°C), the organic vehicle evaporates, and the paste becomes a semisolid-phase mixture of metal and binder powders. In the final curing process (~850°C is recommended for most thick-film products), metal powders form a cohesive film through diffusion, and the inorganic binder molecules migrate to the metal-substrate (e.g., Au-ceramic) interface and form reactive binding chains. Au thin wires can be bonded directly to Au thick-film metallization pads using commercialized wire-bond equipment to provide electrical interconnection in packaging. Some new thick-film materials may be applicable to various ceramic substrates such as alumina (Al_2O_3) [Keusseyyan et al., 1996] and aluminum nitride (AlN)

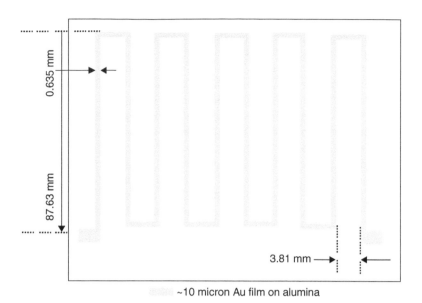

~10 micron Au film on alumina

FIGURE 12.1   Schematic diagram of Au thick-film printed wire for high-temperature tests.

[Chitale, 1994; Shaikh, 1994; Keusseyyan et al., 1996]. Compared with direct thin-film metallization on ceramic substrates, thick-film metallization offers low-cost, simple processing, low electrical resistance, and better adhesion provided by the reactive binders at the metal-substrate interface. Both the electrical and the mechanical properties of Au thick-film materials for applications in hybrid-packaging conventional ICs have been extensively validated at T < 150°C. In order to evaluate Au thick-film materials for 500°C applications, both electrical and mechanical tests at 500°C are necessary. Test results of Au thick-film material-based electrical interconnections (Au thick-film printed wires and thick-film metallization-based Au wire-bond) and a conductive die-attach scheme using Au thick-film as die-attaching material for operation up to 500°C [Chen, 2000c] are reviewed as follows.

12.4.1   Thick-Film Metallization

A Au thick-film printed wire circuit, as shown in Figure 12.1, was screenprinted on a ceramic substrate (AlN or 96% Al_2O_3) and cured at 850°C in air using the curing process recommended by the thick-film manufacturer. The circuit was electrically and thermally tested at 500°C in air for a total of approximately 1500 hours. The electrical resistance of the thick-film circuit was first measured at room temperature by four-probe resistance measurement technique. Afterwards, the temperature was increased to 500°C for approximately 1000 hours without electrical current flow, and the resistance of the wire was measured periodically during this period of time. The resistance fluctuated slightly within ±0.1% dur-ing the 1000-hour test, as shown in Figure 12.2. After testing for 1000 hours without electrical bias, the circuit was biased with 50 mA DC current, and the resistance was again continuously monitored by four-probe resistance measurement technique. The resistance fluctuated slightly within ±0.1% for 500 hours with electrical bias. This very small change in resistance is acceptable for almost all envisioned high-temperature device packaging applications.

As discussed earlier, Au thick-film materials for various substrates have been systematically validated, both electrically and mechanically, for conventional IC packaging; however, in order to be reliable at high temperatures, these Au thick-film material systems must be mechanically evaluated at elevated temperatures in addition to the electrical validation. Figure 12.3 shows the tensile strength of selected Au thick-film metallizations on AlN substrates measured by stud pull tests at room temperature. The tensile strength of Au thick-film metallizations on 96% Al_2O_3 substrates have been tested at room temperature

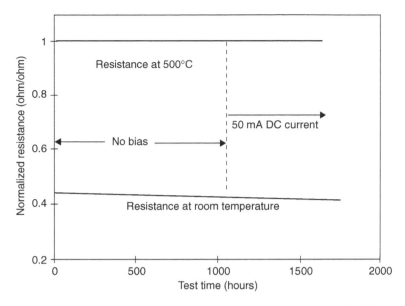

FIGURE 12.2   Normalized resistance of thick-film wire at various temperatures with and without DC bias.

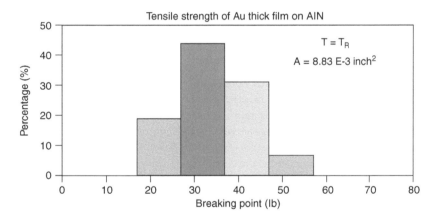

FIGURE 12.3   (**See color insert following page 9-22.**) Tensile strength of adhesion of selected Au thick-film on AlN substrate measured by stud pull tests.

after extended storage at 500°C [Salmon et al., 1998]. In order to examine the mechanical strength and the thermal dynamic stability of the binding system of Au thick-films at high temperatures, the shear strength of selected Au thick-film metallizations on 96% Al_2O_3 substrates were tested at temperatures up to 500°C [Chen et al., 2001]. The shear strength (breaking point) at 500°C reduced by a factor of 0.80 with respect to that at 350°C, whereas the shear strength (measured by breaking point) at 350°C was close to that at room temperature. The shear strength of selected Au thick-films designed for AlN were not as high as those for Al_2O_3, but it was sufficient for application in microsystem packaging operable at 500°C.

12.4.2   Thick Film–Based Wire-Bond

After the qualifications of Au thick-film metallization, the thick-film based wire-bond needs to be evaluated for high-temperature operation. The electrical test circuit including Au thick-film printed wires and pads and multiple thin gold wires bonded to the thick-film pads is illustrated in Figure 12.4. The geometry of the printed thick-film conductor was designed so that the electrical resistance of the test circuit was dominated

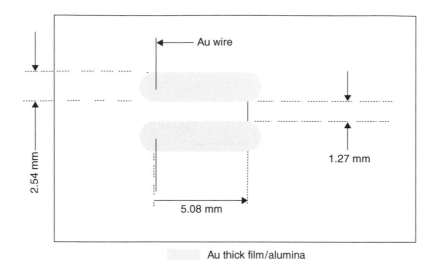

FIGURE 12.4   Schematic diagram of one unit of thick-film metallization based wire-bond test circuit. 0.0254 mm (0.001″) Au wires were bonded to thick film pads using thermal-compression technique.

by the resistance of thin (0.0254 mm diameter) bonded gold wires. The thick-film conductor wires and pads were processed according to standard drying and curing processes [DuPont, 1999] suggested by the material manufacturer. The thin gold wires were bonded to the thick-film pads on the substrate by thermal-compression wire-bonding.

The electrical resistance of a thick-film metallization based wire-bond test circuit (Figure 12.4) includes those of thick-film conductive wires and pads, bonded thin Au wires, and the interfaces of the wire-bonds. The total resistance of 22 units (44 bonds) in series were measured at room temperature and at 500°C vs. accumulated testing time at 500°C. The resistance was first measured at room temperature, after which the temperature was increased to 500°C and the resistance was monitored in air without electrical bias for 670 hours. As shown in Figure 12.5, the temperature was then lowered to room temperature and the resistance was recorded again. After this thermal cycle, the circuit resistance was continuously monitored for a total of 1200 hours at 500°C in air without electrical bias (current flow), followed by another 500 hours at 500°C with 50 mA (DC) current. The resistances under all these conditions were desirably low (less than 0.5 Ω per unit) and decreased slowly and slightly at an average rate of 2.7% over the 1500-hour testing period. The rate of resistance decrease under the DC bias is close to that without electrical bias.

An identical wire-bond sample was electrically tested in a dynamic thermal environment. The same wire-bond circuit was tested in thermal cycles between room temperature and 500°C with an initial temperature rate of 32°C/min for 123 cycles and a dwell time of five minutes at 500°C, then at the temperature rate of 53°C/min. (higher than thermal shock rate) for an additional 100 cycles with 50 mA DC current. The maximum change in electrical resistance, during the thermal cycle test, was 1.5% at room temperature and 2.6% at 500°C [Chen et al., 2001]. The electrical stability of this Au thick-film based wire-bond system should meet most interconnection needs for high-temperature, low-power applications. However, evidence of electro-migration in pure Au wire was observed after the thermal cycling test. Figure 12.6 shows the temperature profiles used for the thermal cycling test, and Figure 12.7 shows the resistance variation of a wire-bond circuit made of 1 mil Au wires upon thermal cycles between room temperature and 500°C.

12.4.3   Conductive Die-Attach

12.4.3.1   SiC Test Dies

Two types of SiC devices were used to demonstrate the packaging material systems. The first SiC high-temperature test device is a metal-semiconductor Schottky diode. The fabrication of a high-temperature

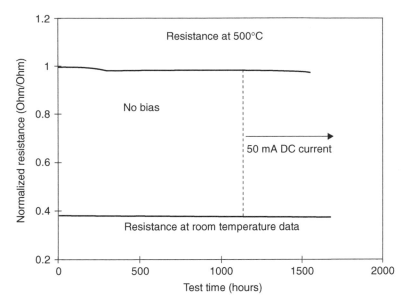

FIGURE 12.5 Normalized resistances of wire-bond test circuit measured at room temperature and at 500°C vs. test time at 500°C with and without DC bias.

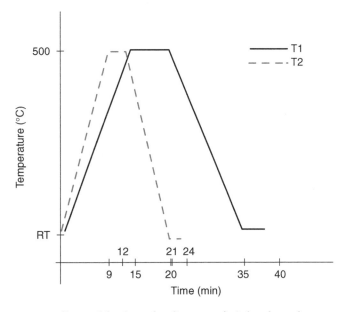

FIGURE 12.6 Temperature profiles used for thermal cycling tests of wirebond samples.

SiC Schottky diode used to test the conductive die-attach scheme was previously reported in detail [Chen et al., 2000b]. An N-type (nitrogen, resistivity less than 0.03 Ω-cm) Si terminated 4H-SiC wafer was used to fabricate the test Schottky diode. The backside of the wafer (unpolished side) was first coated with a nickel (Ni) thin-film by electron beam evaporation. The SiC wafer was then annealed at 950°C in argon in a tube furnace for five minutes, forming an ohmic contact on the backside of the SiC wafer. The device structure on the front side of the SiC wafer was fabricated by electron-beam evaporation of thin titanium (Ti) and thin Au films on the cleaned SiC wafer and then patterned with the liftoff technique.

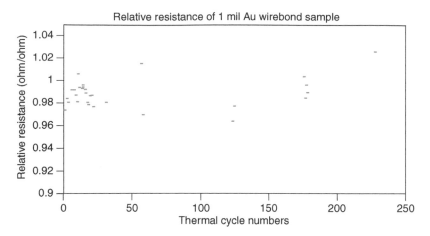

FIGURE 12.7 Resistance variation of 1 mil Au wirebond circuit after thermal cycles between room temperature and 500°C with 50 mA DC bias.

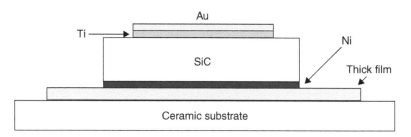

FIGURE 12.8 (**See color insert following page 9-22.**) Schematic diagram of as-fabricated SiC device and die-attach structure.

The second SiC high-temperature test device is a metal-semiconductor field effect transistor (MESFET). The MESFET was fabricated on a commercially purchased off-axis Si-face 6H-SiC epitaxial wafer [Spry et al., 2004]. Triple layers of Pt/Ta$_x$Si$_y$/Ti metallization were used for electrical contact on the top of the device chip. The areas surrounding these contacts were passivated by Si$_3$N$_4$. The wafer was briefly annealed at 600°C after the metallization. The backside of SiC MESFET was metallized with Pt/Ta$_x$Si$_y$/Ti for electrical contact.

12.4.3.2 Conductive Die-Attach

After dicing, the 1 mm × 1 mm SiC diode chips were attached to a ceramic substrate (either AlN or 96% alumina) using selected Au thick-film materials, as shown in Figure 12.8. An optimized thick-film die-attaching process for SiC device results in a low-resistance, conductive die-attach that is very often used for packaging many devices that require backside electrical contact.

The SiC test die with a Ni contact on the back was attached to a ceramic substrate using an optimized two-step Au thick-film processing [Chen et al., 2000b]. A thick film layer was first screenprinted on the substrate and cured at 850°C using the standard process. The SiC die was then attached to the cured thick-film pattern with a minimal amount of subsequent thick-film. A slower drying process (120–150°C) was critical to keeping the thick-film bonding layer uniform and the die parallel to the substrate after the curing process. Following the drying, the attached die was processed at a lower final curing temperature (600°C).

This optimized Au thick-film, die-attach process allows sufficient diffusion of inorganic binders toward the thick-film substrate interface, resulting in a good strength of binding to the ceramic substrate. Meanwhile, it prevents the attached semiconductor chip from being exposed to temperatures above the ultimate operation temperature of SiC devices (600°C) during the die-attach process. The second advantage

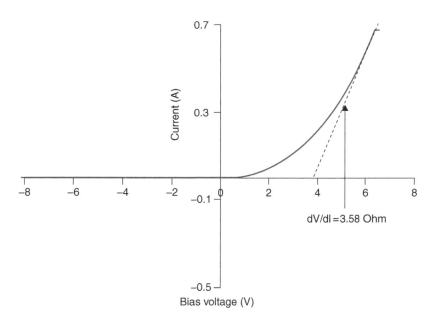

FIGURE 12.9 I-V curve of attached SiC test-diode characterized at 500°C after being tested for 1000 hours at 500°C oxidizing air.

of this die-attach process is that the distribution of the thick-film (after the final curing) between the chip and the substrate can be better controlled, because only a minimal amount of thick-film material is necessary to attach the chip to a cured thick-film pad. Therefore, many potential problems caused by nonuniform thick-film distribution at the chip-substrate interface might be avoided.

A 0.001-in.-diameter Au wire was bonded onto the top of Au thin-film metallization area covered with a Au thick-film overlayer by thermal-compression bonding technique. Thick-film material was also used to reinforce the top Au thin-film for better wire bonding. The Au thin film metallization area was coated with thick-film on the top then dried at 150°C for ten minutes. The thick-film on the device top was cured during the final die-attach process (at 600°C).

A Pt thick-film and glass-based, electrically conductive die-attaching material with low processing (curing) temperature, ~500°C, has been preliminarily developed for various ohmic contacts for backside metallization of high-temperature SiC devices. This material can be used to directly attach SiC die to Au and Pt metallization pads on the packaging substrate with a single curing at temperatures around 500°C. The SiC MESFET was attached to a Au thick-film metallized 96% Al_2O_3 packaging substrate using this material for a long-term, high-temperature test. The die-attaching material was first applied on the substrate and dried between 125 and 150°C. The SiC die was then attached to the dried attaching material and cured at 500°C. Heavily doped 1 mil (diameter) Au wires were used to electrically interconnect the device to the metallization traces on the alumina substrate using thermosonic bonding. Ten mil-diameter Au wires were used to provide the second level of electrical interconnections from the packaging substrate to room temperature terminals for measuring instruments. These Au wires directly extend to the outside of the oven in which the thermal-electrical testing of the device was conducted. This packaging and testing system successfully facilitated continuous electrical testing of the MESFET in 500°C air ambient for over 2000 hours.

12.4.3.3 Electrical Test

The attached SiC test diode (Figure 12.8) was characterized by current–voltage (I–V) measurements at both room temperature and at 500°C for various heating times at 500°C. A minimum dynamic resistance (dV/dI) under forward bias, which is deduced from the I–V curve, was used to estimate the upper limit of resistance of the die-attach structure (both interfaces and materials) and to monitor the resistance stability of the die-attach, as shown in Figure 12.9. This dynamic resistance includes the forward dynamic

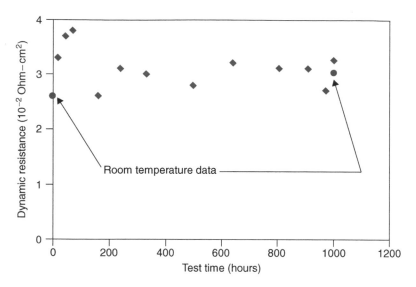

FIGURE 12.10   Minimum specific dynamic resistance (normalized to device area) calculated from I–V data vs. heating time at 500°C. This resistance includes resistances contributed from the Au(Ti)/SiC rectifying interface, SiC wafer, and the die-attach materials and interfaces. Resistances of the bonded wire and the test leads have been subtracted.

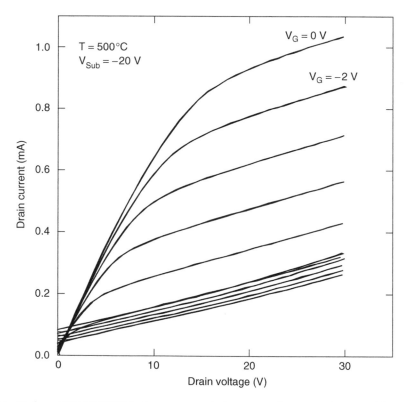

FIGURE 12.11   Packaged SiC MESFET drain current vs. drain-source voltage at various gate biases at 500°C after 558 hours continuous electrical operation in 500°C ambient.

resistance of the AuTi-SiC interface, the SiC wafer bulk resistance, the die-attach materials and interfaces resistance, the bonded wire resistance, and the test leads resistance in series. The resistance contributed from test leads and bonded wire were measured independently and subtracted. The attached device was first characterized by I–V measurements at room temperature. The device exhibited rectifying behavior, and the minimum dynamic resistance after subtracting test-leads and bond-wire resistance (which also applies to all of the following discussion) measured under forward bias was about 2.6 Ω, as shown in Figure 12.10. The temperature was then increased to 500°C (in air) and the diode was in situ characterized periodically by I–V measurement for about 1000 hours. During the first 70 hours at 500°C, the minimum dynamic resistance under forward bias increased slightly from 3.3 to 3.8 Ω. After that, the minimum dynamic resistance decreased slightly and remained at an average of 3.1 Ω. The diode was then cooled down to room temperature and characterized again. The minimum forward dynamic resistance measured at room temperature was 3.3 Ω. It is worth noting that the I–V curve of the device changed somewhat with time during the heat treatment at 500°C. However, the minimum dynamic resistance of the attached diode remained comparatively low over the entire duration of the test and the entire temperature range, indicating a low and relatively stable die-attach resistance.

The packaged SiC MESFET was characterized using a transistor curve tracer at 60 Hz. Electrical testing was conducted after an overnight heat soak in 500°C air ambient. The drain-source I–V curves were continuously swept from 0 to 30 V at gate bias from 0 to −20 V. The substrate bias voltage was either 0 V or −20 V during the test. The channel-substrate diode I–V characteristics were also measured by biasing the substrate while grounding the source, drain, and gate. Figure 12.11 shows the I–V curves of the SiC high-temperature MESFET at various gate biases characterized after 500 hours of continuous electrical operation and test in 500°C air ambient [Neudeck et al., 2004]. The SiC MESFET operated with less than 10% change in operational transistor parameters in the first 500 hours, indicating satisfactory performance of the die-attach.

12.5 Thermomechanical Properties of Die-Attach

By adopting a conventional die-attach structure, in which the die-backside is attached to a substrate using a thin attaching material layer, the die-attach using Au thick-film material discussed in the last section (and illustrated in Figure 12.8) is a typical example of this structure. As discussed previously, the most important issue that concerns the thermomechanical reliability of a packaged high-temperature MEMS is the die-attach thermal stress, i.e., the guideline for material selection, structure design and optimization of the attaching process to minimize the overall die-attach thermal stress effects on the mechanical operation of the device.

We start with the list of static displacement equations governing the thermomechanical behavior of the die-attach structure to analyze the material properties and other factors determining the thermomechanical properties of the structure. Following that, thermomechanical optimization of a Au thick-film based die-attach is discussed using the simulation results of nonlinear finite element analysis (FEA) [Lin and Chen, 2002].

12.5.1 Governing Equations and Material Properties

Assuming that the temperature distribution in the die-attach is static $\left(\frac{\partial T}{\partial t} = 0\right)$ and uniform $\left(\frac{\partial T}{\partial x} = \frac{\partial T}{\partial y} = \frac{\partial T}{\partial z} = 0\right)$, and the external force on the die-attach structure is zero, the general thermomechanical governing equations for the die-attach system are listed as:

$$\frac{\partial^2 u}{\partial x^2} + \frac{\partial^2 v}{\partial x \partial y} + \frac{\partial^2 w}{\partial x \partial z} + (1 - 2v)\nabla^2 u = 0 \tag{12.1}$$

$$\frac{\partial^2 v}{\partial y^2} + \frac{\partial^2 u}{\partial x \partial y} + \frac{\partial^2 w}{\partial y \partial z} + (1 - 2v)\nabla^2 v = 0 \tag{12.2}$$

$$\frac{\partial^2 w}{\partial z^2} + \frac{\partial^2 u}{\partial x \partial z} + \frac{\partial^2 v}{\partial y \partial z} + (1 - 2v)\nabla^2 w = 0 \tag{12.3}$$

where $\nabla^2 = \frac{\partial^2}{\partial x^2} + \frac{\partial^2}{\partial y^2} + \frac{\partial^2}{\partial z^2}$, u, v, and w are the displacements in x, y, and z direction, respectively; v is Poisson's ratio of the material. The normal stress distributions are determined by material properties, the temperature, and the displacements [Lau, 1998]:

$$\sigma_x = \frac{\lambda}{v}\left[(1 - v)\frac{\partial u}{\partial x} + \left(\frac{\partial v}{\partial y} + \frac{\partial w}{\partial z}\right)\right] - \beta(T - T_o) \tag{12.4}$$

$$\sigma_y = \frac{\lambda}{v}\left[(1 - v)\frac{\partial v}{\partial y} + \left(\frac{\partial w}{\partial z} + \frac{\partial u}{\partial x}\right)\right] - \beta(T - T_o) \tag{12.5}$$

$$\sigma_z = \frac{\lambda}{v}\left[(1 - v)\frac{\partial w}{\partial z} + \left(\frac{\partial u}{\partial x} + \frac{\partial v}{\partial y}\right)\right] - \beta(T - T_o) \tag{12.6}$$

Shear stress components are reduced from u, v, and w as:

$$\tau_{xy} = \frac{E}{2 + 2v}\left(\frac{\partial u}{\partial y} + \frac{\partial v}{\partial x}\right) \tag{12.7}$$

$$\tau_{yz} = \frac{E}{2 + 2v}\left(\frac{\partial v}{\partial z} + \frac{\partial w}{\partial y}\right) \tag{12.8}$$

$$\tau_{zx} = \frac{E}{2 + 2v}\left(\frac{\partial w}{\partial x} + \frac{\partial u}{\partial z}\right) \tag{12.9}$$

where $\lambda = \frac{vE}{(1 + v)(1 - 2v)}$, and $\beta = \frac{\alpha E}{1 - 2v}$. E is Young's modulus and α is the coefficient of linear thermal expansion (CTE). Material properties α, E, and v generally vary from the die to the attaching layer and the substrate. If there is no residual stress at a certain temperature, such as the die-attaching temperature, then the mismatch of CTEs of the die-attach is the major source of thermomechanical stress. This condition would become more apparent if the boundary value problems of a die-attach assembly were listed for the die, the attaching layer, and the substrate separately. There would be interface force terms at boundaries between the die and the attaching layer and that between the attaching layer and the substrate, because of the mechanical interactions at these boundaries. These interfacial force terms are generated by the mismatch of material CTEs or the residual stresses, or both. In case there is no residual stress and the CTEs match with each other, these interfacial force terms vanish and solutions for the equations would be trivial.

In the temperature range for conventional IC packaging, all of these materials properties can be approximated as constants. However, in the operation temperature range from room temperature to 500°C, the temperature dependences of these material properties must be quantitatively considered for a precise solution for these equations.

The boundary conditions at the interface between the die and the attaching layer and the interface between the attaching layer and the substrate are largely determined by the bonding quality and the thermomechanical stability of the die/attaching layer interface and the attaching layer/substrate interface. The simplest model of these boundary conditions is the ideal bonding, i.e., the continuity of all displacement components at interfaces in the entire temperature range: $u|_{S_s^+} = u|_{S_s^-}$, $v|_{S_s^+} = v|_{S_s^-}$, $w|_{S_s^+} = w|_{S_s^-}$, $u|_{S_a^+} = u|_{S_a^-}$, $v|_{S_a^+} = v|_{S_a^-}$ and $w|_{S_a^+} = w|_{S_a^-}$. S_s is the interface between the die and the attaching material, and S_a is the interface between the attaching material and the substrate, respectively.

The effects of thermomechanical stress on the performance of a material or component are usually evaluated by either Von Mises Stress, $\sigma_{VM} = 1/2 S_{ij} S_{ij}$ (the deviation stress tensor, $S_{ij} = \sigma_{ij} - 1/3 \sigma_{\alpha\alpha}\delta_{ij}$, where $\delta_{ij} = 1$ for $i = j$, and $\delta_{ij} = 0$ for $i \neq j$), or the maximum principle stress σ_{MP} [Giacomo, 1996].

Assume:

1. The substrate size is much larger than that of the die, and the material is amorphous/polycrystalline (isotropic).
2. One of the basal planes of the chip crystal is parallel to the die-attach interface, and the die material is isotropic in azimuthal planes.
3. Boundary conditions at the interfaces are determined by ideal bonding; all displacement components are continuous at both the interface between the die and the attaching layer and the interface between the attaching layer and the substrate.
4. The die-attach structure is in a uniform temperature field.

Then, the maximum thermal stress distribution, either Von Mises Stress or maximum principal stress (MPS), is basically a function of the thickness of the die (θ_d); the thickness of the attaching material layer

(θ_a); the properties E, α, and ν of the die material (E_d, α_d, and ν_d); the properties of the attaching material (E_a, α_a, and ν_a); the properties of the substrate material (E_s, α_s, and ν_s); and the temperature deviation from the thermal stress relaxing temperature, $T - T_R$, where T_R is the thermal stress relaxing temperature:

$$\sigma = \sigma(x, y, z, l, \theta_d, \theta_a, \theta_s, \overrightarrow{M_d}, \overrightarrow{M_a}, \overrightarrow{M_s}, (T - T_R)) \tag{12.10}$$

where l is the die-size parameter, $\overrightarrow{M}_{d,a,s}$ are material property parameters of the die, and the attaching material and the substrate, respectively. For 500°C operable packaging, because the operation temperature range is, relatively, much wider compared to that for conventional ICs, the temperature dependencies of these material properties and their nonlinearity have to be considered for a precise assessment of the stress and strain distribution. Usually, σ is symmetric about the vertical die central axis if the package dimensions are much larger than those of the die. For example, if the die is in a square shape, σ has approximately lateral fourfold (90°) symmetry about the vertical die's central axis. In case the die surface is off the crystal basal plane, then σ depends on the angle off the basal plan, and the horizontal symmetry of σ reduces from 90° to mirror symmetry. Ideally, if CTEs of the die, attaching material, and the substrate are consistent with each other in the entire operation temperature range and the assembly is relaxed at an initial condition, then at any temperature:

$$\sigma = \sigma(x, y, z, l, \theta_d, \theta_a, \overrightarrow{M_d} = \overrightarrow{M_a} = \overrightarrow{M_s}, (T - T_R)) = 0 \tag{12.11}$$

At the relaxing temperature, T_R, the die-attach assembly is largely relaxed: $\sigma = \sigma(x, y, z, l, \theta_d, \theta_a, \overrightarrow{M_d}, \overrightarrow{M_a}, \overrightarrow{M_s}, (T = T_R)) = 0$. Sometimes the stress (distribution) at relaxing temperature may not be completely relaxed but may reach a minimum; this minimum stress depends on the die-attaching material, the physical and chemical process of the die-attaching, and the thermal experience of the die-attach.

Consider thick-film material based die-attach as an example. T_R is basically the final curing temperature, if the cooling process after the final curing is so rapid that thermal stress does not have a chance to relax through sufficient diffusion at lower temperatures. However, T_R can be lower than the final curing temperature if the cooling process is so slow that a relaxed configuration is reached at lower temperatures. Therefore, T_R is initially determined by die-attach processing temperature, but it may change with the thermal history (temperature vs. time) after the attachment. The dependence of T_R on the thermal history is not desired because this situation may cause a reproducibility problem of device response. The basic physical and chemical attaching process certainly has a significant impact on T_R. For some die-attach materials and attaching processes, such as phase-transition material, T_R may not be the die-attach processing temperature.

In addition to the direct impact from material properties of the attaching layer, the thermomechanical properties of the interface between the die and the attaching layer, and that between the attaching layer and the substrate or also important to the thermal stress or strain configuration of the die-attach structure. The interfacial shear elastic and plastic properties, the interfacial fatigue and creep behavior, and their temperature dependencies are determined by the chemical and physical interactions between the materials composing the interface. If the materials chemically react at the interface, an interphase forms and it may dominate the interfacial thermomechanical properties. The thermomechanical properties of these interphases can certainly be critical to the overall thermomechanical properties of the die-attach structure.

In order to improve the thermomechanical reliability of the die-attach structure, the thermomechanical properties of the die-attach need to be optimized either locally or globally through material selection, structure design, and process control. The guideline for local optimization is to minimize the maximum local thermal stress where the device mechanically operates, which optimizes the device's mechanical operation with the least thermal stress effects. The guideline for global optimization is to minimize the weighted global stress. The parameters that can be adjusted to thermomechanically optimize the die-attach structure are $\overrightarrow{M_a}$, $\overrightarrow{M_s}$, θ_a, and T_R after the determinations of die material and die size.

The thermomechanical governing equations, Equation 12.1 to Equation 12.3, are so complicated that closed nontrivial (analytical) solutions of these equations are very difficult to obtain, even for simple boundary geometry. The nonlinear temperature dependencies of the material properties make it almost impossible to solve the equations with closed analytical solutions; however, numerical computing methods, such as finite element analysis, provide a powerful tool for simulation and optimization of thermomechanical behavior of a die-attach structure through material selection, structure design, and processing control. Numerical methods make it possible to calculate the thermomechanical configuration of the packaging components in a wide temperature range; however, it is still not trivial to optimize a die-attach structure for operation in a wide temperature range using FEA simulation because of the following reasons:

1. The numerical calculations involved in FEA sometimes can be difficult, especially when boundary geometry of the MEMS device is complicated and the dimensions of the mechanical structures are much smaller than the die-attach size [Rudd, 2000].
2. Currently, thermomechanical and fatigue and creep properties of many electronic and packaging materials in such a wide and high temperature range are very often not completely available.
3. If the die-attaching process is based on phase-transition phenomena, then detailed changes in the thermomechanical properties of the attaching material, before and after the phase transition, may not be quantitatively known.
4. Quantitative mechanical modeling of an interfaces which is chemically reactive can be very difficult, especially, when an interphase layer with multiple chemical components forms.

Lin and Zou used a Si stress-monitoring chip to measure surface stress and strain of a die-attach [Lin et al., 1997; Zou et al., 1999]. This unique in situ stress and strain monitoring method can be modified by using a high-temperature-operable SiC stress-monitoring chip to monitor and optimize the die-attach for high-temperature MEMS packaging.

12.5.2 Thermomechanical Simulation of Die-Attach

For high-temperature MEMS packaging, the thermomechanical reliability of a die-attach structure needs to be addressed at two levels: (1) mechanical damages of a die-attach structure resulted from thermomechanical stress, and (2) thermal stress and strain effects on the mechanical operation of the device. Failures at both levels are rooted in thermomechanical stress in the die-attach. In the remaining part of this section, FEA simulation results [Lin and Chen, 2002] of the Au thick-film based SiC die-attach reviewed in the last section are analyzed as an example of die-attach optimization for high-temperature MEMS packaging.

12.5.2.1 High-Temperature Material Properties

The basic thermal and mechanical properties of SiC, Au, AlN, and 96% Al_2O_3 and their temperature dependences needed for FEA simulation are listed in Table 12.1 [Lin and Chen, 2002]. The temperature dependence of Young's modulus of AlN in a wide temperature range has not been reported, so it is extrapolated as a constant from the data at room temperature. The Poisson's ratios of 4H single crystal SiC and AlN are not available, so they were estimated according to those of other carbides and nitrides. The thermal and mechanical properties of 4H SiC were assumed to be isotropic. Because the yield strength of gold thick-film material has not been published either, the simulation was conducted in a range of values, from 650 to 3000 psi. However, the numerical computing using the yield strength at the low end very often diverged, so the calculation was difficult. It is apparent that material properties needed for high-temperature MEMS packaging are not yet fully available.

12.5.2.2 Stress Distribution and Die-Size Effects

Figure 12.12(a) and Figure 12.12(b) show Von Mises stress and maximum principal stress contours of a quarter of SiC die of SiC-Au-AlN die-attach structure at room temperature, assuming the relaxing temperature is 600°C. Horizontally, at the interface with the Au thick-film layer, the stress in the die basically increases with the distance from the center of the die-attach interface, and the stress reaches a maximum at the area close to the die edges, especially at the corner area. This result is due to the dominance of shear

TABLE 12.1   Basic Material Properties of SiC, Au, 96% Alumina, and AlN Used for FEA Simulation for Die-Attach Structure

Temp (deg C)	CTE ($\times$E$-$6/C)	E($\times$E6 psi)	V
AlN Material Properties			
-15	3.14	50.00	0.25
20	3.90	50.00	0.25
105	5.36	50.00	0.25
205	6.51	50.00	0.25
305	7.25	50.00	0.25
405	7.76	50.00	0.25
505	8.25	50.00	0.25
605	8.72	50.00	0.25
705	9.09	50.00	0.25
Au Material Properties			
-15	14.04	11.09	0.44
20	14.24	10.99	0.44
105	14.71	10.83	0.44
205	15.23	10.59	0.44
305	15.75	10.28	0.44
405	16.29	9.92	0.44
505	16.89	9.51	0.44
605	17.58	9.03	0.44
705	18.38	8.50	0.44
96% Alumina Properties			
-15	5.34	44.00	0.21
20	6.20	44.00	0.21
105	7.83	43.58	0.21
205	8.48	43.06	0.21
305	8.89	42.51	0.21
405	9.28	41.93	0.21
505	9.65	41.34	0.21
605	10.00	40.74	0.21
705	10.33	40.13	0.21
SiC Material Properties			
-15	2.93	66.72	0.3
20	3.35	66.72	0.3
105	3.97	66.42	0.3
205	4.23	66.08	0.3
305	4.46	65.74	0.3
405	4.68	65.39	0.3
505	4.89	65.05	0.3
605	5.10	64.71	0.3
705	5.29	64.36	0.3

stress at the boundary, and usually shear stress is the highest at the die edges and corners (highest distance from neutral point, DNP). With these results, it can be predicted that the maximum thermal stress would increase tremendously with the increase in the size of the die or of the attaching area. Vertically, the stress attenuates rapidly with the distance from the interface. At the die's center region, the stress near the die surface attenuates by a factor of 1/80 with respect to the stress at the interface. This picture of the thermal stress distribution indicates that flip-chip bonding would not be recommended for high-temperature MEMS devices if the CTE mismatch is not well controlled, and a thicker die (or a stress buffer layer of the same material as that of the chip) can significantly reduce the thermal stress at the die surface region.

Figure 12.12(c) shows equivalent plastic strain (EPS) in the Au thick-film layer. Because of the same physical mechanism as discussed for the SiC die, the highest EPS in the Au layer is also located at the area

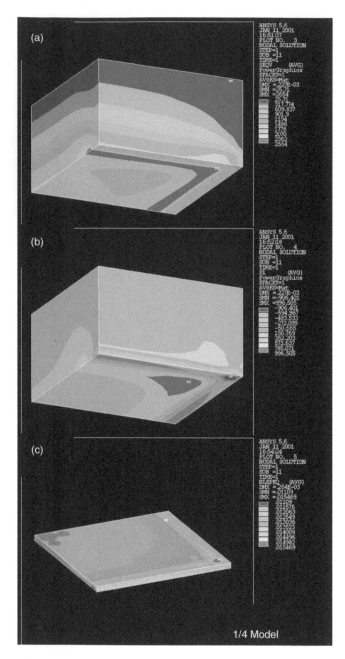

FIGURE 12.12   (**See color insert following page 9-22.**) Thermal stress and strain distribution in SiC die and Au thick-film layer. (a) Von Mises stress distribution in SiC die. (b) Maximum principal stress distribution in SiC die. (c) Equivalent plastic strain in Au thick-film layer.

close to the corner where the DNP is larger. This finding indicates, again, that smaller size of the die or of the attaching area may better satisfy die-attach thermomechanical reliability requirements.

12.5.2.3   Effects of Substrate Material

In comparison with AlN substrate, a 96% Al_2O_3 substrate has relatively higher CTE (~6.2 $\times$ 10^{-6}/°C compare with ~3.9 $\times$ 10^{-6}/°C of AlN at room temperature). In order to evaluate the substrate material's effects on the thermomechanical stress of the die-attach structure, FEA simulation was used to calculate the maximum stresses in the SiC die and the maximum EPSs in the Au thick-film attaching layer for both substrates [Lin and Chen, 2002]. The results indicate that using AlN substrate would result in an improvement of maximum Von Mises Stress in SiC die by a factor of 0.29, an improvement of Von Mises stress in the substrate by a factor of 0.33, and an improvement of EPS in Au thick-film layer by a factor of 0.42, assuming the yield strength of Au thick-film is 3000 psi. This improvement of thermal stress and strain corresponds to an improvement of fatigue lifetime by a factor of 4.3–9.0 (assuming the power law exponent in the Coffin-Mason model, C, is −0.4 and −0.6, respectively). So in terms of the thermomechanical reliability of the die-attach, AlN is suggested (compared to 96% Al$_2$O$_3$) for packaging SiC high-temperature MEMS devices, because the CTE of AlN is closer to that of SiC.

SiC is another candidate substrate material for packaging SiC devices, but currently the cost of single-crystal SiC material is much higher when compared with other ceramics suitable for packaging. The CTEs of α- and β-polycrystalline SiC are very close to those of single-crystal SiC, so thermomechanically they are ideal substrates for packaging large SiC die for high-temperature operation. However, both the dielectric constants and the dissipation factors of these materials are relatively high [Johnson, 1999]. In order to be used as packaging substrates, the surface electrical conductivity of polycrystalline SiC must be reduced, but these materials might still be suitable only for low-frequency application because of the high dissipation factors.

12.5.2.4 Effects of Thickness of the Attaching Layer

The Au attaching layer affects to the thermomechanical behavior of the die-attach structure in two ways:

1. Direct interface effect: The Au layer forms interfaces with both the SiC die and the substrate, so the material properties and the configurations of the Au layer directly influence the thermal stress distributions in both the die and the substrate.
2. Coupling effects: As an interlayer between the die and the substrate, the Au layer couples the die with the substrate mechanically.

When the interlayer is very thick, the substrate and the die are decoupled from each other, thus the second effect vanishes and the first effect dominates; if the thickness of the interlayer → 0, then the die and the substrate are directly coupled with each other, and the influence of the Au material properties vanishes and the second effect dominates.

Figure 12.13(a) and Figure 12.13(b) show maximum Von Mises Stress in the SiC die and the substrates vs. the thickness of the Au attaching layer, θ_a. The maximum stress in the SiC die decreases by a factor of 0.75 with respect to the change in θ_a from 20 to 50 μm for AlN substrate, while the maximum Von Mises stress in the SiC die decreases by a factor of 0.5 with respect to the change in θ_a from 20 to 50 μm for the alumina substrate. The increase of Au layer thickness also significantly reduces the stress in the alumina substrate, as shown in Figure 12.13(b). Figure 12.13(c) shows the maximum Equivalent Plastic Strain (EPS) in the Au layer vs. θ_a. The increase of θ_a from 20 to 50 μm significantly reduces the EPS in the Au layer in both AlN and Al$_2$O$_3$ cases. The maximum stress in the SiC die and the substrate and EPS in the Au attaching layer were calculated against the yield strength of Au thick-film material, as shown in Figures 12.13(d and e).

12.5.2.5 Effects of Relaxing Temperature

In addition to the dependences on material properties of the die, the attaching layer and the substrate, the thermomechanical stress in a die-attach structure also depends on the temperature deviation from the relaxing temperature, T_R, at which the structure is largely relaxed. Therefore, the relaxing temperature of the die-attach structure is another important factor determining thermomechanical configuration (stress and strain) of a die-attach structure at a certain temperature. Generally, the more the temperature deviates from the relaxing temperature, the higher the thermomechanical stress that exists in the die-attach structure. For the type of die-attaching using diffusion-based bonding (e.g., thick-film materials based die-attaching), the relaxing temperature is likely to be close to the processing (curing) temperature. For phase-transition phenomena based die-attach processes, the residual thermal stress could exist at the processing and attaching temperature because of the changes in material properties during the phase

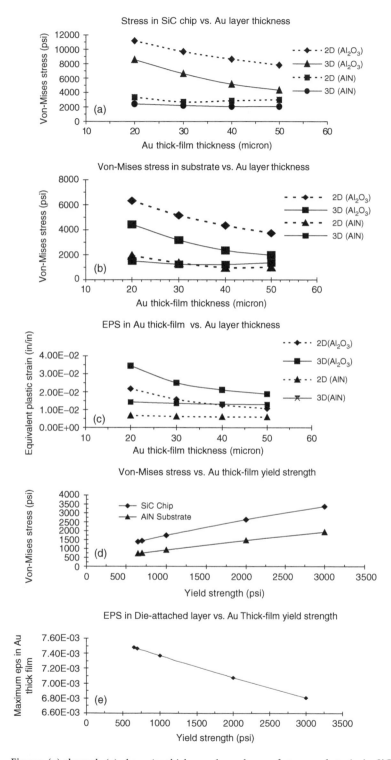

FIGURE 12.13 Figures (a) through (c) show Au thickness dependence of stress and strain in SiC die, substrate, and Au attaching layer. Figures (d) and (e) show stress and strain dependences on the yield strength of Au thick-film material (from Lin and Chen, 2002. with permission).

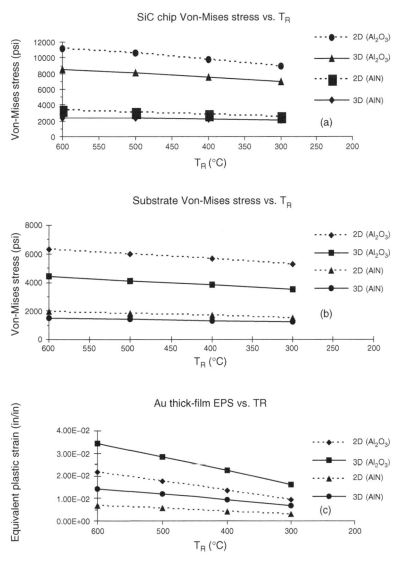

FIGURE 12.14 Relaxing temperature dependence of maximum stress and strain in SiC chip, substrate, and Au attaching layer.

transition. In this case, the die-attach structure reaches a minimum stress configuration at a certain temperature, but it may not be completely relaxed.

In order to assess the relaxing temperature effects on the thermal stress of die-attach structure, the stress distribution of the die-attach at room temperature is simulated by FEA, assuming that the structure is relaxed at various temperatures (from 300–600°C). Figures 12.14(a) through (c) show the maximum Von Mises stress in the die and the substrate, and EPS in Au thick-film layer vs. the relaxing temperature. If the relaxing temperature could be lowered from 600 to 300°C, the maximum Von Mises stress in the die could be reduced by a factor of 0.8, and the maximum EPS in Au thick-film layer could be reduced by a factor of 0.5. The fatigue lifetime corresponding to the stress reduction is improved by a factor of 3 (assuming the exponent in the Coffin–Mason model of fatigue, C, is −0.6). Ideally, the thermomechanical property of the die-attach structure is optimized if the relaxing temperature could be set at the middle of the operation temperature range, but physically this is not always realistic.

Mathematically, the optimization of thermomechanical property of a die-attach is a complicated multiparameter problem. So, even if the numerical computation is possible, both the skill and the number of calculations required for the optimization can be considerable.

12.6   High-Temperature Ceramic Packaging Systems

12.6.1   Chip-Level Packages

Based on the high-temperature packaging material systems introduced in Section 12.4, prototype chip-level packages and printed circuit boards (PCB) have been developed at NASA Glenn Research

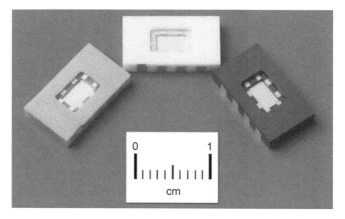

FIGURE 12.15   (See color insert following page 9-22.) AlN (left), 96% Al_2O_3 (top), and 92% Al_2O_3 (right) chip level packages with Au thick-film metallization. These 8-pin (I/O) packages were developed at NASA Glenn Research Center for low-power SiC electronics and sensors for operation up to 500°C.

FIGURE 12.16   (See color insert following page 9-22.) AlN substrate and Au thick-film metallization based printed circuit board designed for the 8-pin (I/O) AlN packages shown in Figure 12.15.

Center. Eight-pin (I/O) 96% Al_2O_3, 92% Al_2O_3, and AlN chip-level packages for low–power, SiC high-temperature electronics and sensors are shown in Figure 12.15. A Au thick-film metallization formula is selected according to the substrate material. Both conductive die-attach materials discussed in Section 12.4.3.2 are compatible to the Au thick-film metallization formulas used for these packages. The Au thick-film metallization pads of these packages are suitable for wire bonding using both pure and doped Au wires.

12.6.2 Printed Circuit Board

PCBs designed for both Al_2O_3 and AlN chip-level packages shown in Figure 12.15 have been developed for the testing of chip-level packages and packaged devices. These PCBs use the same substrate and metallization materials as the chip-level package. The chip level packages are mechanically attached to the PCB with glass or ceramic adhesives. The electrical interconnections between the chip-level packages and a PCB are provided by low curing temperature, conductive adhesives. Figure 12.16 shows an AlN chip-level package mounted on a AlN PCB. Recently, it has been reported that the curing temperature (the temperature at which the material densifies) of metal powders reduces significantly with the particle size of the powders. The increased ratio of surface area to powder volume is important to diffusion-based densifying processes. Au paste with nanometer-size powders is an excellent candidate material for interconnecting thick-film traces at relatively low temperatures.

12.7 Discussion

The combination of versatile functions of MEMS devices and survivability and operability in harsh environments leads to a new generation of microdevices, harsh-environment MEMS, with revolutionary capabilities for aerospace and civil applications. Packaging these revolutionary devices, however, has generated new and challenging research in the device packaging field. In order to meet these challenges, we need to find innovative packaging materials with superior physical and chemical properties suitable for harsh-environment operation; innovative packaging structures and designs to meet the requirements of microstructures and micromechanical operations; and innovative packaging processes to use these innovative packaging materials to fabricate the innovative packaging structures.

12.7.1 Innovative Materials

An ideal die-attaching material suitable for use in a wide temperature range is an illustration of the need for innovative materials to package harsh-environment MEMS devices. In addition to the superior chemical and electrical stabilities in high temperatures and corrosive environments, the die-attaching materials must possess good thermal and electrical conductivity, unique features of metallic material; meanwhile, CTE of such a material should match those of the die and the ceramic substrate. Ideally, if the CTE of the substrate slightly mismatches that of the die, the die-attach material should be able to compensate for the CTE mismatch between the die and the substrate. A material with low Young's modulus and a narrow elastic region certainly would help to absorb thermal strain, thus reducing the stresses in the die and the substrate. However, this condition would reduce the lifetime of the die-attach material layer in a dynamic thermal environment because of the accumulated permanent strain. Comparing the extraordinary material requirements listed above with the features of carbon nanotubes (CNT), we discover that CNT is an ideal die-attach interlayer that meets all of these requirements; CNTs have been reported to have superior electrical and thermal conductivities [Saito et al., 1998; Hone et al., 2000]. The longitudinal mechanical strength (both elastic modulus and breaking point) of CNTs is very high in comparison with steel [Wong et al., 1997; Hernandez et al., 1998; Poncharal et al., 1999; Salvetat et al., 1999; Yu et al., 2000], but the shear modulus of single-walled CNTs is low [Salvetat et al., 1999]. Graphitic bond between neighboring in-wall carbon atoms of each CNT makes the interaction between the neighboring tube walls weak [Girifalco et al., 2000]. Combining all of these features of CNTs allows us to visualize an

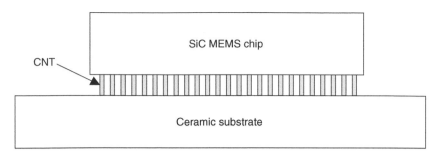

FIGURE 12.17 Schematic diagram of die-attach structure using carbon nanotube as an interlayer with capability to match the CTEs of both the SiC chip and the substrate.

innovative die-attaching interlayer material. If CNT can be vertically grown on the C face of SiC wafers (the C face of SiC is usually used for die-attach because the Si face is favored for fabrication of most electronic devices), it may provide an ideal interlayer with excellent thermal and electrical conductivities, superior mechanical strength, and very low lateral Young's modulus; thus, it may possess the unique capability to manage the CTE mismatch between the die and the substrate. A schematic diagram using CNT as a die-attach interlayer that is expected to decouple the die mechanically from the substrate is illustrated in Figure 12.17.

12.7.2 Innovative Structures

As discussed in previous sections, suppressing thermal stress in the die-attach in a wide temperature range is a common thermomechanical concern for high-temperature MEMS packaging. A bellow structure was invented (for conventional MEMS packaging) to absorb thermal strain and thus release thermal stress in the area where the device operates mechanically [Garcia and Sniegowski, 1995]. The bellow structure can be fabricated between the mechanical operation part and the chip base, so that the mechanical operation part is thermomechanically decoupled laterally from the rest of the device chip. Or the bellow structure can be fabricated between the device chip and the packaging substrate so that the entire chip is thermomechanically decoupled from the packaging environment. A similar structure can also be used for harsh-environment MEMS packaging. Various thermal stress suppression methods for packaging conventional MEMS were summarized by Madou (1997). Some of these methods may be modified for applications in packaging high-temperature MEMS.

Li and Tseng suggested a "four-dot" low-stress die-attach approach for packaging of MEMS accelerometers, in which the die (top) surface strains caused by CTE mismatches were reduced [Li and Tseng, 2001]. FEA simulation results indicated that a significant stress reduction could be achieved when the area of die-bonding "dots," located at the four die corners, was small. This die-attach structure can dramatically reduce or eliminate the transverse normal stress in the central die area. However, this approach may not be optimized to reduce lateral normal stresses in the die, which is important to diaphragm-based piezoresistive MEMS sensors.

As discussed in Section 12.5.2.2, the thermomechanical stress in the SiC die of the die-attach structure discussed in the last section attenuates rapidly with the vertical distance to the interface composed of SiC and Au with different CTEs. Horizontally, the maximum thermal stress increases rapidly with the die size or die-attaching area. These results suggest a simple but effective thermal stress suppression method, lateral stress attenuation die-attach, which partially attaches the die to the packaging base rather than to the complete die. This die-attach scheme allows a lateral distance (on the device chip) from the device mechanical operation area to the direct die-attaching area, and therefore reduces stresses in the unattached die area. Figure 12.18 shows a side view of an example of this low stress die-attach structure, and Figure 12.19(a) and Figure 12.19(b) show top and bottom views of the von Mises stress distribution contours for one-half of the SiC die of the die-attach assembly shown in Figure 12.18. Basically, the stress

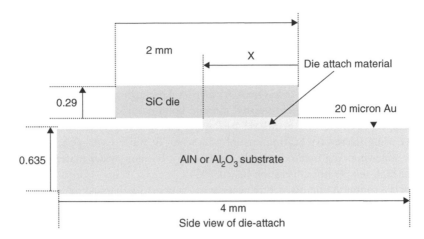

FIGURE 12.18   (See color insert following page 9-22.) Side view of an example of die-attach structure with lateral stress attenuation. Where X (0 mm < X < 2 mm) is a parameter determined by device design and packaging requirements, such as mechanical strength and resonant frequencies.

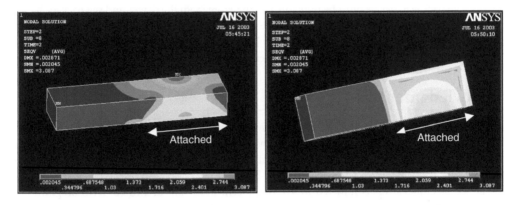

FIGURE 12.19   (See color insert following page 9-22.) FEA simulation of thermal stress of the partially attached SiC die on AlN substrate using 20 μm Au. The attached area is 1 mm × 1 mm. All stress units are MPa. (a) Von-Mises stress contour plot of top half of partially attached SiC die. Attaching area is 1 mm × 1 mm. (b) Von-Mises stress contour plot of bottom half of partially attached SiC die. Attaching area is 1 mm × 1 mm.

generated at the directly attached area attenuate rapidly toward the unattached area, and the overall stress level at the unattached (die) area is low.

In a similar manner, the die can be attached to the substrate using one of the die edges. This low-stress die-attach structure is expected to be especially effective at releasing the thermomechanical stress and strain in the direction vertical to the die-attach interface because the chip has no direct restraint in this direction.

The advantage of this low-stress die-attach structure (lateral stress attenuation die-attach) is that all stress components at the unattached die area are low. The disadvantage of these low-stress die-attach structures is the possibility of vertical cantilever vibration [Chen, 2003].

12.7.3   Innovative Processes

In order to accommodate microsize mechanical features of MEMS devices and support their microlevel mechanical operation, the immediate device packaging environment may have to meet micromechanical requirements in both alignment and assembly. An innovative, organic-solution based, self-assembly method was used to integrate microsized GaAs light emission diodes (LEDs) onto a micromachined Si substrate [Yeh and Smith, 1994]. GaAs LEDs are suspended in ethanol above a micromachined Si wafer.

Macrovibration enables the LEDs to randomly walk in the solution. When a LED hits a vacancy on the Si wafer, it fits in and gains the lowest potential there because of surface–ethanol–surface interaction. This interaction is so strong that perturbation from vibration would not free trapped LEDs. This is a good example indicating how innovative packaging processes at various levels beyond conventional IC packaging technology are introduced to package MEMS devices.

Wafer bonding is one of the key process steps in wafer-level packaging. Madou (1997) summarized oxidation bonding, fusion bonding, field-assisted thermal bonding, and modified field-assisted thermal bonding techniques developed for Si MEMS packaging. Low-pressure and low-temperature hermetic wafer bonding using microwave heating was introduced for Si wafer packaging [Budraa et al., 2000]. Some of these wafer-bonding methods may be modified for bonding wafer materials, such as SiC, that are suitable for high-temperature operation [Chen, 2004].

Microcavity is a common microstructure necessary to support mechanical operation of many MEMS devices such as an absolute pressure sensor or a microresonator [Ikeda et al., 1990; Hanneborg and Øhlckers, 1990]. These cavities are often fabricated at the wafer level using microfabrication processes, instead of being assigned later to conventional packaging that typically is a macrolevel process. This process reflects the trend that micromachining and fabrication processes initially used only for device fabrication now are also used to package MEMS devices at the wafer level. An extreme example of using microfabrication to package MEMS devices would be wireless "bug sensors" for operations in vivo or other harsh environments. The packaged device must be miniaturized, therefore, both the device and the package have to be microfabricated and microassembled. These MEMS devices become "self-packaged," because device fabrication and packaging processes completely merge [Santos, 2001]. These "self-packaged" MEMS devices may have the unique advantage of material consistency for operation in a wide temperature range.

Acknowledgments

One of the authors (L.C.) is very grateful to Dr. Gary W. Hunter for introducing the author to the challenging and vivid research field of high-temperature device packaging. Dr. Lawrence G. Matus is thanked for proofreading the manuscript and for his many suggestions. The authors want to thank Dr. Philip G. Neudeck and David Spry for providing SiC devices for high-temperature packaging research. The FEA modeling of die-attach assemblies was conducted by Dr. Shun-Tien (Ted) Lin at United Technologies Research Center and Prof. McCluskey's research group at the University of Maryland, with support of NASA Electronic Parts and Packaging Program through NASA Glenn Research Center. The authors would like to thank Dr. Lin and Prof. McCluskey for very helpful discussions. Authors want to sincerely thank Dr. Daniel L. Ng for proofreading the first edition of this chapter and for his very helpful suggestions. The harsh-environment microsystems packaging work at NASA Glenn Research Center has been supported by NASA Glennan Microsystems Initiative (GMI) and NASA Electronic Parts and Packaging (NEPP) program.

References

Budraa, K.N., Jakson, H.W., William, T.P., and Mai, J.D. (1999) "Low Pressure and Low Temperature Hermetic Wafer Bonding Using Microwave Heating," in *IEEE Electro Mechanical Systems Technical Digest*, IEEE Catalog Number: 9CH36291C.

Chen, L.-Y., Hunter, G.W., and Neudeck, P.G. (2000a) "Thin and Thick Film Materials Based Interconnection Technology for 500°C Operation," in *Trans. First International AVS Conference on Microelectronics and Interfaces*, 7–11 February, Santa Clara, CA.

Chen, L.-Y., Hunter, G.W., and Neudeck, P.G. (2000b) "Silicon Carbide Die Attach Scheme for 500°C Operation," in *MRS 2000 Spring Meeting Proceedings-Wide-Bandgap Electronic Devices (Symposium T)*, 10–14 April, San Francisco, CA.

Chen, L.-Y., and Neudeck, P.G. (2000c) "Thick and Thin Film Materials Based Chip Level Packaging for High Temperature SiC Sensors and Devices," in *Proc. 5th International High Temperature Electronics Conference (HiTEC)*, 11–16 June, Albuquerque, New Mexico.

Chen, L.-Y., Okojie, R.S., Neudeck, P.G., and Hunter, G.W. (2001) "Material System for Packaging 500°C MicroSystem," in *Proc. 2001 MRS Spring Meeting, Symposium N: Microelectronic, in Optoelectronic, and MEMS Packaging*, 16–20 April, San Francisco.

Chen, L.-Y., McCluskey, P., Meyyappan, K., and Lin, S.-T. (2003) "Low Stress Die-attach Technology for MEMS Packaging," in *Proc. iMAP 5th Topical Workshop on MEMS, Related Microsystems Nanopackaging*, 20–22 November, Boston.

Chen, L.-Y., and Zulueta, P.J. (2004) "MEMS Packaging," *Mechanical Engineering Handbook*, CRC Press, Boca Raton, FL.

Chitale, S.M., Huang, C., and Sten, S.J. (1994) "ELS Thick-Film Materials for AlN," *Adv. Microelectron.*, **21**, 22–23.

DuPont Electronic Materials (1999), *DuPont Processing and Performance Data of Thick Film Materials*, DuPont Electronics Materials, Research Triangle Park, NC.

Garcia, E., and Sniegowski, J. (1995) "Surface Micromachined Microengine," *Sensor. Actuator. A*, **A48**.

Giacomo, G.D. (1996) *Reliability of Electronic Packages and Semiconductor Devices*, McGraw-Hill, New York.

Girifalco, L.A., Hodak, M., and Lee, R.S. (2000) "Carbon Nanotubes, Buckyballs, Ropes, and a Universal Graphitic Potential," *Phys. Rev. B*, **62**, 19, pp. 13104–13110.

Goetz, G.G., and Dawson, W.M. (1996) "Chromium Oxide Protection of High Temperature Conductors and Contacts," in *Trans. Third International High Temperature Electronics Conference (HiTEC)*, 2, 9–14 June, Albuquerque, NM.

Grzybowski, R.R., and Gericke, M. (1999) "Electronics Packaging and Testing Fixture for the 500°C Environment," *High-Temperature Electronics*, R. Kirschman, ed., IEEE Press, Piscataway, NJ.

Hannegorg, A., and Øhlckers, P. (1990) "A Capacitive Silicon Pressure Sensor with Low TCO and High Long-Term Stability," *Sensor. Actuator. A*, **A21**, 151–56.

Harman, G.G. (1999) "Metallurgical Bonding System for High-Temperature Electronics," *High-Temperature Electronics*, R. Kirschman, ed., IEEE Press, Piscataway, NJ.

Hernandez, E., Goze, C., Bernier, P., and Rubio, A. (1998) "Elastic Properties of C and $B_xC_yN_z$ Composite Nanotubes," *Phys. Rev. Lett.*, **80**, 20, pp. 4502–4505.

Hone, J., Batlogg, Z.B., Johnson, A.T., and Fisher, J.E. (2000) "Quantized Phonon Spectrum of Single-Wall Carbon Nanotubes," *Science*, **289**, September 8, pp. 1730–1733. And the news release at the website: www.eurekalert.org/releases/up-iac083000.html.

Hunter, G.W., Liu, C.-C., and Makel, D.B. (2004) "Microfabricated Chemical Sensors for Harsh Environments," *CRC MEMS Handbook*, 2nd ed., CRC Press, Boca Raton, FLA.

Hunter, G.W., Neudeck, P.G., Fralick, G.C. et al. (1999) "SiC-Based Gas Sensors Development," in *Proc. International Conference on SiC and Related Materials*, 10–15 October, Raleigh, NC.

Ikeda, K., Kuwayama, H., Kobayashi, T., Watanabe, T., Nishikawa, T., Yoshida, T., and Harada, K. (1990) "Silicon Pressure Sensor Integrates Resonant Strain Gauge on Diaphragm," *Sensor. Actuator. A*, **A21**, pp. 146–50.

Johnson, R.W. (1999) "Hybrid Materials, Assembly, and Packaging," *High-Temperature Electronics*, R. Kirschman, ed., IEEE Press, Piscataway, NJ.

Keusseyan, R.L., Parr, R., Speck, B.S., Crunpton, J.C., Chaplinsky, J.T., Roach, C.J., Valena, K., and Horne, G.S. (1996) "New Gold Thick Film Compositions for Fine Line Printing on Various Substrate Surfaces," ISHM Symposium.

Kirschman, R. ed. (1999) *High-Temperature Electronics*, IEEE Press, Piscataway, NJ.

Lau, J., Wong, C.P., Prince, J.L., and Nakayama, W. (1998) *Electronic Packaging — Design, Materials, Process, and Reliability*, McGraw-Hill, New York.

Lewis, F.A. (1967) *The Palladium-Hydrogen System*, Academic Press, New York.

Li, G., and Tseng, A.A. (2001) "Low Stress Packaging of a Micromachined Accelerometer," in *IEEE Trans. on Electronics Packaging Manufacturing*, 24.

Lin, S.T., Benoit, J.T., Grzybowski, R.R., Zou, Y., Suhling, J.C., and Jaeger, R.C. (1998) "High Temperature Die-Attach Effects in Die Stresses," in *HiTEC'97 Proceedings*, June 14–18, Albuquerque, NM, pp. 61–67.

Lin, S.T., Han, B., Suhling, J.C., Johnson, R.W., and Evans, J.L. (1997) "Finite Element and Moire Interferometry Study of Chip Capacitor Reliability," in *Proc. InterPack*, Kohala, Hawaii.

Lin, S.T., and Chen, L.-Y. (2002) "Thermomechanical Optimization of a Gold Thick-film based Die-Attach Assembly using Finite Element Analysis," in *Proc. 6th International High Temperature Electronics Conference*, 2–5 June, Albuquerque, NM.

MacKay, C.A. (1991) "Bonding Amalgam and Method Making," United States Patent 5,053,195.

Madou, M. (1997) *Fundamentals of Microfabrication*, CRC Press, Boca Raton, FL.

Martin, T., and Bloom, T. (1999) "High Temperature Aluminum Nitride Packaging," in *High-Temperature Electronics*, R. Kirschman, ed., IEEE Press, Piscataway, NJ.

McCluskey, F.P., Grzybowski, R., and Podlesak, T. (1997) *High Temperature Electronics*, CRC Press, Boca Raton, FL.

McCluskey, P., and Pecht, M. (1999) "Pushing the Limit: The Rise of High Temperature Electronics," in *High-Temperature Electronics*, R. Kirschman, ed., IEEE Press, Piscataway, NJ.

Neudeck, P.G., Beheim, G.M., and Salupo, C. (2000) "600°C Logic Gates Using Silicon Carbide JFETs," in *2000 Government Microcircuit Applications Conference*, 20–23 March, Anaheim, CA. An earlier review article: Davis, R.F., Kelner, G., Shur, M., Palmour, J.W., and Edmond, J.A. (1991) "Thin Film Deposition and Microelectronic and Optoelectronic Device Fabrication and Characterization in Monocrystalline Alpha and Beta Silicon Carbide," in *Proc. IEEE*, 79, March 5, pp. 1513–1516.

Palmer, D.W. (1999) "High–Temperature Electronics Packaging," in *High-Temperature Electronics*, R. Kirschman, ed., IEEE Press, Piscataway, NJ.

Pecht, M.G., Agarwal, R., McCluskey, P., Dishongh, T., Javadpour, S., and Mahajan, R. (1999) *Electronic Packaging — Materials and Their Properties*, CRC Press, Boca Raton, FL.

PIWG (2001), www.piwg.org/PDF/DynamicPressure.pdf.

Poncharal, P., Wang, Z.L., Ugarte, D., and de Heer, W.A. (1999) "Electrostatic Deflections and Electromechanical Resonances of Carbon Nanotubes," *Science*, **283**, March 5, pp. 1513–1516.

Rudd, R.E. (2000) "Coupling of Length Scales in MEMS Modeling: the Atomic Limit of Finite Elements," *Design, Test, Integration, and Packaging of MEMS/MOEMS*, B. Courtois et al., eds., 9–11 May, Paris.

Saito, R., Dresselhaus, G., and Dresselhaus, M.S. (1998) *Physical Properties of Carbon Nanotubes*, Imperial College Press, London.

Salmon, J.S., Johnson, R.W., and Palmer, M. (1998) "Thick-Film Hybrid Packaging Techniques for 500°C Operation," in *Trans. Fourth International High Temperature Electronics Conference (HiTEC)*, 15–19 June, Albuquerque, NM.

Salvetat, J.-P., Briggs, G.A.D., Bonard, J.-M., Bacsa, R.R., Kulik, A.J., Stöckli, T., Burnham, N.A., and Forró, L. (1999) "Elastic and Shear Moduli of Single-Walled Carbon Nanotube Ropes," *Phys. Rev. Lett.*, **82**, 5.

Santos, H.J. D.-L. (1999) *Introduction to Microelectromechanical (MEM) Microwave Systems*, Artech House, Boston, MA.

Shaikh, A. (1994) "Thick-Film Pastes for AlN Substrates," *Adv. Microelectron.*, **21**, pp. 218–221.

Willander, M., and Hartnagel, H.L. (1997) *High Temperature Electronics*, Chapman & Hall, London.

Wong, E.W., Sheehan, P.E., and Lieber, C.M. (1997) "Nanobeam Mechanics: Elasticity, Strength, and Toughness of Nanorods and Nanotubes," *Science*, **277**, p. 26.

Yeh, H.J., and Smith J.S. (1994) "Fluidic Self-Assembly of Microstructures and Its Application to the Integration of GaAs on Si," in *IEEE International Workshop on Microelectromechanical Systems*, Oiso, Japan.

Yu, M.-F., Lourie, O., Dyer, M.L., Moloni, K., Kelly, T.F., and Ruoff R.S. (2000) "Strength and Breaking Mechanism of Multiwalled Carbon Nanotubes Under Tensile Load," *Science*, **287**, p. 28.

Yushin, G.N., Kvit, A.V., Collazo, R. et al. (2003) "SiC to SiC Wafer Bonding," *Silicon Carbide 2002 Materials, Processing and Devices*, S.E. Saddow and D.J. Larkin et al., eds., Proceedings of Mat. Res. Soc Symp. Vol. 742.

Zou, Y., Suhling, J.C., Jaeger, R.C., Lin, S.T., Benoit, J.T., and Grzybowski, R.R. (1999) "Die Surface Stress Variation During Thermal Cycling and Thermal Aging Reliability Tests," in *Proc. Electronic Components & Technology Conference*, San Diego.

13

Fabrication Technologies for Nanoelectro-mechanical Systems

Gary H. Bernstein,
Holly V. Goodson and
Gregory L. Snider
University of Notre Dame

13.1 Introduction

Microelectromechanical systems (MEMS) are typically constructed on the micrometer scale, with some thin layers being perhaps in the nanometer range. As has already been demonstrated by microelectronic circuits, the lateral dimensions of MEMS are being pushed into the nanometer range as well. This advance has been dubbed *nanoelectromechanical systems,* or *NEMS*. The ultimate utility of nanomachining (i.e., the application of robots at the molecular scale that are capable of solving a range of problems) is limitless. Such a regime will likely be attainable only by the "bottom-up" approach, in which atoms are manipulated individually to construct macromolecules or molecular machines. Properties of pure molecules, such as heat conduction, electrical conduction (low power dissipation), speed of performance, and strength, without the limits of boundaries to other molecules and resulting material defects, vastly exceed those of bulk materials.

Drexler wrote about molecular machinery that could be modeled after the ultimate existing nanoelectromechanical system, the biological cell [Drexler, 1981]. He discussed analogs within the cell for such mechanical devices as cables, solenoids, drive shafts, bearings, and so on. Proteins exhibit a remarkable range of functionality and, compared with current MEMS technology, are extremely small. Reasoning that proteins are ideal models upon which to design nanomachines, Drexler envisioned the development of machinery that would allow the artificial production of such nanoscale mechanical components as those just listed. Imagining complex machinery operating at the molecular scale gives rise to images of nanorobots repairing our organs [Langreth, 1993], of smart materials that intelligently conform to our bodies by adjusting a vast number of rigid nanoplates, of robots that scour our clothing for debris to keep them clean [Forrest, 1995], and of nanorobots that build yet more complex nanomachinary. The list of possibilities is endless.

However, at this time, such intriguing visions still reside in the realm of science fiction. If the creation of complex nanosystems of any kind is to be accomplished, it will likely happen soonest by manipulation of biological systems to perform feats of engineering that exploit their existing programming or by chemical synthesis of ever more complex molecules in a beaker. Although the field of molecular electronics is progressing rapidly, molecules are currently combined with relatively large and cumbersome connections in order to study their properties [Reed, 1999], and to our knowledge, no molecular mechanical system has yet been synthesized from its basic atomic constituents. The future is long, though, and progress is unpredictable and often rapid. The advent of scanning probe microscopes, including scanning tunneling and atomic force microscopes, was a turning point in the advancement of nanoscience and technology, because it enabled the manipulation of single atoms [Binnig and Rohrer, 1987]. Even this great advancement is trivial compared with the requirements of assembling thousands of atoms in complex shapes and powering, observing, and controlling them. The problems posed in reaching this capability are staggering.

For the purposes of this chapter, we will define NEMS as an extension of current MEMS technology. However, this chapter includes a discussion of atomic-scale lithography and concludes with a discussion of the synthesis of nanoscale systems by purely biological and molecular synthetic means. Although these latter techniques are in their infancy, it is likely that in the nearer term, using more conventional fabrication techniques, we will be able to exploit new functionalities possible only on the nanoscale. These may include the frictional, thermal, or viscosity properties of materials; interaction with electromagnetic radiation at very small wavelengths; or ultra-high-frequency resonances. These regimes will be approachable by the more conventional "top-down" approach of defining small areas by high-resolution lithography and pattern transfer. In this chapter, we will discuss mainly top-down approaches, except for that of atomic holography and scanning probes based on scanning tunneling or atomic force microscopes, which can be used in either approach. In top-down patterning, several techniques for achieving nanometer resolution of arbitrary shapes have been developed, primarily electron beam lithography (EBL), ion beam lithography (IBL) and X-ray lithography (XRL). Here we will discuss EBL and XRL; due to its expense, complexity, and somewhat lower resolution than EBL, IBL is not as commonly used. We will also discuss two related techniques for achieving a parallel process for high throughput and small features, namely microcontact printing and nanoimprint lithography.

Using these processing techniques, it is certainly possible to machine silicon and related materials in the nanometer regime. Namatsu et al. (1995) used EBL and reactive ion etching (RIE), coupled with etching in KOH/propanol solutions, to create 100-nm vertical walls with a thickness of 6 to 9 nm. Using nanomachined tips (fabricated in a manner similar to contamination resist lithography, discussed later in this chapter), Irmer et al. (1998) used a scanning probe "ploughing" technique to modify thin Al layers, resulting in $100 \times 100 \, nm^2$ Josephson junctions. Others have sought to exploit very high frequency resonances in ultrasmall beams. Erbe et al. (1998) used electromechanical resonators with dimensions of about 150 nm to control physically adjustable tunneling contacts at frequencies up to 73 MHz at room temperature. In a comprehensive review of nanoelectromechanical systems, Roukes (2000) discusses the limits of mechanical resonators made of various semiconducting materials. He reports that at the 10- to 100-nm scale, SiC resonators should oscillate in the 10-GHz range; at molecular scales, nanodevices should exhibit resonant frequencies in the THz range.

Finally, we recognize that biological systems are the ultimate nanomachines. The internal operation of a single cell is more complicated than any human-made factory. In each cell, tens of thousands of different kinds of proteins and enzymes go about their complex tasks in a highly elaborate interplay of activity. We conclude our chapter with a discussion of how some cellular activities are being harnessed as future nanomachines.

13.2 NEMS-Compatible Processing Techniques

13.2.1 Electron Beam Lithography

Electron beam lithography (EBL) is a technique in which the energy imparted by a directed stream of electrons modifies a film on the surface of a substrate. In general, EBL is utilized in the direct writing of patterns on semiconductor substrates, as well as in defining mask patterns for optical and X-ray lithographies. In direct-write mode, a key issue in manufacturing is throughput, a characteristic for which EBL has had difficulty competing with other technologies, such as optical projection lithography. Variations of EBL include exposures by Gaussian-distributed beams, shaped beams in which the pattern of energy deposition is shaped by apertures, and a promising technique for manufacturing called *scattering with angular limitation in projection electron beam lithography*, or *SCALPEL* [Berger et al., 1991]. SCALPEL works by the imaging on a substrate of a pattern of mask features through which electrons are either transmitted without scattering or are scattered at obstructions on the mask. An aperture in the optical column passes preferentially those electrons that are not scattered, so the pattern of unscattered electrons is printed on the wafer. Although shaped beams offer improved throughput compared with Gaussian beams, SCALPEL is superior in this regard and, among the electron beam technologies, offers the most promise for manufacturing. For an excellent discussion of electron beam systems, see McCord and Rook (1997). In this chapter, we are concerned primarily with resolution, as opposed to other aspects of more interest to the mass production of patterns. Although these issues cannot be ignored for a technology to be ultimately viable, we discuss those issues that limit our ability to create nanomachines at the limits of physical possibility.

It is fortuitous that the reliance of nano-EBL on very narrow electron beams is the same as that required of high-resolution scanning electron microscopy (SEM), which now is a mature field. It is common for researchers to modify SEMs [Bazán and Bernstein, 1993], to control the position of the beam, and, in conjunction with a beam blanker that modulates the beam current, to obtain very high resolution patterning. Most commercial EBL systems are optimized for such issues as throughput and placement accuracy, making it less of a priority to achieve resolution in the nanometer range.

13.2.1.1 Resolution Limits

It is useful to understand the dependence of the ultimate size of patterning by EBL on all of the various steps involved. These include creation of the narrowest and brightest beam, choice of electron beam resist, forward- and backscattering of electrons in the resist and substrate, beam–resist interactions, resist development and, finally, pattern transfer to the substrate. Each of these issues will be addressed in turn.

13.2.1.1.1 *Beam Size*
The formation of a narrow, bright beam is critical to the formation of the smallest possible patterns. In all scanning electron optical systems, the column demagnifies an image of the electron source and projects this demagnified image onto the substrate. Electron lenses use magnetic fields created by currents in coils to produce an effect of focusing, much like that of optical lenses. The central function of the column is to demagnify the first crossover point at the source to a small, intense spot on the plane of the wafer. Because the diameter of the beam at the target is some fraction of the diameter of the source, starting with a small, bright source is of central importance. The five types of electron emitters commonly used in electron beam systems can be broadly classified as either thermionic emitters or field emitters. In order of decreasing source size, these are (1) tungsten hairpin filament, or thermionic; (2) lanthanum hexaboride, LaB_6, and the related CeB_6; (3) Schottky emitter; (4) thermal field emitter; and (5) cold cathode field emitter.

Table 13.1 lists the important properties of each emitter, except for the thermal field emitter. Note that these are representative values only and are highly dependent on manufacturers' specifications, beam energy, column design, and so on. Tungsten hairpin filaments are historically the most common and easiest sources to use. These emitters operate by passing current through the filament to raise its temperature to the vicinity of 2700 K, so that electrons can overcome the work function and enter the electron optical column. They require the least stringent vacuum conditions (approximately 10^{-3} Pa), so the overall systems tend to be less expensive. Most importantly, the effective diameter of the emission spot at the

TABLE 13.1   Properties of Various Cathodes Used for Electron Beam Lithography

Type	Vacuum (Pa)	Work Function (eV)	Cathode Temp. (K)	ΔE (eV)	Source Diameter or Tip Radius (μm)	Emission Current Density (A/cm^2)	Brightness (A/cm^2/Sr)	Probe Diameter (nm)	Lifetime (hours)
W	10^{-3}	4.6	2700	2	50	2	105	4	50
LaB$_6$	10^{-5}	2.7	1800	1	10	40	106	3	2000
ZrO/W Schottky	10^{-7}	2.8	1800	0.8	1	500	1010	2	Several thousand
CCFE	10^{-8}	4.6	300	0.3	0.1	105	108	1	Several thousand

crossover below the filament is about 50 μm, resulting in a typical spot size of about 3 to 5 nm. Also important is the brightness of the source, expressed in units of Amperes/cm^2/steradian. Because spot size is reduced through changes in column conditions, a large fraction of electrons is lost by scattering outside of the optical path. In a typical tungsten filament system, 100 μA of current from the source translates to perhaps 10 pA at the sample in obtaining the smallest spot size. This makes imaging at the highest resolution more difficult. In addition to increased throughput at a given spot size, a practical benefit of increased brightness is the ability to form a crisp image during focusing, thereby forming the narrowest possible spot. This translates to better spot size control, and ultimately to higher resolution patterning.

Besides suffering from a large size at the source for tungsten filaments, the thermal energy spread of the beam is about 4.5 eV, which can contribute to an increase in the beam diameter. Although this is a small fraction of the accelerating voltage (typically 30 to 100 kV), it is enough to create noticeable chromatic aberrations, so that focusing of the electrons at a single spot is hampered by the differing focal lengths for electrons with different energies. Because the minimum spot size for each energy is achieved at different focal lengths, the overall minimum spot at any one distance (i.e., at the surface of the substrate) is enlarged.

Closely related to the tungsten hairpin filament is the LaB$_6$, lanthanum hexaboride, emitter, and related CeB$_6$, cerium hexaboride, emitter. These materials often are referred to as *lab-6* and *cebix*, respectively. The emitter is a sharpened piece of crystal that is heated either by passing current directly through it or heating it by a separate heater to about 1800 K. The tip is machined for some optimum shape and produces an effective source diameter, typically 10 μm. Due to its lower work function of 2.7 eV and attending lower energy spread of about 1 eV, it is brighter and produces less chromatic aberration. CeB$_6$ offers a slightly lower work function and greater robustness, but it is not as commonly used. To protect the tip from "poisoning," a better vacuum, typically 10^{-5} Pa, is required of a LaB$_6$ system. In most cases, given the quality of the vacuum system, tungsten hairpin and LaB$_6$ sources can be used interchangeably. One important advantage is that the operating lifetime for LaB$_6$ is up to 2000 hours, versus less than 100 hours for thermionic tungsten.

A significant improvement in resolution and brightness over the previous sources is provided by the Schottky emitter. This source consists of a ZrO-coated W tip sharpened to about 0.5 μm. The reduced work function (from 4.6 eV in W to 2.8 eV) and enhanced electric field, which lowers the barrier (Schottky effect), allow the electrons to be extracted at a temperature of about 1800 K. The advantages of this source are high stability and brightness combined with much higher resolution (smaller spot size) than the thermionic sources. Spot sizes of about 3 nm or better are achievable in these systems. Schottky emitters are becoming the most common in advanced EBL systems because of their combination of beam stability, long lifetime (a few thousand hours), high brightness, and small spot size.

The highest resolution beams for scanning electron microscopy are produced using field emission sources. These consist of a very sharp, single-crystal tungsten tip ($r = 0.01$ to $0.1\,\mu$m) operated under a high electric field. One variation, called *thermal field emission (TFE)*, is to heat the tip; another, called *cold cathode field emission (CCFE)*, is to run the tip at room temperature. The TFE emitter has not commonly been used since the advent of the Schottky emitter. In the CCFE case, a high extraction field renders the potential barrier so thin (about 10 nm) that electrons can tunnel through even at relatively low temperatures. Because of this, thermal energy is reduced to about 0.2 eV, and chromatic aberrations are reduced significantly. Because of the reduced area of emission and reduced chromatic aberrations, probe sizes as small as about 0.5 to 1 nm can be produced. CCFE sources are not commonly used in EBL applications because the beam is inherently unstable due to adsorption of gas layers that affect the emission properties. Heating the tip, as in the case of TFE cathodes, solves this problem. In CCFE SEMs, this instability is dealt with by feeding back a current signal instantaneously to the video monitor to produce a clean image with little or no evidence of current fluctuations. However, for purposes of beam writing, this can lead to fluctuations in instantaneous dose at the wafer, and therefore to poor linewidth control. In spite of this, Dial et al. (1998) have reported excellent lithographic results using a CCFE SEM by building feedback into their exposure rates. Their system has demonstrated a regular array of dots with diameters of 10 nm on a pitch of 25 nm in polymethylmethacrylate (PMMA) at 30 keV.

13.2.1.1.2 Electron Scattering

Formation of a high-quality beam is only the first step in pattern formation. For very narrow beams, a larger contribution to pattern size is the scattering of electrons in the resist and in the substrate. When an electron beam impinges on a resist layer, the electrons scatter through both elastic and inelastic processes. Elastic scattering results in backscattering of the electrons at energies close to the primary beam and in negligible energy transfer to the resist and substrate. Inelastic scattering leads to spreading of the primary beam and generation of low-energy secondary electrons that, in turn, expose the resist. In PMMA, a positive resist, the mechanism of exposure is the scission of bonds in the high-molecular-weight polymer to create regions of lower molecular weight. The lower molecular weight is more soluble in suitable developer solutions, so the resist is removed where exposed. In negative resist, energy deposition results in cross-linking of the polymer, and the resist is rendered less soluble and therefore remains behind after development and removal of unexposed resist.

The spread of the primary beam in the resist layer can be described by:

$$b = 625 \, \frac{Z}{E_0} \, t^{3/2} \left(\frac{\rho}{A} \right)^{1/2} \text{cm} \tag{13.1}$$

where b is the spread of the beam at a distance t into the resist; Z, A, and ρ are the atomic number, atomic weight (g/mol), and density (g/cm³), respectively; and E_0 is the beam energy (keV). For PMMA $((C_5O_2H_8)_n)$, $Z = 3.6$, $A = 6.7$, and $\rho = 1.2$. For a delta function beam (i.e., an ideal beam with zero width) of energy 40 keV, the beam spreads out by 7.5 nm after 100 nm. This leads to a slight undercut profile that helps in lifting off metal patterns, so it is not necessarily undesirable. However, it does suggest that thin resists are preferable for achieving the highest resolution and pattern density. In thicker resists, it is often the case that the width of a lifted-off metal line is wider than the opening at the top surface, because metal can be deposited over the entire width of the line at the wafer surface; therefore, it can sometimes be deceiving to assume that the width of achievable resolution is that observed at the resist surface.

The Bethe retardation law,

$$\frac{dE}{dx} = -7.85 \times 10^4 \left(\frac{Z\rho}{AE_m} \right) \ln \left(\frac{1.66E}{J} \right) \frac{\text{keV}}{\text{cm}} \tag{13.2}$$

(where dE/dX is the energy loss of the beam per unit distance in the solid; E_m is the mean electron energy along the path; Z, A, and ρ are as defined for the previous equation; and J is the mean ionization potential [keV]), tells us that low-energy electrons are more efficient than high-energy electrons at transferring energy to the resist and are in fact responsible for breaking bonds in PMMA. Also, although secondary electrons are generated at energies up to a significant fraction of the beam energy, those at lower energies are again much more effective at exposing resist.

Using the Monte Carlo approach, Joy (1983) has investigated the minimum possible features achievable in PMMA. Because low-energy secondary electrons lose energy more readily than high-energy electrons, they are the dominant agents for transferring energy to the resist. Figure 13.1 shows the absorbed energy due to exposure by a 2-nm beam of 100 keV electrons in 100 nm of PMMA. For finite contrast of the

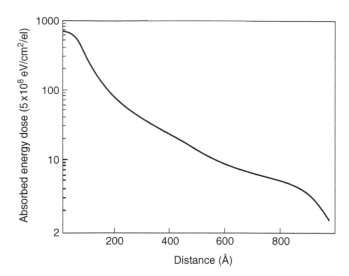

FIGURE 13.1   Absorbed energy due to exposure by a 2-nm beam of 100-keV electrons in 100 nm of PMMA. Energy deposition is primarily due to secondary electrons; minimum feature size is limited to approximately 10 nm. (Adapted from Joy, 1983.)

developer/resist system, the plateau of energy absorption is developed uniformly, and a spatial resolution of approximately 10 nm results. Smaller lines can result from high-contrast developers and very careful processing to take advantage of slight differences in absorbed energy within the plateau region. Chen and Ahmed (1993a) demonstrated linewidths approaching 5 nm with ultrasonic developers and reactive ion etching for pattern transfer. These narrow linewidths are not necessarily developed all the way through the resist, which is necessary to allow pattern transfer to the substrate by lift-off. Chen and Ahmed demonstrated lifted-off lines that were slightly less than 10 nm [Chen and Ahmed, 1993b], in good agreement with Joy's results.

The interaction volume of primary electrons in a substrate can be described by the Kanaya–Okayama (K–O) range [Goldstein, 1981], given by:

$$R_{KO} = 0.0276 \frac{AE_0^{1.67}}{Z^{0.889}\rho} \, \mu m \tag{13.3}$$

where A, E_0, Z, and ρ are as defined earlier. Electrons that are backscattered from the substrate usually are modeled as being Gaussian distributed, with a characteristic width of about half of the K–O range. The effect of backscattered electrons is to expose the resist at some (usually) undesirable background level that depends on the surrounding patterns. This phenomenon is referred to as the *proximity effect*, and much effort has gone into techniques for modifying pattern files to compensate exposure doses. The effect of backscattered electrons on the final resolution of an isolated pattern is less pronounced but can adversely affect the minimum resolution obtainable in dense patterns. Figure 13.2 shows the effect on absorbed energy of exposing lines closer together. Note that the spaces between the lines absorb more energy, and the difference in effective exposure dose between the intended patterns and spaces is reduced. This makes the job of developing only the lines a more difficult one for the developer to accomplish, as explained in more detail below.

A common technique for reducing the effects of backscattered electrons is to expose patterns on thin membranes (typically silicon nitride) formed by patterning the backside of a silicon wafer and etching through to expose the membrane over a usable area. Although not very practical for making electronic devices, this technique has been used to demonstrate very high resolution lithography and to allow direct imaging of the results in a transmission electron microscope.

13.2.1.1.3   *Conventional Resists and Developers*
The quality of a developer for a particular resist is defined in terms of the sensitivity and contrast. A typical dose/exposure plot showing normalized thickness of the resist after development for positive resist is shown

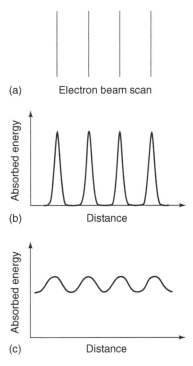

FIGURE 13.2   Absorbed energy in resist due to proximity effect. (a) Line scans, (b) energy deposition without proximity effect, and (c) total energy deposition including proximity effect. As the space between the line scans absorbs
more energy, it is more difficult for the developer to resolve the patterns, and linewidths increase. (Adapted from
Huang et al., 1993.)

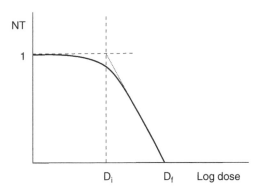

FIGURE 13.3   Dose/exposure plot of normalized positive resist thickness after development.

in Figure 13.3. (A curve for negative resist is zero at low doses and maximum at high doses.) The threshold
dose for positive resist can be defined as that which results in clearing of the resist in the exposed areas. It
is generally desirable for the threshold dose to be as small as possible for purposes of throughput, but if it is
too small, then statistical effects due to the number of electrons required to expose a single pixel can become
important. In general, resists that exhibit the highest resolution also are the slowest (least sensitive).

The contrast of a resist (using a particular developer under a particular set of conditions) is defined as:

$$\gamma = \log\left(\frac{D_i}{D_f}\right)^{-1} \tag{13.4}$$

where (for positive resist) D_i is the dose at the steepest part of the contrast curve extrapolated to full thickness, and D_f is that dose extrapolated to zero thickness. The role of contrast is central to obtaining the

finest pattern dimensions. As discussed earlier, each exposure results in some spatially varying energy deposition. Contrast (for positive resist) is the ability of the developer to selectively remove resist that has absorbed more than a critical value and ignore that resist that has absorbed less. For an infinite contrast, the contrast curve would be vertical; resist above a sharp value of absorbed energy would be perfectly developed, and resist below that value would be totally unaffected. In this case, we could arbitrarily approach the peak in Figure 13.1 and obtain patterns equal to the beam width. In addition, we could make the lines arbitrarily close together, because the lines could be perfectly selectively removed. In reality, the slope of a real dose/exposure curve is finite, so a range of absorbed energy contours is removed, and it is difficult to achieve much less than the 10 nm discussed by Joy.

Typical unexposed, high-resolution PMMA has a molecular weight of about 10^6 Da, which is reduced to a few thousand Da after exposure [Harris, 1973]. The role of the developer is to selectively remove the lower molecular mass polymer components that result from chain scission of larger ones. The two most common developers of PMMA are methyl isobutyl ketone (MIBK) mixed with IPA in a ratio of 1:3 and 2-ethoxyethanol (Cellosolve) mixed with MeOH in a ratio of 3:7. The γ values provided by these mixtures are about -6 to -9 [Bernstein et al., 1992]. Bernstein et al. reported that the addition of about 1% methyl ethyl ketone (MEK) increases the contrast of these developers to greater than -12. This is ascribed to enhanced removal of high-molecular-weight components that are not sufficiently broken at lower doses, without significantly affecting the solubility of the low-molecular-weight components.

As mentioned earlier, Chen and Ahmed (1993a) found that developing with ultrasonic agitation offers benefits to resolution, and using this technique demonstrated 5- to 7-nm lines. They assert that for exposures smaller than 10 nm, the resulting exposed PMMA fragments are trapped in a potential well created by the adjoining unexposed molecules. By exposing lines wider than 10 nm, the developer is able to dissolve the exposed resist but is not able to remove them in the narrower lines. In the case of narrower lines, ultrasonic agitation can provide additional energy by which the exposed fragments can escape this potential well in the presence of the developer. (They used the Cellosolve solution mentioned earlier.) Although lift-off was not demonstrated, the patterns were transferred to a Si substrate by reactive ion etching. Even smaller features have been demonstrated by allowing regions of exposure by secondary electrons to overlap. Using this technique, Cumming et al. (1996) have created metal lines by lift-off as small as 3 nm. These are perhaps the smallest features ever created by exposure of PMMA.

Finally, upon exposure to very high doses (i.e., greater than about 10^{-3} C/cm^2), PMMA will cross-link to act as a negative resist. Tada and Kanayama (1995) report that with careful control, highly dosed PMMA developed in acetone can yield features as small as 10 nm. As the dose is increased, resistance to dry etching increases as well, to about twice that of unexposed PMMA. Using an electron cyclotron resonance (ECR) etcher, they created Si pillars with a width of 10 nm and a height of 95 nm.

The process of nanomachining can benefit from the use of high-resolution negative resists as well. One of the highest resolution (conventional) negative resists is Shipley's SAL 601-ER7, a member of the Shipley ANR series of resists. It has been demonstrated to provide features as small as 50 nm [Bernstein et al., 1988]. This resist is one example of a chemically amplified resist, consisting of three components: an acid generator, a cross-linker, and a resin matrix. Exposure by the electron beam generates an acid that upon heating catalyzes the cross-linking of the linker and the resin matrix, leading to very high sensitivity. Typical threshold exposure doses are about an order of magnitude less than for PMMA. Other negative resists that exhibit nearly similar resolution include IBM APEX-E and Shipley UVIIHS.

Because many deep ultraviolet optical resists also are sensitive to electron beams, a plethora of choices exists for resists that maximize sensitivity, resolution, dry-etch resistance, or other properties. These include PBS (Mead), RE-4200N (Hitachi), APEX-E and KRS (IBM), UVIIHS and UV-5 (Shipley), AZ-PF514 and AZ-PN114 (Hoechst Celanese), and ZEP-520 (Nippon Zeon). Several of these are capable of achieving sub-100-nm resolution.

13.2.1.1.4 *Unconventional Resists*

Many other alternative resist schemes for defining patterns with an electron beam have been investigated. One common technique is the condensation of carbon compounds from the vapors present in electron beam vacuum systems. This is referred to as *contamination resist* lithography because it utilizes the same mechanism that often causes contamination and degradation of imaging quality in electron microscopes. As early as 1964, Broers (1964), using a 10-nm electron beam, obtained a resolution of 50 nm after ion milling a contamination resist mask. It is generally accepted that the mechanism of deposition is the cracking by secondary electrons of adsorbed hydrocarbons at the sample surface. As with PMMA, the resolution of this technique benefits from the use of membranes for supporting the patterns. Contamination resist cannot be used in a lift-off mode, but it does perform reasonably well as an etch mask. Broers (1995) later demonstrated 5-nm Au–Pd lines created by deposition on a membrane followed by ion milling.

In an effort to merge with molecular electronics, self-assembled monolayers (SAMs) have more recently been investigated as electron beam resists. Because these layers are only a few nanometers thick, it is thought that the resolution achieved in this system will ultimately be higher than with PMMA. Also, very low energies can be used to expose thin layers, which can be beneficial to backscattering issues and also allow the use of exposure by low-energy scanning probe tips. SAMs consist of an ordered layer of single molecules adsorbed to a surface. Examples include *n*-octadecyltrichlorosilane (OTS), *n*-octadecanethiolate (ODT) [Lercel et al., 1993], octadecylsiloxanes (ODS) [Lercel et al., 1996], and dodecanethiol (DDT). Their value lies in the presence of alkyl groups at their outer surface, which gives SAMs hydrophobic properties and resistance to wet etching. Exposure by an electron beam (or ultraviolet radiation) removes the alkyl groups, rendering them hydrophilic, so exposed areas can be etched. They are, therefore, positive tone in nature. Using ODS SAMs, Lercel et al. exposed individual dots with diameters of approximately 6 nm. These are roughly the size of individual molecules that could be anchored for NEMS systems.

Another technique employing an electron beam for lithography is that of direct sublimation of material. This is often referred to as simply *drilling* through a layer of inorganic material. This is quite literally a nanomachining process. Materials that are amenable to drilling include Al_2O_3, LiF, NaCl, CaF_2, MgO, and AlF_3. Mochel et al. (1983) has demonstrated 1-nm holes drilled in Al_2O_3, which is perhaps the highest resolution ever demonstrated with a high-energy beam of electrons. In a related technique, Hiroshima et al. (1995) created lines on a 15-nm pitch by exposure of SiO_2. Development was performed in a buffered HF solution. Such a technique may have applications in nanomachining.

13.2.2 X-Ray Lithography

X-ray lithography (XRL) is basically an extension of optical proximity printing that uses much higher energy X-ray photons, with wavelengths in the range of about 0.1 to 10 nm. It was first described by Spears and Smith (1972). For MEMS technology, it has been extended to the LIGA process for creating microscale machined pieces such as gears and other mechanical parts. Because the use of X-rays is so common in micromachining applications, it is worthwhile to investigate the limits of resolution for nanomachined systems. An excellent review of XRL is provided by Cerrina and Rai-Choudhury (1997).

Figure 13.4 schematically shows a basic X-ray exposure configuration. The main issue in achieving workable X-ray systems in manufacturing is the development of suitable X-ray sources, masks, and alignment techniques. The problems involved in developing bright, collimated X-ray sources are, in general, shared with the LIGA community, but are even more pronounced for nanolithography. For example, the geometrical considerations of a finite-sized source, such as penumbral blurring and runout produced by an X-ray tube or plasma source, would be more severe (as a fraction of their total size) for nanometer-scale features. Because synchrotron sources produce nearly parallel beams, these would be most advantageous for printing ultrasmall patterns. In the late 1980s, IBM installed a minisynchrotron source [Andrews and Wilson, 1989] capable of servicing more than a dozen steppers. The cost of such an X-ray source today would be on the order of about $20M [Cerrina, 2000] and might compete economically with advanced optical steppers.

For proximity printing in general, resolution is limited by diffraction effects, because the minimum linewidth is approximately:

$$W = k \sqrt{G\lambda} \qquad (13.5)$$

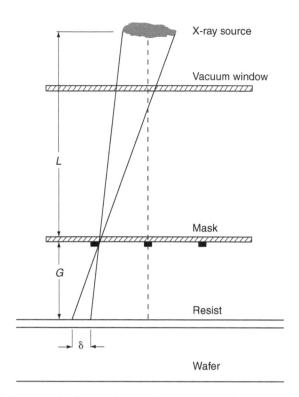

FIGURE 13.4 Schematic diagram of basic X-ray lithography exposure configuration. *d* is the amount of penumbral blurring due to a finite-sized source.

where *k* is a process-dependent parameter in the range of 0.6 to 1 [Cerrina, 1996], *G* is the gap, and λ is the wavelength. The *k* parameter depends on the processing conditions, including the type of mask, resist, and other conditions. With the use of phase-shift mask technology, 40-nm features at 20-μm gaps have been achieved [Chen et al., 1997].

With conventional masks, for an λ of 1 Å and a relatively narrow gap of 5 μm, we can expect to achieve feature sizes of a few tens of nanometers. This is well within the useful range of nanomachining. Better results can be achieved for even smaller gaps. However, there are other potential limitations to the maximum resolution. Feldman and Sun (1992) and Chlebek et al. (1987) have discussed several potential contributions to resolution degradation, shown in Figure 13.5. First, photoelectrons and Auger electrons that are energetic enough to expose the resist are released from the mask. Second, photons that pass along the edge of the absorber create photoelectrons in the resist that expose regions at the edge and under the absorber shadow. Third, strong photoabsorption at the substrate creates additional photoelectrons that can lead to under-cutting of the resist. According to Feldman and Sun, the high-energy photoelectrons do not contribute significantly to the limits to resolution and, according to Chlebek et al., neither do fluorescence photons.

It appears, then, that the limits to XRL resolution in PMMA are nearly identical to those determined by Joy (1983), as discussed earlier; the source of energy, either X-ray photons or electrons, is not a significant factor. In fact, Flanders (1980) performed experiments demonstrating very high resolution XRL of PMMA. The masks used in this experiment were nanomachined (partially employing angled shadow evaporation) to have alternating, vertical, 100-nm-thick bands of 17.5-nm-wide tungsten and carbon layers. As seen vertically by the X-ray source, this resulted in a mask of alternating regions of differing transmission values. In fact, successful contact printing of this pattern demonstrated that resolution of XRL in PMMA is comparable with that of high-resolution EBL. Using a similar technique, Early et al. (1990) demonstrated 30-nm lines independent of wavelengths ranging from 0.83 to 4.5 nm. The limit to useful printing scales in practical XRL systems will depend on other factors, such as patterning accuracy and mask distortions. Even including all of these effects, XRL will likely be an important technology for future NEMS applications.

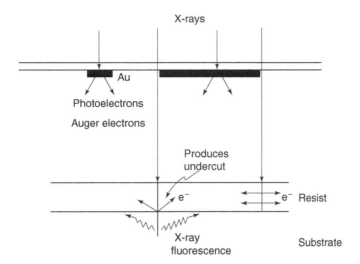

FIGURE 13.5 Contributions by scattered energy deposition in resist for X-ray lithography, including photoelectrons and Auger electrons from the mask and photoelectrons from the resist and substrate.

13.2.3 Other Parallel Nanoprinting Techniques

Two other printing methods deserve mention due to their simplicity, potential high-resolution capability, and potentially high throughput, which will be necessary for a commercially feasible technology. These are *microcontact printing*, developed by Whitesides, and *nanoimprint lithography*, developed by Chou. Both of these techniques function by physical contact between a patterned stamp and a substrate.

In the first case, the stamp is a pliable, elastomeric material, namely polydimethylsiloxane (PDMS), that is coated with a SAM (typically one of those mentioned earlier) and then pressed onto a substrate, resulting in a printed monolayer of SAM molecules. The stamps are formed by first preparing a master pattern etched into a hard substrate typically made of glass, resists, polymers, or silicon. The master is made using standard photolithography and etching techniques. The elastomer is poured onto the master, cured, peeled off, and then treated with an appropriate SAM "ink." Finally, the stamp is pressed directly onto the target material by hand or with some other simple pressing system. Using this technique, remarkably high-quality patterns and processing applications are possible. St. John and Craighead (1995) have demonstrated the use of OTS as ink, followed by plasma etching, to demonstrate 5-μm patterns in silicon. Kumar and Whitesides (1993) used a variety of thiol-based SAMs to pattern Au layers, which were used for further processing of substrates. They demonstrated features as small as 200 nm using hexadecanethiolate as the ink and wet etching of underlying gold layers using KCN solutions. Additional techniques demonstrated were selective-area plating of nickel and selective-area condensation of water vapor in hydrophilic regions for use as optical diffraction gratings.

The second stamping technique utilizes a rigid stamp in contact with a viscoelastic polymer, typically PMMA. The stamp is pressed into a resist layer that is heated above the glass transition temperature, T_g. Polymers such as PMMA are mixtures of amorphous and crystalline morphologies. T_g is an important parameter, because it is the temperature at which the polymer chains begin to slip in their amorphous arrangement and the material can flow easily when subjected to pressure. (Note that relaxation of the crystalline portion of the arrangement is referred to as *melting*, which is not important to this discussion.) T_g for bulk, heterotactic PMMA is approximately 105°C. According to Kleideiter et al. (1999), T_g can vary depending on the thickness of the film and the substrate upon which it is cast. It is important for a successful nanoimprinting resist to have a low glass-transition temperature, low viscosity, low sticking forces to the stamp, high stability, and high dry-etch resistance. Typical imprint temperatures are about 90°C greater than T_g.

Chou et al. (1997) have demonstrated 10-nm holes on a 40-nm pitch in PMMA. To ensure a clean surface after imprinting, the residual PMMA at the bottom of the patterns is etched away using an oxygen plasma. These patterns are then suitable for lift-off of metal patterns. By additional etching of the stamp features, 6-nm metal dots have been lifted off. Schulz et al. (2000) have investigated alternative resists, including thermoplastic polymers based on aromatic polymethacrylates and thermoset polymers based on multifunctional allylesters. These exhibit overall lower T_g (as low as 39°C in the case of the allylesters before imprinting), improved dry-etch resistance, and imprint performance comparable with PMMA.

13.2.4 Achieving Atomic Resolution

The lithographic techniques just discussed are not capable of achieving atomic resolution; however, proximal, or scanning, probes offer a path for scaling fabrication down to atomic dimensions. The two basic types of scanned probes used for lithography are atomic force microscopes (AFMs) and scanning tunneling microscopes (STMs). With its proven atomic resolution, STMs have been used to pattern substrates with the highest resolution reported to date, approximately 0.5 nm. AFM patterning has yet to attain this level of resolution, but the introduction of carbon nanotube tips promises to increase the resolution toward 5 nm.

13.2.4.1 STM Patterning

The most successful technique demonstrated for STM patterning has been the passivation and selective depassivation of silicon. In this method, pioneered by Lyding et al. (1997), a silicon surface is prepared and hydrogen passivated in an ultra-high-vacuum (UHV) environment. Surface preparation is, of course, crucial to this process. The sample must be cleaned carefully before inserting it into the UHV chamber, which must achieve pressures on the order of 10^{-8} Pa to preserve the cleanliness of the surface. The sample is flash heated to a temperature of 1200 to 1300°C for 1 minute, while maintaining a pressure of 4×10^{-8} Pa, to drive off any surface contamination and oxides. At a temperature of 650°C, hydrogen is introduced into the chamber to bring the pressure to $\sim 10^{-4}$ Pa. The hydrogen molecules are cracked using a tungsten filament, and hydrogen atoms attach to the dangling bonds on the silicon surface. The result for a (100) Si surface is a 2×1 reconstructed surface passivated by hydrogen.

To perform lithography, an STM tip is brought close to the surface (still in UHV), and a tip voltage of approximately -5 V relative to the sample is applied. Electrons streaming from the tip to the substrate transfer energy to the hydrogen atoms and break bonds between the hydrogen and silicon. Approximately 106 electrons are needed to remove each hydrogen atom. As the tip is scanned across the surface, lines of hydrogen atoms are removed, leaving depassivated Si atoms, as shown in Figure 13.6. Lyding et al. (1997) have demonstrated that through proper control of the depassivation process, individual hydrogen atoms can be removed, yielding true atomic resolution.

The depassivated silicon atoms are far more chemically active than when bonded to hydrogen, and this difference can be used to transfer the pattern into the substrate by selective chemistry. Perhaps the simplest method is to dose the surface with O_2 molecules at 10^{-4} Pa. These molecules will react with the depassivated silicon to form a thin SiO_2 layer but leave the passivated silicon unchanged. Although the

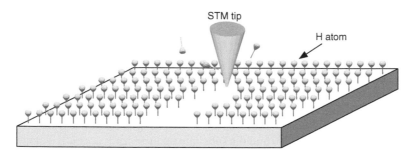

FIGURE 13.6 Depassivation of a silicon surface using an STM tip. Hydrogen atoms are knocked off the surface by electrons tunneling from the tip to the substrate.

oxide layers obtained by dosing with O_2 at low pressures are submonolayer, thicker ($\sim$4-nm) layers have been formed with oxygen at higher pressures, although with higher defect levels [Shen et al., 1995]. Thin nitride layers can be formed by similar treatments with NH_3. Although these layers are very thin, layers of similar thickness have been used to transfer patterns into silicon [Snow et al., 1996].

The formation of nanoscale metal lines by STM also is being pursued. All future applications of nanofabrication, both electronic and nanomechanical, will require very small lines of metal to connect circuits and subsystems or to connect the nanosystem to the outside world. Work in this area using STM has been attempted with limited success. One attempted technique was the field evaporation of gold from a gold tip, which yielded acceptable metal dots but gave only discontinuous lines [Mamin et al., 1990]. STM-assisted deposition of metals from organometallic precursors also has been attempted, but the resulting lines were contaminated with impurities [Rubel et al., 1994]. Selective metal deposition also has been investigated. The most successful method uses thermal chemical vapor deposition (CVD) of metal on selectively depassivated substrates. Deposition can be performed at a temperature of $\sim$400°C, which is low enough to preserve the hydrogen passivation. Proper choice of a precursor produces deposition on the depassivated areas and leaves the passivated areas unaffected. Using trimethylphosphine-Au, a selective deposition was achieved on depassivated areas of Si.

While STM has demonstrated atomic-level patterning on silicon substrates, many challenges remain before it can be incorporated into the fabrication of NEMS and nanomachines. First, the patterns produced are very fragile. It will be a significant challenge to a pattern-transfer process to retain this resolution while producing the three-dimensional structures often needed for NEMS. Throughput, which applies to all scanned-probe fabrication, also is an issue. Because the probe must form each feature sequentially, complex patterns will take a very long time to write. Work by Quate [Wilder et al., 1999] on multitip systems offers the possibility of high-throughput lithography systems. Scanned-probe lithography can also be used to form the masks for other high-resolution, but more parallel, fabrication techniques, such as nanoimprint lithography.

13.2.4.2 AFM Patterning

AFMs can also be used for ultra-high-resolution lithography. Approaches investigated range from the use of the AFM tip as scriber to locally activated oxidation of silicon or metals. AFM lithography in general has a much lower resolution than that of STM lithography. Whereas STMs achieve atomic resolution by tunneling electrons out of only a tiny area of an already very sharp tip, AFM resolution is limited by the shape and size of the full tip. Indeed, a trade-off must be made between resolution and tip durability in many applications, especially those that require a mechanical modification of the surface by the tip. In scribing lithography, the tip is used to scratch the substrate surface. The substrate typically is covered with a soft material, such as PMMA, which can be displaced by the tip without causing excessive wear of the tip. A limitation of this technique is that the soft film becomes stressed and distorted as material is moved, and the displaced material has to go somewhere. Because this displaced material can intrude on nearby features, the packing density is limited. The resolution is influenced strongly by the size and shape of the tip. Work has been done using electron-beam-grown "super-tips," which are formed from contamination, as discussed earlier; however, these ultra-sharp tips are more prone to breakage and wear-induced degradation.

Anodic oxidation induced by an AFM tip is a technique that, although having a resolution lower than that of STM, produces patterns that are more durable. To perform the anodic oxidation of a silicon substrate, the substrate is first cleaned using standard silicon cleaning techniques and then finished with an HF dip and water rinse, leaving the surface hydrogen passivated. The sample is then placed in a controlled-humidity atmosphere (humidity = 20–80%) at ambient pressure and an AFM tip is brought to the surface. A meniscus of water forms between the tip and sample, as shown in Figure 13.7, and the surface oxidizes when a voltage of -2 to -10 V is applied to the tip. If no voltage is applied, the surface does not oxidize, thus the surface can be imaged independently of the oxidation. The resulting oxide is approximately 1.5 nm thick [Snow et al., 2000]. Although this is a very thin oxide, it is sufficiently thick for transfer into the underlying substrate. Both wet- and dry-etching techniques have been used. Wet etching is attractive because its lack of a physical sputtering component allows a very high selectivity with the proper choice

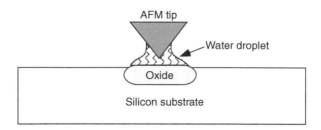

FIGURE 13.7   Anodic oxidation of a silicon surface using an AFM tip. A water droplet forms between the tip and substrate. The substrate is oxidized when a voltage is applied between tip and substrate.

of etchant. Snow et al. (1996) used a 1-nm oxide mask to etch through the 0.3-µm-thick top silicon layer of an SOI wafer using 70% hydrazine. Wet etching has limitations due to the isotropic or crystallographic nature of the etch, which can degrade the resolution of the pattern transfer. Dry etching can give very anisotropic etch profiles, but the physical component of the etch reduces the selectivity of the etch relative to the mask, a serious concern when the mask is only 1 to 1.5 nm thick. For this reason, dry etching is best carried out in high-density plasma systems, such as electron cyclotron resonance (ECR) or inductively coupled plasma (ICP), where the physical component is controlled separately and can be kept low. ECR etching has been investigated as a pattern-transfer mechanism [Snow et al., 1995]. Silicon structures 10 nm wide and 30 nm deep were etched using a Cl_2/O_2 plasma. Oxidation of metallic films has also been demonstrated and may be useful for the fabrication of fine metallic lines and the electronic devices associated with a nanomechanical device. Ti, Al, and Nb have been selectively oxidized in this fashion [Snow and Campbell, 1995; Schmidt et al., 1998; Matsumoto, 1999], and the technique has been used to fabricate a room-temperature, single-electron transistor [Matsumoto, 2000].

The resolution of anodic oxidation is largely limited by the size of the AFM tip. Using conventional tips, features as small as 10 nm have been demonstrated. The development of carbon nanotube tips opens the way for dramatic improvements in resolution, and features as small as 5 nm may be possible. As with all scanning probe techniques, AFM patterning is a serial technique and is therefore slow, but as with STM lithography, multitip systems may improve throughput. AFM lithography has some advantages over STM; because it is an ambient-pressure technique, the experimental setup is much simpler and less expensive.

13.2.4.3   Atomic Holography

It also is possible to use a variation of holography to directly control the placement of individual atoms as they deposit on a surface. Holography is a familiar technique that uses optical waves to create three-dimensional images by utilizing both the amplitude and phase information carried by the light. However, holography can be applied to any wave, including matter waves [Sone et al., 1999]. In matter waves, the "image" produced is made up of atoms and, if a substrate is placed in the image plane, the atoms will deposit on the surface. In this way, it is possible to create very small patterns. Whereas optical holography uses a photographic plate to manipulate the optical wavefront, in atomic holography a nanofabricated membrane is used to manipulate the atomic wavefront. The design of this hologram is critical to the formation of the desired image and requires intensive calculations. Using the pattern that is to be created on the substrate, one must calculate back to find what wavefront must leave the membrane. The hologram then becomes a series of holes etched into the surface of the membrane, which allows portions of the matter front to pass through in the required areas. The interference of the atomic wave from each hole as it propagates toward the surface produces the desired image. The membrane typically is 100-nm-thick silicon nitride, and the holes etched through the membrane have a typical dimension of 200 nm.

The key to atomic holography is to produce a mono-energetic beam of atoms (one wavelength) analogous to the coherent light used in optical holography. This is done by producing a beam of ionized atoms in a dc discharge, extracting the atoms, and then choosing a particular species with a mass selector. The atoms are then passed through a Zeeman Slower and caught in an atom trap, a magneto-optical

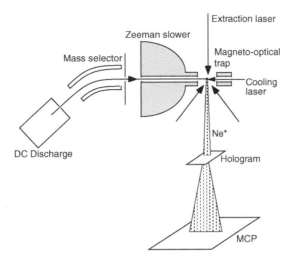

FIGURE 13.8 Schematic of an atomic holography system. (Adapted from Sone et al., 1999.)

trap with four laser beams, as shown in Figure 13.8. The cooled atoms are removed from the trap by the transfer laser and fall vertically under the gravitational force. The atoms are passed through the etched membrane hologram, where the beam is diffracted to form a reconstructed image at the substrate. In initial demonstrations, this substrate is a microchannel-plate (MCP) electron multiplier that detects the impact of atoms and produces a corresponding image on a fluorescent screen. This allows an image to be directly viewed and recorded by computer. The spatial resolution of the image is therefore limited by the resolution of the MCP, which is typically a few tens of microns. The actual resolution of the pattern is limited by the wavelength of the atomic beam, which is typically on the nanometer scale, in contrast to optical beams, for which the wavelength is typically more than 200 nm.

A challenge to atomic holography is the deposition of useful materials. Initial demonstrations have used atoms of limited utility, such as neon and sodium [Behringer et al., 1996; Matsui and Fujita, 1998]. These atoms were chosen due to properties that make them attractive for atomic beam manipulations rather than for their usefulness as materials. In order for this technique to become widely applied, depositions of materials such as aluminum or copper must be demonstrated.

13.3   Fabrication of Nanomachines: The Interface with Biology

The preceding pages have described the current state and predicted future of MEMS manufacturing based on real and projected advances in electronics, materials science, and chemistry. As discussed, it is expected that many of the most important advances will come as a result of cross-fertilizations between these fields. However, there is a growing consensus that the greatest advances, indeed the future of nanotechnology itself, lie in the integration of biological tools and systems into nanotechnological design, function, and manufacturing.

The basis for this statement is the obvious fact that nature has performed feats of nanotechnological engineering that we are only beginning to describe or understand. Modern approaches to biological research have revealed molecular machines that are remarkable in their beauty, surprising in their efficiency, and sometimes astounding in their self-assembled, three-dimensional complexity. Examples of molecular machines include "simple" machines, such as the bacterial flagellar motor, which powers the movement of organisms such as *Escherichia coli* through viscous media (Figure 13.9; see DeRosier, 1998, for an electron micrograph). The striking beauty of this rotating assembly (complete with a drive shaft, bushing, and a universal joint) is all the more remarkable when one realizes that it is only 50 nm in diameter, is capable of rotating at 300 revolutions per second, and produces a torque of 550 pN-nm [DeRosier, 1998]. Another example is bacteriophage T4, a virus (thankfully specific to bacteria) that bears a remarkable

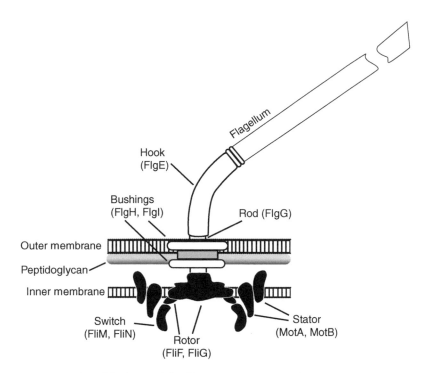

FIGURE 13.9   Representation of the bacterial flagellar motor. Gene names for proteins comprising individual components are given in parentheses. Protons pass through MotA to drive rotation of the propeller (the flagellum). Figure is drawn approximately to scale (width of motor = 50 nm). (Adapted from Berg, 1995, and Mathews, 2000.)

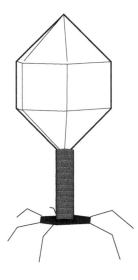

FIGURE 13.10   Representation of bacteriophage T4 (a bacterial virus). The phage attaches to the bacterium with specific receptors on its "legs" and injects the DNA stored in its "head" through the tailpiece and into the bacterium to initiate the process of infection. The head is approximately 85 to 115 nm.

resemblance to a lunar lander (Figure 13.10). For electron micrographs, see, for example, Nelson and Cox (2000) or http://www.asmusa.org/division/m/foto/T4Mic.html. This nanosyringe (only 85 nm in diameter) lands on and attaches to the surface of a susceptible bacterium and then uses its tail projection to inject its DNA into the bacterium. Once the DNA is inside the cell, it essentially reprograms the bacterium, converting it into a factory for making more viruses. Though these examples exist on the nanoscale, the existence of our brains and bodies demonstrates that biological nanotechnology can be integrated and coordinated into three-dimensional structures more than seven orders of magnitude larger.

13.3.1   Inspiration from Biology

Producing an engineered version of a system as complex as the human brain is obviously still in the realm of science fiction, but nanotechnological engineers are finding inspiration from a number of biological sources. The most obvious sources are existing biological molecules. Biological molecules are finding application both as functional components and as structural components. Some molecules have natural functions that engineers hope to harness directly. These include molecular motors, such as the flagellar motor just described [DeRosier, 1998; Mehta et al., 1999; Vale and Milligan, 2000] (see also below), as well as electric switches or ion pumps, such as the voltage-gated ion channels found in nerve cells. More complicated systems, such as the photosynthetic apparatus, also are key targets [Gust and Moore, 1994]. One example of proteins that are naturally useful in a structural way is bacterial "S proteins." These pore-forming proteins found on the extracellular surface of bacteria can be extracted and then experimentally reassembled (on a variety of surfaces) into matrices with specific characteristics as variable as the number of different species of bacteria [Sleytr et al., 1997; Sleytr and Beveridge, 1999]. One could imagine that naturally assembling filamentous proteins, such as cytoskeletal proteins, could be utilized in similar ways.

Some of the most exciting developments may come from application of the natural characteristics of biological molecules in completely novel ways. For example, DNA has been used as a "rope" of defined length that can be positioned in specific places by virtue of its sequence-specific "sticky" cohesive ends and then used as a scaffold on which silver wires are deposited [Braun et al., 1998]. The self-organizing capacity of these sequence-specific "sticky" ends also has been utilized to assemble simple mixtures of cross-linked DNA units into well-organized two-dimensional "DNA crystals" with repeating surface features of various widths [Winfree et al., 1998]. Other DNA molecules have been used to construct more complicated three-dimensional structures, including cubes and truncated octahedrons [Seeman, 1999]. The fact that specific positions in these DNA molecules can be derivatized in a large variety of ways suggests that DNA may become an important structural tool for the design and manufacture of molecular electronic and other nanotechnological devices [Winfree et al., 1998].

Though biological molecules are useful, it is the harnessing of biological ideas, such as self-assembly and evolution by natural selection, that may in the end have the greatest impact. Two obvious problems for nanotechnology are (1) how does one design a molecule with the activity desired and (2) how does one assemble it. As noted earlier, nature has solved both of these problems to create "machines" of daunting complexity and remarkable ability. It has been argued that there is no reaction for which Nature has not developed an enzyme. This array of well-honed activities has been achieved by random mutation followed by reproductive selection of the more efficient variants. In design-based engineering, we are limited by our imagination and our ability to model. With mutation/selection-based engineering, one is limited only by the number of variants that can be tested and the efficiency with which improved variants can be selected. The difficult part of applying this idea is figuring out how to identify and isolate or reproduce the molecules with desired characteristics. Combinatorial chemists are actively using this approach to identify new drugs [Floyd et al., 1999]; biologists have used it to develop entirely new enzymatic activities, either in vivo (bioremediation) [Chen et al., 1999] or in vitro [Wilson and Szostak, 1999]. The breadth of practical future applications for "directed evolution" is difficult to predict, but it is clear that the idea is powerful.

The problem of three-dimensional self-assembly is closely related to that of structure and design. Indeed, natural machines have been selected by evolution to self-assemble (or to assemble with the assistance of other natural machines, called *chaperones*; see Nelson [2000], for an introduction to this process). For example, an enzyme is a linear chain of amino acids that folds into a complex and reproducible three-dimensional structure. The precise array of amino acid side chains endows it with specific affinities for other molecules and for a specific chemical activity. To properly place these side chains artificially in three-dimensional space would be difficult, if not impossible, with existing technology. However, many proteins spontaneously fold into the correct three-dimensional structure in seconds. These proteins can be assembled into larger machines of remarkable complexity and function. One beautiful example is the ribosome, a biological machine that functions (according to the analogy of Peter Moore) rather like a programmable loom [Ban et al., 2000; Nissen et al., 2000]. The ribosome "reads" the sequence of a given messenger

RNA molecule and strings together amino acids according to the mRNA program, thus synthesizing the prescribed protein. The ribosome is indeed a programmable, positional, assembly fabrication device existing on the nanometer scale, thus demonstrating that such nanofabrication devices are possible [Merkle, 1999].

The problem of how to predict the three-dimensional structure of a protein from the linear sequence of amino acids is one of the remaining "Holy Grails" of molecular biology. If one could figure out how to predict three-dimensional structure from a sequence, it seems likely that one could figure out how to design a sequence necessary to produce a particular structure (see DeGrado et al. [1999] and Koehl and Levitt [1999a; 1999b] for discussions of current attempts to achieve this aim). However, there is the very real problem of whether this thermodynamically designed protein fold could be achieved in a kinetically realistic amount of time. One possibility is to model new proteins on old (evolutionarily selected) folds [Koehl and Levitt, 1999a; 1999b]. It also seems likely that engineers could use in vitro evolution to optimize the self-assembly (folding ability) of these designed protein machines. If single proteins could be designed de novo, why not entire complexes? If (when) this aim of de novo protein design is achieved, the machines of nanotechnology could very well start assembling themselves.

13.3.2 Practical Fabrication of Biological Nanotechnology

To begin to understand how devices based on biological molecules might be fabricated now, it is necessary to review some general techniques for the production of biological molecules. Many of the DNA-based applications just described require small DNA molecules (<100 basepairs), and small DNA molecules are relatively easy to make in large amounts either chemically (available commercially; 0.1 μmol of a 20 mer oligonucleotide can be made for less than $1/base, 10 μmol for less than $20/base by Sigma Genosys, for example) or by biological replication. One advantage of the synthetic approach is that the nucleotides at specific sites can be chemically derivatized in a wide variety of ways, limited only by the compatibility of the derivatization with the chemical process of DNA synthesis. Another advantage of chemical synthesis is that the engineer defines the sequence. For pieces of DNA larger than 100 bases (and up to chromosome size), biological manufacture often is preferable. Biological manufacture of DNA is performed by replicating an existing molecule, and it is done either in vitro or in vivo. In vitro replication is performed by putting the starting DNA and the necessary purified components and enzymes into a test tube and then letting the enzymes do their job. Polymerase chain reaction (PCR) is one particularly useful enzymatic approach for producing large numbers of replicas of a DNA fragment in the range of 0.1 to 10 kilobasepairs. To replicate a desired piece of DNA in vivo, the DNA fragment is ligated into a plasmid (a different piece of DNA that can replicate in bacteria independently of the chromosome), the plasmid is grown in bacteria, and then the plasmid (containing the DNA of interest) is purified out of the bacteria. The desired DNA can then be cut out of the plasmid using sequence-specific restriction enzymes and purified away from the rest of the plasmid DNA. In this process, the amount of DNA produced is limited only by the quantity of bacteria grown.

Proteins also can be made either chemically or biologically, but most useful proteins are too large to be easily made in reasonable amounts chemically. Therefore, they are made biologically by inserting a piece of DNA encoding the desired protein into a cell type suitable for making the protein. Bacteria usually are the most productive hosts for protein production, but many human proteins cannot be grown in bacteria, either because they are not folded properly into their three-dimensional shape or because they are not modified properly (proteins can require phosphorylations, sugar additions, etc.). Yeast cells, insect cells, and human cells are (in that order) more likely to make active proteins, but also are less productive and (considerably) more expensive.

After the problem of how to express the active protein is solved, the protein must be purified away from other cellular proteins. This problem can be made considerably easier by engineering an *affinity tag* into the DNA (and therefore the protein). This tag is a specific amino acid sequence that enables the engineered protein to stick specifically to a particular affinity matrix (several types exist), allowing the tagged protein to be retained on the matrix and the vast majority of undesired proteins to be washed away. In addition, a variety of tags or modifications can be chemically added to the protein after it is purified. A tag of one form or another often is necessary to allow positioning of the protein in the desired place and in the desired orientation; it also may be necessary to harness the protein to other mechanical devices (e.g., in the case of a molecular motor). For more details on the processes of working with DNA or protein, see Ausubel et al. (2000) and Coligan et al. (2000).

To understand these issues in more practical terms, it is useful to describe the approaches used by two engineers to try to harness the work of the molecular motor enzyme F1 ATPase. F1 ATPase is a rotary motor present in essentially every living cell. Its natural purpose is to synthesize the energy carrier molecule adenosine triphosphate (ATP) by harnessing the flow of protons down an electrochemical gradient. During ATP synthesis, the central subunit rotates in a clockwise direction. However, if given a supply of ATP, the motor will utilize the ATP to drive rotation in a counterclockwise direction. Montemagno and Bachand (1999) have produced and purified a tailored form of this molecule that they have attached specifically and in a specific orientation to a substrate. They attached the central subunit specifically to a latex bead and used it to drive the rotation of the latex bead.

In order to do this, Montemagno and Bachand first isolated the pieces of DNA encoding the subunits of the ATP synthase using standard molecular biology procedures. They spliced these pieces of DNA into a bacterial plasmid that was engineered so that production of the ATPase proteins could be turned on or off in response to an externally applied chemical signal (the sugar analog IPTG). They then tailored the DNA encoding the ATPase subunits to provide solutions to the other problems. For example, they modified the DNA of one of the subunits at the base of the ATPase by fusing it to a His tag, which is simply a string of 6 to 10 histidines that allows any protein (or complex of proteins) to bind specifically to nickel. This His tag allowed the ATPase complex to be purified on a column of nickel resin. It also allowed the ATPase complex to bind to a nickel substrate with a defined orientation. As a final bit of genetic engineering, Montemagno and Bachand changed one of the amino acids at the tip of the central rotating subunit (the "driveshaft") to cysteine. Cysteine can be chemically cross-linked to a variety of different molecules, and this engineered cysteine allowed them to cross-link the driveshaft to biotin. Because biotin binds streptavidin specifically and tightly, the biotin-linked driveshaft bound specifically to a streptavidin-coated bead. This bead was therefore coupled to the rotating driveshaft, enabling the motor to be harnessed to move the bead.

Although these experiments are just a demonstration of principle, it is easy to imagine replacing the latex bead with some form of molecular drivetrain. Thus far, the integration of biology into nanotechnology has yet to live up to expectations. However, with the breathtaking advances in biotechnology and the enormous breadth of biochemical biodiversity, the expectations are great indeed.

13.4 Summary

This chapter has attempted to provide an overview of technologies, from established to nascent, that may be used to create nanostructures suitable for future nanoelectromechanical systems. Many visions exist as to what is possible in the near and further future for ultrasmall mechanical devices, so it is difficult to tell at this time which of the various fabrication technologies available today may become the most useful in the future. It is likely that some combination of many technologies will find applications. In order to address this, we discussed the most common nanofabrication technologies, such as electron beam and X-ray lithographies, and what limits their resolution in achieving the smallest possible pattern dimensions. We also discussed an alternative method, microcontact printing, for achieving high throughput at very small dimensions. We showed that these techniques were capable of achieving 10-nm resolution in a straightforward manner. Using exotic techniques, much smaller features are attainable.

Scanning probe techniques have proven to be viable technologies in attaining true atomic resolution. We discussed the use of scanning tunneling microscopy and atomic force microscopy in defining Angstrom-scale patterns and pointed out a method of achieving high throughput with this technique. We also discussed atomic holography, which someday may be used to build NEMS devices by direct deposition of the constituent atoms.

We concluded this chapter with a discussion of how biological systems qualify as NEMS and how future technologists might exploit existing cellular structures to create artificial nanomachines. We discussed specific cellular functions that are similar in function to potential nanomachines, including flagellar motors, voltage switches, ion pumps, cytoskeletal proteins for structural support, and even DNA for use as a patterning template. We discussed how biologists succeed in customizing and fabricating batch quantities of these potential NEMS structural elements.

Acknowledgments

GHB is grateful to F. Cerrina for helpful discussions. This work was supported in part by ONR/DARPA and the Indiana 21st Century Research and Technology Fund.

References

Andrews, D.E., and Wilson, M.N. (1989) "High-Energy Lithography Illumination by Oxford's Synchrotron: A Compact Superconducting Synchrotron X-ray Source," *J. Vac. Sci. Technol. B* 7, pp. 1696–1701.

Ausubel, F., Brent, M., Kingston, R.E., Moore, D.D., Seidman, J.G., and Struhl, K., eds. (2000) *Current Protocols in Molecular Biology*, John Wiley & Sons, New York.

Ban, N., Nissen, P., Hansen, J., Moore, P.B., and Steitz, T.A. (2000) "The Complete Atomic Structure of the Large Ribosomal Subunit at 2.4 Å Resolution," *Science* 289, pp. 905–20.

Bazán, G., and Bernstein, G.H. (1993) "Electron Beam Lithography Over Very Large Scan Fields," *J. Vac. Sci. Technol. A* 11, pp. 1745–52.

Behringer, R.E., Natarajan, V., and Timp, G. (1996) "Laser Focused Atomic Deposition: A New Lithography Tool," *Appl. Phys. Lett.* 68, pp. 1034–36.

Berger, S.D., Gibson, J.M., Camarda, R.M., Farrow, R.C., Huggins, H.A., Kraus, J.S., and Liddle, J.A. (1991) "Projection Electron-Beam Lithography: A New Approach," *J. Vac. Sci. Technol. B* 9, pp. 2996–99.

Bernstein, G.H., Liu, W.P., Khawaja, Y.N., Kozicki, M.N., Ferry, D.K., and Blum, L. (1988) "High-Resolution Electron Beam Lithography with Negative Organic and Inorganic Resists," *J. Vac. Sci. Technol. B* 6, pp. 2298–2302.

Bernstein, G.H., Hill, D.A., and Liu, W.P. (1992) "New High-Contrast Developers for Poly(Methyl Methacrylate) Resist," *J. Appl. Phys.* 71, pp. 4066–75.

Binnig, G., and Rohrer, H. (1987) "Scanning Tunneling Microscopy — From Birth to Adolescence," *Rev. Mod. Phys.* 59, pp. 615–25.

Braun, E., Eichen, Y., Sivan, U., and Ben-Yoseph, G. (1998) "DNA-Templated Assembly and Electrode Attachment of a Conducting Silver Wire," *Nature* 391, pp. 775–78.

Broers, A.N. (1964) "Micromachining by Sputtering Through a Mask of Contamination Laid Down by an Electron Beam," in *Proc. 1st Int. Conf. on Electron and Ion Beam Science and Technology* (R. Bakish, ed.), pp. 191–204, Wiley, New York.

Broers, A.N. (1995) "Fabrication Limits of Electron Beam Lithography, and of UV, X-ray, and Ion-Beam Lithographies," *Phil. Trans. R. Soc. London A* 353, pp. 291–311.

Cerrina, F. (1996) "The Limits of Patterning in X-ray Lithography," *Mat. Res. Soc. Symp. Proc.* 380, pp. 173–77.

Cerrina, F. (2000) Private communication.

Cerrina, F., and Rai-Choudhury, P., eds. (1997) *Handbook of Microlithography, Micromachining, and Microfabrication. Vol. 1: Microlithography*, SPIE Optical Engineering Press, Bellingham, WA.

Chen, W., and Ahmed, H. (1993a) "Fabrication of 5–7 nm Wide Etched Lines in Silicon Using 100 KeV Electron-Beam Lithography and Polymethylmethacrylate Resist," *Appl. Phys. Lett.* 62, pp. 1499–1501.

Chen, W., and Ahmed, H. (1993b) "Fabrication of Sub-10 nm Structures by Lift-Off and by Etching After Electron-Beam Exposure of Poly(methylmethacrylate) Resist on Solid Substrates," *J. Vac. Sci. Technol. B* 11, pp. 2519–23.

Chen, W., Bruhlmann, F., Richins, R.D., and Mulchandani, A. (1999) "Engineering of Improved Microbes and Enzymes for Bioremediation," *Curr. Opin. Biotechnol.* **10**, pp. 137–41.

Chen, Z.G., Leonard, Q.T., Khan, M., Cerrina, F., and Seeger, D.E. (1997) "X-ray Phase-Mask: Nanostructures," *Proc. SPIE, Emerging Lithographic Technol.* **3048**, pp. 183–92.

Chlebek, J., Betz, H., Heuberger, A., and Huber, H.-L. (1987) "The Influence of Photoelectrons and Fluorescence Radiation on Resolution in X-ray Lithography," *Microelectron. Eng.* **6**, pp. 221–26.

Chou, S.Y., Krauss, P.R., Wei-Zhang, Lingjie-Guo, Lei-Zhuang (1997) "Sub-10 nm Imprint Lithography and Applications" *J. Vac. Sci. Technol. B* **15**, pp. 2897–2904.

Coligan, J.E., Dunn, B.M., Ploegh, H.L., Speiche, Da.W., and Wingfield, P.T., eds. (2000) *Current Protocols in Protein Science*, John Wiley & Sons, New York.

Cumming, D.R.S., Thoms, S., Beaumont, S.P., and Weaver, J.M.R. (1996) "Fabrication of 3 nm Wires Using 100 keV Electron Beam Lithography and Poly(methyl methacrylate) Resist," *Appl. Phys. Lett.* **68**, pp. 322–24.

DeGrado, W.F., Summa, C.M., Pavone, V., Nastri, F., and Lombardi, A. (1999) "De Novo Design and Structural Characterization of Proteins and Metalloproteins," *Ann. Rev. Biochem.* **68**, pp. 779–819.

DeRosier, D.J. (1998) "The Turn of the Screw: The Bacterial Flagellar Motor," *Cell* **93**, pp. 17–20.

Dial, O., Cheng, C.C., and Scherer, A. (1998) "Fabrication of High-Density Nanostructures by Electron Beam Lithography," *J. Vac. Sci. Technol. B* **16**, pp. 3887–90.

Drexler, K.E. (1981) "Molecular Engineering: An Approach to the Development of General Capabilities for Molecular Manipulation," *Proc. Natl. Acad. Sci. USA* **78**, pp. 5275–78.

Early, K., Schattenburg, M.L., and Smith, H.L. (1990) "Absence of Resolution Degradation in X-ray Lithography for λ from 4.5 nm to 0.83 nm," *Microelectron. Eng.* **11**, pp. 317–21.

Erbe, A., Blick, R.H., Kriele, A., and Kotthaus, J.P. (1998) "A Mechanically Flexible Tunneling Contact Operating at Radio Frequencies," *Appl. Phys. Lett.* **73**, pp. 3751–53.

Feldman, M., and Sun, J. (1992) "Resolution Limits in X-ray Lithography," *J. Vac. Sci. Technol. B* **10**, pp. 3173–76.

Flanders, D.C. (1980) "Replication of 175-Å Lines and Spaces in Polymethylmethacrylate Using X-ray Lithography," *Appl. Phys. Lett.* **36**, pp. 93–96.

Floyd, C.D., Leblanc, C., and Whittaker, M. (1999) "Combinatorial Chemistry as a Tool For Drug Discovery," *Prog. Med. Chem.* **36**, pp. 91–168.

Forest, D.R. (1995) "The Future Impact of Molecular Nanotechnology on Textile Technology and on the Textile Industry," in *Proc. of Discover Expo '95: Industrial Fabric & Equipment Exposition*, PAGES, DATE, Charlotte, NC.

Goldstein, J.I., Newbury, D.E., Echlin, P., Joy, D.C., Fiori, C., and Lifshin, E. (1981) *Scanning Electron Microscopy and X-ray Microanalysis*, Plenum Press, New York.

Gust, D., and Moore, T. (1994) "Photosynthesis Mimics as Molecular Electronic Devices," *IEEE Eng. Med. Biol.* **13**, pp. 58–66.

Harris, R.A. (1973) "Polymethyl Methacrylate as an Electron Sensitive Resist," *J. Electrochem. Soc. Solid-State Sci. Technol.* **120**, pp. 270–74.

Hiroshima, H., Okayama, S., Ogura, M., Komuro, M., Nakazawa, H., Nakagawa, Y., Ohi, K., and Tanaka, K. (1995) "Nanobeam Process System: An Ultrahigh Vacuum Electron Beam Lithography System with 3-nm Probe Size," *J. Vac. Sci. Technol. B* **13**, pp. 2514–17.

Huang, X., Bernstein, G.H., Bazán, G., and Hill, D.A. (1993) "Spatial Density of Lines Exposed in Poly (methylmethacrylate) by Electron Beam Lithography," *J. Vac. Sci. Technol. A* **11**, pp. 1739–44.

Irmer, B., Blick, R.H., Simmel, F., Godel, W., Lorenz, H., and Kotthaus, J.P. (1998) "Josephson Junctions Defined by a Nanoplough," *Appl. Phys. Lett.* **73**, pp. 2051–53.

Joy, D.C. (1983) "The Spatial Resolution Limit of Electron Lithography," *Microelectron. Eng.* **1**, pp. 103–19.

Kleideiter, G., Prucker, O., Bock, H., and Frank, C.W. (1999) "Polymer Thin Film Properties as a Function of Temperature and Pressure," *Macromol. Symp.* **145**, pp. 95–102.

Koehl, P., and Levitt, M. (1999a) "De Novo Protein Design. I. In Search of Stability and Specificity," *J. Mol. Biol.* **293**, pp. 1161–81.

Koehl, P., and Levitt, M. (1999b) "De Novo Protein Design. II. Plasticity in Sequence Space," *J. Mol. Biol.* **293**, pp. 1183–93.

Kumar, A., and Whitesides, G.M. (1993) "Features of Gold Having Micrometer to Centimeter Dimensions Can Be Formed Through a Combination of Stamping with an Elastomeric Stamp and an Alkanethiol "Ink" Followed by Chemical Etching," *Appl. Phys. Lett.* **63**, pp. 2002–04.

Langreth, R. (1993) "Molecular Marvels," *Pop. Sci.* pp. 91–94.

Lercel, M.J., Tiberio, R.C., Chapman, P.F., Craighead, H.G., Sheen, C.W., Parikh, A., and Allara, D.L. (1993) "Self-Assembled Monolayer Electron-Beam Resists on GaAs and SiO2," *J. Vac. Sci. Technol. B* **11**, pp. 2823–28.

Lercel, M.J., Craighead, H.G., Parikh, A.N., Seshadri, K., and Allara, D.L. (1996) "Sub-10 nm Lithography with Self-Assembled Monolayers," *Appl. Phys. Lett.* **68**, pp. 1504–06.

Lyding, J.W. (1997) "UHV STM Nanofabricatrion: Progress, Technology Spin-Offs, and Challenges," *Proc. IEEE* **85**, pp. 589–600.

Mamin, H.J., Guethner, P.H., and Rugar, D. (1990) "Atomic Emission from a Gold Scanning-Tunneling-Microscope Tip," *Phys. Rev. Lett.* **65**, pp. 2418–21.

Matsui, S., and Fujita, J.I. (1998) "Material-Wave Nanotechnology: Nanofabrication Using a de Broglie Wave," *J. Vac. Sci. Technol. B* **16**, pp. 2439–43.

Matsumoto, K. (1999) "Room Temperature Operated Single Electron Transistor Made by a Scanning Tunnelling Microscopy/Atomic Force Microscopy Nano-oxidation Process," *Int. J. Electron.* **86**, pp. 641–62.

Matsumoto, K. (2000) "Room-Temperature Single Electron Devices by Scanning Probe Process," *Int. J. High Speed Electron. Sys.* **10**, pp. 83–91.

McCord, M.A., and Rook, M.J. (1997) "Electron Beam Lithography," in *Handbook of Microlithography, Micromachining, and Microfabrication, Vol. 1: Microlithography* (P. Rai-Choudhury, ed.), pp. 139–250. SPIE Press, Bellingham, WA.

Mehta, A.D., Rief, M., Spudich, J.A., Smith, D.A., and Simmons, R.M. (1999) "Single-Molecule Biomechanics with Optical Methods," *Science* **283**, pp. 1689–95.

Merkle, R.C. (1999) "Biotechnology as a Route to Nanotechnology," *Trends Biotechnol.* **17**, pp. 271–74.

Mochel, M.E., Humphreys, C.J., Mochel, J.M., and Eades, J.A. (1983) "Cutting of 20 Å Holes and Lines in Metal-Aluminas," in *Proc. 41st Annual Meeting of the Electron Microscopy Society of America*, pp. 100–101.

Montemagno, C., and Bachand, G. (1999) "Constructing Nanomechanical Devices Powered by Biomolecular Motors," *Nanotechnology* **10**, pp. 225–31.

Namatsu, H., Nagase, M., Kurihara, K., Iwadate, K., Furuta, T., and Murase, K. (1995) "Fabrication of Sub-10-nm Silicon Lines with Minimum Fluctuation," *J. Vac. Sci. Technol. B* **13**, pp. 1473–76.

Nelson, D.L., and Cox, M.M. (2000) *Lehninger Principles of Biochemistry*, Worth, New York.

Nissen, P., Hansen, J., Ban, N., Moore, P.B., and Steitz, T.A. (2000) "The Structural Basis of Ribosome Activity in Peptide Bond Synthesis," *Science* **289**, pp. 920–30.

Reed, M.A. (1999) "Molecular-Scale Electronics," *Proc. IEEE* **87**, pp. 652–58.

Roukes, M.L. (2000) "Nanoelectromechanical Systems," in *Proc. Solid-State Sensor and Actuator Workshop*, pp. 367–76, 4–8 June, Hilton Head Island, SC.

Rubel, S., Trochet, M., Ehrichs, E.E., Smith, W.F., and de-Lozanne, A.L. (1994) "Nanofabrication and Rapid Imaging with a Scanning Tunneling Microscope," *J. Vac. Sci. Technol. B* **12**, pp. 1894–97.

Schmidt, T., Martel, R., Sandstrom, R.L., and Avouris, P. (1998) "Current-Induced Local Oxidation of Metal Films: Mechanism and Quantum-Size Effects," *Appl. Phys. Lett.* **73**, pp. 2173–75.

Schulz, H., Scheer, H.-C., Hoffmann, T., Sotomayor Torres, C.M., Pfeiffer, K., Bleidiessel, G., Grutzner, G., Cardinaud, Ch., Gaboriau, F., Peignon, M.-C., Ahopelto, J., and Heidari, B. (2000) "New Polymer Materials for Nanoimprinting," *J. Vac. Sci. Technol. B* **18**, pp. 1861–65.

Seeman, N.C. (1999) "DNA Engineering and Its Application to Nanotechnology," *Trends Biotechnol.* **17**, pp. 437–43.

Shen, T.C., Wang, C., Lyding, J.W., and Tucker, J.R. (1995) "Nanoscale Oxide Patterns on Si(100) Surfaces," *Appl. Phys. Lett.* **66**, pp. 976–78.

Sleytr, U.B., and Beveridge, T.J. (1999) "Bacterial S-Layers," *Trends Microbiol.* **7**, pp. 253–60.

Sleytr, U.B., Bayley, H., Sara, M., Breitwieser, A., Kupcu, S., Mader, C., Weigert, S., Unger, F.M., Messner, P., Jahn-Schmid, B., Schuster, B., Pum, D., Douglas, K., Clark, N.A., Moore, J.T., Winningham, T.A., Levy, S., Frithsen, I., Pankovc, J., Beale, P., Gillis, H.P., Choutov, D.A., and Martin, K.P. (1997) "Applications of S-Layers," *FEMS Microbiol. Rev.* **20**, pp. 151–75.

Snow, E.S., and Campbell, P.M. (1995) "AFM Fabrication of Sub-10-nanometer Metal-Oxide Devices with In Situ Control of Electrical Properties," *Science* **270**, pp. 1639–41.

Snow, E.S., Juan, W.H., Pang, S.W., and Campbell, P.M. (1995) "Si Nanostructures Fabricated by Anodic Oxidation with an Atomic Force Microscope and Etching with an Electron Cyclotron Resonance Source," *Appl. Phys. Lett.* **66**, pp. 1729–31.

Snow, E.S., Campbell, P.M., and McMarr, P.J. (1996) "AFM-Based Fabrication of Free-Standing Si Nanostructures," *Nanotechnology* **7**, pp. 434–37.

Snow, E.S., Jernigan, G.G., and Campbell, P.M. (2000) "The Kinetics and Mechanism of Scanned Probe Oxidation of Si," *Appl. Phys. Lett.* **76**, pp. 1782–84.

Sone, J., Fujita, J., Ochiai, Y., Manako, S., Matsui, S., Nomura, E., Baba, T., Kawaura, H., Sakamoto, T., Chen, C.D., Nakamura, Y., and Tsai, J.S. (1999) "Nanofabrication Toward Sub-10 nm and Its Application to Novel Nanodevices," *Nanotechnology* **10**, pp. 135–41.

Spears, D., and Smith, H.I. (1972) "High-Resolution Pattern Replication Using Soft X-rays," *Electron. Lett.* **8**, pp. 102–04.

St. John, P.M., and Craighead, H.G. (1995) "Microcontact Printing and Pattern Transfer Using Trichlorosilanes on Oxide Substrates," *Appl. Phys. Lett.* **68**, pp. 1022–24.

Tada, T., and Kanayama, T. (1995) "Fabrication of Silicon Nanostructures with a Poly(methylmethacry-late) Single-Layer Process," *J. Vac. Sci. Technol. B* **13**, pp. 2801–04.

Vale, R.D., and Milligan, R.A. (2000) "The Way Things Move: Looking Under the Hood of Molecular Motor Proteins," *Science* **288**, pp. 88–95.

Wilder, K., Soh, H.T., Atalar, A., and Quate, C.F. (1999) "Nanometer-Scale Patterning and Individual Current-Controlled Lithography Using Multiple Scanning Probes," *Rev. Sci. Inst.* **70**, pp. 2822–27.

Wilson, D.S., and Szostak, J.W. (1999) "In Vitro Selection of Functional Nucleic Acids," *Annu. Rev. Biochem.* **68**, pp. 611–47.

Winfree, E., Liu, F., Wenzler, L.A., and Seeman, N.C. (1998) "Design and Self-Assembly of Two-Dimensional DNA Crystals," *Nature* **394**, pp. 539–544.

14

Molecular Self-Assembly: Fundamental Concepts and Applications

Jill A. Miwa and
Federico Rosei
University of Quebec

14.1 Introduction

In the past decade, research efforts in nanoscience have demonstrated that atoms [Eigler and Schweizer, 1990], molecules [Heinrich et al., 2002], and clusters and nanoparticles [Kiely et al., 1998; Brust et al., 1995; Brust et al., 1994] can be used as functional building blocks for fabricating advanced and new phases of condensed matter on the nanometer length scale. The optimal size of such components depends on the particular property to be optimized: by altering the dimensions of the building blocks and controlling their surface geometry, chemistry, and assembly, it is now possible to tailor functionalities in unprecedented ways.

At the start of the new millennium, we are thus confronted with the desire to investigate in greater detail the atomic scale structure of matter. Besides the interest in basic science, the ultimate goal in this context is to develop tomorrow's functional devices so that they can operate on the nanoscale. The obsessive trend toward miniaturization is driven partly by an analogous trend in the semiconductor industry, which aims at developing ever smaller and faster devices (e.g., computers, mobile telephones, and other portable equipment) and partly by the desire to develop new instruments for other purposes (e.g., microelectromechanical systems [MEMS] and nanoelectromechanical systems [NEMS], gas and chemical sensors, or even biomedical equipment). For these reasons, it is expected that nanotechnology will have a much greater impact on modern society than even the silicon integrated circuit (which led to the electronic revolution of the 20th century), since it will be applicable to many different fields of human activity, including medicine, security, and telecommunications.

Proper tools must be used to study the properties of materials and surfaces on the nanometer length scale. In fact, the development of this multidisciplinary field was undoubtedly accelerated by the advent of relatively recent technologies that allow the visualization, design, characterization, and manipulation of nanoscale systems. The birth of scanning probe microscopy (SPM) techniques, in particular scanning tunneling microscopy (STM) [Binnig et al., 1982; Binnig and Rohrer 1999], and atomic force microscopy (AFM) [Binnig et al., 1986] is largely considered responsible for the emergence of nanoscience as a novel field of research. Although it does not have strong chemical sensitivity, the STM has proven to be a powerful tool in the study of adsorbate–surface interactions and of elementary surface processes in general [Besenbacher, 1996; Neddermeyer, 1996; Rosei and Rosei, 2002; Rosei et al., 2003; Rosei, 2004].

The motivation to integrate organic molecules into existing MEMS/NEMS technology stems from a projected limitation that is well exemplified by Moore's heuristic. In 1965, an Intel cofounder by the name of Gordon Moore predicted that the number of transistors per integrated circuit would double every 18 months [Moore 1965]. His empirical observation has held true for almost four decades. Unfortunately, in the years to come, fabrication costs will be formidable, and more importantly, silicon chip manufacturing processes will reach fundamental physical limits. If this down-sizing pace is to be maintained, new roads toward the miniaturization of devices must be explored.

Organic molecules offer a host of compelling and tunable properties [Wolkow, 1999; Rosei, 2004], and thus they have stimulated considerable interest as an alternative to top-down methodologies (i.e., photolithography, etc.). Molecules represent the definitive limit for electronic devices and can be regarded as prefabricated building blocks [Schunack, 2002]. But, how will a multitude of these tiny molecular blocks arrange into an ordered device?

Currently, there are two active strategies to attack this problem: atom-by-atom engineering and self-assembly processes. The idea of manipulating individual atoms and molecules on the nanoscale was simply a dream just 20 years ago. The first demonstration of controlled STM manipulations was published less than a decade after the invention of the STM itself: individual Xe atoms were deposited on Ni(110) at 4 K and were positioned to write IBM's company logo [Eigler and Schweizer, 1990]. However, at present, this approach is far too cumbersome and tedious to play a dominant role in the semiconductor (or any other) industry. On the other hand, the latter trend is far more amenable to current chip manufacturing technologies. The concept of self-assembly was introduced by Whitesides [Whitesides et al., 1991], and is defined as the spontaneous alignment of molecules into stable and well-defined aggregates by noncovalent forces. This phenomenon is very advantageous because it carries out the most difficult task of nano-fabrication, namely atomic-level modification or surface functionalization, in a parallel fashion. Furthermore, self-assembled structures tend toward thermodynamic equilibrium and therefore produce nearly defect-free or self-healing structures. But what are the fundamental principles driving this process? And how can self-assembly be harnessed and actually used to fabricate devices?

The governing forces for self-assembly are not yet well understood. They can be divided into two main categories: intermolecular interactions and molecule–substrate interactions. The organization of molecules on a surface is guided by a delicate balance between these two competing interactions. Noncovalent bonding (e.g., hydrogen bonding) and electrostatic forces (e.g., van der Waals) between molecules can lend a hand in the creation of complex structures. Strong or weak coupling with the underlying surface can also play a crucial role in the development of the adlayer. If the adsorbate–surface relationship is weak (physisorption), the substrate can be envisioned as a checkerboard that only offers an array of adsorption sites for the molecules. A stronger interaction and greater binding energy (chemisorption) between adsorbate and substrate may result in the surface dictating the direction and boundaries of growth.

In this chapter we draw examples from the recent literature, describing experiments whereby self-assembly mechanisms and molecule–surface interactions are demonstrated. Although this field is still far from a commercial application, we envisage that these concepts will soon be used to develop functionalized surfaces leading to new generations of nanoscale systems and sensors. Examples will include biological and chemical sensors that are able to detect the presence of minute amounts of substance and to labs on a chip that will perform reliable blood testing and similar operations.

14.2 Molecule–Molecule Interactions

The manner in which molecular building blocks fit together when deposited on a substrate largely depends on the geometry and chemical nature of their functional groups. Different end groups lead to different types of interaction and binding mechanisms between molecules. A clear understanding of these different types of connectivities should lead to the control of self-assembly mechanisms. Moreover, organic molecules are flexible building blocks. They can be endowed with a vast assortment of functional groups; thus, the possible self-assembled motifs that can we can create is largely limited only by our imagination. In this section, we describe recent experimental reports that demonstrate how molecule–molecule interactions can be used to create ordered molecular arrangements on flat metal and semiconductor surfaces.

14.2.1 Hydrogen Bonding and van der Waals Interactions

George Flynn and his group at Columbia University have investigated the packing order of organic molecules by STM at the solution–graphite interface. Through the use of chemical markers (i.e. single atoms or molecules that exhibit high contrast in STM images) they have been able to directly observe the exact orientation and position of molecules in a self-assembled monolayer (SAM) and reveal the driving forces that underpin this process [Wintgens et al., 2003; Giancarlo and Flynn, 2000].

By choosing a long chain hydrocarbon (i.e. 15-hydroxypentadecanoic acid) that is flanked by a carboxylic acid head group and a hydroxyl moiety, Flynn and co-workers are able to simultaneously study van der Waals interactions and different types of hydrogen bonding connectivities. The molecules lie in an all-trans configuration on the graphite surface and form broad lamellae segregated by dark troughs that correspond to the chemical marker groups. The individual molecules within each lamella arrange in such a way that van der Waals interactions between the hydrogen atoms of the carbon backbone are maximized. These forces, in turn, help to stabilize the film.

The two types of markers are easily distinguished in STM images (Figures 14.1a and 14.1b). The wider troughs are composed of carboxylic acid moieties, while the narrower troughs are composed of the smaller alcohol functional groups. The troughs can be distinguished further by the characteristic angle between the molecular axes of adjoining lamellae. Hydrogen bonding induces the formation of interdigitated dimers between the carboxylic acid moieties of neighboring molecules, and thus the angle between the molecular axes is 180°. Hydrogen bonding through alcohol groups results in an angle of 120° between molecules in adjacent lamellae. These competing forces give rise to an overall super-herringbone pattern, as directly revealed by STM.

Interestingly, by changing the length of this molecule by a single carbon atom, a substantially different global pattern emerges. Similar to the previous case, 16-hydroxyhexadecanoic acid molecules lie in an all-trans configuration on the surface, but now molecular axes in abutting lamellae are fully aligned with one another (Figures 14.1c and 14.1d). The terminations of the molecules come together to form four different types of tetramers; each consists of two hydroxyl and two carboxylic acid groups hydrogen bonded together. The tetramers appear as large dark spots in the STM image.

From these experiments, it is evident that adsorbate–adsorbate interactions are far from straightforward and that nuances in the geometry and chemical composition of the adsorbed molecule can significantly influence the long-range ordering of the SAM, with clear implications on its potential functionality.

14.2.2 Metal–Ligand Bonding

Significant progress has been made in exploiting hydrogen bonding and electrostatic intermolecular coupling for creating extended and highly organized molecular networks. Some of the patterns created are not simply elegant but also functional. Unfortunately, hydrogen and electrostatic type bonds are weak,

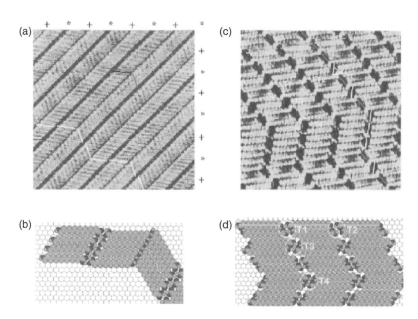

FIGURE 14.1 (See color insert following page 9-22.) (a) 15-hydroxypentadecanoic acid at the interface of a 1-nonanol solution and the basal plane of graphite, imaged by STM. Scan size: 15 × 15 nm². A single molecular length is indicated by a black or a blue bar. The asterisks (∗) mark troughs that are composed of carboxylic acid dimers, and the pluses (+) mark troughs that are composed of alcohol functional groups. The yellow bars depict the global pattern, which can be described as a super-herringbone structure. (b) Ball model of 15-hydroxypentadecanoic acid on a graphite surface: carbon atoms are green, oxygen atoms are red, hydrogen atoms are grey. The molecules are positioned in registry with the underlying graphite lattice. (c) STM images of 16-hydroxyhexadecanoic acid diluted in a 1-hexanol solution on graphite. Molecular lengths are indicated by the blue bars. The blue bars are collinear, indicating that the fatty acids occupy the same graphite lattice rows. The red and the yellow bars depict the two different orientations observed for the dark spots. Image size: 10 × 10 nm². (d) Ball model of 16-hydroxyhexadecanoic acid on a graphite surface. (Reprinted with permission from Wintgens, D. et al. [2003] "Packing of HO(CH$_2$)$_{14}$COOH and HO(CH$_2$)$_{15}$COOH on Graphite at the Liquid–Solid Interface Observed by Scanning Tunneling Microscopy: Methylene Unit Direction of Self-Assembly Structures," *J. Phys. Chem. B.* **107**, pp. 173–79.)

and thus in adverse circumstances (e.g. elevated temperatures) major instabilities within the molecular framework can arise. A stronger type of interaction is the metal–ligand bond [Holliday and Mirkin, 2001]. This type of bonding mechanism is proving to be a strong candidate for organizing molecular building blocks.

Kern's group in Stüttgart has recently focused on metal–ligand bonding of trimesic acid (tma) and copper atoms [Lin et al., 2002; Dmitriev et al., 2002]. Their investigations are conducted with an ultra-high vacuum (UHV) compatible STM. Trimesic acid, which consists of a phenyl ring and three carboxy end groups, is deposited onto a Cu(100) surface. The substrate is held at room temperature so that copper adatoms move about freely and cannot be resolved by the microscope. The tma molecules, however, appear as bright equilateral triangles in STM images. This adsorption geometry indicates that the molecule is bound with its phenyl ring lying parallel to the underlying metal substrate.

At the copper surface the tma undergoes a deprotonation process. Four of these now negatively charged ion groups coalesce around a single copper atom; a cloverleaf configuration with a bright central protrusion evolves. The copper atom is thus trapped and can be imaged by STM. If a sufficient coverage of tma molecules is deposited, a long ranging array of cloverleafs can be seen (Figure 14.2).

Upon closer inspection of the cloverleaf topography, thermal rotations of individual tma molecules are revealed in a sequence of STM images over a time period of a few minutes. If the carboxylate ligands are rotated such that the formation of a metal–ligand bond is *not* favorable, the central metal atom will diffuse

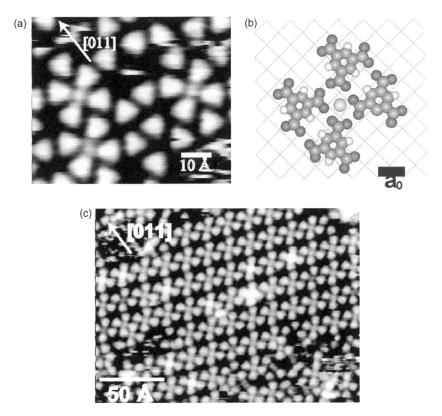

FIGURE 14.2 (See color insert following page 9-22.) (a) STM image of trimesic acid molecules adsorbed on a Cu(100) surface at room temperature. The triangular shape reflects a flat-lying adsorption geometry. Cloverleaf-shaped arrangements of four tma molecules with a central protrusion represent $[Cu(tma)_4]^{n-}$ coordination compounds. (b) Model for the $[Cu(tma)_4]^{n-}$ configuration with a central Cu atom coordinated by four carboxylate ligands. Gray: carbon atoms; red: oxygen atoms; white: hydrogen atoms; yellow: Cu adatoms. The Cu substrate is represented by the orange lattice where $a_0 = 3.61$ Å. In agreement with the tunneling data the tma triangles do not point straight to the central site, which indicates a unidentate coordination of the Cu adatom. (c) Regular array of $[Cu(tma)_4]^{n-}$ complexes. (Reprinted with permission from Lin, N. et al. [2002] "Real-Time Single-Molecule Imaging of the Formation and Dynamics of Coordination Compounds," *Angew. Chem. Int. Ed.* **41**, pp. 4779–83).

away. The corresponding STM image shows a cloverleaf with a dark hole at its center indicating the absence of the copper atom. When the tma triangles are properly oriented, the four ligands will capture one of the diffusing copper adatoms. In this latter scenario, the STM image shows a cloverleaf with a bright central protrusion. In short, a system of molecular rotors that work together to create a dynamical atom trap has been developed. Moreover, the studies indicate that the lifetime of a trap (i.e. how long the copper atom remains locked into position by the surrounding ligands) is highly dependent on the presence of neighboring clovers and the surface morphology of the substrate.

This experiment has demonstrated that two reacting species can be specifically positioned by means of a simple self-assembly process. Even though the trapping process is not controlled, the results of this experiment represent a promising start and will have an impact on scientists who wish to use metal–organic coordination to fit different molecular building blocks together to form a rich variety of rigid geometries.

In more recent developments, the same group demonstrated that organic molecules such as terephthalic acid (TPA), trimellitic acid (TMLA), and 4,1′,4′,1″-terphenyl-1,4″-dicarboxlic acid (TDA) along with iron atoms can arrange into extended trellises on a Cu(100) surface [Stepanow et al., 2004]. The molecules link in a linear fashion via iron–ligand bonding to produce robust 2-D ladder arrays. These structures are further stabilized by hydrogen bonding between the carboxylate end groups that do not participate in the coordination bond and from interactions between adjacent phenyl rings. The dimension of the resulting rectangular cavities of the network can be tuned by choosing the appropriate length of organic linker molecule. Moreover, TMLA contains an additional side group that can be used to chemically functionalize the cavity.

These systems represent ideal host sites for guest molecules because each cavity is geometrically and chemically well-defined. Kern and co-workers investigated this latter idea by dosing the trellises with fullerenes. The C_{60} molecules preferentially adsorbed to sterically compatible cavities. The Fe–TPA arrays were able to accommodate single C_{60} molecules, while the larger cavities of the Fe–TDA networks could hold up to three fullerenes. This is an interesting study since it nicely illustrates the potential for using these arrays in molecular recognition applications. These types of metal–organic coordination networks (MOCNs) are likely to find their way into viable technological applications.

14.3   Molecule–Substrate Interactions

Thus far, this chapter has focused on intermolecular interactions (i.e. van der Waals forces, hydrogen bonding, and metal–ligand bonds). The present section is dedicated to molecule–substrate interactions. Two types of surfaces are discussed, metal and semiconductor.

A substrate can play a passive or active role in the formation of a SAM and in general in the interaction with complex molecules. Metal surfaces often act as static checkerboards that only offer an array of adsorption sites for molecules, while semiconductor surfaces have been known to greatly influence the growth of the molecular adlayer. However, this is a rather naive vision of molecule–substrate interactions. Here we review a system in which the adsorption of molecules on a metal surface is completely dominated by molecule–substrate interactions. Surfaces that are intrinsically sensitive only to certain molecules and adsorbates that cause reconstruction of the underlying surface can add greater complexity and scope to the issue at hand. A few examples of these are briefly discussed.

14.3.1   Adsorption of Molecules on Metal Substrates

Thiols on Au(111) are the archetypal SAM system. Thiols are easily prepared, cost efficient, and well-ordered, making them well compatible with current microfabrication techniques. In particular, alkanethiols are the simplest type of thiols. A traditional route for preparing an alkanethiol SAM is to simply dip the Au substrate into a solution of ethanol with micromolar to millimolar concentrations of the alkanethiol.

The alkanethiol consists of a thiol (SH) headgroup and a carbon backbone. The S atom strongly bonds to the Au surface, and is it is this reaction that is responsible for the creation of the SAM. It is believed that this type of linkage could be the nanoscale equivalent of an alligator clip. Due to the characteristic nature of the bond, the carbon chain extends up and away from the surface. In fact, the chains are slightly tilted with respect to the surface normal. The amount of tilt is dependent on the length of the backbone and presence of end groups that could potentially interact with neighboring alkanethiols. It is a very flexible setup. Simply changing the termination of the thiol can lead to an entirely different ordering of the SAM and functionalization of a surface. OH– terminated, COOH– terminated, and fluorinated alkanethiols as well as a host of other types of thiols have been studied in depth [Schreiber, 2000]. Thiols on Au beautifully illustrate how surfaces can be intrinsically sensitive only to certain molecules. This model system shows enormous potential for applications in NEMS/MEMS technology.

A novel scheme for functionalizing a gold surface with thiols is dip-pen nanolithography (DPN). This is a positive printing technique that delivers molecules to a surface using an AFM tip. It is a new take on conventional lithographic methods. The ultimate resolution of DPN is still uncertain because there are many factors to contend with, such as the shape of the AFM tip, the size of the molecules, the writing speed, and the grain size of the substrate. In particular, Piner et al. (1999) have written 30 nm wide lines with alkanethiols on a gold thin film. The process works by first coating an AFM tip with molecules. Upon exposure to air, a water meniscus forms on and between the tip and the substrate. The molecules are then transported to the surface by capillary action. The deposited molecules are stabilized by the strong coupling between the thiol headgroup and the Au surface.

The major advantage of DPN is its simplicity. The process does not require the use of resists, stamps, complex procedures, or expensive equipment. Single dots, individual lines, arrays, and grids have been generated using DPN. It is a powerful technique. Presently, DPN could be used for precise functionalization of devices that are created by conventional microfabrication techniques. Of course, writing with a single AFM tip is relatively slow, but one can envision arrays of tips writing out patterns in parallel on nanoscale devices.

An excellent example of how molecule–substrate interactions can be dominant was recently reported by Weckesser et al. (1999), who studied the adsorption properties of 4-[*trans*-2-(pyrid-4-yl-vinyl)] benzoic acid (PVBA) on the (110) surface of palladium (Pd). PVBA is a rod-like and planar molecule. It has been the subject of intense investigation because it was specifically designed for applications in nonlinear optics. It is comprised of a pyridyl head group joined to a benzoic acid moiety via a vinyl group. The molecular subunits are intended to form strong hydrogen bonds; thus in a PVBA/Pd(110) system, appreciable lateral interactions between individual PVBA molecules are expected. This is not the case, however, and adsorbate–substrate interactions completely determine the growth of the film.

The interaction of the aromatic rings and the underlying surface is optimized when the molecule lies flat and diagonally across three adjacent Pd atomic rows. This particular binding geometry of PVBA to a clean Pd(110) surface, is aptly referred to as a dog-bone configuration. At low coverages, the molecules are randomly distributed with the molecular axis of the molecules always oriented by ±35° with respect to the [1–10] direction (Figure 14.3). This observation suggests that molecule–substrate forces have a greater influence over the growth of the adlayer than intermolecular interactions.

Similar conclusions were drawn when higher PVBA coverages and higher substrate temperatures were investigated. At increased temperatures, the molecules exhibit movement only along the [1–10] direction. The activation barrier for diffusion is suitably high indicating a strong correlation between the π-orbitals of the aromatic groups and the Pd surface. Lateral interactions do not come into play during diffusion, and it is proposed that the dog-bone adsorption geometry is more energetically favorable than the energy gained from hydrogen bonding. Moreover, the surface-induced orientation of the molecules is not conducive

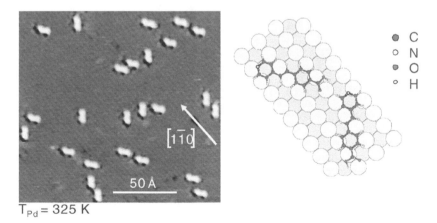

FIGURE 14.3 (**See color insert following page 9-22.**) Left: PVBA molecules randomly distributed on the Pd(110) surface at low coverage ($\theta = 0.018$ ML). The molecules were adsorbed and imaged at a substrate temperature of 325 K and deposited with a rate of 3×10^{-5} ML/s. The atomic rows of the Pd substrate and the dog-bone internal molecular structure are resolved (image size: 19×18 nm^2: $U_t = 1.04$ V, $I_t = 1$ nA). Right: Ball model for unrelaxed PVBA molecules on Pd(110). Large light and dark grey circles: Pd atomic rows of the surface; small dark grey circles: carbon atoms; yellow: nitrogen atoms; red: oxygen atoms; and white: hydrogen atoms. The molecular axis is oriented by ±35.3° with respect to the [1–10] Pd rows providing optimal coordination of surface Pd and molecular subunits. (Reprinted with permission from Weckesser, J. et al. [1999] "Binding and Ordering of Large Organic Molecules on an Anisotropic Metal Surface: PVBA on Pd(110)," *Surf. Sci.* **431**, pp. 168–73.)

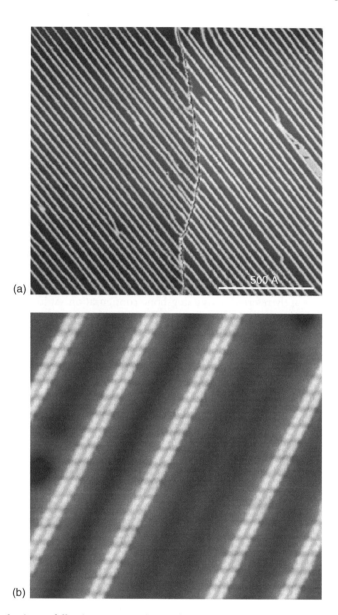

FIGURE 14.4 **(See color insert following page 9-22.)** One-dimensional supramolecular PVBA super-grating by self-assembly mediated by H-bond formation on an Ag(111) surface at 300 K (measured at 77 K). (a) An STM topograph of a single domain extending over two terraces demonstrates ordering at the micron length scale. (b) A close-up image of the self-assembled twin chains reveals that they consist of coupled rows of PVBA molecules. (Reprinted with permission from Barth, J.V. et al. [2000] "Building Supramolecular Nanostructures at Surfaces by Hydrogen Bonding," *Angew. Chem. Int. Ed.* **39**, pp. 1230–3.)

for the formation of H-bonding between abutting molecules. Due to the architecture of PVBA, the angle between the molecular axes would have to be 180° in order to favor H-bonding. On the Pd surface, adjacent PVBA molecules form an angle of 110° with each other. It is also presumed that the weaker van der Waals forces are assuaged by the surface electrons of the metal surface, thereby further reducing any lateral coupling.

Despite intentions to use a molecule that strongly favors intermolecular covalent bonding, the growth and bonding geometry of PVBA on Pd(110) is ruled by adsorbate–substrate interactions. Weckesser et al.

have made considerable advances in understanding how large organic molecules bond, order, and self-assemble onto metal surfaces. The scope of this incredible work includes studies of PVBA on Ag(111), Au(111), and Cu(111) surfaces and STM observations of a similar molecule, 4-[(pyrid-4-yl-ethynyl)] benzoic acid (PEBA), on Ag(111). The comparative study of PVBA and PEBA on Ag(111) is particularly impressive since the intermolecular and molecule–substrate interactions are well balanced, and what becomes strikingly evident is that the subtle difference in structure of the two molecules results in radically different self-assembled nanostructures. The PVBA molecules form chiral H-bonded twin chains that extend over to the micron scale and are regularly spaced (Figure 14.4) whereas PEBA molecules align to form vast 2-D islands that have straight, kink-free, borders (Figure 14.5) [Weckesser, 2000; Barth et al., 2000; Barth et al., 2002; Barth et al., 2003].

In some cases, the strength of molecule–surface interactions can be so intense as to induce a reconstruction of the underlying surface. An example of such a phenomenon is found by depositing Lander $(C_{90}H_{98})$ molecules onto an ordinary Cu(110) surface [Rosei et al., 2002; Schunack et al., 2002]. At room temperature, the molecules adsorb to the surface and diffuse toward the highly mobile step edges. If the temperature of the system is lowered to approximately 150 K, the movement of the molecules and steps can be frozen out.

To investigate the way in which the Landers are anchored to the surface, the STM tip can be used to gently push the molecules away from the now static (frozen) step edges. What was discovered was that the molecules created facets at the steps (Figure 14.6). These facets are tooth-like in structure and mirror the dimensions of the Landers! This is an intriguing result because it illustrates how a surface can be modified by cooperative adsorbate–surface interactions. Normally, molecules are arranged on a surface in order to create a new surface, but in this scenario the adsorbed molecules are used to modify the existing surface. This is a unique self-fabrication process for altering surfaces at the nanoscale. Using appropriately designed molecules, it is envisioned that metal surfaces may be reshaped or even engineered to the desired atomic-scale specifications.

Molecule–substrate interactions also can be attenuated or even screened by fabricating an appropriate substrate template. This concept was demonstrated in recent work from Besenbacher's group at the University of Aarhus (DK). They reported a method for organizing molecules into a predetermined design by modulating molecule–substrate interactions. To this end, Otero et al. (2004) have controlled the oxidation of a Cu(110) surface, forming an array of alternating copper oxide (Cu–O) and copper (Cu) gratings. The size of these lines can be altered by adjusting the oxygen exposure time and the substrate temperature during oxidation. An increase in temperature will result in an increase in the width of the stripes, while increasing the exposure time gives rise to narrower gratings. The dimensions of the tunable template are impressive. The length of the stripes extends over several hundred nanometers and the smallest grating sizes are 2 nm wide Cu troughs separated by periodic 5 nm wide Cu–O stripes.

The Single Lander (SL, $C_{90}H_{98}$) molecule seemed an ideal candidate to deposit onto the template surface. The molecule's shape resembles a table whose four legs extend equally above and below the tabletop. What seems like poor carpentry makes for ingenious science. The tabletop consists of a highly conductive backbone and can be envisioned as a short molecular wire. The legs act as bulky spacers that separate the conducting backbone from the substrate surface. Ideally, upon deposition one would like to see a string of SL molecules confined to one of the gratings so as to generate a long conducting wire.

As hoped, the SL molecules preferentially adsorb only onto the bare Cu troughs. It is proposed that the main molecule–substrate interaction is an attractive van der Waals force between the backbone of the SL molecule and the copper surface. The oxide layer reduces this interaction, and thus the Cu–O stripes remain bare. Unfortunately, comparison with theoretical calculations shows that the SL molecules assemble such that the wire segment is aligned perpendicular to the direction of the grating (Figure 14.7). The reason is that the length of the SL is approximately 1.7 nm, which is compatible with the 2 nm width of the Cu troughs. Furthermore, the SL is inclined to align its backbone along the [001] direction of the copper surface. Incidentally, this direction is perpendicular to the direction of the stripes.

To remedy the situation, the longer Violet Lander (VL, $C_{108}H_{104}$) was deposited onto the gratings. This molecule has a structure similar to the SL except it is approximately 0.5 nm longer and slightly wider. This

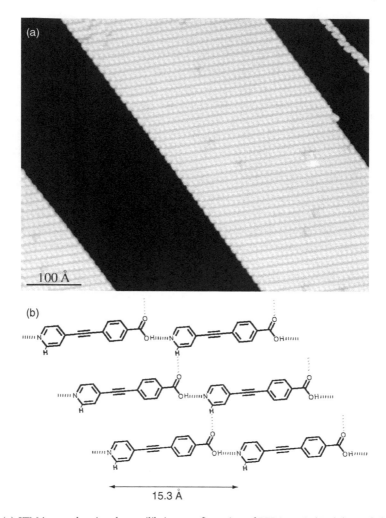

FIGURE 14.5 (a) STM image showing the equilibrium configuration of PEBA on Ag(111) (annealed to room temperature following adsorption at 150 K and measured at 77 K). 2-D islands of PEBA form on large terraces in which the molecules are aligned strictly parallel. The rod-like shape of the flat-lying molecules is clearly visible. (b) Schematic model for the PEBA supramolecular structure with OH … N head-to-tail and weak lateral CH … OC hydrogen bonds indicated. For closer proximity of the functional groups and the 2-D interconnection of the rows, the molecular axis is slightly rotated with respect to the orientation of the superstructure. (Reprinted with permission from Barth, J.V. et al. [2000], "Building Supramolecular Nanostructures at Surfaces by Hydrogen Bonding," *Angew. Chem. Int. Ed.* **39**, pp. 1230–34.)

slight change in dimensions would cause a part of the VL to sit on top of an oxidized region. However, the molecule adopts a more energetically favorable configuration where the entire molecule resides only on the bare Cu area and the backbone is thus aligned in the direction of the grating (Figure 14.8). When the width of the Cu troughs is slightly increased, the VL molecules tilt in an attempt to realign their backbone with the [110] direction of the bare copper surface, further demonstrating that molecule–substrate interactions are the driving forces behind this self-assembly process.

The real beauty of this experiment is that it takes advantage of molecule–substrate interactions so as to enhance the self-assembly process of molecules into predetermined architectures. A patterned surface was created that favored molecular adsorption only in specific regions, and more importantly, the size of these sites can be tailored to direct the orientation of the adsorbed molecules in the template.

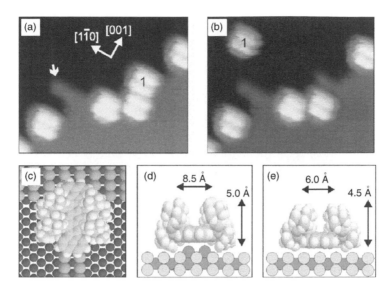

FIGURE 14.6 (See color insert following page 9-22.) (a, b) Manipulation sequence of a Lander molecule labeled 1 from a step edge on Cu(110). A nanostructure from a previous manipulation experiment is visible and indicated by an arrow. ($13.6 \times 11.2\,nm^2$, V $= -1.768$ V, I $= -37$ nA, T $= 95$ K). (b) After the manipulation of the molecule along the [1–10] direction, the nanostructure underneath becomes visible (tunneling parameters for manipulation: V $= -0.055$ V, I $= -1.05$ nA). (c) Model of the Lander on the nanostructure, showing that the board is parallel to the nanostructure. For clarity, the length of the nanostructure is extended beyond the molecule. Cross-sectional side view of the Lander (d) on a nanostructure and (e) on a flat terrace. The distances between and heights of the spacer groups (from calculated STM images) are stated. (Reprinted with permission from Schunack, M. et al. [2002] "Adsorption Behavior of Lander Molecules on Cu(110) Studied by Scanning Tunnelling Microscopy," *J. Chem. Phys.* 117, pp. 6259–65.)

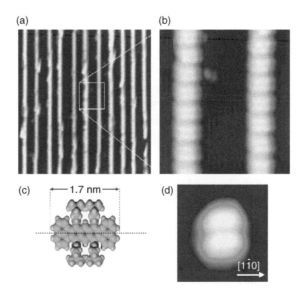

FIGURE 14.7 (See color insert following page 9-22.) (a) $60 \times 60\,nm^2$ STM image showing the molecular chains formed after the deposition of SL molecules on a nanopatterned Cu–O surface. The molecules adsorb exclusively on bare Cu stripes. (b) $14 \times 14\,nm^2$ high-resolution STM image. The individual SL molecules can be individually resolved, and their conformation can be extracted. (c) Filled-space model of the single Lander (SL) molecule. (d) Typical $3 \times 3\,nm^2$ STM image of the SL molecule on a clean Cu(110) surface revealing the SL as four lobes corresponding to the spacer legs. (Reprinted with permission from Otero, R. et al. [2004] "One Dimensional Assembly and Selective Orientation of Lander Molecules on a O–Cu Template," *Angew. Chem. Int. Ed.* 43, pp. 2092–95.)

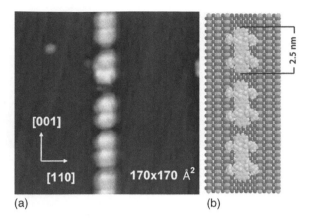

FIGURE 14.8   (See color insert following page 9-22.) (a) $17 \times 17 \, nm^2$ STM image showing the molecular chains formed after the deposition of VL on the nanopatterened O–Cu surface. The molecules adsorb exclusively on bare Cu stripes, but their orientation is no longer perpendicular to the direction of the chain. The molecules are found either perfectly aligned with the direction of the chain or forming an angle of 20° with respect to that direction. (b) Ball model of the VL adsorbed in a Cu trench. (Reprinted with permission from Otero, R. et al. [2004] "One Dimensional Assembly and Selective Orientation of Lander Molecules on a O–Cu Template," *Angew. Chem. Int. Ed.* **43**, pp. 2092–95.)

14.3.2 Adsorption of Molecules on Semiconductor Surfaces

Molecules on semi-conductor surfaces have been the subject of intense research [Wolkow, 1999]. The goal is to incorporate organic molecules into existing technologies and to create novel molecular devices. Organic molecules offer a range of properties that complement those of silicon, and thus they will enrich the current state of the silicon industry. There are many challenges to be faced in fabricating hybrid organic-molecule–silicon devices, one of which is a detailed understanding of molecular linkages to semiconductor surfaces.

Silicon surfaces are highly reactive. Studies are often carried out under UHV conditions in order to add an element of control to the experiment. This of course represents a problem because such an environment is unrealistic for devices that will be created for consumer use. To circumvent this problem, some scientists have aspired to work on passivated silicon surfaces. Hydrogen terminated silicon surfaces can be easily prepared by dry or wet processes and have shown to be relatively stable in air. The following examples involve H-terminated Si(100). These experiments were carried out under UHV conditions.

It has been reported that the diffusion of molecules on silicon surfaces is frustrated and thus presents a problem for the formation of SAMs. In spite of that, Lopinski et al. (2000) have presented an approach for growing styrene lines on a H–Si(100) surface. The growth process is a chemical one, and the silicon lattice specifies the direction of the lines and the spacing between the molecules within the line. Styrene consists of a phenyl ring and alkene moiety. These molecules are completely unreactive with the hydrogen terminated portion of the surface. Line growth is initiated at exposed single Si dangling bonds. Upon interaction with the dangling bond the π-bond of the alkene breaks up to form a Si–C bond and a C radical. The phenyl group stabilizes the radical giving it the necessary time to abstract a hydrogen atom from an adjacent surface site. A new Si dangling bond site is now exposed. Another styrene molecule can react with this site, and the entire mechanism is repeated creating a chain reaction (Figure 14.9).

Line growth is often terminated at defect sites. Interestingly, "unzipping" of the lines has been observed for styrene lines terminated at a Si dangling bond. The dangling bond can abstract a hydrogen from an end styrene and induce desorption of the molecule. The chain reaction is reversed and the line is unzipped. This event is rare, but even so, it has been suggested that the lines could be capped with a protecting molecule in order to prevent further reactions from taking place.

Another interesting example is TEMPO (2,2,6,6-tetramethylpiperidinyloxy), a molecule that is able to passivate residual silicon dangling bonds [Pitters et al., 2003]. This molecule can be used to refine the

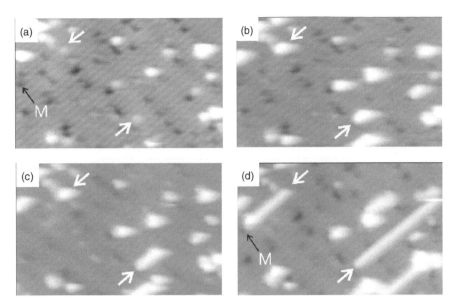

FIGURE 14.9 Growth of styrene lines on a H-terminated Si(100) surface with a dilute concentration of single Si dangling bonds. The figure shows a sequence of STM images (size: 25 × 14 nm^2, −2.1 V, 47 pA) corresponding to an increasing exposure to styrene. (a) 3 L, (b) 28 L, (c) 50 L, (d) 105 L. White arrows denote two particular dangling-bond sites that lead to the growth of long styrene lines. The missing dimer defect (M) marked in the figure terminates the growth of the line in the upper left-hand corner of the image. (Reprinted with permission from Lopinski, G.P. et al. [2000] "Self-Directed Growth of Molecular Nanostructures on Silicon," *Nature* **406**, pp. 48–51.)

position, orientation, and extent of styrene line growth in order to create a predetermined surface pattern. TEMPO is an ideal candidate for the job because it is a nitroxide that can form a stable, radical–radical link with a silicon dangling bond via its lone electron that resides on the oxygen atom. Molecules that dative bond to the dangling bond are too weak and would desorb at elevated temperatures, and of course, alkenes and aldehydes form carbon radicals that fail to pacify the site. TEMPO does not react with the H-terminated portions of the surface. Together, these experiments [Lopinski et al., 2000; Pitters et al., 2003] offer a promising approach for functionalizing silicon surfaces in a controlled manner to create complex surface patterns.

14.4 Applications of Functionalized Surfaces

The manner in which molecules assemble on a surface influences their growth and thus the final properties of the adlayer. This is why a proper understanding of intermolecular and molecule–surface interactions is imperative. Understanding self-assembly processes will make it possible to harness these driving forces for technological applications. The previous examples highlighted some of the amazing work that has been done in this exciting field of research. They also exemplify the complex nature of these interactions and offer a number of approaches to functionalizing surfaces.

Bestowing MEMS/NEMS devices with organic molecules may serve to: (1) revolutionize current devices, (2) fabricate novel devices, and (3) create new materials. In the remaining section we discuss current achievements in organic molecular functionalization of MEMS/NEMS devices, specifically the notion of endowing microfabricated silicon cantilevers with organic molecules for sensor-type applications. We then conclude with the recent work of Jackson et al. (2004) on constructing unique functionalized nanoparticles that shows incredible potential for biosensor applications.

14.4.1 Sensors

Fritz et al. (2000) have taken advantage of and derived inspiration from the facts that biomolecules are a favorite for molecular recognition purposes and that the lateral tension of a lipid bilayer can be altered by interactions with membrane molecules to create a so-called nanomechanical nose [Lang et al., 1999]. This is an ultrasensitive device that can be used to detect reactions and assay molecules. This breakthrough device does not involve fluorescence or radioactive tags. It is reusable and it is compatible with today's silicon technology.

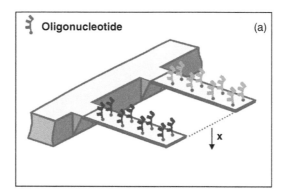

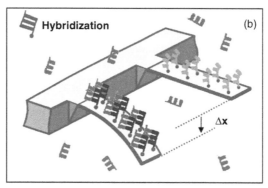

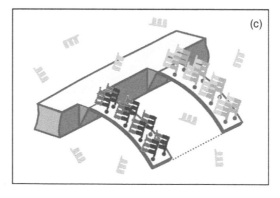

FIGURE 14.10   (See color insert following page 9-22.) Simple scheme illustrating the hybridization experiment. Each cantilever is functionalized on one side with a different oligonucleotide base sequence (marked red or blue). (a) The differential signal is set to zero. (b) After injection of the first complementary oligonucleotide (green), hybridization occurs on the cantilever that provides the matching sequence (red), increasing the differential signal. (c) Injection of the second complementary oligonucleotide (yellow) causes the cantilever functionalized with the second oligonucleotide (blue) to bend. (Reprinted with permission from Fritz, J. et al. [2000] "Translating Biomolecular Recognition into Nanomechanics," *Science* 288, pp. 316–18.)

The foundation of the nanomechanical nose is a microfabricated silicon cantilever array. One side of the cantilevers is coated with a gold film. Thiol-modified DNA strands are covalently bonded to the gold side of the cantilevers. The interaction between the thiol headgroup and the Au surface is strong enough to stabilize the adlayer. Each lever is functionalized with either 12-mer oligonucleotides or 16-mer oligonucleotides. These two strands have different base sequences.

The array is housed in a liquid cell. Solutions of complementary 16-mer oligonucleotides and 12-mer olignonucleotides are injected in sequence into the cell. The DNA strands in solution match up and anchor to their counterparts on the cantilever. The difference in surface stresses between the bare silicon side and the functionalized gold side results in a deflection of the cantilever (Figure 14.10). An optical beam deflection technique is used to monitor the bending motions of the cantilevers in the array.

The nanomechanical nose is a quantitative and qualitative detection device. It provides a clear yes or no response in molecular recognition surveys, and it is sensitive to picomolar concentrations of assayed molecules. Moreover, this cantilever sensor array incorporates the cooperation of intermolecular interactions (i.e., DNA hybridization) and molecule–substrate interactions (i.e., functionalized gold surface). It marries MEMS techniques with nanoscience to create a unique sensing device.

Another take on the nanomechanical nose is the nanomechanical oscillator [Craighead, 2000; Ilic et al., 2004]. This sensor also makes use of a silicon cantilever array, but in this case the cantilever is coated with antibodies that provide a binding site for specific bacteria. The presence of a bonded pathogen is detected by a shift in the resonant frequency of the cantilever (Figure 14.11). This device is capable of subattogram sensitivity under vacuum conditions.

These two types of sensors demonstrate the rich opportunities that molecules can provide when they are incorporated into MEMS/NEMS technology. This initial work is already making its mark in laboratories around the world and will continue to do so in the future.

14.4.2 Functional New Materials

The development of new classes of nanostructured nanomaterials through the functionalization of surfaces by organic molecules is a major achievement. An example of such is a monolayer-protected metal nanoparticle (MPMN). Material properties can be created or improved by restricting the size of a material to the nanoscale and/or by controlling the placement of surface adsorbates with atomic precision.

A MPMN consists of a nanosized metallic core enclosed by a ligand shell. Very recently, Stellacci's group at MIT has observed that mixtures of ligands will self-assemble into well defined domains onto a gold nanoparticle [Jackson et al., 2004]. Notably, the domains are phase-separated by an unprecedented length scale of 5 Å. These nanoparticles can be easily synthesised by a one-step method that allows for

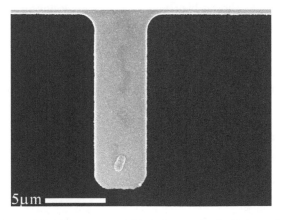

FIGURE 14.11   **(See color insert following page 9-22.)** SEM micrograph of a single *E. coli* bacterium on an antibody-coated silicon nitride cantilever oscillator. (Reprinted with permission from Craighead, H.G. [2000] "Nanoelectro-mechanical Systems," *Science* **290**, pp. 1532–35.)

FIGURE 14.12   (**See color insert following page 9-22.**) Decanethiol/MPA (2:1 molar ratio) self-assembled onto gold nanoparticles. The molecules separate into alternating domains that wrap around the gold core. Left: These domains appear as "ripples" in the STM images. Right: Schematic model of the MPMN. (Reprinted with permission from Jackson, A.M. et al. [2004] "Spontaneous Assembly of Subnanometre–Ordered Domains in the Ligand Shell of Monolayer-Protected Nanoparticles," *Nature Materials* 3, pp. 330–36.)

control over the size of the metallic core and the ligand pattern of the outer shell. Through detailed and controlled studies, it has been determined that the morphology of the molecular coating can be altered by changing the core size or the stoichiometry of the ligand mixture.

Jackson et al. (2004) created MPMNs that consisted of gold cores and ligand mixtures of octanethiol (OT) and mercaptopropionic acid (MPA). They discovered via STM, XRD, and TEM measurements that the ligands form domains that align into parallel ripples around the outside of the core (Figure 14.12). Nanoparticles that are functionalized with a single type of ligand do not exhibit these ripples; instead, they show a hexagonally packed arrangement of headgroups. By varying the reagents in the ligand solution, the difference in height of the peaks and troughs of the ripples can be altered. Moreover, the domain spacing can be tuned by changing the size of the core. It is the radius of curvature of the sphere that is responsible for this phenomenon. Therefore, by increasing the diameter of the core, the distance between the domains is decreased. The spacing can also be modified by changing the ratios of the molecules in the solution.

Interestingly, these MPMNs show a high aptitude for avoiding nonspecific adsorption of proteins, making this material an ideal candidate for biosensor applications. This phenomenon is a result of the small domain sizes. The ripples of the ligand shell are composed of alternating hydrophobic and hydrophilic regions. Proteins are extremely varied and versatile and can either adsorb to a hydrophobic or a hydrophilic region. Since the modulations of the ripples are typically smaller than the dimensions of a protein, a single protein will feel an alternation of attractive and repulsive forces. The resulting effect is a zero net attraction, making protein adsorption unfavorable. After 24 hours of exposure to concentrated protein solutions, Jackson et al. (2004) observed that the rippled, OT/MPA-covered nanoparticles showed no evidence of protein adsorption.

14.5   Conclusions and Perspectives

Organically functionalized surfaces offer a range of properties that can be tailored and therefore are of interest for numerous and varied fields and applications. However, a detailed understanding of molecule–molecule and molecule–substrate interactions is essential. These forces determine the manner in which molecules adsorb and assemble onto a surface, thereby mediating the resulting properties of the molecular adlayer.

In this chapter, we reviewed a number of surprising studies on intermolecular linkages that emphasize the effect of these interactions on the formation of a SAM. They also illustrate how van der Waals forces, directional bonding, and robust metal–ligand links potentially can be used to fabricate novel devices and how nuances in the geometry and composition of an adsorbate can result in drastically different self-assembled configurations.

Another objective of this work was to illustrate the importance of molecule–substrate interactions in SAMs and how substrates can play a passive or active role in the adsorption of molecules. The archetypal system of thiols on Au(111) surfaces was discussed so as to show how surfaces can be intrinsically sensitive only to certain molecules. Other examples focused on molecules that are strongly adsorbed to a surface and how this can result in a reconstruction of the underlying surface or even completely determine the arrangement of adducts on a surface. These investigations shed light on new methods for functionalizing surfaces for technological applications such as sensors.

The final section of this chapter highlighted accomplished work that incorporates organic molecules and MEMS/NEMS technology for sensor applications and reviewed recent work that employs SAMs for the generation of a new class of nanosized and nanostructured materials.

A deeper understanding of the underlying forces and mechanisms for adsorption processes is still required. With this knowledge, scientists will be able to harness these interactions to create more complex nanostructured designs for integration into the Si industry. This is an exciting time and a lot of promising work lies ahead.

References

Barth, J.V., Weckesser, J., Cai, C., Günter, P., Bürgi, L., Jeandupeaux, O., and Kern, K. (2000) "Building Supramolecular Nanostructures at Surfaces by Hydrogen Bonding," *Angew. Chem. Int. Ed.* **39**, pp. 1230–34.

Barth, J.V., Weckesser, J., Lin, N., Dmitriev, A., and Kern K. (2003) "Supramolecular Architectures and Nanostructures at Metal Surfaces," *Appl. Phys. A* **76**, pp. 645–52.

Barth, J.V., Weckesser, J., Trimarchi, G., Vladimirova, M., De Vita, A., Cai, C., Brune, H., Günter, P., and Kern K. (2002) "Stereochemical Effects in Supramolecular Self-Assembly at Surfaces: 1-D vs. 2-D Enantiomorphic Ordering for PVBA and PEBA on Ag(111)," *J. Am. Chem. Soc.* **124**, pp. 7991–8000.

Besenbacher, F. (1996) "Scanning Tunnelling Microscopy Studies of Metal Surfaces," *Rep. Prog. Phys.* **59**, pp. 1737–1802.

Binnig, G., Quate, C.F., and Gerber, Ch. (1986) "Atomic Force Microscope," *Phys. Rev. Lett.* **56**, pp. 930–933.

Binnig, G., and Rohrer H. (1999) "In Touch with Atoms," *Rev. Mod. Phys.* **71**, S324–S330.

Binnig, G., Rohrer, H., Gerber, Ch., and Weibel E. (1982) "Surface Studies by Scanning Tunneling Microscopy," *Phys. Rev. Lett.* **49**, pp. 57–61.

Brust, M., Fink, J., Bethell, D., Schiffrin, D.J., and Kiely C. (1995) "Synthesis and Reactions of Functionalised Gold Nanoparticles," *J. Chem. Soc., Chem. Commun.* **16**, pp. 1655–56.

Brust, M., Walker, M., Bethell, D., Schiffrin, D.J., and Whyman R. (1994) "Synthesis of Thiol-Derivatised Gold Nanoparticles in a Two-Phase Liquid–Liquid System," *J. Chem. Soc., Chem. Commun.* **7**, pp. 801–2.

Craighead, H.G. (2000) "Nanoelectromechanical Systems," *Science* **290**, pp. 1532–35.

Dmitriev, A., Lin, N., Weckesser, J., Barth, J.V., and Kern, K. (2002) "Supramolecular Assemblies with Trimesic Acid at a Cu(100) Surface," *J. Phys. Chem. B* **106**, pp. 6907–12.

Eigler, D.M., and Schweizer E.K. (1990) "Positioning Single Atoms with a Scanning Tunnelling Microscope," *Nature* **344**, pp. 524–26.

Fritz, J., Baller, M.K., Lang, H.P., Rothuizen, H., Vettiger, P., Meyer, E., Güntherodt, H.-J., Gerber, Ch., and Gimzewski, J.K. (2000) "Translating Biomolecular Recognition into Nanomechanics," *Science* **288**, pp. 316–18.

Giancarlo, L.C., and Flynn, G.W. (2000) "Raising Flags: Applications of Chemical Marker Groups to Study Self-Assembly, Chirality, and Orientation of Interfacial Films by Scanning Tunneling Microscopy," *Acc. Chem. Res.* **33**, pp. 491–501.

Heinrich, A.J., Lutz, C.P., Gupta, J.A., and Eigler, D.M. (2002) "Molecule Cascades," *Science* **298**, pp. 1381–87.

Holliday, B.J., and Mirkin, C.A. (2001) "Strategies for the Construction of Supramolecular Compounds through Coordination Chemistry," *Angew. Chem. Int. Ed.* **40**, pp. 2022–43.

Ilic, B., Craighead, H.G., Krylov, S., Senaratne, W., Ober, C., and Neuzil, P. (2004) "Attogram Detection Using Nanoelectromechanical Oscillators," *J. App. Phys.* **95**, pp. 3694–703.

Jackson, A.M., Myerson, J.W., and Stellacci, F. (2004) "Spontaneous Assembly of Subnanometre–Ordered Domains in the Ligand Shell of Monolayer-Protected Nanoparticles," *Nat. Mater.* **3**, pp. 330–36.

Kiely, C.J., Fink, J., Brust, M., Bethell, D., and Schiffrin, D.J. (1998) "Spontaneous Ordering of Bimodal Ensembles of Nanoscopic Gold Clusters," *Nature* **396**, pp. 444–46.

Lang, H.P., Baller, M.K., Battiston, F.M., Fritz, J., Berger, R., Ramseyer, J.-P., Fornaro, P., Meyer, E., Günterodt, H.-J., Brugger, J., Drechsler, U., Rothuizen, H., Despont, M., Vettiger, P., Gerber, Ch., and Gimzewski, J.K. (1999) "The Nanomechancial Nose," *Technical Digest of MEMS '99, 12th IEEE International Micro Electro Mechanical Systems Conference*, pp. 9–13, Jan. 17–21, Orlando, FL, USA.

Lin, N., Dmitriev, A., Weckesser, J., Barth, J.V., and Kern, K. (2002) "Real-Time Single-Molecule Imaging of the Formation and Dynamics of Coordination Compounds," *Angew. Chem. Int. Ed.* **41**, pp. 4779–83.

Lopinski, G.P., Wayner, D.D.M., and Wolkow, R.W. (2000) "Self-Directed Growth of Molecular Nanostructures on Silicon," *Nature* **406**, pp. 48–51.

Moore, G.E. (1965) "Cramming More Components onto Integrated Circuits," *Electronics* **38**, pp. 114–18.

Neddermeyer, H. (1996) "Scanning Tunnelling Microscopy of Semiconductor Surfaces," *Rep. Prog. Phys.* **59**, pp. 701–69.

Otero, R., Naitoh, Y., Rosei, F., Jiang, P., Thostrup, P., Gourdon, A., Laegsgaard, E., Stensgaard, I., Joachim, C., and Besenbacher, F. (2004) "One Dimensional Assembly and Selective Orientation of Lander Molecules on a O–Cu Template," *Angew. Chem. Int. Ed.* **43**, pp. 2092–95.

Piner, R.D., Zhu, J., Xu, F., Hong, S., and Mirkin, C.A. (1999) "Dip–Pen Nanolithography," *Science* **283**, pp. 661–63.

Pitters, J.L., Piva, P.G., Tong, X., and Wolkow, R.A. (2003) "Reversible Passivation of Silicon Dangling Bonds with the Stable Radical TEMPO," *Nano Lett.* **3**, pp. 1431–35.

Rosei, F. (2004) "Nanostructured Surfaces: Challenges and Frontiers in Nanotechnology," *J. Phys.: Condens. Matter* **16**, pp. S1373–S1436.

Rosei, F., and Rosei R. (2002) "Atomic Description of Surface Processes: Diffusion and Dynamics," *Surf. Sci.* **500**, pp. 395–413.

Rosei, F., Schunack, M., Jiang, P., Gourdon, A., Laegsgaard, E., Stensgaard, I., Joachim, C., and Besenbacher, F. (2002) "Organic Molecules Acting as Templates on Metal Surfaces," *Science* **296**, pp. 328–31.

Rosei, F., Schunack, M., Naitoh, Y., Jiang, P., Gourdon, A., Laegsgaard, E., Stensgaard, I., Joachim, C., and Besenbacher, F. (2003) "Properties of Large Organic Molecules on Metal Surfaces," *Prog. Surf. Sci.* **71**, pp. 95–146.

Schreiber, F. (2000) "Structure and Growth of Self-Assembling Monolayers," *Prog. Surf. Sci.* **65**, pp. 151–256.

Schunack, M. (2002) "Scanning Tunneling Microscopy Studies of Organic Molecules on Metal Surfaces," Ph.D. thesis, University of Aarhus.

Schunack, M., Rosei, F., Naitoh, Y., Jiang, P., Gourdon, A., Laegsgaard, E., Stensgaard, I., Joachim, C., and Besenbacher, F. (2002) "Adsorption Behavior of Lander Molecules on Cu(110) Studied by Scanning Tunnelling Microscopy," *J. Chem. Phys.* **117**, pp. 6259–65.

Stepanow, S., Lingenfelder, M., Dmitriev, A., Spillmann, H., Delvigne, E., Lin, N., Deng, X., Cai, C., Barth, J.V., and Kern, K. (2004) "Steering Molecular Organization and Host–Guest Interactions Using Two-Dimensional Nanoporous Coordination Systems," *Nat. Mater.* **3**, pp. 229–33.

Weckesser, J. (2000) "Atomic Scale Observations of Large Organic Molecules at Metal Surfaces: Bonding, Ordering and Supramolecular Self-assembly," Ph.D. thesis, Ecole Polytechnique Federale de Lausanne.

Weckesser, J., Barth, J.V., Cai, Ch., Müller, B., and Kern, K. (1999) "Binding and Ordering of Large Organic Molecules on an Anisotropic Metal Surface: PVBA on Pd(110)," *Surf. Sci.* **431**, pp. 168–73.

Weckesser, J., Barth, J.V., and Kern, K. (1999) "Direct Observation of Surface Diffusion of Large Organic Molecules at Metal Surfaces: PVBA on Pd(110)," *J. Chem. Phys.* **110**, pp. 5351–54.

Whitesides, G.M., Mathias, J.P., and Seto, C.T. (1991) "Molecular Self-Assembly and Nanochemistry: A Chemical Strategy for the Synthesis of Nanostructures," *Science* **254**, pp. 1312–19.

Wintgens, D., Yablon, D.G., and Flynn, G.W. (2003) "Packing of $HO(CH_2)_{14}COOH$ and $HO(CH_2)_{15}COOH$ on Graphite at the Liquid–Solid Interface Observed by Scanning Tunneling Microscopy: Methylene Unit Direction of Self-Assembly Structures," *J. Phys. Chem. B.* **107**, pp. 173–79.

Wolkow, R.A. (1999) "Controlled Molecular Adsorption on Silicon: Laying a Foundation for Molecular Devices," *Annu. Rev. Phys. Chem.* **50**, pp. 413–41.

Index